T0343176

Mathematics for Physical Science and Engineering

Mathematics for Physical Science and Engineering

Symbolic Computing Applications in Maple and Mathematica

Frank E. Harris

University of Utah, Salt Lake City, UT
and University of Florida, Gainesville, FL

AMSTERDAM • BOSTON • HEIDELBERG • LONDON • NEW YORK • OXFORD
PARIS • SAN DIEGO • SAN FRANCISCO • SYDNEY • TOKYO
Academic Press is an imprint of Elsevier

Academic Press is an imprint of Elsevier
225 Wyman Street, Waltham, MA 02451, USA
The Boulevard, Langford Lane, Kidlington, Oxford, OX5 1GB, UK

British Library Cataloguing in Publication Data
A catalogue record for this book is available from the British Library

Library of Congress Cataloging-in-Publication Data
A catalog record for this book is available from the Library of Congress

ISBN: 978-0-12-801000-6

For information on all Academic Press publications
visit our web site at store.elsevier.com

14 15 16 17 18 10 9 8 7 6 5 4 3 2 1

CONTENTS

PREFACE

This text is designed to provide an understanding of the mathematical concepts that form a basis for our understanding of physics and engineering, and to introduce students to the symbolic computation tools that have become essential for the use of that mathematics. The book is at a level of sophistication and rigor appropriate to students in the latter part of a standard undergraduate curriculum in physics or engineering. It should be an easy read for well-prepared graduate students or professionals whose main need is for a good knowledge of symbolic computing.

There are many books that contain the mathematical topics that are covered in the present text, but this author knows of no others that tie that material in an effective and comprehensive way to symbolic computation. Such a connection is of importance, in part because understanding is enhanced by using symbolic techniques to apply the mathematics and visualize its results. In addition, many important mathematical methods become cumbersome when applied to real-world problems, and symbolic computation methods are often superior to the purely numerical approaches that until recently dominated in practical applications.

Today's physical science and engineering students are adept in using computers for information retrieval and some are skilled in digital computer programming, but too many have little or no experience in symbolic computing. This lack of experience is particularly unfortunate because symbolic computing can aid greatly in the visualization of concepts, can eliminate drudgery (and errors) from otherwise complicated computations, and can even facilitate studies of types that were previously completely impractical. But despite the rapid advances in symbolic computing and the increasing extent of its use in the physical and engineering sciences, very little of the instructional material in this area is designed for students of advanced mathematical methods in science and engineering.

In determining how best to incorporate symbolic computation into a presentation of the mathematics, the issue of the choice of symbolic language immediately arises. One could take the view that a noncommercial open-access language should be chosen, but no widely accepted such language presently exists. The alternative of discussing symbolic computation in an artificial "pseudocode" (a device sometimes used to present numerical computational procedures) does not provide students with the necessary practice or experience. It seemed inevitable that one must choose among the two symbolic languages that presently dominate, MAPLE and MATHEMATICA. Each of these languages has its passionate defenders and equally vocal critics, and each has its individual strengths and weaknesses. Moreover, as a practical matter the choice of language depends upon the language that is actually available to the student, and that is often determined by the instructor or the educational institution.

This book approaches the language dilemma by developing both languages in parallel, thereby giving both the students and their instructor the opportunity to make language choices that fit their current instructional situation. This approach therefore has features that extend its value to a large population of practicing physicists and engineers: it enables those fluent in one of the two languages rapidly to gain proficiency in the other. In addition, the comparative discussion of the two languages helps to identify areas where one has advantages over the other.

It is important to realize that skill in symbolic computing cannot be a substitute for an understanding of the related mathematics, and the exposition of this text starts each topic with a discussion of the mathematics involved. To reinforce this point, the Exercises include many that focus on the analytic properties of the quantities involved, and not just on using computers to provide answers. We have in addition placed some emphasis on the use of symbolic computation to explore the content hidden in the underlying mathematics, and have also encouraged students to use symbolic methods to check results obtained by other means. In some areas (e.g., matrix eigenvalue problems, evaluation of inverse Laplace transforms) symbolic computation permits the exploration of a much wider collection of practical examples than would otherwise be possible.

Because this is not a reference book or an advanced treatise, we have omitted proofs of some theorems when we believe they are not essential to a basic understanding of the issues involved. This decision enables us to create a text that falls within the time constraints of instructional programs while still including a wide range of useful topics. Readers desiring treatment of topics at greater depth may find what they seek in the more comprehensive work by Arfken, Weber, and Harris, *Mathematical Methods for Physicists*, 7th edition (New York: Academic Press, 2013).

While a different choice might have been more mathematically elegant, the text has been constructed in a manner that does not require the use of complex variable theory (in particular, the residue theorem and contour integration) in the discussion of special functions, especially the gamma function. We do, however, present early in the book and use throughout it the algebra of complex functions, including Euler's formula for $e^{i\theta}$, de Moivre's theorem, multiple-valued functions, and formulas for inverse trigonometric functions.

There is one technical issue which arises so often that it should be addressed here: The coding that illustrates symbolic computing methods has been written assuming that the user has not defined quantities or procedures that can interfere with the problem under discussion. The presence of potentially interfering definitions is probably the most frequent source of puzzling errors in symbolic computation, and both the instructor and the student need to be cognizant of the need to make computations in a "clean" symbolic environment, obtained either by opening a new computing session or by being vigilant in undefining quantities that are no longer needed or relevant.

In general, this text has been prepared in a way such that the background needed for each chapter has been presented in earlier chapters. However, one can omit or delay the study of Chapter 8 (Tensor Analysis), Chapter 12 (Fourier Series), Chapter 13 (Integral Transforms), Chapter 16 (Calculus of Variations), or Chapter 17 (Complex Variable Theory) without much impact on the remainder of the book. On the other hand, it is recommended not to be dismissive of the material in the Appendices. Much of that material is there, not because it is parenthetical, unimportant, or too specialized, but because it also applies in contexts far removed from that which represented its first use. That material deserves detailed discussion at appropriate points in an instructional program.

An examination of the introductory chapter will reveal that the discussion of symbolic computing assumes that the student has already gotten the symbolic system running on the computer to be used. An instructor may need to supplement this text by providing instructions as to how to access the symbolic system available to his/her students, how to obtain computer output, and how to preserve and reuse workspaces.

The author has benefited from the advice and help of a number of people. Professor Nelson H. F. Beebe of the University of Utah provided invaluable counsel regarding symbolic computing languages, in which he is a recognized expert. At Elsevier, substantial assistance was provided at all stages of the publication process by Editorial Project Manager Jessica Vaughan. The author also gratefully acknowledges the support and encouragement of his friend and partner Sharon Carlson. Without her, he might not have had the energy and sense of purpose needed to bring this project to a timely fruition.

Additional Information

Qualified instructors: See www.textbooks.elsevier.com for the in-depth Instructor's Guide

Chapter 1

COMPUTERS, SCIENCE, AND ENGINEERING

Digital computers are revolutionizing the ways in which scientists and engineers can solve numerical problems of practical interest. This fact is well appreciated both in academic institutions and in governmental and industrial laboratories, and over the past half-century research and development organizations have invested tremendous amounts of money and computer personnel into ever larger and faster digital computing environments. These resources permit the solution of problems for which the relevant science is known, but which are too complicated to solve either by formal analysis or by hand computation. Typical examples of problems whose study has been made possible by digital computation include:

- Engineering problems such as the design of aircraft (requiring the application of compressible fluid mechanics [air] to complicated geometric shapes),
- Physics instrumentation problems such as the design of nuclear reactors (requiring analysis of heat flow and neutron transport through diverse materials in difficult geometries as well as modeling the nuclear reactions involved),
- Modeling of chemical reactions (involving solution of the Schrödinger equation for the quantum states of the relevant species and study of the time-dependent dynamics of the reactants and products),
- Modeling of the physical properties of materials (involving the simulation of processes such as cracking or fragmentation, time evolution of defects or grain boundaries, and other changes in morphology),
- Prediction of electronic or optical properties in complex systems (involving studies of the effects of basic composition and impurities, nonlinear response to incoming signals, and other effects),
- Analysis of the time evolution of the spatial configuration of a large molecular system (examples include biological systems such as the folding processes in proteins, but also the formation or destruction of inanimate structures such as carbon cages or nanotubes).

What these processes have in common is that they involve numerical computations that are so lengthy or complicated that they cannot be done without using machinery to perform the computations and store their results.

Mathematics for Physical Science and Engineering.
http://dx.doi.org/10.1016/B978-0-12-801000-6.00001-8

More recently, accelerating into prominence within the past 25 years, computer software has emerged for carrying out mathematical analyses that are not basically numerical. Software for this purpose, sometimes referred to as **computer algebra** or **symbolic computing**, can assist human investigators in exploring the application of more advanced mathematics. Again, we illustrate with a short list:

- Conversion of the form of algebraic expressions. These conversions include expansions of products of polynomials, solution of algebraic equations or systems of such equations, and insertion of the explicit form of defined quantities,
- Expansions into power series, trigonometric series, asymptotic series, or other types of expansions,
- Integration or differentiation of algebraic forms,
- Solution of differential or integral equations,
- Evaluation and manipulation of special functions,
- Plotting and visualization of functions of interest,
- Reduction of expressions to numerical form at user-chosen levels of numerical precision.

These capabilities are of great importance for both the developers and the users of mathematical methods in science and engineering. Symbolic methods are inherently slower computationally than the purely numerical methods that they supplement, but they are greatly superior for obtaining qualitative understanding of functions and mathematical processes. They also enable the processing of algebraic operations that are too cumbersome or error-prone to be carried out by hand.

It is important to realize that the availability of symbolic computing cannot be a substitute for a knowledge of basic mathematical analysis. A computer algebra program only provides answers to the specific mathematical questions asked by the user. At least for the foreseeable future, the choice of mathematical methods for a science or engineering problem remains the job of the scientist or engineer. It turns out that this task usually cannot even be turned over to a mathematician. A knowledge of the essence of the scientific problem is ordinarily a prerequisite to its successful solution.

This book is not primarily about numerical methods for solving problems in applied mathematics, as that area has reached great sophistication and deserves study as a topic of its own. Rather, our objective here is to introduce the reader to symbolic computation tools that will be extremely useful in gaining understanding of the mathematical processes and functions that we introduce, while at the same time providing access to standard methods for carrying out numerical calculations. The application of advanced mathematical methods to science and engineering is greatly facilitated by this multifaceted approach, and such an approach is becoming an essential part of the education of scientists and engineers. Accordingly, the mission of this introductory chapter is to introduce and develop familiarity with some ideas relevant to symbolic computation, thereby preparing the reader for the use of such systems in the remainder of the book and beyond.

1.1 COMPUTING: HISTORICAL NOTE

Although mechanical machines to do digital computation date back as far as Babbage (1791–1871) and continued to develop until the mid-20th century (with prominent manufacturers then being Friden, Monroe, and others), one of the key advances that made sophisticated computing practical was the development of electronic circuits that permitted an output that depended upon the logical state of an electronically-stored quantity. The development of logic gates and of elements that could be set to either of two states (on vs. off, or 0 vs. 1) made it possible to store data digitally and to manipulate it to produce desired outcomes. The early logic gates used electronic tubes. Data was represented by voltage levels and subsequently by the magnetization states of so-called magnetic cores, and these elements could be combined to represent numeric data (as binary numbers), to perform arithmetic operations on the numeric data, and to store the result.

To make use of these new capabilities, engineers designed electronic circuitry that could, in steps called **cycles**, carry out the successive operations of a defined process on arbitrary data, thereby for the first time creating a need for what is now known as computer software. To use this software, it of course needed to be loaded into a computer and stored. Initially this was done (laboriously, and with many potential errors) from a typewriter-like keyboard or from punched paper tape. It was not long before punched cards came into use for this purpose; error correction then only required replacement of the offending card. Punched cards also had the desirable property that IBM and others had developed the technology for preparing and reading such cards, copying them, and printing their content. It thereby became convenient to convert software to human-readable form and to distribute it to other users.

Initially, software was produced by entering the binary codes that controlled the computer logic; this procedure was not at all user-friendly, and rapidly became supplanted by software called **assemblers**, which would take the successive commands in a user-friendly form (e.g., words like ADD, MULT, STORE) and with the locations of stored data given names instead of their binary addresses. Even this advance could be improved upon. Before long, computer languages were invented to allow the writing of expressions that more closely resemble mathematical statements. Special computer programs, called **compilers**, convert those languages into assembly language and thence into computer-usable form. Such higher-level languages improved the convenience by which successions of numerical steps could be carried out on arbitrary data. They included commands that conditionally (or unconditionally) caused the sequence of commands being executed to continue from a different location in the program. The language that became dominant was called FORTRAN; despite criticism of its design from practitioners of the new field called computer science, FORTRAN rapidly became the language of choice in science and engineering, and has retained that status up to the present day. Because of FORTRAN's dominant position, computer manufacturers and others have continually worked to improve the efficiency of the code produced by its compilers, and FORTRAN continues to be a highly efficient platform for the carrying out of computations that are primarily numeric in nature. Another language that provides good access to computer features that are not limited to those most important in numerical computation is called C. A well-equipped computer environment supports C as well as FORTRAN.

Neither FORTRAN nor C are optimum for support of our study of the mathematical methods that are the subject of the present book. For that purpose we turn to the symbolic computing systems that were developed in a period starting in the 1960s and 1970s. Two of the efforts started in that time frame dominate (as of 2012) the commercial symbolic computing software market: MAPLE, a product of MapleSoft (headquartered in Waterloo, Ontario, Canada), and MATHEMATICA, produced by Wolfram Research (headquarters Champaign, IL, USA). Both MAPLE and MATHEMATICA support a wide variety of areas in symbolic computation and provide interfaces to print, plots, and other forms of visual output. Both are fully equipped to handle all the special functions mentioned in this book, plus many more. Symbolic software has now matured to the point where it is an indispensable tool for serious students of mathematical methods in science and engineering, and we will use such software both to support the mathematical topics in this book and to prepare the student for his/her future professional endeavors.

1.2 BASICS OF SYMBOLIC COMPUTING

In this section we review some of the basic features of symbolic computation and identify the ways in which they appear in the two dominant symbolic computation systems, MAPLE and MATHEMATICA. As we encounter opportunities to use symbolic computation we will use these languages for illustrative purposes. This section will not be sufficient to make its readers highly skilled in the use of either MAPLE or MATHEMATICA, but should provide a base from which simple computations can be carried out.

Because many operations that are needed for mathematics convert expressions containing variables into other related forms, it is necessary to maintain a clear distinction between symbolic expressions and the numerical (or symbolic) values they may acquire when evaluated. For example, a symbolic operation that takes the derivative of an expression with respect to x will not know what to do if x has already been replaced by something else. As another simple example, the factoring of a polynomial in x has no meaning if a numerical value has already been assigned to x and the numerical values of the polynomial's terms have been combined into a single number. In both MAPLE and MATHEMATICA, if a variable has been assigned a value (either symbolic or numeric), the value is usually substituted **before** any commands involving it are processed. This means that the language must not only permit the assignment of values to symbols, but it must also permit the "unassignment" of values.

Even though symbolic computation systems are built to handle general algebraic quantities, they also permit symbolic quantities to be evaluated to numerical values, but with the feature that the numerical precision and accuracy can be manipulated by the user rather than being dictated by the precision level of the computer being used. This means that symbolic computing can be used to study the numerical stability of proposed computational processes, and it permits the graphing of functions and numerical solutions to mathematical problems.

An important feature of both MAPLE and MATHEMATICA is that they include the definition of an extremely wide variety of special functions, with much information about their properties and the relationships they satisfy. This not only enables the use of those functions where appropriate, but also permits them to be graphed and thereby visualized.

Because the commercial symbolic computation systems are designed to have all the features needed for use by professionals, it is totally impractical to provide in this text a description of all (or even most) of their properties. To become an "advanced user," one needs to examine the documentation provided by the distributor and possibly to study some of the many books that present these languages at varying degrees of readability and detail. See the Additional Readings at the end of this chapter. We do provide here enough introductory material to make it reasonable for the reader to gain further symbolic computation proficiency through self study.

The following two subsections deal with the individual systems MAPLE and MATHEMATICA. The reader need only study the material referring to the system he/she intends to use.

MAPLE

Processing in MAPLE normally occurs by executing commands in a workspace MAPLE calls a **worksheet**. It is possible to run MAPLE as an ordinary command in a terminal window, but novice users may find that approach a difficult starting point. MAPLE supports two types of worksheets, of which the newer (and that obtained by default on many computers) is designed also to permit document preparation. The older worksheet type, called "classic" by MAPLE, focuses only on the mathematical computations and their graphic display, and corresponds directly with the summary presented here. It is not difficult to adapt this summary to the document-preparation worksheet. Accordingly, we assume the user has clicked on an icon or typed whatever is needed to open a classic MAPLE worksheet.

Readiness to accept user input is indicated by the issuance of the MAPLE prompt `>`. In response to the prompt, the user can type a **statement**, which is then executed by MAPLE. Although the material presented here assumes that a worksheet is constructed sequentially, the user can (using the cursor and toolbar) insert and execute material in any order. MAPLE generates results that correspond to the order of execution (not location in the worksheet).

The simplest types of statements available to the user are arithmetic expressions, which may involve variables (they start with a letter and may continue with letters or numbers—letters are case-sensitive), and/or constants. Constants can be numbers or they may be represented by symbols that have been predefined by MAPLE and cannot be changed (e.g., `Pi`, which is $3.14159\cdots$).

The arithmetic operations are illustrated by `a+b`, `a-b`, `a*b` (multiplication), `a/b`, `a^b` (a^b). One may also use the FORTRAN notation for the exponentiation operator: `a**b` also produces a^b. Multiplication and division are left-associative, so `a/b*c` stands for $(a/b)c$ and `a/b/c` is $(a/b)/c = a/bc$. Parentheses can be used to control grouping, and must be used, e.g., for `a^(b^c)` as opposed to `(a^b)^c`. Parentheses must also be used to surround the argument lists of functions or commands. Other grouping symbols such as `[ ]` or `{ }` have special meanings and cannot be used in place of `( )`.

A statement may only end with `;` (in which case MAPLE prints its value), or `:` (printing is suppressed but the statement is nevertheless executed). A statement can consist of just an expression (in which case it is not retained long-term by MAPLE), or it may be an **assignment**, in which case a variable is assigned a value (which may itself be symbolic, depending on other variables). The assignment operator is `:=` (the equals sign by itself is used as a relational operator and causes no assignment). Once a variable has been assigned, its occurrence in all **subsequently processed** statements (no matter where they physically occur in the worksheet) will be replaced by the assigned value.

When a statement is complete (including the final ; or :) it is communicated to MAPLE by pressing `Enter`. It is permissible to have more than one statement on an input line (each properly terminated). Then none of the statements is processed by MAPLE until `Enter` is pressed. After MAPLE processes an input line, if printing the result has not been suppressed it prints the value(s) of the statement(s), in a different font and color than the input, usually centered on a new line. Then, on a new line, MAPLE issues a prompt for the next statement. Often it is useful to refer to the output of the most recently executed statement. MAPLE gives it the name `%`; the two statements executed before that are designated `%%` and `%%%`.

Comments (material that is not processed by MAPLE) can be introduced using the `#` **(sharp)** character; the comment runs from sharp to the end of the input line.

Here are a few lines of MAPLE input and output:

```
> # This entire line is a comment.
```

```
> x1 := z + 5;
```

$$x1 := z + 5$$

Note that the assignment operator is := (not =)

```
> x1;
```

$$z + 5$$

```
> % + ymax + 3;
```

$$ymax + z + 8$$

% denotes value of x1

```
> sin(Pi/2);
```

$$1$$

MAPLE knows to insert exact value

```
>
```

Notice that if a symbol has already been defined when a statement is executed, the value is used when the result is evaluated. Undefined quantities remain as symbols. To undefine a symbol x to which something has been assigned, write `x := 'x';` the single quotes indicate that the quantity they enclose is **not** to be evaluated, so x then has only the symbolic value x. If the single quotes were omitted, this statement would cause x to be evaluated and then assigned to x; that would cause the value of x to remain unchanged. Thus, continuing the above session,

```
> x1 := x1;
```

$$x1 := z + 5$$

```
> x1 := 'x1';
```

$$x1 := x1$$

The notion of assignment in MAPLE gives meaning to statements that seem nonsensical in conventional mathematics. For example,

```
> t := 3;
```

$$t := 3$$

```
> t := t + 1;
```

$$t := 4$$

showing that the assignment operator is not just a glorified equals sign. If t had not been defined at the time the assignment was attempted MAPLE would have reported an error because an unevaluated t would then have been on both sides of the assignment statement and MAPLE's attempt to resolve the ambiguity would generate an infinite amount of recursion.

A need that often arises is to evaluate an expression for some value of a variable contained in the expression. For example, we may need to calculate $x^2 - 2x + 3$ for $x = 1$. We could accomplish this by the MAPLE statements

```
> y := x^2 - 2*x + 3:
> x := 1:
> y;
```

$$2$$

but this has the undesirable feature that x has now been assigned a value, and that may cause us trouble later unless we remember to undefine x before we assume it to be available again as a variable. Moreover, if we (carelessly) repeat the command `y := x^2-2*x+3` while x is still set to 1, we will lose the functional form of y and be left with `y:= 2`. MAPLE provides a good alternative: the command `subs(x=1,y)` causes the substitution of 1 for x when evaluating the expression y without affecting the status of either x or y. Thus, a better sequence of commands for evaluating $x^2 - 2x + 3$ for $x = 1$ is

```
> y := x^2 - 2*x + 3:
> subs(x=1, y);
                                        2
```

MAPLE input is regarded as consisting of **execution groups**, marked by tie lines (drawn by MAPLE at the left margin of the worksheet). Unless the user takes explicit action to group statements, each use of `Enter` creates a separate execution group. Execution groups are significant because the pressing of `Enter` anywhere within a complete execution group causes the entire group to be executed. This behavior means that one cannot use `Enter` to open a blank line within an execution group for the insertion of new material (not even if the execution group consists of only a single statement). The user will either have to type the new material without carriage returns (letting MAPLE wrap any resulting long lines) or use MAPLE's `edit` menu to `split` the input, insert a new line, and then recombine the old and the new material using the edit feature `join`.

Numbers without decimal points are regarded as exact, so fractions are reduced to lowest terms but not converted to decimal form. Numbers with decimal points are rendered to `Digits` significant figures, where MAPLE initially sets `Digits:=10`; the user can change the precision to any desired level (e.g., 20) by a statement like `Digits:=20`. If a statement suffers loss of significant figures (e.g., by differencing errors), MAPLE will usually report answers to correspondingly lower precision. Sometimes MAPLE does not automatically convert quantities to decimal form; for example, `Pi` is ordinarily left as the symbol π. Evaluation in decimal form can be forced by the command `evalf` ("floating-point evaluation"). This feature is also useful when we want decimals instead of fractions or if we need a numerical approximation to an integral in place of an analytical form. The command `evalf` accepts an optional second argument that specifies the number of decimal digits of precision (without changing the global variable `Digits`. Here is some sample code:

```
> Digits;
                          10
```
Reads current value of `Digits`

```
> evalf(sin(Pi/4));
                     0.7071067810
```
Uses `evalf` to force numerical evaluation

```
> evalf(Pi, 20);
               3.1415926535897932385
```
Evaluates to 20 digits but does not change `Digits`

```
> Digits := 16;
                     Digits := 16
```
Assigns new value to `Digits`

```
> evalf(sin(Pi/4));
               0.7071067811865475
```

MAPLE is designed to be useful for computations involving imaginary and complex quantities. The symbol `I` is defined as the square root of -1 (upper case so i remains available as a subscript). No symbol has been defined to be e (base of natural logarithms); it must be accessed using the function `exp`.

Special functions are accessed by their MAPLE names. For example, `sin(x)` produces $\sin x$; note that in MAPLE the function argument is **always** within ordinary parentheses. Functions with both arguments and indices are rendered using only arguments; an example is `BesselJ(n,x)` for the Bessel function $J_n(x)$.

A major problem with symbolic computing is the escalation in complexity that often occurs. This makes it important to be able to simplify expressions or convert them to more convenient forms. Because what is most convenient or simple is not a universal and well-defined feature, it is necessary to be familiar with commands that can effect simplifications or changes in form. Commands that are useful include: `simplify`, which produces MAPLE's opinion of simplification; `normal`, which converts expressions to a single fraction with factored numerator and denominator (canceling where possible); `factor`, which does not always yield the same result as `simplify` or `normal`; `expand` (which sometimes expands more than desired); and `collect`, which enables collection in powers of an indicated variable. Finally, there is a catch-all command `convert` that provides a large number of ways expressions can be changed in form.

Much of the power of MAPLE lies in its ability to carry out differentiations and integrations. The commands `diff` and `int` do just that. MAPLE knows the differentiation rules for all elementary and many special functions, and is pretty good at integration. If MAPLE cannot evaluate an integral analytically, it returns the unevaluated integral as an expression involving an integral sign. The user can still get a numerical value for the integral using `evalf`.

```
> diff(sin(x)^2,x);
```

$$2\sin(x)\cos(x)$$

Derivative. Must write `sin(x)^2`, not `sin^2(x)`.

```
> diff(sin(x)^2,x,x);
```

$$2\cos(x)^2 - 2\sin(x)^2$$

Second derivative

Can integrate analytically

```
> int(exp(-x^2)*cosh(x)^(1/2)/x, x=1 .. 2);
```

$$\int_1^2 \frac{e^{-x^2}\sqrt{\cosh(x)}}{x}\,dx$$

MAPLE can't integrate this analytically

```
> evalf(%);
```

$$0.1495664907$$

`%` denotes previous result. Numerical evaluation succeeds.

Notice that the integration range takes the form `x=`*lower limit* `..` *upper limit*. This construction to indicate a range is also used in other contexts where a range is to be specified. For an indefinite integral in the variable `x` the range argument is simply replaced by `x`.

MAPLE is also able to evaluate summations through use of the command `sum`; here "evaluation" means finding a closed form corresponding to a finite or infinite summation—i.e., the discrete analog of an integral evaluation. MAPLE's implementation of the `sum` command leads to difficulties (which we won't discuss here) when the objects being summed involve indexed quantities, and MAPLE provides the alternative command `add`, which is more flexible and should always be used when a formal evaluation is not the objective. Thus,

```
> f := 1/n^2:
```

Define `f`

```
> sum(f, n = 1 .. infinity);
```

Evaluate $\sum_1^\infty 1/n^2$

$$\frac{1}{6}\pi^2$$

```
> add(f, n = 1 .. infinity);
```

Error...

```
> u[1] := 1: u[2] := -1/2:  u[3] := 1/3:
```

Define indexed quantities

```
> sum(u[i], i = 1 .. 3);
```

Error...

```
> add[u[i], i = 1 .. 3);
```

`add` works properly

$$\frac{5}{6}$$

In this book we make extensive use of MAPLE's ability to draw graphs. The plotting system gives the user the capability to select or adjust many features of a function plot (for both ordinary two-dimensional plots and for representations of three-dimensional objects). It requires an extremely complicated set of options to provide this degree of control; we present here only the most basic plotting features. The techniques described here are adequate for making a plot in which function values are plotted against its argument, but there are other ways of generating plots, several of which are discussed in some detail in Appendix A. That appendix also includes a review of some of the more useful plotting options. The types of plots reviewed in the appendix include those for functions described parametrically, contour plots, and plots of discrete point sets. It is also shown how to combine plots of the same or different types in a single graph.

Two elementary 2-D plotting commands for the graphing of $f(x)$ as a function of x for the horizontal range $a \leq x \leq b$ are

```
> plot(f(x), x = a .. b);                    Use here =, not :=.
> plot(f(x), x = a .. b, c .. d);
```

The second of these commands limits the vertical range to $[c, d]$; if the vertical range is not given, MAPLE chooses it to be sufficient to display all, or almost all, the relevant values of $f(x)$. If only one range is given, MAPLE assumes it to be the horizontal range. The horizontal range can sometimes be replaced by giving the name of the independent variable, in which case MAPLE chooses a default range. The quantity being plotted, $f(x)$, must evaluate to a number for all relevant values of x, but the variable x must be **undefined**.

Sometimes we want MAPLE to print material other than a statement's result; examples are (1) to attach a label or explanatory matter to a numerical value, or (2) to make a table of function values. A command to accomplish this is `print`. One can print values of MAPLE variables or arbitrary **strings** of characters (placed within opening and closing double quotes). A string is **not** a MAPLE variable, though a MAPLE variable can have a string as its value. We illustrate:

```
> A := "The result is";
                    A := "The result is"
> x := 6: print(A,x);
                    "The result is", 6
> "The result is" := 6;
Error, invalid left hand side of assignment
```

Detailed control of printing is possible and is discussed briefly in Section 1.5. That section includes examples that show how to print tables of numerical data.

Many of the features discussed above are illustrated in the worksheet depicted in Table 1.1. To save space we have placed the output of each statement to its right instead of centered below it on a new line. Procedures for simplification or conversion between equivalent forms are illustrated in Table 1.2.

MAPLE provides an extensive run-time help facility. The command ?*topic* (with no following ; or :) accesses the help page for *topic*. One easily finds in this way the names and definitions of special functions and the properties and syntax of any MAPLE command.

It is useful to annotate worksheets with comments that inform actual or potential users as to what is going on. In addition to the possibility of using the sharp character

Table 1.1: A MAPLE worksheet. Commands are executed in the order shown.

Input	Output	Comment
`> 12/3/4;`	1	This is (12/3)/4
`> 2^2*3/4;`	3	This is $(2^2)3/4$
`> 5*x + 7;`	$5x + 7$	This isn't stored
`> y = 5*x + 7;`	$y = 5x + 7$	Neither is this!
`> y := 5*x + 7;`	$y := 5x + 7$	y now set to $5x + 7$
`> subs(x=1, y);`	12	y evaluated for $x = 1$
`> x := 3:`		Output print suppressed
`> %;`	3	% is most recent output
`> x;`	3	It did get stored in x
`> 4*Pi;`	4π	π is not evaluated
`> evalf(4*Pi);`	12.56637062	Evaluation now forced
`> Digits := 8;`	$Digits := 8$	Precision changed
`> %%;`	12.56637062	Second previous output
`> evalf(4*Pi);`	12.566371	Recomputed at 8 digits
`> Digits := 10:`		Output suppressed
`> y;`	22	Uses current value of x
`> I^2;`	-1	I is $\sqrt{-1}$
`> sin(Pi/2);`	1	This value is known
`> x := 'x':`		single quotes prevent evaluation
`> x;`	x	x is now undefined
`> ln(e);`	$\ln e$	`e` is just a variable
`> ln(exp(1));`	1	Now we get $\ln e = 1$
`> diff(x^2 - 3*x*z + 2, x);`	$2x - 3z$	Derivative
`> diff(x^2 - 3*x*z + 2, x, z);`	-3	`z` could be another `x`
`> int(sin(x)^2, x = 0 .. 2*Pi);`	π	Range of x is 0 to 2π
`> int(6*x^2, x);`	$2x^3$	Indefinite integral
`> sum(1/x^2, x = 1 .. infinity;`	$\frac{1}{6}\pi^2$	Formal evaluation
`> add(1/n^2, n = 1 .. 3);`	$\frac{49}{36}$	Use for most finite sums
`> plot(sin(y), y = -Pi .. Pi);`	(error)	Bad args: `sin(22), 22=-Pi..Pi`
`> plot(sin(x), x = -Pi .. Pi);`	(see Fig. 1.1)	OK because `x` is undefined
`> print("y=", y);`	"y=", 22	The print statement has no value but causes this to print

Table 1.2: Elementary form conversions in MAPLE.

```
> S := 8 - 2/(2*x-1) + 1/(x-1);
```

$$S := 8 - \frac{2}{2x-1} + \frac{1}{x-1}$$

(For this S, `simplify` and `normal` produce the same result; `factor` does not)

```
> simplify(S);
```

$$\frac{16x^2 - 24x + 9}{(x-1)(2x-1)}$$

```
> normal(S);
```

$$\frac{16x^2 - 24x + 9}{(x-1)(2x-1)}$$

```
> factor(S);
```

$$\frac{(4x-3)^2}{(x-1)(2x-1)}$$

```
> expand(S);
```

$$\frac{16x^2}{(x-1)(2x-1)} - \frac{24x}{(x-1)(2x-1)} + \frac{9}{(x-1)(2x-1)}$$

(`expand` does not really bring back the original form)

```
> convert(%,parfrac);
```

$$8 - \frac{2}{2x-1} + \frac{1}{x-1}$$

(Here `convert` applies a partial fraction decomposition; look at `?convert`)

```
> T := exp(a+ln(b*exp(c)));
```

$$T := e^{(a + \ln(be^c))}$$

(Here `simplify` and `normal` produce different results)

```
> normal(T);
```

$$e^{(a + \ln(be^c))}$$

```
> simplify(T);
```

$$b\,e^{(a+c)}$$

```
> V := 8*x^3*y^3 + 24*x^2*y^2 + 24*x*y + 8*y^3 + x + y - 5 + 2*y^2;
```

$$V := 8x^3y^3 + 24x^2y^2 + 243xy + 8y^3 + x + y - 5 + 2y^2$$

```
> collect(V,x);
```

$$8x^3y^3 + 24x^2y^2 + (1 + 24y)x + y - 5 + 8y^3 + 2y^2$$

```
> collect(V,y);
```

$$(8x^3 + 8)y^3 + (2 + 24x^2)y^2 + (1 + 24x)y + x - 5$$

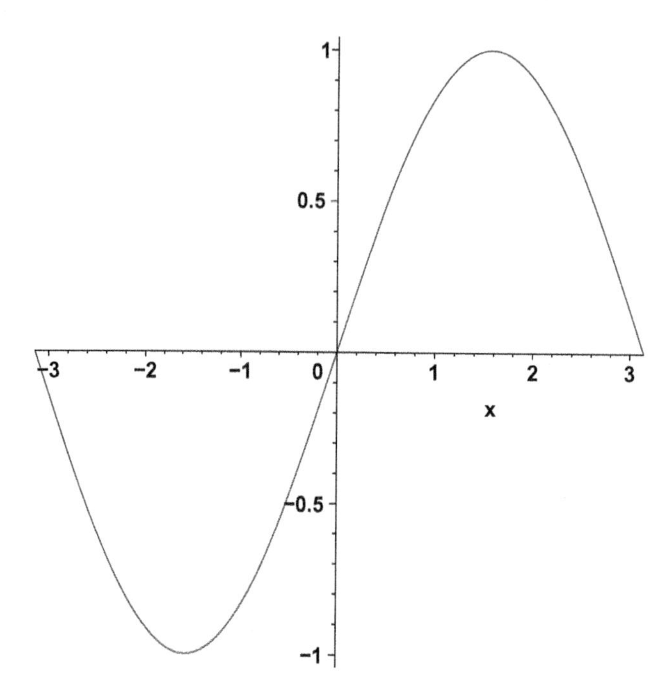

Figure 1.1: Plot of $\sin x$ (similar in both MAPLE and MATHEMATICA). Coding in Tables 1.1 and 1.3.

to initiate a comment, the pull-down menu item `insert text` permits the insertion of comments, causing them to be rendered in different fonts and/or colors than real MAPLE input. Liberal use of comments is highly recommended.

Worksheets can be named and saved for reuse (use MAPLE's pull-down menu). When a worksheet is reopened in a new MAPLE session, none of its statements are automatically executed, so the new MAPLE session has not generated or stored any of its data. The user may execute any statement from the worksheet by positioning the cursor on it and pressing `Enter`. A pull-down menu permits execution of the entire worksheet or selected portions thereof.

MATHEMATICA

Processing in MATHEMATICA occurs by executing commands in a workspace called a **notebook**. We assume the user has clicked on a icon or typed whatever is needed to cause MATHEMATICA to open a blank notebook.

A notebook will not exhibit a prompt, but when a MATHEMATICA command is entered it will be processed. Although the material presented here assumes that a notebook is constructed sequentially, the user can (using the cursor) insert and execute material in any order. MATHEMATICA generates results that correspond to the order of execution (not location in the notebook). When a complete command has been entered, MATHEMATICA assigns it a sequence number n, prefaces it by the notation `In[`n`]:=`, and writes the statement's output preceded by `Out[`n`]=`.

The simplest types of commands available to the user are ordinary arithmetic expressions, which may involve variables (they start with a letter and may continue with letters or numbers—letters are case-sensitive), and/or constants. Constants can be numbers or they may be represented by symbols that have been predefined by MATHEMATICA and cannot be changed (e.g., `Pi`, which is $3.14159\cdots$).

The arithmetic operations are illustrated by `a+b`, `a-b`, `a b` or `a*b` (multiplication), `a/b`, `a^b` (a^b). The FORTRAN notation for exponentiation, `a**b`, is **not** recognized by MATHEMATICA. Multiplication and division are left-associative, so `a/b*c` stands for $(a/b)c$ and `a/b/c` is $(a/b)/c = a/bc$. When no operator symbol is used to indicate multiplication, symbolic operands (e.g., `a` and `b`) must be separated by at least one space; otherwise they would be interpreted as a single symbol (e.g., `ab`). A better practice, followed in most of the coding examples in this book, is to always use the `*` symbol for multiplication. Parentheses can be used to control grouping, and must be used when the grouping is not that obtained in MATHEMATICA by default. For example, `a^b^c` is interpreted by MATHEMATICA as a^{b^c}; if the user wants $(a^b)^c = a^{bc}$, he/she must write `(a^b)^c`. Argument lists for functions or commands are not surrounded by parentheses, but by square brackets, `[ ]`. Other grouping symbols or combinations thereof, such as `{ }`, have special meanings and cannot be used in place of `( )` or `[ ]`.

Commands are ended with `Shift/Enter` (keep the shift key depressed while pressing `Enter`). `Enter` (by itself) simply permits a continuation of the input on the next line of the notebook. If `;` is the last character before `Shift/Enter`, printing of the output (but not its generation) is suppressed. A command can consist of just an expression, or it may be an **assignment**, in which case a variable is assigned a value (which may itself be symbolic, depending on other variables). The assignment operator is `=` and once a variable has been assigned, its occurrence in all **subsequently processed** statements (no matter where they physically occur in the notebook) will be replaced by the assigned value. All output that has been generated in a MATHEMATICA session remains available for the duration of the session, and may be referred to as `%` (most recently produced output), `%%` (next most recent output), `%%...%` (k times) (the kth previous output), or `%`n (the output line `Out[`n`]`). Using `%`n may cause problems if a notebook is saved and reused, because the output numbering depends on the session history. Here are a few lines of MATHEMATICA input and output. The user does not type `In[..]:=`; that is supplied by MATHEMATICA when the statement is processed.

`In[1]:= x1 = z + 5`	Assigns $z + 5$ as value of `x1`
`Out[1]=` $z + 5$	
`In[2]:= x1`	Checks that assignment was done
`Out[2]=` $z + 5$	
`In[3]:= % + ymax + 3`	`%` denotes `Out[2]`;
`Out[3]=` $ymax + z + 8$	addends are combined
`In[4]:= Sin[Pi/2]`	Note style: `S` and `[ ]`
`Out[4]=` 1	MATHEMATICA knows the value

Notice that if a symbol has already been defined when a statement is executed, the value is used when the result is evaluated. Undefined quantities remain as symbols. To undefine a symbol `x` to which something has been assigned, write `Clear[x]`.

A need that often arises is to evaluate an expression for some value of a variable contained in the expression. For example, we may need to evaluate `x1=z+5` for $z = 1$.

We could accomplish this by the statements

```
x1 = z + 5;
z = 1;
y                                    6
```

but this has the undesirable feature that `z` has now been assigned a value, and that may cause us trouble later unless we remember to undefine `z` before we assume it to be available again as a variable. Moreover, if we (carelessly) repeat the command `x1=z+5` while `z` is still set to 1, we will lose the functional form of `x1` and be left with `x1=6`. MATHEMATICA provides a good alternative: the command `x1/.z -> 1` (`->` is two key strokes, `-` followed by `>`) causes the substitution of 1 for `z` when evaluating the expression `x1` without affecting the status of either `x1` or `z`.

The evaluation procedure in MATHEMATICA continues iteratively if a statement redefines any of its operands; this feature causes constructions like `x=x+1` (with `x` not defined) to generate an infinite sequence of substitutions. If `x` **has** been defined, the right-hand `x` is replaced by its value when evaluation begins and there is then no problem. Continuing the above session, we illustrate these points.

```
In[5]:= x1/.z -> 1              x1 is evaluated for z = 1
Out[5]= 6
In[6]:= x1 = x1                 Attempt to undefine x1 fails
Out[6]= z + 5
In[7]:= Clear[x1]               This produces no output
In[8]:= x1 = x1
Out[8]= x1                      x1 is now undefined
In[9]:= x1 = x1 + 1
                                Attempt to increment x1 fails
(error message)
In[10]:= x1 = 3                 Give x1 a value
Out[10]= 3
In[11]:= x1 = x1 + 1
Out[11]= 4                      Increment now works
In[12]:= x1 = %
Out[12]= 4
```

Numbers without decimal points are regarded as exact, so fractions are reduced to lowest terms but not converted to decimal form. In the absence of appropriate action by the user (see below), numbers with decimal points are generated at machine precision for the computer in use, though they are often displayed at lower precision (typically six decimal digits). An output number at its full stored precision can be exhibited by placing the cursor on it and then pressing `Enter` (without `Shift`). The full-precision numerical value is marked with a grave accent. Fractions and symbols representing numbers can be converted to decimal form (at machine precision) using the operator `N`, as in `N[Pi]` or `N[2/3]`. To set the precision of the conversion of *expr* to *n* decimal digits, write `N[`*expr*`,`*n*`]`. Arbitrary-precision arithmetic is used by MATHEMATICA when the operands have been specified at higher than machine precision.

`In[1]:= N[2/3]`	Evaluates at machine precision
`Out[1]=` 0.666667	Displays only six digits
0.6666666666666666`	Pressing `Enter` reveals more Accent means at machine precision
`In[2]:= N[2/3,20]`	Evaluation demanded at 20 figures
`Out[2]=` 0.66666666666666666667	Thus, 20 figures are shown
0.666666666666666 666666666666670874`20	This is what MATHEMATICA keeps The final 20 means 20 figures are good

MATHEMATICA is designed to be useful for computations involving imaginary and complex quantities. The symbol `I` is defined as the square root of -1 (upper case so i remains available as a subscript). The symbol `E` denotes the base of natural logarithms.

Special functions are accessed by their MATHEMATICA names, which begin with an upper-case letter even when that is not the usual notation. For example, `Sin[x]` produces $\sin x$; note that in MATHEMATICA function arguments are **always** enclosed in square brackets. Functions with both arguments and indices are rendered using only arguments; an example is `BesselJ[n,x]` for the Bessel function $J_n(x)$.

Internally, MATHEMATICA processes every command as a function call, so for example when the user writes `a+b`, MATHEMATICA changes it internally to `Plus[a,b]`. Functions defined by MATHEMATICA are always given names that start with an upper-case letter (suggesting that some errors can be avoided if user-defined quantities are given names that start with lower-case letters). An understanding of MATHEMATICA's use of function calls may help the reader to remember to capitalize their initial letters and to enclose their arguments in square brackets `[ ]`.

A major problem with symbolic computing is the escalation in complexity that often occurs. This makes it important to be able to simplify expressions or convert them to more convenient forms. MATHEMATICA does quite a bit of spontaneous expression simplification, but the process is not infallible, in part because convenience and simplification do not have precise definitions. It is therefore necessary to be familiar with commands that can effect simplifications or changes in form.

The MATHEMATICA commands that are useful for manipulation of expressions include `Simplify`, which produces MATHEMATICA's opinion of simplification. Also helpful are `Together`, which combines terms over a common denominator; `Apart`, which carries out a partial fraction decomposition; `Expand` (which sometimes expands more than desired); and `Collect`, which enables collection in powers of an indicated variable. One can `Factor` an entire expression or, using `FactorTerms`, extract common factors from its individual terms without forcing them to be combined.

Much of the power of MATHEMATICA lies in its ability to carry out differentiations and integrations. The commands `D` and `Integrate` do just that. MATHEMATICA knows the differentiation rules for all elementary and many special functions, and is pretty good at integration. If MATHEMATICA cannot evaluate an integral analytically, it returns the unevaluated integral as an expression involving an integral sign. However, the user can still get a numerical value for the integral using the command `N`.

`In[1]:= D[Sin[x]^2,x]`	Derivative. Must write `Sin[x]^2`, not `Sin^2[x]`.
`Out[1]=` $2\,\mathrm{Cos}[x]\,\mathrm{Sin}[x]$	
`In[2]:= D[Sin[x]^2,x,x]`	Second derivative
`Out[2]=` $2\,\mathrm{Cos}[x]^2 - 2\,\mathrm{Sin}[x]^2$	

```
In[3]:= Integrate[Exp[-x^2]*Cosh[x]^(1/2)/x,{x,1,2}]
```

Out[3]= $\int_1^2 \frac{e^{-x^2}\sqrt{\text{Cosh}[x]}}{x}\,dx$ MATHEMATICA can't integrate this analytically

`In[4]:= N[%]` % denotes previous result.

Out[4]= 0.149566 Numerical evaluation succeeds.

0.1495664907063264` Pressing **Enter** when cursor is on result gives it to full precision

Note that the integration variable and range are enclosed in braces; this is MATHEMATICA's general way of indicating ranges. For an indefinite integral with integration variable `x` the entire brace construction would be replaced by `x`.

Summations can be carried out using the command `Sum`. MATHEMATICA will try to bring infinite series into closed forms. Thus,

```
In[1]:= Sum[1/n^2 ,{n,1,Infinity}]
```

Out[1]= $\frac{\pi^2}{6}$

In the present book we will make extensive use of MATHEMATICA's ability to draw graphs. The plotting system gives the user the capability to select or adjust many features of a function plot (both for ordinary two-dimensional plots and for representations of three-dimensional objects). It requires an extremely complicated set of options to provide this degree of control; we present here only the most basic plotting features. The techniques described here are adequate for making a plot in which function values are plotted against its argument, but there are other ways of generating plots, several of which are discussed in some detail in Appendix A. That appendix also includes a review of some of the more useful plotting options. The types of plots reviewed in the appendix include those for functions described parametrically, contour plots, and plots of discrete point sets. It is also shown how to combine plots of the same or different types in a single graph.

Two elementary 2-D plotting commands for the graphing of $f(x)$ as a function of x for the horizontal range $a \leq x \leq b$ are

```
Plot[f(x),{x,a,b}]

Plot[f(x),{x,a,b}, PlotRange->{c,d}]
```

The second of these commands limits the vertical range to $[c, d]$; if the vertical range is not given, MATHEMATICA chooses it to be sufficient to display all (or almost all) of the relevant values of $f(x)$. The variable (x) used in `Plot` is regarded as local to that command so it does not matter whether or not it has been defined and any previous definition remains unchanged.

Sometimes we want MATHEMATICA to print material other than a statement's result; examples are (1) to attach a label or explanatory matter to a numerical value, or (2) to make a table of function values. A command to accomplish this is `Print`. One can print values of MATHEMATICA variables or arbitrary **strings** of characters (placed within opening and closing double quotes). A string is **not** a MATHEMATICA variable, though a MATHEMATICA variable can have a string as its value. We illustrate (the material following `In[..]:=` was typed in):

```
In[1]:= A = "The result is "
Out[1]= The result is
In[2]:= x=6; Print[A,x]
Out[2]= The result is 6
In[3]:= "The result is " = 6
     Set :: setraw :
     Cannot assign to raw object The result is .
Out[3]= 6
```

Notice that the `Print` command has no value (and hence does not generate an `Out` statement, but it caused printing to occur.

Many of the features discussed above are illustrated in the notebook depicted in Table 1.3. The sequence numbering shows that the notebook entries are processed in order. Procedures for simplification or conversion between equivalent forms are illustrated in Table 1.4.

MATHEMATICA provides an extensive run-time help facility, accessed from an entry in the menu at the top of the notebook. In addition, a large number of tutorials (available free as PDF files or for purchase as hard copy) could, as of this book's publication date, be reached from MATHEMATICA's web pages.

It is useful to annotate notebooks with comments that inform actual or potential users as to what is going on. To insert a comment, type

(* *any desired text, which may extend over*
  multiple lines, continuing until an instance of *)

Notebooks can be named and saved for reuse (use MATHEMATICA's pull-down menu). When a notebook is reopened in a new MATHEMATICA session, none of its statements is automatically executed, so the new MATHEMATICA session has not generated or stored any data. The user may execute all or part of the notebook (use the pull-down menu); that causes the executed commands and their output to be numbered consecutively (with numbers unrelated to those assigned when the notebook was originally created).

Exercises

Carry out these exercises using the symbolic computation system of your choice.

1.2.1. Reduce the following polynomial in x and z to fully factored form:

$$3x^2 + 3x^4 - 4xz + 3x^2z + 2x^3z + 6x^5z + z^2 - 4xz^2 - 7x^2z^2 + 6x^3z^2$$
$$-8x^4z^2 + z^3 + 2xz^3 - 5x^2z^3 + 2x^3z^3 - 2xz^4 + 6x^3z^4 + z^5 - 8x^2z^5 + 2xz^6$$

1.2.2. Because the polynomial in Exercise 1.2.1 can be factored, symbolic computer systems can easily find the values of x that are its roots. Calling the polynomial in its original form `poly`, use one of the following:

`solve(poly=0,x)` or `Solve[poly==0,x]`

Note. Less convenient forms than that found here are usually obtained as the solutions to more general root-finding problems.

Table 1.3: A MATHEMATICA notebook. In actual notebooks the output is placed below the input, as in the main text.

Input	Output	Comments
`In[1] := 12/3/4`	`Out[1]` $= 1$	Evaluated as $(12/3)/4$
`In[2] := 2^2*3/4`	`Out[2]` $= 3$	Could write `(2^2)3/4`
`In[3] := y = 5*x + 7`	`Out[3]` $= 7 + 5x$	y contains symbolic data
`In[4] := x = 3`	`Out[4]` $= 3$	Does not change y
`In[5] := y`	`Out[5]` $= 22$	Evaluates with $x = 3$
`In[6] := Clear[x]`		Undefines x (no output)
`In[7] := y`	`Out[7]` $= 7 + 5x$	
`In[8] := %5`	`Out[8]` $= 22$	Just copies `Out[5]`
`In[9] := y/. x -> 1`	`Out[9]` $= 12$	Evaluates y for $x = 1$
`In[10] := 4*Pi`	`Out[10]` $= 4\pi$	π stays symbolic
`In[11] := N[%]`	`Out[11]` $= 12.5664$	At machine precision
To see more digits we press **Enter**:	12.566370614359172`	` is the full-precision marker
`In[12] := 22/7`	`Out[12]` $= \frac{22}{7}$	Left as fraction
`In[13] := 22./7`	`Out[13]` $= 3.14286$	Force decimal conversion
`In[14] := I^2`	`Out[14]` $= -1$	I is $\sqrt{-1}$
`In[15] := Log[E]`	`Out[15]` $= 1$	Log is ln; E is base for ln
`In[16] := D[x^2 - 3*x*z + 2*x,x]`	`Out[16]` $= 2x - 3z$	Derivative
`In[17] := D[x^2 - 3*x*z + 2*x,x,z]`	`Out[17]` $= -3$	z could be another x
`In[18] := Integrate[Sin[x]^2, {x, 0, 2*Pi}]`		
	`Out[18]` $= \pi$	Range of x is 0 to 2π
`In[19] := Integrate[6*x^2,x]`	`Out[19]` $= 2x^3$	Indefinite integral
`In[20] := Sum[x^n, {n, 0, Infinity}]`	`Out[20]` $= \frac{1}{1-x}$	Evaluates sum
`In[21] := Plot[Sin[y], {y, -Pi, Pi}]`		OK even if y is defined
	(see Fig. 1.1)	Plot but no other output
`In[22] := Print["y = ", y]`		Prints $y = 7 + 5x$

Table 1.4: Elementary form conversions in MATHEMATICA.

`In[1] :=` `S = 8 - 2/(2x-1) + 1/(x-1)`

`Out[1] =` $8 + \frac{1}{-1+x} - \frac{2}{-1+2x}$

(For this S, **Simplify**, **Factor**, and **Together** all give different results)

`In[2] :=` `Simplify[S]`

`Out[2] =` $8 + \frac{1}{-1+x} - \frac{2}{-1+2x}$

`In[3] :=` `Factor[S]`

`Out[3] =` $\frac{(-3+4x)^2}{(-1+x)(-1+2x)}$

`In[4] :=` `Together[S]`

`Out[4] =` $\frac{9-24x+16x^2}{(-1+x)(-1+2x)}$

(The result of **Together** can be further simplified)

`In[5] :=` `SS = Simplify[%]`

`Out[5] =` $\frac{(3-4x)^2}{(-1+x)(-1+2x)}$

(**Apart** returns to the original form of this S; **Expand** does not)

`In[6] :=` `Apart[SS]`

`Out[6] =` $8 + \frac{1}{-1+x} - \frac{2}{-1+2x}$

`In[7] :=` `Expand[SS]`

`Out[7] =` $\frac{9}{(-1+x)(-1+2x)} - \frac{24x}{(-1+x)(-1+2x)} + \frac{16x^2}{(-1+x)(-1+2x)}$

`In[8] :=` `T = Exp[a + Log[b*Exp[c]]]`

`Out[8] =` $e^{a + \text{Log}\left[be^{c}\right]}$

`In[9] :=` `Simplify[%]`

`Out[9] =` $b\,e^{a+c}$

`In[10] :=` `V = 8x^3y^3 + 24x^2y^2 + 24x y + 8y^3 + x + y - 5 + 2y^2`

`Out[10] =` $-5 + x + y + 24xy + 2y^2 + 24x^2y^2 + 8y^3 + 8x^3y^3$

(`Collected` powers are not necessarily sorted)

`In[11] :=` `Collect[V,x]`

`Out[11] =` $-5 + y + 2y^2 + 24x^2y^2 + 8y^3 + 8x^3y^3 + x(1+24y)$

`In[12] :=` `Collect[V,y]`

`Out[12] =` $-5 + x + (1+24x)y + (2+24x^2)y^2 + (8+8x^3)y^3$

1.2.3. (a) Starting with $f(x) = 1 + 2x + 5x^2 - 3x^3$, obtain an expansion of $f(x)$ in powers of $x + 1$ by carrying out the following steps: (1) Define a variable `s` and assign to it the value of the polynomial representing $f(x)$; (2) Define x to have the value $z - 1$ and recompute `s`; (3) Expand `s` in powers of z; (4) Define z to have the value $x + 1$ and recompute `s`. You will need to be careful to have x and/or z **undefined** at various points in this process.

(b) Expand your final expression for `s` and verify that it is correct.

Note. Your algebra system may combine the linear and constant terms in a way that causes $(x + 1)$ not to be explicitly visible.

1.2.4. Verify the trigonometric identity $\sin(x + y) = \sin x \cos y + \cos x \sin y$ by simplifying $\sin(x + y) - \sin x \cos y - \cos x \sin y$ to the final result zero. Use the commands for simplifying and expanding (try both even if the first works). These two commands do not always give the same results.

1.2.5. Obtain numerical values, precise to 30 decimal digits, for π^2 and $\sin 0.1\pi$.

Note. Obtain these 30-digit results even if your computer does not normally compute at this precision.

1.2.6. (a) Verify that your computer system knows the special function corresponding to the following integral:

$$\text{erf}(x) = \frac{2}{\sqrt{\pi}} \int_0^x e^{-t^2} dt .$$

In MAPLE, it is `erf(x)`; in MATHEMATICA, `Erf[x]`.

(b) Write symbolic code to evaluate the integral defining $\text{erf}(x)$, and check your code by comparing its output with calls to the function in your symbolic algebra system. Check values for $x = 0$, $x = 1$, and $x = \infty$.

Note. Infinity is `infinity` in MAPLE, `Infinity` in MATHEMATICA.

1.2.7. Plot the function defined in Exercise 1.2.6 for various ranges of x. Use enough ranges to get some experience in using your algebra system's plotting utility.

1.2.8. Obtain the value of $\text{erf}(\pi)$ to 30 decimal places and print out the message "erf(pi)= $\cdots$ ".

Note. The value of erf must be supplied directly by the computer; you should not obtain an answer by typing it in.

1.2.9. Obtain the sixth derivative with respect to x of $1/\sqrt{x^2 + z^2}$. Evaluate it numerically for $x = 0$, $z = 1$.

1.2.10. (a) Evaluate the indefinite integral $\displaystyle\int \frac{dz}{(z^2 + 4)^{7/2}}$.

(b) Evaluate the above integral numerically for the range $(0, \pi)$.

1.2.11. The amplitude A of the light of wavelength λ that is transmitted through a slit of width d is in the approximation of **Fraunhofer diffraction** given as $A = A_0 \sin x / x$, where $x = \pi d \sin\theta / \lambda$, with θ the angle from the normal to the slit; see Fig. 1.2.

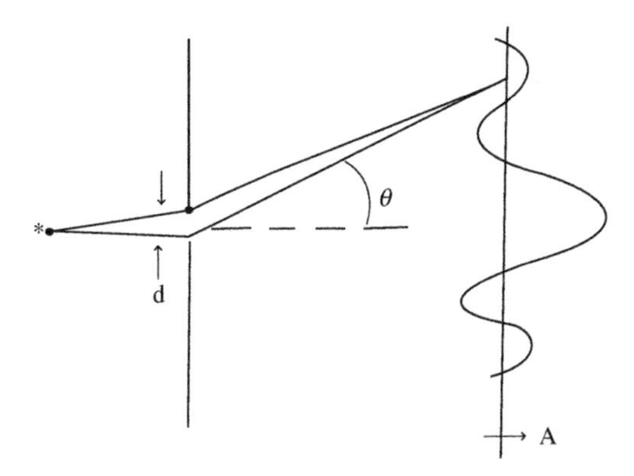

Figure 1.2: Fraunhofer diffraction.

(a) Plot A over a range of x that is symmetric about $x = 0$ and contains 11 extrema.

(b) The places where $A = 0$ occur are at $x = \pm\pi$, $\pm 2\pi$, etc., and the extrema lie between (but not exactly half-way between) these zeros. Using the plot facilities of your symbolic computing system, change the range of your plot in ways that permit each extremum to be located to an accuracy of 0.001 in x.

(c) Returning now to the formula for A, compute the integral of A^2 for each region bounded by adjacent zeros of A, and thereby determine the ratio of the intensity for the region containing $x = 0$ (referred to as the principal maximum) to the intensity in one of the immediately adjacent regions.

1.2.12. Consider the function $f(x) = x^6 - 2x^5 + x^4 - 1.783x^3 + 1.234x^2 + 0.877x - 0.434$. Find all the roots of $f(x)$ to three decimal places by making plot(s) over a suitable range or ranges.

1.2.13. Given the function $f(x) = a^3x^3 - x^2 + 3x - 4a$, find by making suitable plots a value of a such that $f(2) = 0$. Is your value of a unique?

1.3 SYMBOLIC COMPUTATION PROGRAMS

Although an ability to evaluate and graph expressions can have great value, the power of symbolic computation systems can be increased almost without limit by using the tools they provide to undertake tasks that are too complex to be carried out in a practical fashion by the evaluation of a single statement. The key features that are needed to support programming are (1) commands that control conditional or repeated execution, and (2) the capability to create, save, and reuse command sequences. Both MAPLE and MATHEMATICA have these features.

The level of programming expertise needed in connection with the use of this book is fairly limited, and accordingly we present here only a minimal number of programming commands and strategies. The full sets of programming capabilities of our symbolic computation systems are far more extensive than the present survey.

RELATIONAL AND LOGICAL OPERATORS

Relational operators (in conventional mathematical notation, $>$, $\geq$, $=$, $\neq$, $<$, $\leq$) are needed to indicate whether a statement or series of statements is to be executed, with the execution to take place only if some relation is satisfied. In a computer program, a relational statement such as $a < b$ is not a report that a is less than b, but is a construction that will either be **true** or **false**, depending upon the values of a and b when the statement $a < b$ is evaluated. It therefore makes sense to discuss a variable C **defined** to have the value **true** if $a < b$ and the value **false** otherwise, and to indicate that fact by writing (using the syntax of our symbolic computing system) something equivalent to $C = (a < b)$. Because the possible values if C are **true** and **false**, C is a logical-valued (and not a numeric or algebraic) variable. This will enable us to write program elements that are executed, for example, only if the value of C is **true**.

It is also essential to be able to construct variables that have the value **true** only if multiple or more complicated conditions are satisfied; for example, if $C = a < b$ and $C' = d < a$, then a logical quantity that is **true** only if $d < a < b$ will take a form corresponding to C **and** C', meaning that C **and** C' is **true** only if C and C' are both **true**. On the other hand, we may need to specify a quantity that will have the value **true** if either or both of C and C' are **true**. This situation corresponds to C **or** C'.[1] Finally, we may encounter a need for an expression that is **true** only if C is **false** and vice versa; this operation corresponds to **not** C.

In MAPLE, the relational operators are indicated by the symbols `<`, `<=`, `=` (note the absence of the `:`), `<>` (this is $\neq$ and is formed from the two keystrokes $<$ and $>$), `>`, and `>=`. The logical operators are easy to remember; they are `and`, `or`, and `not`.

In MATHEMATICA, the relational operators are indicated by `<`, `<=`, `==` (a single $=$ sign will not do), `!=` (this is $\neq$), `>`, and `>=`. The logical operators are taken from the notation of C, in which **and** is `&&`, while **or** is `||` (this is two keystrokes), and **not** is `!` (the exclamation point).

Example 1.3.1. Specifying Conditions

Suppose that we want to execute some statements when x is either in the range $0 < x < 1$ or in the range $2 < x \leq 10$. A logical statement that is true only within these ranges can take the form

$$A = [(0 < x) \text{ and } (x < 1)] \text{ or } [(2 < x) \text{ and } (x \leq 10)] ,$$

and this translates into

MAPLE
```
A := (0 < x) and (x < 1) or (2 < x) and (x <= 10)
```

MATHEMATICA
```
A = (0 < x) && (x < 1) || (2 < x) && (x <= 10)
```

Because `and` has higher precedence than `or`, just like $\times$ has precedence over $+$, e.g., $2y + 5$ means $(2y) + 5$, not $2(y + 5)$, it is not necessary to use parentheses around the `and` constructions.

An alternative specification of A with the same truth value is

$$A = [(0 < x) \text{ and } (x \leq 10)] \text{ and not}[(1 \leq x) \text{ and } (x \leq 2)] ,$$

[1]This is the accepted mathematical meaning of **or**. An operation that yields **true** only if exactly one of C and C' is **true** is called **exclusive or** and is sometimes abbreviated **xor**.

corresponding to

MAPLE
```
A := (0 < x) and (x <= 10) and not((1 <= x) and (x <= 2))
```

MATHEMATICA
```
A = (0 < x) && (x <= 10) && !((1 <= x) && (x <= 2))
```

Here the parentheses showing the range of application of `not` are essential, as otherwise `not`, which has higher precedence than `and`, would apply only to `1 <= x`.

Note that in all these MAPLE and MATHEMATICA examples, the groupings are effected using ordinary parentheses, and not other delimiters, such as [ ], which are often used in (noncomputer) mathematics to improve legibility.

■

CONDITIONAL EXECUTION

Now that we can specify a condition, we are ready to show how to arrange for statements to be executed subject to that condition. In MAPLE, the basic program construct for this purpose is

> `if` *condition* `then` *then-statement* `else` *else-statement* `end if;`

This command will cause *then-statement* to be executed if *condition* is true, with *else-statement* executed otherwise. If nothing is to be executed when *condition* is false, the word `else` and the *else-statement* can be omitted; the command must still terminate with `end if`. In some early computer languages a conditional command that began with `if` was ended by `fi`; MAPLE permits `fi` to be used as an alternative to `end if`. However, `fi` is regarded as obsolete and will eventually disappear from the MAPLE language, so don't use it! The quantities *then-statement* and *else-statement* can each actually be a series of statements separated by ; characters, so the alternatives for execution are not intrinsically limited in extent. If MAPLE is unable to determine whether *condition* is true, an error message is produced.

In MATHEMATICA, there is also an `If` function,

`If[`*condition*`,` *then-statement*`,` *else-statement*`]`

If *then-statement* or *else-statement* is actually a series of statements, the individual statements must be separated using the ; character; commas may **not** be used for that purpose. However, commas **must** be used to separate *condition, then-statement,* and *else-statement.* If MATHEMATICA cannot determine whether *condition* is true or false, neither the *then-* nor *else-statement* is executed, and the input is returned as the command output. If the *else-statement* is not needed, it and the comma that precedes it may be omitted.

Example 1.3.2. A Conditional Computation

We need symbolic code to do the following:

> Given a defined numerical value of x and a defined integer value of $m : m = 0,\ m = \pm 1,\ m = \pm 2,\ \ldots,$ return as output the numerical value of $f(x)$, defined as
>
> $$f(x) = \begin{cases} \cos mx, & \text{if } m \geq 0, \\ \sin |m| x, & \text{if } m < 0. \end{cases}$$
>
> Also print a line identifying which formula was used for $f(x)$.

In MAPLE, our code can take the form

```
> if (m >= 0) then
>    print("Case m >= 0");
>    R := cos(m*x)
> else
>    print("Case m < 0");
>    R := sin(-m*x)
> end if;
> R;
```

Result saved in R; then R is reiterated as last statement so the value of $f(x)$ will be printed.

Because the `if` statement has a null value this strategy is needed to output the value of $f(x)$.

When the user types the above into a terminal, MAPLE will make all of it except the last line a single execution group, and if the user forces a break-up of the seven-line `if` command into more than one execution group an error will be created. If we want to include the final R in the same execution group (as shown here), we will need to use MAPLE's `join` edit option. When the above code is executed with m previously set to 2 and x set to 0.54, the terminal will display the following:

```
"Case m >= 0"
 0.4713283642
```

This decimal number is the value of cos(1.08) (argument in radians).

Corresponding code for the same example in MATHEMATICA can be

```
If[ m >= 0,
      Print["Case m >= 0"];
        Cos[m*x],
      Print["Case m < 0"];
        Sin[-m*x]
  ]
```

See comments to MAPLE code

It is important to write the coding in a way that helps to avoid errors. In this example that was accomplished by line breaking and indentation. Note carefully the use of comma and semicolon characters.

In MATHEMATICA, `If` is a function whose value is the last command executed and is therefore either $\sin|m|x$ or $\cos mx$. When the MATHEMATICA code is executed with $m = 2$ and $x = 0.54$, there results

```
          Case m >= 0
Out[..]= 0.471328
```

If the cursor is placed on this number and `Enter` pressed, we get

```
0.47132836417373997`
```

■

CONTROL LOOPS

Repetition of a series of commands (often referred to as executing a **control loop**) is an important computational tool. We present here only the simplest loop commands. Consider first the following loop constructs, which operate similarly in both our symbolic computation languages:

MAPLE	`for` *j* `from` *a* `to` *b* `by` *s* `do` *do-statement* `end do`
MATHEMATICA	`Do[`*do-statement*`, {`*j*, *a*, *b*, *s*`} ]`

Here j, a, b, and s are arbitrary, but a, b, and s must evaluate to numerical values which need not be integers. When this construct is executed, j, whether or not it has previously been defined, is set to a, and then (assuming $s > 0$) the following three steps are taken repeatedly until the loop is exited:

(1) If $j > b$, exit the loop,

(2) Evaluate *do-statement*, using the current value of j,

(3) Increment j by s.

This prescription causes *do-statement* to be evaluated at the j values a, $a+s \cdots$ until j is larger than b (with no steps at all if $a > b$). If $s < 0$, *do-statement* is evaluated at a decreasing set of j values starting at a and continuing until $j < b$ (with no steps at all if $a < b$). One may omit "`by` s" or "`,` s", in which case s defaults to unity. Notice that while noninteger values of a, b, and s are permitted, extreme care should be exercised to ensure that floating-point round-off cannot affect the range of the control loop.

In MAPLE, after the loop execution is complete, j has the value of the first step **not taken**. In MATHEMATICA, j is a variable local to the `Do` procedure, so any definition previously given j is not affected. In MAPLE, an obsolete alternative to `end do` is the whimsical `od`.

A second loop construct, also of similar effect in both symbolic languages, is

MAPLE `while` *condition* `do` *do-statement* `end do`

MATHEMATICA `While[`*condition* `,` *do-statement*`]`

In this loop construct, the following two steps are taken until exit occurs:

(1) If *condition* is not `true`, exit the loop,

(2) Evaluate the *do-statement.*

Obviously this prescription is meaningful only if *do-statement* has the potential to alter the truth value of *condition*. Note that the test of *condition* takes place **before** each iteration, so *do-statement* will not be executed at all if *condition* is `false` when execution of the loop begins.

The *do-statement* in these constructions can actually be a sequence containing an unlimited number of statements. In both MAPLE and MATHEMATICA, the statements are separated by the ; character. Commas may not be used for that purpose.

It turns out to be very useful to have a means for immediate exit from a control loop or for proceeding immediately to its next iteration. Modern computer languages tend not to support commands such as `go to` (*any location in the program*). The command for immediate exit from a control loop,

`break` (MAPLE), `Break[ ]` (MATHEMATICA),

causes execution to continue from the first statement following the loop. The command for proceeding immediately to the next iteration of the loop is

`next` (MAPLE), `Continue[ ]` (MATHEMATICA).

Example 1.3.3. Some Simple Program Loops

As an initial example, suppose we want to compute

$$S_n = \sum_{j=1}^{n} j^2$$

for various values of n and did not happen to know that $S_n = n(2n+1)(n+1)/6$. A MAPLE PROGRAM to do this could be the following (assuming n has been defined to have a positive integer value):

```
> S := 0;                         Initialize accumulation cell S
> for j from 1 to n do            Take n steps
>    S := S + j^2 end do:         Increment S; terminate with : to avoid
                                  getting output at every step
> S;                              Final value is now output
```

We have caused all this code to be in a single execution group.

MATHEMATICA code for the same problem can take the form

```
In[1]:= s = 0;                    Use lower-case accumulation cell name
                                  Use ; character so zero will not print
In[2]:= Do[s = s + j^2, {j,1,n}]
                                  Do command has no value;
In[3]:= s                         repeat s to print result
```

A second example is the evaluation of

$$\sum_{n=1}^{\infty} \frac{1}{n^4}$$

in the approximation that we cut off the sum at the smallest n value such that $1/n^4$ is smaller than a small number t. If summed to infinity, the exact value of this summation is $\pi^4/90$.

MAPLE code for this example, assuming that t has previously been assigned a numerical value, could be

```
> S := 0:                         Initialize S and n
> n := 1:
> while (1/n^4) >= t do           Iterate while condition satisfied
>    S := S + 1./n^4;             Increment S (as decimal) and n
>    n := n + 1 end do:           Use colon so steps will not print

> print("Largest n used",n-1):    n was incremented after last use
> S;                              Final result will print
```

In MATHEMATICA, this example can be coded

```
In[1]:= s = 0;                    Use lower-case variables
In[2]:= n = 1;                    Initialize s and n
In[3]:= While[(1/n^4) >= t,       Iterate while condition satisfied
              s = s + 1./n^4;     Note: s forced to be decimal
              n = n + 1
        ]                         While has no output value
In[4]:= Print["Largest n used ", n-1]   n was incremented after last use
In[5]:= s                         Final result will be in Out[5]
```

The only output (other than the print line) will be Out[5].

Let's now repeat this second example, but this time using the break command instead of the while construction. Possible MAPLE code (again with t previously assigned a numerical value) is

```
> S := 0:
> for n from 1 to 1000000 do
>   a := 1./n^4;
>   if (a < t) then break end if
>   S := S + a  end do
> if (n > 1000000) then print("Series did not converge")
>   else print("Largest n used ",n-1);  S  end if;
```

In MATHEMATICA, with t previously assigned a numerical value,

```
In[1]:= s = 0;
In[2]:= Do[a = 1./n^4;  If[a < t, Break[ ] ];  s = s + a,
            {n,1,1000000}
           ]
In[3]:= If[n > 1000000, Print["Series did not converge"],
              Print["Largest n used ", n-1];  s
           ]
```

This alternative coding provides an upper limit (1,000,000) to the number of terms taken in an attempt to reach convergence.

Finally, we illustrate use of the command next or Continue. Given the definition

$$a_n = \int_0^\pi \sin nx \, e^{x/\pi} \, dx \, ,$$

we wish to know how many of the first 1000 coefficients a_n exceed (in absolute value) the threshold value 0.01. One way to code this in MAPLE is

```
> m := 0:                                        initialize counter
> for n from 1 to 1000 do                        loop over values of n
>   V := evalf(int(sin(n*x)*exp(x/Pi),           compute a_n; call it V
>              x=0 .. Pi));
>   if(abs(V) < 0.01) then next end if;          test V; if too small, go to next n
>   m := m + 1    end do:                        if V large enough, increment m
> m;                                             print m after looping complete
```

Here abs(V) is the absolute value of V.

In MATHEMATICA, this problem can be coded

```
In[1]:= m = 0;                                     initialize counter
In[2]:= Do[v = Integrate[Sin[n*x]*Exp[x/Pi],       loop over values of n
                    {x,0,Pi}];                       compute a_n, call it v
          If[Abs[N[v]] < 0.01, Continue[ ] ];      test v, if small go to next n
          m = m + 1,                                 if v not small increment m
          {n,1,1000}                               loop is for n from 1 to 1000
         ]                                         Do command has no output
In[3]:= m                                          print result
```

Here Abs[v] is the absolute value of v.

■

Exercises

Carry out these exercises using the symbolic computation system of your choice.

1.3.1. Code the problem of Example 1.3.2 and execute it for a variety of values of m and x. Verify that your code performs as expected.

1.3.2. Write code that examines the values of two numerical quantities x and y, and produces as output one of the strings `Case I` or `Case II`, whichever applies. Case I is when $x + y > 10$ and $0 < xy < 3$; Case II occurs otherwise.

1.3.3. Code the first two problems of Example 1.3.3 and execute them with a variety of input parameters (n and t). For the first problem, verify that your code yields the exact values given by the formula that was presented as its solution. For the second problem, see how rapidly the finite summations converge toward the analytical result.

Note. The fact that individual terms get small is not a guarantee that a series converges. In the present case, the series is convergent.

1.3.4. Determine what happens if, in the second problem of Example 1.3.3, the decimal point is removed from the $1./n^4$ in the fourth line of the coding. Explain the reason for what you observe.

1.3.5. Write code that adds members of the series

$$1 + \frac{1}{2} + \frac{1}{3} + \frac{1}{4} + \cdots$$

until the sum has reached a value exceeding 10, then giving as output the number of addends used.

1.3.6. Write code that, by testing successive integers, finds the largest integer whose square root is less than or equal to an input number `n`:

(a) Using the construction `while` or `While`,

(b) Using the command `break` or `Break`.

1.3.7. Given the definition $f(n) = \sin(0.27n)$, write code using the command `next` or `Continue` that will determine the number of integer n values in the range $[1, 1000]$ for which $|f(n)| > 0.4$.

1.4 PROCEDURES

Programming can be carried out either by writing sequences of commands into a workspace or by the definition of **procedures** which define command sequences to be carried out when called. A procedure is a programming construct that can have **arguments** and which produces specific results that depend upon the values its arguments possess when the procedure is executed. Functions are procedures which generate as output the value of the function for the current input; for example, the procedure `sin` (MAPLE) or `Sin` (MATHEMATICA), which has a single argument, produces the value of $\sin x$ when the expression `sin(x)` (in MAPLE) or `Sin[x]` (in MATHEMATICA) is executed. However, procedures can be more general than mathematical functions; a procedure call can cause the generation of multiple items of data, the output of newly created or previously existent data, the generation of plots or other graphic objects, or the carrying out of any action supported by the capabilities of the computer system. Many procedures for general use have been written by symbolic software developers, but procedures can also be written by individual users to encapsulate programming units for their own use or for sharing among interested parties.

It is important to realize that definition of a procedure does not cause any computation to take place. The definition simply establishes a rule by which computations take place when the procedure is called. Our symbolic computation languages enable procedures to be defined by the following coding:

MAPLE
```
> proc-name := proc(proc-args) local local-vbls;
>    proc-body
> end proc;
```

MATHEMATICA
```
proc-name[proc-args] := Module[{local-vbls},proc-body]
```

Here *proc-name* is a name that meets the symbolic system's naming conventions, and *proc-args* is a comma-separated list of the procedure's arguments. The notation *proc-body* denotes a sequence of statements, separated by the `;` character if there is more than one statement; commas may **not** be used for this purpose. Ordinarily *proc-body* will depend upon the procedure's arguments, but may also depend upon other variables, identified as *local-vbls*, that are defined and used only within the procedure. By listing the local variables, we assure that they have meaning only within the procedure and are not identified with variables that may have the same name(s) that are used elsewhere. This feature permits procedures to be written independently of the programs within which they may be used. The local variables are written as a comma-separated list (in any order); if no local variables are needed, we may omit `local` *local-vbls* (but not the `;` that follows) from the MAPLE procedure definition or, in the MATHEMATICA definition, replace the entire `Module` construct by (*proc-body*) (the enclosing parentheses are obligatory here if *proc-body* is more than one statement). Finally, only in MATHEMATICA, each argument in *proc-args* must have an underscore attached at its end; no such attachment is applied when the arguments occur within *proc-body*.

It is important to realize that executing a command such as

`> f(x) := x*sin(x):` (MAPLE) or `f[x] = x*Sin[x];` (MATHEMATICA)

does not establish $f(x)$ as a procedure defining a function. In both languages the above command treats f as a tabulated quantity with an entry identified by x. Thus `f(y)` or `f[y]` has no meaning unless it has also been defined separately.

The somewhat detailed specifications for procedures are now illustrated with several examples.

Example 1.4.1. A Procedure

Let's turn the first problem of Example 1.3.3 into a procedure. We decide to name our procedure `squaresum`; it will have one argument, `n`. It does not matter whether there is a variable called n in the coding within which `squaresum` will be used; the only purpose of the argument is to show how to use the actual argument when the procedure is executed. The value returned when a procedure is called is the value (if any) of the last statement executed within the procedure. Depending upon the program flow, this may or may not be the last item in the procedure coding. Note also that once a procedure has been defined, it remains available for use until the end of the MAPLE or MATHEMATICA session unless its name has been undefined or defined to be something else.

In MAPLE, our procedure is

```
> squaresum := proc(n) local j,S;
>    S= 0;
>    for j from 1 to n do
>       S= S + j^2 end do;
>    S;
> end proc;
```

squaresum must start with a letter, may continue with letters or numbers. j and S are defined as local to avoid collisions with other programming. Otherwise, same as in Example 1.3.3

Notice that MAPLE puts the entire procedure into a single execution group. This means that modification of a procedure definition will require the steps needed to make changes within an execution group. An error will result if this grouping is altered. Because we terminated the procedure definition with a semicolon, MAPLE will print its program listing; to save space we have omitted this output.

To use the present procedure (for $n = 4$), we simply type (anywhere a symbol can be used)

```
> squaresum(4);
                    30
> 4*squaresum(4)/squaresum(2);
                    24
```

In MATHEMATICA, we write, noting that the equals sign in the procedure definition is `:=`, not just `=`,

```
In[1]:= squaresum[n_] :=
            Module[{j,s},
              s = 0;
              Do[s=s+j^2,{j,1,n}];
              s  ]
```

Procedure name starts lower-case
Underscore added to n on left but not on right. Made j and s local. Otherwise, like Example 1.3.3

In MATHEMATICA, a procedure definition produces no output. To use this procedure, we simply type its name, with the desired argument (**without the underscore**). For example,

```
In[2]:= squaresum[4]
Out[2]= 30
In[3]:= 4*squaresum[4]/squaresum[2]
Out[3]= 24
```

■

Example 1.4.2. A Procedure of Convenience

One may want to define a procedure that uses neither conditional execution nor control loops just to avoid the repeated typing of complicated input. For example, a student may have a frequent need to use a function $g(x)$, defined on the range

$-1 \le x \le 1$, given by the formula

$$g(x) = e^{-x^2}\,\frac{x^2 + \cos(\pi x/2) - 0.05\,x + 1}{1 + \cos^2(\pi x/2)} \,.$$

To avoid having to retype or copy the right-hand side of this equation repeatedly for various values of x, she defines (in MAPLE)

```
g := proc(x) local C;
        C := cos(Pi*x/2);
        exp(-x^2)*(x^2 + C - 0.05*x + 1)/(1 + C^2);
end proc:
```

Use of a colon to terminate the procedure keeps MAPLE from printing its program listing.

We can use our procedure g whether or not its argument has been assigned a numerical value. Examples:

```
> g(0);
```

$$1$$

```
> g(x+1);
```

$$e^{-(x+1)^2}\,\frac{(x+1)^2 + \cos(\pi(x+1)/2) - 0.05\,x + 0.95}{1 + \cos(\pi(x+1)/2)^2}$$

If our student was a MATHEMATICA user, she would have defined

```
In[1]:= g[x_] := Module[{C},C = Cos[Pi x/2];
                        Exp[-x^2]*(x^2 + C - 0.05x + 1) / (1+C^2) ]
```

Execution of this command produces no output line. Notice that the two statements in the procedure body are separated by a semicolon, not a comma.

Use of this procedure is illustrated here:

```
In[2]:= g[0]
```

Out[2] $= 1$

```
In[3]:= g[x+1]
```

Out[3] $= e^{-(1+x)^2}\,\dfrac{1 - 0.05(1+x) + (1+x)^2 + \mathrm{Cos}[\pi(1+x)/2]}{1 + \mathrm{Cos}[\pi(1+x)/2]^2}$

■

Example 1.4.3. Double Factorial Notation

For positive integers, the notation $n!$ (called **n factorial**) stands for the product of the first n integers (for example, $3! = 1 \cdot 2 \cdot 3$). For reasons that we discuss later, $0!$ is defined to be unity and the factorials of negative integers are undefined. The notation $n!$ is known to both MAPLE and MATHEMATICA.

Sometimes it is useful to introduce symbols which stand for the products of either even or odd integers; in particular

$$(2n)!! \equiv 2 \cdot 4 \cdots (2n) \,, \tag{1.1}$$

$$(2n+1)!! \equiv 1 \cdot 3 \cdots (2n+1) \,. \tag{1.2}$$

We supplement the above with $0!! = (-1)!! = 1$. These notations are often referred to as **double factorials**, and are **not** known to MAPLE or MATHEMATICA. The objective of this Example is to present a procedure `dfac(n)` which will work for all integer $n \geq -1$, both even and odd.

In MAPLE, we may write

```
> dfac := proc(n) local s,j;
>   s := 1; j:= n;
>   while (j > 0) do
>     s := s*j;  j := j - 2;  end do;
>   s;
> end proc;
```

MAPLE users should input the above and make checks for both even and odd choices of n.

In MATHEMATICA,

```
In[1]:= dfac[n_]:= Module[{s,j,t},
                   s = 1;  j = n;
                   While[j > 0, s = s * j; j = j - 2]; s]
```

MATHEMATICA users should input and test this procedure.

These procedure implementations are not regarded as ideal from a programming viewpoint, as they do not guard against situations in which the procedure is called with noninteger n or with $n < -1$. Readers desiring more robust procedure definitions can insert additional statements that test whether n is an integer (your symbolic language documentation shows how to do this).

■

There may be circumstances where one wants to define a procedure that has no arguments. In MAPLE, one leaves the argument list empty, and writes *procname* `:= proc( );`. The empty argument list must be included when the procedure is called, e.g.,

> *result* := *procname*( );

In MATHEMATICA, one omits the entire construction [*proc-args*] from the procedure definition and calls the procedure as *procname* only.

```
In[1]:= y = 4;
In[2]:= silly := 3*y          Procedure definitions produce no output
In[3]:= silly
Out[3]= 12
```

PROCEDURES INVOLVING RECURSION

Both MAPLE and MATHEMATICA permit recursively defined procedures, meaning those in which a procedure contains code referring to itself (but, of course, with different arguments). We give a simple example.

Example 1.4.4. Recursion for Factorial

One of the main features of the factorial function $n!$ is that it satisfies the **functional relation** $n! = n(n-1)!$. Because this relation is recursive, meaning that $n!$ can be defined in terms of $(n-1)!$, we can write a procedure for $n!$ that makes use of the recursivity.

Calling our procedure `fac`, in MAPLE, and assuming that input to `fac` will always be a nonnegative integer, we write

```
> fac := proc(n) local s;
>    if n=0 then s:=1 else s:=n*fac(n-1) end if;
> s; end proc:
```

In MATHEMATICA, with the same assumptions,

```
In[1]:= fac[n_] := Module[{s},If[n == 0, s=1, s=n*fac[n-1]]; s]
```

An important thing to notice here is that the recursive process does not continue indefinitely; it terminates when `fac` is evaluated with argument zero.

We note also that although the recursive definition may simplify the coding, it does not necessarily simplify the sequence of individual computational steps; the computer will still need to evaluate $n!$ by multiplying together all the integers from 1 through n. But simplifying the coding and clarifying the logic promotes understanding and has a value that should not be minimized.

This example is of value only as an illustration. Use of the built-in notation `n!` is far more convenient than calling a procedure explicitly.

■

Here is another example illustrating a less trivial application of recursion.

Example 1.4.5. Legendre Polynomials

Later in this book we will encounter a set of polynomials of successive degrees n, called the Legendre polynomials and usually denoted P_n. We find there that successive $P_n(x)$ satisfy a **recurrence relation** of the form

$$(n+1)P_{n+1}(x) - (2n+1)xP_n(x) + nP_{n-1}(x) = 0 .$$

Special cases of the P_n include $P_0(x) = 1$ and $P_1(x) = x$. It is clear that, starting from the given values of P_0 and P_1, we can use the recurrence relation to get $P_2(x)$, then P_3, and so on. As we go to larger n the algebra involved in the process gets more complicated, and it is natural to turn to a symbolic-algebra program to assist in the work.

Rearranging our recurrence relation to the form

$$P_n(x) = \frac{(2n-1)xP_{n-1}(x) - (n-1)P_{n-2}(x)}{n} ,$$

and allowing for the fact that our recurrence formula (in this form) will only be useful here when $n \geq 2$, we write the following MAPLE procedure:

```
> legenP := proc(n,x) local s;
>   if n=0 then s:=1 end if;
>   if n=1 then s:=x end if;
>   if n>1 then s:=expand(((2*n-1)*x*legenP(n-1,x)
>                         -(n-1)*legenP(n-2,x))/n) end if;
>    s; end proc:
```

The simplifying command **expand** was inserted to prevent an undesired escalation in complexity as the recursion is repeated. In MATHEMATICA, the same procedure takes the form

```
In[1]:= legenP[n_,x_] :=
            Module[ {s},
                    If[n==0,s=1];
                    If[n==1,s=x];
                    If[n>1,s=Expand[((2*n-1)*x*legenP[n-1,x]
                                     -(n-1)*legenP[n-2,x])/n]];
                    s          (* function return value *)
                  ]
```

Again, note that the recursion is finite in extent, extending for n steps. Use of the above procedure is left as an exercise.

■

After a procedure has been written and tested, the workspace containing it can be saved for reuse in future symbolic computing sessions. The procedure definition will need to be executed again in the current session to make it available for use there.

The developers of both MAPLE and MATHEMATICA have written procedures that define a wide variety of mathematical functions, and the definitions of these built-in procedures are already available at the opening of any symbolic computing session.

Exercises

Carry out these exercises using the symbolic computation system of your choice.

1.4.1. Enter the code for the procedure `dfac` of Example 1.4.3 and verify that it gives correct results for all integer n from -1 to 8.

1.4.2. Convert the second problem of Example 1.3.3 into a procedure and test its application for several precision levels t.

1.4.3. Enter the code for the procedure `legenP` of Example 1.4.5.

(a) Verify that it gives correct results by checking that

$$P_4(x) = \frac{35}{8}\,x^4 - \frac{15}{4}\,x^2 + \frac{3}{8}\,,$$

and that (for several n, some even and some odd), $P_n(1) = 1$.

(b) Remove the simplifying command **expand** or **Expand** and notice how the results are changed.

1.4.4. To see how the use of recursion in procedure definitions can make the coding cleaner, consider a procedure that computes the Legendre polynomial $P_n(x)$ without using the recursive strategy illustrated in Example 1.4.5.

Referring to that example for definitions, write a procedure `altLegP` that computes the Legendre polynomial $P_n(x)$ in the following way:

1. Letting P be the polynomial P_j in Step j, and assuming that Q and R contain the polynomials of the two previous steps, write for P the formula $P = [(2j-1)xQ - (j-1)R]/j$.
2. After making P, update Q and R by moving the contents of Q into R and those of P into Q.
3. You will need starting values of Q and R to initiate this process and must organize the coding to deal with special cases and control the number of steps to be taken.

1.4.5. Generate formulas that give $\sin nx$ and $\cos nx$ in terms of $\sin x$ and $\cos x$ by defining procedures based on the formulas

$$\sin(n+1)x = \sin nx \cos x + \cos nx \sin x \,,$$

$$\cos(n+1)x = \cos nx \cos x - \sin nx \sin x \,.$$

1.5 GRAPHS AND TABLES

Both MAPLE and MATHEMATICA permit detailed formatting of numerical data, and, as already mentioned, support plotting of functions (both as ordinary plots and as views of three-dimensional objects). For the purposes of the present text it is useful to provide a small amount of additional detail on the use of these symbolic computing capabilities.

PLOTTING MULTIPLE FUNCTIONS

Sometimes the solutions of an equation can be better understood if its left- and right-hand sides are presented on the same plot, making it obvious where the two sides coincide. As a simple example, consider the equation

$$(x-2)^2 = \frac{\tanh x}{x} \,. \tag{1.3}$$

In MAPLE, we can get both $(x-2)^2$ and $\tanh x/x$ to plot on the same graph by typing

```
> plot({(x-2)^2,tanh(x)/x},x=-3 .. 3);
```

The output of this plot command is rendered in black and white in Fig. 1.3.

The essential features of this plot would be easier to see if we had more control of the vertical range of the plot. We can accomplish this objective by using the alternate form of the plot command that includes the vertical range as a third argument. For example, if we want our vertical range to be from 0 to 2, we write

```
> plot({(x-2)^2,tanh(x)/x}, x=-3 .. 3, 0 .. 2);
```

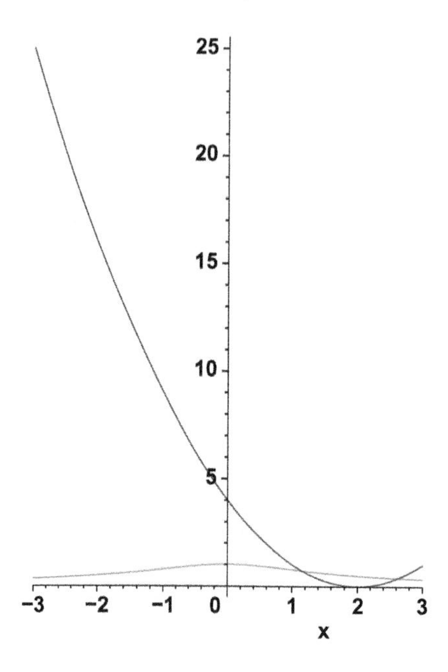

Figure 1.3: Plot of $(x-2)^2$ and $\tanh x/x$.

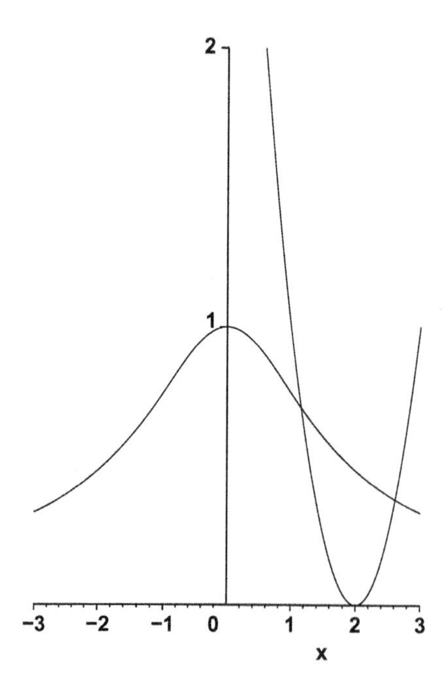

Figure 1.4: Portion of Fig. 1.3.

The result is in Fig. 1.4. It is clear that Eq. (1.3) has two solutions, both falling within the range $0 < x < 3$.

In MATHEMATICA, quantities to be included in the same plot are also enclosed in braces, with the vertical range controlled as shown:

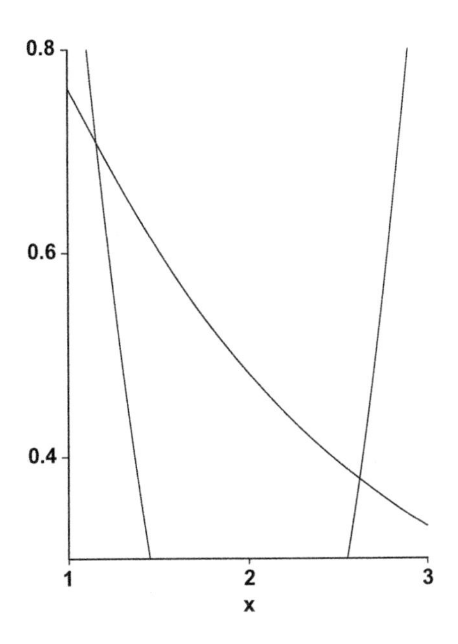

Figure 1.5: Portion of Fig. 1.4.

```
Plot[{(x-2)^2,Tanh[x]/x},{x,-3,3}, PlotRange -> {0, 2}]
```

This command produces a plot similar to that in Fig. 1.4.

Example 1.5.1. Zeroing in on Solutions

With ability to control both the horizontal and vertical range of a plot, it is now easier to get accurate values of the values x that solve Eq. (1.3). Noting from Fig. 1.4 that its two curves cross twice in the range $1 < x < 3$, we change our plot specification to

MAPLE
```
plot({(x-2)^2,tanh(x)/x}, x=1 .. 3, 0.3 .. 0.8);
```

MATHEMATICA
```
Plot[{(x-2)^2,Tanh[x]/x},{x,1,3}, PlotRange -> {0.3, 0.8}]
```

and obtain a plot similar to that in Fig. 1.5. Further range changes focused on the individual crossing points can enable their location to any desired precision.

■

TABULATING FUNCTION VALUES

Sometimes it will be useful to make tables of function values. For readability it is desirable to obtain output with consistent alignment of the data, and this is not always achieved by straightforward use of the symbolic system's print command. We provide here basic procedures in MAPLE and MATHEMATICA that will print a tabulation of an arbitrary function for a user-chosen range of its argument. These procedures and a discussion that will be sufficient to enable users to modify and generalize them are presented in Appendix B. To use the procedure for a chosen symbolic language without modification, the user needs only to paste the procedure code for that language into a worksheet or notebook, execute it to activate the procedure, and call it with the prescribed argument sequence.

Example 1.5.2. A Table of Function Values

Appendix B contains a procedure `MakeTable`, with first argument `func`, the name of the function to be tabulated. The function `func` can be either a built-in function of your symbolic system or a user-written procedure that accepts a single argument `x` and returns the corresponding function value. The remaining arguments of `MakeTable` are `x1` and `x2`, the first and last values of `x` in the tabulation, and `delta`, the spacing of the `x` values.

We illustrate this procedure by applying it to the error function (which was defined in Exercise 1.2.6 and is a function known to both MAPLE and MATHEMATICA). First, copy the code for `MakeTable` in your chosen symbolic language from Appendix B, paste it into a worksheet or notebook, and execute it (thereby causing `MakeTable` to be available for use). Then construct a table of the error function by invoking the following command:

In MAPLE, `> MakeTable(erf, 0, 4, 0.8):`

This command should produce the following output:

```
Values of erf(x) for x from 0 to 4 in steps of .8

      x           erf(x)
    0.000       0.000000E+00
    0.800       7.421010E-01
    1.600       9.763484E-01
    2.400       9.993115E-01
    3.200       9.999940E-01
    4.000       1.000000E+00
```

In MATHEMATICA, `MakeTable[Erf, 0, 4, 0.8]`

This command should produce the following output:

Values of Erf[x] for x from 0 to 4 in steps of 0.8

x	Erf[x]
0.000	0.000000
0.800	7.421010×10^{-1}
1.600	9.763484×10^{-1}
2.400	9.993115×10^{-1}
3.200	9.999940×10^{-1}
4.000	1.000000

■

Exercises

Carry out these exercises using the symbolic computation system of your choice.

1.5.1. Using the procedure for generating a table of function values (Example 1.5.2), make a table giving values of the function erfc(x) for $0 \le x \le 10$ in unit steps. This function is defined as $1 - \text{erf}(x)$ and is a known function in both MAPLE and MATHEMATICA.

1.5.2. Using the procedure `dfac` developed in Example 1.4.3, make a table of double factorials for the range $-1 \le n \le 20$.

1.5.3. By graphical methods, find the three smallest positive roots of the transcendental equation $\cot x - 0.1x = 0$ and estimate the precision of your answers.

1.5.4. Figure out how to plot three functions on the same graph, and then make a plot of $\sin x$, $2\sin 2x$, and $3\sin 3x$ on the range $0 \le x \le \pi$.

Note. By including the coefficients 2 and 3, it may make it easier to tell which plot corresponds to each function. Most versions of MAPLE and MATHEMATICA by default plot different functions in different colors, but there is nothing universal about the color assignment.

1.6 SUMMARY: SYMBOLIC COMPUTING

It is obvious from the material already presented that it will be useful for the reader to have a library of symbolic computing procedures and other pieces of symbolic code that may assist in carrying out analyses of types that will arise later in this book. The exercises at the end of this section provide a prescription for the start of such a library. The reader should make appropriate additions to his/her library as procedures of general utility are identified in later chapters. Ways of preserving and reusing coding are discussed later in this section.

When dealing with arrays of data, there may be situations where it will be useful to understand how your symbolic computation system organizes the data and provides mechanisms for accessing or changing individual data items. A survey of this topic and some commands that may be useful for dealing with data structures can be found in Appendix C.

Inside the front and back covers of this book are lists of almost all the commands used in this text in both MAPLE and MATHEMATICA. This summary by itself cannot provide everything one needs to know about either language, and it does not include all features or options of the commands it identifies. Despite its limitations it may be helpful in recalling language features that the reader cannot remember accurately.

SAVING AND REUSING SYMBOLIC CODE

Both MAPLE and MATHEMATICA provide (via their pull-down menus) mechanisms for saving entire workspaces. While these capabilities are useful (and in fact essential) for managing work that is in progress, they do not constitute the only way in which symbolic coding can be preserved for archiving or later reuse. It is also possible to move material from a workspace to a text file. Even though each symbolic system uses its own proprietary data format, material that is identified for copying (using a mouse and the cursor) is in plain text and remains in that form when it is pasted into any text file.

It is recommended that you save code by pasting into a text file rather than only saving a workspace, because your text file will remain readable even when you do not have access to the relevant symbolic system, and you are thereby also insured against the relatively remote possibility that the symbolic system's proprietary format becomes unsupported.

Once you have saved symbolic code in an external text file, two approaches are available for using it in subsequent symbolic computations. One possibility is simply to copy and then paste it into a symbolic computing session. A second approach is to use the symbolic system's ability to read the external file you want to include in your workspace.

We illustrate for MAPLE with a file named `maple_input.txt`. That file contains:

```
# Maple input for double factorial procedure.  Makes n!!
dfac := proc(n) local s,j;
  s := 1; j:= n;
  while (j > 0) do
  s := s*j; j := j - 2; end do;
  s;
end proc:
print("Results for 2, 3, and 4:", dfac(2), dfac(3), dfac(4));
```

In a MAPLE session, we write

```
> read("C:/Users/harrisfe/Desktop/maple_input.txt");
```

"Results for 2, 3, and 4"

There are several things to note about the above. First, it is necessary to provide MAPLE with an unambiguous path to the text file, which by default is assumed to be in the directory containing the MAPLE session. If that session has not been saved to memory or if it is not where the text file actually is, it will be necessary to provide a complete specification of the location of `maple_input.txt`. Next, note that even if the user is in an environment where the backslash character is used in file and folder specification (e.g., Windows), MAPLE accepts only the forward slash (solidus) for this purpose. In addition, note that the command `read` causes all the material that is read to be executed, while displaying the output (if any) but not the input. This behavior is different than the result of pasting the file contents into the worksheet; if that is done, the copied material is shown, but the code it represents is not automatically executed.

In MATHEMATICA, we illustrate with a file named `mathematica_input.txt`, containing

```
(* Mathematica input for double factorial procedure.  Makes n!! *)
dfac[n_]:= Module[{s,j,t},
s = 1; j = n;
While[j > 0, s = s * j; j = j - 2]; s]
Print["Results for 2, 3, and 4: ", dfac[2]," ",dfac[3]," ",dfac[4]]
```

In a MATHEMATICA session, we write

```
<< "C:/Users/harrisfe/Desktop/mathematica_test_input.txt"
```

Results for 2, 3, and 4: 2 3 8

There are several things to note about the above. First, it is necessary to provide MATHEMATICA with a full path to the text file; there is no useful default. Next, note that even if the user is in an environment where the backslash character is used in file and folder specification (e.g., Windows), MATHEMATICA accepts only the forward slash (solidus) for this purpose. In addition, note that the command `<<` causes all the material that is read to be executed, while displaying the output (if any) but not the input. This behavior is different than the result of pasting the file contents into the worksheet; if that is done, the copied material is shown, but the code it represents is not automatically executed.

Exercises

1.6.1. Using the symbolic computing system of your choice, open a new worksheet or notebook and, going through the earlier sections of this chapter, enter pieces of code that you anticipate might have future use. For each piece of code you add to your workspace, include comments that remind you of its purpose. When the code is that of a complete procedure, your comments should include the identification of each procedure argument and any information needed to help in reaching an understanding of the output. Once the correctness and appropriacy of a piece of code has been verified and documented, preserve it for future use. It is recommended that you do this by using copy-and-paste techniques to place the coding into a text file that can be loaded as input into a symbolic computing session. You may wish to keep a set of these text files together in a single folder (directory), each with a name that makes its content easily identifiable.

It is also possible to preserve these codings in workspaces. The main limitation of such an approach is that workspaces are written in the proprietary format of the symbolic system and will only be available when you are online in that system.

1.6.2. If you have not already included it in response to Exercise 1.6.1, add the procedure `dfac` to your preserved collection of symbolic code.

1.6.3. If you have not already included it in response to Exercise 1.6.1, add the procedure `MakeTable` to your preserved collection of symbolic code.

Additional Readings

Don, E. (2009). *Schaum's outline of Mathematica* (2nd ed.). New York: McGraw-Hill (A typical offering from Schaum's Outline Series; presents what users **must** know with examples and just enough explanation).

Garvan, F. G. (2001). *The Maple book.* New York: Chapman and Hall/CRC (Comprehensive, organized by level and subject area of mathematics, starting with high school algebra and continuing through number theory and beyond. Usable both as an introduction to MAPLE and as a reference).

Geddes, K., Labahn, G., & Monagan M. (2008). *Maple 12 introductory programming guide*, also *Maple 12 advanced programming guide*. Waterloo, ON, Canada: Maplesoft.

Meade, D. B. (2009). *Getting started with Maple* (3rd ed.). New York: Wiley.

Shingareva, I. K., & Lizárraga-Celaya, C. (2009). *Maple and Mathematica: A problem solving approach for mathematics* (2nd ed.). New York: Springer (Not recommended as a preliminary introduction, but useful for more experienced users. Contains many examples of coding in both symbolic computing languages).

Torrence, B. F., & Torrence, E. A. (2009). *The student's introduction to Mathematica: A handbook for precalculus, calculus, and linear algebra* (2nd ed.). Cambridge, UK: Cambridge University Press (A thorough, readable elementary introduction. Recommended).

Wolfram, S. (2003). *The Mathematica book* (5th ed.). Champaign, IL: Wolfram Media (General reference).

Chapter 2

INFINITE SERIES

Perhaps the most widely used technique in the toolbox of a scientist or engineer is the use of **infinite series** (i.e., sums consisting formally of an infinite number of terms) to represent functions, to bring them to forms facilitating further analysis, or even as a prelude to numerical evaluation. The acquisition of skill in creating and manipulating series expansions is therefore an absolutely essential part of the training of one who seeks competence in the mathematical methods used in modern science and engineering, and it is therefore the first mathematical topic in this text. An important part of this skill set is the ability to recognize the functions represented by commonly encountered expansions, and it is also of importance to understand issues related to the convergence of infinite series.

2.1 DEFINITION OF SERIES

The usual way of assigning a meaning to the sum of an infinite number of terms is by introducing the notion of partial sums. If we are given an infinite sequence of terms u_1, u_2, u_3, u_4, u_5,..., the ith partial sum of these terms is defined as

$$s_i = \sum_{n=1}^{i} u_n \, . \tag{2.1}$$

This is a finite summation, so its convergence is not a relevant mathematical question. We define an infinite series to be **convergent** if its partial sums s_i converge to a finite limit S as $i \to \infty$,

$$\lim_{i\to\infty} s_i = S \, . \tag{2.2}$$

The infinite series $\sum_{n=1}^{\infty} u_n$ is then assigned the value S. Note that a necessary condition for convergence to a limit is that $\lim_{n\to\infty} u_n = 0$. But this condition is not sufficient to guarantee convergence. The essential requirement is that beyond a large enough i **all** the partial sums approach arbitrarily closely a finite limit.

Some series **diverge** because the sequence of partial sums approaches $\pm\infty$; others diverge because their partial sums oscillate between two values, as for example

$$\sum_{n=1}^{\infty} u_n = 1 - 1 + 1 - 1 + 1 - \cdots - (-1)^n + \cdots \, .$$

Mathematics for Physical Science and Engineering.
http://dx.doi.org/10.1016/B978-0-12-801000-6.00002-X

It is important to be able to determine whether, or under what conditions, a series we would like to use is convergent.

Example 2.1.1. The Geometric Series

The geometric series, starting with $u_0 = 1$ and with further terms formed by successive multiplications by a fixed number r, is

$$1 + r + r^2 + r^3 + \cdots + r^{n-1} + \cdots .$$

Its nth partial sum s_n(that of the first n terms) is

$$s_n = \frac{1 - r^n}{1 - r} . \tag{2.3}$$

A simple way to establish Eq. (2.3) is to multiply s_n by $1 - r$:

$$\begin{aligned}(1 - r)s_n &= (1 - r)(1 + r + r^2 + \cdots + r^{n-1})\\ &= 1 + r + r^2 + \cdots + r^{n-1}\\ &\quad - r - r^2 - \cdots - r^{n-1} - r^n\\ &= 1 - r^n .\end{aligned}$$

Restricting attention to $|r| < 1$, so that for large n, r^n approaches zero, s_n possesses the limit

$$\lim_{n\to\infty} s_n = \frac{1}{1 - r} , \tag{2.4}$$

showing that for $|r| < 1$, the geometric series converges. It clearly diverges for $|r| \geq 1$, as the individual terms do not then approach zero at large n.

■

Example 2.1.2. The Harmonic Series

The **harmonic series** has definition

$$\sum_{n=1}^{\infty} \frac{1}{n} = 1 + \frac{1}{2} + \frac{1}{3} + \frac{1}{4} + \cdots + \frac{1}{n} + \cdots . \tag{2.5}$$

The terms approach zero for large n, i.e., $\lim_{n\to\infty} 1/n = 0$, but that is not sufficient to guarantee convergence. If we group the terms (without changing their order) as

$$1 + \frac{1}{2} + \left(\frac{1}{3} + \frac{1}{4}\right) + \left(\frac{1}{5} + \frac{1}{6} + \frac{1}{7} + \frac{1}{8}\right) + \left(\frac{1}{9} + \cdots + \frac{1}{16}\right) + \cdots ,$$

each pair of parentheses encloses p terms of the form

$$\frac{1}{p+1} + \frac{1}{p+2} + \cdots + \frac{1}{p+p} > \frac{p}{2p} = \frac{1}{2} .$$

Forming partial sums by taking increasing numbers of parenthesized groups, we find

$$s_1 = 1, \quad s_2 = \frac{3}{2}, \quad s_3 > \frac{4}{2}, \quad s_4 > \frac{5}{2}, \ldots, \quad s_n > \frac{n+1}{2},$$

and we are forced to the conclusion that the harmonic series diverges.

Although the harmonic series diverges, its partial sums have relevance among other places in number theory, where $H_n = \sum_{m=1}^{n} m^{-1}$ are sometimes referred to as **harmonic numbers**.

■

SYMBOLIC COMPUTATION OF SUMS

If f is an expression whose value for integer m reduces to the **numerical value** of the term u_m of the summation

$$S = \sum_{m=n}^{N} u_m ,$$

then S is given in MAPLE as `sum(f,m = n..N)` or `add(f,m = n..N)` and in MATHEMATICA as `Sum[f,{m,n,N}]`. In MAPLE, one should use `sum` if one seeks the formal evaluation of an infinite series; otherwise it is better to use `add`. If the upper limit of a sum is infinity, N is replaced by `infinity` (MAPLE) or `Infinity` (MATHEMATICA). Those languages will return the value of the summation if they know how to obtain it, including a report that the value is infinite when that is known to be the case. If the summation can neither be evaluated nor determined to be divergent, an unevaluated expression involving a summation sign is returned.

If the sum of a finite number of terms is requested and decimal evaluation is not forced, the result may appear as a fraction in which the numerator and denominator might be such large integers that the meaning of the result may be rather obscure.

Example 2.1.3. Symbolic Coding

The output of each symbolic statement is shown to its right. Expressions involving `sum` or `add` are in MAPLE; those involving `Sum` are in MATHEMATICA.

$\sum_{n=1}^{\infty} \frac{(-1)^{n+1}}{n}$ ⟹ `sum((-1)^(n+1)/n,n=1..infinity);` $\ln(2)$

⟹ `Sum[(-1)^(n+1)/n,{n,1,Infinity}]` Log[2]

$\sum_{n=1}^{40} \frac{(-1)^{n+1}}{n}$ ⟹ `Sum[(-1)^(n+1)/n,{n,1,40}]` $\frac{3637485804655193}{5342931457063200}$

⟹ `add((-1.)^(n+1)/n,n=1 .. 40);` 0.6808033818

$\sum_{n=1}^{\infty} \sin(n\sqrt{2})$ ⟹ `Sum[Sin[Sqrt[2]*n],{n,1,Infinity}]` $\sum_{n=1}^{\infty}$ Sin[$\sqrt{2}$ n]

■

Exercises

2.1.1. A repeating decimal is equivalent to a geometric series. For example, $0.909090\cdots = 0.90+(0.01)(0.90)+(0.01)^2(0.90)+\cdots$. Using this information, find the fractions equivalent to

(a) $0.909090\cdots$ (b) $0.243243243\cdots$ (c) $0.538461538461\cdots$

2.1.2. We may encounter series that are summable to give an analytical result but we do not recognize them as known summations. Sometimes a symbolic computation system can evaluate the sum.

Write code that will form $\sum_{n=1}^{\infty} u_n$, and see which of the following summations your symbolic computation system can evaluate.

(a) $1 - \frac{1}{3} + \frac{1}{5} - \frac{1}{7} + \cdots$ (d) $1 - \frac{1}{2} + \frac{1}{3} - \frac{1}{4} + \cdots$

(b) $1 + \frac{1}{4} + \frac{1}{9} + \frac{1}{16} + \cdots$ (e) $1 + \frac{1}{3} + \frac{1}{5} + \frac{1}{7} + \cdots$

(c) $1 - \frac{1}{4} + \frac{1}{9} - \frac{1}{16} + \cdots$ (f) $1 - \frac{1}{4} + \frac{1}{7} - \frac{1}{10} + \cdots$

2.1.3. Obtain (to six significant figures) a numerical value for $\sum_{n=1}^{\infty} \frac{\sin n}{n^3}$.

Note. Here (and in most mathematical contexts) angles are in radians.

Hint. Do not panic if your answer is an unfamiliar function or something that is not useful. If all else fails, approach your answer through large finite N.

2.1.4. Partial fraction decompositions (see Appendix E) are sometimes helpful in evaluating summations. Apply the technique to show that the following summation converges and has the indicated value.

$$\sum_{n=1}^{\infty} \frac{1}{n(n+1)} = 1\,.$$

2.1.5. Another technique for establishing the value of a summation (most useful if you know or suspect the value and need to confirm it) is the method of **mathematical induction** (see Appendix F). Prove that

$$\sum_{n=1}^{\infty} \frac{1}{(2n-1)(2n+1)} = \frac{1}{2}$$

in the following way: first, use mathematical induction to show that the partial sum (through $n = m$) has the value $s_m = m/(2m+1)$. Then take the limit of s_m as $m \to \infty$.

Note. The result is easily obtained from a partial fraction decomposition, but this problem is specifically an exercise in the use of mathematical induction.

2.2 TESTS FOR CONVERGENCE

We now turn to a more detailed study of the convergence and divergence of series, considering here series all of whose terms are positive. Series with terms of both signs are treated later.

COMPARISON TEST

If term-by-term a series of terms u_n satisfies $0 \le u_n \le a_n$, where the a_n form a convergent series, then the series $\sum_n u_n$ is also convergent.

Letting s_i and s_j be partial sums of the u series, with $j > i$, the difference $s_j - s_i$ is $\sum_{n=i+1}^{j} u_n$, and this is smaller than the corresponding quantity for the a series, thereby proving convergence.

If term-by-term a series of terms v_n satisfies $0 \le b_n \le v_n$, where the b_n form a divergent series, then $\sum_n v_n$ is also divergent.

We already have two series that can be used in comparison tests. For the convergent series a_n we can use the geometric series, while the harmonic series can be used as a divergent comparison series b_n. We later identify additional convergent and divergent series that may then be used as known series for comparison tests.

Example 2.2.1. A Divergent Series

Test for convergence $\sum_{n=1}^{\infty} u_n$, where $u_n = n^{-p}$, $p = 0.999$. Let's compare this series with the harmonic series, which is divergent and with terms $b_n = n^{-1}$. We note that for all n $u_n > b_n$, so that the comparison test indicates $\sum_n n^{-0.999}$ is divergent. In fact, $\sum_n n^{-p}$ is divergent for all $p \le 1$.

■

RATIO TEST

If $u_{n+1}/u_n \le r < 1$ for all sufficiently large n and r is independent of n, then $\sum_n u_n$ is convergent. If $u_{n+1}/u_n \ge 1$ for all sufficiently large n, then $\sum_n u_n$ is divergent.

The convergence case of this test can be established by direct comparison with the geometric series $(1 + r + r^2 + \cdots)$. For the divergence case, $u_{n+1} \ge u_n$ and the terms do not approach zero at large n. The ratio test is one of the easiest to apply and is widely used. An alternate statement of the ratio test is in the form of a limit:

$$\text{If} \quad \lim_{n\to\infty} \frac{u_{n+1}}{u_n} \begin{cases} < 1, & \text{convergence}, \\ > 1, & \text{divergence}, \\ = 1, & \text{indeterminate}. \end{cases} \tag{2.6}$$

Because of the indeterminacy at $u_{n+1}/u_n \to 1$, the ratio test often fails for important series, and other tests then become necessary. Note that the convergence case implies that there exists a constant r such that for large enough n, $u_{n+1}/u_n \le r < 1$. This is a stronger condition than $u_{n+1}/u_n < 1$ for all **finite** n. We already have, in the harmonic series, $u_{n+1}/u_n < 1$, but there is no $r < 1$ such that for all finite n $u_{n+1}/u_n < r$. The harmonic series, which is divergent, is therefore a case where the ratio test fails to establish convergence or divergence.

Example 2.2.2. Ratio Test

Test $\sum_{n=1}^{\infty} n/2^n$ for convergence. Applying the ratio test,

$$\frac{u_{n+1}}{u_n} = \frac{(n+1)/2^{n+1}}{n/2^n} = \frac{1}{2}\,\frac{n+1}{n}\,.$$

Since

$$\frac{u_{n+1}}{u_n} \le \frac{3}{4} \quad \text{for } n \ge 2,$$

we have convergence.

Note that for $n = 1$ we have $u_2/u_1 = 1$, but this is irrelevant since the convergence or divergence of the series depends upon the limiting behavior for large n, not on the behavior for any finite number of initial terms.

■

Example 2.2.3. Analyzing Series

Here are some examples that illustrate techniques for establishing the limiting behavior of successive terms of a series.

1. Consider the series $\sum_{n=1}^{\infty} u_n = \sum_{n=1}^{\infty} \frac{n^2}{2^n}$.

The first few u_n are $u_1 = 1/2$, $u_2 = 1$, $u_3 = 9/8$, which might suggest that the terms are increasing in size. A better approach would be to compare the logarithms of the numerator and denominator: $\ln n^2 = 2 \ln n$, while $\ln 2^n = n \ln 2$. Since it is obvious that in the limit of large n we will have $n \gg \ln n$, we see that the terms of this series tend toward zero. To determine convergence, we proceed to the ratio of successive terms:

$$\frac{u_{n+1}}{u_n} = \frac{1}{2}\left(\frac{n+1}{n}\right)^2 .$$

If $n \ge 4$, $[(n+1)/n]^2 \le 25/16$, which is less than 2. The series therefore converges.

2. Now consider $\sum_{n=1}^{\infty} u_n = \sum_{n=1}^{\infty} \frac{\sqrt{n^2+n-1}}{n^2-n+5}$.

When n becomes large, the numerator of u_n approaches $\sqrt{n^2} \approx n$, while the denominator approaches n^2. This suggests (but does not prove) that the series might diverge. Based on that observation, we look for a comparison series whose terms are smaller than u_n. We notice that dropping the $n-1$ within the numerator of u_n decreases the size of the numerator for all $n > 1$, and that dropping the $-n+5$ in the denominator would, for $n > 5$, increase the denominator. Each of these changes is in the direction of making the resulting term (which we call v_n) smaller than u_n. If we make both changes we would have a series $v_n = 1/n$ with the property that, for $n > 5$, $v_n < u_n$. Since the v_n describe the divergent harmonic series, the u_n series must also diverge.

■

CAUCHY INTEGRAL TEST

This test compares a series with an integral. The test compares the area of a series of unit-width rectangles with the area under a curve.

Choose $f(x)$ to be a continuous, **monotonic decreasing function** in which $f(n) = u_n$ and $f(0)$ is finite. The Cauchy integral test states that

$\sum_n u_n$ converges if $\int_1^\infty f(x)\,dx$ is finite and diverges if the integral is infinite.

To prove the validity of the test, we start by noting that the ith partial sum of the series is

$$s_i = \sum_{n=1}^{i} u_n = \sum_{n=1}^{i} f(n)\ .$$

But, because $f(x)$ is monotonic decreasing, see Fig. 2.1(a),

$$s_i \geq \int_1^{i+1} f(x)\,dx\ . \tag{2.7}$$

On the other hand, as shown in Fig. 2.1(b),

$$s_i \leq \int_0^{i} f(x)\,dx\ . \tag{2.8}$$

Taking the limit as $i \to \infty$, we have

$$\int_1^\infty f(x)\,dx \leq \sum_{n=1}^{\infty} u_n \leq \int_0^\infty f(x)\,dx\ . \tag{2.9}$$

This equation shows that the infinite series converges or diverges as the corresponding integral converges or diverges.

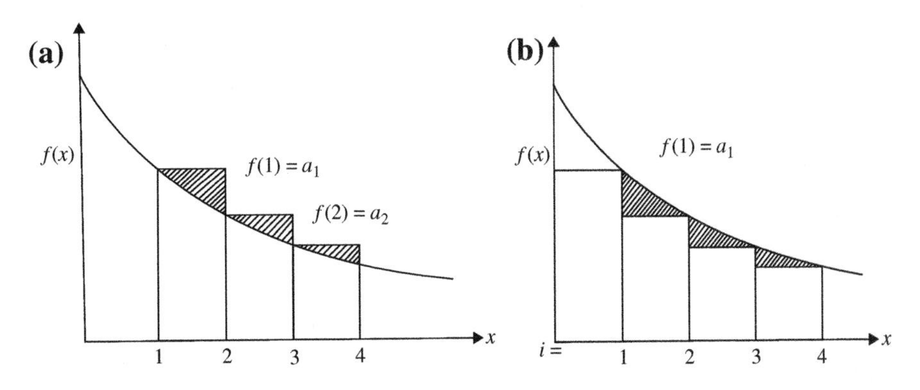

Figure 2.1: (a) Comparison of integral and sum-blocks leading. (b) Comparison of integral and sum-blocks lagging.

If $f(x)$ is singular or not monotonic for a finite range of x, the integral test can be applied to the portion of a series beyond some starting value n_0. The behavior of any finite number of leading terms is irrelevant to the convergence of the series; convergence or divergence is determined entirely by the behavior in the limit of large n. This integral test is often used to set upper and lower bounds on the remainder of a series after some number of initial terms have been summed. That is,

$$\sum_{n=1}^{\infty} u_n = \sum_{n=1}^{N} u_n + \sum_{n=N+1}^{\infty} u_n \,, \tag{2.10}$$

and

$$\int_{N+1}^{\infty} f(x)\, dx \le \sum_{n=N+1}^{\infty} u_n \le \int_{N}^{\infty} f(x)\, dx \,. \tag{2.11}$$

Example 2.2.4. Riemann Zeta Function

The **Riemann zeta function** has definition

$$\zeta(p) = \sum_{n=1}^{\infty} n^{-p} \tag{2.12}$$

for all p such that the series converges. Let's apply the Cauchy integral test to determine the range of p for which Eq. (2.12) applies. Take $f(x) = x^{-p}$, and note that

$$\begin{aligned} \int_1^{\infty} x^{-p}\, dx &= \left.\frac{x^{-p+1}}{-p+1}\right|_{x=1}^{\infty}, \qquad & p \ne 1, \\ &= \ln x \Big|_{x=1}^{\infty}, \qquad & p = 1 \,. \end{aligned}$$

The integral and therefore also the series are divergent for $p \le 1$, and convergent for $p > 1$. The present analysis also shows that the harmonic series ($p = 1$) diverges logarithmically.

■

The fact that the harmonic series diverges logarithmically suggests that the combination

$$\gamma = \lim_{n\to\infty} \left(\sum_{m=1}^{n} m^{-1} - \ln n \right) \tag{2.13}$$

might approach a definite limit. This is indeed the case, and the expression designated as γ in Eq. (2.13) is an important quantity known as the **Euler-Mascheroni constant.**

The constant γ is known to both MAPLE and MATHEMATICA:

γ :   `gamma` (MAPLE),   `EulerGamma` (MATHEMATICA).   (2.14)

To see the numerical value of this constant, invoke decimal evaluation:

```
> gamma;                          γ
> evalf(gamma);                   0.5772156649
N[EulerGamma,20]                  0.57721566490153286061
```

Example 2.2.5. Euler-Mascheroni Constant

Consider a sequence of partial sums s_n of the form

$$s_n = \sum_{m=1}^{n} m^{-1} - \ln n\,. \tag{2.15}$$

We wish to show that the s_n approach a limit, by which we mean that if integers p and q have arbitrary values larger than a sufficiently large n, the difference $s_p - s_q$ can be made arbitrarily small.

Assuming $p > q$, we form

$$s_p - s_q = \sum_{m=q+1}^{p} m^{-1} - \ln p + \ln q$$

and apply the inequalities of Eqs. (2.7) and (2.8) to the summation:

$$\int_{q+1}^{p+1} x^{-1}\,dx - \ln p + \ln q \le s_p - s_q \le \int_{q}^{p} x^{-1}\,dx - \ln p + \ln q\,. \tag{2.16}$$

Evaluating the integrals, Eq. (2.16) reduces to

$$\ln\left(\frac{p+1}{p}\right) - \ln\left(\frac{q+1}{q}\right) \le s_p - s_q \le 0\,; \tag{2.17}$$

the left member of this inequality approaches zero as p and q become large (even if $p - q$ is also large), thereby showing that s_p and s_q become equal for large p and q and that the limit γ exists.

The limit is approached exceedingly slowly if we attempt to reach it by direct evaluation of Eq. (2.13) for various values of n. An accurate value of γ, obtained by other means, is $0.57721\ 56649\cdots$, while $s_{10,000}$ of Eq. (2.15) has the value $0.57726\cdots$, and billions of terms would be needed to obtain γ to ten significant figures!

■

Example 2.2.6. A Slowly Diverging Series

Consider now the series

$$S = \sum_{n=2}^{\infty} \frac{1}{n \ln n}\,.$$

We form the integral

$$\int_2^{\infty} \frac{1}{x \ln x}\,dx = \int_{x=2}^{\infty} \frac{d \ln x}{\ln x} = \ln \ln x \Big|_{x=2}^{\infty},$$

which diverges, indicating that S is divergent. Notice that the lower limit of the integral is in fact unimportant so long as it does not introduce any spurious singularities, as it is the large-x behavior that determines the convergence. Because $n \ln n > n$, the divergence is slower than that of the harmonic series. But because $\ln n$ increases more slowly than n^ε, where ε can have an arbitrarily small positive value, we have divergence even though the series $\sum_n n^{-(1+\varepsilon)}$ converges.

■

Additional and more powerful methods for determining the convergence of infinite series are known. For further discussion of such methods, the reader is referred to Arfken et al. in the Additional Readings.

Exercises

2.2.1. Prove that

(a) If a series $\sum_{n=1}^{\infty} u_n$ converges, so also does $\sum_{n=1}^{\infty} k u_n$, where k is any constant.

(b) If a series $\sum_{n=1}^{\infty} u_n$ diverges, so also does $\sum_{n=1}^{\infty} k u_n$, where k is any constant.

(c) If a series $\sum_{n=n_0}^{\infty} u_n$ converges, so also does $\sum_{n=n_1}^{\infty} u_n$, where $n_1 > n_0$.

2.2.2. Prove the following **limit tests** for a series of terms u_n:

(a) If, for some $p > 1$, $\lim_{n\to\infty} n^p u_n < \infty$, then $\sum_n^{\infty} u_n$ converges.

(b) If $\lim_{n\to\infty} n u_n > 0$, then $\sum_n^{\infty} u_n$ diverges.

2.2.3. (a) Show that the series $\displaystyle\sum_{n=2}^{\infty} \frac{1}{n\,(\ln n)^2}$ converges.

(b) Write a symbolic computation program that will evaluate

$$\sum_{n=2}^{1000} \frac{1}{n\,(\ln n)^2},$$

and, using Eq. (2.11), make an estimate of the sum of the infinite series. Determine the accuracy you can guarantee for your estimate.

2.2.4. Determine whether each of the following series converges.

(a) $\displaystyle\sum_{n=2}^{\infty} (\ln n)^{-1}$ (b) $\displaystyle\sum_{n=0}^{\infty} \frac{1}{n+3}$ (c) $\displaystyle\sum_{n=0}^{\infty} \frac{1}{2n-1}$

(d) $\displaystyle\sum_{n=2}^{\infty} [n(n-1)]^{-1/2}$ (e) $\displaystyle\sum_{n=2}^{\infty} \frac{\sqrt{n^2+1}}{n^2-1}$ (f) $\displaystyle\sum_{n=1}^{\infty} \frac{1}{2n(2n-1)}$

(g) $\displaystyle\sum_{n=1}^{\infty} \frac{n!}{10^{n+1}}$ (h) $\displaystyle\sum_{n=0}^{\infty} \frac{2^{5n}}{5^{2n}}$ (i) $\displaystyle\sum_{n=1}^{\infty} \frac{\arctan(n)}{n^2}$.

2.2.5. Determine whether each of the following series converges.

(a) $\sum_{n=1}^{\infty} \frac{1}{n(n+1)}$ (b) $\sum_{n=1}^{\infty} \frac{1}{n2^n}$ (c) $\sum_{n=0}^{\infty} e^{-n}$

(d) $\sum_{n=2}^{\infty} \frac{1}{n \ln n}$ (e) $\sum_{n=1}^{\infty} \frac{1}{n \cdot n^{1/n}}$ (f) $\sum_{n=0}^{\infty} \frac{(2n)!\,n!}{(3n)!}$

(g) $\sum_{n=1}^{\infty} \ln\left(1 + \frac{1}{n}\right)$ (h) $\sum_{n=1}^{\infty} \frac{\mathrm{arccot}(n)}{n^{1/2}}$ (i) $\sum_{n=0}^{\infty} \mathrm{arccot}(n)$.

2.2.6. Try to evaluate the summations of the two preceding exercises using MAPLE or MATHEMATICA. This approach will show that some of the divergent series are confirmed as divergent and will give values of some of the convergent series. Notice that symbolic computation is not yet as accomplished as skilled humans in identifying series that are divergent.

2.2.7. Write a symbolic computation program to evaluate $\sum_{n=1}^{1000} n^{-1}$, and use the result to obtain upper and lower bounds on the Euler-Mascheroni constant.

2.3 ALTERNATING SERIES

The previous section dealt only with series of positive terms. We now turn to infinite series in which the signs alternate. It turns out that the partial cancellation due to alternating signs makes convergence more rapid and easier to identify. The key result to be established in this section is the **Leibniz criterion**, a general condition for the convergence of an alternating series.

The Leibniz criterion applies to series of the form

$$S = u_1 - u_2 + u_3 - u_4 + \cdots ,$$

with all $u_n > 0$. It states:

If for all sufficiently large n u_n *is* **monotonically decreasing** *and* $\lim_{n\to\infty} u_n = 0$, *then the series* $\sum_{n=1}^{\infty}(-1)^{n+1}u_n$ *converges.*

To prove this theorem, notice that the remainder R_{2n} of the series beyond s_{2n}, the partial sum of the first $2n$ terms, can be written in two alternate ways:

$$R_{2n} = (u_{2n+1} - u_{2n+2}) + (u_{2n+3} - u_{2n+4}) + \cdots$$

$$= u_{2n+1} - (u_{2n+2} - u_{2n+3}) - (u_{2n+4} - u_{2n+5}) - \cdots .$$

Since the u_n are decreasing, the first of these equations implies $R_{2n} > 0$, while the second implies $R_{2n} < u_{2n+1}$, so

$$0 < R_{2n} < u_{2n+1} .$$

Thus, R_{2n} is positive but bounded, and the bound can be made arbitrarily small by taking larger values of n. An argument similar to that made above for the remainder R_{2n+1} after an odd number of terms would show that it is negative and bounded by $-u_{2n+2}$. Thus, it is generally true that the remainder after truncating an alternating series with monotonically decreasing terms is smaller in magnitude and of the same sign as the first term dropped.

The Leibniz criterion depends for its applicability on the presence of strict sign alternation. Less regular sign changes present more challenging problems for convergence determination.

ABSOLUTE AND CONDITIONAL CONVERGENCE

An infinite series is **absolutely** convergent if the absolute values of its terms form a convergent series. If it converges, but not absolutely, it is termed **conditionally** convergent. An example of a conditionally convergent series is the alternating harmonic series,

$$\sum_{n=1}^{\infty}(-1)^{n-1}n^{-1} = 1 - \frac{1}{2} + \frac{1}{3} - \frac{1}{4} + \cdots + \frac{(-1)^{n-1}}{n} + \cdots . \tag{2.18}$$

This series is convergent, based on the Leibniz criterion. It is clearly not absolutely convergent; if all terms are taken with + signs, we have the harmonic series, which we already know to be divergent. The tests described earlier in this section for series of positive terms are, then, tests for absolute convergence.

Exercises

2.3.1. Determine whether each of these series is convergent, and if so, whether it is absolutely convergent:

(a) $\frac{\ln 2}{2} - \frac{\ln 3}{3} + \frac{\ln 4}{4} - \frac{\ln 5}{5} + \frac{\ln 6}{6} - \cdots ,$

(b) $\frac{1}{1} + \frac{1}{2} - \frac{1}{3} - \frac{1}{4} + \frac{1}{5} + \frac{1}{6} - \frac{1}{7} - \frac{1}{8} + \cdots ,$

(c) $1 - \frac{1}{2} - \frac{1}{3} + \frac{1}{4} + \frac{1}{5} + \frac{1}{6} - \frac{1}{7} - \frac{1}{8} - \frac{1}{9} - \frac{1}{10} + \frac{1}{11} \cdots + \frac{1}{15} - \frac{1}{16} \cdots - \frac{1}{21} + \cdots .$

2.3.2. **Catalan's constant**, sometimes designated $\beta(2)$, is defined by the series

$$\beta(2) = \sum_{k=0}^{\infty}(-1)^k(2k+1)^{-2} = \frac{1}{1^2} - \frac{1}{3^2} + \frac{1}{5^2} - \cdots .$$

Using symbolic computations, calculate $\beta(2)$ to six-digit accuracy.

Hint. The rate of convergence is enhanced by pairing the terms:

$$(4k-1)^{-2} - (4k+1)^{-2} = \frac{16k}{(16k^2-1)^2} .$$

If you kept enough digits in your summation, $\sum_{1\le k\le N} 16k/(16k^2-1)^2$, additional significant figures may be obtained by setting upper and lower bounds on the tail of the series, $\sum_{k=N+1}^{\infty}$. These bounds may be set by comparison with integrals, as in the Cauchy integral test.

Check your work against the built-in value of Catalan's constant in your symbolic computation system; its name is `Catalan` (the same in both MAPLE and MATHEMATICA).

2.4 OPERATIONS ON SERIES

We need to know what operations can legitimately be performed on the terms of an infinite series. In this connection, the establishment of absolute convergence is important, because it can be shown that

- If an infinite series is absolutely convergent, the series sum is independent of the order in which the terms are added.
- An absolutely convergent series may be added termwise to, or subtracted termwise from, or multiplied termwise with another absolutely convergent series, and the resulting series will also be absolutely convergent.

No similar properties can be assured for conditionally convergent series, though some relevant statements can be made if only one of the series to be combined is conditionally convergent. The danger of applying the above properties indiscriminately to conditionally convergent series is illustrated by the following example.

Example 2.4.1. Rearrangement, Alternating Harmonic Series

Writing the alternating harmonic series as

$$1-\frac{1}{2}+\frac{1}{3}-\frac{1}{4}+\cdots=1-\left(\frac{1}{2}-\frac{1}{3}\right)-\left(\frac{1}{4}-\frac{1}{5}\right)-\cdots, \tag{2.19}$$

it is clear that

$$\sum_{n=1}^{\infty}(-1)^{n-1}n^{-1}<1\,.$$

However, if we rearrange the order of the terms, we can make this series converge to any desired value or even diverge (a situation that can be brought about by taking all the positive terms first). Let's suppose we wish to make the alternating harmonic series sum to 3/2. We start by taking positive terms (in the order in which they occur in the original series) until their partial sum exceeds 3/2. Then we take negative terms (in order) until the partial sum has been reduced below 3/2. We continue with positive terms until 3/2 has again been exceeded, negative terms until the partial sum is less than 3/2, continuing this process indefinitely. The first few steps of this procedure correspond to

$$\left(1+\frac{1}{3}+\frac{1}{5}\right)-\left(\frac{1}{2}\right)+\left(\frac{1}{7}+\frac{1}{9}+\frac{1}{11}+\frac{1}{13}+\frac{1}{15}\right)-\left(\frac{1}{4}\right)$$
$$+\left(\frac{1}{17}+\cdots+\frac{1}{25}\right)-\left(\frac{1}{6}\right)+\left(\frac{1}{27}+\cdots+\frac{1}{35}\right)-\left(\frac{1}{8}\right)+\cdots. \tag{2.20}$$

Letting s_i denote the partial sum through the ith group of terms, we obtain

$$\begin{aligned} s_1 &= 1.5333 & s_2 &= 1.0333\\ s_3 &= 1.5218 & s_4 &= 1.2718\\ s_5 &= 1.5143 & s_6 &= 1.3476\\ s_7 &= 1.5103 & s_8 &= 1.3853\\ s_9 &= 1.5078 & s_{10} &= 1.4078\,. \end{aligned}$$

From this tabulation of s_n, we see that we are indeed getting convergence to $\frac{3}{2}$. As the series extends to infinity, all original terms will eventually appear, and (at least in principle) we can continue until all further partial sums will be arbitrarily close to $\frac{3}{2}$.

■

The statement that an alternating series can be rearranged so as to sum to any result is sometimes called **Riemann's theorem**. It is clear that it is not legitimate to undertake arbitrary rearrangements of the terms of an alternating series.

IMPROVEMENT OF CONVERGENCE

Thus far the emphasis of our discussion has focused on the treatment of convergence as a mathematical property. But in the practical use of series expansions, the **rate** of convergence is clearly of interest and importance. A method for improving convergence, due to Kummer, is to form a linear combination of our slowly converging series and one or more series whose sum is known, with the most slowly convergent part of our series replaced by the series with a known sum. Kummer suggested the following series for his method:

$$\alpha_1 = \sum_{n=1}^{\infty} \frac{1}{n(n+1)} = 1\,,$$

$$\alpha_2 = \sum_{n=1}^{\infty} \frac{1}{n(n+1)(n+2)} = \frac{1}{4}\,,$$

$$\alpha_3 = \sum_{n=1}^{\infty} \frac{1}{n(n+1)(n+2)(n+3)} = \frac{1}{18}\,,$$

$$\cdots$$

$$\alpha_p = \sum_{n=1}^{\infty} \frac{1}{n(n+1)\cdots(n+p)} = \frac{1}{p\,p!}\,. \tag{2.21}$$

These sums can be evaluated via partial fraction expansions, and are the subject of Exercise 2.4.2.

The series we wish to sum and one or more known series (multiplied by coefficients) are combined term-by-term. The coefficients in the linear combination are chosen to cancel the most slowly converging terms.

Example 2.4.2. Riemann Zeta Function $\zeta(3)$

From the definition in Eq. (2.12), we identify $\zeta(3)$ as $\sum_{n=1}^{\infty} n^{-3}$. Noticing that α_2 of Eq. (2.21) has a large-n dependence $\sim n^{-3}$ and that α_2 has the value $1/4$, we consider the linear combination

$$\sum_{n=1}^{\infty} n^{-3} + a\alpha_2 = \zeta(3) + \frac{a}{4}\,. \tag{2.22}$$

Table 2.1: Riemann Zeta Function.

s	$\zeta(s)$
2	1.64493 40668
3	1.20205 69032
4	1.08232 32337
5	1.03692 77551
6	1.01734 30620
7	1.00834 92774
8	1.00407 73562
9	1.00200 83928
10	1.00099 45751

We did not use α_1 because it converges more slowly than $\zeta(3)$. Combining the two series on the left-hand side termwise, we obtain

$$\sum_{n=1}^{\infty}\left[\frac{1}{n^3}+\frac{a}{n(n+1)(n+2)}\right]=\sum_{n=1}^{\infty}\frac{n^2(1+a)+3n+2}{n^3(n+1)(n+2)}\,.$$

If we choose $a=-1$, we remove the leading term from the numerator; then, setting this equal to the right-hand side of Eq. (2.22) and solving for $\zeta(3)$,

$$\zeta(3)=\frac{1}{4}+\sum_{n=1}^{\infty}\frac{3n+2}{n^3(n+1)(n+2)}\,. \tag{2.23}$$

This series converges as n^{-4}, faster than n^{-3}. A more convenient form with even faster convergence is introduced in Exercise 2.4.1. There, the symmetry leads to convergence as n^{-5}.

■

Sometimes it is helpful to use the Riemann zeta function in a way similar to that illustrated for the α_p in the foregoing example. That approach is practical because the zeta function has been tabulated (see Table 2.1).

Example 2.4.3. Convergence Improvement

The problem is to evaluate the series $\sum_{n=1}^{\infty} 1/(1+n^2)$. Expanding $(1+n^2)^{-1} = n^{-2}(1+n^{-2})^{-1}$ by direct division, we have

$$(1+n^2)^{-1}=n^{-2}\left(1-n^{-2}+n^{-4}-\frac{n^{-6}}{1+n^{-2}}\right)$$

$$=\frac{1}{n^2}-\frac{1}{n^4}+\frac{1}{n^6}-\frac{1}{n^8+n^6}.$$

Therefore

$$\sum_{n=1}^{\infty}\frac{1}{1+n^2}=\zeta(2)-\zeta(4)+\zeta(6)-\sum_{n=1}^{\infty}\frac{1}{n^8+n^6}.$$

The remainder series converges as n^{-8}. Clearly, the process can be continued as desired. You make a choice between how much algebra you will do and how much arithmetic the computer will do.

■

Exercises

2.4.1. The convergence improvement of Example 2.4.2 may be carried out more effectively if we rewrite α_2, from Eq. (2.21), into a more symmetric form: replacing n by $n-1$ and changing the lower summation limit to $n=2$, we get

$$\alpha_2' = \sum_{n=2}^{\infty} \frac{1}{(n-1)n(n+1)} = \frac{1}{4}.$$

Show that by combining $\zeta(3)$ and α_2' we can obtain convergence as n^{-5}.

2.4.2. The formula for α_p, Eq. (2.21), is a summation of the form $\sum_{n=1}^{\infty} u_n(p)$, with

$$u_n(p) = \frac{1}{n(n+1)\cdots(n+p)}.$$

Applying a partial fraction decomposition to the first and last factors of the denominator, i.e.,

$$\frac{1}{n(n+p)} = \frac{1}{p}\left[\frac{1}{n} - \frac{1}{n+p}\right],$$

show that $u_n(p) = \dfrac{u_n(p-1) - u_{n+1}(p-1)}{p}$ and that $\sum_{n=1}^{\infty} u_n(p) = \dfrac{1}{p\,p!}$.

Hint. Consider using mathematical induction. In that connection it is useful to note that $u_1(p-1) = 1/p!$.

2.5 SERIES OF FUNCTIONS

We now consider infinite series in which each term u_n may be a function of some variable, $u_n = u_n(x)$. The partial sums then become functions of the variable x,

$$s_n(x) = u_1(x) + u_2(x) + \cdots + u_n(x), \tag{2.24}$$

and we define the series sum as the limit of the partial sums:

$$\sum_{n=1}^{\infty} u_n(x) = S(x) = \lim_{n\to\infty} s_n(x). \tag{2.25}$$

In principle we may encounter series in which the successive terms are not closely functionally related, but it is more usual that successive terms involve a recognizable set of functions. Here are examples of typical types of series:

$$S(x) = a_0 + a_1 x + a_2 x^2 + \cdots + a_n x^n + \cdots,$$

$$S(x) = a_0 + a_1 \cos x + a_2 \cos 2x + \cdots + a_n \cos nx + \cdots,$$

$$S(x) = a_0 J_0(x) + a_1 J_1(x) + a_2 J_2(x) + \cdots + a_n J_n(x) + \cdots.$$

The first of these is known as a **power series**, and will be a subject for discussion here. The second is a trigonometric, or **Fourier series**. Fourier series are of sufficient importance to be the topic of a later chapter in this book. The third series illustrates the fact that we often encounter series in which the individual terms are members of a set of special functions; here $J_n(x)$ denotes a **Bessel function** and the series can be referred to as a **Bessel series**; these are also treated in some detail in a later chapter.

Irrespective of the way in which the terms of a series are formed, we have the same concerns that we have previously investigated, but now with the additional feature that the properties of the series may depend upon the value of the series variable, x. In particular, we may now inquire as to the range of x for which a series converges; we may ask whether a series represents the expansion of a known function or we may wish to find series expansions of given functions; and we may ask what operations (including integration or differentiation) can be applied termwise to our series, and with what results.

We begin our study of these questions by considering the extremely useful expansions we identified as **power series**.

POWER SERIES

The first question we ask is: how can we represent a known function $f(x)$ as a power series, i.e., as a series of the form

$$f(x) = a_0 + a_1 x + a_2 x^2 + a_3 x^3 + a_4 x^4 + \cdots . \tag{2.26}$$

Assuming that the desired power series exists, our task is to find the coefficients a_i. To start, we can set $x = 0$, and Eq. (2.26) then reduces to $f(0) = a_0$. Next, we differentiate the expansion of $f(x)$, reaching

$$f'(x) = a_1 + 2a_2 x + 3a_3 x^2 + 4a_4 x^3 + \cdots ,$$

from which, setting $x = 0$, we read out $f'(0) = a_1$. Differentiating again, we reach

$$f''(x) = 2a_2 + 2 \cdot 3a_3 x + 3 \cdot 4a_4 x^2 + \cdots .$$

Again setting $x = 0$, we find $f''(0) = 2a_2$, or $a_2 = f''(0)/2$. Differentiating a third time, we find

$$f^{(3)}(x) = 2 \cdot 3a_3 + 2 \cdot 3 \cdot 4a_4 x + \cdots ,$$

where the notation $f^{(3)}$ stands for the third derivative of f; the parentheses tell us that the "3" is not an exponent. Setting $x = 0$, we obtain $f^{(3)}(0) = 2 \cdot 3a_3 = 3!\, a_3$, or $a_3 = f^{(3)}/3!$.

Continuing this process indefinitely, we find $a_n = f^{(n)}(0)/n!$, reaching the final result

$$f(x) = f(0) + f^{(1)}(0)\, x + f^{(2)}(0)\, \frac{x^2}{2!} + \cdots + f^{(n)}(0)\, \frac{x^n}{n!} + \cdots . \tag{2.27}$$

This is the power-series expansion of $f(x)$, also known as its **Maclaurin series**.

Although the above derivation appears unambiguous it cannot be concluded that every function which is finite and differentiable at $x = 0$ will have a power-series expansion, or that the expansion will be convergent. A case in point is the function $\exp(-1/x^2)$; it and all its derivatives vanish at $x = 0$, which would suggest a series that is identically zero. But clearly $\exp(-1/x^2)$ is nonzero for all nonzero x. We will see later that it is useful to consider this as a situation in which the interval in x where the power series converges to the function consists only of the point $x = 0$.

UNIQUENESS OF POWER SERIES

Our procedure for finding the coefficients in the power series expansion of a function shows that each coefficient has a specific relationship to a derivative of the function being expanded, so the expansion is necessarily unique. The uniqueness enables us to obtain the power series expansion of an unknown function by equating coefficients of individual powers of x in an equation in which the unknown function appears.

The uniqueness also reflects the fact that different powers of x are linearly independent, meaning that no power is (over a range of x) equal to a linear combination of other powers.

REMAINDER OF POWER SERIES EXPANSION

If the power-series expansion of a function $f(x)$ is truncated after $n+1$ terms (the term proportional to x^n), the difference between the truncated series and $f(x)$ is called the **remainder** of the series and denoted R_n; thus

$$R_n(x) = f(x) - \left[f(0) + f^{(1)}(0)\,x + \cdots + f^{(n)}(0)\frac{x^n}{n!} \right] . \tag{2.28}$$

It can be shown that if $f(x)$ is sufficiently differentiable (whether or not the power series converges) the remainder is given by

$$R_n(x) = f^{(n+1)}(\xi)\frac{x^{n+1}}{(n+1)!} , \tag{2.29}$$

where ξ is an (unknown) point between 0 and x.

In principle we can determine the convergence of a power series by seeing whether R_n as given by Eq. (2.29) tends to zero as n is increased, but it is usually easier just to apply the ratio test to the series, as discussed in the next subsection. However, the remainder formula helps us to understand unusual situations such as that we illustrate below, which we encounter when we try to expand $\exp(-1/x^2)$ in powers of x.

Example 2.5.1. Remainder Formula

Consider the remainder $R_n(x)$ when the power series expansion of $f(x) = \exp(-1/x^2)$ is truncated after $n+1$ terms. To use Eq. (2.29) we need the $(n+1)$th derivative of $f(x)$; for our present purposes it suffices to identify the contribution to this derivative which has the smallest negative power of x; by analysis or by inspection of computer-generated derivatives, we find

$$f^{(n+1)}(x) = \frac{(-1)^n (n+2)!}{x^{n+3}}\, e^{-1/x^2} \pm \cdots .$$

Inserting this into Eq. (2.29), we find

$$R_n(x) = \frac{(-1)^n (n+2)}{\xi^2} \left(\frac{x}{\xi}\right)^{n+1} e^{-1/x^2} \pm \cdots ,$$

where ξ is between zero and x. This is not the largest contribution to R_n, but its behavior suffices to suggest that $R_n(x)$ cannot be expected to approach zero as n is increased.

■

CONVERGENCE OF POWER SERIES

If we apply the ratio test to a power series, we see that it will converge absolutely if

$$\lim_{n\to\infty}\left|\frac{a_{n+1}x^{n+1}}{a_n x^n}\right| < 1\,, \quad \text{or} \quad |x| < R = \lim_{n\to\infty}\left|\frac{a_n}{a_{n+1}}\right|\,. \tag{2.30}$$

In other words, our power series will exhibit absolute convergence if $-R < x < R$. It will certainly diverge if $|x| > R$ because successive terms then increase in magnitude. The ratio test fails for the cases $x = \pm R$, so the end points of the interval of convergence will require special attention. Note that (except possibly for behavior at the end points) the interval of convergence is symmetric about $x = 0$.

EXAMPLES OF POWER SERIES

We present here several power series expansions that are of sufficient importance that they should become familiar to all readers of this book.

Example 2.5.2. Exponential Function

Let $f(x) = e^x$. Differentiating, then setting $x = 0$, we have

$$f^{(n)}(0) = 1$$

for all n, $n = 1,\ 2,\ 3, \ldots$. Then, applying Eq. (2.27), we obtain

$$e^x = 1 + x + \frac{x^2}{2!} + \frac{x^3}{3!} + \cdots = \sum_{n=0}^{\infty}\frac{x^n}{n!}\,. \tag{2.31}$$

This is the series expansion of the exponential function.

Although this series is clearly convergent for all x, as may be verified using the ratio test, it is instructive to check the remainder term, R_n. From Eq. (2.29) we get

$$R_n(x) = \frac{x^{n+1}}{(n+1)!}\,f^{(n+1)}(\xi) = \frac{x^{n+1}}{(n+1)!}\,e^{\xi}\,,$$

where ξ is between 0 and x. Irrespective of the sign of x,

$$|R_n(x)| \le \frac{|x|^{n+1}e^{|x|}}{(n+1)!}\,;$$

No matter how large $|x|$ may be, a sufficient increase in n will cause the denominator of this form for R_n to dominate over the numerator, and $\lim_{n\to\infty} R_n = 0$. Thus, the Maclaurin expansion of e^x converges absolutely over the entire range $-\infty < x < \infty$.

■

Example 2.5.3. Sine and Cosine

Consider next $f(x) = \sin x$. Since $f'(x) = \cos x$ and $f''(x) = -\sin x$, $f^{(3)}(x) = -\cos x$, $f^{(4)}(x) = \sin x$, so these values repeat indefinitely for the higher derivatives. Evaluating at $x = 0$, we have

$$f^{(2n)}(0) = 0, \qquad f^{(2n+1)}(0) = (-1)^n, \qquad n = 0,\ 1,\ 2,\ \cdots\,.$$

Inserting these values into the Maclaurin expansion, we get

$$\sin x = x - \frac{x^3}{3!} + \frac{x^5}{5!} - \frac{x^7}{7!} + \cdots = \sum_{n=0}^{\infty} \frac{(-1)^n x^{2n+1}}{(2n+1)!} \,. \tag{2.32}$$

The same values of the derivatives occur (but shifted in position) for $f(x) = \cos x$, with

$$f^{(2n)}(0) = (-1)^n, \qquad f^{(2n+1)}(0) = 0, \qquad n = 0,\, 1,\, 2,\, \cdots,$$

and the resulting Maclaurin series is

$$\cos x = 1 - \frac{x^2}{2!} + \frac{x^4}{4!} - \frac{x^6}{6!} + \cdots = \sum_{n=0}^{\infty} \frac{(-1)^n x^{2n}}{(2n)!} \,. \tag{2.33}$$

Applying the ratio test, we find that both these series converge absolutely for all x, but it is obvious that the rate of convergence will become slower as $|x|$ increases. Since the functions $\sin x$ and $\cos x$ are actually periodic, it will be most efficient to use Eqs. (2.32) and (2.33) for $|x| \le \pi$.

■

Example 2.5.4. Logarithm

For another Maclaurin expansion, let $f(x) = \ln(1+x)$. Differentiating, we reach

$$f'(x) = (1+x)^{-1} \,,$$

$$f^{(n)}(x) = (-1)^{n-1}\,(n-1)!\,(1+x)^{-n} \,. \tag{2.34}$$

Thus,

$$f(0) = \ln 1 = 0 \,, \quad f^{(1)}(0) = 1 \,, \quad f^{(2)}(0) = -(1!) \,, \quad f^{(n)}(0) = (-1)^{n-1}(n-1)! \,.$$

Equation (2.27) then yields

$$\ln(1+x) = x - \frac{x^2}{2} + \frac{x^3}{3} - \frac{x^4}{4} + \cdots$$

$$= \sum_{p=1}^{\infty} (-1)^{p-1}\, \frac{x^p}{p} \,. \tag{2.35}$$

Applying the ratio test, absolute convergence is guaranteed if

$$\lim_{p\to\infty} \left| \frac{u_{p+1}}{u_p} \right| = \lim_{p\to\infty} \left(\frac{p}{p+1} \right) |x| < 1 \,.$$

This condition is met for $-1 < x < 1$. Irrespective of the sign of x, successive terms increase in magnitude if $|x| > 1$, so the series then diverges. The ratio test fails at $x = \pm 1$, but at $x = 1$ we have an alternating series whose convergence follows from the Leibniz criterion. At $x = -1$ we have (minus) the harmonic series, which we know to be divergent. Thus, our power series for the logarithm converges on the interval $-1 < x \le 1$.

Returning to the value at $x = 1$, we have the interesting result

$$\ln 2 = 1 - \frac{1}{2} + \frac{1}{3} - \frac{1}{4} + \frac{1}{5} - \cdots = \sum_{n=1}^{\infty} (-1)^{n-1}\, n^{-1} . \tag{2.36}$$

■

UNIFORM CONVERGENCE

Some important properties of an infinite series expansion of a function $f(x)$ depend upon the rate at which it converges, viewed as a function of the variable x. The key concept here is uniform convergence. Loosely speaking, a series is uniformly convergent on a range of x if for that entire range (including the end points) it converges at least as rapidly as some convergent expansion that is independent of x. It can be shown that in the absence of uniform convergence it is possible for a series of individually continuous terms to add to a discontinuous result. Uniform convergence is also generally a requirement which, if satisfied, permits interchange of the order of summations and integrations.

Summarizing, if a series $\sum_n u_n(x)$ is uniformly convergent in $[a, b]$ and the individual terms $u_n(x)$ are continuous,

1. The series sum $f(x) = \sum_{n=1}^{\infty} u_n(x)$ is also continuous.

2. The series may be integrated term-by-term. The sum of the integrals is equal to the integral of the sum:

$$\int_a^b f(x)\, dx = \sum_{n=1}^{\infty} \int_a^b u_n(x)\, dx . \tag{2.37}$$

3. The derivative of the series sum $f(x)$ equals the sum of the individual-term derivatives:

$$\frac{d}{dx} f(x) = \sum_{n=1}^{\infty} \frac{d}{dx} u_n(x) , \tag{2.38}$$

providing the following additional conditions are satisfied:

$$\frac{du_n(x)}{dx} \text{ is continuous in } [a, b],$$

$$\sum_{n=1}^{\infty} \frac{du_n(x)}{dx} \text{ is uniformly convergent in } [a, b].$$

Power series are uniformly convergent on any interval **interior** to their range of convergence. Thus, if a power series is convergent on $-R < x < R$, it will be uniformly convergent on any interval $-S \le x \le S$, where $S < R$. This is so because the convergence of the series is slowest at either S or $-S$, and that is a fixed comparison series that can be used for the entire range $|x| \le S$. It is necessary that $S < R$ because we need an x-independent convergence rate **at** the end points of our interval of uniform convergence and generally this will not be possible at $|x| = R$.

Since differentiation of a power series produces a new power series with the same range of convergence as the original series, the differentiated series will also be uniformly convergent on the same range as the original series. Therefore, convergent power series have the following useful properties:

Within their region of uniform convergence,

1. They describe continuous functions.
2. They may be integrated term-by-term as often as desired.
3. They may be differentiated term-by-term as often as desired.

Those properties, which are much less restrictive than those applying to series in general, make power series extremely useful.

TAYLOR SERIES

Once we have a power series in a variable x, we can replace x by $x - a$ providing we rewrite the coefficients so they correspond to derivatives evaluated at $x - a = 0$, i.e., at $x = a$. This yields an expansion known as a Taylor series, of the form

$$f(x) = f(a) + f^{(1)}(a)(x-a) + f^{(2)}(a)\frac{(x-a)^2}{2!} + \cdots + f^{(n)}(a)\frac{(x-a)^n}{n!} + \cdots . \quad (2.39)$$

Our Taylor series specifies the value of a function at one point, x, in terms of the value of the function and its derivatives at a reference point a. It is an expansion in powers of the **change** in the variable, namely $x - a$. This idea can be emphasized by writing Taylor's series in an alternate form in which we replace x by $x + h$ and a by x:

$$f(x+h) = \sum_{n=0}^{\infty} \frac{h^n}{n!} f^{(n)}(x) . \quad (2.40)$$

FINDING POWER SERIES EXPANSIONS

While Eq. (2.27) provides a general formula for the power series expansion of a general function, there are often easier ways to obtain the expansions. Here are some of the possibilities.

Example 2.5.5. Integrate a Known Series

Let's suppose that we remember the form of the geometric series

$$\frac{1}{1-x} = 1 + x + x^2 + x^3 + x^4 + \cdots ,$$

in which we replace x by $-x$:

$$\frac{1}{1+x} = 1 - x + x^2 - x^3 + x^4 - \cdots .$$

Integrating this term-by-term, we get

$$\int_0^x \frac{dx}{1+x} = \ln(1+x) = x - \frac{x^2}{2} + \frac{x^3}{3} - \frac{x^4}{4} + \frac{x^5}{5} - \cdots . \tag{2.41}$$

■

Example 2.5.6. Differentiate a Known Series

We recognize that

$$\frac{1}{(1-x)^2} = \frac{d}{dx}\left(\frac{1}{1-x}\right).$$

We may therefore expand it as

$$\frac{1}{(1-x)^2} = \frac{d}{dx}\Big[1 + x + x^2 + x^3 + \cdots\Big] = 1 + 2x + 3x^2 + \cdots . \tag{2.42}$$

■

Example 2.5.7. Substitute for Variable

The power series expansion of e^{-x^2} is easily obtained by replacing x by $-x^2$ in the expansion of e^x:

$$e^{-x^2} = 1 - x^2 + \frac{(-x^2)^2}{2!} + \frac{(-x^2)^3}{3!} + \cdots = 1 - x^2 + \frac{x^4}{2!} - \frac{x^6}{3!} + \cdots . \tag{2.43}$$

■

Example 2.5.8. Multiply Series by Function

The power series expansion of $x/(1-x)$ is obtained with virtually no effort by multiplying the expansion of $1/(1-x)$ through by x. It is considerably more work to take successive derivatives of $x/(1-x)$ and use Eq. (2.27) to form the expansion.

■

Example 2.5.9. A More Complicated Example

If we notice that $\arctan x$ (sometimes written $\tan^{-1} x$, which **does not** mean $1/\tan x$) satisfies the equation

$$\arctan x = \int_0^x \frac{dx}{1+x^2},$$

we see that we can expand the integrand in a power series and integrate term-by-term:

$$\arctan x = \int_0^x \Big[1 - x^2 + x^4 - x^6 + \cdots\Big]\,dx = x - \frac{x^3}{3} + \frac{x^5}{5} - \cdots . \tag{2.44}$$

■

SYMBOLIC COMPUTATION OF POWER SERIES

Both MAPLE and MATHEMATICA have built-in procedures that develop power series without the user having to compute derivatives and put together the expansion. In MAPLE, the relevant command is

```
series(expr, eqn,n);
```

Here *eqn* defines the point about which *expr* is to be expanded, with *eqn* just being the name of the power-series variable or an equation of the form `x=a` to generate a series in powers of $x - a$. If *eqn* has the form `x=infinity` the result is an expansion in powers of $1/x$. The quantity `n`, which is optional, controls (but is not always equal to) the number of terms kept in the expansion.

In MATHEMATICA, a series expansion is generated by the procedure

```
Series[expr,{x,a,n}]
```

Here `x` is the expansion variable, `a` is the point about which the expansion takes place, and `n` (which is not optional) is the highest power retained in the expansion. If `a` is given the value `Infinity`, we get an expansion in powers of $1/x$.

In both MAPLE and MATHEMATICA the expansion variable `x` must be an undefined name; the expansion point may either be defined or undefined. In both symbolic systems, the result of the expansion is in a special form that is not suitable for ordinary computations using the truncated series, but can be converted into an ordinary polynomial by use of

`convert(`*series*`,polynom);` (MAPLE) or `Normal[`*series*`]` (MATHEMATICA)

These conversions generate expressions at the truncation level of the series being converted.

While the series forms cannot be used in most computations, they can be differentiated or integrated term-by-term, using the usual operators (`diff`, `int`, `D`, or `Integrate`). The operators `diff` and `D` operate on series in the usual fashion. However, due to the special form of the series construction, the integration operators `int` and `Integrate` can **only** be applied as **indefinite integrals**, i.e., using commands of the form `int(`*series*`,x)` or `Integrate[`*series*`,x]`.

In both symbolic systems the series procedure often works even when a Maclaurin or Taylor series does not exist because the expansion requires some negative or fractional powers. These more general expansions may be useful in certain contexts.

Example 2.5.10. Symbolic Coding of Series Expansions

Here are some series generated by the symbolic procedures:

`A:=series(tan(x),x);`
$$A := x + \frac{1}{3}\,x^3 + \frac{2}{15}\,x^5 + O(x^6)$$

`B:=series(cot(x),x,6);`
$$B := x^{-1} - \frac{1}{3}\,x - \frac{1}{45}\,x^3 - \frac{2}{945}\,x^5 + O(x^7)$$

`series(arctan(x),x=infinity,6);`
$$\frac{\pi}{2} - \frac{1}{x} + \frac{1}{3x^3} - \frac{1}{5x^5} + O\left(\frac{1}{x^6}\right)$$

`C=Series[1/(1+x),{x,a,3}]`
$$\frac{1}{1+a} - \frac{x-a}{(1+a)^2} + \frac{(x-a)^2}{(1+a)^3} + O[x-a]^3$$

```
F=Series[1/(x+Sqrt[x]),{x,0,2}]
```

$$\frac{1}{\sqrt{x}} - 1 + \sqrt{x} - x + x^{3/2} - x^2 + O[x]^{5/2}$$

Now let's try to evaluate the expansion of tan(0.1) by substituting 0.1 for x in the series **A**:

```
subs(x=0.1,A);
```

Error, invalid substitution...

We get better results if we first convert A into a polynomial:

```
AA:=convert(A,polynom);
```

$$AA := x + \frac{1}{3}x^3 + \frac{2}{15}x^5$$

The disappearance of the term $O(x^6)$ is an indication that the conversion to an ordinary expression has occurred. Now,

```
subs(x=0.1,AA);
```

0.1003346666

A similar conversion is needed in MATHEMATICA.

```
FF=Normal[F]
```

$$-1 + \frac{1}{\sqrt{x}} + \sqrt{x} - x + x^{3/2} - x^2$$

FF is an ordinary expression, as evidenced by the removal of the term $O[x]^{5/2}$.

■

Exercises

2.5.1. Show that

$$\text{(a)} \ \sin x = \sum_{n=0}^{\infty} (-1)^n \frac{x^{2n+1}}{(2n+1)!},$$

$$\text{(b)} \ \cos x = \sum_{n=0}^{\infty} (-1)^n \frac{x^{2n}}{(2n)!}.$$

2.5.2. (a) Show that $f(x) = x^{1/2}$ has no Maclaurin expansion.

(b) Show that the Taylor series for $x^{1/2}$ about $x = a > 0$ is

$$x^{1/2} = a^{1/2} + \sum_{n=1}^{\infty} u_n (x-a)^n, \quad \text{with} \quad u_n = \frac{(-1)^{n-1}(2n-2)!}{2^{2n-1}\,n!\,(n-1)!\,a^{n-1/2}}.$$

(c) Determine the range over which this Taylor series converges.

2.5.3. (a) Use MAPLE or MATHEMATICA to obtain the power series expansions of $\tan x$ and $\tan^2 x$. Arrange to keep terms at least as far as x^{10}.

(b) Differentiate the series for $\tan x$ term-by-term; to what trigonometric function does your new series correspond?

(c) Verify that the power series found in parts (a) and (b) are consistent with the trigonometric identity $1 + \tan^2 x = \sec^2 x$.

2.5.4. (a) Use MAPLE or MATHEMATICA to obtain the Taylor series expansion of $1/\sin x$ about the point $x = \pi/2$. Keep terms at least as far as $(x - \pi/2)^{10}$.

(b) Integrate this series term-by-term. Notice that the variable of integration should be specified as `x`, not `x-Pi/2`. Thus, the form of the termwise integration command is

`int(`*series*`,x);` (MAPLE) or `Integrate[`*series*`,x]` (MATHEMATICA).

(c) The integral of $1/\sin x$ from $x = \pi/2$ to $x = x$ can be written in the form $\ln\tan(x/2)$. Note that at $x = \pi/2$ we have $\tan(x/2) = 1$, giving the required value of zero for this reference value of x. Expand $\ln\tan(x/2)$ about $x = \pi/2$, and compare your result with that obtained in part (b).

2.5.5. (a) Using the expansion of $\ln(1+x)$ given in Eq. (2.41), show that, for sufficiently large n,

$$\text{(i)}\ \frac{1}{n} + \ln\left(\frac{n-1}{n}\right) < 0\,, \qquad \text{(ii)}\ \frac{1}{n} - \ln\left(\frac{n+1}{n}\right) > 0\,.$$

(b) Use these inequalities to show that the limit defining the Euler-Mascheroni constant, Eq. (2.13), is finite.

2.5.6. Show that $\lim_{x\to 0}\left[\dfrac{\sin(\tan x) - \tan(\sin x)}{x^7}\right] = -\dfrac{1}{30}$.

2.5.7. A power series T converges for $-R < x < R$.

(a) Show that a series obtained by termwise integration of T converges for any range $-S < x < S$ such that $|S| < |R|$.

(b) Show that termwise differentiation of T also produces a series that converges on all ranges interior to the convergence range of T.

2.5.8. Show that the integral $\displaystyle\int_0^1 \tan^{-1} t\,\frac{dt}{t}$ has a value equal to Catalan's constant.

Note. The definition and numerical computation of Catalan's constant was addressed in Exercise 2.3.2.

2.6 BINOMIAL THEOREM

An extremely important application of the Maclaurin expansion is the derivation of the binomial theorem.

Let $f(x) = (1+x)^m$, in which m may be either positive or negative and is not limited to integral values. Writing the Maclaurin series, Eq. (2.27),

$$(1+x)^m = 1 + mx + \frac{m(m-1)}{2!}x^2 + \cdots\,. \tag{2.45}$$

In the present case the remainder is

$$R_n = \frac{x^n}{n!}(1+\xi)^{m-n}\,m(m-1)\cdots(m-n+1)\,, \tag{2.46}$$

with ξ between 0 and x. Restricting attention for now to $x \ge 0$, we note that for $n > m$, $(1+\xi)^{m-n}$ is a maximum for $\xi = 0$, so for positive x and sufficiently large n,

$$|R_n| \le \frac{x^n}{n!}\,|m(m-1)\cdots(m-n+1)|\,, \tag{2.47}$$

with $\lim_{n\to\infty} R_n = 0$ when $0 \le x < 1$. Because the radius of convergence of a power series is the same for positive and for negative x, the binomial series converges for $-1 < x < 1$. Convergence at the limit points ± 1 is not addressed by the present analysis, and depends upon m.

Summarizing, we have established the **binomial expansion**,

$$(1+x)^m = 1 + mx + \frac{m(m-1)}{2!}x^2 + \frac{m(m-1)(m-2)}{3!}x^3 + \cdots , \tag{2.48}$$

convergent for $-1 < x < 1$. It is important to note that Eq. (2.48) applies whether or not m is integral, and for both positive and negative m. If m is a nonnegative integer, R_n for $n > m$ vanishes for all x, corresponding to the fact that under those conditions $(1+x)^m$ is a finite sum with x^m as its largest power of x.

Because the binomial expansion is of frequent occurrence, the coefficients appearing in it, which are called **binomial coefficients**, are given the special symbol

$$\binom{m}{n} = \frac{m(m-1)\cdots(m-n+1)}{n!} , \tag{2.49}$$

and the binomial expansion assumes the general form

$$(1+x)^m = \sum_{n=0}^{\infty} \binom{m}{n} x^n . \tag{2.50}$$

In evaluating Eq. (2.49), notice that when $n = 0$, the product in its numerator is empty (starting from m and **descending** to $m+1$); in that case the convention is to assign the product the value unity. In addition, the empty product corresponding to $n! = 1 \cdot 2 \cdots n$ with $n = 0$ (i.e., 0!) is also defined to be unity, so for any m we have the trivial value

$$\binom{m}{0} = 1 .$$

In the special case that m is a positive integer, we may write our binomial coefficient in terms of factorials:

$$\binom{m}{n} = \begin{cases} \dfrac{m!}{n!\,(m-n)!} , & n = 0, 1, \cdots, m, \\ 0, & \text{otherwise.} \end{cases} \tag{2.51}$$

In writing Eq. (2.51) we have taken account of the fact that these binomial coefficients vanish if $n > m$.

For positive integer m, the $\binom{m}{n}$ also arise in combinatorial theory, being the number of different ways n out of m objects can be selected. That, of course, is consistent with the coefficient set if $(1+x)^m$ is expanded. The term containing x^n has a coefficient that corresponds to the number of ways one can choose the "x" from n of the factors $(1+x)$ and the 1 from the $m-n$ other $(1+x)$ factors.

For negative integer m, we can still use the special notation for binomial coefficients, but their evaluation is more easily accomplished if we set $m = -p$, with p a positive integer, and write

$$\binom{-p}{n} = (-1)^n \frac{p(p+1)\cdots(p+n-1)}{n!} = \frac{(-1)^n\,(p+n-1)!}{n!\,(p-1)!} = (-1)^n \binom{p+n-1}{n} . \tag{2.52}$$

These binomial coefficients remain nonzero for all n, so we have an infinite series, and not a polynomial.

For nonintegral m, it is convenient to use the **Pochhammer symbol**, defined for general a and nonnegative integer n and given the notation $(a)_n$, as

$$(a)_0 = 1, \quad (a)_1 = a, \quad (a)_{n+1} = a(a+1)\cdots(a+n), \quad (n \geq 1) \,. \tag{2.53}$$

For both integral and nonintegral m, the binomial coefficient formula can be written

$$\binom{m}{n} = \frac{(m-n+1)_n}{n!} \,. \tag{2.54}$$

There is a rich literature on binomial coefficients and relationships between them and on summations involving them. We mention here only one such formula that arises if we evaluate $1/\sqrt{1+x}$, i.e., $(1+x)^{-1/2}$. The binomial coefficient

$$\begin{aligned}\binom{-\frac{1}{2}}{n} &= \frac{1}{n!}\left(-\frac{1}{2}\right)\left(-\frac{3}{2}\right)\cdots\left(-\frac{2n-1}{2}\right) \\ &= (-1)^n \frac{1\cdot 3\cdots(2n-1)}{2^n\, n!} = (-1)^n \frac{(2n-1)!!}{(2n)!!} \,,\end{aligned} \tag{2.55}$$

where the "double factorial" notation indicates products of even or odd positive integers as follows:

$$\begin{aligned} 1\cdot 3\cdot 5\cdots(2n-1) &= (2n-1)!! \\ 2\cdot 4\cdot 6\cdots(2n) &= (2n)!! \,. \end{aligned} \tag{2.56}$$

These are related to the regular factorials by

$$(2n)!! = 2^n\, n! \qquad \text{and} \qquad (2n-1)!! = \frac{(2n)!}{2^n\, n!} \,. \tag{2.57}$$

Notice that these relations include the special cases $0!! = (-1)!! = 1$.

Example 2.6.1. Relativistic Energy

The total relativistic energy of a particle of mass m and velocity v is

$$E = mc^2\left(1 - \frac{v^2}{c^2}\right)^{-1/2} , \tag{2.58}$$

where c is the velocity of light. Let's introduce a binomial expansion to see how this expression behaves when v/c is small compared to unity. Using Eq. (2.50) with $m = -1/2$ and $x = -v^2/c^2$, and evaluating the binomial coefficients using Eq. (2.55), we have

$$\begin{aligned} E &= mc^2\left[1 - \frac{1}{2}\left(-\frac{v^2}{c^2}\right) + \frac{3}{8}\left(-\frac{v^2}{c^2}\right)^2 - \frac{5}{16}\left(-\frac{v^2}{c^2}\right)^3 + \cdots\right] \\ &= mc^2 + \frac{1}{2}mv^2 + \frac{3}{8}mv^2\left(\frac{v^2}{c^2}\right) + \frac{5}{16}mv^2\left(\frac{v^2}{c^2}\right)^2 + \cdots \,. \end{aligned} \tag{2.59}$$

The first term, mc^2, is the rest-mass energy. The remainder, which we identify as the kinetic energy, is

$$E_{\text{kinetic}} = \frac{1}{2}mv^2 \left[1 + \frac{3}{4}\frac{v^2}{c^2} + \frac{5}{8}\left(\frac{v^2}{c^2}\right)^2 + \cdots\right]. \tag{2.60}$$

For particle velocity $v \ll c$, the expression in the brackets approaches unity and we see that the kinetic energy then reduces to the nonrelativistic result.

■

The binomial expansion can be generalized for positive integer n to polynomials:

$$(a_1 + a_2 + \cdots + a_m)^n = \sum \frac{n!}{n_1!n_2!\cdots n_m!} a_1^{n_1} a_2^{n_2} \cdots a_m^{n_m}, \tag{2.61}$$

where the summation includes all different combinations of nonnegative integers $n_1, n_2, \ldots, n_m$ with $\sum_{i=1}^{m} n_i = n$. This generalization finds considerable use in statistical mechanics.

In everyday analysis, the combinatorial properties of the binomial coefficients make them appear often. For example, Leibniz's formula for the nth derivative of a product of two functions, $u(x)v(x)$, can be written

$$\left(\frac{d}{dx}\right)^n \Big(u(x)\,v(x)\Big) = \sum_{i=0}^{n} \binom{n}{i} \left(\frac{d^i\,u(x)}{dx^i}\right)\left(\frac{d^{n-i}\,v(x)}{dx^{n-i}}\right). \tag{2.62}$$

Example 2.6.2. Application of Binomial Expansion

Sometimes the binomial expansion provides a convenient indirect route to the Maclaurin series when direct methods are difficult. We consider here the power series expansion

$$\arcsin x = \sum_{n=0}^{\infty} \frac{(2n-1)!!}{(2n)!!}\,\frac{x^{2n+1}}{(2n+1)} = x + \frac{x^3}{6} + \frac{3x^5}{40} + \cdots. \tag{2.63}$$

Starting from $\sin y = x$, we find $dy/dx = 1/\sqrt{1-x^2}$, and write the integral

$$\arcsin x = y = \int_0^x \frac{dt}{(1-t^2)^{1/2}}.$$

We now introduce the binomial expansion of $(1-t^2)^{-1/2}$ and integrate term-by-term. The result is Eq. (2.63).

■

SYMBOLIC COMPUTATION OF BINOMIAL COEFFICIENTS

Both MAPLE and MATHEMATICA support procedures that compute binomial coefficients. We have

$\binom{n}{m}$: `binomial(n,m)` (MAPLE), `Binomial[n,m]` (MATHEMATICA).

In MAPLE binomial coefficients are not always evaluated automatically. To force evaluation, use **expand**.

Example 2.6.3. Binomial Coefficients

Here is a MAPLE session:

```
> binomial(5,2),binomial(2,5);
```

$$10,\ 0$$

```
> binomial(a,3);
```

binomial(a,3)

```
> expand(%);
```

$$\frac{1}{6}a^3 - \frac{1}{2}a^2 + \frac{1}{3}a$$

```
> factor(%);
```

$$\frac{a(a-1)(a-2)}{6}$$

Notice that we had to force the evaluation when the result was not numeric.

In MATHEMATICA the evaluation is obtained more directly:

```
Binomial[a,3]
```

$$\frac{1}{6}(-2+a)(-1+a)a$$

■

Exercises

2.6.1. Write $(a+b)^n$ as a binomial series which is convergent for all n when $|a| < |b|$ and as another binomial series that converges when $|a| > |b|$. Which of these series can be used when $|a| > |b|$ and n is a positive integer?

2.6.2. Using MAPLE or MATHEMATICA,

(a) Generate (as exact fractions) the binomial coefficients

$$\binom{1/2}{0}, \quad \binom{1/2}{1}, \quad \binom{1/2}{2},$$

and then generate a list of $\binom{1/2}{n}$ (as decimal quantities) for $n = 0$ through $n = 10$.

2.6.3. Show that for integral $n \geq 0$, $\displaystyle\frac{1}{(1-x)^{n+1}} = \sum_{m=n}^{\infty} \binom{m}{n} x^{m-n}$.

2.6.4. For positive integer m, show that $\displaystyle(1+x)^{-m/2} = \sum_{n=0}^{\infty} (-1)^n \frac{(m+2n-2)!!}{2^n n!(m-2)!!} x^n$.

2.6.5. Write the function $(1-x)^{-n-1}$ as a series (keeping terms through x^5). Use MAPLE or MATHEMATICA to verify that the coefficient of each power of x corresponds to the formula in Exercise 2.6.3.

2.6.6. Use symbolic computation to verify correctness of the coefficients in the formula of Exercise 2.6.4. It suffices to check four values of n for each of three m values.

2.6.7. Prove the identities (for positive integer n)

(a) $\sum_{m=0}^{n} \binom{n}{m} = 2^n$,

(b) $\sum_{m=0}^{n} (-1)^m \binom{n}{m} = 0.$

2.6.8. Show that

$$\ln\left(\frac{1+x}{1-x}\right) = 2\left(x + \frac{x^3}{3} + \frac{x^5}{5} + \cdots\right), \qquad -1 < x < 1.$$

2.7 SOME IMPORTANT SERIES

There are a few series that arise so often that all students should recognize them. Here is a short list that is worth committing to memory.

$$\exp(x) = \sum_{n=0}^{\infty} \frac{x^n}{n!} = 1 + x + \frac{x^2}{2!} + \frac{x^3}{3!} + \frac{x^4}{4!} + \cdots, \qquad -\infty < x < \infty, \tag{2.64}$$

$$\sin(x) = \sum_{n=0}^{\infty} \frac{(-1)^n x^{2n+1}}{(2n+1)!} = x - \frac{x^3}{3!} + \frac{x^5}{5!} - \frac{x^7}{7!} + \cdots, -\infty < x < \infty, \tag{2.65}$$

$$\cos(x) = \sum_{n=0}^{\infty} \frac{(-1)^n x^{2n}}{(2n)!} = 1 - \frac{x^2}{2!} + \frac{x^4}{4!} - \frac{x^6}{6!} + \cdots, \qquad -\infty < x < \infty, \tag{2.66}$$

$$\frac{1}{1-x} = \sum_{n=0}^{\infty} x^n = 1 + x + x^2 + x^3 + x^4 + \cdots, \qquad -1 \le x < 1, \tag{2.67}$$

$$\ln(1+x) = \sum_{n=1}^{\infty} \frac{(-1)^{n-1} x^n}{n} = x - \frac{x^2}{2} + \frac{x^3}{3} - \frac{x^4}{4} + \cdots, \ -1 < x \le 1, \tag{2.68}$$

$$(1+x)^p = \sum_{n=0}^{\infty} \binom{p}{n} x^n = \sum_{n=0}^{\infty} \frac{(p-n+1)_n}{n!} x^n, \qquad -1 < x < 1. \tag{2.69}$$

Reminder: The notation $(a)_n$ is the Pochhammer symbol: $(a)_0 = 1$, $(a)_1 = a$, and for integers $n > 1$, $(a)_n = a(a+1)\cdots(a+n-1)$. It is not required that a, or p in Eq. (2.69), be positive or integral.

2.8 SOME APPLICATIONS OF SERIES

An obvious place where series are useful is in the numerical evaluation of functions that do not have convenient closed forms. With the nearly universal availability of computers of all sizes ranging from hand-held devices to the largest high-performance equipment, the work of numerical evaluation has largely been automated, but scientists and engineers will still encounter functions that are too specialized or nonstandard to have preprogrammed evaluations.

Often it will be desired to have a functional form which is a relatively simple approximation to a complicated function; here the leading term (or terms) of a series expansion can be extremely useful. Sometimes an exact analytical expression for a quantity of interest may be so numerically ill-conditioned that it is difficult to evaluate directly. Again, a series expansion may provide a route toward evaluation.

Finally, there are analytical processes which are facilitated by the use of expansions. A good example is the evaluation of indeterminate forms. These types of applications are illustrated by the following examples.

Example 2.8.1. Leading Terms of Expansion

Suppose we require $f(x) = \dfrac{1-\sin x}{1-x}$ for small x. Either by inserting the expansions of $1-\sin x$ and $1/(1-x)$ and multiplying them together or by using symbolic computing, we may find that

$$f(x) = 1 + \frac{1}{6}x^3 + \frac{1}{6}x^4 + \frac{19}{120}x^5 + \cdots,$$

showing us that the departure of $f(x)$ from unity has a leading dependence proportional to x^3, and that for $|x| < 0.1$ the first three terms of the series yield a result that is good to about five significant figures.

■

Example 2.8.2. Ill-Conditioned Expression

Sometimes analytical cancellations can reduce an ill-conditioned expression to a more easily evaluated form. This can often be accomplished by the introduction of series expansions.

Suppose we need to evaluate $\sin(\tan x) - \tan(\sin x)$ for $x = 0.01$. On some ten-digit calculators we get the answer zero; on others 0.1×10^{-10}. The correct answer is approximately -0.333×10^{-15}. The reason for the discrepancy is that

$$\sin(\tan 0.01) = 0.0100001666641661 2099,$$

$$\tan(\sin 0.01) = 0.01000016666416645436,$$

and a numerical evaluation loses about 14 significant figures in the subtraction.

However, power series expansions of these functions, which we can easily carry out using either MAPLE or MATHEMATICA, take the forms

$$\sin(\tan x) = x + \frac{x^3}{6} - \frac{x^5}{40} - \frac{55x^7}{1008} - \cdots,$$

$$\tan(\sin x) = x + \frac{x^3}{6} - \frac{x^5}{40} - \frac{107x^7}{5040} - \cdots,$$

and the subtraction of these series leaves us with

$$\sin(\tan x) - \tan(\sin x) = -\frac{x^7}{30} - \cdots.$$

This analysis is similar to that of Exercise 2.5.6.

The leading contribution to our function for $x = 0.01$ is $-10^{-14}/30$, in qualitative agreement with the high-precision numerical data.

■

Example 2.8.3. Indeterminate Form

The power-series representation of functions is often useful in evaluating indeterminate forms, and is the basis of **l'Hôpital's rule**, which states that if the ratio of two differentiable functions $f(x)$ and $g(x)$ becomes indeterminate, of the form 0/0, at $x = x_0$, then

$$\lim_{x \to x_0} \frac{f(x)}{g(x)} = \lim_{x \to x_0} \frac{f'(x)}{g'(x)}. \tag{2.70}$$

Equation (2.70) can be proved in general by inserting the Taylor series expansions about x_0 of $f(x)$ and $g(x)$. Because $f(x_0) = g(x_0) = 0$, the respective leading terms of the two series are $f'(0)(x-x_0)$ and $g'(0)(x-x_0)$. Dividing numerator and denominator of f/g by $x - x_0$ we find that at $x = x_0$ the numerator and denominator respectively reduce to $f'(x_0)$ and $g'(x_0)$, thereby establishing Eq. (2.70).

Sometimes it is easier just to introduce power-series expansions than to evaluate the derivatives that enter l'Hôpital's rule. Here is an example of this strategy. We wish to evaluate

$$\lim_{x \to 0} \frac{1 - \cos x}{x^2} . \tag{2.71}$$

Replacing $\cos x$ by its Maclaurin-series expansion, Exercise 2.5.1, we obtain

$$\frac{1 - \cos x}{x^2} = \frac{1 - (1 - \frac{1}{2!}x^2 + \frac{1}{4!}x^4 - \cdots)}{x^2} = \frac{1}{2!} - \frac{x^2}{4!} + \cdots .$$

Letting $x \to 0$, we have

$$\lim_{x \to 0} \frac{1 - \cos x}{x^2} = \frac{1}{2} . \tag{2.72}$$

One reason why this procedure may be convenient is that we need the derivatives only in the limit $x = x_0$, and that is what we get from the series expansions.

■

SYMBOLIC COMPUTATION OF LIMITS

MAPLE and MATHEMATICA both support the computation of limits. The command for taking the limit of *expr* as x approaches x_0 is

`limit(`*expr*, x=x_0`)` (MAPLE) or `Limit[`*expr*, x->x_0`]` (MATHEMATICA).

If in a strict mathematical sense the limit does not exist but can be unambiguously characterized as $+\infty$ or $-\infty$, then that value is reported. If the symbolic system cannot supply an unambiguous value for a limit, it may return "*undefined*" or specify an interval of uncertainty as to the result. It is possible to request a limit at $x_0 = \infty$ or $-\infty$ (where ∞ is designated `infinity` in MAPLE, `Infinity` in MATHEMATICA).

For functions such that the limit is the same if approached from either direction, both symbolic systems behave equivalently. But if the limit when x_0 is approached from $x < x_0$ (referred to as from **below** or from the **left**) differs from the limit from **above** (also referred to as from the **right**), the two symbolic systems behave differently. In MAPLE, the limit is then reported as *undefined*, while in MATHEMATICA, the limit is by default taken from the right.

A single-sided limit in which x_0 is approached from below can be obtained by writing

`limit(`*expr*, x=x_0, `left)` or `Limit[`*expr*, x->x_0, `Direction->1]` .

To get a one-sided limit in MAPLE with approach from $x > x_0$, we write instead

`limit(`*expr*, x=x_0, `right)` .

One could explicitly specify a one-sided limit from the right in MATHEMATICA by inserting `Direction->-1` in the `limit` command, but that is not really necessary because it is the default. We note that the symbolic systems can usually find single-sided limits even though their existence means that power-series expansions are not available.

Example 2.8.4. Limits

Consider the limits of $f(x) = \sin x/x$ and $g(x) = 1/x$ as $x \to 0$. Issuing limit commands in various ways, we get

MAPLE:	`>limit(sin(x)/x, x=0);`	1
	`> limit(1/x, x=0);`	*undefined*
	`> limit(1/x, x=0, left);`	$-\infty$
	`> limit(1/x, x=0, right);`	∞
MATHEMATICA:	`Limit[Sin[x]/x, x->0]`	1
	`Limit[1/x, x->0]`	∞
	`Limit[1/x, x->0, Direction -> 1]`	$-\infty$
	`Limit[1/x, x->0, Direction -> -1]`	∞

■

Exercises

Find the following limits. Use symbolic computation to check any limits you find by hand.

2.8.1. $\lim_{x\to 0} \frac{\cos^2 x - 1}{x^2}$

2.8.2. $\lim_{x\to \pi} \frac{x^2 \sin x}{x - \pi}$

2.8.3. $\lim_{x\to 0} \frac{\ln x}{x}$

2.8.4. $\lim_{x\to 0} x^p \ln x$.

Here p is an arbitrary positive real number; give the limit as a function of p.

2.9 BERNOULLI NUMBERS

A large number of common functions have power-series expansions whose coefficients can be expressed in terms of a set of numbers first identified by Jacques Bernoulli. These so-called **Bernoulli numbers** also appear in other mathematical contexts; they are found in formulas for the Riemann zeta functions of even integer argument, and they enter in the Euler-Maclaurin summation formula (which relates the integral of a function to a sum over discrete values of the same function). They are also of interest from a pedagogical viewpoint; their definition and properties illustrate the concept and use of **generating functions** and **recurrence formulas**.

GENERATING FUNCTION

The Bernoulli numbers B_n are defined by the Maclaurin expansion

$$\frac{t}{e^t - 1} = \sum_{n=0}^{\infty} \frac{B_n t^n}{n!} \,. \tag{2.73}$$

The function on the left-hand side of this formula is known as a **generating function** (here, "the generating function for the B_n"), meaning that its expansion "generates" the set of quantities B_n as the coefficients of t^n.

Before developing the properties of the Bernoulli numbers, we should point out that not all writers use Eq. (2.73) as their definition. Some authors omit the $n!$ from the denominator, while others modify the definition in ways motivated by the properties we shall shortly examine. It is therefore essential to notice the exact definition used in any work from which the reader will take formulas or values of the B_n.

Continuing now from Eq. (2.73), we can, at least in principle, find the Bernoulli numbers by developing the Maclaurin series of the generating function:

$$B_n = \left[\frac{d^n}{dt^n} \left(\frac{t}{e^t - 1} \right) \right]_{t=0} \,. \tag{2.74}$$

The direct use of Eq. (2.74) rapidly becomes cumbersome because the generating function is at $t = 0$ an indeterminate form. However, it is not too much work to find B_0 and B_1:

$$B_0 = \lim_{t=0} \left(\frac{t}{e^t - 1} \right) = \lim_{t=0} \left(\frac{t}{t + t^2/2 + \cdots} \right) = 1 \,, \tag{2.75}$$

$$B_1 = \lim_{t=0} \left[\frac{d}{dt} \left(\frac{t}{e^t - 1} \right) \right] = \lim_{t=0} \left[\frac{1}{e^t - 1} - \frac{te^t}{(e^t - 1)^2} \right] = -\frac{1}{2} \,. \tag{2.76}$$

We are now in a position to identify a perhaps unexpected symmetry of the Bernoulli expansion. Rewriting Eq. (2.73) with the B_1 term of the summation moved to the left-hand side,

$$\begin{aligned} \frac{t}{e^t - 1} + \frac{t}{2} &= \sum_{n \neq 1} \frac{B_n t^n}{n!} \\ &= \frac{t}{2} \frac{e^t + 1}{e^t - 1} = \frac{t}{2} \frac{e^{t/2} + e^{-t/2}}{e^{t/2} - e^{-t/2}} \,. \end{aligned} \tag{2.77}$$

The second line of Eq. (2.77) shows that the each side of its top line is an even function of t, meaning that all the B_n of odd n (other than B_1) must vanish.

Table 2.2: Bernoulli Numbers.

n	B_n	B_n
0	1	1.000000000
1	$-\frac{1}{2}$	−0.500000000
2	$\frac{1}{6}$	0.166666667
4	$-\frac{1}{30}$	−0.033333333
6	$\frac{1}{42}$	0.023809524
8	$-\frac{1}{30}$	−0.033333333
10	$\frac{5}{66}$	0.075757576

Further values are easily found using symbolic computing methods.

COMPUTATION OF BERNOULLI NUMBERS

In principle, either of the following finite summations can be used to find B_{2N} when B_{2n} is known for all $n < N$:

$$N - \frac{1}{2} = \sum_{n=1}^{N} B_{2n} \binom{2N+1}{2n} ,$$
$$N = \sum_{n=1}^{N} B_{2n} \binom{2N+2}{2n} . \tag{2.78}$$

Since the Riemann zeta function for large n can be computed efficiently directly from its defining sum, one may then also compute B_{2n} as

$$B_{2n} = \frac{(-1)^{n-1} 2(2n)!}{(2\pi)^{2n}} \zeta(2n) , \qquad n = 1, 2, 3, \cdots . \tag{2.79}$$

Derivation of these formulas can be found in Arfken et al., Additional Readings. The first few Bernoulli numbers are given in Table 2.2.

SYMBOLIC COMPUTATION

Both MAPLE and MATHEMATICA support the generation of Bernoulli numbers. In MAPLE the Bernoulli number B_n is obtained from the procedure `bernoulli(n)`. In MATHEMATICA, the corresponding command is `BernoulliB[n]`.

Exercises

2.9.1. Write a symbolic computing procedure `bern(p)` or `bern[p]` that will generate the Bernoulli number B_p (as an exact fraction) recursively using one of Eqs. (2.78). Your code should give correct results for all nonnegative integers `p`, both even and odd.

Hint. You may want to consult Appendix D to see how to manage the data for this problem (here, intermediate Bernoulli numbers).

2.9.2. Here are some series expansions that use Bernoulli numbers. Write procedures that evaluate these series (keeping terms through x^{15}), using your symbolic system's Bernoulli number function. Then compare the accuracy of your procedures with the results of direct function evaluations.

(a) $\cot x = \sum_{n=0}^{\infty} \frac{(-1)^n 2^{2n} B_{2n}}{(2n)!} x^{2n-1}$,

(b) $\tan x = \sum_{n=1}^{\infty} \frac{(-1)^{n-1} 2^{2n}(2^{2n}-1) B_{2n}}{(2n)!} x^{2n-1}$,

(c) $\csc x = \sum_{n=0}^{\infty} \frac{(-1)^{n-1} 2(2^{2n-1}-1) B_{2n}}{(2n)!} x^{2n-1}$,

(d) $\ln\cos z = \sum_{n=1}^{\infty} \frac{(-1)^n 2^{2n-1}(2^{2n}-1) B_{2n}}{n(2n)!} x^{2n}$.

2.10 ASYMPTOTIC SERIES

Sometimes we wish to describe a function $f(x)$ as a series in inverse powers of x, so as to have a result in which the terms (other than perhaps the leading term) will become negligible as x becomes very large. In favorable cases we can simply write the Maclaurin expansion, using $1/x$ as the expansion variable:

$$e^{-1/x} = 1 - \frac{1}{x} + \frac{1}{2!}\frac{1}{x^2} - \frac{1}{3!}\frac{1}{x^3} + \cdots .$$

This expansion converges for all nonzero x.

Often a simple approach becomes impractical, and other routes to an expansion must be pursued. We illustrate with the function

$$f(x) = \int_0^{\infty} \frac{e^{-xv}}{1+v^2}\, dv . \tag{2.80}$$

We write the denominator of the integrand in Eq. (2.80) as the first n terms of the binomial expansion plus the remainder:

$$\frac{1}{1+v^2} = 1 - v^2 + \cdots + (-1)^n v^{2n} - \frac{(-1)^n v^{2n+2}}{1+v^2} . \tag{2.81}$$

Notice that because n is finite, this is an ordinary closed expression that is not only exact but also introduces no convergence issues. Inserting Eq. (2.81) into Eq. (2.80), making the substitution $v = u/x$ and integrating, we obtain

$$f(x) = \frac{0!}{x} - \frac{2!}{x^3} + \cdots + \frac{(-1)^n (2n)!}{x^{2n+1}} - \int_0^{\infty} \frac{(-1)^n e^{-xv} v^{2n+2}}{1+v^2}\, dv . \tag{2.82}$$

We cannot interpret the large-n limit of Eq. (2.82) as an infinite series because it will diverge for all x; however, if we fix n at some finite value, the n-term series will be an approximation to $f(x)$ with an error dependent on the size of the remainder integral. If for fixed n we make x sufficiently large, the error in our n-term series can be made as small as we like. Thus, even though the infinite series for $f(x)$ in powers of $1/x$ diverges, its truncation to n terms is still useful as an estimate of $f(x)$ for sufficiently large x. Series of this type are called **asymptotic** or **semiconvergent**, and we distinguish them notationally from convergent series by writing $\sim$ in place of the equals sign:

$$f(x) \sim \frac{0!}{x} - \frac{2!}{x^3} + \cdots + \frac{(-1)^n (2n)!}{x^{2n+1}} .$$

Summarizing, we identify a series as convergent (for a particular range of x) if the partial sums (as $n \to \infty$) converge to a finite limit; a series is asymptotic (semiconvergent) if it diverges but approaches a limit if n is fixed and x is increased toward infinity. If a series for $f(x)$ is semiconvergent, its error in the estimation of $f(x)$ for a given x may initially become smaller as n is increased, but after a certain n value the estimates of $f(x)$ become poorer with increasing n.

Another frequently useful way to obtain asymptotic series for functions defined as integrals is by repeated integrations by parts. The special function $E_1(x)$ (called an **exponential integral**) has the definition

$$E_1(x) = \int_x^\infty \frac{e^{-t}}{t}\, dt .$$

Integrating by parts, with $1/t$ differentiated and e^{-t} integrated, we find

$$E_1(x) = \frac{-e^{-t}}{t}\bigg|_x^\infty - \int_x^\infty \frac{-e^{-t}}{-t^2}\, dt = \frac{e^{-x}}{x} - \int_x^\infty \frac{e^{-t}}{t^2}\, dt .$$

A second integration by parts brings us to

$$E_1(x) = \frac{e^{-x}}{x} - \frac{e^{-x}}{x^2} + \int_x^\infty \frac{2e^{-t}}{t^3}\, dt ,$$

and further integrations by parts lead to

$$E_1(x) \sim e^{-x}\left[\frac{0!}{x} - \frac{1!}{x^2} + \frac{2!}{x^3} + \cdots + (-1)^n \frac{n!}{x^{n+1}}\right] . \tag{2.83}$$

It is instructive to examine the behavior of the partial sums from Eq. (2.83). Taking for illustrative purposes $x = 8$, we can find, using MAPLE or MATHEMATICA, that the accurate value of $e^8 E_1(8)$ is 0.1123; the partial sums of this quantity for $3 \le n \le 15$ are plotted in Fig. 2.2. Because of the sign alternation in the series, successive partial sums alternate between values larger and smaller than the exact value. We note an apparent convergence through the seventh and eighth partial sums; thereafter the partial sums diverge from the exact value. We see that at $x = 8$, we cannot reduce the error below one or two units in the fourth decimal place. In this respect, an asymptotic series differs from the convergent expansions we have heretofore studied in that we cannot (for given x) improve the accuracy beyond a certain point by including more terms. Despite this limitation, asymptotic series can be useful, and can give essentially quantitative results when the expansion variable is sufficiently large.

Example 2.10.1. Symbolic Computation

The symbolic algebra systems produce expansions in powers of $1/x$ if the command `series` (`Series`) is called with the expansion point set to `infinity` (`Infinity`), yielding a series that may be either convergent or semiconvergent, depending upon the function being expanded. This is a good topic for illustrating the limitations of the symbolic systems, as attempts to obtain fairly simple asymptotic expansions

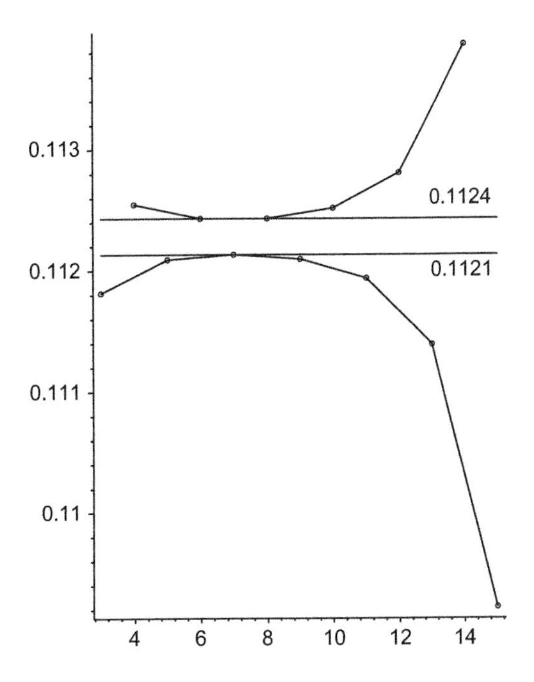

Figure 2.2: Partial sums of $e^x E_1(x)\,|_{x=8}$.

sometimes lead to simplification problems that the reader may find challenging. We consider here the first asymptotic expansion of the current section, that of $f(x)$ as defined by Eq. (2.80). In MAPLE, the function $f(x)$ can be defined as

```
F := int(exp(-v*x)/(1+v^2),v=0 .. infinity):
```

An attempt to obtain an asymptotic expansion fails unless MAPLE is informed that x is to be restricted to positive values; this can be accomplished by issuing the command `assume(x>0);`. We find, to our dismay, that

```
> assume(x>0):
> series(F,x=infinity,10);
```

produces several screenfuls of output which does not get reduced in complexity by application of the command `simplify`. However, we note that the MAPLE output contains both the trigonometric functions `sin(x)`, `cos(x)` and complex exponentials such as `exp(I*x)`. This suggests that simplification may occur if all the trigonometric functions are converted to complex exponentials (or the reverse). We therefore apply to the result of the foregoing operations the commands

```
> convert(%,exp):
> simplify(%):
> expand(%);
```

(this converts sin and cos to exponentials)

$$\frac{1}{x} - \frac{2}{x^3} + \frac{24}{x^5} - \frac{720}{x^7} + \frac{40320}{x^9} + O\left(\frac{1}{x^{10}}\right)$$

This is the expected result, but the reader still needs to recognize the numerators as the factorials of the even integers.

MATHEMATICA does a bit better with this problem. We need the commands

```
f = Integrate[Exp[-x*v]/(1 + v^2),{v,0,Infinity}]
Series[f, {x,Infinity,10}]
```

This produces a lot of output, but it simplifies:

`Simplify[%]` $\left(\frac{1}{x} - \frac{2}{x^3} + \frac{24}{x^5} - \frac{720}{x^7} + \frac{40320}{x^9} + O\left[\frac{1}{x}\right]^{11}\right) + O\left[\frac{1}{x}\right]^{11} \text{Sin}[2x]$

Both these symbolic procedures require more knowledge (and fiddling) than simply expanding the denominator and integrating term-by-term.

■

Exercises

2.10.1. The complementary Gauss error function is defined as

$$\text{erfc}(x) = \frac{2}{\sqrt{\pi}} \int_x^\infty e^{-t^2}\, dt\,.$$

Show, applying repeated integrations by parts, that erfc(x) has the asymptotic expansion

$$\text{erfc}(x) \sim \frac{e^{-x^2}}{\sqrt{\pi}\, x}\left[1 - \frac{1}{2x^2} + \frac{1 \cdot 3}{(2x^2)^2} - \frac{1 \cdot 3 \cdot 5}{(2x^2)^3} + \cdots + (-1)^n \frac{(2n-1)!!}{(2x^2)^n}\right].$$

See if you can obtain this asymptotic expansion from your symbolic computing system. The complementary error function is called `erfc(x)` or `Erfc[x]`.

2.11 EULER-MACLAURIN FORMULA

If n_1 and n_2 are integers, with $n_1 < n_2$, the integral of a function $f(x)$ from n_1 to n_2 can be related to a summation of $f(n)$, with n taking on values from n_1 to n_2. This relation, known as the **Euler-Maclaurin formula**, takes the form

$$\begin{aligned}\int_{n_1}^{n_2} f(x)\, dx &= \frac{1}{2} f(n_1) + f(n_1+1) + \cdots + f(n_2-1) + \frac{1}{2} f(n_2) \\ &\quad - \frac{B_2}{2!}\left[f'(n_2) - f'(n_1)\right] - \frac{B_4}{4!}\left[f^{(3)}(n_2) - f^{(3)}(n_1)\right] \\ &\quad - \cdots - \frac{B_{2p}}{(2p)!}\left[f^{(2p-1)}(n_2) - f^{(2p-1)}(n_1)\right] - \cdots\,. \end{aligned} \tag{2.84}$$

Here f' stands for the first derivative of f, with $f^{(2p-1)}$ designating higher derivatives. Notice that this equation gives the trapezoid-rule formula as the leading contribution to the integral. The remaining terms contain derivatives of f at the ends of the integration interval; these terms can be considered as corrections to the trapezoid rule. Note the occurrence of the Bernoulli numbers. For the derivation of Eq. (2.84), see Arfken et al., Additional Readings.

In the oft-occurring case where the upper limit of the integral is infinity, with $f(x)$ and all its derivatives vanishing there, Eq. (2.84) simplifies to

$$\int_n^\infty f(x)\,dx = \frac{1}{2}\,f(n) + \sum_{j=n+1}^{\infty} f(j) + \sum_{p=1}^{P} \frac{B_{2p}}{(2p)!}\,f^{(2p-1)}(n)\,. \tag{2.85}$$

In many applications of interest the Euler-Maclaurin expansion does not converge, but in such cases it is usually a semiconvergent series. Then one may use the procedure outlined in the section on asymptotic series: we can approximate an integral by keeping terms only up to the point where they begin to increase in magnitude.

Although Eq. (2.85) is in a form that expresses an integral as a summation plus corrections, that equation may also be interpreted as a recipe for evaluating a summation as an integral plus corrections. This approach can be of interest because we have more general methods for evaluating integrals than are available for infinite series.

Finally, we note that convergence (or weaker divergence) of the Euler-Maclaurin series can be achieved by choosing the integration/summation interval in a way that decreases the magnitudes of the end-point derivatives. For example, if we are using the expansion for a convergent sum with range $(0,\infty)$, we may get better results by keeping several explicit terms, then applying the Euler-Maclaurin expansion for a range (n,∞) in which $f(x)$ is varying less rapidly.

Example 2.11.1. Zeta Function

We consider here the use of the Euler-Maclaurin formula for the evaluation of the Riemann zeta function; the strategy is to write the summation defining the zeta function as an integral plus correction terms involving the Bernoulli numbers. Accordingly, we specialize Eq. (2.85) to the present situation as follows:

$$\begin{aligned}\zeta(s) &= \sum_{n=1}^{\infty}\frac{1}{n^s} \approx \frac{1}{2}\\ &+ \left[\int_x^\infty \frac{dt}{t^s} - \sum_{p=1}^{P}\frac{B_{2p}}{(2p)!}\,\frac{(-1)^{2p-1}s(s+1)\cdots(s+2p-2)}{x^{s+2p-1}}\right]_{x=1}\,.\end{aligned} \tag{2.86}$$

Here the "1/2" allows for the fact that the summation we wish to evaluate has a full contribution at $n=1$ while the first two terms on the right-hand side of Eq. (2.85) only give half-weight to the $n=1$ term. The expression multiplying $B_{2p}/(2p)!$ is the $(2p-1)$th derivative of $1/x^s$.

An attempt to use Eq. (2.86) for $\zeta(6)$ fails; the p summation is strongly divergent. However, we can weaken the divergence and obtain a useful semiconvergent series by summing the first few terms of $\zeta(s)$ explicitly and then applying the Euler-Maclaurin formula only to the remaining infinite set of terms. Removal of the first few terms causes the derivatives in the Bernoulli summation to be smaller, thereby delaying the onset of divergence. The equation corresponding to Eq. (2.86) then takes the form, with the Euler-Maclaurin formula only for the terms from k to ∞,

$$\begin{aligned}\zeta(s) &\approx \sum_{n=1}^{k-1}\frac{1}{n^s}+\frac{1}{2k^s}\\ &+\left[\int_k^\infty \frac{dt}{t^s}-\sum_{p=1}^{P}\frac{B_{2p}}{(2p)!}\,\frac{(-1)^{2p-1}s(s+1)\cdots(s+2p-2)}{x^{s+2p-1}}\right]_{x=k}. \end{aligned}\tag{2.87}$$

The expression in square brackets can be simplified; the integral can be written $B_0\,k^{1-s}/(s-1)$ and (because $2p$ is even) the terms with nonzero p become

$$+\frac{B_{2p}\,(s+2p-2)!}{(s-1)!(2p)!\,k^{s+2p-1}}=\frac{B_{2p}}{s-1}\binom{s+2p-2}{2p}\frac{1}{k^{s+2p-1}}\,.$$

We now recognize that the integral is the $p=0$ case of the Bernoulli sum; that is not surprising if we think of it as the (-1)th derivative. Thus, our final form of Eq. (2.87) is

$$\zeta(s)\approx\sum_{n=1}^{k-1}\frac{1}{n^s}+\frac{1}{2k^s}+\frac{1}{s-1}\sum_{p=0}^{P}\binom{s+2p-2}{2p}\frac{B_{2p}}{k^{s+2p-1}}\,. \tag{2.88}$$

Using Eq. (2.88) we now find that there are choices of k and P (the range of the Bernoulli sum) that permit highly accurate evaluation of $\zeta(s)$ with only modest amounts of work.

A MAPLE procedure that enables test of Eq. (2.88) is the following:

```
> Ztest := proc(s, k, P) local p;                    approximation to ζ(s)
>    add(1/p^s, p=1 .. (k-1) )                       first k − 1 terms are explicit
>   +(1/2)/k^s                                       adjustment to k term
>   +1/(s-1)*add(binomial(s+2*p-2, 2*p)              p = 0 term is the integral
>              *bernoulli(2*p)/k^(s+2*p-1),          the factors combine into
>              p = 0 .. P);                          a binomial coefficient
>   evalf(%)  end proc;                              want floating-point output
```

A MATHEMATICA procedure based on Eq. (2.88) is

```
ztest[s_, k_, p_] :=                                 approximation to ζ(s)
Module[{q,u}, u = Sum[1/q^s, {q, 1, k-1}]            first k − 1 terms are explicit
      + (1/2)/k^s                                    adjustment to k term
      + 1/(s-1)*Sum[Binomial[s+2*q-2, 2*q]           q = 0 term is the integral
                *BernoulliB[2*q]/k^(s+2*q-1),        the factors combine into
                {q, 0, p}];                          a binomial coefficient
      N[u]                                           want floating-point output
     ]
```

■

Exercises

2.11.1. (a) Verify the algebraic steps leading to Eq. (2.88).

(b) Enter MAPLE or MATHEMATICA code for Example 2.11.1 into your computer and test its function for a wide variety of values of the arguments s, k and P (or p).

2.11.2. The Euler-Maclaurin integration formula may be used for the evaluation of finite series:

$$\sum_{m=1}^{n} f(m) = \int_1^n f(x)\,dx + \frac{1}{2}f(1) + \frac{1}{2}f(n) + \frac{B_2}{2!}\Big[f'(n) - f'(1)\Big] + \cdots .$$

Show that

(a) $\sum_{m=1}^{n} m = \frac{1}{2}\,n(n+1)$,

(b) $\sum_{m=1}^{n} m^2 = \frac{1}{6}\,n(n+1)(2n+1)$,

(c) $\sum_{m=1}^{n} m^3 = \frac{1}{4}\,n^2(n+1)^2$,

(d) $\sum_{m=1}^{n} m^4 = \frac{1}{30}\,n(n+1)(2n+1)(3n^2+3n-1)$.

2.11.3. Use the Euler-Maclaurin formula to compute the Euler-Mascheroni constant as follows:

(a) Show that $\gamma = \sum_{s=1}^{n} s^{-1} - \int_1^n \frac{dx}{x} + \lim_{N\to\infty}\left[\sum_{s=n+1}^{N} s^{-1} - \int_n^N \frac{dx}{x}\right]$

(b) Using Eq. (2.85), show that the equation of part (a) can be rearranged to

$$\gamma \approx \sum_{s=1}^{n} s^{-1} - \ln n - \frac{1}{2n} + \sum_{k=1}^{k_{\rm max}} \frac{B_{2k}}{(2k)n^{2k}}\,.$$

(c) Write a symbolic procedure `EulerMasch` that will compute approximations to γ as a function of n and $k_{\rm max}$ and find values of these parameters that will give a value of γ that is accurate to 15 decimal places.

Note. Remember to do your symbolic computations at sufficiently high accuracy, and compare your results to an authentic value of γ obtained as a built-in constant of your symbolic computing system.

Additional Readings

Abramowitz, M., & Stegun, I. A. (Eds.) (1972). *Handbook of mathematical functions with formulas, graphs, and mathematical tables (AMS-55).* Washington, DC: National Bureau of Standards (Reprinted, Dover (1974). Contains power series expansions of a large number of elementary and special functions).

Arfken, G. B., Weber, H. J., & Harris, F. E. (2013). *Mathematical methods for physicists* (7th ed.). New York: Academic Press.

Galambos, J. (1976). *Representations of real numbers by infinite series.* Berlin: Springer.

Gradshteyn, I. S., & Ryzhik, I. M. In A. Jeffrey, & D. Zwillinger (Eds.) (2007). *Table of integrals, series, and products* (7th ed.). New York: Academic Press. (Corrected and enlarged).

Hansen, E. (1975). *A table of series and products.* Englewood Cliffs, NJ: Prentice-Hall (A tremendous compilation of series and products).

Jeffrey, A., & Dai, H.-H. (2008). *Handbook of mathematical formulas and integrals* (4th ed.). New York: Academic Press.

Knopp, K. (1971). *Theory and application of infinite series.* (2nd ed.). London/New York: Blackie and Son/Hafner (Reprinted: A. K. Peters Classics (1997). This is a thorough, comprehensive, and authoritative work that covers infinite series and products. Proofs of almost all the statements about series not proved in this Chapter will be found in this book).

Mangulis, V. (1965). *Handbook of series for scientists and engineers.* New York: Academic Press (A most convenient and useful collection of series. Includes algebraic functions, Fourier series, and series of the special functions: Bessel, Legendre, and so on).

Olver, F. W. J., Lozier, D. W., Boisvert, R. F., & Clark, C. W. (Eds.) (2010). *NIST handbook of mathematical functions.* Cambridge, UK: Cambridge University Press (Update of AMS-55 (Abramowitz and Stegun, 1972), but links to computer programs are provided instead of tables of data).

Rainville, E. D. (1967). *Infinite series.* New York: Macmillan (A readable and useful account of series constants and functions).

Sokolnikoff, I.S., & Redheffer, R. M. (1966). *Mathematics of physics and modern engineering* (2nd ed.). New York: McGraw-Hill (A long Chapter 2 (101 pages) presents infinite series in a thorough but very readable form. Extensions to the solutions of differential equations, to complex series, and to Fourier series are included).

Chapter 3

COMPLEX NUMBERS AND FUNCTIONS

3.1 INTRODUCTION

Complex numbers and analysis based on complex-variable theory have become extremely important and valuable tools for the mathematical analysis of physical theory. Although the results of the measurement of physical quantities must, we firmly believe, ultimately be described by real numbers, there is ample evidence that successful theories predicting the results of those measurements require the use of complex numbers and analysis.

In electrical engineering, complex numbers are used to deal with the fact that the current through a circuit element such as a capacitor or inductor is not in phase with the voltage across it. In elementary quantum mechanics, time-dependent wave functions are described by complex-valued functions, and many other phenomena (e.g., dielectric constants and dielectric loss) can be represented as the real and imaginary parts of the same complex-valued quantity. In all these and other areas, physical phenomena are derived from a complex-valued formulation by recipes that assign real values to observable (i.e., measurable) quantities.

In elementary mathematics, complex numbers arise because simple algebraic equations often have solutions that cannot be expressed in terms of real numbers alone. One of the situations early encountered in a school classroom is the problem of finding solutions to a quadratic equation, of the form

$$az^2 + bz + c = 0\,. \tag{3.1}$$

As is well known, the values of z that solve this equation are given by the **quadratic formula**,

$$z = \frac{-b \pm \sqrt{b^2 - 4ac}}{2a}\,, \tag{3.2}$$

which exposes the feature that there is no real number z that solves Eq. (3.1) if the **discriminant** $b^2 - 4ac$ (the quantity whose square root appears in the solution) is negative.

The response of mathematicians to this dilemma was to expand the number system by including another set of numbers that are **defined** to have negative squares. Because these new numbers must be distinguished from ordinary numbers, they must

Mathematics for Physical Science and Engineering.
http://dx.doi.org/10.1016/B978-0-12-801000-6.00003-1

be labeled distinctively, and because of their nontraditional property (the negative square) they were termed **imaginary**. All imaginary numbers can be written as multiples of a unit imaginary defined to be the square root of -1. Euler (and later Gauss) chose i as a symbol standing for the imaginary unit, so that, for example, $\sqrt{-4}$ was written as $2i$, and **complex numbers** containing both real and imaginary terms were written in forms like $z = 3 + 2i$. General algebraic quantities were found to behave consistently if one added (or subtracted) complex numbers as though they were linear forms containing the "variable" i. Multiplication was defined by assigning i^2 the value -1. But this state of affairs left many mathematicians uneasy.

The choice of i to denote the imaginary unit has become universal in mathematics. However, in engineering, where the symbol i is customarily used for current, the imaginary unit is often given the name j. This book uses i.

We now know that a relatively straightforward and rigorous way of introducing complex numbers is to define them as ordered pairs of real numbers with specified rules for operating on these ordered pairs. From this viewpoint, one may define a complex number as nothing more than an ordered pair of two real numbers (a, b). Similarly, a complex variable is an ordered pair of two real variables,

$$z \equiv (x, y) . \tag{3.3}$$

The ordering is significant. In general (a, b) is not equal to (b, a) and (x, y) is not equal to (y, x). As usual, we continue writing a real number $(x, 0)$ simply as x, and we give $(0, 1)$ the name i. All of complex analysis can be developed in terms of ordered pairs of numbers, variables, and functions $(u(x, y), v(x, y))$.

We now define **addition** of complex numbers in terms of their components as

$$z_1 + z_2 = (x_1, y_1) + (x_2, y_2) = (x_1 + x_2, y_1 + y_2) . \tag{3.4}$$

Multiplication of complex numbers is defined as

$$z_1 z_2 = (x_1, y_1) \cdot (x_2, y_2) = (x_1 x_2 - y_1 y_2, x_1 y_2 + x_2 y_1) . \tag{3.5}$$

Using Eq. (3.5) we verify that $i^2 = (0, 1) \cdot (0, 1) = (-1, 0) = -1$, so we can also identify $i = \sqrt{-1}$ and further rewrite Eq. (3.3) as

$$z = (x, y) = (x, 0) + (0, y) = x + (0, 1) \cdot (y, 0) = x + iy . \tag{3.6}$$

Notice that the addition and multiplication rules we just defined for complex numbers have the properties we expect: they are consistent with those for ordinary arithmetic with the additional property that $i^2 = -1$. For addition, this is obvious. For multiplication,

$$(x_1 + iy_1)(x_2 + iy_2) = x_1 x_2 + i^2 y_1 y_2 + i(x_1 y_2 + y_1 x_2) = (x_1 x_2 - y_1 y_2) + i(x_1 y_2 + y_1 x_2),$$

in agreement with Eq. (3.5).

Some additional definitions and properties include the following:

Complex Conjugation: Like all complex numbers, i has an inverse under addition, denoted $-i$; in two-component form $(0, -1)$. Given a complex number $z = x + iy$, it is useful to define another complex number, $z^* = x - iy$, which we call the **complex conjugate** of z.[1] Notice specifically that the complex conjugate of i is $-i$, and that complex conjugation leaves real numbers unchanged. Forming

$$zz^* = (x + iy)(x - iy) = x^2 + y^2 , \tag{3.7}$$

[1]The complex conjugate of z is often denoted $\overline{z}$ in the mathematical literature.

we see that zz^* is real; we define the **absolute value** or **magnitude** of z, denoted $|z|$, as $\sqrt{zz^*}$.

Division: Consider now the division of two complex numbers: z_1/z. We need to manipulate this quantity to bring it to the complex number form $u + iv$ (with u and v real). We may do so as follows:

$$\frac{z_1}{z} = \frac{z_1 z^*}{zz^*} = \frac{(x_1 + iy_1)(x - iy)}{x^2 + y^2},$$

or

$$\frac{x_1 + iy_1}{x + iy} = \frac{xx_1 + yy_1}{x^2 + y^2} + i\,\frac{xy_1 - x_1 y}{x^2 + y^2}. \tag{3.8}$$

Inverse: If in Eq. (3.8) we set $x_1 = 1$ and $y_1 = 0$, we have a formula for $1/z$ (which we can also write as z^{-1}):

$$\frac{1}{z} = \frac{x - iy}{x^2 + y^2}, \tag{3.9}$$

a result that is easily verified if we note that the denominator can be written in the factored form $(x - iy)(x + iy)$. The existence of this formula makes it clear that every nonzero z (having a positive and nonzero value of $x^2 + y^2$) has a finite and nonzero inverse.

Real and Imaginary Parts: From time to time we refer to the real and imaginary parts of a complex variable. If that variable is $z = x + iy$, we identify x, the real part of z, by the notation $\mathfrak{Re}\, z$ and y, its imaginary part, by $\mathfrak{Im}\, z$. Notice that $\mathfrak{Im}\, z = y$, not iy. Thus, $\mathfrak{Re}\, z$ and $\mathfrak{Im}\, z$ are the two **real** variables that together define the **complex** variable $z = x + iy = \mathfrak{Re}\, z + i\mathfrak{Im}\, z$.

Complex Equations: It is obvious from the definitions that the real and imaginary units are linearly independent, meaning that neither is a linear function of the other. This fact means that an equation connecting two complex quantities is equivalent to two equations in the real domain: one for their real parts and another for their imaginary parts.

SYMBOLIC COMPUTATION

Both MAPLE and MATHEMATICA support the use of complex numbers by defining the symbol `I` to stand for the imaginary unit; the use of upper case permits lower-case i to be available as an index, a frequent and traditional use.

Numeric quantities that can be reduced to the form $\mathfrak{Re} + i\,\mathfrak{Im}$ are automatically presented in that form in both symbolic languages; for example, the expression $1/(1 + 3i)$ processes as follows (MAPLE to left, MATHEMATICA to right):

`> 1/(1+3*I);` $\longrightarrow \quad \frac{1}{10} - \frac{3}{10}I$ or `1/(1+3*I)` $\longrightarrow \quad \frac{1}{10} - \frac{3i}{10}$.

Real and imaginary parts, complex conjugates, and absolute values are also available in the symbolic systems. Assuming z to have previously been defined as $4 - 3i$, we have

`Re(z);`	4	`Re[z]`	4
`Im(z);`	-3	`Im[z]`	-3
`conjugate(z);`	$4+3I$	`Conjugate[z]`	$4+3i$
`abs(z);`	5	`Abs[z]`	5

Exercises

3.1.1. Verify the formula for $1/z$, Eq. (3.9).

3.1.2. Find the complex conjugates of (a) $\dfrac{z+z^*}{2i}$, (b) $\dfrac{z-z^*}{2}$.

3.1.3. Find the magnitude of (a) $3+4i$, (b) $1-i$.

3.1.4. Find both the roots of $z^2+2z+5=0$. Show that they are complex conjugates.

3.1.5. (a) Find the real and imaginary parts of $W = \dfrac{(3+i)^2(3-2i)(1-7i)}{(2-i)^3(5-i)(4+i)^2}$.

(b) Find the magnitude of W.

(c) Verify that if W^*, the complex conjugate of W, is obtained by changing the sign of i everywhere in the original expression for W, the result reduces to $\mathfrak{Re}\, W - i\,\mathfrak{Im}\, W$.

3.2 FUNCTIONS IN THE COMPLEX DOMAIN

Because a complex variable z contains two real variables, $x = \mathfrak{Re}\, z$ and $y = \mathfrak{Im}\, z$, the most general possibility for a function of z, which we call $w(z)$, can be written

$$w(z) = w(x+iy) = u(x,y) + iv(x,y)\,, \tag{3.10}$$

where $u(x,y) = \mathfrak{Re}\, w$ is the real part of w and $v(x,y) = \mathfrak{Im}\, w$ is its imaginary part. Note that $u(x,y)$ and $v(x,y)$ are real. Functions also have complex conjugates:

$$[w(z)]^* = u(x,y) - iv(x,y)\,. \tag{3.11}$$

For a completely arbitrary function it is not true that $w(z^*) = [w(z)]^*$, but if the function is given explicitly in terms of the elementary arithmetic operations (including exponentiation) we can form its complex conjugate simply by taking the complex conjugates of all the quantities that appear in it. For example, assuming that z is a complex variable and that x and y are real,

$$(z^2 - i)^* = (z^*)^2 + i\,, \quad (x^2 - iy^2)^* = x^2 + iy^2\,, \quad (z - x^2)^* = z^* - x^2\,,$$

$$(3xyz)^* = 3xyz^*\,, \quad \left(\frac{x+iy}{x-iy}\right)^* = \frac{x-iy}{x+iy}\,, \quad (e^{iz})^* = e^{-iz^*}\,.$$

Simple functions of a complex variable z can be analyzed simply by inserting in them $z = x+iy$. For example,

$$z^2 = (x+iy)^2 = x^2 + 2ixy + (iy)^2 = (x^2 - y^2) + 2ixy\,.$$

Since the fundamental operations in the complex domain obey the same rules as those for arithmetic in the space of real numbers, it is natural to define trigonometric, exponential, and other functions so that their real and complex incarnations are similar, and specifically so that the complex and real definitions agree when both are applicable. This means, among other things, that if a function is represented by a power series, we should, within the region of convergence of the power series, be able to use the same series with complex values of the expansion variable. This notion is called **permanence of the algebraic form**.

POWER SERIES

Let's consider a function $f(z)$ for which we have a power series expansion:

$$f(z) = a_0 + a_1 z + a_2 z^2 + \cdots . \tag{3.12}$$

The question we now ask is, "When will it converge?" A fairly good answer to this question is provided by the ratio test. We have convergence when, for large n,

$$R = \left| \frac{a_{n+1} z^{n+1}}{a_n z^n} \right| < 1 .$$

Rearranging to the form

$$R = \left| \frac{a_{n+1}}{a_n} \right| |z| < 1 ,$$

we see that the region of convergence depends only upon the absolute value of z, and is for all z such that

$$|z| < R_\infty = \lim_{n\to\infty} \left| \frac{a_n}{a_{n+1}} \right| .$$

For $|z| > R_\infty$ the power series diverges. The ratio test fails when $|z| = R_\infty$; the extent of convergence there (if of interest) needs to be established in other ways.

The concepts of absolute and uniform convergence continue to apply when the elements of a series are complex; just as for real power series, a complex power series is absolutely and uniformly convergent for all $|z| < S$ where $0 < S < R_\infty$. Thus, subject to this range limitation, the terms of a power series can be reordered at will without changing its value, and we may invoke the integration and differentiation properties of uniformly convergent series.

It is of course usually possible to represent a function by a Taylor series with an expansion point b other than $z = 0$. In that event the region of convergence would be for a region such that $|z - b| < R_\infty$.

EXPONENTIAL

As a first example of complex power series, let's look at the exponential function e^z. Using the power series expansion we found in Chapter 2, we define

$$e^z = 1 + z + \frac{1}{2!} z^2 + \frac{1}{3!} z^3 + \frac{1}{4!} z^4 + \cdots . \tag{3.13}$$

This expansion converges for all finite z, since the ratio of the nth term to the $(n-1)$th term is z/n, which has an absolute value smaller than unity for all terms n with $n > |z|$.

As a first step toward understanding the behavior of e^z, note that

$$e^z = e^{x+iy} = e^x\, e^{iy}\,, \tag{3.14}$$

and the only unfamiliar part of this expression is the factor e^{iy}. Writing e^{iy} as its power series expansion, we have

$$\begin{aligned} e^{iy} &= 1 + iy + \frac{1}{2!}(iy)^2 + \frac{1}{3!}(iy)^3 + \frac{1}{4!}(iy)^4 + \cdots \\ &= \left[1 - \frac{1}{2!}y^2 + \frac{1}{4!}y^4 - \cdots\right] + i\left[y - \frac{1}{3!}y^3 + \frac{1}{5!}y^5 - \cdots\right]. \end{aligned} \tag{3.15}$$

For reasons that will soon become apparent we have regrouped the terms of the expansion so as to separate those that are real from those that are imaginary. The regrouping is permissible for arbitrary y because the original series is absolutely convergent for all y.

We now make the key observation that the bracketed expansions in the last line of Eq. (3.15) are the power series expansions of $\cos y$ and $\sin y$ (which are also convergent for all y), so we have the extremely valuable result (known as **Euler's formula**)

$$e^{iy} = \cos y + i \sin y\,. \tag{3.16}$$

Although our motivation for the above analysis was the separation of the real and imaginary terms of the expansion, Euler's formula continues to be valid even if y is made complex, but it is most useful when y is real.

Combining Eqs. (3.14) and (3.16), we have the general result

$$e^z = e^x(\cos y + i \sin y)\,. \tag{3.17}$$

SYMBOLIC COMPUTATION

Our symbolic computation systems do not automatically reduce functions of complex variables to the form $\mathfrak{Re} + i\,\mathfrak{Im}$. However, we can force an attempt to obtain this reduction (which often produces results that are more complicated than what we started from). In fact, readers without a firm grasp of the properties of elementary and special functions may be surprised by the results that sometimes appear when fairly simple expressions are separated into their real and imaginary parts.

The commands involved here are `evalc` (MAPLE; this stands for "complex evaluation") and `ComplexExpand` (MATHEMATICA). The commands `expand` and `Expand` sometimes work, but do not collect the real and imaginary parts and do not know anything about the behavior of various functions.

Both `evalc` and `ComplexExpand` assume that the variables in the expressions to which these commands are applied are real (although advanced users can tell the algebra system to consider them complex). These commands also know how to handle elementary functions with complex arguments. We illustrate with a simple and two more complicated examples.

Example 3.2.1. Simple Real/Imaginary Separation

Let's assume that z has been defined as $x + iy$. Then

`> evalc(z^3);` $\longrightarrow$ $x^3 - 3xy^2 + (3x^2y - y^3)I\,,$

`ComplexExpand[z^3]` $\longrightarrow$ $x^3 - 3xy^2 + i(3x^2y - y^3)$.

If, for example, we had used **expand**, we would have gotten

`> expand(z^3);` $\longrightarrow$ $x^3 + 3Ix^2y - 3xy^2 - y^3I$.

■

Example 3.2.2. More Complicated Real/Imaginary Separations

Again we assume that z has been defined as $x + iy$. Now

`> evalc(z*exp(z));` $\longrightarrow$ $xe^x \cos(y) - ye^x \sin(y) + (ye^x \cos(y) + xe^x \sin(y))I$,

`ComplexExpand[z*E^z]` $\longrightarrow$ $e^x x\,\mathrm{Cos}[y] - e^x y\,\mathrm{Sin}[y] + i(e^x y\,\mathrm{Cos}[y] + e^x x\,\mathrm{Sin}[y])$.

This time the use of **expand** or **Expand** is less successful:

`Expand[z*E^z]` $\longrightarrow$ $e^{x+iy}x + ie^{x+iy}y$.

As a final example, let's ask for a separation of $\tan z$.

`> evalc(tan(z));` $\longrightarrow$ $\dfrac{\sin(x)\cos(x)}{\cos(x)^2 + \sinh(y)^2} + \dfrac{\sinh(y)\cosh(y)\,I}{\cos(x)^2 + \sinh(y)^2}$.

`ComplexExpand[Tan[z]]` $\longrightarrow$ $\dfrac{\mathrm{Sin}[2x]}{\mathrm{Cos}[2x] + \mathrm{Cosh}[2y]} + \dfrac{i\,\mathrm{Sinh}[2y]}{\mathrm{Cos}[2x] + \mathrm{Cosh}[2y]}$.

These results, which appear different, can be shown to be equal. Our two symbolic systems also behave differently if we compare the results of using **expand** and **Expand**. `Expand[Tan[z]]` does nothing; it returns `Tan[x+iy]`; **expand** does a partial expansion:

`expand(tan(z));` $\longrightarrow$ $\dfrac{\tan(x) + \tanh(y)\,I}{1 - \tan(x)\tanh(y)I}$.

■

Exercises

3.2.1. Separate into two real equations (but do not try to solve them):

(a) $z^3 = \dfrac{(7-4i)^3}{(1-i)^2}$,

(b) $e^{1/z} = 1 - 5i$,

(c) $z^7 = 1$.

3.2.2. Separate into real and imaginary parts:

(a) e^{z^5},

(b) $\cos^2(1/z)$.

3.2.3. (a) Separate $\sin z = \sin(x + iy)$ into real and imaginary parts.

(b) Find the two real equations equivalent to $\sin z = 0$.

(c) Plot $\cosh y$ and $\sinh y$ over a range sufficient to convince yourself that there is no real value of y such that $\cosh y = 0$, and that the only real value of y for which $\sinh y = 0$ is $y = 0$.

(d) Based on your solutions to parts (b) and (c), determine all the values of $z = x + iy$ for which $\sin z = 0$.

Note. This exercise shows how judicious use of a symbolic computing system can help you identify properties of the mathematical quantities involved in a problem.

3.3 THE COMPLEX PLANE

Real numbers can be identified with points on a line; because complex numbers have two components, they can be identified with points on a plane. Given $z = x + iy$, it is customary to plot z as a point whose horizontal coordinate is x and whose vertical coordinate is y. The plane on which complex numbers are plotted is formally known as an **Argand diagram**; it is often just referred to as the **complex plane**, and the horizontal and vertical axes are usually called the **real** and **imaginary** axes. See Fig. 3.1.

Once we have located $z = x+iy$ on a complex plane, it is natural to notice that the point corresponding to z can alternatively be expressed in polar coordinates. Letting r be the distance from the origin and taking θ as the angle measured counterclockwise from the real axis, we have

$$x = r\cos\theta, \quad y = r\sin\theta\,, \qquad \text{or} \qquad r = \sqrt{x^2+y^2}, \quad \theta = \tan^{-1} y/x\,. \tag{3.18}$$

The arctan function $\tan^{-1}(y/x)$ is multiple valued; the value of θ needs to be consistent with the individual values of x and y.

The Cartesian and polar representations of a complex number can also be related by writing

$$x + iy = r(\cos\theta + i\sin\theta)\,. \tag{3.19}$$

From Eq. (3.19) we see that $r = |z|$; given values of $r \geq 0$ and θ identify a unique complex number. In complex-variable theory r is also called the **modulus** of z and θ is termed the **argument** or the **phase** of z.

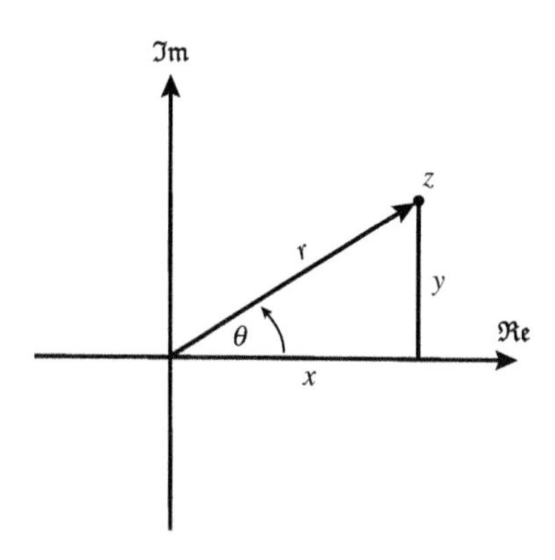

Figure 3.1: Argand diagram, showing location of $z = x + iy = re^{i\theta}$.

Based on our earlier exploration of e^{iy}, we also notice that the polar representation of $x + iy$ can be written

$$x + iy = r(\cos\theta + i\sin\theta) = re^{i\theta} \; . \tag{3.20}$$

We remind the reader that r is the magnitude of z and we note that $e^{i\theta}$ is a complex function whose magnitude (for all θ) is unity; an elementary proof of this result is obtained by invoking the well-known identity $\cos^2\theta + \sin^2\theta = 1$. Another simple proof proceeds as follows:

$$\left|e^{i\theta}\right|^2 = e^{i\theta}\left(e^{i\theta}\right)^* = e^{i\theta}e^{-i\theta} = e^0 = 1 \; .$$

Complex numbers of the same $r = |z|$ lie on the complex plane at equal distances from the origin, so the locus of all z with the same $|z|$ is a circle. In contrast, all z with the same argument θ fall on a ray that starts at the origin and makes an angle θ with the positive real axis.

Returning now to the convergence condition $|z| < R_\infty$ of a power series expansion, we note that the region of convergence is a circular disk in the complex plane whose boundary is a circle of radius R_∞ centered at $z = 0$. This region is sometimes referred to as the **disk of convergence** of our power series, and the radius of the circle bounding that disk is known as the **radius of convergence**. Remember that convergence of our power series is guaranteed only **within** (but not on) the radius of convergence.

Since addition of complex numbers corresponds to separate addition of their real and imaginary parts, the sum of two complex numbers $z_1 = x_1 + iy_1$ and $z_2 = x_2 + iy_2$ (each represented by an arrow in the complex plane) can be formed by placing the first arrow (at its proper orientation) with its tail at the origin, then placing the second with its tail at the head of the first arrow, so that the resultant, $z = z_1 + z_2$, has $x = x_1 + x_2$ and $y = y_1 + y_2$. See Fig. 3.2.

Given the location of z on an Argand diagram, where will we find its complex conjugate z^* and its additive inverse $-z$? Since both z and z^* have the same value of x but with y values of the same magnitude but opposite sign, z and z^* are related by reflection about the real axis. See Fig. 3.3. This symmetry makes it obvious that no matter what z we choose, the sum $z + z^*$ lies on the real axis and therefore be real. Again see Fig. 3.3.

Since $-z = -x - iy$, its location in the complex plane is opposite to that of z, as also shown in the figure. We also see that $z - z^*$ lies on the imaginary axis and is therefore pure imaginary. Algebraically, these relations correspond to

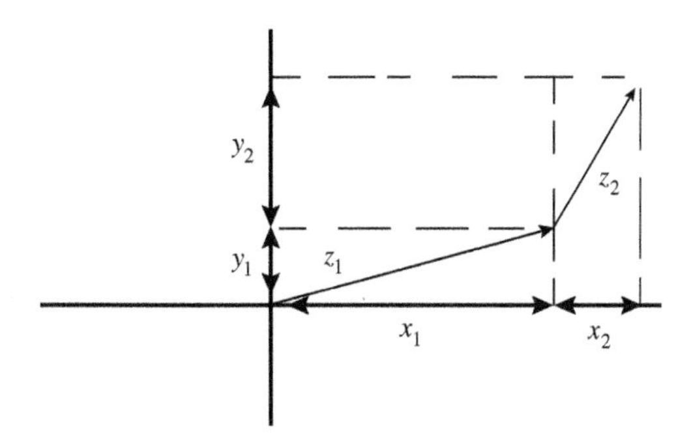

Figure 3.2: Graphical addition of two complex numbers.

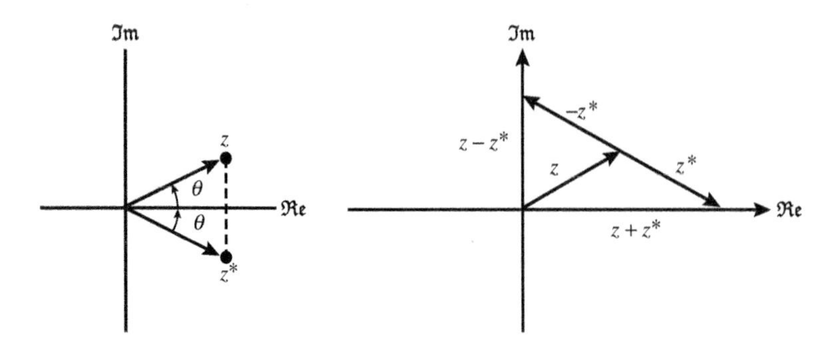

Figure 3.3: Left: Relation of z and z^*; Right: $z + z^*$ and $z - z^*$.

$z+z^* = (x+iy)+(x-iy) = 2x$ and $z-z^* = (x+iy)-(x-iy) = 2iy$, equivalent to

$$\mathfrak{Re}\, z = \frac{z+z^*}{2}\,, \qquad \mathfrak{Im}\, z = \frac{z-z^*}{2i}\,. \tag{3.21}$$

We emphasize: $z + z^*$ is real; $z - z^*$ is pure imaginary.

COMPLEX NUMBERS OF UNIT MAGNITUDE

We start from our previous results that

$$e^{i\theta} = \cos\theta + i\sin\theta \tag{3.22}$$

and that

$$\left|e^{i\theta}\right| = 1\,, \tag{3.23}$$

showing that $e^{i\theta}$ has magnitude unity. The points $\exp(i\theta)$ therefore lie in the complex plane on the **unit circle**, at polar angles θ. This observation makes obvious a number of relations that could in principle also be deduced from Eq. (3.22). For example, if θ has the special value $\pi/2$, the corresponding z value will be at 90° from the real axis and will therefore be at $+1$ on the imaginary axis, which is the point i. A similar analysis for $\theta = \pi$ gives us the result -1, while the choice $\theta = 3\pi/2$ leads to the result $-i$. Summarizing,

$$e^0 = 1, \quad e^{i\pi/2} = i, \quad e^{i\pi} = -1, \quad e^{3i\pi/2} = -i. \tag{3.24}$$

Particularly the result $e^{i\pi} = -1$ has been cited for its apparently mystical connection between the fundamental constants e and π. Values of z on the unit circle are illustrated in Fig. 3.4.

We also see that $\exp(i\theta)$ is periodic, with period 2π, so for example

$$e^{\pm 2i\pi} = e^{\pm 4i\pi} = \cdots = 1\,, \qquad e^{\pm i\pi} = e^{\pm 3i\pi} = \cdots = -1\,, \qquad e^{3i\pi/2} = e^{-i\pi/2} = -i\,. \tag{3.25}$$

The periodicity, together with the fact that the real and imaginary parts of $e^{i\theta}$ are sinusoidal functions, make complex exponentials useful in describing oscillatory motion. For example, the real part of

$$e^{i(\omega t+\delta)} = \cos(\omega t + \delta) + i\sin(\omega t + \delta)$$

describes an oscillation whose angular frequency is ω and whose phase (relative to that of an oscillation with a maximum at $t = 0$) is δ. The imaginary part of $\exp(i\omega t)$ also describes oscillation, but displaced in phase from the real part by 90°.

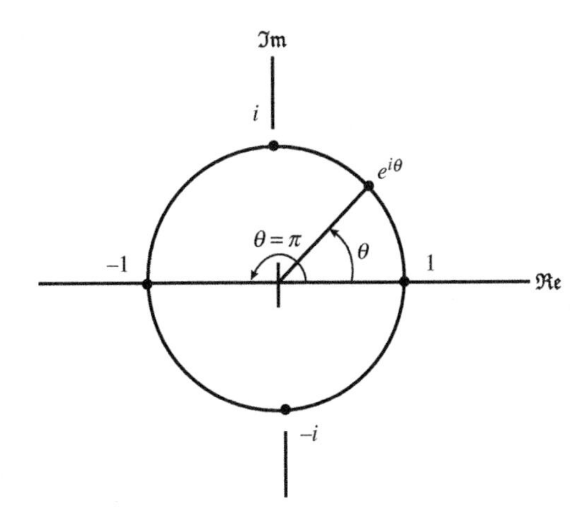

Figure 3.4: Some values of z on the unit circle.

Note that when using complex exponentials to describe angles and phases, the formulas require the use of radian measure. One may of course speak of degrees, but one must remember that

$$\theta_{\text{radians}} = \theta_{\text{degrees}} \times \frac{\pi}{180} \,. \tag{3.26}$$

PRODUCTS IN POLAR FORM

If we have two complex numbers, z and z_1, in polar form, their product zz_1 can be written

$$zz_1 = (re^{i\theta})(r_1 e^{i\theta_1}) = (rr_1)e^{i(\theta+\theta_1)} \,, \tag{3.27}$$

showing that the location of the product in an Argand diagram has argument (polar angle) equal to the sum of the polar angles of the factors, and has a magnitude that is the product of the factor magnitudes. Conversely, the quotient z/z_1 has magnitude r/r_1 and argument $\theta - \theta_1$. These relationships should aid in getting a qualitative understanding of complex multiplication and division. This discussion also shows that multiplication and division are easier in the polar representation, whereas addition and subtraction have simpler forms in Cartesian coordinates.

Where in the complex plane is the reciprocal of $z = re^{i\theta}$? Since $z^{-1} = r^{-1}e^{-i\theta}$, it must have an argument which is the negative of that of z and a magnitude which is the reciprocal of that of z. Thus, if z is within the unit circle, z^{-1} will be outside, and vice versa. If z has unit magnitude, both z and z^{-1} are on the unit circle.

PLOTTING FUNCTIONS

We can use an Argand diagram to plot values of a function $w(z)$ as well as just z itself, in which case we could label the axes u and v, referring to the real and imaginary parts of w. In that case, we can think of the function $w(z)$ as providing a **mapping** from the xy plane to the uv plane, with the effect that any curve in the xy (sometimes called z) plane is mapped into a corresponding curve in the uv $(= w)$ plane. In fact, as a nearly trivial example, the unit circle in Fig. 3.4 is the map of e^z when $z = i\theta$, with θ varied over a real-valued range of length 2π.

SYMBOLIC COMPUTATION

Conversion between Cartesian and polar representations of complex quantities is probably more important for gaining understanding than for doing computer-assisted computations, in part because the difference in complexity of various uses of the two representations is not significant to a computer. Nevertheless, we provide here the tools for converting between these two representations.

To convert a complex quantity *expr* from Cartesian to polar form, we need its modulus and argument; in MAPLE, these are obtained as `abs(`*expr*`)` and `argument(`*expr*`)`. In MATHEMATICA, these commands are `Abs[`*expr*`]` and `Arg[`*expr*`]`. To convert from polar to Cartesian form, one simply applies `evalc` or `ComplexExpand` to *expr* written in polar form.

Example 3.3.1. Cartesian—Polar Conversions

```
> C:=4 - 3*I;
```
$$4-3I$$
```
> abs(C);
```
$$5$$
```
> argument(C);
```
$$-\arctan\left(\frac{3}{4}\right)$$
```
> evalc(5*exp(-I*arctan(3/4)))
```
$$4-3I$$
```
> polar(5,-arctan(3/4))
```
$$4-3I$$

Notice that MAPLE provides `polar` as an alternative to the explicit construction of the $r\,e^{i\theta}$ polar form. In MATHEMATICA,

```
F=4 - 3*I
```
$$4-3i$$
```
Abs[F]
```
$$5$$
```
Arg[F]
```
$$-\text{ArcTan}\left[\frac{3}{4}\right]$$
```
ComplexExpand[5*Exp[-I*ArcTan[3/4]]]
```
$$4-3i$$

■

Plots in the complex plane are often best produced as **parametric** plots, in which the real and imaginary parts of $z(t) = x(t) + iy(t)$ assume a range of values controlled by a **real** parameter t. In MAPLE, the syntax for obtaining such a plot is

> `plot([`$x(t)$, $y(t)$, $t = t_1\,..\,t_2$`])`

where $x(t)$ and $y(t)$ are expressions giving the current values of x and y and the range of t is from t_1 to t_2. The square brackets `[ ]` and the range of t are required elements of the command.

In MATHEMATICA, a similar parametric plot is achieved by the command

`ParametricPlot[`$\{x[t],\ y[t]\},\ \{t, t_1, t_2\}$`]`

In both languages, multiple parametric plots can be specified in a single command; in MAPLE, they must be listed within an outer set of square brackets, as in

$[\ [x_1(t),\ y_1(t),\ t = t_1..t_2],\ [x_2(t),\ y_2(t),\ t = t_3..t_4]\ ];$

in MATHEMATICA, they must be grouped within outer braces and be **for the same parameter range**, as in

{ $\{x_1[t],\ y_1[t]\}$, $\{x_2[t],\ y_2[t]\}$ }, $\{t, t_1, t_2\}$.

More detail on parametric plots can be found in Appendix A.

Example 3.3.2. Function Plots

As a first example, assume that we want to evaluate the function $w(z) = 1/(4-z+z^2)$ on the upper half of the unit circle, on which $y = \sqrt{1-x^2}$. In MAPLE, we define

```
> y := proc(x); sqrt(1-x^2) end proc;
> w := proc(z); 1/(4-z+z^2) end proc;
```

and we can plot the trajectory of z and that of $w(z)$ as follows:

```
> plot([x,y(x),x=-1 .. 1]);
> plot([Re(w(x+I*y(x))),Im(w(x+I*y(x))),x=-1 .. 1]);
```

We may wish to force the vertical and horizontal scales to be the same; if so, insert in the above commands an additional argument, as illustrated by

```
> plot([x,y(x),x=-1 .. 1],scaling=constrained);
```

The plot output (with constrained scaling) is presented in Fig. 3.5. The left panel shows the curve in the z-plane on which w is evaluated; the right panel are the values of w calculated on that curve (its "map in the w-plane").

This same example, in MATHEMATICA, can be produced by the following code:

```
y[x_] := Sqrt[1-x^2]
f[z_] := 1/(4-z+z^2)
ParametricPlot[{Re[f[x+I*y[x]]],Im[f[x+I*y[x]]]}, {x,-1,1}]
```

A more involved example is to plot the curves identifying the values of x and y that satisfy the equation $|z^2-1| = s$, where s is a positive-valued parameter. Writing first $|z^2-1|^2 = (z^2-1)(z^{*2}-1) = s^2$ and inserting $z = x+iy$, we reach

$$(z^2-1)(z^{*2}-1) = (x^2-y^2-1)^2 + 4x^2y^2 = (x^2+y^2+1)^2 - 4x^2 = s^2\,,$$

which can be solved to yield $y = \pm\sqrt{\sqrt{s^2+4x^2} - x^2 - 1}$.

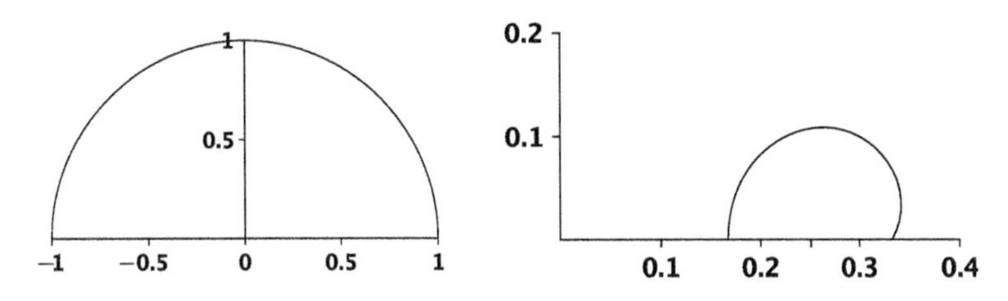

Figure 3.5: First function plot, Example 3.3.2. Left: Curve in the z-plane. Right: Its map in the w-plane.

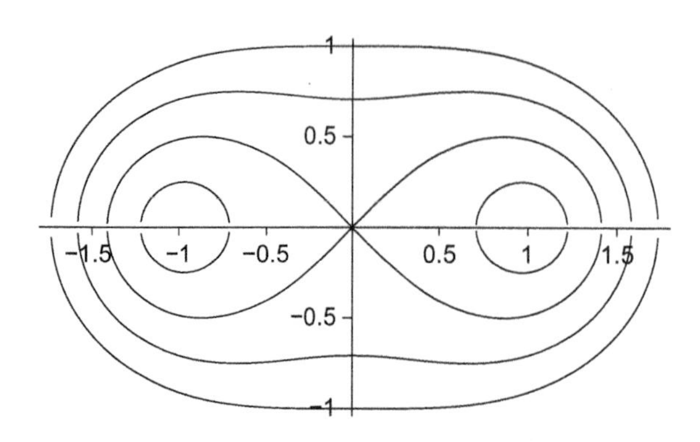

Figure 3.6: Second function plot, Example 3.3.2. Outer curves correspond to larger s values.

Since we will make plots for several values of s, we define (in MAPLE)

```
F:=proc(s); sqrt(sqrt(s^2+4*x^2)-x^2-1) end proc;
```

and then make plots for both the $+y$ and $-y$ branches for $s = 0.5$, 1.0, 1.5, and 2.0:

```
> plot([[x,F(0.5),x=-2 .. 2], [x,F(1.),x=-2 .. 2],

  [x,F(1.5),x=-2 .. 2],   [x,F(2.),x=-2 .. 2], [x,-F(0.5),x=-2 .. 2],

  [x,-F(1.),x=-2 .. 2],  [x,-F(1.5),x=-2 .. 2], [x,-F(2.),x=-2 .. 2]],

  scaling=constrained);
```

Note that it does no harm for the range of x to be larger than the range for which F is real. Portions of curves for which F is not real are ignored by MAPLE's plotting program. The plot produced by the above code is shown in Fig. 3.6.

The corresponding MATHEMATICA code is

```
f[s_] := Sqrt[Sqrt[s^2+4*x^2]-x^2-1]

ParametricPlot[{{x,f[0.5]}, {x,f[1.]}, {x,f[1.5]}, {x,f[2.]},

      {x,-f[0.5]}, {x,-f[1.]}, {x,-f[1.5]}, {x,-f[2.]}}, {x,-2,2}]
```

■

Exercises

3.3.1. Locate the following quantities on the complex plane.

(a) $1 + i$ (b) $(1 + i)^*$ (c) $-(1 + i)$

(d) $4e^{i\pi/4}$ (e) $4e^{-i\pi/2}$ (f) $e^{i\pi/4}/4$.

3.3.2. Locate the following quantities on the complex plane, given that $z = 1 + i$.

(a) $\dfrac{z}{z}$ (b) $\dfrac{z}{z+2}$ (c) $\dfrac{1}{z-2}$

(d) $\dfrac{1}{z+i}$ (e) $e^{\pi z/5}$ (f) $e^{-i\pi z/5}$.

3.3.3. For each of the following complex numbers, compute its absolute value and then multiply it by a real number such that the resulting complex number has absolute value unity.

$$\text{(a) } 3+4i \qquad \text{(b) } \frac{1}{4-3i} \qquad \text{(c) } \frac{1+3i}{3-i}\,.$$

3.3.4. Find the reciprocal of $x + iy$, working in polar form but expressing the final result in Cartesian form.

3.3.5. Let $z_1 = 3 - 4i$ and $z_2 = 4 + 3i$.

(a) Find the polar forms of z_1 and z_2. Express the arguments as angles given both in degrees and in radians.

(b) Form $z_1 + z_2$ and $z_1 - z_2$, expressing each of these results both in Cartesian and in polar form.

(c) Form $z_1 z_2$ and z_1/z_2, expressing each of these results both in Cartesian and in polar form.

3.3.6. Plot the function $e^{(i-a)t}$ in the complex plane for $a = 0.05$, 0.10, and 0.15, in each case for the range $0 \le t \le 10$.

Hint. Make parametric plots.

3.4 CIRCULAR AND HYPERBOLIC FUNCTIONS

Euler's formula, Eq. (3.22), enables us to obtain interesting formulas for the sine and cosine. Taking the sum and difference of $\exp(+i\theta)$ and $\exp(-i\theta)$, we have

$$\cos\theta = \frac{e^{i\theta}+e^{-i\theta}}{2}\,, \qquad \sin\theta = \frac{e^{i\theta}-e^{-i\theta}}{2i}\,. \tag{3.28}$$

The circular functions of argument θ are the x and y values of a point on the unit circle (equation $x^2+y^2=1$) reached from the point $x=1,\ y=0$ by counterclockwise travel along the arc of the circle a distance θ. The corresponding hyperbolic functions of argument μ are the x and y values of a point on a hyperbola (equation $x^2-y^2=1$) reached by upward travel along the arc a distance μ starting from $(1,0)$. It can be shown that

$$\cosh\mu = \frac{e^{\mu}+e^{-\mu}}{2}\,, \qquad \sinh\mu = \frac{e^{\mu}-e^{-\mu}}{2}\,. \tag{3.29}$$

Equation (3.29) can be regarded as the **definition** of the hyperbolic functions $\cosh\mu$ and $\sinh\mu$.

While the preceding discussion dealt explicitly with real values of the quantities θ and μ, the relationships we have identified extend to arbitrary complex numbers z. To emphasize this point, we rewrite Eqs. (3.28) and (3.29) in the more general notation

$$\begin{aligned} \cos z &= \frac{e^{iz}+e^{-iz}}{2}\,, & \sin z &= \frac{e^{iz}-e^{-iz}}{2i}\,. \\ \cosh z &= \frac{e^{z}+e^{-z}}{2}\,, & \sinh z &= \frac{e^{z}-e^{-z}}{2}\,, \end{aligned} \tag{3.30}$$

Comparing the members of this set of equations, it is possible to establish the formulas

$$\cosh iz = \cos z \,, \qquad \sinh iz = i \sin z \,. \tag{3.31}$$

Proof is left to Exercise 3.4.2.

The equation showing that $\exp(in\theta)$ can be written in the two equivalent forms

$$e^{in\theta} = \cos n\theta + i \sin n\theta = (\cos\theta + i\sin\theta)^n \tag{3.32}$$

is known as **de Moivre's Theorem**. By expanding the right member of Eq. (3.32), we easily obtain trigonometric multiple-angle formulas, of which the simplest examples are the well-known results

$$\cos 2\theta = \cos^2\theta - \sin^2\theta \,, \qquad \sin 2\theta = 2\sin\theta\cos\theta \,.$$

INTEGRALS

For integer n, we consider now the integral

$$\int_0^{2\pi} e^{in\theta}\, d\theta = \int_0^{2\pi} (\cos n\theta + i \sin n\theta)\, d\theta = \begin{cases} 2\pi & n = 0, \\ 0 & n \neq 0. \end{cases} \tag{3.33}$$

This result is obvious because (for nonzero n) the sine and cosine are periodic, and in the interval 2π each undergoes n complete oscillations. When $n = 0$, the integral is just that of $d\theta$, which integrates to the length of the interval, namely 2π. Equation (3.33) is useful for establishing a variety of trigonometric integrals by writing them using Eq. (3.30); see Exercise 3.4.6.

Other integrals whose evaluation is made easier using Euler's formula are of forms like

$$\int e^{ax} \cos bx\, dx \qquad \text{or} \qquad \int e^{ax} \sin bx\, dx \,.$$

These integrals are the real and imaginary parts of the complex integral

$$\int e^{(a+ib)x}\, dx = \frac{e^{(a+ib)x}}{a+ib} \,. \tag{3.34}$$

Incidentally, Eq. (3.34) illustrates the fact that the appearance of a complex parameter in a mathematical expression does not prevent us from using established operational techniques. Of course, a complex parameter value may affect the convergence of sums or integrals in which it appears.

The use of Eq. (3.34) is illustrated in Exercise 3.4.7.

SYMBOLIC COMPUTATION

The trigonometric functions are known to the symbolic systems by the names sin, cos, tan, cot, sec, csc, with (only in MATHEMATICA) the initial letter capitalized. The hyperbolic functions are designated sinh, cosh, tanh, coth, sech, and csch (also with the initial letter capitalized in MATHEMATICA). Both symbolic systems automatically evaluate these functions when special values of their arguments make it possible. Although both systems know how to simplify trigonometric expressions, neither does so unless asked by a simplification or form-modification command.

Some of these functions have definite values when infinity is approached from a specific direction in the complex plane. The symbolic systems interpret the arguments `infinity` (`Infinity`) and `-infinity` (`-Infinity`) as an approach from real values of the indicated sign.

Example 3.4.1. Circular and Hyperbolic Functions

Maple		Mathematica	
`> sin(Pi/2);`	1	`Sin[Pi/2]`	1
`> cosh(I*Pi);`	-1	`Cosh[I*Pi]`	-1
`> tanh(-infinity);`	-1	`Tanh[-Infinity]`	-1
`> sinh(-infinity);`	$-\infty$	`Sinh[-Infinity]`	$-\infty$
`> sin(infinity);`	*undefined*	`Sin[Infinity]`	Interval[{−1, 1}]
`> tan(0.1);`	0.1003346721	`Tan[0.1]`	0.100335

■

Exercises

3.4.1. Evaluate, using symbolic computation if helpful:

(a) $\sin(0.1)$, $\sin(1)$, $\sin 3\pi/2$, $\sin \pi i$, $\sin 8\pi$,

(b) $\sinh(0.1)$, $\sinh(1)$, $\sinh \pi i/3$, $\sinh \pi$, $\sinh 8\pi$,

(c) $\cos(0)$, $\cos(0.1)$, $\cos(\frac{1}{2}\pi + 0.1)$, $\cos \pi i$, $\cos 2\pi/3$, $\cos 8\pi$,

(d) $\cosh(0)$, $\cosh(0.1)$, $\cosh(\frac{1}{2}\pi+0.1)$, $\cosh \pi i$, $\cosh 2\pi/3$, $\cosh 8\pi$,

(e) $\tan(0)$, $\cot(0)$, $\tan \pi/4$, $\cot \pi/4$, $\tan 8\pi i$, $\cot 8\pi i$,

(f) $\tanh(0)$, $\coth(0)$, $\tanh \pi$, $\coth \pi$, $\tanh \pi i/4$, $\coth \pi i/4$,

(g) $\tanh(\infty)$, $\coth(\infty)$, $\tanh(-\infty)$, $\coth(-\infty)$.

3.4.2. Assume that the trigonometric functions and the hyperbolic functions are defined for complex argument by the appropriate power series. If you cannot remember these power series, use MAPLE or MATHEMATICA to obtain them. Then show that

$$i \sin z = \sinh iz, \quad \sin iz = i \sinh z,$$

$$\cos z = \cosh iz, \quad \cos iz = \cosh z.$$

3.4.3. Using the identities

$$\cos z = \frac{e^{iz} + e^{-iz}}{2}, \qquad \sin z = \frac{e^{iz} - e^{-iz}}{2i},$$

show that

(a) $\sin(x + iy) = \sin x \cosh y + i \cos x \sinh y,$

$\cos(x + iy) = \cos x \cosh y - i \sin x \sinh y,$

(b) $|\sin z|^2 = \sin^2 x + \sinh^2 y, \quad |\cos z|^2 = \cos^2 x + \sinh^2 y.$

These results show that $|\sin z|$ and $|\cos z|$ are not bounded by unity when z is permitted to assume complex values.

3.4.4. From the identities in Exercises 3.4.2 and 3.4.3 show that

$$\text{(a)}\ \sinh(x+iy) = \sinh x \cos y + i\ \cosh x \sin y,$$
$$\cosh(x+iy) = \cosh x \cos y + i \sinh x \sin y,$$
$$\text{(b)}\ |\sinh z|^2 = \sinh^2 x + \sin^2 y, \quad |\cosh z|^2 = \cosh^2 x + \sin^2 y.$$

3.4.5. Show that

$$\text{(a)}\ \tanh\frac{z}{2} = \frac{\sinh x + i\sin y}{\cosh x + \cos y}\,, \qquad \text{(b)}\ \coth\frac{z}{2} = \frac{\sinh x - i\sin y}{\cosh x - \cos y}\,.$$

3.4.6. By inserting Euler's formula for the trigonometric functions and applying Eq. (3.33), show that

$$\text{(a)}\ \int_0^{2\pi} \cos 2x \cos 3x\, dx = 0\,, \qquad \text{(b)}\ \int_0^{2\pi} \sin 2x \sin 3x\, dx = 0\,,$$

$$\text{(c)}\ \int_0^{2\pi} \sin 2x \cos 3x\, dx = 0\,, \qquad \text{(d)}\ \int_0^{2\pi} \sin 2x \cos 2x\, dx = 0\,,$$

$$\text{(e)}\ \int_0^{2\pi} \cos^2 2x\, dx = \pi\,, \qquad \text{(f)}\ \int_0^{2\pi} \sin^2 3x\, dx = \pi\,.$$

3.4.7. Using Eq. (3.34), show that

$$(a)\ \int e^{ax}\cos bx\, dx = \frac{(a\cos bx + b\sin bx)e^{ax}}{a^2+b^2}\,,$$

$$(b)\ \int e^{ax}\sin bx\, dx = \frac{(a\sin bx - b\cos bx)e^{ax}}{a^2+b^2}\,.$$

3.5 MULTIPLE-VALUED FUNCTIONS

A multiple-valued function is one which can have more than one value for the same argument; most students had their first encounter with such functions when they learned that the square root of a positive number can have either sign (e.g., $\sqrt{4} = +2$ but also -2). This type of situation normally arises when we consider the inverse of a function for which more than one argument leads to the same function value; in the example at hand the function $f(x) = x^2$ yields the same result for $-x$ as it does for $+x$, so an attempt to invert it for given x^2 has two equally legitimate results.

In the world of complex variables, the most fundamental multiple-valued function is the logarithm, whose behavior stems from the fact that different quantities, when exponentiated, can yield the same result. For example, $e^0 = e^{2\pi i} = e^{4\pi i} = \cdots$, and in general $e^z = e^{z+2\pi i} = e^{z+4\pi i} = \cdots$.

LOGARITHM

The logarithm (to the base e, which is that most natural in mathematical analysis, and therefore sometimes called the **natural logarithm**) is defined as the function inverse to the exponential. Using the notation **ln** (thereby keeping **log** available for logarithms to base 10, even though they have been made obsolete by the proliferation of hand-held calculators), the basic definition is the implicit equation

$$z = e^{\ln z} \,, \tag{3.35}$$

which leads to the basic property

$$zz_1 = e^{\ln z}\, e^{\ln z_1} = e^{\ln z + \ln z_1} \,,$$

showing that multiplication of numbers corresponds to addition of their logarithms. Since this basic result remains true if e is replaced by 10, one can use **common**, or **Briggsian** logarithms and have the practical advantage that integer powers of 10 are notationally easy to use.

Returning to natural logarithms, we note that Eq. (3.35) remains valid if we change it to

$$z = e^{\ln z + 2n\pi i} \,, \qquad n = 0, \pm 1, \pm 2, \cdots \,, \tag{3.36}$$

so all the quantities $\ln z + 2n\pi i$ are equally valid as the logarithm of z. If all we plan to do is to use logarithms to carry out multiplications, this multiple-valuedness is of little concern, because when we take the antilogarithm (i.e., exponential), the fact that $e^{2n\pi i} = 1$ causes us to recover a unique result.

Let's move on to consider the logarithms of complex numbers in polar form. Writing $z = re^{i\theta}$ and writing the real quantity r as $e^{\ln r}$, Eq. (3.36) becomes

$$z = re^{i\theta} = e^{\ln r + i\theta} = e^{\ln z + 2n\pi i} \,; \tag{3.37}$$

equating exponents of the last two members of Eq. (3.37), we get

$$\ln z = \ln r + i(\theta + 2n\pi) \,. \tag{3.38}$$

That is a key result that we will use repeatedly in complex analysis.

If we now apply Eq. (3.38) to zz_1, we get

$$\ln zz_1 = \ln z + \ln z_1 = \ln r + \ln r_1 + i(\theta + \theta_1 + 2n\pi) \,,$$

showing that the real part of $\ln zz_1$ is the logarithm of the magnitude of zz_1 and that the imaginary part of $\ln zz_1$ is the phase of zz_1 (plus an arbitrary multiple of 2π).

We can interpret Eq. (3.38) relative to its simpler form with $n = 0$, namely

$$\ln z = \ln r + i\theta$$

by noting that $\theta + 2n\pi$ corresponds on an Argand diagram to making n circuits of the polar angle about $z = 0$ before moving it an additional amount θ. This procedure brings us to the same z, no matter how many circuits we make, but different numbers of circuits yield different values for the logarithm.

Often we suppress explicit discussion of the multiple values of a function by reporting a single value that is relatively simple in form, designating this as the **principal value** of the function. For the logarithm, it is usual to choose the principal value to be real when z is real, and for complex z to specify the value of θ that falls in a convenient range of length 2π, such as either $-\pi < \theta \le \pi$ or perhaps $0 \le \theta < 2\pi$. With either of these principal-value definitions, we would then report $\ln(-1) = \pi i$, and $\ln i = \pi i/2$. The first definition would cause us to report $\ln(-i) = -\pi i/2$, while the second corresponds to $\ln(-i) = 3\pi i/2$.

Summarizing, the logarithm is infinitely multiple-valued, with the values differing by amounts 2π in their imaginary parts. The complex-variable definition of the logarithm gives meaning to the logarithms of negative real numbers, with the principal value of $\ln(-x)$ (where $x > 0$) equal to $\ln x + \pi i$, and the logarithm is defined for all complex numbers except $z = 0$.

POWERS AND ROOTS

The polar form is very convenient for expressing powers and roots of complex numbers. For integer powers, the result is obvious and unique:

$$z = re^{i\theta}, \qquad z^n = r^n e^{in\theta} .$$

However, if the power, which we now call p, is not an integer, it is convenient to introduce the logarithm of z, writing

$$z^p = \left(e^{\ln z}\right)^p = e^{p \ln z} . \tag{3.39}$$

If we now insert for $\ln z$ the formula $\ln r + i(\theta + 2n\pi)$, where n can be any integer of either sign, our expression for z^p becomes

$$z^p = e^{p \ln r + ip\theta + 2np\pi i} = r^p \, e^{ip\theta} \, e^{p(2n\pi i)} , \tag{3.40}$$

and we see that the final exponential does not reduce to unity because p has been assumed nonintegral. The number of different values we can obtain for z^p depends upon whether p is rational; if it is not, the exponential $e^{p(2n\pi i)}$ can assume **any** value of magnitude unity.

A more interesting and practical case is when p is a small rational fraction; consider first $p = 1/2$. Then $e^{p(2n\pi i)}$ can assume only one of two values; when n is even, it evaluates to unity; if n is odd, we get -1. This means that if z is written as $re^{i\theta}$, the two possible values of $z^{1/2}$ are $+r^{1/2}e^{i\theta/2}$ and $-r^{1/2}e^{i\theta/2}$. The significance of the $+$ and $-$ is that the two values of this square root differ in phase by half a revolution, i.e., π (180°).

The cube root exposes us to the fuller generality of the complex-variable power/root process. With $p = 1/3$, we have, taking steps as illustrated in Eqs. (3.39) and (3.40),

$$z^{1/3} = r^{1/3} \, e^{i\theta/3} \, e^{(2n\pi/3)i} . \tag{3.41}$$

As n is varied, the final exponential in Eq. (3.41) assumes three distinct values: 1, $e^{(2\pi/3)i}$, and $e^{-(2\pi/3)i}$. These three complex values are of magnitude unity and are separated in direction by 120°. This pattern means that, irrespective of the value of θ, at least two of the three cube roots of any number are complex. Perhaps this is why you may not have encountered them some years ago.

In general, $z^{1/n}$ is n-valued, with successive values having arguments that differ by $2\pi/n$. Figure 3.7 illustrates the multiple values of $1^{1/3}$, $i^{1/3}$, and $(-1)^{1/3}$. Powers like $z^{m/n}$ are also n-valued; one way to see this is to realize that $z^{m/n}$ is an integer power (m) of $z^{1/n}$.

One could designate a particular one of the multiple values of a power or root as its **principal value**. In general this choice is not particularly useful, except that it is customary for the square root of a positive number x to associate the symbol $\sqrt{x}$ with that of positive sign; the square root of a negative number $-x$ is often assigned the principal value $i\sqrt{-x}$.

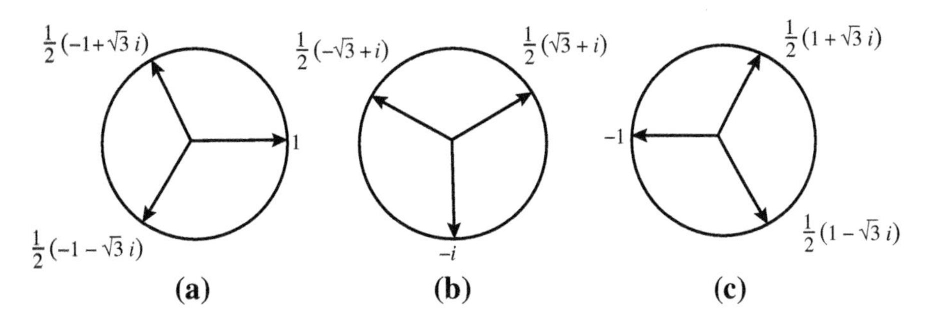

Figure 3.7: Cube roots: (a) $1^{1/3}$; (b) $i^{1/3}$; (c) $(-1)^{1/3}$.

Example 3.5.1. A Helpful Hint

The device of writing a quantity as the exponential of its logarithm can be helpful in the analysis of seemingly obscure forms. For example,

$$i^i = \left(e^{\ln i}\right)^i = e^{i \ln i} = e^{i(i\pi/2 + 2n\pi i)} = e^{-\pi(2n+1/2)} \quad \text{(all integer } n\text{)},$$

an infinite set of real numbers!

In general, powers of complex quantities can be reduced in this way. If a is a complex number whose polar form is $re^{i\theta}$, and m is also complex, of the form $x + iy$, then

$$a^m = e^{m \ln a} = e^{(x+iy)(\ln r + i\theta + 2n\pi i)} = e^{x \ln r - y(\theta + 2n\pi)}\, e^{i(y \ln r + x[\theta + 2n\pi])} ,$$

showing that a^m has an infinite set of values that lie on a spiral (each n changes the magnitude as well as the phase).

■

INVERSE CIRCULAR AND HYPERBOLIC FUNCTIONS

You probably already know that the inverse circular functions are multiple-valued; this must be so because many angles have the same value of the sine or cosine.

If we multiply the $\sin\theta$ formula of Eq. (3.28) by $e^{i\theta}$ and rearrange it to

$$\left(e^{i\theta}\right)^2 - 2i \sin\theta \left(e^{i\theta}\right) - 1 = 0 ,$$

we can identify it as a quadratic equation in $e^{i\theta}$ which we can then solve, using the quadratic formula, to reach

$$e^{i\theta} = i \sin\theta + \sqrt{-\sin^2\theta + 1} . \tag{3.42}$$

We have chosen to take the plus sign for the radical.

If we now set $\sin\theta = z$ and $\theta = \arcsin z = \sin^{-1} z$, and then take the logarithm of both sides of Eq. (3.42), we will have succeeded in inverting the equation for $\sin\theta$:

$$\sin^{-1} z = -i \ln\left[iz + \sqrt{1 - z^2}\right] .$$

The appearance of the logarithm causes this formula to be multiple-valued; if we add $2\pi i$ to the logarithm we see that it increases $\sin^{-1} x$ by 2π, an expected result because the sine has period 2π. Within each interval of length 2π there are two angles with the same value of the sine; that not given above is obtained by taking the minus sign for the radical in the quadratic formula.

Similar analyses can be carried out for all the circular and hyperbolic functions. Some such results, in forms consistent with the usual designations of principal value ($-\pi/2 \le \sin^{-1} x \le \pi/2$, $-\pi/2 < \tan^{-1} x < \pi/2$), and with all functions real for real arguments, are

$$\sin^{-1} z = -i \ln\left[iz + \sqrt{1-z^2}\right], \qquad \sinh^{-1} z = \ln\left[z + \sqrt{1+z^2}\right],$$

$$\tan^{-1} z = \frac{i}{2} \ln\left(\frac{1-iz}{1+iz}\right), \qquad \tanh^{-1} z = \frac{1}{2} \ln\left(\frac{1+z}{1-z}\right).$$

SYMBOLIC COMPUTATION

Fractional powers can be entered into the symbolic computation systems by the usual notation for exponentiation, but alternatively the square root can be accessed as `sqrt(`*expr*`)` or `Sqrt[`*expr*`]`.

In MAPLE the natural logarithm is denoted `ln`, but the alternative `log` (a synonym for `ln`) is also recognized. In MATHEMATICA, the natural logarithm is uniquely denoted `Log`.

The inverse circular and hyperbolic functions are identified by prefixing the corresponding function by `arc` (MAPLE) or `Arc` (MATHEMATICA).

Often these multiple-valued functions yield results that can be simplified using `evalc` or `ComplexExpand`; in all cases when simplification is complete a single **principal value** is obtained.

Example 3.5.2. Multiple-Valued Functions

MAPLE input	Output	MATHEMATICA input	Output
`> (-1)^(1/3);`	$(-1)^{(1/3)}$	`(-2)^(-1/3)`	$\frac{(-1)^{2/3}}{2^{1/3}}$
`> evalc(%);`	$\frac{1}{2} + \frac{1}{2} I\sqrt{3}$	`ComplexExpand[%]`	$\frac{1}{2\,2^{1/3}} - \frac{i\sqrt{3}}{2\,2^{1/3}}$
`> ln(I);`	$\frac{1}{2} I\pi$	`Log[(1+I)/Sqrt[2]]`	$\text{Log}\left[\frac{1+i}{\sqrt{2}}\right]$
`> arccos(-1/sqrt(2));`	$\frac{3\pi}{4}$	`ComplexExpand[%]`	$\frac{i\pi}{4}$
`> arctan(-infinity);`	$-\frac{\pi}{2}$	`ArcSinh[1]`	ArcSinh[1]
`> arccsch(1);`	$\ln(1+\sqrt{2})$	`ComplexExpand[%]`	$\text{Log}[1+\sqrt{2}]$
`> arcsech(1.5);`	$0.8410686706I$	`ArcCoth[Infinity]`	0

■

Exercises

3.5.1. Show that complex numbers have square roots and that the square roots are contained in the complex plane. What are the square roots of i?

3.5.2. Explain why $e^{\ln z}$ always equals z, but $\ln e^z$ does not always equal z.

3.5.3. By comparing series expansions, show that $\tan^{-1} z = \frac{i}{2} \ln\left(\frac{1-iz}{1+iz}\right)$.

3.5.4. Find the Cartesian form for **all values** of

(a) $(-8)^{1/3}$,
(b) $i^{1/4}$,
(c) $e^{i\pi/4}$.

3.5.5. Find the polar form for **all values** of

(a) $(1+i)^3$,
(b) $(-1)^{1/5}$.

3.5.6. Bring the following to $x + iy$ form. It suffices to identify a principal value for each of these quantities. Check your work using MAPLE or MATHEMATICA.

(a) $\cos^{-1}(3i)$ (b) $\cosh^{-1}(1+i)$ (c) $\cosh^{-1}\pi$
(d) $\sin^{-1}(1-i)$ (e) $\sin^{-1} i\pi$ (f) $\sinh^{-1}(i)$
(g) $\tan^{-1}(i-1)$ (h) $\tanh^{-1}(2i)$ (i) $\cot^{-1}(\pi+i)$.

3.5.7. For each of the multiple-valued quantities in Exercise 3.5.6, indicate how all its other values are related to the principal value found as a solution to that exercise.

3.5.8. In Exercise 3.2.1 we decomposed the equation

$$z^3 = \frac{(7-4i)^3}{(1-i)^2} \tag{3.43}$$

into two real equations which appeared difficult to solve. We can now obtain solutions simply by taking the cube root of the right-hand side of the equation. Obtain all solutions to Eq. (3.43).

Note. This shows how the use of complex variables can sometimes cause an apparently difficult problem to assume a simpler form.

Additional Readings

Abramowitz, M., & Stegun, I. A. (Eds.) (1972). *Handbook of mathematical functions with formulas, graphs, and mathematical tables (AMS-55).* Washington, DC: National Bureau of Standards (Reprinted, Dover (1974). Contains power series expansions of a large number of elementary and special functions).

Galambos, J. (1976). *Representations of real numbers by infinite series.* Berlin: Springer.

Gradshteyn, I. S., & Ryzhik, I. M. (2007). In A. Jeffrey, & D. Zwillinger (Eds.), *Table of integrals, series, and products* (Corrected and enlarged 7th ed.). New York: Academic Press.

Hansen, E. (1975). *A table of series and products.* Englewood Cliffs, NJ: Prentice-Hall (A tremendous compilation of series and products).

Jeffrey, A. (1995). *Handbook of mathematical formulas and integrals.* San Diego: Academic Press.

Knopp, K. (1971). *Theory and application of infinite series* (2nd ed.). London/New York: Blackie and Son/Hafner (Reprinted: A. K. Peters Classics (1997). This is a thorough, comprehensive, and authoritative work that covers infinite series and products. Proofs of almost all the statements about series not proved in this Chapter will be found in this book).

Mangulis, V. (1965). *Handbook of series for scientists and engineers.* New York: Academic Press (A most convenient and useful collection of series. Includes algebraic functions, Fourier series, and series of the special functions: Bessel, Legendre, and so on).

Olver, F. W. J., Lozier, D. W., Boisvert, R. F., & Clark, C. W. (Eds.) (2010). *NIST handbook of mathematical functions.* Cambridge, UK: Cambridge University Press (Update of AMS-55 (Abramowitz and Stegun, 1972), but links to computer programs are provided instead of tables of data).

Rainville, E. D. (1967). *Infinite series.* New York: Macmillan (A readable and useful account of series constants and functions).

Sokolnikoff, I. S., & Redheffer, R. M. (1966). *Mathematics of physics and modern engineering* (2nd ed.). New York: McGraw-Hill (A long Chapter 2 (101 pages) presents infinite series in a thorough but very readable form. Extensions to the solutions of differential equations, to complex series, and to Fourier series are included).

Chapter 4

VECTORS AND MATRICES

Many physical phenomena have the property that contributions to them from different sources enter additively. An example is the electrostatic potential at a point produced by an array of charges: the total potential is the sum of the potentials arising from the individual charges. Another example is that the overall force on an object can be treated as the sum of the individual forces acting on it (with the understanding that we will use vector addition to form the sum). The superposition of waves provides yet another illustration of this additivity; the total amplitude of a collection of waves at any point at a given time is the sum of the amplitudes of the individual waves passing through that point at the specified time. Specific illustrations of additivity are provided by sound waves, water waves, or wave patterns on a vibrating string or membrane. More esoteric examples are provided by wave functions in quantum mechanics; if a system is described by a superposition of basis wave functions, the overall wave function is their sum.

The additive property all these phenomena have in common is termed **linearity**, and the phenomena can be described either by linear algebraic equations, linear differential equations, or both. The handling of these linear problems is systematized and made much easier by the introduction of appropriate mathematical objects, in particular vectors (one-dimensional arrays) and matrices[1] (two-dimensional arrays). For this reason the present chapter, which could have been called "An introduction to linear algebra" is given the more specific title "Vectors and matrices."

Linearity is very helpful in reducing the difficulty of mathematical analysis. By the end of the 19th century, formal methods for handling linear problems had reached a high level of sophistication. In fact, by the 1890s many prominent physicists believed that physical theory was more or less complete and all that remained to be done was further refinement in methods of numerical calculation.

The discoveries that changed the face of physics in the early twentieth century (particularly quantum mechanics and special relativity) turned out to be expressible as linear problems, reaffirming the importance of linearity in physics. World War II and the advent of large-scale digital computation spurred the development of efficient methods for the numerical solution of linear problems, and symbolic computation systems have also given linear problems a great deal of attention.

[1]**Matrices** (pronounced **may**-trih-seeze) is the plural of **matrix**.

Mathematics for Physical Science and Engineering.
http://dx.doi.org/10.1016/B978-0-12-801000-6.00004-3

It should be pointed out, however, that not all physics is linear, and some of today's important questions (e.g., turbulent fluid flow, other chaotic phenomena, problems in general relativity) are inherently nonlinear. Nonlinearity also results in wave-propagation problems when the medium supporting the wave is limited in its ability to accept large amplitudes (saturation effects). The methods that have been highly developed for linear problems are in general not fully adaptable to nonlinear problems, with the result that the latter remain difficult.

We begin our study of linear problems with a short discussion of vectors.

4.1 BASICS OF VECTOR ALGEBRA

In an elementary physics course, you have probably heard a **vector** described as a quantity possessing magnitude and direction. While that is certainly true, it is by no means a complete specification. Central to the notion of a vector is that it possesses several properties, which we now introduce and discuss.

VECTORS

In two-dimensional space, which we often abbreviate as **2-D** (e.g., the xy-plane), we identify a vector as an object that can correspond to an arrow in the plane, oriented in some direction, and with a length indicating its magnitude. However, an arrow is not regarded mathematically as a vector unless it is subject to the operations that define a **vector algebra**. In this book we will normally identify a vector by a letter in a bold-faced roman font, such as **A** or **z**. If our space is three-dimensional (**3-D**), our vector corresponds to an arrow of arbitrary length and any orientation in the 3-D space.

As already stated, a vector may be geometrically represented by an arrow with length proportional to the magnitude. The direction of the arrow indicates the direction of the vector, the positive sense of direction being indicated by the point. In this representation, vector addition

$$\mathbf{C} = \mathbf{A} + \mathbf{B} \tag{4.1}$$

consists of placing the tail of vector **B** at the point of vector **A** (head to tail). Vector **C** is then represented by the arrow drawn from the tail of **A** to the point of **B**. This procedure, the triangle law of addition, assigns meaning to Eq. (4.1) and is illustrated in the left panel of Fig. 4.1. If in the same figure we draw the representation of the addition **B** + **A** (see the right panel of Fig. 4.1), we obtain a parallelogram, showing that **B** + **A** also yields **C** as a result: thus,

$$\mathbf{C} = \mathbf{A} + \mathbf{B} = \mathbf{B} + \mathbf{A} \,. \tag{4.2}$$

In words, we find that **vector addition is commutative.**

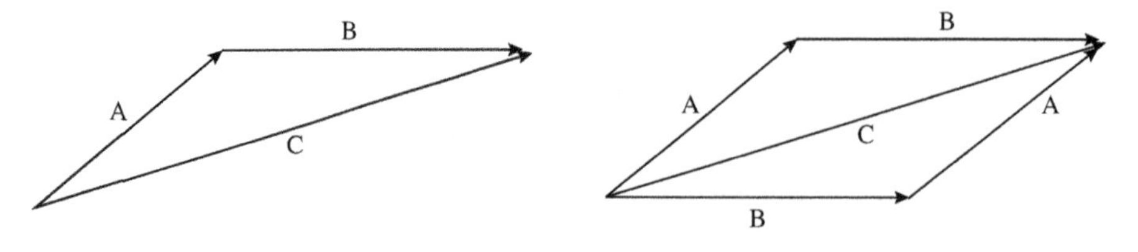

Figure 4.1: Vector Addition: left: $\mathbf{A}+\mathbf{B}$; right: also $\mathbf{B}+\mathbf{A}$, illustrating parallelogram law.

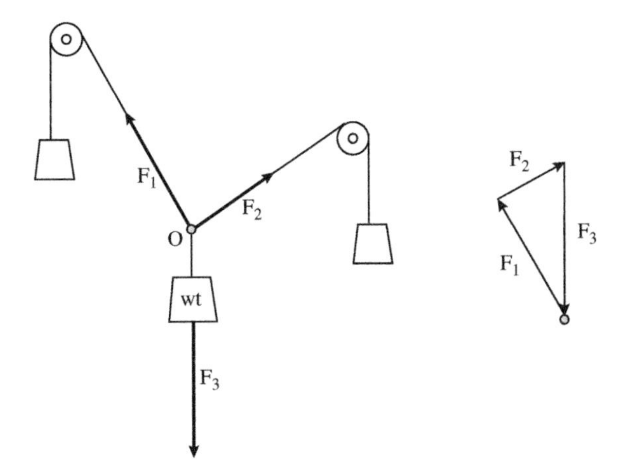

Figure 4.2: A mass at static equilibrium. Right panel shows vector addition of forces.

If we add three vectors, first adding **A** and **B**, and then adding **C** to that result, we get the same result as when we first add **B** and **C**, then adding **A**. These observations correspond to the equation

$$\mathbf{D} = (\mathbf{A} + \mathbf{B}) + \mathbf{C} = \mathbf{A} + (\mathbf{B} + \mathbf{C}) . \tag{4.3}$$

We say that **vector addition is associative**; from a practical viewpoint, this means that we do not need to use parentheses to indicate the order in which vector additions are to be carried out.

Vector subtraction is handled by defining the negative of a vector as a vector of the same magnitude but with reversed direction. Then

$$\mathbf{A} - \mathbf{B} = \mathbf{A} + (-\mathbf{B}) . \tag{4.4}$$

The rules we have just introduced are found to correspond to the ways in which forces add in mechanical systems. For example, look at the apparatus shown in Fig. 4.2. For this system to be in mechanical equilibrium, the forces at point O of the diagram must add to zero. The three forces at O are in the respective directions of the strings, and have magnitudes given by the three weights. Experiment shows that at mechanical equilibrium the vector sum of $\mathbf{F}_1$ and $\mathbf{F}_2$ is equal in magnitude and opposite in direction to $\mathbf{F}_3$. This (and similar experiments in electrostatics) show that our vector addition law corresponds in general to the addition of forces.

Vectors are also used to describe **velocities** or mass-weighted velocities (i.e., **momenta**). The law of conservation of momentum states that the vector sum of all the individual momenta in an isolated many-particle system is constant. To work with momenta we need to give meaning to the multiplication of a vector by an ordinary number, such as a mass. These ordinary numbers are called **scalars**, and the operation we are considering is called **multiplication (of a vector) by a scalar**. If **A** is a vector and $k \geq 0$ a scalar, we define $k\mathbf{A}$ as a vector in the same direction as **A** but with length kA; if $k < 0$, consistency demands that $k\mathbf{A}$ be of length $|k|A$ but in the direction of $-\mathbf{A}$.

Many physical phenomena involve the notion of a vector that is defined for each point in space. Such a collection of vectors is called a **vector field**; an important example is the **electric field**, whose value at each point is the vector describing the electric force a unit positive electric charge would experience if it were placed there.

In other words, a vector field is a function whose value at each point is a vector. The variation from point to point of a vector field provides an opportunity to consider derivatives and integrals involving vector fields. These important topics, which are at the heart of what is called **vector analysis**, are taken up later in a separate chapter.

Example 4.1.1. Medians of Triangle Intersect at a Point

The geometric and pictorial properties of vectors suffice to solve various problems. This example deals with the **medians** of a triangle; medians are lines that connect the vertices of a triangle to the midpoints of the opposite sides.

The task under consideration here is to show that the medians of a triangle intersect at a point. We also want to show that this intersection is at 2/3 the length of each median, measured from the vertex at which it starts.

Let's place one vertex of our triangle at an origin point $\mathbf{O}$, with the other vertices at points defined by vectors $\mathbf{A}$ and $\mathbf{B}$ starting from $\mathbf{O}$. The midpoint of the OA side of the triangle is at the point $\mathbf{A}/2$; the midpoint of the OB side is at $\mathbf{B}/2$. See Fig. 4.3 for the geometry.

Two geometric features of the figure are important: (1) the line connecting $\mathbf{A}/2$ and $\mathbf{B}/2$ is parallel to the triangle side AB, and therefore (2) for each median, the distance from its vertex to the point $\mathbf{M}$ where the medians cross is the same fraction k of that median's total length. We may now use this information to compute the location of $\mathbf{M}$ in two ways: taking a path from $\mathbf{O}$ to $\mathbf{A}$ and then to $\mathbf{M}$, we find $\mathbf{M}$ to be displaced from $\mathbf{O}$ by

$$\mathbf{M} = \mathbf{A} + k\left(\frac{\mathbf{B}}{2} - \mathbf{A}\right) .$$

However, if we reach $\mathbf{M}$ by going first to $\mathbf{B}$, we find

$$\mathbf{M} = \mathbf{B} + k\left(\frac{\mathbf{A}}{2} - \mathbf{B}\right) .$$

Equating these expressions for $\mathbf{M}$ and simplifying, we reach

$$(\mathbf{A} - \mathbf{B})\left(1 - \frac{3k}{2}\right) = 0 .$$

If we actually have a triangle, $\mathbf{A} \neq \mathbf{B}$, so it is necessary that $1 - 3k/2 = 0$, equivalent to $k = 2/3$.

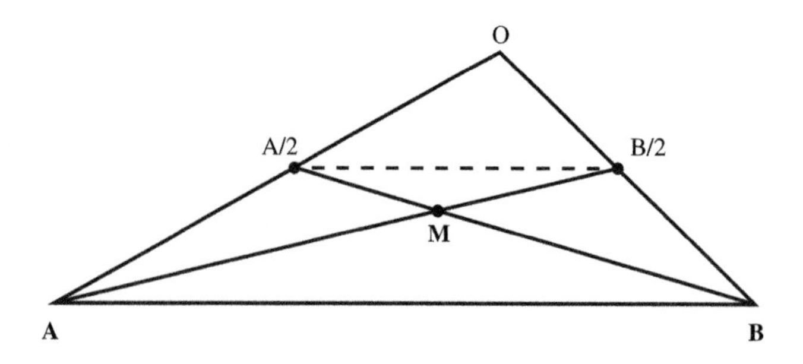

Figure 4.3: A triangle and two of its medians.

Since k has a fixed value independent of the specific shape of the triangle, it must also apply to intersections involving the remaining median (that starting from **O**), so we may conclude that all three medians intersect at a point.

■

VECTOR COMPONENTS

Often we want to identify a vector in ways that do not require the drawing of a diagram depicting arrows in 2-D or 3-D space; it is then useful to describe a vector by the Cartesian coordinates of its head when its tail is placed at the origin of the coordinate system. Introduction of these coordinates will permit us to solve problems involving vectors using algebraic methods when they are more convenient or powerful than the geometric method used in Example 4.1.1. It is both obvious and important that a 2-D vector will be described by the values of **two** coordinates; in fact that is what we really mean when we say a space is 2-D. A 3-D vector will require three coordinates for its description, corresponding to its existence in a 3-D space.

The coordinates describing a vector can for convenience be placed in an array; the form we usually choose for this purpose is illustrated here for a 2-D vector we have named **A**:

$$\mathbf{A} = \begin{pmatrix} A_x \\ A_y \end{pmatrix}. \tag{4.5}$$

When we write **A** this way we sometimes call it a **column vector**. The coordinates describing vector **A** in Eq. (4.5) are referred to as its **components**.

The law we have adopted for vector addition enables us to give physical meaning to the components of a vector. Our vector **A** can be regarded as the sum of a vector of length A_x in the x direction of the Cartesian system and a vector of length A_y in the y direction. See Fig. 4.4.

The length or magnitude of a vector **A**, denoted $|\mathbf{A}|$ or sometimes just A, can be computed from the values of its components. Because Cartesian axes are mutually perpendicular, we can use the Pythagorean theorem to find

$$|\mathbf{A}|^2 = (A_x)^2 + (A_y)^2 \quad \text{(2-D)}, \qquad |\mathbf{A}|^2 = (A_x)^2 + (A_y)^2 + (A_z)^2 \quad \text{(3-D)}. \tag{4.6}$$

Notice that A or its equivalent $|\mathbf{A}|$ is **not** a vector. It consists of a single number that is independent of the direction of **A** and is thus an ordinary number, a **scalar**.

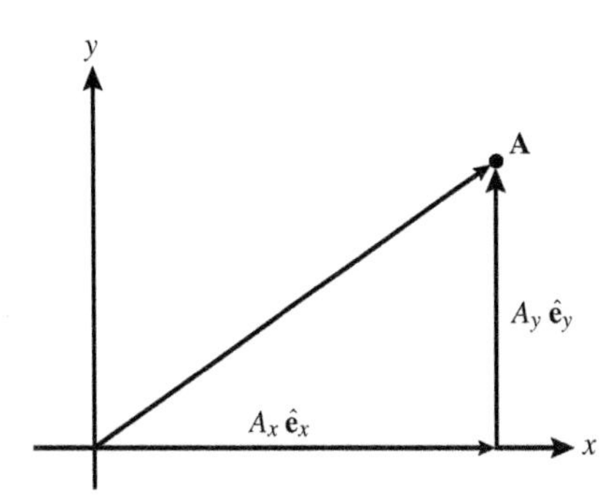

Figure 4.4: Vector as sum of contributions in coordinate directions.

All vectors (not only those in 2-D or 3-D space, but also those in nonphysical spaces with arbitrary numbers of dimensions) can be written as column vectors. While there are some special properties unique to 3-D space that we will explore later in great detail, our mission here is to introduce the operations that **all** quantities called vectors must possess. These are the basic operations of vector algebra.

Finally, we call attention to a very important vector that identifies points in space. It is called the **radius vector**, is usually denoted $\mathbf{r}$, and its components are the coordinates of the point it designates. Thus, in 3-D,

$$\mathbf{r} = \begin{pmatrix} x \\ y \\ z \end{pmatrix} . \tag{4.7}$$

It is obvious that $r = |\mathbf{r}| = \sqrt{x^2 + y^2 + z^2}$.

DIRECTION COSINES

It is often of interest to know the angles a particular vector makes with the coordinate directions. In 2-D, the angle relative to the x axis is usually called α, while the angle from the y axis is denoted β. In 3-D, there will also be an angle γ from the z axis. Then, applying elementary trigonometry, it is easy to show that, for a 3-D vector $\mathbf{A}$, of magnitude A,

$$A_x = A\cos\alpha\ , \qquad A_y = A\cos\beta\ , \qquad A_z = A\cos\gamma\ . \tag{4.8}$$

See Fig. 4.5. The quantities $\cos\alpha$, $\cos\beta$, $\cos\gamma$ are called the **direction cosines** of the vector to which they refer. Since $(A_x)^2 + (A_y)^2 + (A_z)^2 = A^2$, substitution from Eq. (4.8) yields

$$\cos^2\alpha + \cos^2\beta + \cos^2\gamma = 1\ , \tag{4.9}$$

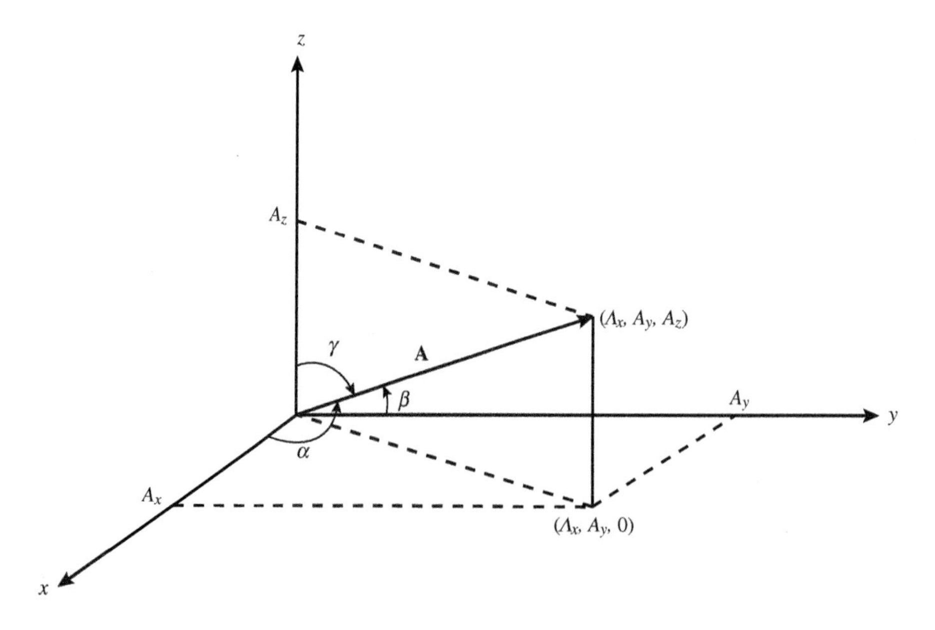

Figure 4.5: Direction cosines of **A**.

a result connecting the direction cosines of any 3-D vector. In 2-D, we have a similar result without the γ term. In 2-D, it is also clear that $\cos\beta = \pm\sin\alpha$, so the direction-cosine identity reduces to $\cos^2\alpha + \sin^2\alpha = 1$. The 3-D result, Eq. (4.9), is a bit less obvious.

Example 4.1.2. Direction Cosines

Let $\mathbf{A}$ be the column vector $\begin{pmatrix} 4 \\ 2 \\ -4 \end{pmatrix}$.

To find its direction cosines, we first compute the magnitude of $\mathbf{A}$. We get $A^2 = 4^2 + 2^2 + (-4)^2 = 36$, so $A = 6$. Then, using Eq. (4.8),

$$\cos\alpha = \frac{4}{6} = \frac{2}{3}\,, \qquad \cos\beta = \frac{2}{6} = \frac{1}{3}\,, \qquad \cos\gamma = \frac{-4}{6} = -\frac{2}{3}\,.$$

Checking Eq. (4.9), we have

$$\left(\frac{2}{3}\right)^2 + \left(\frac{1}{3}\right)^2 + \left(-\frac{2}{3}\right)^2 = \frac{4}{9} + \frac{1}{9} + \frac{4}{9} = 1\,.$$

■

VECTOR OPERATIONS USING COMPONENTS

We have already noted that a 2-D vector $\mathbf{A}$ can be thought of as the vector sum of a vector of length A_x in the x direction and a vector of length A_y in the y direction; making similar observations for another 2-D vector $\mathbf{B}$, we can use the fact that vector addition is commutative to group the x components together, finding that the x component of $\mathbf{C} = \mathbf{A} + \mathbf{B}$ must be $C_x = A_x + B_x$. Making a similar observation for the y components, we have

$$\mathbf{C} = \mathbf{A} + \mathbf{B} = \begin{pmatrix} A_x \\ A_y \end{pmatrix} + \begin{pmatrix} B_x \\ B_y \end{pmatrix} = \begin{pmatrix} A_x + B_x \\ A_y + B_y \end{pmatrix}. \tag{4.10}$$

Notice that when written in component form, the rule for addition is seen to be similar in form to that used to add complex numbers. Equation (4.10) makes it obvious that vector addition must be both commutative and associative.

The use of components makes simpler many problems in mechanics. Here is an example.

Example 4.1.3. Mass in Static Equilibrium

We are now ready to solve the static equilibrium problem associated with Fig. 4.2. Since the weight being supported has mass M, the gravitational force $\mathbf{F}_3$ has magnitude Mg, where g is the acceleration of gravity. Our task here is to determine the magnitudes of $\mathbf{F}_1$ and $\mathbf{F}_2$. Since the sum of the forces acting on the junction point, labeled O, must be zero, the vertical components of $\mathbf{F}_1$ and $\mathbf{F}_2$ must add to Mg, while their horizontal components cancel (i.e., add algebraically to zero). Using the direction cosine formulas or just applying our knowledge of trigonometry, we have the following two equations for the magnitudes of $\mathbf{F}_1$ and $\mathbf{F}_2$:

$$F_1 \cos\theta_1 + F_2 \cos\theta_2 = Mg\ , \qquad F_1 \sin\theta_1 = F_2 \sin\theta_2\ .$$

Equilibrium is possible for a wide range of the angles θ_1 and θ_2 of the figure. Let's get a numerical solution for the case $\theta_1 = 30°$, $\theta_2 = 60°$, so $\sin\theta_1 = 1/2$, $\cos\theta_1 = \sqrt{3}/2$, and $\sin\theta_2 = \sqrt{3}/2$, $\cos\theta_2 = 1/2$. Then our equations reduce to

$$\frac{\sqrt{3}}{2}\,F_1 + \frac{1}{2}\,F_2 = Mg\ ,$$

$$\frac{1}{2}\,F_1 - \frac{\sqrt{3}}{2}\,F_2 = 0\ .$$

The solution to these simultaneous equations is $F_1 = (\sqrt{3}/2)Mg$, $F_2 = Mg/2$. Notice that the magnitudes of $\mathbf{F}_1$ and $\mathbf{F}_2$ add to a result greater than Mg. These two forces are pulling against each other in addition to combining to hold up the mass M.

■

MULTIPLICATION BY A SCALAR

If we add a vector $\mathbf{A}$ to itself, the resulting vector will have components twice that of $\mathbf{A}$; it is natural to call this vector $2\mathbf{A}$. That observation motivates the definition that if k is an arbitrary ordinary number (a scalar), the multiplication of $\mathbf{A}$ by k will (in 2-D) have components kA_x and kA_y, or

$$\mathbf{A} = \begin{pmatrix} A_x \\ A_y \end{pmatrix}; \qquad k\mathbf{A} = \begin{pmatrix} kA_x \\ kA_y \end{pmatrix}. \tag{4.11}$$

If $k = -1$, Eq. (4.11) produces the vector $-\mathbf{A}$, and it is clear that $\mathbf{A} + (-\mathbf{A}) = \mathbf{A} - \mathbf{A} = 0$.

The rules for addition and multiplication by a scalar satisfy the **distributive laws** of ordinary algebra, i.e., we have

$$(k + k')\mathbf{A} = k\mathbf{A} + k'\mathbf{A} \quad \text{and} \quad k(\mathbf{A} + \mathbf{B}) = k\mathbf{A} + k\mathbf{B}\ .$$

This means that we can give meaning to linear forms involving vectors, such as (in 2-D)

$$\mathbf{A} + 2\mathbf{B} - 3\mathbf{C} = \begin{pmatrix} A_x \\ A_y \end{pmatrix} + \begin{pmatrix} 2B_x \\ 2B_y \end{pmatrix} + \begin{pmatrix} -3C_x \\ -3C_y \end{pmatrix} = \begin{pmatrix} A_x + 2B_x - 3C_x \\ A_y + 2B_y - 3C_y \end{pmatrix}.$$

These rules extend straightforwardly to vectors in 3-D and in higher-dimensional spaces.

Example 4.1.4. Addition of Momenta

A system consists of three particles of masses $m_1 = 1.5$ kg, $m_2 = 2$ kg, and $m_3 = 3$ kg. The particles are moving in the xy-plane, at the respective velocities $v_1 = 4$ m/s in the $+x$ direction; $v_2 = 5$ m/s at an angle with direction cosines $\alpha = -\sqrt{3}/2$, $\beta = +1/2$; and $v_3 = 2.5$ m/s in the $-y$ direction. Notice that the v_i are **magnitudes**

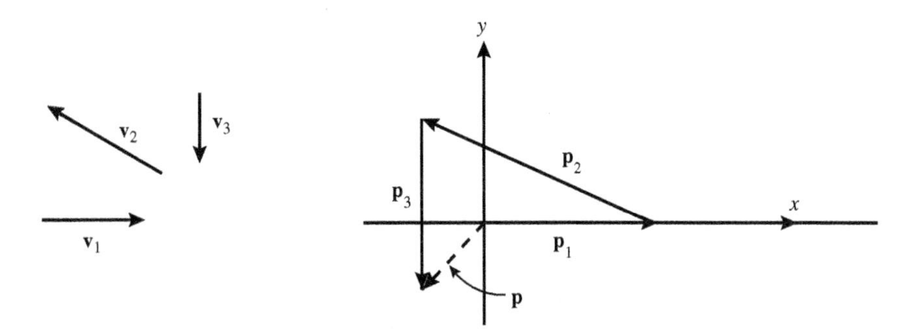

Figure 4.6: Velocities and momenta of the particles in Example 4.1.4.

of the velocities, and are therefore scalars. Find the magnitude and direction cosines of the total momentum of this system. See Fig. 4.6.

The total momentum $\mathbf{p}$ is given by the formula $\mathbf{p} = \sum_i m_i \mathbf{v}_i$, so we have, writing each velocity as a column vector and determining its components,

$$\mathbf{p} = \begin{pmatrix} p_x \\ p_y \end{pmatrix} = 1.5 \begin{pmatrix} 4 \\ 0 \end{pmatrix} + 2 \begin{pmatrix} -5\sqrt{3}/2 \\ 5/2 \end{pmatrix} + 3 \begin{pmatrix} 0 \\ -5/2 \end{pmatrix} = \begin{pmatrix} -2.66 \\ -2.5 \end{pmatrix}.$$

Thus the momentum has components $p_x = -2.66$ kg-m/s, $p_y = -2.5$ kg-m/s; its magnitude is therefore

$$p = \sqrt{(-2.66)^2 + (-2.5)^2} = 3.65 \text{ kg-m/s},$$

and its direction cosines are

$$\cos\alpha = -2.66/3.65 = -0.729, \qquad \cos\beta = -2.5/3.65 = -0.685.$$

To check the consistency of the direction cosines, we can compute $(-0.729)^2 + (-0.685)^2 = 1.00$.

■

Exercises

4.1.1. Given the two vectors $\mathbf{A}+\mathbf{B}$ and $\mathbf{A}-\mathbf{B}$, show (using a drawing) how to find the individual vectors $\mathbf{A}$ and $\mathbf{B}$.

4.1.2. Two airplanes are moving (relative to the ground) at the respective velocities $\mathbf{v}_A$ and $\mathbf{v}_B$. Their relative velocity, $\mathbf{v}_{\text{rel}}$, is defined by the equation $\mathbf{v}_{\text{rel}} = \mathbf{v}_A - \mathbf{v}_B$. Determine $\mathbf{v}_{\text{rel}}$ if

$$\mathbf{v}_A = 150 \text{ mi/hr west},$$
$$\mathbf{v}_B = 200 \text{ mi/hr north}.$$

4.1.3. If a motorboat can travel at a speed of 10 mi/hr (relative to the water) and is in a river that flows due north at 5 mi/hr, what compass heading should the boat take to make its net travel due east? What will be the speed of the boat relative to the river banks?

4.1.4. The vertices A, B, and C of a triangle are given by the points $(-2,0,1)$, $(0,2,2)$, and $(1,-2,0)$, respectively. Find the point D that makes the figure $ABCD$ a plane parallelogram.

4.1.5. A triangle is defined by the points of three vectors $\mathbf{A}, \mathbf{B}$, and $\mathbf{C}$ that start at the origin. Show that the **vector** sum of the successive sides of the triangle $(AB + BC + CA)$ is zero, where the side AB stands for the vector $\mathbf{B} - \mathbf{A}$, etc.

4.1.6. A sphere of radius a is centered at a point $\mathbf{r}_1$.

(a) Give the algebraic equation for the sphere.

(b) Using vectors, describe the locus of the points of the sphere.

4.1.7. Using formulas for vector components, show that the diagonals of a parallelogram bisect each other.

4.1.8. A corner reflector is formed by three mutually perpendicular reflecting surfaces. Show that a ray of light incident upon the corner reflector (and sufficiently near the intersection of its three surfaces) is reflected back along a line parallel to the line of incidence.

Hint. A reflection reverses the sign of the vector component perpendicular to the reflecting surface.

4.1.9. Consider a unit cube with one corner at the origin and its three sides lying along Cartesian coordinates axes.

(a) Show that there are four body diagonals, each with length $\sqrt{3}$. Representing these as vectors, find their components.

(b) Show that the faces of the cube have diagonals of length $\sqrt{2}$ and determine their components.

UNIT VECTORS AND BASES[2]

The notion that vectors can be represented by their components can be expressed formally if we introduce the concept of **unit vector**, namely a vector of unit length in some relevant direction. Letting $\hat{\mathbf{e}}_x$ be a unit vector in the x direction of a coordinate system, we can write our vector in the x direction with length A_x as $A_x\hat{\mathbf{e}}_x$. Similarly, $A_y\hat{\mathbf{e}}_y$ is a vector of length A_y in the y direction. Written as column vectors, these unit vectors are

$$\hat{\mathbf{e}}_x = \begin{pmatrix} 1 \\ 0 \end{pmatrix}, \qquad \hat{\mathbf{e}}_y = \begin{pmatrix} 0 \\ 1 \end{pmatrix}, \tag{4.12}$$

and therefore

$$A_x\hat{\mathbf{e}}_x = \begin{pmatrix} A_x \\ 0 \end{pmatrix}, \qquad A_y\hat{\mathbf{e}}_y = \begin{pmatrix} 0 \\ A_y \end{pmatrix},$$

and it is obvious that a 2-D vector $\mathbf{A}$ with components A_x and A_y can be represented as the vector sum

$$\mathbf{A} = A_x\hat{\mathbf{e}}_x + A_y\hat{\mathbf{e}}_y = \begin{pmatrix} A_x \\ 0 \end{pmatrix} + \begin{pmatrix} 0 \\ A_y \end{pmatrix} = \begin{pmatrix} A_x \\ A_y \end{pmatrix}. \tag{4.13}$$

[2]**Bases**, pronounced "base-eeze," is the plural of **basis**.

Because any vector in our 2-D space can be written as a linear combination of $\hat{\mathbf{e}}_x$ and $\hat{\mathbf{e}}_y$, we refer to these two vectors as a **basis** for our 2-D vector space, where **vector space** refers to the set of all vectors that can exist under the conditions we have chosen to impose (here, that they correspond to points in an infinite plane). Our basis has two members, corresponding to the dimensionality of the space, and our space is **closed** under the operations of addition and multiplication by a scalar, meaning that these operations on any vector(s) in our space produces a vector that is also in our space. In order for this to be true, our space must include a **zero** element (a vector of length zero), since that element can be reached by addition $\mathbf{A}+(-\mathbf{A})$ or by multiplication of any vector by $k = 0$.

We have chosen to give the unit vector in the x direction the name $\hat{\mathbf{e}}_x$, using $\mathbf{e}$ in part to emphasize that it is a basis vector of our space, and using the circumflex accent ($\hat{\ }$) to indicate that it is of unit length. Later we will sometimes call this unit vector $\hat{\mathbf{e}}_1$, to associate it with the first member of our basis (this will be particularly useful if that member is not in the x direction). In engineering, the unit vectors in the x, y, and z directions are often given the names $\mathbf{i}$, $\mathbf{j}$, and $\mathbf{k}$.

Notice that unit vectors are not restricted to the orientations of the coordinate axes. We can find a unit vector in the direction of any vector $\mathbf{A}$ simply by forming $\mathbf{A}/|\mathbf{A}|$, and will indicate that it is a unit vector by placing a hat (circumflex) over it. For example, a unit vector in the direction of the radius vector $\mathbf{r}$ can be obtained as $\hat{\mathbf{r}} = \mathbf{r}/r$.

The notion of **basis** is more general than identification of unit vectors in the coordinate directions. For example, any two vectors that are not collinear form a basis for the 2-D space. But bases are far easier to use if they are vectors that are perpendicular to each other and of unit length.

Exercises

4.1.10. Calculate the components of a unit vector that makes equal angles with the positive directions of the x-, y-, and z-axes.

4.2 DOT PRODUCT

We now introduce a rule for the multiplication of two vectors, of a type that produces as its result an ordinary number (not another vector). It is called the **dot product**. We want the dot product of two vectors $\mathbf{A}$ and $\mathbf{B}$ to be independent of the orientation of the vectors relative to the coordinate system, and to depend only upon the magnitudes of $\mathbf{A}$ and $\mathbf{B}$ and upon the angle θ between their directions. Using a central dot as the symbol for this operation, its definition takes the form

$$\mathbf{A}\cdot\mathbf{B} = AB\cos\theta\,. \tag{4.14}$$

The angle θ is identified in Fig. 4.7.

Notice that Eq. (4.14) makes the dot product a commutative operation,

$$\mathbf{A}\cdot\mathbf{B} = \mathbf{B}\cdot\mathbf{A}\,, \tag{4.15}$$

and that the dot product is zero when $\mathbf{A}$ and $\mathbf{B}$ are at right angles to each other, as $\cos\theta$ is then zero. Vectors which are perpendicular are called **orthogonal**; if they are also of unit length, they are referred to as **orthonormal** (a vector of unit length is sometimes called **normalized**).

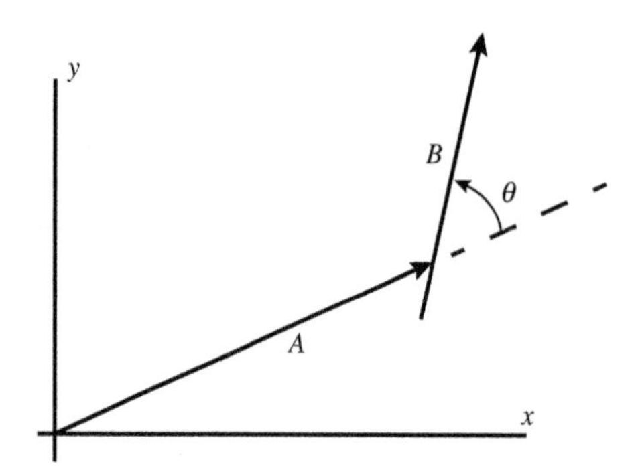

Figure 4.7: Angle between vectors **A** and **B**.

It can be shown that the dot product is linear in **A** and **B**, which means that in the dot-product formula we can replace **A** and **B** with

$$\mathbf{A} = A_x\hat{\mathbf{e}}_x + A_y\hat{\mathbf{e}}_y + A_z\hat{\mathbf{e}}_z \ , \qquad \mathbf{B} = B_x\hat{\mathbf{e}}_x + B_y\hat{\mathbf{e}}_y + B_z\hat{\mathbf{e}}_z \ ,$$

and expand the result. We get

$$\begin{aligned}\mathbf{A}\cdot\mathbf{B} = A_xB_x\hat{\mathbf{e}}_x\cdot\hat{\mathbf{e}}_x + A_yB_y\hat{\mathbf{e}}_y\cdot\hat{\mathbf{e}}_y + A_zB_z\hat{\mathbf{e}}_z\cdot\hat{\mathbf{e}}_z + (A_xB_y + A_yB_x)\hat{\mathbf{e}}_x\cdot\hat{\mathbf{e_y}} \\ +(A_xB_z + A_zB_x)\hat{\mathbf{e}}_x\cdot\hat{\mathbf{e_z}} + (A_yB_z + A_zB_y)\hat{\mathbf{e}}_y\cdot\hat{\mathbf{e_z}} \ . \qquad (4.16)\end{aligned}$$

Equation (4.16) can be simplified by using the properties of the unit vectors in the coordinate directions. Since these directions are mutually perpendicular, and all the unit vectors are (by definition) of unit length, application of Eq. (4.14) yields

$$\hat{\mathbf{e}}_x\cdot\hat{\mathbf{e}}_x = \hat{\mathbf{e}}_y\cdot\hat{\mathbf{e}}_y = \hat{\mathbf{e}}_z\cdot\hat{\mathbf{e}}_z = 1 \quad \text{and} \quad \hat{\mathbf{e}}_x\cdot\hat{\mathbf{e}}_y = \hat{\mathbf{e}}_x\cdot\hat{\mathbf{e}}_z = \hat{\mathbf{e}}_y\cdot\hat{\mathbf{e}}_z = 0 \ . \qquad (4.17)$$

These relationships are statements that the Cartesian unit vectors are **orthonormal**. Using Eq. (4.17), Eq. (4.16) reduces to

$$\mathbf{A}\cdot\mathbf{B} = A_xB_x + A_yB_y + A_zB_z \ . \qquad (4.18)$$

To make an equation like Eq. (4.18) apply to vectors in spaces with other than three dimensions, we write that formula in a more general form. Using $1, 2, \ldots$ to identify the basis unit vectors, the dot product takes this form for a space of dimension n:

$$\mathbf{A}\cdot\mathbf{B} = \sum_{i=1}^{n} A_iB_i \ . \qquad (4.19)$$

Looking back at the definition, we see that this formula is applicable only if our coordinate system is **orthogonal** (i.e., axes mutually perpendicular), as is the case when we use Cartesian coordinates. This is not a severe limitation, as we only use oblique coordinate systems under duress (e.g., for discussions involving a crystal whose symmetry axes are not at right angles to each other).

Our first use of the dot product will be to obtain a formula for the magnitude of a vector using its components. Letting A or $|\mathbf{A}|$ stand for the magnitude of a vector **A**, we have (again in 2-D), compare Eq. (4.6),

$$\mathbf{A}\cdot\mathbf{A} = (A_x)^2 + (A_y)^2 = A^2 = |\mathbf{A}|^2 \ . \qquad (4.20)$$

We therefore have yet another way of writing the square of the magnitude of a vector **A**, namely $\mathbf{A}\cdot\mathbf{A}$, frequently shortened to $\mathbf{A}^2$. Notice that $\mathbf{A}^2$ is not a vector but is an ordinary (nonnegative) number; the notation is to be interpreted as meaning that the vectors **A** are combined by taking their dot product.

It is of interest to consider the dot product $\mathbf{C}\cdot\mathbf{C}$, where we set $\mathbf{C} = \mathbf{A}+\mathbf{B}$. Expanding, we have

$$\mathbf{C}\cdot\mathbf{C} = (\mathbf{A}+\mathbf{B})\cdot(\mathbf{A}+\mathbf{B}),$$

which expands to

$$\mathbf{C}^2 = \mathbf{A}^2 + \mathbf{B}^2 + 2\mathbf{A}\cdot\mathbf{B}\,.$$

Solving for $\mathbf{A}\cdot\mathbf{B}$, we reach

$$\mathbf{A}\cdot\mathbf{B} = \frac{C^2 - A^2 - B^2}{2} = AB\cos\theta\,, \tag{4.21}$$

which we identify as the law of cosines, familiar from trigonometry. Note that the minus sign in the law of cosines as usually presented refers to the (interior) angle of a triangle ABC opposite the side of length C. Here, as indicated in Fig. 4.7, we have defined θ as an exterior angle, so that $\theta = 0$ corresponds to **B** being in the same direction as **A**. This choice of θ explains the lack of a minus sign on the right-hand side of Eq. (4.21).

Because the dot product is independent of overall orientation, it is often also called a **scalar product**. We are now using the term **scalar** to mean more than "not a vector"; we are defining it to mean a quantity that is independent of our choice of coordinate system. Emphasizing this point, we have noted that although a rotation of either the vectors or the coordinate system changes the values of vector components, the combination of those components that forms the dot product evaluates to a rotation-independent result. It is this invariance that is associated with the term **scalar**.

Since $\cos\theta$ can assume all values in the range $(-1,+1)$, we see that the formula $\mathbf{A}\cdot\mathbf{B} = AB\cos\theta$ will have values that range from $+AB$ (when the vectors are in the same direction) to $-AB$ (when they are oppositely directed), with $\mathbf{A}\cdot\mathbf{B} = 0$ when **A** and **B** are at right angles (orthogonal). In mathematical contexts, **orthogonal** refers to quantities with a zero scalar product even when they are more abstract and there is no obvious physical basis for the language.

In 2-D, two vectors that are orthogonal have components that satisfy the equation

$$A_xB_x + A_yB_y = 0\,, \quad \text{which we can rearrange to} \quad \frac{A_y}{A_x}\frac{B_y}{B_x} = -1\,.$$

Here A_y/A_x is the slope corresponding to the direction of **A**, while B_y/B_x is the slope corresponding to **B**. The fact that the product of these slopes is -1 is a well-known indicator that they are perpendicular.

Example 4.2.1. Angle Between Two Vectors

We desire the angle between the two vectors

$$\mathbf{A} = \begin{pmatrix} 4 \\ -2 \\ 4 \end{pmatrix} \quad \text{and} \quad \mathbf{B} = \begin{pmatrix} 2 \\ 4 \\ -4 \end{pmatrix}.$$

To find it we need values of A, B, and $\mathbf{A}\cdot\mathbf{B}$. We get

$$A = \sqrt{4^2 + (-2)^2 + 4^2} = 6, \quad B = \sqrt{2^2 + 4^2 + (-4)^2} = 6,$$

$$\mathbf{A}\cdot\mathbf{B} = 4(2) + (-2)(4) + 4(-4) = -16 .$$

From these data we compute

$$\cos\theta = \frac{\mathbf{A}\cdot\mathbf{B}}{AB} = \frac{-16}{36} = -\frac{4}{9}.$$

Taking the arccosine, we find $\theta = 2.031$ radians, or about 116°.

■

PROJECTIONS

The simplest examples of what are called **projections** of a vector correspond to the removal of one or more (but not all) of its components. For example, consider a 3-D vector whose tail is at (x_1, y_1, z_1) and whose head is at (x_2, y_2, z_2). If we suppress (i.e., remove) its z component, it will then be in a 2-D region (the xy-plane) with ends at $(x_1, y_1, 0)$ and $(x_2, y_2, 0)$. Because this is a result that corresponds to the shadow (i.e., image) of the vector on the xy plane if illuminated from the z direction, it is reasonable to call this the **projection** of our vector onto the xy plane.

Alternatively, we might choose to remove both the y and z components of a 3-D vector. What will then remain is a vector along the x axis from x_1 to x_2. This is also called a projection, but now onto a line (the x axis).

The dot product gives us a formal way of finding projections of arbitrary vectors. If we want the projection of a vector $\mathbf{A}$ on the x axis (the result is obviously $A_x\hat{\mathbf{e}}_x$), and we write $\mathbf{A} = A_x\hat{\mathbf{e}}_x + A_y\hat{\mathbf{e}}_y + A_z\hat{\mathbf{e}}_z$, it is clear that the result we seek can be obtained if we can eliminate the $\hat{\mathbf{e}}_y$ and $\hat{\mathbf{e}}_z$ terms from the expression for $\mathbf{A}$. The recipe for accomplishing this is to take the dot product of $\mathbf{A}$ with $\hat{\mathbf{e}}_x$. This will work because the basis unit vectors are orthonormal, with the relationships given in Eq. (4.17). Therefore,

$$\hat{\mathbf{e}}_x \cdot \mathbf{A} = \hat{\mathbf{e}}_x \cdot (A_x\hat{\mathbf{e}}_x + A_y\hat{\mathbf{e}}_y + A_z\hat{\mathbf{e}}_z) = A_x , \tag{4.22}$$

and

$$\text{Projection of } \mathbf{A} \text{ on } x \; = (\hat{\mathbf{e}}_x \cdot \mathbf{A})\hat{\mathbf{e}}_x . \tag{4.23}$$

Notice that the projection is not just A_x, but a **vector** in the x direction of length A_x.

A formula of the above type is still applicable (but less trivial) if we want a projection onto a direction other than a coordinate axis. Suppose we want to obtain the projection of $\mathbf{A}$ on the direction of another vector, $\mathbf{B}$. Then we need a unit vector in the direction of $\mathbf{B}$, namely $\hat{\mathbf{B}} = \mathbf{B}/B$, and form our projection as

$$\text{Projection of } \mathbf{A} \text{ on } \mathbf{B} \; = \left[\left(\frac{\mathbf{B}}{B}\right)\cdot\mathbf{A}\right]\hat{\mathbf{B}} = \frac{(\mathbf{B}\cdot\mathbf{A})}{B}\,\hat{\mathbf{B}} .$$

This equation also shows that the magnitude of the projection of $\mathbf{A}$ onto $\mathbf{B}$ is $\mathbf{B}\cdot\mathbf{A}/B$, equivalent to the observation that $\mathbf{A}\cdot\mathbf{B}$ is equal to the product of the magnitudes of $\mathbf{B}$ and the projection of $\mathbf{A}$ on $\mathbf{B}$. Because the dot product is symmetric in $\mathbf{A}$ and $\mathbf{B}$, this observation can be generalized to state that the dot product $\mathbf{A}\cdot\mathbf{B}$ is equal to the product of the magnitude of either factor multiplied by the magnitude of the projection on it of the other factor. When $\mathbf{A}\cdot\mathbf{B}$ is written as $AB\cos\theta$, $A\cos\theta$ is the magnitude of the projection of $\mathbf{A}$ on $\mathbf{B}$, while $B\cos\theta$ is the magnitude of the projection of $\mathbf{B}$ on $\mathbf{A}$.

Exercises

4.2.1. Find the angle between the vectors $3\hat{\mathbf{e}}_x - 2\hat{\mathbf{e}}_y + \hat{\mathbf{e}}_z$ and $\hat{\mathbf{e}}_x + 2\hat{\mathbf{e}}_y - 3\hat{\mathbf{e}}_z$.

4.2.2. Given $\mathbf{A} = 2\hat{\mathbf{e}}_x - 3\hat{\mathbf{e}}_z$ and $\mathbf{B} = 3\hat{\mathbf{e}}_x - \hat{\mathbf{e}}_y$, find $\mathbf{A} \cdot \mathbf{B}$.

4.2.3. Show that $\hat{\mathbf{e}}_x - 2\hat{\mathbf{e}}_y - 4\hat{\mathbf{e}}_z$ and $2\hat{\mathbf{e}}_x + 5\hat{\mathbf{e}}_y - 2\hat{\mathbf{e}}_z$ are perpendicular, and find a third vector perpendicular to both.

4.2.4. Show that $|B|\mathbf{A} + |A|\mathbf{B}$ is perpendicular to $|B|\mathbf{A} - |A|\mathbf{B}$.

4.2.5. The vector $\mathbf{r}$, starting at the origin, terminates at and specifies the point in space (x, y, z). If $\mathbf{a}$ is an arbitrary constant vector and $\mathbf{r}$ is allowed to take on all values that satisfy one of the following conditions, characterize the surface defined by the tip of $\mathbf{r}$ if

(a) $(\mathbf{r} - \mathbf{a}) \cdot \mathbf{a} = 0$,

(b) $(\mathbf{r} - \mathbf{a}) \cdot \mathbf{r} = 0$.

4.2.6. A pipe comes diagonally down the south wall of a building, making an angle of φ with the vertical. Coming into a corner, the pipe turns and continues diagonally down a west-facing wall, still making an angle of φ with the vertical. Given that the angle between the south-wall and west-wall sections of the pipe is 120°, find the angle φ.

4.3 SYMBOLIC COMPUTING, VECTORS

To do symbolic computing involving vectors, we must first learn how to enter them into the symbolic system. The command to define `A` as a 2-D vector with components A_x, A_y is

MAPLE: `> A := Vector( [`A_x`,`A_y`] );`   MATHEMATICA: `A = {`A_x`,`A_y`}`

For vectors with more than two components these commands would have an appropriate number of additional arguments.

Sometimes we will want to work with a vector $\mathbf{A}$ whose components have values that are the output of a computation. We then need to define $\mathbf{A}$ in a way that causes the symbolic system to know that $\mathbf{A}$ is a vector of a specified dimension, and we also need a way to load computed data into $\mathbf{A}$ and to read or change the values of its components. In MAPLE, the command `A := Vector(`n`)` defines $\mathbf{A}$ to be a vector of dimension n (initialized to zero), and the ith component of $\mathbf{A}$ has the name `A[i]`, which can be used to read out or change the value of A_i. In MATHEMATICA, the command `Array[A,n]` causes definition of an array named $\mathbf{A}$ of dimension n, with the elements A_i defined only as symbols. MATHEMATICA identifies the components A_i by the names `A[[i]]`. Thus,

MAPLE: `> A := Vector(3):`		MATHEMATICA: `A = Array[a,3];`	
`A[1];`	0	`A[[1]]`	`A[[1]]`
`A[1] := 4;`	$A_1 := 4$	`A[[1]] = 4`	4
`A[2] := 2;`	$A_2 := 2$	`A[[2]] = 2`	2
`A;`	$\begin{pmatrix} 4 \\ 2 \\ 0 \end{pmatrix}$	`A`	`{4, 2, A[[3]]}`

More detail on the MAPLE and MATHEMATICA data structures can be found in Appendix C. Moreover, both symbolic systems have additional ways of providing vectors as input; we leave these for exploration by advanced users.

In both symbolic systems vector addition and multiplication by a scalar are obtained using the same notations that the system would use for addition and multiplication of ordinary numbers. However, an error condition results if it is attempted to add vectors that do not conform (i.e., that have different numbers of components). In neither MAPLE nor MATHEMATICA can one write $\mathbf{A}^2$ as a synonym for $\mathbf{A} \cdot \mathbf{A}$. One must use the dot-operator form, as explained in detail in the next subsection.

In MAPLE, the output of a vector is displayed in column-vector form, providing the number of its components (its dimension) is not too large (>10). However, in MATHEMATICA, vector output is written in the same notation as vector input. To obtain `A` displayed as a column vector in MATHEMATICA, one can use the command `MatrixForm`:

```
A = {1,2};   MatrixForm[A]
```

$$\begin{pmatrix} 1 \\ 2 \end{pmatrix}$$

DOT OPERATOR

In both MAPLE and MATHEMATICA there are multiple ways to obtain dot products of vectors. Perhaps the least complicated way to obtain a dot product is to write the vectors to be combined with a dot (typed as a period) between them. In both symbolic systems the dot is used not only as a decimal point but also to represent the multiplication of various different kinds of objects, and it is sometimes necessary to use parentheses to avoid ambiguity. Unexpected results or errors may occur if it has not been made clear that the operands of the "dot" are vectors and that the dot is not a decimal point. Thus, if `A` and `B` are vectors, instead of `A.3*B` write `A.(3*B)`.

The discussion presented here for the dot product assumes that all the vectors involved are real. Symbolic computations for vectors with components that may be complex present issues that are not addressed here; those issues are reviewed in Chapter 11.

Example 4.3.1. Symbolic Operations with Vectors

In MAPLE,

```
A := Vector([1,2]);
```

$$A := \begin{bmatrix} 1 \\ 2 \end{bmatrix}$$

```
B := Vector([2,-1]);
```

$$B := \begin{bmatrix} 2 \\ -1 \end{bmatrix}$$

```
C := Vector([2,1,4]);
```

$$C := \begin{bmatrix} 2 \\ 1 \\ 4 \end{bmatrix}$$

```
A - 2*B;
```

$$\begin{bmatrix} -3 \\ 4 \end{bmatrix}$$

```
A + C;
```

(error condition)

```
A^2;
```

(error condition)

```
A . A;
```

5

```
A . B;
```

0

In MATHEMATICA,

`A = {1,2}`	$\{1,2\}$
`B = {2,-1}`	$\{2,-1\}$
`C = {2,1,4}`	$\{2,1,4\}$
`A - 2*B`	$\{-3,0\}$
`MatrixForm[%]`	$\begin{pmatrix} -3 \\ 4 \end{pmatrix}$
`A + C`	(error condition)
`A^2`	Answer isn't $\mathbf{A}\cdot\mathbf{A}$!
`A . A`	5
`A . B`	0

■

Exercises

4.3.1. Using a symbolic computing system, define vectors $\mathbf{A} = A_x\hat{\mathbf{e}}_x + A_y\hat{\mathbf{e}}_y + A_z\hat{\mathbf{e}}_z$ and $\mathbf{B} = B_x\hat{\mathbf{e}}_x + B_y\hat{\mathbf{e}}_y + B_z\hat{\mathbf{e}}_z$ and then form $(\mathbf{A}+\mathbf{B})\cdot(\mathbf{A}-\mathbf{B})$. Show that your result is equal to $\mathbf{A}^2 - \mathbf{B}^2$.

Note. In MAPLE, the components of **A** and **B** are assumed complex, but this exercise assumes them to be real. Tell MAPLE this with `assume(Ax,real)` and similar commands for all other components of **A** and **B**. This issue does not arise in MATHEMATICA.

4.3.2. Use your symbolic computing system to verify the result of Exercise 4.2.1. Go from the cosine of an angle to the angle itself using the function `arccos` or `ArcCos`.

4.3.3. Using your symbolic computing system, define the four vectors

$$\mathbf{A} = \begin{pmatrix} 5 \\ 3 \\ 1 \end{pmatrix}, \quad \mathbf{B} = \begin{pmatrix} 3 \\ -5 \\ 0 \end{pmatrix}, \quad \mathbf{C} = \begin{pmatrix} 1 \\ 0 \\ -5 \end{pmatrix}, \quad \mathbf{D} = \begin{pmatrix} 0 \\ -1 \\ 3 \end{pmatrix}.$$

(If any of these symbols have a fixed meaning in your symbolic system and therefore cannot represent user-defined quantities, change them as needed to avoid errors—e.g., to lowercase letters.)

(a) Calculate the dot products for all pairs of the above vectors (including those with both members the same).

(b) Based on the results of part (a), what can you say about the directions of **B**, **C**, and **D**?

(c) Based on your answer to part (b), find a relation connecting the vectors **B**, **C**, and **D**.

4.3.4. Continuing with the vectors defined in Exercise 4.3.3, use your symbolic system to compute $(\mathbf{A}-3\mathbf{B})\cdot\mathbf{C}$ and $(\mathbf{A}-3\mathbf{C})\cdot\mathbf{B}$.

4.4 MATRICES

Matrices are important in linear algebra because they have been defined in a way that makes them useful for working with systems of linear equations. This point can be illustrated by the following short introduction to linear equation systems.

LINEAR EQUATION SYSTEMS

Let's look at the following set of three simultaneous linear equations:

$$\begin{aligned} a_1x_1 + a_2x_2 + a_3x_3 &= h_1 , \\ b_1x_1 + b_2x_2 + b_3x_3 &= h_2 , \\ c_1x_1 + c_2x_2 + c_3x_3 &= h_3 . \end{aligned} \tag{4.24}$$

Here the a_i, b_i, c_i, and h_i are known quantities, while the x_i are unknowns which we may be able to find by solving the equations. We now observe that if we identify (a_1, a_2, a_3) as the components of a vector $\mathbf{a}$, with $(b_1, b_2, b_3) = \mathbf{b}$, $(c_1, c_2, c_3) = \mathbf{c}$, and $(x_1, x_2, x_3) = \mathbf{x}$, the left-hand side of each of Eqs. (4.24) has the form of a dot product, and our three equations can then be written

$$\mathbf{a} \cdot \mathbf{x} = h_1 , \qquad \mathbf{b} \cdot \mathbf{x} = h_2 , \qquad \mathbf{c} \cdot \mathbf{x} = h_3 . \tag{4.25}$$

The properties of the equation system in Eqs. (4.24) depend crucially upon whether the forms constituting their left-hand sides are **linearly dependent**. A set of forms is called linearly dependent if at least one of the forms is a linear combination of others. If the three forms in Eq. (4.24) are linearly independent, they provide three independent equations for x_1, x_2, x_3, and we can solve those equations to obtain unique values of the x_i. On the other hand, if the three forms in Eq. (4.24) exhibit linear dependence, those equations have fewer than three independent left-hand sides and, depending on the values of the h_i, they either reduce to only one or two independent equations (in which case the solution is not unique) or they can be combined to cause the same left-hand side to be equal to two different right-hand sides. In that case the equation set is said to be **inconsistent**, as it has no solution.

Since linear dependence is a property of the coefficients on the left-hand side of our equation system, it is useful to adopt a notation that contains only that information. We do so by placing the coefficients in an array called a **matrix**, with the coefficients in positions corresponding to their locations in Eqs. (4.24). Based on that equation set, and giving our matrix the name A, we write

$$\mathsf{A} = \begin{pmatrix} a_1 & a_2 & a_3 \\ b_1 & b_2 & b_3 \\ c_1 & c_2 & c_3 \end{pmatrix} . \tag{4.26}$$

A deeper analysis of linear equation systems now leads us in two directions, both of which will be elaborated in this chapter: (1) we want to know how we can determine whether a matrix A describes a linearly dependent set of linear forms, and (2) we need rules for working with sets of linear equations involving the coefficients in a matrix A.

As an initial foray into the second direction, we choose to write Eqs. (4.24) in the symbolic form

$$\mathsf{A}\mathbf{x} = \mathbf{h} , \qquad \text{or} \qquad \begin{pmatrix} a_1 & a_2 & a_3 \\ b_1 & b_2 & b_3 \\ c_1 & c_2 & c_3 \end{pmatrix} \begin{pmatrix} x_1 \\ x_2 \\ x_3 \end{pmatrix} = \begin{pmatrix} h_1 \\ h_2 \\ h_3 \end{pmatrix} . \tag{4.27}$$

Since we have already made the observation that the left-hand sides of the three equations of our current example could be written $\mathbf{a}\cdot\mathbf{x}$, $\mathbf{b}\cdot\mathbf{x}$, $\mathbf{c}\cdot\mathbf{x}$, Eq. (4.27) could alternatively be written as

$$\begin{pmatrix} a_1 & a_2 & a_3 \\ b_1 & b_2 & b_3 \\ c_1 & c_2 & c_3 \end{pmatrix} \begin{pmatrix} x_1 \\ x_2 \\ x_3 \end{pmatrix} = \begin{pmatrix} \mathbf{a}\cdot\mathbf{x} \\ \mathbf{b}\cdot\mathbf{x} \\ \mathbf{c}\cdot\mathbf{x} \end{pmatrix} = \begin{pmatrix} h_1 \\ h_2 \\ h_3 \end{pmatrix}. \tag{4.28}$$

In order for Eqs. (4.27) and (4.28) to have a mathematical, as well as a symbolic meaning, we need to assume that the arrays $\mathbf{x}$ and $\mathbf{h}$ can be interpreted as column vectors and that the expression $\mathsf{A}\mathbf{x}$ corresponds to an operation on the matrix A and the vector $\mathbf{x}$ that we will call **matrix multiplication**, and which produces the result shown as the central member of Eq. (4.28). We then interpret the equality of the last two members of Eq. (4.28) as that of two column vectors, so we recover the three equations previously given as Eqs. (4.25).

The preceding discussion can be regarded as motivation for the formal definition of matrices and matrix operations.

MATRICES: DEFINITION AND OPERATIONS

We define a **matrix** to be a rectangular array of numbers (called its **elements**) for which various operations are defined. The elements are arranged in horizontal **rows** and vertical **columns**; if a matrix has m rows and n columns, it is referred to as an $m \times n$ matrix. In spoken language, an $m \times n$ matrix is usually called an "m by n matrix." If the number of rows is equal to the number of columns, a matrix is identified as **square**. In this book a matrix will usually be denoted by an upper-case character in a sans-serif font; an example is A. The elements of matrix A are often referred to as A_{ij} or a_{ij}; both conventions are in common use, and both mean the element in row i and column j. Notice that the row index is always given first. Note also that if $i \neq j$, A_{ij} and A_{ji} refer to different positions in the array and there is in general no reason to expect that $A_{ij} = A_{ji}$. For later use, we note that elements A_{ij} with $i = j$ (hence A_{ii}) are called the **diagonal** elements of the matrix, and a line through these diagonal elements is called the **principal diagonal**. The array of numbers (or symbols) constituting a matrix is conventionally enclosed in ordinary parentheses, not curly braces { } or vertical lines | |. Matrix algebra is defined in a way such that a matrix with only one column is synonymous with a column vector; matrices with only one row are called **row vectors**. Some matrices are exhibited in Fig. 4.8. It is important to realize that (like a vector) a matrix is not a single number and cannot be reduced to a single number; it is an array.

$$\begin{pmatrix} u_1 \\ u_2 \\ u_3 \\ u_4 \end{pmatrix} \quad \begin{pmatrix} 4 & 2 \\ -1 & 3 \\ 0 & 1 \end{pmatrix} \quad \begin{pmatrix} 6 & 7 & 0 \\ 1 & 4 & 3 \end{pmatrix} \quad \begin{pmatrix} 0 & 1 \\ 1 & 0 \end{pmatrix} \quad \begin{pmatrix} a_{11} & a_{12} \end{pmatrix}.$$

Figure 4.8: From left to right, matrices of dimension 4×1 (column vector), 3×2, 2×3, 2×2 (square), 1×2 (row vector).

EQUALITY

If A and B are matrices, $\mathsf{A} = \mathsf{B}$ only if $a_{ij} = b_{ij}$ for all values of i and j. A necessary but not sufficient condition for equality is that both matrices have the same dimensions; we then may say that they have the same **shape**.

MATRIX ADDITION

To add two matrices, we add corresponding elements. As an example,

$$\mathsf{A} + \mathsf{B} = \begin{pmatrix} a_{11} & a_{12} \\ a_{21} & a_{22} \end{pmatrix} + \begin{pmatrix} b_{11} & b_{12} \\ b_{21} & b_{22} \end{pmatrix} = \begin{pmatrix} a_{11} + b_{11} & a_{12} + b_{12} \\ a_{21} + b_{21} & a_{22} + b_{22} \end{pmatrix} . \tag{4.29}$$

To express this relationship in general, we can write

$$\mathbf{C} = \mathbf{A} + \mathbf{B} \qquad \text{if, for all } i \text{ and } j, \qquad c_{ij} = a_{ij} + b_{ij} \; . \tag{4.30}$$

In order for addition to be defined, the matrices to be added must have the same shape (i.e., both must have the same number of rows and both must have the same number of columns). They need not be square.

It is clear from the definition that matrix addition is commutative and associative:

$$\mathsf{A} + \mathsf{B} = \mathsf{B} + \mathsf{A} \, , \qquad (\mathsf{A} + \mathsf{B}) + \mathsf{C} = \mathsf{A} + (\mathsf{B} + \mathsf{C}) \, . \tag{4.31}$$

MULTIPLICATION BY A SCALAR

The multiplication of matrix A by the scalar k yields a matrix B of the same shape as A, according to

$$\mathsf{B} = k\mathsf{A} \, , \qquad \text{with } b_{ij} = k\, a_{ij} \text{ for all } i \text{ and } j. \tag{4.32}$$

This operation is commutative, with $k\,\mathsf{A} = \mathsf{A}\,k$.

The rules for matrix addition and multiplication by a scalar give unambiguous meaning to linear forms involving matrices of conforming dimensions. Note also that these rules correspond exactly with the corresponding rules for vectors, confirming that it is consistent to call a matrix with only one column a column vector.

Example 4.4.1. Elementary Matrix Operations

Given the matrices

$$\mathsf{A} = \begin{pmatrix} 1 & 2 \\ 3 & 4 \\ 5 & 6 \end{pmatrix}, \qquad \mathsf{B} = \begin{pmatrix} -1 & 0 \\ 2 & -3 \\ 0 & 6 \end{pmatrix}, \qquad \mathsf{C} = \begin{pmatrix} 1 & 2 \\ 3 & 4 \end{pmatrix},$$

then

$$\mathsf{A} + 2\mathsf{B} = \begin{pmatrix} -1 & 2 \\ 7 & -2 \\ 5 & 18 \end{pmatrix}, \qquad \mathsf{A} - \mathsf{B} = \begin{pmatrix} 2 & 2 \\ 1 & 7 \\ 5 & 0 \end{pmatrix}, \qquad 3\mathsf{C} = \begin{pmatrix} 3 & 6 \\ 9 & 12 \end{pmatrix}.$$

Matrices to be added must have conforming dimensions: here $\mathsf{A} + \mathsf{C}$ is not defined.

■

MATRIX MULTIPLICATION

The most useful definition for **matrix multiplication** is not an element-by-element product, but is instead one that is motivated by Eq. (4.28). The product of matrices A and B, written $\mathsf{A\,B}$, is defined as

$$\mathsf{C} = \mathsf{A\,B}\,, \qquad \text{with} \qquad c_{ij} = \sum_k a_{ik} b_{kj}\,. \tag{4.33}$$

This definition causes the ij element of C to be the dot product of the ith row of $\mathbf{A}$ with the jth column of $\mathbf{B}$. Obviously Eq. (4.33) is meaningful only if A has the same number of columns as B has rows. Moreover, the product matrix, C, will have the same number of rows as A and the same number of columns as B. One way to visualize Eq.(4.33) is to think of matrix A as composed of row vectors $\mathbf{a}_i$, and to consider B as composed of column vectors $\mathbf{b}_j$. Then the matrix product AB has the form

$$\begin{pmatrix} \mathbf{a}_1 \\ \hline \mathbf{a}_2 \\ \hline \mathbf{a}_3 \\ \hline \cdots \end{pmatrix} \begin{pmatrix} \mathbf{b}_1 \,\Big|\, \mathbf{b}_2 \,\Big|\, \mathbf{b}_3 \,\Big|\, \cdots \end{pmatrix} = \begin{pmatrix} \mathbf{a}_1\cdot\mathbf{b}_1 & \mathbf{a}_1\cdot\mathbf{b}_2 & \mathbf{a}_1\cdot\mathbf{b}_3 & \cdots \\ \mathbf{a}_2\cdot\mathbf{b}_1 & \mathbf{a}_2\cdot\mathbf{b}_2 & \mathbf{a}_2\cdot\mathbf{b}_3 & \cdots \\ \mathbf{a}_3\cdot\mathbf{b}_1 & \mathbf{a}_3\cdot\mathbf{b}_2 & \mathbf{a}_3\cdot\mathbf{b}_3 & \cdots \\ \cdots & \cdots & \cdots & \cdots \end{pmatrix}.$$

Because the roles of the two operands in a matrix multiplication are different (the first is processed by rows, the second by columns), the operation is in general not commutative, that is, $\mathsf{A\,B} \neq \mathsf{B\,A}$. In fact, $\mathsf{A\,B}$ may even have a different shape than $\mathsf{B\,A}$. Matrix multiplication is, however, **associative**, meaning that $\mathsf{(AB)C{=}A(BC)}$. Proof of this statement is the topic of Exercise 4.4.6.

If A and B are square and of the same dimension, it is useful to define the **commutator** of A and B,

$$[\mathsf{A}, \mathsf{B}] = \mathsf{A\,B} - \mathsf{B\,A}\,, \tag{4.34}$$

which, as stated above, will in many cases be nonzero.

If matrices A and B are not square, $\mathsf{A\,B}$ and $\mathsf{B\,A}$ will not even have the same shape and therefore cannot be added or subtracted. In fact, these products will not even be defined unless the number of columns of the first factor is equal to the number of rows of the second factor. The following are examples of meaningless expressions:

$$\begin{pmatrix} 1 \\ 3 \\ 5 \end{pmatrix} \begin{pmatrix} 1 & 2 & 3 \\ 4 & 5 & 6 \\ 7 & 8 & 9 \end{pmatrix}, \qquad \begin{pmatrix} 1 & 2 \\ 3 & 4 \\ 5 & 6 \end{pmatrix} \begin{pmatrix} 1 & 2 \\ 3 & 4 \\ 5 & 6 \end{pmatrix}.$$

Example 4.4.2. Multiplication, Pauli Matrices

These three 2×2 matrices, which occurred in early work in quantum mechanics by Pauli, are encountered frequently in physics contexts, so a familiarity with them is highly advisable. They are

$$\sigma_1 = \begin{pmatrix} 0 & 1 \\ 1 & 0 \end{pmatrix}, \quad \sigma_2 = \begin{pmatrix} 0 & -i \\ i & 0 \end{pmatrix}, \quad \sigma_3 = \begin{pmatrix} 1 & 0 \\ 0 & -1 \end{pmatrix}. \tag{4.35}$$

Let's form $\sigma_1\sigma_2$. The $1,1$ element of the product involves the first **row** of σ_1 and the first **column** of σ_2; these are shaded and lead to the indicated computation:

$$\begin{pmatrix} 0 & 1 \\ 1 & 0 \end{pmatrix} \begin{pmatrix} 0 & -i \\ i & 0 \end{pmatrix} \quad \to \quad 0(0) + 1(i) = i\,.$$

Continuing, we have

$$\sigma_1\sigma_2 = \begin{pmatrix} 0(0)+1(i) & 0(-i)+1(0) \\ 1(0)+0(i) & 1(-i)+0(0) \end{pmatrix} = \begin{pmatrix} i & 0 \\ 0 & -i \end{pmatrix} . \tag{4.36}$$

In a similar fashion, we can compute

$$\sigma_2\sigma_1 = \begin{pmatrix} 0 & -i \\ i & 0 \end{pmatrix}\begin{pmatrix} 0 & 1 \\ 1 & 0 \end{pmatrix} = \begin{pmatrix} -i & 0 \\ 0 & i \end{pmatrix} . \tag{4.37}$$

It is clear that σ_1 and σ_2 do not commute. We can construct their commutator:

$$[\sigma_1, \sigma_2] = \sigma_1\sigma_2 - \sigma_2\sigma_1 = \begin{pmatrix} i & 0 \\ 0 & -i \end{pmatrix} - \begin{pmatrix} -i & 0 \\ 0 & i \end{pmatrix}$$

$$= 2i\begin{pmatrix} 1 & 0 \\ 0 & -1 \end{pmatrix} = 2i\sigma_3 . \tag{4.38}$$

Notice that not only have we verified that σ_1 and σ_2 do not commute, we have even evaluated and simplified their commutator.

■

MULTIPLYING MATRICES WITH VECTORS

The rule for matrix multiplication continues to apply if one or both factors are of dimension 1 in one direction (thereby being equivalent to a row or column vector). However, when an array consists of only one row or column, its elements are usually identified by a single index rather than an index pair. Correct results are obtained from matrix multiplications if a second index, always set to 1, is considered as inserted (for a column vector) after the variable index, or (for a row vector) before the variable index. Here are two examples, in which A is a 2×3 matrix, $\mathbf{u}$ is a column vector of dimension 3, and $\mathbf{v}$ is a row vector of dimension 2.

$$\mathsf{A}\mathbf{u} = \begin{pmatrix} a_{11} & a_{12} & a_{13} \\ a_{21} & a_{22} & a_{23} \end{pmatrix}\begin{pmatrix} u_1 \\ u_2 \\ u_3 \end{pmatrix} = \begin{pmatrix} \sum_i a_{1i}u_i \\ \sum_i a_{2i}u_i \end{pmatrix},$$

$$\mathbf{v}\,\mathsf{A} = \begin{pmatrix} v_1 & v_2 \end{pmatrix}\begin{pmatrix} a_{11} & a_{12} & a_{13} \\ a_{21} & a_{22} & a_{23} \end{pmatrix} = \begin{pmatrix} \sum_i v_i a_{i1} & \sum_i v_i a_{i2} & \sum_i v_i a_{i3} \end{pmatrix}.$$

Attempts to form $\mathbf{u}\mathsf{A}$ or $\mathsf{A}\mathbf{v}$ will fail, as those products do not have conforming dimensions.

Here is an example in which both matrices are vectors.

Example 4.4.3. Multiplication, Row, and Column Matrices

Consider

$$\mathsf{A} = \begin{pmatrix} 1 \\ 2 \\ 3 \end{pmatrix}, \qquad \mathsf{B} = \begin{pmatrix} 4 & 5 & 6 \end{pmatrix}.$$

Let us form $\mathsf{A}\,\mathsf{B}$ and $\mathsf{B}\,\mathsf{A}$:

$$\mathsf{A}\mathsf{B} = \begin{pmatrix} 4 & 5 & 6 \\ 8 & 10 & 12 \\ 12 & 15 & 18 \end{pmatrix}, \qquad \mathsf{B}\mathsf{A} = \begin{pmatrix} 4\times 1 + 5\times 2 + 6\times 3 \end{pmatrix} = (32)\,.$$

The results speak for themselves. Often when a matrix operation leads to a 1×1 matrix, the parentheses are dropped and the result is written as an ordinary number or function.

■

Exercises

4.4.1. Given the matrices

$$\mathsf{A} = \begin{pmatrix} 3 & 4 & -1 & 3 \\ 1 & 2 & 0 & 4 \end{pmatrix}, \quad \mathsf{B} = \begin{pmatrix} 2 & 3 \\ -1 & 1 \\ 4 & 1 \end{pmatrix}, \quad \mathsf{C} = \begin{pmatrix} 2 & 3 & 1 \\ -1 & -2 & 3 \\ 1 & 0 & -1 \end{pmatrix},$$

evaluate or identify as meaningless each of the following matrix products: A^2, AB, BA, B^2, AC, CA, BC, CB, C^2.

4.4.2. Given the matrices

$$\mathsf{A} = \begin{pmatrix} 1 \\ 4 \\ 2 \\ 3 \end{pmatrix}, \qquad \mathsf{B} = \begin{pmatrix} 4 & 3 & 2 & 1 \end{pmatrix},$$

evaluate or identify as meaningless each of the following expressions: $2\mathsf{AB}$, $3\mathsf{BA}$, $[\mathsf{A},\mathsf{B}]$, A^2, B^2.

4.4.3. Evaluate $(2\ \ 5)\begin{pmatrix} 1 & 2 \\ 3 & -1 \end{pmatrix}\begin{pmatrix} 2 \\ 1 \end{pmatrix}$.

4.4.4. Evaluate $(x\ \ y)\begin{pmatrix} 1 & 2 \\ 3 & -1 \end{pmatrix}\begin{pmatrix} x \\ y \end{pmatrix}$.

4.4.5. A matrix A with elements $a_{ij} = 0$ for $j < i$ is called upper triangular. This nomenclature is appropriate because the elements below the principal diagonal vanish.

Show that the product of two upper triangular matrices is also an upper triangular matrix.

4.4.6. By writing the matrices in terms of their elements, show that matrix operations satisfy

(a) The distributive laws $\mathsf{A}(\mathsf{B}+\mathsf{C}) = \mathsf{AB} + \mathsf{AC}$ and $(\mathsf{B}+\mathsf{C})\mathsf{A} = \mathsf{BA} + \mathsf{CA}$.

(b) The associative law $\mathsf{A}(\mathsf{BC}) = (\mathsf{AB})\mathsf{C}$.

4.4.7. Show that if and only if A and B commute will it be true that $(\mathsf{A}+\mathsf{B})(\mathsf{A}-\mathsf{B}) = \mathsf{A}^2 - \mathsf{B}^2$.

4.4.8. Show that the matrices

$$\mathsf{A} = \begin{pmatrix} 0 & 1 & 0 \\ 0 & 0 & 0 \\ 0 & 0 & 0 \end{pmatrix}, \quad \mathsf{B} = \begin{pmatrix} 0 & 0 & 0 \\ 0 & 0 & 0 \\ 1 & 0 & 0 \end{pmatrix}, \quad \mathsf{C} = \begin{pmatrix} 0 & 0 & 0 \\ 0 & 0 & 0 \\ 0 & 1 & 0 \end{pmatrix}$$

satisfy the commutation relations

$$[\mathsf{B}, \mathsf{A}] = \mathsf{C}, \qquad [\mathsf{A}, \mathsf{C}] = 0, \qquad [\mathsf{B}, \mathsf{C}] = 0.$$

4.4.9. The three Pauli spin matrices are

$$\sigma_1 = \begin{pmatrix} 0 & 1 \\ 1 & 0 \end{pmatrix}, \quad \sigma_2 = \begin{pmatrix} 0 & -i \\ i & 0 \end{pmatrix}, \quad \sigma_3 = \begin{pmatrix} 1 & 0 \\ 0 & -1 \end{pmatrix}.$$

Show that

(a) $\sigma_i^2 = \begin{pmatrix} 1 & 0 \\ 0 & 1 \end{pmatrix}, \quad i = 1, 2, 3.$

(b) $\sigma_1\sigma_2 = i\sigma_3, \quad \sigma_2\sigma_3 = i\sigma_1, \quad \sigma_3\sigma_1 = i\sigma_2.$

(c) $\sigma_i\sigma_j + \sigma_j\sigma_i = 0, \quad i \neq j.$

Property (c) indicates that different σ_i **anticommute** (this means that commutation changes the sign of the matrix product).

ZERO AND UNIT MATRICES

By direct matrix multiplication, it is possible to show that a square matrix with elements of value unity on its **principal diagonal** (the elements (i, j) with $i = j$), and zeros everywhere else, will leave unchanged any matrix with which it can be multiplied. Notationally, the situation encountered here is traditionally indicated by introducing a quantity called the **Kronecker delta** and defined as

$$\delta_{ij} = \begin{cases} 1, & i = j, \\ 0, & i \neq j. \end{cases} \tag{4.39}$$

Now a unit matrix, to which we give the name **1**, can be said to have elements

$$\mathbf{1}_{ij} = \delta_{ij} \,. \tag{4.40}$$

For example, the 3×3 unit matrix has the form

$$\begin{pmatrix} 1 & 0 & 0 \\ 0 & 1 & 0 \\ 0 & 0 & 1 \end{pmatrix};$$

note that it is **not** a matrix all of whose elements are unity. The defining property of a unit matrix is that it satisfies the following equation:

$$\mathbf{1}\,\mathsf{A} = \mathsf{A}\,\mathbf{1} = \mathsf{A} \,. \tag{4.41}$$

In interpreting this equation, we must keep in mind that unit matrices, which are square and therefore of dimensions $n \times n$, exist for all n; the n values for use in

Eq. (4.41) must be those consistent with the applicable dimension of A. So if A is $m \times n$, the unit matrix in $\mathbf{1}\mathsf{A}$ must be $m \times m$, while that in $\mathsf{A}\mathbf{1}$ must be $n \times n$.

A matrix with all elements zero, sometimes written O, will have the properties expected of zero. Zero matrices can have whatever dimensions are needed for matrix addition or multiplication. When chosen to have dimensions making the operations defined, we have for all A

$$\mathsf{O}\,\mathsf{A} = \mathsf{A}\,\mathsf{O} = \mathsf{O} \quad \text{and} \quad \mathsf{A} + \mathsf{O} = \mathsf{A}\,. \tag{4.42}$$

Many authors write an ordinary zero, 0, in place of O.

DIAGONAL MATRICES

If a matrix D has nonzero elements d_{ij} only for $i = j$, it is said to be **diagonal**; a 3×3 example is

$$\mathsf{D} = \begin{pmatrix} 1 & 0 & 0 \\ 0 & 2 & 0 \\ 0 & 0 & 3 \end{pmatrix}.$$

If we multiply two square diagonal matrices together, we get the following result:

$$\begin{pmatrix} a_{11} & 0 & 0 \\ 0 & a_{22} & 0 \\ 0 & 0 & a_{33} \end{pmatrix} \begin{pmatrix} b_{11} & 0 & 0 \\ 0 & b_{22} & 0 \\ 0 & 0 & b_{33} \end{pmatrix} = \begin{pmatrix} a_{11}b_{11} & 0 & 0 \\ 0 & a_{22}b_{22} & 0 \\ 0 & 0 & a_{33}b_{33} \end{pmatrix}.$$

From the result we see that the product is also diagonal, and its symmetry (between a and b) shows that the matrices commute. However, unless proportional to a unit matrix, diagonal matrices will not commute with nondiagonal matrices containing arbitrary elements.

TRANSPOSE, COMPLEX CONJUGATE, ADJOINT

The **transpose** of a matrix is obtained by interchanging its row and column indices, and is indicated either by placing a tilde symbol over the matrix name or by appending a superscript T. Thus,

$$\tilde{\mathsf{A}}_{ij} = \left(\mathsf{A}^T\right)_{ij} = a_{ji}, \quad \text{all } i \text{ and } j. \tag{4.43}$$

Note that transposition changes an $n \times m$ matrix into a $m \times n$ matrix, so if $n \neq m$ the operation changes the matrix shape. In particular, transposition converts a column vector into a row vector, and vice versa. Here are a few examples:

$$\mathsf{A} = \begin{pmatrix} 1 & 2 & 3 \\ 4 & 5 & 6 \end{pmatrix}, \quad \tilde{\mathsf{A}} = \mathsf{A}^T = \begin{pmatrix} 1 & 4 \\ 2 & 5 \\ 3 & 6 \end{pmatrix}, \quad \mathbf{x} = \begin{pmatrix} x_1 \\ x_2 \\ x_3 \end{pmatrix}, \quad \tilde{\mathbf{x}} = \mathbf{x}^T = (x_1\ x_2\ x_3)\,.$$

A matrix which is equal to its transpose is said to be **symmetric**. In terms of its elements, this corresponds to

$$\text{If } \mathsf{A} \text{ is symmetric,} \quad \mathsf{A}^T = \mathsf{A}, \text{ and } a_{ji} = a_{ij}\,. \tag{4.44}$$

It is obvious that a symmetric matrix must be square, but that a square matrix may or may not be symmetric.

The **complex conjugate** of a matrix A, indicated A^*, has elements which are complex conjugates of the elements of A:

$$(\mathsf{A}^*)_{ij} = a_{ij}^*, \quad \text{all } i \text{ and } j. \tag{4.45}$$

A matrix is called **real** if all its elements are real, and in that case $\mathsf{A}^* = \mathsf{A}$. Matrices with elements that may be complex are encountered, among other places, in quantum mechanics.

The **adjoint** of a matrix A, denoted $\mathsf{A}^\dagger$, is obtained from A by taking both its transpose and its complex conjugate. Some authors identify the adjoint by the more specific name **Hermitian adjoint**. In terms of elements, this means

$$\mathsf{A}^\dagger = \left(\mathsf{A}^T\right)^* = (\mathsf{A}^*)^T \; ; \qquad \left(\mathsf{A}^\dagger\right)_{ij} = a_{ji}^* \, . \tag{4.46}$$

A matrix which is equal to its adjoint is termed **Hermitian**. Thus,

$$\text{If } \mathsf{A} \text{ is Hermitian,} \quad \mathsf{A}^\dagger = \mathsf{A}, \text{ and } a_{ji}^* = a_{ij} \, . \tag{4.47}$$

Note that if a matrix is both real and symmetric, it is also Hermitian.

DOT PRODUCT AGAIN

Because vectors have been identified as matrices consisting of a single row or column, we can rewrite the dot product of two vectors as a matrix multiplication if we arrange for the first vector to be a row vector, with the second a column vector. This situation was illustrated in Example 4.4.3. Thus, if $\mathbf{a}$ and $\mathbf{b}$ are real column vectors,

$$\mathbf{a} \cdot \mathbf{b} = \sum_i a_i b_i = \mathbf{a}^T \mathbf{b} \, , \qquad (\mathbf{a} \text{ and } \mathbf{b} \text{ real}). \tag{4.48}$$

A discussion of dot products of complex vectors is deferred to Chapter 11.

INVERSE OF A MATRIX

It will often be the case that, given a square matrix A, there will be a square matrix B such that $\mathsf{A}\mathsf{B} = \mathsf{B}\mathsf{A} = 1$. A matrix B with this property is called the **inverse** of A and is given the name A^{-1}. If A^{-1} exists, it must be unique. The proof of this statement is simple: if B and C are both inverses of A, then

$$\mathsf{A}\mathsf{B} = \mathsf{B}\mathsf{A} = \mathsf{A}\mathsf{C} = \mathsf{C}\mathsf{A} = \mathbf{1} \, .$$

We now look at

$$\mathsf{C}\mathsf{A}\mathsf{B} = (\mathsf{C}\mathsf{A})\mathsf{B} = \mathsf{B} \, , \quad \text{but also} \quad \mathsf{C}\mathsf{A}\mathsf{B} = \mathsf{C}(\mathsf{A}\mathsf{B}) = \mathsf{C} \, .$$

This shows that $\mathsf{B} = \mathsf{C}$.

Every nonzero real (or complex) number α has a nonzero multiplicative inverse, often written $1/\alpha$. The corresponding property does **not** hold for matrices; there exist nonzero matrices that do not have inverses. To demonstrate this, consider the following;

$$\mathsf{A} = \begin{pmatrix} 1 & 1 \\ 0 & 0 \end{pmatrix}, \quad \mathsf{B} = \begin{pmatrix} 1 & 0 \\ -1 & 0 \end{pmatrix}, \quad \text{and} \quad \mathsf{A}\mathsf{B} = \begin{pmatrix} 0 & 0 \\ 0 & 0 \end{pmatrix} .$$

If A has an inverse, we can multiply the equation $\mathsf{A}\,\mathsf{B} = \mathsf{O}$ **on the left** by A^{-1}, thereby obtaining

$$\mathsf{AB} = \mathsf{O} \quad\longrightarrow\quad \mathsf{A}^{-1}\mathsf{AB} = \mathsf{A}^{-1}\mathsf{O} \quad\longrightarrow\quad \mathsf{B} = \mathsf{O}\,.$$

Since we started with a matrix B that was nonzero, this is an inconsistency, and we are forced to conclude that A^{-1} does not exist. A matrix without an inverse is said to be **singular**, so our conclusion is that A is singular. Notice that in our derivation, we had to be careful to multiply both members of $\mathsf{A}\,\mathsf{B} = \mathsf{O}$ from the left, because multiplication is noncommutative. Alternatively, assuming B^{-1} to exist, we could multiply this equation **on the right** by B^{-1}, obtaining

$$\mathsf{AB} = \mathsf{O} \quad\longrightarrow\quad \mathsf{ABB}^{-1} = \mathsf{OB}^{-1} \quad\longrightarrow\quad \mathsf{A} = \mathsf{O}\,.$$

This is inconsistent with the nonzero A with which we started; we conclude that B is also singular. Summarizing, there are nonzero matrices that do not have inverses and are identified as singular.

While there exist formal methods for obtaining the inverse of a matrix, we do not address them in detail, as they are somewhat cumbersome and in practice it is usually far easier to obtain a matrix inverse numerically or by symbolic computation methods. However, a full discussion of the conditions under which a matrix has no inverse (i.e., is singular) **is** of current importance and will be discussed later.

Some matrices have special relationships with their inverses. If the transpose of a matrix is also its inverse, the matrix is termed **orthogonal**. A generalization of this terminology, applicable when matrices can be complex, is that a matrix is called **unitary** when its Hermitian adjoint is also its inverse. Because these definitions are important, we set them forth in display format:

$$\begin{aligned} &\text{A matrix } \mathsf{V} \text{ is orthogonal if } \ \mathsf{V}^{\mathrm{T}} = \mathsf{V}^{-1}\,, \\ &\text{A matrix } \mathsf{U} \text{ is unitary if } \ \mathsf{U}^{\dagger} = \mathsf{U}^{-1}\,. \end{aligned} \tag{4.49}$$

RELATIONS AMONG MATRIX OPERATIONS

We often encounter situations in which one of the matrix operations (transpose, adjoint, etc.) is applied to a matrix product. It is useful to note how such expressions can be reduced. Taking the transpose first, the relevant relation is

$$(\mathsf{A}\,\mathsf{B})^{\mathrm{T}} = \mathsf{B}^{\mathrm{T}}\mathsf{A}^{\mathrm{T}}\,. \tag{4.50}$$

Equation (4.50) is easily proved by writing it in terms of its elements:

$$\left[(\mathsf{A}\,\mathsf{B})^{\mathrm{T}}\right]_{ij} = \sum_k a_{jk}b_{ki} = \sum_k b_{ki}a_{jk} = \sum_k \left(\mathsf{B}^{\mathrm{T}}\right)_{ik}\left(\mathsf{A}^{\mathrm{T}}\right)_{kj} = \left(\mathsf{B}^{\mathrm{T}}\mathsf{A}^{\mathrm{T}}\right)_{ij}\,.$$

A similar formula applies to a product of adjoints, since that formula is simply the complex conjugate of Eq. (4.50):

$$(\mathsf{A}\,\mathsf{B})^{\dagger} = \mathsf{B}^{\dagger}\mathsf{A}^{\dagger}\,. \tag{4.51}$$

Note, however, that the formula for complex conjugation of a product does not involve reversal of the order of the factors:

$$(\mathsf{A}\,\mathsf{B})^{*} = \mathsf{A}^{*}\mathsf{B}^{*}\,. \tag{4.52}$$

The inverse of a product of nonsingular matrices satisfies a relation similar to Eqs. (4.50) and (4.51):

$$(\mathsf{A}\,\mathsf{B})^{-1} = \mathsf{B}^{-1}\mathsf{A}^{-1}. \tag{4.53}$$

Here the proof is quite direct; we simply form the product of $\mathsf{A}\,\mathsf{B}$ and its alleged inverse:

$$(\mathsf{A}\,\mathsf{B})(\mathsf{A}\,\mathsf{B})^{-1} = \mathsf{A}\,\mathsf{B}\mathsf{B}^{-1}\mathsf{A}^{-1} = \mathbf{1}\,,$$

as may be seen by first combining B with B^{-1} and then A with A^{-1}.

Exercises

These exercises refer to the following matrices:

$$\mathsf{A} = \begin{pmatrix} \frac{1}{2} & \frac{\sqrt{3}}{2} \\ -\frac{\sqrt{3}}{2} & \frac{1}{2} \end{pmatrix}, \quad \mathsf{B} = \begin{pmatrix} \frac{1}{2} & -\frac{\sqrt{3}}{2} \\ \frac{\sqrt{3}}{2} & \frac{1}{2} \end{pmatrix}, \quad \mathsf{C} = \begin{pmatrix} \frac{i}{2} & \frac{\sqrt{3}}{2} \\ \frac{\sqrt{3}}{2} & -\frac{i}{2} \end{pmatrix},$$

$$\mathsf{D} = \begin{pmatrix} 0 & 1 \\ 2 & 3 \end{pmatrix}, \quad \mathsf{F} = \begin{pmatrix} 3 & -4i \\ 4i & 2 \end{pmatrix}, \quad \mathsf{G} = \begin{pmatrix} 0 & i \\ 0 & 4 \end{pmatrix}, \quad \mathsf{H} = \begin{pmatrix} 3 & i \\ 0 & 0 \end{pmatrix}.$$

4.4.10. (a) Compute AB, BA, and their commutator $[\mathsf{A},\mathsf{B}] \equiv \mathsf{AB} - \mathsf{BA}$.
(b) Repeat for AC, CA, and $[\mathsf{A},\mathsf{C}]$.
(c) Repeat for BC, CB, and $[\mathsf{B},\mathsf{C}]$.

4.4.11. Which of the matrices A through H are (1) symmetric, (2) orthogonal, (3) unitary, (4) Hermitian?

4.4.12. Given that $\mathbf{a}$ is a real column vector and $\mathbf{b}$ is a real row vector, write a matrix product that corresponds to the dot product $\mathbf{a}\cdot\mathbf{b}$.

4.4.13. Prove that matrices G and H are singular without using the formulas for the determinant or matrix inverse that are introduced in later sections of this book.

4.4.14. By carrying out the indicated matrix operations, verify that $(\mathsf{DF})^T = \mathsf{F}^T\mathsf{D}^T$, that $(\mathsf{DF})^* = \mathsf{D}^*\mathsf{F}^*$, and that $(\mathsf{DF})^\dagger = \mathsf{F}^\dagger\mathsf{D}^\dagger$.

4.4.15. Show that if matrices U and V are unitary, so are the matrix products UV and VU.

4.5 SYMBOLIC COMPUTING, MATRICES

The following examples illustrate how matrices are defined in MAPLE and MATHEMATICA; note that the finest-level (inner) grouping is that of the elements in a **row** of the matrix. Do not use these definitions for one-dimensional arrays; use those given for vectors.

MAPLE:
```
> Matrix( [ [1,2,3],[4,5,6] ] );
```
$$\begin{bmatrix} 1 & 2 & 3 \\ 4 & 5 & 6 \end{bmatrix}$$

MATHEMATICA: `{ {1,2,3},{4,5,6} }` → `{ {1,2,3}, {4,5,6} }`

`MatrixForm[%]` → $\begin{pmatrix} 1 & 2 & 3 \\ 4 & 5 & 6 \end{pmatrix}$

MATHEMATICA makes no distinction between column and row vectors; it interprets vectors as being either column or row, whichever is needed to make valid the expression in which the vector occurs. However, MAPLE distinguishes between row and column vectors. The MAPLE command **Vector** produces by default a column vector. To get a row vector in MAPLE, use the following syntax:

`> Vector[row]([1,2,3]);` → $[1, 2, 3]$

Just as for vectors, it is sometimes useful to define a matrix of specified dimensions and then load computed data into its elements. In MAPLE, an $m \times n$ matrix A (initialized to zero) is produced by the command `A := Matrix(m,n)`. In MATHEMATICA, the corresponding command is `Array[A,{m,n}]`. The individual elements of the matrix can be accessed as `A[i,j]` (MAPLE) or `A[[i,j]]` (MATHEMATICA). In MATHEMATICA, this definition of A does not set the matrix elements to zero but leaves them defined only as symbols `A[[i,j]]`. More detail concerning the matrix data structures of both symbolic languages can be found in Appendix C.

Example 4.5.1. Hilbert Matrix

The **Hilbert matrix** H is often used in the testing of numerical methods in matrix algebra. It is a family of square matrices of dimensions n with elements given as $H_{ij} = 1/(i + j - 1)$. With n previously defined, MAPLE code to generate a Hilbert matrix can take the form

```
> H := Matrix(n,n):
> for i from 1 to n do for j from 1 to n do
>   H[i,j] := 1/(i+j-1)
> end do: end do:
```

Corresponding MATHEMATICA code is

```
Array[H, {n,n}];
Do[Do[H[[i,j]]=1/(i+j-1), {i,n}], {j,n}];
```

■

ADDITION, MULTIPLICATION BY SCALAR

In both symbolic systems, addition, subtraction, and multiplication by a scalar can be indicated using ordinary algebraic notation. Errors are generated if these operations are used to add or subtract matrices of different shapes. Matrix multiplication, however, requires a more specific notation, as described in the following subsection.

DOT OPERATOR

In both symbolic systems, the dot operator can be used to indicate matrix multiplication, including matrix-matrix, matrix-vector, and vector-matrix products. If both factors A and B are matrices, the notation `A.B` produces the matrix product A B or an error message, depending upon whether or not the dimensions of the matrices are consistent with the indicated matrix product.

If one factor of a product (but not both) are vectors, the two symbolic systems exhibit different behavior. MAPLE has a strict orientation requirement. If A is a $n \times n$ matrix ($n > 1$) and **u** is a **column** vector of dimension n (the default orientation for a vector in MAPLE), then the product `A.u` is valid (and is a column vector), but `u.A` generates an error. However, if **u** is a **row** vector, then MAPLE accepts `u.A` (with the result a row vector), while `A.u` is not valid. On the other hand, in MATHEMATICA the lack of distinction between row and column vectors causes both `A.u` and `u.A` to be accepted, with `A.u` yielding the result obtained when **u** is assumed to be a column vector, and with `u.A` evaluated assuming **u** to be a row vector.

When both factors are vectors (so the product is of the form `u.v`), MATHEMATICA assumes them to be in the orientations that make their matrix product the dot product as given by Eq. (4.19). That is, MATHEMATICA treats `u` as a row vector and `v` as a column vector. This choice corresponds to what was presented in Section 4.3.

In MAPLE, `u.v` is processed in ways that depend upon the orientations of `u` and `v`. Assuming `u` and `v` to be real (the complex case is treated in Chapter 11), the orientation combinations `(row).(column)` and `(column).(row)` are treated as ordinary matrix multiplications, causing `(row).(column)` to reduce to a scalar, while `(column).(row)` becomes a square matrix. The other two orientations, `(column).(column)` and `(row).(row)`, are processed by MAPLE as though they were `(row).(column)`, in both cases thereby generating a scalar result. These choices permit operations in which matrices are not in play to proceed as naively expected, yielding the vector dot product when `u` and `v` are both in the default orientation (as was tacitly assumed in Section 4.3).

Note that MAPLE processes the orientation combination `(column).(row)` as the actual matrix product (compare Example 4.4.3), while MATHEMATICA cannot use the dot operator to obtain the matrix product from that orientation combination. There is, however, an alternative for MATHEMATICA which we illustrate here:

```
Outer[Times,{a,b},{c,d}]; MatrixForm[%]
```

$$\begin{pmatrix} ac & ad \\ bc & bd \end{pmatrix}$$

The name `Outer` reflects the fact that this operation is also known as an outer product; `Times` appears because MATHEMATICA also uses this construct in contexts other than multiplication.

It is important to note that if `A` and `B` have been defined as matrices or vectors, incorrect results or error indications are obtained if they are combined using the notations for ordinary multiplication. Thus, one must avoid constructions like `A*B`.

OTHER MATRIX OPERATIONS

In MAPLE, additional commands that may be applied to a matrix A include

`Transpose`(A), `HermitianTranspose`(A) (= adjoint), `MatrixInverse`(A).

Those commands are part of MAPLE's Linear Algebra package and must be preceded by the command `with(LinearAlgebra)`. MAPLE's complex conjugate command does not work for matrices; in the infrequent case where one wants complex conjugation without transposition, one can successively apply `HermitianTranspose` and `Transpose` (the latter to "undo" the transposition).

In MATHEMATICA, the corresponding commands are

`Transpose[A]`, `ConjugateTranspose[A]`, `Inverse[A]`.

We note that the command `Conjugate[A]` **does** work in MATHEMATICA.

Example 4.5.2. Symbolic Computations in MAPLE

Let's start by defining two real matrices and a real vector:

```
> M := Matrix([[1,-1],[0,4]]);
```
$$M := \begin{pmatrix} 1 & -1 \\ 0 & 4 \end{pmatrix}$$

```
> Unit := Matrix([[1,0],[0,1]]);
```
$$Unit := \begin{pmatrix} 1 & 0 \\ 0 & 1 \end{pmatrix}$$

```
> a := Vector([1,-1]);
```
$$a := \begin{pmatrix} 1 \\ -1 \end{pmatrix}$$

These operations use $+$, $-$, $*$:

```
> 2*M - Unit;
```
$$\begin{pmatrix} 1 & -2 \\ 0 & 7 \end{pmatrix}$$

But these require use of the dot operator:

```
> M . M;
```
$$\begin{pmatrix} 1 & -5 \\ 0 & 16 \end{pmatrix}$$

```
> M . a;
```
$$\begin{pmatrix} 2 \\ -4 \end{pmatrix}$$

The dot operator is associative and integer powers of matrices are interpreted as successive dot products, so

```
> M . M . M;
```
$$\begin{pmatrix} 1 & -21 \\ 0 & 64 \end{pmatrix}$$

```
> M^3;
```
$$\begin{pmatrix} 1 & -21 \\ 0 & 64 \end{pmatrix}$$

Some operations on M:

```
> Transpose(M);
```
(returns input unprocessed)

```
> with(LinearAlgebra):
```
(this was needed)

```
> Transpose(M);
```
$$\begin{pmatrix} 1 & 0 \\ -1 & 4 \end{pmatrix}$$

```
> MX := MatrixInverse(M);
```
$$MX := \begin{pmatrix} 1 & \frac{1}{4} \\ 0 & \frac{1}{4} \end{pmatrix}$$

```
> M . MX;
```
$$\begin{pmatrix} 1 & 0 \\ 0 & 1 \end{pmatrix}$$

Now look at operations illustrating MAPLE's treatment of real vectors in dot products. The vector a which we defined above is a column vector. Let's make two row vectors: the transpose of a and another that we will call b.

```
> ax := Transpose(a);
```
$ax := (1 \quad -1)$

```
> b := Vector[row]([1,-3]);
```
$b := (1 \quad -3)$

MAPLE's strict orientation requirement causes the results

```
> a . M;
```
(error)

```
> ax . M;
```
$(1 \quad -13)$

MAPLE gives the same result for the two following expressions because its convention is to take the transpose of a column vector in the left member of a dot product of two column vectors:

```
> a . a;
```
2

```
> ax . a;
```
2

For the following expression MAPLE takes the transpose of the row vector in the right member of a dot product of two row vectors:

```
> b . b;
```
10

No adjustments are made to the following expressions:

```
> b . a;
```
4

```
> a . b;
```
$$\begin{pmatrix} 1 & -3 \\ -1 & 3 \end{pmatrix}$$

The peculiarities in MAPLE's handling of dot products of vectors can lead to strange behavior, such as the following:

```
> a . M . a;
```
(error)

```
> a . ( M . a);
```
6

In the first of those two examples MAPLE detects an illegal (column vector).(matrix) product; in the second example the parenthesized subexpression is first reduced to a column vector so MAPLE sees a (column).(column) vector product for which processing is defined. One way to avoid problems from these peculiarities of MAPLE is to always use vectors in mathematically appropriate orientations, taking transposes or adjoints as necessary to represent the mathematics properly. Thus, the two expressions given above would have been better written as `ax.M.a`, where the reader may recall that `ax` was defined to be the transpose of `a`.

■

Example 4.5.3. Symbolic Computations in MATHEMATICA

Let's start by defining two real matrices and a real vector:

```
M = {{1,-1},{0,4}}                    {{1, -1}, {0, 4}}
Unit = {{1,0},{0,1}}                  {{1, 0}, {0, 1}}
a = {1,-1}                            {1, -1}
```

We can see the above output in regular matrix form by using `MatrixForm`:

```
MatrixForm[M]
```
$$\begin{pmatrix} 1 & -1 \\ 0 & 4 \end{pmatrix}$$

These operations use +, −, ∗:

`2*M - Unit`	`{{1, -2}, {0, 7}}`

These require use of the dot operator:

`M . M`	`{{1, -5}, {0, 16}}`
`M . a`	`{2, -4}`

It is important to note that in MATHEMATICA `M^2` does not mean `M.M`:

`MatrixForm[M . M ]`	$\begin{pmatrix} 1 & -5 \\ 0 & 16 \end{pmatrix}$
`MatrixForm[M^2]`	$\begin{pmatrix} 1 & 1 \\ 0 & 16 \end{pmatrix}$

Some operations on M:

`MatrixForm[Transpose[M]]`	$\begin{pmatrix} 1 & 0 \\ -1 & 4 \end{pmatrix}$
`MX = Inverse[M]`	$\left\{\left\{1, \frac{1}{4}\right\}, \left\{0, \frac{1}{4}\right\}\right\}$
`MatrixForm[M . MX]`	$\begin{pmatrix} 1 & 0 \\ 0 & 1 \end{pmatrix}$

Now look at operations involving MATHEMATICA's treatment of vectors in vector-matrix products. Because MATHEMATICA does not distinguish between row and column vectors, row/column conversions do not work:

`Transpose[a]`	(error, no processing)
`ConjugateTranspose[a]`	(error, no processing)

To create a mathematical expression involving the adjoint of a vector `c`, we need to first complex conjugate `c` and then use it where a row vector is expected. Thus, define

`c = {1,3*I}`	`{1, 3i}`
`cx = Conjugate[c]`	$\{1, \; -3i\}$
`cx . M`	$\{1, \; -1-12i\}$

Also legal, but useful only if it is what is intended:

`MatrixForm[M . cx]`	$\begin{pmatrix} 1+3i \\ -12i \end{pmatrix}$

■

Exercises

4.5.1. Using a symbolic computing system, form the two matrices

$$\mathsf{X} = \begin{pmatrix} 0 & 1 \\ -1 & 0 \end{pmatrix} \quad \text{and} \quad \mathsf{Y} = \begin{pmatrix} 1 & 2 \\ 3 & 4 \end{pmatrix}.$$

Then form X^{-1}, Y^{-1}, X^{T}, Y^{T}, $(\mathsf{XY})^{-1}$, $(\mathsf{XY})^{\mathrm{T}}$, and verify that $(\mathsf{XY})^{-1} = \mathsf{Y}^{-1}\mathsf{X}^{-1}$ and that $(\mathsf{XY})^{\mathrm{T}} = \mathsf{Y}^{\mathrm{T}}\mathsf{X}^{\mathrm{T}}$.

4.5.2. `HermitianTranspose`, `MatrixInverse`, and `ConjugateTranspose` are inconveniently long commands. Write procedures for your symbolic computing system that use `Adj` instead of `HermitianTranspose` or `ConjugateTranspose`, with `Conj` standing for complex conjugation (both MAPLE and MATHEMATICA) and `Inv` for `MatrixInverse` or `Inverse`. Present examples to show that these shortened names work properly.

4.5.3. Repeat Exercise 4.4.10 using symbolic computation.

4.5.4. Verify that matrices G and H of Exercise 4.4.13 are singular using symbolic computation.

4.5.5. Repeat Exercise 4.4.14 using symbolic computation.

4.5.6. Given $\mathsf{X} = \begin{pmatrix} 1 & 0 & 0 \\ 0 & 0 & 1 \\ 0 & -1 & 0 \end{pmatrix}$,

use symbolic computation to find $\mathsf{Y} = \mathsf{X}^{-1}$, and check that $\mathsf{XY} = \mathsf{YX} = \mathbf{1}$.

Is X Hermitian, orthogonal, or unitary (note that these designations are not mutually exclusive)?

4.5.7. Define $\mathsf{A} = \begin{pmatrix} 0 & 0 & 0 \\ 0 & 0 & -i \\ 0 & i & 0 \end{pmatrix}$ and $\mathsf{B} = \begin{pmatrix} 1 & 0 & 0 \\ 0 & 0 & 0 \\ 0 & 0 & 0 \end{pmatrix}$.

Using symbolic computation,

(a) Verify that A is singular, but that A + B is nonsingular.

(b) Form $(\mathsf{A}+\mathsf{B})^{-1} - \mathsf{B}$. Check that its lower-right 2×2 block is the inverse of the lower-right 2×2 block of A.

4.6 SYSTEMS OF LINEAR EQUATIONS

We now return to systems of linear equations, the problem that motivated the development of matrix algebra. Our objective is to understand the conditions under which equation systems have unique solutions. We start by considering the simplest possible case of this sort, the set of two equations

$$\begin{aligned} a_{11}x_1 + a_{12}x_2 &= h_1 \,, \\ a_{21}x_1 + a_{22}x_2 &= h_2 \,. \end{aligned} \tag{4.54}$$

If both a_{11} and a_{21} vanish these equations cannot have a unique solution (the value of x_1 is then completely arbitrary). We therefore assume that the equations are written in an order such that $a_{11} \neq 0$. We proceed by eliminating the x_1 term from the second equation; we do so by making the replacement

$$\text{(new 2nd equation)} = \text{(original 2nd equation)} - \frac{a_{21}}{a_{11}} \times \text{(1st equation)}. \tag{4.55}$$

Our two equations then become

$$\begin{aligned} a_{11}x_1 + a_{12}x_2 &= h_1 \,, \\ a'_{22}x_2 &= h'_2 \,, \end{aligned} \tag{4.56}$$

with

$$a'_{22} = a_{22} - \frac{a_{21}a_{12}}{a_{11}} \qquad \text{and} \qquad h'_2 = h_2 - \frac{a_{21}}{a_{11}}\, h_1 \,.$$

Viewed as a matrix equation, our problem has now assumed the form

$$\mathsf{A}'\mathbf{x} = \mathbf{h}' \,, \qquad \text{or} \qquad \begin{pmatrix} a_{11} & a_{12} \\ 0 & a'_{22} \end{pmatrix} \begin{pmatrix} x_1 \\ x_2 \end{pmatrix} = \begin{pmatrix} h_1 \\ h'_2 \end{pmatrix} .$$

If $a'_{22} \neq 0$ we can solve the second equation for x_2, after which (because we chose to have $a_{11} \neq 0$) we can then solve the first equation for x_1. Combining all the foregoing analysis, we see that we will have a unique solution for x_1 and x_2 only if both a_{11} and a'_{22} are nonzero, i.e., only if $a_{11}a'_{22} \neq 0$. This product, which evaluates to

$$a_{11}a'_{22} = a_{11}a_{22} - a_{21}a_{12} \,,$$

is called the **determinant** of our two equations, or, since it depends only upon the a_{ij}, it can also be called the determinant of the matrix A whose elements are the a_{ij}. Since the determinant is an object of importance, we identify the following notations for it:

$$\det \mathsf{A} \equiv |\mathsf{A}| \equiv \begin{vmatrix} a_{11} & a_{12} \\ a_{21} & a_{22} \end{vmatrix} = a_{11}a_{22} - a_{21}a_{12} \,. \tag{4.57}$$

In the notation $|\mathsf{A}|$ the vertical bars do not stand for absolute value; this does not cause confusion because A is presumably known to stand for a square array. In the case where the matrix elements are given explicitly, notice that they are enclosed by vertical lines, not ( ), [ ], or, { }.

Before leaving this nearly trivial example, note that when the determinant is nonzero, we get x_1 and x_2 as expressions that are linear combinations of the h_i. If the determinant is zero, our second equation becomes $0 = h'_2$, and we get no solution at all unless $h'_2 = 0$. But if $h'_2 = 0$, our second equation is $0 = 0$, and we then have only one equation in x_1 and x_2, and there are solutions, but they are not unique. The important conclusion to be reached here is that if the matrix A defining a pair of linear equations has a zero determinant, (1) the forms defining the left-hand sides of the equations are linearly dependent, and (2) the equations do not have a unique solution (they either have many solutions or no solution at all, depending on the values of their right-hand sides).

Consider now a system of three equations,

$$\begin{aligned} a_{11}x_1 + a_{12}x_2 + a_{13}x_3 &= h_1 \,, \\ a_{21}x_1 + a_{22}x_2 + a_{23}x_3 &= h_2 \,, \\ a_{31}x_1 + a_{32}x_2 + a_{33}x_3 &= h_3 \,. \end{aligned} \tag{4.58}$$

Making replacements similar to that in Eq. (4.55), we can eliminate the x_1 term from the second and third equations, arriving at an equation set that, in matrix notation, has the form

$$\mathsf{A}'\mathbf{x} = \mathbf{h}' \,, \qquad \text{or} \qquad \begin{pmatrix} a_{11} & a_{12} & a_{13} \\ 0 & a'_{22} & a'_{23} \\ 0 & a'_{32} & a'_{33} \end{pmatrix} \begin{pmatrix} x_1 \\ x_2 \\ x_3 \end{pmatrix} = \begin{pmatrix} h_1 \\ h'_2 \\ h'_3 \end{pmatrix} . \tag{4.59}$$

Here

$$a'_{22} = a_{22} - \frac{a_{21}a_{12}}{a_{11}}\,, \qquad a'_{23} = a_{23} - \frac{a_{21}a_{13}}{a_{11}}\,,$$

$$a'_{32} = a_{32} - \frac{a_{31}a_{12}}{a_{11}}\,, \qquad a'_{33} = a_{33} - \frac{a_{31}a_{13}}{a_{11}}\,.$$

We next eliminate x_2 from the third equation of this set by subtracting from the third equation an appropriate multiple of the second equation, reaching, in matrix form,

$$\mathsf{A}''\mathbf{x} = \mathbf{h}''\,, \qquad \text{with} \qquad \mathsf{A}'' = \begin{pmatrix} a_{11} & a_{12} & a_{13} \\ 0 & a'_{22} & a'_{23} \\ 0 & 0 & a''_{33} \end{pmatrix}. \tag{4.60}$$

Here

$$a''_{33} = a'_{33} - \frac{a'_{32}a'_{23}}{a'_{22}}\,.$$

The process illustrated above whereby zeros are introduced into the coefficient matrix is called **row reduction**, and its result is a modified coefficient matrix which is **upper triangular**, meaning that all its elements below the principal diagonal are zero. Because row reduction only involves the formation of linear combinations of equations each of which is individually valid, the equation set corresponding to the modified coefficient matrix will have the same solutions as the original set of equations.

The equation set represented by Eq. (4.60) will have a unique solution only if a_{11}, a'_{22}, and a''_{33} are all nonzero. We therefore define the determinant for the three equations in Eq. (4.58) as the product $a_{11}a'_{22}a''_{33}$. Evaluating this product in terms of the original matrix elements a_{ij}, we find (after some effort)

$$\det \mathsf{A} = a_{11}a_{22}a_{33} - a_{21}a_{12}a_{33} - a_{31}a_{22}a_{13} - a_{11}a_{32}a_{23} + a_{21}a_{32}a_{13} + a_{31}a_{12}a_{23}\,. \tag{4.61}$$

The determinants we have identified for matrices of dimensions 2 and 3 can be generalized to matrices of arbitrary size, and we shall consider that matter in the next section of this chapter. The importance of the determinant is that

If a square matrix A *of any dimension has a zero determinant,*

(1) the left-hand sides of the linear equations defined by the rows of A *are linearly dependent, and*
(2) the equations will not have a unique solution.

Exercises

For each of the following equation sets:

(a) Compute the determinant of the coefficients, using Eq. (4.61).

(b) Row-reduce the coefficient matrix to upper triangular form and either obtain the most general solution to the equations or explain why no solution exists.

(c) Confirm that the existence and/or uniqueness of the solutions you found in part (b) correspond to the zero (or nonzero) value you found for the determinant.

4.6.1.

$$\begin{aligned} x - \ y + 2z &= \ \ 5\,, \\ 2x \ + \ z &= \ \ 3\,, \\ x + 3y - \ z &= -6\,. \end{aligned}$$

4.6.2.

$$\begin{aligned} 2x + 4y + 3z &= 4\,, \\ 3x + \ y + 2z &= 3\,, \\ 4x - 2y + \ z &= 2\,. \end{aligned}$$

4.6.3.

$$\begin{aligned} 2x + 4y + 3z &= \ \ 2\,, \\ 3x + \ y + 2z &= \ \ 1\,, \\ 4x - 2y + \ z &= -1\,. \end{aligned}$$

4.6.4.

$$\begin{aligned} x - y + 2z &= 0\,, \\ 2x + z &= 0\,, \\ x + 3y - z &= 0\,. \end{aligned}$$

4.6.5.

$$\begin{aligned} \sqrt{6}\,x + 2\sqrt{3}\,y + 3\sqrt{2}\,z &= 0\,, \\ \sqrt{2}\,x \ + \ 2y \ + \ \sqrt{6}\,z &= 0\,, \\ (1/\sqrt{2})\,x \ + \ y + \sqrt{1.5}\,z &= 0\,. \end{aligned}$$

4.6.6.

$$\begin{aligned} \sqrt{6}\,x + 2\sqrt{3}\,y + 3\sqrt{2}\,z &= \sqrt{3}\,, \\ \sqrt{2}\,x \ + \ 2y \ + \ \sqrt{6}\,z &= 2\,, \\ (1/\sqrt{2})\,x \ + \ y + \sqrt{1.5}\,z &= 1\,. \end{aligned}$$

4.7 DETERMINANTS

If we examine the forms we obtained for the determinants of orders 2 and 3 (meaning that they are from matrices of dimensions 2 and 3), we may guess that a determinant of order n is a linear combination of products of n matrix elements, such that (1) each product in the determinant of A contains one element from each row of A, chosen in such a way that every factor in the product comes from a different column of A. Each term in the determinant exhibits a different way of choosing a product in which each row and column is represented. Note also that half the products are assigned negative signs. These features provide the symmetry needed to give determinants their useful properties.

DEFINITIONS

Before defining the determinant, we need to introduce some related concepts and definitions.

- Starting from a set of n objects numbered from 1 through n and arranged in numerical order, we can make a **permutation** of them to some other order; the total number of distinct permutations that are possible is $n!$ (choose the first object n ways, then choose the second in $n-1$ ways, etc.).
- Every permutation of n objects can be reached from the original numerical order by a succession of pairwise interchanges (e.g., $1234 \to 4132$ can be reached by the successive steps $1234 \to 1432 \to 4132$). Although the number of pairwise interchanges needed for a given permutation depends on the path (compare the above example with $1234 \to 1243 \to 1423 \to 4123 \to 4132$), for a given permutation the number of interchanges will always either be **even** or **odd**. Thus a permutation can be identified as having either even or odd **parity**.
- It is convenient to introduce the **Levi-Civita symbol**, which for an n-object system is denoted by $\varepsilon_{ij\ldots}$, where ε has n subscripts, each of which identifies one of the objects. This Levi-Civita symbol is defined to be $+1$ if $ij\ldots$ represents an even permutation of the objects from numerical order; it is defined to be -1 if $ij\ldots$ represents an odd permutation of the objects, and zero if $ij\ldots$ does not represent a permutation of the objects (e.g., contains an entry duplication). Because this is an important definition, we set it out in a display format:

$$\begin{aligned} \varepsilon_{ij\ldots} &= +1\,, \quad ij\ldots \text{ an even permutation,} \\ &= -1\,, \quad ij\ldots \text{ an odd permutation,} \\ &= \;\;0\,, \quad ij\ldots \text{ not a permutation.} \end{aligned} \tag{4.62}$$

We now write a determinant of **order** n as an $n \times n$ square array of numbers (or functions), with the array conventionally written within vertical bars (not parentheses, braces, or any other type of brackets), as follows:

$$D_n = \begin{vmatrix} a_{11} & a_{12} & \cdots & a_{1n} \\ a_{21} & a_{22} & \cdots & a_{2n} \\ a_{31} & a_{32} & \cdots & a_{3n} \\ \cdots & \cdots & \cdots & \cdots \\ a_{n1} & a_{n2} & \cdots & a_{nn} \end{vmatrix} . \tag{4.63}$$

The determinant D_n has a value that is obtained from its array by

1. Forming all $n!$ products that can be formed by choosing one entry from each row in such a way that one entry comes from each column.
2. Assigning each product a sign that corresponds to the parity of the sequence in which the columns were used (assuming the rows were used in numerical order).
3. Adding (with the assigned signs) the products.

More formally, the determinant in Eq. (4.63) is defined to have the value

$$D_n = \sum_{ij\ldots} \varepsilon_{ij\ldots} a_{1i} a_{2j} \cdots . \tag{4.64}$$

The summations in Eq. (4.64) need not be restricted to permutations, but can be assumed to range independently from 1 through n; the presence of the Levi-Civita symbol will cause only the index combinations corresponding to permutations to actually contribute to the sum.

As indicated earlier, we can indicate the determinant of a square matrix A either as $\det(\mathsf{A})$ or as $|\mathsf{A}|$. We reiterate a point made earlier but often not fully appreciated: a matrix has no value other than the array of quantities that represent it. Its determinant is not its value; the determinant is the value of a useful number that can be obtained from the matrix.

Example 4.7.1. Determinants of Orders 2 and 3

To make the definition more concrete, we illustrate first with a determinant of order 2. The Levi-Civita symbols needed for this determinant are $\varepsilon_{12} = +1$ and $\varepsilon_{21} = -1$ (note that $\varepsilon_{11} = \varepsilon_{22} = 0$), leading to

$$D_2 = \begin{vmatrix} a_{11} & a_{12} \\ a_{21} & a_{22} \end{vmatrix} = \varepsilon_{12} a_{11} a_{22} + \varepsilon_{21} a_{12} a_{21} = a_{11} a_{22} - a_{12} a_{21} \,.$$

We see that this determinant expands into $2! = 2$ terms, and the result agrees with the definition introduced earlier, in Eq. (4.57).

Determinants of order 3 expand into $3! = 6$ terms. The relevant Levi-Civita symbols are $\varepsilon_{123} = \varepsilon_{231} = \varepsilon_{312} = +1$, $\varepsilon_{213} = \varepsilon_{321} = \varepsilon_{132} = -1$; all other index combinations have $\varepsilon_{ijk} = 0$, so

$$D_3 = \begin{vmatrix} a_{11} & a_{12} & a_{13} \\ a_{21} & a_{22} & a_{23} \\ a_{31} & a_{32} & a_{33} \end{vmatrix} = \sum_{ijk} \varepsilon_{ijk} a_{1i} a_{2j} a_{3k}$$

$$= a_{11}a_{22}a_{33} - a_{11}a_{23}a_{32} - a_{13}a_{22}a_{31} - a_{12}a_{21}a_{33} + a_{12}a_{23}a_{31} + a_{13}a_{21}a_{32} \,.$$

This expression is also in agreement with that introduced earlier, in Eq. (4.61).

■

It may be worth pointing out that many students have learned simple methods for reading out determinants using signed diagonals of 2×2 or 3×3 arrays. This type of scheme does not work for larger determinants; they need to be evaluated from the basic formula or using methods presented later in this text.

PROPERTIES OF DETERMINANTS

The symmetry properties of the Levi-Civita symbol translate into a number of symmetries exhibited by determinants. For simplicity, we illustrate with determinants of order 3. The interchange of any two columns of a determinant (they need not be adjacent) causes the Levi-Civita symbol multiplying each term of the expansion to change sign; the same is true if any two rows are interchanged. Moreover, the roles of rows and columns may be interchanged without changing the value of the determinant. In other words a matrix and its transpose have equal determinants. Summarizing:

> *Interchanging any two rows (or any two columns) changes the sign of the value of a determinant. Transposition does not alter its value.*

Thus,

$$\begin{vmatrix} a_{11} & a_{12} & a_{13} \\ a_{21} & a_{22} & a_{23} \\ a_{31} & a_{32} & a_{33} \end{vmatrix} = - \begin{vmatrix} a_{12} & a_{11} & a_{13} \\ a_{22} & a_{21} & a_{23} \\ a_{32} & a_{31} & a_{33} \end{vmatrix} = \begin{vmatrix} a_{11} & a_{21} & a_{31} \\ a_{12} & a_{22} & a_{32} \\ a_{13} & a_{23} & a_{33} \end{vmatrix} . \tag{4.65}$$

Further consequences of the definition in Eq. (4.64) are:

(1) Multiplication of all members of a single column (or a single row) by a constant k causes the value of the determinant to be multiplied by k.

(2) If the elements of a column (or row) are actually sums of two quantities, the determinant can be decomposed into a sum of two determinants.

Thus,

$$k \begin{vmatrix} a_{11} & a_{12} & a_{13} \\ a_{21} & a_{22} & a_{23} \\ a_{31} & a_{32} & a_{33} \end{vmatrix} = \begin{vmatrix} ka_{11} & a_{12} & a_{13} \\ ka_{21} & a_{22} & a_{23} \\ ka_{31} & a_{32} & a_{33} \end{vmatrix} = \begin{vmatrix} ka_{11} & ka_{12} & ka_{13} \\ a_{21} & a_{22} & a_{23} \\ a_{31} & a_{32} & a_{33} \end{vmatrix} , \tag{4.66}$$

$$\begin{vmatrix} a_{11}+b_1 & a_{12} & a_{13} \\ a_{21}+b_2 & a_{22} & a_{23} \\ a_{31}+b_3 & a_{32} & a_{33} \end{vmatrix} = \begin{vmatrix} a_{11} & a_{12} & a_{13} \\ a_{21} & a_{22} & a_{23} \\ a_{31} & a_{32} & a_{33} \end{vmatrix} + \begin{vmatrix} b_1 & a_{12} & a_{13} \\ b_2 & a_{22} & a_{23} \\ b_3 & a_{32} & a_{33} \end{vmatrix} . \tag{4.67}$$

These properties and/or the basic definition mean that

1. Any determinant with two rows equal, or two columns equal, has the value zero. To prove this, interchange the two identical rows or columns; the determinant both remains the same and changes sign, and therefore must be zero.
2. An extension of the above is that if two rows (or columns) are proportional, the determinant is zero.
3. The value of a determinant is unchanged if a multiple of one row is added (column by column) to another row or if a multiple of one column is added (row by row) to another column. Applying Eq. (4.67), we see that the addition does not contribute to the value of the determinant.
4. If every element in a row or every element in a column is zero, the determinant has the value zero.

DETERMINANTS AND ROW REDUCTION

Row reduction refers to the process whereby zeros are introduced into the coefficient matrix A describing a system of linear equations, i.e., the process used in Section 4.6. Note the following:

1. Because of item 3 in the list of properties, the determinant of a coefficient matrix will remain unchanged under row reduction. When the row reduction is complete, the coefficient matrix will have been brought to upper triangular form (meaning that all matrix elements below the principal diagonal are zero).
2. The only contribution to the determinant of an upper triangular matrix will be the product of its diagonal elements. Every other term in this determinant must include at least one matrix element below the diagonal and therefore must vanish.

3. The determinant will then be nonzero only if all the diagonal elements of the upper triangular matrix are nonzero, corresponding to a linearly independent set of linear forms.

We therefore see that

The determinant as formally defined in the present section is in fact the same identifier for linear dependence that we previously generated via the row-reduction process.

DETERMINANT PRODUCT THEOREM

The **determinant product theorem** is an important result that relates the determinants of two square matrices of the same dimension, A and **B**, to the determinant of their product:

$$\det(\mathsf{A}\,\mathsf{B}) = \det(\mathsf{A})\,\det(\mathsf{B})\,. \tag{4.68}$$

For a proof of this theorem, the reader is referred to Arfken *et al*, Additional Readings.

The determinant product theorem has some useful consequences, one of which is that its form indicates that $\det(\mathsf{A}\,\mathsf{B}) = \det(\mathsf{B}\,\mathsf{A})$ whether or not A and B commute.

Example 4.7.2. Determinant of Matrix Product

Consider the two matrices

$$\mathsf{A} = \begin{pmatrix} -1/2 & -\sqrt{3}/2 \\ \sqrt{3}/2 & -1/2 \end{pmatrix}, \qquad \mathsf{B} = \begin{pmatrix} -1 & \sqrt{3} \\ -\sqrt{3} & -1 \end{pmatrix}.$$

We have $\det(\mathsf{A}) = 1/4+3/4 = 1$, $\det(\mathsf{B}) = 1+3 = 4$, and $\det(\mathsf{A})\det(\mathsf{B}) = 4$, whereas the matrix product is

$$\mathsf{A}\,\mathsf{B} = \begin{pmatrix} -1/2 & -\sqrt{3}/2 \\ \sqrt{3}/2 & -1/2 \end{pmatrix}\begin{pmatrix} -1 & \sqrt{3} \\ -\sqrt{3} & -1 \end{pmatrix} = \begin{pmatrix} 2 & 0 \\ 0 & 2 \end{pmatrix}.$$

From the matrix product we also get $\det(\mathsf{A}\,\mathsf{B}) = 4$, confirming this case of the determinant product theorem.

■

The determinant product theorem provides a route toward the discussion of singularity in matrices: the determinants of a matrix A and its inverse A^{-1} (assuming it exists) are related by the equation

$$\det(\mathsf{A})\det(\mathsf{A}^{-1}) = \det(\mathsf{A}\,\mathsf{A}^{-1}) = 1\,. \tag{4.69}$$

Any matrix with finite elements will have a determinant; we know this because we have an explicit rule for its construction. If the determinant is nonzero, with value D, the inverse of the matrix must exist, because Eq. (4.69) tells us that it has a determinant of value D^{-1}. But if the determinant D of a matrix is zero, then the determinant of its inverse does not exist, meaning that the inverse matrix does not exist either. We thus have the important result

A matrix is singular (meaning that it does not have an inverse) if and only if its determinant is zero.

SOLVING LINEAR EQUATION SYSTEMS

We are now in a position to sharpen our earlier discussion of linear equation systems.

> *A linear equation system, written in matrix form as* $\mathsf{A}\mathbf{x} = \mathbf{h}$, *has a unique solution if and only if* $\det(\mathsf{A}) \neq 0$ *(meaning also that* A *is nonsingular); the solution can be obtained by left-multiplying this matrix equation by* A^{-1}, *thereby reaching* $\mathbf{x} = \mathsf{A}^{-1}\mathbf{h}$. *If* $\det(\mathsf{A}) = 0$, *the solution, if it exists, will not be unique.*

LAPLACIAN DEVELOPMENT BY MINORS

The fact that a determinant of order n expands into $n!$ terms means that it is important to identify efficient means for determinant evaluation. One approach is to expand in terms of **minors**. The minor corresponding to a_{ij}, denoted M_{ij}, or $M_{ij}(\mathsf{A})$ if we need to identify M as coming from the a_{ij}, is the determinant (of order $n-1$) produced by striking out row i and column j of the original determinant. When we expand into minors, the quantities to be used are the **cofactors** of the (ij) elements, defined as $(-1)^{i+j}M_{ij}$. The expansion can be made for any row or column of the original determinant. If, for example, we expand the determinant of Eq. (4.63) using row i, we have

$$D_n = \sum_{j=1}^{n} a_{ij}(-1)^{i+j}M_{ij} \,. \tag{4.70}$$

This expansion reduces the work involved in evaluation if the row or column selected for the expansion contains zeros, as the corresponding minors need not be evaluated.

Example 4.7.3. Expansion in Minors

Consider the determinant (arising in Dirac's relativistic electron theory)

$$D \equiv \begin{vmatrix} a_{11} & a_{12} & a_{13} & a_{14} \\ a_{21} & a_{22} & a_{23} & a_{24} \\ a_{31} & a_{32} & a_{33} & a_{34} \\ a_{41} & a_{42} & a_{43} & a_{44} \end{vmatrix} = \begin{vmatrix} 0 & 1 & 0 & 0 \\ -1 & 0 & 0 & 0 \\ 0 & 0 & 0 & 1 \\ 0 & 0 & -1 & 0 \end{vmatrix} .$$

Expanding across the top row, only one 3×3 matrix survives:

$$D = (-1)^{1+2}a_{12}M_{12}(a) = (-1)\cdot(1)\begin{vmatrix} -1 & 0 & 0 \\ 0 & 0 & 1 \\ 0 & -1 & 0 \end{vmatrix} \equiv (-1)\begin{vmatrix} b_{11} & b_{12} & b_{13} \\ b_{21} & b_{22} & b_{23} \\ b_{31} & b_{32} & b_{33} \end{vmatrix} .$$

Expanding now across the second row, we get

$$D = (-1)(-1)^{2+3}b_{23}M_{23}(b) = \begin{vmatrix} -1 & 0 \\ 0 & -1 \end{vmatrix} = 1 \,.$$

When we finally reached a 2×2 determinant, it was simple to evaluate it without further expansion.

■

Minors also appear in the general formula for the inverse of a matrix. An explicit, but cumbersome formula for the inverse of a matrix A has the form

$$\left(\mathsf{A}^{-1}\right)_{ij} = \frac{(-1)^{i+j} M_{ji}}{\det(\mathsf{A})} \,. \tag{4.71}$$

The presence of det(A) in the denominator shows that this formula will fail when $\det(\mathsf{A}) = 0$, a result that is expected because the inverse does not then exist. We present Eq. (4.71) mainly to show the existence of an explicit formula for matrix inversion. The reader will generally find it more advisable to invert matrices using symbolic computation.

SYMBOLIC COMPUTATION

The methods presented earlier for the evaluation of determinants are useful for understanding their properties, but application to practical problems usually demands machine-assisted computation. Programs for the evaluation of determinants given as arrays of numbers are efficient and widely available. Our focus here, however, is the use of symbolic computation systems which can handle not only determinants specified numerically, but also those given in terms of algebraic expressions.

In MAPLE, the determinant of a square matrix M is obtained by the command `Determinant(M)`. This command is not a basic MAPLE command, but must be preceded by a call to make available MAPLE's Linear Algebra package (the capitalization of Determinant is important because MAPLE still supports an obsolete linear algebra package called `linalg`, in which `determinant` is a command). For example,

```
> M := Matrix( [[1,2,3],[2,0,-1],[-1,3,2]] );
```

$$M := \begin{bmatrix} 1 & 2 & 3 \\ 2 & 0 & -1 \\ -1 & 3 & 2 \end{bmatrix}$$

```
> with(LinearAlgebra):
```

(using ; gives the entire list of Linear Algebra commands)

```
> Determinant(M);
```

15

In MATHEMATICA, the determinant of M is obtained without fanfare by the command `Det[M]`.

```
M = { {1,2,3}, {2,0,-1}, {-1,3,2} }; Det[M]
```

15

Exercises

Using symbolic computation, obtain values for the following determinants:

4.7.1. $\begin{vmatrix} 3 & 1 & -2 & 2 \\ -1 & 0 & 1 & 2 \\ 2 & 1 & -1 & 0 \\ 1 & 3 & 1 & 2 \end{vmatrix}$.

4.7.2. $\begin{vmatrix} 3 & 4 & 0 & 1 & -2 & 2 \\ -2 & 0 & -2 & -1 & 1 & 2 \\ 2 & 1 & 1 & -1 & 0 & 1 \\ 1 & 3 & 0 & -2 & 1 & 2 \\ -2 & 1 & 1 & 3 & 2 & 1 \\ 0 & -4 & 2 & -1 & -3 & -3 \end{vmatrix}$.

4.7.3. $\begin{vmatrix} x-3 & 3 & 4 \\ 3 & x-2 & 1 \\ 4 & 1 & x-1 \end{vmatrix}$.

4.7.4. $\begin{vmatrix} x-1 & 1 & 0 & 2 \\ 1 & x-1 & 3 & 0 \\ 0 & 3 & x-2 & 4 \\ 2 & 0 & 4 & x-3 \end{vmatrix}$.

4.8 APPLICATIONS OF DETERMINANTS

Determinants can be used to systematize the solution of systems of linear equations.

SOLVING LINEAR EQUATION SYSTEMS

Suppose we have the simultaneous equations

$$\begin{aligned} a_1x_1 + a_2x_2 + a_3x_3 &= h_1 \,, \\ b_1x_1 + b_2x_2 + b_3x_3 &= h_2 \,, \\ c_1x_1 + c_2x_2 + c_3x_3 &= h_3 \,. \end{aligned} \tag{4.72}$$

To use determinants to help solve this equation system, we define

$$D = \begin{vmatrix} a_1 & a_2 & a_3 \\ b_1 & b_2 & b_3 \\ c_1 & c_2 & c_3 \end{vmatrix} \,. \tag{4.73}$$

Starting from $x_1\,D$, we manipulate it by (1) moving x_1 to multiply the entries of the first column of D, then (2) adding to the first column x_2 times the second column and x_3 times the third column (neither of these operations change the value). The result is the first line of the equation below. We then reach the second line of the equation by substituting the right-hand sides of Eqs. (4.72). Thus:

$$\begin{aligned} x_1\,D = \begin{vmatrix} a_1\,x_1 & a_2 & a_3 \\ b_1\,x_1 & b_2 & b_3 \\ c_1\,x_1 & c_2 & c_3 \end{vmatrix} &= \begin{vmatrix} a_1\,x_1 + a_2\,x_2 + a_3\,x_3 & a_2 & a_3 \\ b_1\,x_1 + b_2\,x_2 + b_3\,x_3 & b_2 & b_3 \\ c_1\,x_1 + c_2\,x_2 + c_3\,x_3 & c_2 & c_3 \end{vmatrix} \\ &= \begin{vmatrix} h_1 & a_2 & a_3 \\ h_2 & b_2 & b_3 \\ h_3 & c_2 & c_3 \end{vmatrix} \,. \end{aligned} \tag{4.74}$$

If $D \neq 0$, Eq. (4.74) may now be solved for x_1:

$$x_1 = \frac{1}{D} \begin{vmatrix} h_1 & a_2 & a_3 \\ h_2 & b_2 & b_3 \\ h_3 & c_2 & c_3 \end{vmatrix} . \tag{4.75}$$

Analogous procedures starting from $x_2 D$ and $x_3 D$ give the parallel results

$$x_2 = \frac{1}{D} \begin{vmatrix} a_1 & h_1 & a_3 \\ b_1 & h_2 & b_3 \\ c_1 & h_3 & c_3 \end{vmatrix} , \qquad x_3 = \frac{1}{D} \begin{vmatrix} a_1 & a_2 & h_2 \\ b_1 & b_2 & h_2 \\ c_1 & c_2 & h_3 \end{vmatrix} . \tag{4.75}$$

We see that the solution for x_i is $1/D$ times a numerator obtained by replacing the ith column of D by the right-hand-side coefficients, a result that can be generalized to an arbitrary number n of simultaneous equations. This scheme for the solution of linear equation systems is known as **Cramer's rule**.

The above construction of the x_i is definitive and unique, showing that when $D \neq 0$ there will be exactly one solution to the equation set. Cramer's rule obviously becomes inapplicable when $D = 0$.

Example 4.8.1. Electric Circuit

Here is a practical problem involving simultaneous linear equations. Consider a multi-loop electric circuit containing batteries V_i and resistances R_i across which, according to Ohm's law, there is a voltage drop IR_i, where I is the current through the resistance. We can find the steady-state current in such a circuit by applying Kirchoff's law, which states that the current is continuous at the circuit junctions and that the net voltage change around any loop is zero. To apply this analysis to the circuit in Fig. 4.9, we assume that the current in the upper loop (in a clockwise sense) has the value I_2 while the current in the lower loop (also clockwise) is I_1. This means that the total current in the circuit branch containing V_3 and R_3 (flowing from right to

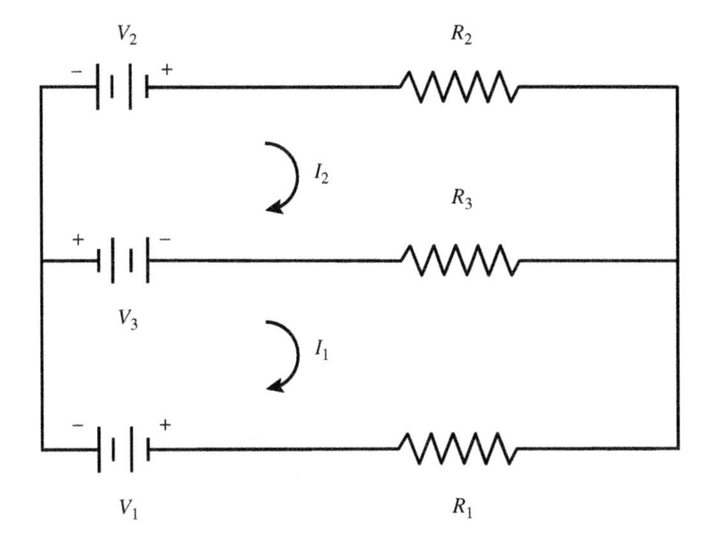

Figure 4.9: Battery-powered circuit.

left) is $I_2 - I_1$. Our assumption about the directions of current flow is not a physical limitation; if our solution to the problem indicates that either current flows in the direction opposite from our assumption, we will simply get a negative value for that current.

Our assignment of currents I_1 and I_2 and their superposition in the central branch of the circuit automatically produces continuity in the total current flow. To compute the voltage change around the upper loop, we must note carefully the assumed direction of the current flow and the polarities of the batteries. Taking the upper loop first, starting at its upper left and moving in the assumed current direction, the potential is first **increased** by V_2 as we pass through Battery 2, then decreased by I_2R_2, then decreased by $(I_2 - I_1)R_3$, and finally increased by V_3. Since the overall result must combine to give zero, we have the linear equation

$$V_2 - I_2R_2 - (I_2 - I_1)R_3 + V_3 = 0\ .$$

From the lower loop, starting at its upper left and moving in the direction of I_1 (note that in this direction the current in the central branch is $I_1 - I_2$), the voltage changes are $-V_3$, $-(I_1 - I_2)R_3$, $-I_1R_1$, and $-V_1$. Setting the sum of these to zero,

$$-V_3 - (I_1 - I_2)R_3 - I_1R_1 - V_1 = 0\ .$$

Writing these two equations in standard form, we have

$$-R_3I_1 + (R_2 + R_3)I_2 = V_2 + V_3\ ,$$

$$-(R_1 + R_3)I_1 + R_3I_2 = V_1 + V_3\ .$$

Solving these two equations for I_1 and I_2 by application of Cramer's Rule, we have

$$I_1 = \frac{D_1}{D},\quad I_2 = \frac{D_2}{D},\quad D = \begin{vmatrix} -R_3 & (R_2 + R_3) \\ -(R_1 + R_3) & R_3 \end{vmatrix},$$

$$D_1 = \begin{vmatrix} V_2 + V_3 & R_2 + R_3 \\ V_1 + V_3 & R_3 \end{vmatrix},\quad D_2 = \begin{vmatrix} -R_3 & V_2 + V_3 \\ -(R_1 + R_3) & V_1 + V_3 \end{vmatrix}.$$

Expanding the determinants and canceling where possible, we get

$$I_1 = \frac{(V_2 - V_1)R_3 - (V_1 + V_3)R_2}{R_1R_2 + R_1R_3 + R_2R_3}\ ,$$

$$I_2 = \frac{(V_2 - V_1)R_3 + (V_2 + V_3)R_1}{R_1R_2 + R_1R_3 + R_2R_3}\ .$$

For the specific values $R_1 = 4$ ohm, $R_2 = 3$ ohm, $R_3 = 2$ ohm, $V_1 = 1$ volt, $V_2 =$ 2 volt, $V_3 = 3$ volt, the above formulas reduce to

$$I_1 = \frac{(1)\cdot 2 - 4\cdot 3}{4\cdot 3 + 4\cdot 2 + 3\cdot 2} = -\frac{5}{13}\ \text{amp},\qquad I_2 = \frac{(-1)\cdot 2 + 5\cdot 4}{4\cdot 3 + 4\cdot 2 + 3\cdot 2} = \frac{11}{13}\ \text{amp}.$$

The negative value of I_1 shows that the current in the lower loop flows in the direction opposite to that assumed in setting up the problem.

■

SYMBOLIC SOLUTION OF LINEAR EQUATIONS

If one only wanted the determinant as a step toward solving a linear equation system, both symbolic systems provide a more direct route to the solution. In MAPLE, the equations to be solved are most conveniently given as named quantities, such as

```
> Eq1 := 2*x1+4*x2+5*x3+2*x4 = 6:
```

Note that equality in an equation is `=`, not `:=`, and that the `:=` causes the entire equation to be assigned the name `Eq1`. If the `= 6` had been omitted, MAPLE would have put in its place `= 0`. Assuming additional equations `Eq2`, `Eq3`, `Eq4` have been defined, the command for solving the equation set for `x1, x2, x3, x4` is

```
> solve( [Eq1, Eq2, Eq3, Eq4], [x1, x2, x3, x4] );
```

If the linear forms are independent, the equation set has a unique solution, and MAPLE returns (this is fictitious data)

$$\left[\left[x1 = \frac{5}{16},\ x2 = \frac{-3}{32},\ x3 = \frac{3}{4},\ x4 = -2\right]\right].$$

If the linear forms are dependent and the equations are inhomogeneous, there may be no solution, and MAPLE then returns `[ ]`. In the homogeneous case (and some inhomogeneous cases) the solution may contain one or more free parameters, and MAPLE returns output like

$$\left[\left[x1 = 2\,x2,\ x2 = x2,\ x3 = \frac{3}{4}\,x2,\ x4 = 12\,x2\right]\right].$$

Here the entry $x2 = x2$ tells us that we have a set of solutions parameterized by the arbitrary value of x_2.

In MATHEMATICA, the situation is similar but the notation is slightly different. A named equation has a form like

```
Eq1 = 2*x1 + 4*x2 + 5*x3 + 2*x4 == 6
```

Note that the construction indicating equality in an equation is `==`. The `== 6` cannot be absent to leave a default value `== 0`. The command for solving the equations is

```
Solve[ {Eq1, Eq2, Eq3, Eq4}, {x1, x2, x3, x4} ]
```

and if the linear forms are independent the output has a form like

$$\left\{\left\{x1 \to \frac{5}{16},\ x2 \to \frac{-3}{32},\ x3 \to \frac{3}{4},\ x4 \to -2\right\}\right\}.$$

If the linear forms are dependent and there is no solution, MATHEMATICA returns $\{\ \}$. If there is a parameterized set of solutions, it writes a warning message and output like

$$\left\{\left\{x1 \to \frac{x4}{8},\ x2 \to \frac{x4}{16},\ x3 \to -\frac{x4}{2}\right\}\right\}.$$

It is to be understood that in this construction x_4 is arbitrary.

Exercises

4.8.1. Using symbolic computation, write a procedure that will evaluate (in decimal form) the **Hilbert determinant** H of order n, whose elements are

$h_{ij} = (i + j - 1)^{-1}$. When $n = 3$, this determinant has the form

$$\begin{vmatrix} 1 & 1/2 & 1/3 \\ 1/2 & 1/3 & 1/4 \\ 1/3 & 1/4 & 1/5 \end{vmatrix} .$$

Get results for $n = 1$ through $n = 8$.

Note. This determinant becomes notoriously ill-conditioned as n increases, with a resultant value that is far smaller than any of its elements.

4.8.2. Solve the following equation system (keeping results to at least six decimal places):

$$\begin{aligned}
1.0x_1 + 0.9x_2 + 0.8x_3 + 0.4x_4 + 0.1x_5 &= 1.0 \,, \\
0.9x_1 + 1.0x_2 + 0.8x_3 + 0.5x_4 + 0.2x_5 + 0.1x_6 &= 0.9 \,, \\
0.8x_1 + 0.8x_2 + 1.0x_3 + 0.7x_4 + 0.4x_5 + 0.2x_6 &= 0.8 \,, \\
0.4x_1 + 0.5x_2 + 0.7x_3 + 1.0x_4 + 0.6x_5 + 0.3x_6 &= 0.7 \,, \\
0.1x_1 + 0.2x_2 + 0.4x_3 + 0.6x_4 + 1.0x_5 + 0.5x_6 &= 0.6 \,, \\
0.1x_2 + 0.2x_3 + 0.3x_4 + 0.5x_5 + 1.0x_6 &= 0.5 \,.
\end{aligned}$$

4.8.3. Using a symbolic computing system, solve the set of equations

$$x + 3y + 3z = 3 \,, \qquad x - y + z = 1 \,, \qquad 2x + 2y + 4z = 4 \,.$$

What can you conclude from the results?

4.8.4. For the electric circuit shown in Fig. 4.10, find the currents I_1 to I_6 that flow through the respective resistors R_1 to R_6. Take $V_1 = 2$ volt, $V_2 = 5$ volt,

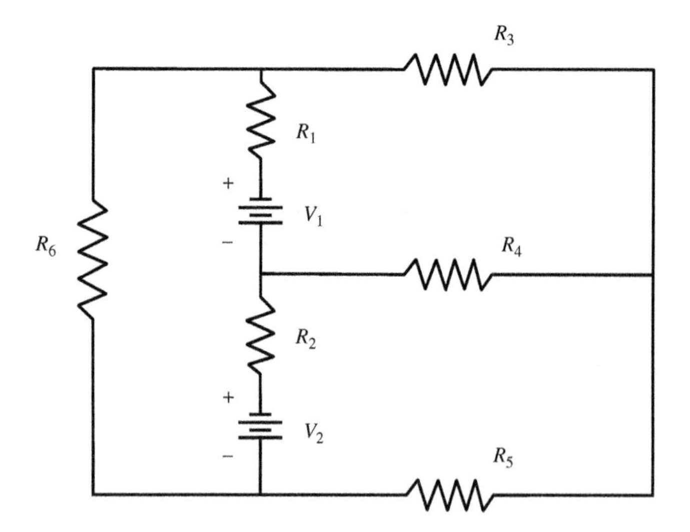

Figure 4.10: Battery-powered circuit for Exercise 4.8.4.

$R_1 = 2$ ohm, $R_2 = 4$ ohm, $R_3 = 6$ ohm, $R_4 = 8$ ohm, $R_5 = 3$ ohm, $R_6 = 5$ ohm. For the current through each resistor indicate its direction of flow as one of "up," "down," "left," or "right."

Hint. There are only three independent current loops.

HOMOGENEOUS LINEAR EQUATIONS

A set of linear equations is called **homogeneous** if all its terms are linear in the unknowns; in matrix form it corresponds to the equation $\mathbf{Ax} = \mathbf{0}$. If A is nonsingular, Cramer's rule gives us, unambiguously, the unique solution $\mathbf{x} = \mathbf{0}$. If, however, A is singular, then the equation set exhibits linear dependence and its left-hand sides can be manipulated to bring at least one of them to zero. Because the right-hand sides of a set of homogeneous equations are all zero, these manipulations will always retain the zero right-hand sides, so the linear dependence will produce equation(s) of the form $0 = 0$. Our conclusion, then, is that

> *A homogeneous equation set, of the form* $\mathsf{A}\mathbf{x} = \mathbf{0}$*, will, if* A *is nonsingular, have only the solution* $\mathbf{x} = \mathbf{0}$*; if* A *is singular, it will have a family of nonzero solutions.*

Example 4.8.2. Linearly Dependent Homogeneous Equations

Consider the equation set

$$\begin{aligned} 2x_1 + 4x_2 + 5x_3 &= 0 \,, \\ x_1 - 2x_2 + 2x_3 &= 0 \,, \\ 3x_1 + 2x_2 + 7x_3 &= 0 \,. \end{aligned}$$

The coefficient matrix is of the form

$$\mathsf{A} = \begin{pmatrix} 2 & 4 & 5 \\ 1 & -2 & 2 \\ 3 & 2 & 7 \end{pmatrix}.$$

The reader can easily verify that $\det \mathsf{A} = 0$, so these equations exhibit linear dependence.

In this equation set, we note that the third equation is the sum of the first two, so it adds no new information and we may discard it. Solving the first two equations for x_1 and x_2 in terms of x_3, we find

$$x_1 = -\frac{9}{4}\,x_3 \,, \qquad x_2 = -\frac{1}{8}\,x_3 \,.$$

Thus, we may choose **any** value for x_3; after doing so, the above formulas tell us the values of x_1 and x_2 which, along with the chosen x_3, constitute a solution to our original equation system. If we regard the x_i as a column vector, our solutions are of the form

$$\mathbf{x} = C \begin{pmatrix} -9/4 \\ -1/8 \\ 1 \end{pmatrix},$$

where C is arbitrary (it is the value we assign to x_3).

It is not surprising that our solution is arbitrary as to scale. The fact that the equations are homogeneous means that any multiple of a solution should also be a solution.

■

LINEAR INDEPENDENCE OF FUNCTIONS

We often need to know whether a set of functions is **linearly independent**, meaning that no member of the set is expressible as a linear combination of others. A formal definition of linear dependence, therefore, is that the functions φ_i, $i = 1,\ldots,n$ are linearly dependent if there exist constants $k_1,\ k_2,\ldots,k_n$, not all zero, such that

$$k_1\varphi_1(x) + k_2\varphi_2(x) + \cdots + k_n\varphi_n(x) \equiv 0\,. \tag{4.76}$$

Note that Eq. (4.76) only implies linear dependence of the φ_i if that equation is satisfied for all x relevant to the function set.

It is important to realize that linear dependence is not the same as any dependence whatsoever. For example, there is an obvious relation between x and x^2 that is unique for any region such that $x > 0$, and (for $0 \le x \le \pi/2$) $\sin x$ and $\cos x$ are related (i.e., either can be determined from the other).

The existence of linear dependence of a function set may or may not be obvious under casual inspection, but determinants provide a way to test for linear dependence. The relevant theorem is the following:

If the functions $\varphi_1(x)$, $\varphi_2(x)$, $\ldots$, $\varphi_n(x)$ have derivatives of order $n-1$, then they are linearly independent if the determinant

$$W = \begin{vmatrix} \varphi_1(x) & \varphi_2(x) & \cdots & \varphi_n(x) \\ \varphi_1'(x) & \varphi_2'(x) & \cdots & \varphi_n'(x) \\ \varphi_1''(x) & \varphi_2''(x) & \cdots & \varphi_n''(x) \\ \cdots & \cdots & \cdots & \cdots \\ \varphi_1^{(n-1)}(x) & \varphi_2^{(n-1)}(x) & \cdots & \varphi_n^{(n-1)}(x) \end{vmatrix} \tag{4.77}$$

is not identically equal to zero. This determinant is called the **Wronskian** *of the function set, and is of particular importance in differential equation theory.*

Example 4.8.3. Wronskian: Linearly Independent Functions

Let's check that the functions x, $\sin x$, $\cos x$ are linearly independent. We form their Wronskian:

$$W = \begin{vmatrix} x & \sin x & \cos x \\ 1 & \cos x & -\sin x \\ 0 & -\sin x & -\cos x \end{vmatrix} = -x\,.$$

W is not identically zero; the functions are linearly independent.

■

Example 4.8.4. Wronskian: Linearly Dependent Functions

Let's check for linear dependence the functions $\sin 2x$, $\cos 2x$, $\sin x \cos x$. Forming their Wronskian,

$$W = \begin{vmatrix} \sin 2x & \cos 2x & \sin x \cos x \\ 2\cos 2x & -2\sin 2x & \cos^2 x - \sin^2 x \\ -4\sin 2x & -4\cos 2x & -4\sin x \cos x \end{vmatrix} = 0\,.$$

The zero value of W is obvious because its first and third rows are proportional. We were able to show that these functions are linearly dependent without even having to use the formula $2\sin x \cos x = \sin 2x$.

■

Exercises

For each of the following equation sets,

(a) Obtain by hand computation its general solution.

(b) Check your answer using symbolic computation.

4.8.5.

$$\begin{aligned} 3x \quad + y + 2z &= 0, \\ 4x - 2y \quad + z &= 0, \\ 2x + 4y + 3z &= 0. \end{aligned}$$

4.8.6.

$$\begin{aligned} 2x + y + 2z &= 0, \\ 4x + y \quad + z &= 0, \\ 2x \qquad - 3z &= 0. \end{aligned}$$

Determine whether each of the following sets of functions exhibits linear dependence.

4.8.7. $\sin x, \quad \sin 2x, \quad \sin 3x.$

Hint. Can you find any value of x for which the Wronskian is easily evaluated and has a nonzero value?

4.8.8. $1, \quad x, \quad (3x^2 - 1)/2.$

4.8.9. $\cosh x, \quad \sinh x, \quad e^x.$

4.8.10. $\sin x, \quad \cos x, \quad e^{ix}.$

4.8.11. $\sin x, \quad \cos x, \quad \sin x \cos x.$

4.8.12. $\sin^2 x, \quad \cos^2 x, \quad \sin 2x, \quad \cos 2x.$

Additional Readings

Aitken, A. C. (1956). *Determinants and matrices.* New York: Interscience (Reprinted, Greenwood (1983). A readable introduction to determinants and matrices).

Arfken, G. B., Weber, H. J., & Harris, F. E. (2013). *Mathematical methods for physicists* (7th ed.). New York: Academic Press.

Barnett, S. (1990). *Matrices: methods and applications.* Oxford: Clarendon Press.

Bickley, W. G., & Thompson, R. S. H. G. (1964). *Matrices—their meaning and manipulation.* Princeton, NJ: Van Nostrand (A comprehensive account of matrices in physical problems, their analytic properties, and numerical techniques).

Brown, W. C. (1991). *Matrices and vector spaces.* New York: Dekker.

Gilbert, J., & Gilbert, L. (1995). *Linear algebra and matrix theory.* San Diego: Academic Press.

Golub, G. H., & Van Loan, C. F. (1996). *Matrix computations* (3rd ed.). Baltimore: JHU Press (Detailed mathematical background and algorithms for the production of numerical software, including methods for parallel computation. A classic computer science text).

Heading, J. (1958). *Matrix theory for physicists.* London: Longmans, Green and Co. (A readable introduction to determinants and matrices, with applications to mechanics, electromagnetism, special relativity, and quantum mechanics).

Vein, R., & Dale, P. (1998). *Determinants and their applications in mathematical physics.* Berlin: Springer.

Watkins, D.S. (1991). *Fundamentals of matrix computations.* New York: Wiley.

Chapter 5

MATRIX TRANSFORMATIONS

When vectors are used to represent physical quantities such as velocities or forces, their individual components have values that depend upon the orientation of the coordinate system. However, the connections between related quantities cannot depend upon the choice of coordinate system, because that would be inconsistent with our experimental experience that space is isotropic (i.e., has no preferred direction). This idea has profound consequences, some of which we now study by examining how vectors, and linear equations connecting vectors, transform under transformations of the coordinate system. Since linear equations involving vectors can be written as matrix equations, our study necessarily includes the way in which matrix equations must transform in order to remain consistent under coordinate transformations.

5.1 VECTORS IN ROTATED SYSTEMS

We consider here what happens when we rotate the coordinate system used to describe a vector. Working in two dimensions (2-D) for simplicity, consider a fixed vector $\mathbf{A}$, with components A_x, A_y in our original coordinate system defined by unit vectors $\hat{\mathbf{e}}_x$, $\hat{\mathbf{e}}_y$. We now wish to express $\mathbf{A}$ in terms of its components $A_{x'}$, $A_{y'}$ in new coordinates whose unit vectors $\hat{\mathbf{e}}_{x'}$, $\hat{\mathbf{e}}_{y'}$ are rotated relative to the original coordinates through an angle θ, with counterclockwise rotation corresponding to positive values of θ. This situation is sometimes referred to as a **passive** rotation or transformation, to distinguish it from an **active** transformation in which the coordinate system remains fixed and the vector is rotated. In this book we will generally deal with passive rotations; formulas for active rotations will differ from those for passive rotations by a change in sign(s) of the rotation angle(s).

A simple way to analyze the effect of coordinate rotations is to start by finding out what happens to $A_x\hat{\mathbf{e}}_x$ and $A_y\hat{\mathbf{e}}_y$ when the coordinates are rotated. Continuing in 2-D, and looking at Fig. 5.1, we see that a vector $A_x\hat{\mathbf{e}}_x$ in the original coordinate system will have component $A_x \cos\varphi$ in the new x direction (which we call x') and component $-A_x \sin\varphi$ in the new y direction (called y'). Figure 5.1 also shows that the vector $A_y\hat{\mathbf{e}}_y$ will have component $+A_y \sin\varphi$ in the x' direction and component

Mathematics for Physical Science and Engineering.
http://dx.doi.org/10.1016/B978-0-12-801000-6.00005-5

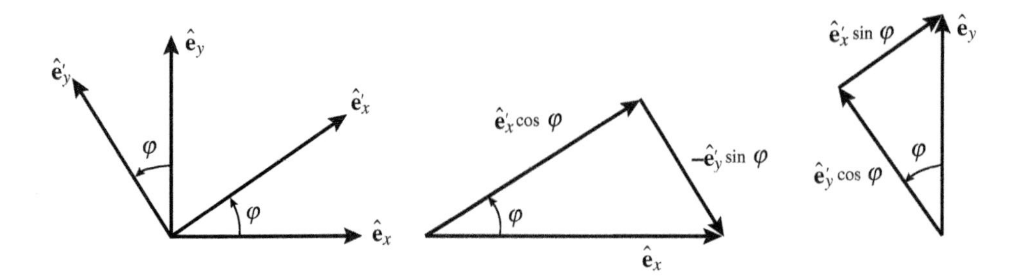

Figure 5.1: Coordinate rotation. Left, rotated coordinates; center and right, decomposition of $\hat{\mathbf{e}}_x$ and $\hat{\mathbf{e}}_y$ in the rotated system.

$A_y \cos\varphi$ in the y' direction. Putting these relations together, we see that

$$\begin{aligned} A_x\hat{\mathbf{e}}_x &= A_x \cos\theta\, \hat{\mathbf{e}}_{x'} - A_x \sin\theta\, \hat{\mathbf{e}}_{y'}\ , \\ A_y\hat{\mathbf{e}}_y &= A_y \cos\theta\, \hat{\mathbf{e}}_{y'} + A_y \sin\theta\, \hat{\mathbf{e}}_{x'}\ , \end{aligned} \tag{5.1}$$

which is equivalent to

$$\mathbf{A} = A_x\hat{\mathbf{e}}_x + A_y\hat{\mathbf{e}}_y = (A_x\cos\theta + A_y\sin\theta)\,\hat{\mathbf{e}}_{x'} + (-A_x\sin\theta + A_y\cos\theta)\,\hat{\mathbf{e}}_{y'}\ . \tag{5.2}$$

We can write Eq.(5.2) as $\mathbf{A} = A_{x'}\hat{\mathbf{e}}_{x'} + A_{y'}\hat{\mathbf{e}}_{y'}$ with

$$\begin{aligned} A_{x'} &= A_x\cos\theta + A_y\sin\theta\ , \\ A_{y'} &= -A_x\sin\theta + A_y\cos\theta\ , \end{aligned}$$

which in matrix notation takes the form

$$\begin{pmatrix} A_{x'} \\ A_{y'} \end{pmatrix} = \begin{pmatrix} \cos\theta & \sin\theta \\ -\sin\theta & \cos\theta \end{pmatrix} \begin{pmatrix} A_x \\ A_y \end{pmatrix}. \tag{5.3}$$

If we let $\mathbf{a}$ and $\mathbf{a}'$ represent the column vectors containing the components of $\mathbf{A}$ in the original and primed coordinates, and let $\mathsf{U}(\theta)$ stand for the matrix connecting $\mathbf{a}$ and $\mathbf{a}'$, we have

$$\mathbf{a}' = \mathsf{U}(\theta)\,\mathbf{a}\ , \qquad \text{with}\quad \mathsf{U}(\theta) = \begin{pmatrix} \cos\theta & \sin\theta \\ -\sin\theta & \cos\theta \end{pmatrix}, \tag{5.4}$$

Note that the same matrix $\mathsf{U}(\theta)$ describes the effect of the coordinate transformation on any vector $\mathbf{a}$.

ORTHOGONALITY OF ROTATION MATRICES

It is important to understand the properties of the rotational transformation matrix U. Physically, the transformation must preserve the magnitude of any vector to which it is applied, so $\mathbf{a}' = \mathsf{U}\mathbf{a}$ must have the same magnitude as $\mathbf{a}$. Moreover, if we change the sign of the rotation angle θ in U, we must obtain a transformation that is the inverse of that described by U. The first of these requirements can be checked by setting the scalar product $\mathbf{a}'\cdot\mathbf{a}'$ equal to $\mathbf{a}\cdot\mathbf{a}$, which in matrix notation corresponds to

$$\mathbf{a}'\cdot\mathbf{a}' = (\mathsf{U}\,\mathbf{a})^{\mathrm{T}}\mathsf{U}\,\mathbf{a} = \mathbf{a}^{\mathrm{T}}\mathsf{U}^{\mathrm{T}}\mathsf{U}\,\mathbf{a} = \mathbf{a}^{\mathrm{T}}\mathbf{a}\ .$$

This equation can be valid for all $\mathbf{a}$ only if

$$\mathsf{U}^{\mathrm{T}}\mathsf{U} = \mathbf{1}\,, \tag{5.5}$$

which is a severe requirement on U.

Let's check that our U satisfies this requirement. We have, taking U and its transpose from Eq. (5.4),

$$\begin{pmatrix} \cos\theta & -\sin\theta \\ \sin\theta & \cos\theta \end{pmatrix}\begin{pmatrix} \cos\theta & \sin\theta \\ -\sin\theta & \cos\theta \end{pmatrix} = \begin{pmatrix} 1 & 0 \\ 0 & 1 \end{pmatrix} = \mathbf{1}\,,$$

as required. We note also that changing the sign of θ indeed converts U into U^{-1}.

Another way of expressing Eq. (5.5) is to observe that it states that $\mathsf{U}^{\mathrm{T}} = \mathsf{U}^{-1}$. We previously noted that matrices with this property are called **orthogonal**; we now see that this designation arises because such matrices describe coordinate transformations that preserve the orthogonality of the coordinate axes. Replacing U by U^{T}, we see that if U is orthogonal, so also is U^{T}. Moreover, U and U^{T} commute:

$$\mathsf{U}^{\mathrm{T}}\mathsf{U} = \mathsf{U}\,\mathsf{U}^{\mathrm{T}} = \mathbf{1}\,. \tag{5.6}$$

Further insight into orthogonal matrices can be obtained if we look at the elements of the matrix product $\mathsf{U}^{\mathrm{T}}\,\mathsf{U} = \mathbf{1}$. Taking the general case, where U and U^{T} are of dimension $n \times n$, we think of U as an array of column vectors $\mathbf{u}_i$, which in turn means that we consider U^{T} as an array of row vectors $\mathbf{u}_i^{\mathrm{T}}$. That is,

$$\mathsf{U} = \left(\begin{array}{c|c|c|c} \mathbf{u}_1 & \mathbf{u}_2 & \mathbf{u}_3 & \cdots \end{array}\right), \qquad \mathsf{U}^{\mathrm{T}} = \left(\begin{array}{c} \mathbf{u}_1^{\mathrm{T}} \\ \hline \mathbf{u}_2^{\mathrm{T}} \\ \hline \mathbf{u}_3^{\mathrm{T}} \\ \hline \cdots \end{array}\right).$$

Remember that $\mathbf{u}_i^{\mathrm{T}}$ are not just arbitrary vectors that form the rows of U^{T}; each $\mathbf{u}_i^{\mathrm{T}}$ is the transpose of the corresponding $\mathbf{u}_i$.

Now, forming $\mathsf{U}^{\mathrm{T}}\,\mathsf{U}$, and identifying the matrix multiplication in terms of scalar products, we reach

$$\mathsf{U}^{\mathrm{T}}\mathsf{U} = \left(\begin{array}{c} \mathbf{u}_1^{\mathrm{T}} \\ \hline \mathbf{u}_2^{\mathrm{T}} \\ \hline \mathbf{u}_3^{\mathrm{T}} \\ \hline \cdots \end{array}\right)\left(\begin{array}{c|c|c|c} \mathbf{u}_1 & \mathbf{u}_2 & \mathbf{u}_3 & \cdots \end{array}\right) = \begin{pmatrix} \mathbf{u}_1\cdot\mathbf{u}_1 & \mathbf{u}_1\cdot\mathbf{u}_2 & \mathbf{u}_1\cdot\mathbf{u}_3 & \cdots \\ \mathbf{u}_2\cdot\mathbf{u}_1 & \mathbf{u}_2\cdot\mathbf{u}_2 & \mathbf{u}_2\cdot\mathbf{u}_3 & \cdots \\ \mathbf{u}_3\cdot\mathbf{u}_1 & \mathbf{u}_3\cdot\mathbf{u}_2 & \mathbf{u}_3\cdot\mathbf{u}_3 & \cdots \\ \cdots & \cdots & \cdots & \cdots \end{pmatrix},$$

$$= \begin{pmatrix} 1 & 0 & 0 & \cdots \\ 0 & 1 & 0 & \cdots \\ 0 & 0 & 1 & \cdots \\ \cdots & \cdots & \cdots & \cdots \end{pmatrix}.$$

We conclude that

$$\mathbf{u}_i \cdot \mathbf{u}_j = \delta_{ij}. \tag{5.7}$$

Here δ_{ij} is the Kronecker delta, defined in Eq. (4.39). Equation (5.7) tells us that the columns of an orthogonal matrix form an orthonormal set of vectors.

Now, what are these vectors that form the columns of U? Each column contains the components of one of the original unit vectors when expressed in terms of the primed coordinates. To see that this is so, return to the 2-D case in Eqs. (5.1), there setting $A_x = A_y = 1$. Taking scalar products with $\hat{\mathbf{e}}_{x'}$ and $\hat{\mathbf{e}}_{y'}$, we find

$$\hat{\mathbf{e}}_x \cdot \hat{\mathbf{e}}_{x'} = \hat{\mathbf{e}}_y \cdot \hat{\mathbf{e}}_{y'} = \cos\theta \,, \qquad \hat{\mathbf{e}}_y \cdot \hat{\mathbf{e}}_{x'} = -\hat{\mathbf{e}}_x \cdot \hat{\mathbf{e}}_{y'} = \sin\theta \,. \tag{5.8}$$

Using these expressions to rewrite the matrix elements of U, we get

$$\mathsf{U} = \begin{pmatrix} \hat{\mathbf{e}}_x \cdot \hat{\mathbf{e}}_{x'} & \hat{\mathbf{e}}_y \cdot \hat{\mathbf{e}}_{x'} \\ \hat{\mathbf{e}}_x \cdot \hat{\mathbf{e}}_{y'} & \hat{\mathbf{e}}_y \cdot \hat{\mathbf{e}}_{y'} \end{pmatrix}, \tag{5.9}$$

thereby confirming that the first column of U contains the new components of $\hat{\mathbf{e}}_x$, while the second column contains the new components of $\hat{\mathbf{e}}_y$. Alternatively, we can identify each row of U as containing the components of one of the rotated unit vectors when expressed in terms of the original coordinates. This is to be expected since the rows of U are the columns of U^{-1}.

The most important idea to be gained from the above discussion is the following:

The columns of an orthogonal matrix give the components of an orthonormal set of unit vectors when expressed in a rotated coordinate system. Because the transpose of an orthogonal matrix is also its inverse, it is also the case that the rows of an orthogonal matrix give the components of the transformed unit vectors in terms of the original unit vector set.

Finally, we note that because $\det(\mathsf{U}) = \det(\mathsf{U}^T)$, the determinant product theorem tells us that $[\det(\mathsf{U})]^2 = 1$, so that (because U is real), $\det(\mathsf{U}) = \pm 1$. In other words, the only values possible for $\det(\mathsf{U})$ are $+1$ and -1. However, because U must vary continuously with θ and $\det(\mathsf{U}) = +1$ when $\theta = 0$ (no rotation), we must have $\det(\mathsf{U}) = +1$ for **all** rotations. That means that though the definition of an orthogonal matrix admits matrices with determinant -1, such matrices cannot represent rotations. Thus,

The orthogonal matrices that represent rotations have determinant $+1$.

SUCCESSIVE ROTATIONS

If U is the matrix for 2-D rotation through an angle θ_1 and V describes a rotation through an angle θ_2, the matrix product $\mathsf{U}\,\mathsf{V}$ corresponds to a rotation through θ_2 followed by one through θ_1. The result is a rotation through $\theta_1 + \theta_2$, and it should also be described by an orthogonal matrix. We can check this:

$$(\mathsf{U}\,\mathsf{V})(\mathsf{U}\,\mathsf{V})^T = \mathsf{U}\,\mathsf{V}\,\mathsf{V}^T\mathsf{U}^T = \mathbf{1}\,,$$

verifying that $\mathsf{U}\,\mathsf{V}$ is also orthogonal. In 2-D, the same result is obtained if the rotations are carried out in either order; that implies that U and V commute.

SUMMARY

Our illustration of orthogonal matrices and rotation was restricted to a 2-D example. Nevertheless, the concepts extend to three-dimensional (3-D) space (and formally, to spaces of arbitrary dimensionality). In 3-D space, for example, a rotation carries the x, y, and z axes (conventionally defined as a **right-handed** system) into a right-handed set of orthogonal unit vectors at another orientation, and the effect of the rotation will be encapsulated in a 3×3 orthogonal matrix whose rows give the components of the rotated unit vectors in terms of the original coordinates. However, in three and higher dimensions, rotations about different axes do not commute, and the commutation of 2-D rotation matrices does not extend to general dimensionality. For more detail regarding 3-D rotations and their description using what are known as their Euler angles, see Arfken et al in the Additional Readings. Despite the fact that the situation in 3-D is more complicated, rotations in 3-D (and in spaces of higher dimension) are described by orthogonal matrices with determinant +1.

Example 5.1.1. A 3-D Rotation

Apply $\mathsf{U} = \begin{pmatrix} 0 & 1 & 0 \\ 0 & 0 & 1 \\ 1 & 0 & 0 \end{pmatrix}$ to the vectors $\mathbf{a} = \begin{pmatrix} 3 \\ 1 \\ 2 \end{pmatrix}$ and $\mathbf{b} = \begin{pmatrix} 1 \\ 2 \\ 0 \end{pmatrix}$.

Working first in MAPLE, we determine what happens by finding the vectors $\mathbf{a}' = \mathsf{U}\mathbf{a}$ and $\mathbf{b}' = \mathsf{U}\mathbf{b}$:

```
>U := Matrix([[0,1,0],[0,0,1],[1,0,0]]);
```

$$U := \begin{pmatrix} 0 & 1 & 0 \\ 0 & 0 & 1 \\ 1 & 0 & 0 \end{pmatrix}$$

```
> a := Vector([3,2,1]);
```

$$a := \begin{pmatrix} 3 \\ 1 \\ 2 \end{pmatrix}$$

```
> b := Vector([1,2,0]);
```

$$b := \begin{pmatrix} 1 \\ 2 \\ 0 \end{pmatrix}$$

```
> a1 := U . a;
```

$$a1 := \begin{pmatrix} 1 \\ 2 \\ 3 \end{pmatrix}$$

```
> b1 := U . b;
```

$$b1 := \begin{pmatrix} 2 \\ 0 \\ 1 \end{pmatrix}$$

Alternatively, in MATHEMATICA,

```
u = { {0,1,0), {0,0,1}, {1,0,0} }        { {0,1,0), {0,0,1}, {1,0,0} }
a = {3,1,2}                               { {3,1,2} }
b = {1,2,0}                               { {1,2,0} }
a1 = U . a                                { {1,2,3} }
b1 = U . b                                { {2,0,1} }
```

We check that $\mathbf{a}'$ and $\mathbf{b}'$ have the same magnitudes as $\mathbf{a}$ and $\mathbf{b}$:

MAPLE		MATHEMATICA	
`> a . a;`	14	`a . a`	14
`> a1 . a1;`	14	`a1 . a1`	14
`> b . b;`	5	`b . b`	5
`> b1 . b1;`	5	`b1 . b1`	5

and we compute the dot products $\mathbf{a} \cdot \mathbf{b}$ and $\mathbf{a}' \cdot \mathbf{b}'$:

```
> a . b;          5          a . b      5
> a1 . b1;        5          a1 . b1    5
```

These are necessary (but not sufficient) conditions that U describes a rotation; a complete demonstration would require a check on at least one more pair of vectors. We can verify that U is a rotation matrix by showing that it is orthogonal, with determinant +1:

MAPLE		MATHEMATICA
`> with{LinearAlgebra):`		
`> Transpose(U) . U;`	$\begin{pmatrix} 1 & 0 & 0 \\ 0 & 1 & 0 \\ 0 & 0 & 1 \end{pmatrix}$	`Transpose[U] . U` `{ {1,0,0}, {0,1,0}, {0,0,1} }`
`> Determinant(U);`	1	`Det[U]` 1

To more fully understand what this U does, notice what happens when we apply it to unit vectors in the three coordinate directions:

`> U . Vector([1,0,0]);`	$\begin{pmatrix} 0 \\ 0 \\ 1 \end{pmatrix}$	`U . {1,0,0}`	`{0,0,1}`

Note that the first column of U contains the components of the rotated x-axis; the second and third columns hold the rotated y- and z-axes:

`> U . Vector([0,1,0]);`	$\begin{pmatrix} 1 \\ 0 \\ 0 \end{pmatrix}$	`U . {0,1,0}`	`{1,0,0}`
`> U . Vector([0,0,1]);`	$\begin{pmatrix} 0 \\ 1 \\ 0 \end{pmatrix}$	`U . {0,0,1}`	`{0,1,0}`

Thus, U is a rotation such that $\hat{\mathbf{e}}_x \to \hat{\mathbf{e}}_z$, $\hat{\mathbf{e}}_y \to \hat{\mathbf{e}}_x$, and $\hat{\mathbf{e}}_z \to \hat{\mathbf{e}}_y$.

■

Exercises

5.1.1. Consider the 3-D rotation matrix $\mathsf{U} = \begin{pmatrix} -0.8 & 0 & 0.6 \\ 0 & 1 & 0 \\ -0.6 & 0 & -0.8 \end{pmatrix}$.

(a) Verify that the vector whose components are the first column of U is orthogonal to the vector described by the third column of U, and verify that each of these columns describes a vector of unit magnitude.

(b) Because all the columns of U are mutually orthogonal, we can conclude that U is an orthogonal matrix. Explain why this is so, and carry out matrix operations confirming that the matrix U is orthogonal.

(c) What is the orientation of $\hat{\mathbf{e}}_y$ in the rotated coordinates?

(d) Describe in words the rotation corresponding to U. Be specific about the sense of the rotation (assume a right-handed coordinate system).

5.1.2. Find the matrix V that produces the rotation inverse to that caused by the matrix U of Exercise 5.1.1. Verify that U and V are mutually inverse by applying them in succession to the vector which (in row form) is

$$\mathbf{a} = (1.2345, -0.7507, 0.6733)\,.$$

5.1.3. Apply the U of Exercise 5.1.1 three times in succession to the vector **a** given in Exercise 5.1.2. Verify that the result has the same magnitude as **a**.

5.1.4. In 4-D space consider the matrix $\mathsf{U} = \begin{pmatrix} 0.5 & 0.5 & 0.5 & 0.5 \\ 0.5 & 0.5 & -0.5 & -0.5 \\ 0.5 & -0.5 & 0.5 & -0.5 \\ -0.5 & 0.5 & 0.5 & -0.5 \end{pmatrix}$.

Does this matrix describe a rotation of the 4-D coordinate system? How do you know?

5.1.5. Find the matrix U that corresponds to a rotation of 45° about the y-axis of a right-handed 3-D coordinate system. The positive sense of such a rotation tilts the z-axis toward the original position of the x-axis (can you see why?).

5.2 VECTORS UNDER COORDINATE REFLECTIONS

It is sometimes useful to subject a coordinate system (in 3-D space) to a transformation that can be described as a reflection. The two such transformations that are usually considered are **inversion**, in which the direction of each of the three coordinates is reversed, and a simple reflection, which corresponds to viewing the coordinate axes through a plane mirror (this causes reversal of the coordinate direction normal to the mirror). Both these coordinate transformations change the usual right-handed coordinates to a left-handed system. However, the equations of physics must predict the same physical phenomena irrespective of the handedness of the coordinate system used for their description.

Our current interest is in the matrix representation of these reflection transformations. Letting **i** designate the inversion operation and $\boldsymbol{\sigma}_z$ a reflection through the xy-plane (the normal to which is in the z-direction), these reflections correspond to

$$\mathsf{U}(\mathbf{i}) = \begin{pmatrix} -1 & 0 & 0 \\ 0 & -1 & 0 \\ 0 & 0 & -1 \end{pmatrix}, \qquad \mathsf{U}(\boldsymbol{\sigma}_z) = \begin{pmatrix} 1 & 0 & 0 \\ 0 & 1 & 0 \\ 0 & 0 & -1 \end{pmatrix}. \tag{5.10}$$

Each of these matrices is orthogonal. Moreover, each is its own inverse, and therefore satisfies $\mathsf{U}^2 = \mathbf{1}$. Note that each of these matrices has $\det(\mathsf{U}) = -1$, showing that this value of the determinant signals a reflection operation. If we now carry out both a reflection and an arbitrary rotation, the result will be described by an arbitrary orthogonal matrix U with determinant -1. Our conclusion is that, in 3-D space, a U of determinant -1 corresponds to conversion between a right-handed and a left-handed coordinate system.

Example 5.2.1. Reflection Transformations

Using matrix multiplication, let's find the matrix corresponding to a reflection of the coordinates through the xy-plane followed by rotating the coordinates 36.87° about the current position of the y-axis. These two operations have the respective matrix representations $\mathsf{U}(\boldsymbol{\sigma}_z)$ and $\mathsf{U}(R_y)$, and because these matrices are to operate on some vector to their right, they should be applied in the order $\mathsf{U}(R_y)\mathsf{U}(\boldsymbol{\sigma}_z)$. We therefore get (note that $\cos 36.87° = 0.80$ and $\sin 36.87° = 0.60$)

$$\mathsf{U}(R_y\boldsymbol{\sigma}_z) = \mathsf{U}(R_y)\mathsf{U}(\boldsymbol{\sigma}_z) = \begin{pmatrix} 0.80 & 0 & -0.60 \\ 0 & 1 & 0 \\ 0.60 & 0 & 0.80 \end{pmatrix} \begin{pmatrix} 1 & 0 & 0 \\ 0 & 1 & 0 \\ 0 & 0 & -1 \end{pmatrix}$$

$$= \begin{pmatrix} 0.80 & 0 & 0.60 \\ 0 & 1 & 0 \\ 0.60 & 0 & -0.80 \end{pmatrix}.$$

A simple hand computation reveals that this matrix has determinant -1.

Note. If the signs of the quantities 0.60 in $\mathsf{U}(R_y)$ are puzzling, look at Exercise 5.1.5.

Suppose now that instead of the above we rotate the coordinates about the y-axis **before** carrying out the coordinate reflection. We get

$$\mathsf{U}(\boldsymbol{\sigma}_z R_y) = \mathsf{U}(\boldsymbol{\sigma}_z)\mathsf{U}(R_y) = \begin{pmatrix} 1 & 0 & 0 \\ 0 & 1 & 0 \\ 0 & 0 & -1 \end{pmatrix} \begin{pmatrix} 0.80 & 0 & -0.60 \\ 0 & 1 & 0 \\ 0.60 & 0 & 0.80 \end{pmatrix}$$

$$= \begin{pmatrix} 0.80 & 0 & -0.60 \\ 0 & 1 & 0 \\ -0.60 & 0 & -0.80 \end{pmatrix}.$$

The determinant is again -1 but the matrix U is different than before. These two operations do not commute.

■

Exercises

5.2.1. Does the matrix $\mathsf{U} = \begin{pmatrix} 0.408248 & 0.816497 & 0.408248 \\ -0.707107 & 0 & 0.707107 \\ -0.577350 & 0.577350 & -0.577350 \end{pmatrix}$

describe a rotation, a rotation and reflection, or neither (to within the precision of its specification)?

5.2.2. Find the matrix U that corresponds to the following three successive operations: (1) Reflection through the xy-plane; (2) Rotation about the current x axis by 90° (the positive sense of rotation tilts $\hat{\mathbf{e}}_y$ toward $\hat{\mathbf{e}}_z$); (3) Reflection through the current xy-plane. Is this overall operation a rotation, a rotation and reflection, or neither?

5.3 TRANSFORMING MATRIX EQUATIONS

We now proceed to consider how matrix equations connecting physical quantities are affected by coordinate transformations. The situation with which we are concerned can be illustrated by the equation connecting the induced dipole moment $\mathbf{p}$ when an anisotropic polarizable object is placed in an electric field $\mathbf{E}$. That equation has the form $\mathbf{p} = \mathsf{X}\mathbf{E}$, where X (often written χ in the research literature) is called the electric susceptibility matrix; we have a matrix equation because the anisotropy may cause the induced moment to be in a different direction than $\mathbf{E}$. Another example is the equation connecting the angular velocity $\boldsymbol{\omega}$ and the angular momentum $\mathbf{L}$ of a rotating anisotropic rigid body: $\mathbf{L} = \mathsf{I}\boldsymbol{\omega}$. Because an anisotropic object has different moments of inertia for rotations about different axes, its angular momentum may not be in the same direction as its angular velocity; hence these two quantities are connected by a matrix equation.

Writing a generic equation of this sort as $\mathbf{f} = \mathsf{A}\mathbf{g}$, we note that if we rotate the coordinate system and thereby change the specific form of $\mathbf{f}$ and $\mathbf{g}$, we must also change the form of A in a way that preserves the physical content of the equation. In other words (letting primes denote the transformed quantities), a coordinate rotation must yield $\mathbf{f}' = \mathsf{A}'\mathbf{g}'$. From a mathematical viewpoint, the rotation of the coordinate system corresponds to a change of the basis in terms of which we are describing not only $\mathbf{f}$ and $\mathbf{g}$, but also A.

Introducing U as the orthogonal matrix describing a coordinate rotation, we note that $\mathbf{f}' = \mathsf{U}\mathbf{f}$ and $\mathbf{g}' = \mathsf{U}\mathbf{g}$, so

$$\mathbf{f}' = \mathsf{A}'\mathbf{g}' \quad \longrightarrow \quad \mathsf{U}\mathbf{f} = \mathsf{A}'\mathsf{U}\mathbf{g}\,.$$

Multiplying both sides of the resulting equation on the left by U^{-1}, we reach

$$\mathbf{f} = \mathsf{U}^{-1}\mathsf{A}'\mathsf{U}\mathbf{g}\,.$$

Comparing with the original equation $\mathbf{f} = \mathsf{A}\mathbf{g}$, we see that A must transform in a way such that

$$\mathsf{U}^{-1}\mathsf{A}'\mathsf{U} = \mathsf{A}\,, \quad \text{equivalent to} \quad \mathsf{A}' = \mathsf{U}\mathsf{A}\mathsf{U}^{-1}\,. \tag{5.11}$$

Equation (5.11) is the rule for transforming a matrix when the coordinates are rotated. Notice that U enters the formula twice, in contrast to its single appearance in the formula for transforming a vector. This transformation rule is often referred to as a **similarity transformation**; the name recognizes the similarity between A and A'. Both A and A' must contain the same physics; we shall shortly see what these matrices have in common.

Incidentally, an instructive alternate way of obtaining Eq. (5.11) is simply to insert unity (in the form $\mathsf{U}^{-1}\mathsf{U}$) into the equation $\mathbf{f} = \mathsf{A}\mathbf{g}$ and then multiply on the left by U. Thus,

$$\mathbf{f} = \mathsf{A}\mathsf{U}^{-1}\mathsf{U}\mathbf{g} \quad \longrightarrow \quad \mathsf{U}\mathbf{f} = \mathsf{U}\mathsf{A}\mathsf{U}^{-1}(\mathsf{U}\mathbf{g})\,.$$

Finally, note that since we are considering orthogonal transformations, $\mathsf{U}^{-1} = \mathsf{U}^{\mathrm{T}}$, so we can also write

$$\mathsf{A}' = \mathsf{U}\mathsf{A}\mathsf{U}^{\mathrm{T}}\,. \tag{5.12}$$

Example 5.3.1. Transforming a Matrix Equation

Consider the matrix equation $\mathbf{g} = \mathsf{A}\mathbf{f}$, with

$$\mathsf{A} = \begin{pmatrix} 1 & 2 & 3 \\ 2 & 6 & 0 \\ 3 & 0 & 4 \end{pmatrix}, \quad \mathbf{f} = \begin{pmatrix} 2 \\ -1 \\ 1 \end{pmatrix}, \quad \mathbf{g} = \begin{pmatrix} 1 & 2 & 3 \\ 2 & 6 & 0 \\ 3 & 0 & 4 \end{pmatrix} \begin{pmatrix} 2 \\ -1 \\ 1 \end{pmatrix} = \begin{pmatrix} 3 \\ -2 \\ 10 \end{pmatrix}.$$

Applying the matrix transformation $\mathsf{U} = \begin{pmatrix} 0.408248 & 0.816497 & 0.408248 \\ -0.707107 & 0 & 0.707107 \\ 0.577350 & -0.577350 & 0.577350 \end{pmatrix}$,

$$\mathsf{A}' = \mathsf{UAU}^T = \begin{pmatrix} 7.16667 & -0.28868 & -0.23570 \\ -0.28868 & -0.50000 & -2.04124 \\ -0.23570 & -2.04124 & 4.33333 \end{pmatrix},$$

$$\mathbf{f}' = \mathsf{U}\mathbf{f} = \begin{pmatrix} 0.40825 \\ -0.70711 \\ -2.30940 \end{pmatrix}, \qquad \mathbf{g}' = \mathsf{U}\mathbf{g} = \begin{pmatrix} 3.67423 \\ 4.94975 \\ -8.66025 \end{pmatrix}.$$

We may now check that $\mathsf{A}'\mathbf{f}' = \mathbf{g}'$:

$$\mathsf{A}'\mathbf{f}' = \begin{pmatrix} 7.16667 & -0.28868 & -0.23570 \\ -0.28868 & -0.50000 & -2.04124 \\ -0.23570 & -2.04124 & 4.33333 \end{pmatrix} \begin{pmatrix} 0.40825 \\ -0.70711 \\ -2.30940 \end{pmatrix} = \begin{pmatrix} 3.76422 \\ 4.94974 \\ -8.66024 \end{pmatrix}.$$

You may want to verify the above using your symbolic computing system.

Note that if we had mistakenly computed $\mathsf{A}' = \mathsf{U}^T\mathsf{AU}$, we would not have obtained consistent results. It is necessary to use the correct definitions for similarity transformations.

■

Exercises

5.3.1. Using $\mathsf{H} = \begin{pmatrix} 1 & 2 \\ 2 & 4 \end{pmatrix}$, $\mathsf{U} = \begin{pmatrix} 0.6 & -0.8 \\ 0.8 & 0.6 \end{pmatrix}$, and $\mathbf{x} = \begin{pmatrix} 3 \\ -1 \end{pmatrix}$,

(a) Verify that U describes a rotation.

(b) Find the vector $\mathbf{y} = \mathsf{H}\mathbf{x}$.

(c) Transform the equation of part (b) by appropriate application of U. Call the results $\mathbf{x}'$, $\mathbf{y}'$, and H'.

(d) Verify that $\mathbf{y}' = \mathsf{H}'\mathbf{x}'$ and that $|\mathbf{x}'| = |\mathbf{x}|$ and $|\mathbf{y}'| = |\mathbf{y}|$.

(e) Verify that the angle between $\mathbf{x}$ and $\mathbf{y}$ is the same as that between $\mathbf{x}'$ and $\mathbf{y}'$.

Note. The results of parts (d) and (e) are consistent with the notion that the original and transformed equations describe the same phenomenon in different coordinate systems.

5.3.2. Repeat part (b) of Exercise 5.3.1 for the same H and U but with $\mathbf{x} = \begin{pmatrix} 2 \\ -1 \end{pmatrix}$.

Can you predict the value of $\mathbf{y}'$ before transforming with U?

5.4 GRAM-SCHMIDT ORTHOGONALIZATION

Early in this chapter we considered 2-D coordinate systems, assuming that before we applied a rotational transformation they corresponded to a basis that consisted of the unit vectors $\hat{\mathbf{e}}_x = \begin{pmatrix} 1 \\ 0 \end{pmatrix}$ and $\hat{\mathbf{e}}_y = \begin{pmatrix} 0 \\ 1 \end{pmatrix}$. After rotating the axes, the new basis vectors (which are mutually orthogonal) are given in terms of the original coordinates as the rows of the transformation matrix U. In our 2-D example we derived expressions for the new basis vectors by analyzing the trigonometry of a specified rotation. However, the analysis we carried out would have been equally valid if we had chosen a new orthogonal basis in any other way and then used it to construct the transformation matrix U.

There are a number of circumstances (some of which we will encounter later in this chapter) in which we will want to generate an orthogonal set of unit vectors that do not necessarily correspond to the directions of our original coordinate system. Such a set is referred to as **orthonormal**, meaning orthogonal and normalized (i.e., of unit length). A typical situation of this sort arises when we are given two or more vectors that are not orthonormal and which we would like to make orthonormal so that they can be part of an orthogonal unit-vector basis. There are a number of ways to accomplish this; perhaps the orthonormalization procedure that is conceptually simplest is the **Gram-Schmidt** process.

We start with a set of vectors $\mathbf{a}_i$ that are not orthogonal and are not necessarily of unit length; our mission to take linear combinations of the $\mathbf{a}_i$ in such a way that the vectors thereby produced are mutually orthogonal unit vectors. Since we want a method that will work for a space with an arbitrary number of dimensions, we do not want to use any properties that are unique to two- or three-dimensional space.

We start by taking the first vector from our input set, $\mathbf{a}_1$. If it is not of unit length, we normalize it by dividing it by $\sqrt{\mathbf{a}_1^T\mathbf{a}_1}$, designating the result as our first orthonormal vector $\hat{\mathbf{b}}_1$:

$$\hat{\mathbf{b}}_1 = \frac{\mathbf{a}_1}{\sqrt{\mathbf{a}_1^T\mathbf{a}_1}} . \tag{5.13}$$

We now obtain a vector that is orthogonal to $\hat{\mathbf{b}}_1$ by taking $\mathbf{a}_2$ and subtracting from it its projection on $\hat{\mathbf{b}}_1$. Since we are assuming our set of $\mathbf{a}_i$ are linearly independent, there will be a contribution from $\mathbf{a}_2$ remaining after we remove its $\hat{\mathbf{b}}_1$ projection; this projection, given by Eq. (4.23), is $(\hat{\mathbf{b}}_1^T\mathbf{a}_2)\hat{\mathbf{b}}_1$. Therefore a vector orthogonal to $\hat{\mathbf{b}}_1$ is

$$\mathbf{b}_2 = \mathbf{a}_2 - (\hat{\mathbf{b}}_1^T\mathbf{a}_2)\hat{\mathbf{b}}_1 ,$$

which after normalization is

$$\hat{\mathbf{b}}_2 = \frac{\mathbf{b}_2}{\sqrt{\mathbf{b}_2^T\mathbf{b}_2}} . \tag{5.14}$$

A third orthonormal vector can now be obtained, starting from $\mathbf{a}_3$, by removing its projections on $\hat{\mathbf{b}}_1$ and $\hat{\mathbf{b}}_2$,

$$\mathbf{b}_3 = \mathbf{a}_3 - (\hat{\mathbf{b}}_1^T\mathbf{a}_3)\hat{\mathbf{b}}_1 - (\hat{\mathbf{b}}_2^T\mathbf{a}_3)\hat{\mathbf{b}}_2 ,$$

and then normalizing:

$$\hat{\mathbf{b}}_3 = \frac{\mathbf{b}_3}{\sqrt{\mathbf{b}_3^T\mathbf{b}_3}} . \tag{5.15}$$

The generalization to the nth orthonormal vector is

$$\mathbf{b}_n = \mathbf{a}_n - \sum_{i=1}^{n-1}(\hat{\mathbf{b}}_i^T\mathbf{a}_n)\hat{\mathbf{b}}_i , \qquad \hat{\mathbf{b}}_n = \frac{\mathbf{b}_n}{\sqrt{\mathbf{b}_n^T\mathbf{b}_n}} . \tag{5.16}$$

Notice that because the process is sequential, the result obtained will depend upon the order in which the vectors $\mathbf{a}_i$ are arranged. For that reason the Gram-Schmidt process is also called **serial orthonormalization**.

Example 5.4.1. Vector Orthonormalization

In 3-D space, we are given the vectors

$$\mathbf{a}_1 = \begin{pmatrix} 2 \\ 1 \\ 3 \end{pmatrix}, \qquad \mathbf{a}_2 = \begin{pmatrix} 3 \\ 1 \\ 2 \end{pmatrix}, \qquad \mathbf{a}_3 = \begin{pmatrix} 1 \\ -1 \\ 1 \end{pmatrix}.$$

Applying the Gram-Schmidt process, we wish to use the $\mathbf{a}_i$ to form a set of three orthonormal basis vectors.

Our first basis vector $\hat{\mathbf{b}}_1$ is a normalized version of $\mathbf{a}_1$:

$$\hat{\mathbf{b}}_1 = \frac{\mathbf{a}_1}{\sqrt{\mathbf{a}_1^T \mathbf{a}_1}} = \frac{\mathbf{a}_1}{\sqrt{14}} = \frac{1}{\sqrt{14}} \begin{pmatrix} 2 \\ 1 \\ 3 \end{pmatrix}.$$

Noting that $\hat{\mathbf{b}}_1^T \mathbf{a}_2 = 13/\sqrt{14}$, we compute the second basis vector, before normalization, as

$$\mathbf{b}_2 = \mathbf{a}_2 - (\hat{\mathbf{b}}_1^T \mathbf{a}_2)\hat{\mathbf{b}}_1 = \begin{pmatrix} 3 \\ 1 \\ 2 \end{pmatrix} - \frac{13}{\sqrt{14}} \frac{1}{\sqrt{14}} \begin{pmatrix} 2 \\ 1 \\ 3 \end{pmatrix} = \frac{1}{14} \begin{pmatrix} 16 \\ 1 \\ -11 \end{pmatrix}.$$

The normalized version of $\mathbf{b}_2$ is

$$\hat{\mathbf{b}}_2 = \frac{1}{\sqrt{378}} \begin{pmatrix} 16 \\ 1 \\ -11 \end{pmatrix}.$$

Continuing to $\mathbf{b}_3$, we need $\hat{\mathbf{b}}_1^T \mathbf{a}_3 = 4/\sqrt{14}$ and $\hat{\mathbf{b}}_2^T \mathbf{a}_3 = 4/\sqrt{378}$, so

$$\mathbf{b}_3 = \begin{pmatrix} 1 \\ -1 \\ 1 \end{pmatrix} - \frac{4}{\sqrt{14}} \frac{1}{\sqrt{14}} \begin{pmatrix} 2 \\ 1 \\ 3 \end{pmatrix} - \frac{4}{\sqrt{378}} \sqrt{\frac{1}{378}} \begin{pmatrix} 16 \\ 1 \\ -11 \end{pmatrix} = \frac{7}{27} \begin{pmatrix} 1 \\ -5 \\ 1 \end{pmatrix}.$$

Normalizing $\mathbf{b}_3$,

$$\hat{\mathbf{b}}_3 = \frac{1}{\sqrt{27}} \begin{pmatrix} 1 \\ -5 \\ 1 \end{pmatrix}.$$

The complexity of the computations increases as we proceed. That observation suggests that this is a place where symbolic computation might be rather valuable. ■

Exercises

5.4.1. Check that the vectors $\hat{\mathbf{b}}_i$ of Example 5.4.1 are orthonormal.

SYMBOLIC COMPUTATION

The tasks that we now want to perform with symbolic computation involve input and output sets of vectors, and in order to work effectively with such sets it is useful to understand the way our symbolic systems organize data and provide for access to the individual data items. Information about the data structures used in MAPLE and MATHEMATICA is presented in Appendix C; the reader should consult that appendix if the examples given here do not seem sufficiently clear or if it is desired to manipulate data arrays in ways not explained in this subsection.

In MAPLE, an ordered set of vectors $\mathbf{A}$ with members $\mathbf{a}_1, \cdots, \mathbf{a}_n$ to be orthonormalized by the Gram-Schmidt process can be represented as `A:=[a1,···,an]`, where `a1` etc. are MAPLE vectors (i.e., formed using the `Vector` command). Notice that in MAPLE an ordered set of vectors is **not** the same as a matrix. Then the command

```
> B := GramSchmidt(A,normalized);
```

produces a serially orthonormalized set of vectors $\mathbf{B}$ as the ordered set `B:=[b1,···,bn]`. If "`,normalized`" is omitted, the output set of vectors will be at arbitrary individual scales. The vectors to which the Gram-Schmidt process is applied must be of the same dimension and orientation, and the orthonormalized vectors will be in the same orientation as the input set.

In MATHEMATICA, an ordered set of vectors $\mathbf{A}$ takes the form `A={a1,···,an}`, where `a1` etc. are vectors all of the same dimension, i.e., consisting of components enclosed within braces, with each pair of braces enclosing the same number of elements. Note that a set of vectors in MATHEMATICA is identical to a matrix with those vectors as rows. Then the command

```
Orthogonalize[A]
```

produces from its input the Gram-Schmidt orthonormalization as a set of output vectors (or equivalently a matrix with the orthonormal vectors as its rows).

It is, of course, possible to invoke the Gram-Schmidt process for a set of input vectors that turns out to be linearly dependent. This will certainly be the case if the number of input vectors exceeds the dimension of the vector space, and may be the case under other circumstances. In MAPLE, the output of `GramSchmidt` will consist of only the orthogonal (and therefore surely linearly independent) vectors that can be constructed; in MATHEMATICA, `Orthogonalize` places a zero vector into the output whenever the computation reaches an input vector that is linearly dependent upon those that precede it in the input; this action permits the user to identify the source of the linear dependence.

Example 5.4.2. Symbolic Orthonormalization

Consider again the construction of a set of orthonormal vectors from the vectors $\mathbf{a}_1$, $\mathbf{a}_2$, $\mathbf{a}_3$ that were introduced in Example 5.4.1. Using MAPLE, we need to start by defining the $\mathbf{a}_i$ and assembling them into a list of vectors:

```
> a1 := Vector([2,1,3]):
> a2 := Vector([3,1,2]):
> a3 := Vector([1,-1,1]):
> A := [a1, a2, a3];
```

$$A := \left[\begin{bmatrix} 2 \\ 1 \\ 3 \end{bmatrix}, \begin{bmatrix} 3 \\ 1 \\ 2 \end{bmatrix}, \begin{bmatrix} 1 \\ -1 \\ 1 \end{bmatrix} \right]$$

Then we call `GramSchmidt`:

```
B := GramSchmidt(A,normalized);
```

$$B := \left[\left[\begin{array}{c} \frac{\sqrt{14}}{7} \\ \frac{\sqrt{14}}{14} \\ \frac{3\sqrt{14}}{14} \end{array}\right], \left[\begin{array}{c} \frac{8\sqrt{42}}{63} \\ \frac{\sqrt{42}}{126} \\ -\frac{11\sqrt{42}}{126} \end{array}\right], \left[\begin{array}{c} \frac{\sqrt{3}}{9} \\ -\frac{5\sqrt{3}}{9} \\ \frac{\sqrt{3}}{9} \end{array}\right]\right].$$

Note that B is not in a form permitting its use as a transformation matrix. However, to assemble these vectors as the columns of a transformation matrix, we execute the command

```
U = Matrix(B);
```

$$U := \left[\begin{array}{ccc} \frac{\sqrt{14}}{7} & \frac{8\sqrt{42}}{63} & \frac{\sqrt{3}}{9} \\ \frac{\sqrt{14}}{14} & \frac{\sqrt{42}}{126} & -\frac{5\sqrt{3}}{9} \\ \frac{3\sqrt{14}}{14} & -\frac{11\sqrt{42}}{126} & \frac{\sqrt{3}}{9} \end{array}\right].$$

Note that if it were our intention to transform vectors or matrices from the basis $(\hat{\mathbf{e}}_x, \hat{\mathbf{e}}_y, \hat{\mathbf{e}}_z)$ to the basis we have just found, we would need the transpose of U, so as to place the components of the new basis (in terms of the old) as **rows** of the transformation matrix.

In MATHEMATICA, the code for this Gram-Schmidt orthonormalization is

```
A = { {2,1,3}, {3,1,2}, {1,-1,1} };

B = Orthogonalize[A]; MatrixForm[B]
```

$$\left(\begin{array}{ccc} \sqrt{\frac{2}{7}} & \frac{1}{\sqrt{14}} & \frac{3}{\sqrt{14}} \\ \frac{8\sqrt{2/21}}{3} & \frac{1}{3\sqrt{42}} & -\frac{11}{3\sqrt{42}} \\ \frac{1}{3\sqrt{3}} & -\frac{5}{3\sqrt{3}} & \frac{1}{3\sqrt{3}} \end{array}\right).$$

We see that MATHEMATICA produces each orthonormal vector as a **row** of B. The output is ready for use as a transformation matrix.

■

Exercises

5.4.2. Verify that the transformation matrix found in Example 5.4.2 is orthogonal.

5.5 MATRIX EIGENVALUE PROBLEMS

In many problems of importance for physics, a matrix equation can be brought to a simpler form by applying a suitable rotation to the coordinate system. For example, when an electric field is applied to an anisotropic object, there will be certain directions in which the polarization and the applied field are collinear. A coordinate rotation that aligns the coordinate axes along these directions separates this three-dimensional problem into three one-dimensional problems. It is also found that there are directions in which the angular velocity and angular momentum of a rigid body become collinear; the use of coordinates aligned in these directions also permits meaningful simplification. These features are typical of a large class of problems involving matrix equations; these problems are (for reasons that will soon become apparent) known as **eigenvalue problems**.

A PRELIMINARY EXAMPLE

We start our study of eigenvalue problems by considering the following situation. A particle slides frictionlessly in an ellipsoidal basin (see Fig. 5.2). If we release the particle (initially at rest) at an arbitrary point in the basin, it will start to move downhill in the (negative) gradient direction, which in general will not aim directly at the potential minimum at the bottom of the basin. The particle's overall trajectory will then be a complicated path, as sketched in the right-hand panel of Fig. 5.2. Our objective is to find the positions, if any, from which the trajectories will aim at the potential minimum, and will therefore represent simple one-dimensional oscillatory motion.

To make the problem more specific, we take the potential to be of the form

$$V(x, y) = ax^2 + bxy + cy^2, \tag{5.17}$$

with parameters a, b, c in ranges that describe an ellipsoidal basin with a minimum at $x = y = 0$. Then the forces on our particle when at (x, y) will be

$$F_x = -\frac{\partial V}{\partial x} = -2ax - by\,, \qquad F_y = -\frac{\partial V}{\partial y} = -bx - 2cy\,.$$

For most values of x and y $F_x/F_y \neq x/y$, so the force will not be directed toward the minimum at $x = y = 0$.

Our first step toward analyzing this problem is to write the equations for the force components as a matrix equation,

$$\begin{pmatrix} F_x \\ F_y \end{pmatrix} = \begin{pmatrix} -2a & -b \\ -b & -2c \end{pmatrix} \begin{pmatrix} x \\ y \end{pmatrix}, \quad \text{or} \quad \mathbf{f} = \mathsf{H}\mathbf{r}\,,$$

where $\mathbf{f}$, H, and $\mathbf{r}$ are defined as indicated. Now the condition $F_x/F_y = x/y$ is equivalent to the statement that $\mathbf{f}$ and $\mathbf{r}$ are proportional, corresponding to the matrix equation

$$\mathsf{H}\mathbf{r} = \lambda\mathbf{r}\,. \tag{5.18}$$

Here H is known but both λ and $\mathbf{r}$ are to be determined. Equations of this type are known as **eigenvalue equations** because **eigen** is German for "[its] own." The

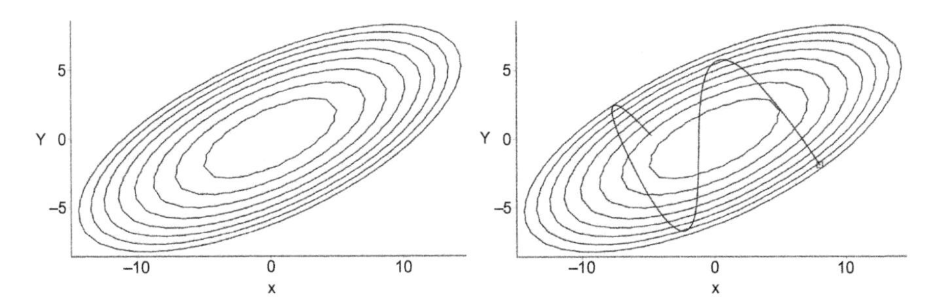

Figure 5.2: Left: Contour lines of basin potential $V = x^2 - \sqrt{5}\,xy + 3y^2$. Right: Trajectory of sliding particle of unit mass starting from rest at $(8.0, -1.92)$.

column vectors $\mathbf{r}$ that are solutions to Eq. (5.18) are called **eigenvectors**, and the associated values of λ are called **eigenvalues**. An interesting feature of eigenvalue equations is that solutions typically exist only for specific eigenvalues.

Equation (5.18) is a homogeneous linear equation system, as becomes more obvious if written

$$(\mathsf{H} - \lambda \mathbf{1})\mathbf{r} = 0 \,, \tag{5.19}$$

and we know from our analysis in Section 4.8 that it will have the unique solution $\mathbf{r} = 0$ unless $\det(\mathsf{H} - \lambda\mathbf{1}) = 0$. However, the value of λ is at our disposal, so we can search for values of λ that cause this determinant to vanish. Proceeding symbolically, we look for λ such that

$$\det(\mathsf{H} - \lambda\mathbf{1}) = \begin{vmatrix} h_{11} - \lambda & h_{12} \\ h_{21} & h_{22} - \lambda \end{vmatrix} = 0.$$

Expanding the determinant, which is sometimes called a **secular determinant** (the name arising from early applications in celestial mechanics), we have an algebraic equation, the **secular equation**,

$$(h_{11} - \lambda)(h_{22} - \lambda) - h_{12}h_{21} = 0 \,, \tag{5.20}$$

which can be solved for λ. The left-hand side of Eq. (5.20) is also called the **characteristic polynomial** (in λ) of H, and Eq. (5.20) is for that reason also known as the **characteristic equation** of H.

Equation (5.20) is of second degree in λ, and it will have two roots. For each value of λ which solves Eq. (5.20), we can return to the homogeneous equation system, Eq. (5.19), and solve it for the vector $\mathbf{r}$. This can be repeated for all λ that are solutions to the secular equation, thereby giving a set of eigenvalues and the associated eigenvectors. These eigenvectors define the coordinate directions that simplify our problem.

The outline for the solution of matrix eigenvalue problems is now complete, but a good operational understanding of the process requires that we look at a number of specific examples, which we now proceed to do.

Example 5.5.1. Ellipsoidal Basin

We return to the example involving the potential given in Eq. (5.17), with $a = 1$, $b = -\sqrt{5}$, $c = 3$. Then our matrix H has the form

$$\mathsf{H} = \begin{pmatrix} -2 & \sqrt{5} \\ \sqrt{5} & -6 \end{pmatrix},$$

and the secular equation is

$$\det(\mathsf{H} - \lambda\mathbf{1}) = \begin{vmatrix} -2 - \lambda & \sqrt{5} \\ \sqrt{5} & -6 - \lambda \end{vmatrix} = \lambda^2 + 8\lambda + 7 = 0 \,.$$

Because $\lambda^2 + 8\lambda + 7 = (\lambda + 1)(\lambda + 7)$, we see that the secular equation has as solutions the eigenvalues $\lambda = -1$ and $\lambda = -7$.

To get the eigenvector corresponding to $\lambda = -1$, we return to Eq. (5.19), which, written in great detail, is

$$(\mathsf{H} - \lambda \mathbf{1})\mathbf{r} = \begin{pmatrix} -2-(-1) & \sqrt{5} \\ \sqrt{5} & -6-(-1) \end{pmatrix} \begin{pmatrix} x \\ y \end{pmatrix} = \begin{pmatrix} -1 & \sqrt{5} \\ \sqrt{5} & -5 \end{pmatrix} \begin{pmatrix} x \\ y \end{pmatrix} = 0 \, ,$$

which expands into a linearly dependent pair of equations:

$$-x + \sqrt{5}\, y = 0 \, ,$$

$$\sqrt{5}\, x - 5y = 0 \, .$$

This is, of course, the intention associated with the secular equation, because if these equations were linearly independent they would inexorably lead to the solution $x = y = 0$. Instead, from either equation, we have $x = \sqrt{5}\, y$, so we have the eigenvalue/eigenvector pair

$$\lambda_1 = -1 \, , \qquad \mathbf{r}_1 = C \begin{pmatrix} \sqrt{5} \\ 1 \end{pmatrix} \, ,$$

where C is a constant that can assume any value. Thus, there is an infinite number of x, y pairs that define a **direction** in the 2-D space, with the magnitude of the displacement in that direction arbitrary. The arbitrariness of scale is a natural consequence of the fact that the equation system was homogeneous; any multiple of a solution of a linear homogeneous equation set will also be a solution. This eigenvector corresponds to trajectories that start from the particle at rest anywhere on the line defined by $\mathbf{r}_1$. A trajectory of this sort is illustrated in the left panel of Fig. 5.3.

We have not yet considered the possibility $\lambda = -7$. This leads to a different eigenvector, obtained by solving

$$(\mathsf{H} - \lambda \mathbf{1})\mathbf{r} = \begin{pmatrix} -2+7 & \sqrt{5} \\ \sqrt{5} & -6+7 \end{pmatrix} \begin{pmatrix} x \\ y \end{pmatrix} = \begin{pmatrix} 5 & \sqrt{5} \\ \sqrt{5} & 1 \end{pmatrix} \begin{pmatrix} x \\ y \end{pmatrix} = 0 \, ,$$

corresponding to $y = -x\sqrt{5}$. This defines the eigenvalue/eigenvector pair

$$\lambda_2 = -7 \, , \qquad \mathbf{r}_2 = C' \begin{pmatrix} -1 \\ \sqrt{5} \end{pmatrix} \, .$$

A trajectory of this sort is shown in the right panel of Fig. 5.3.

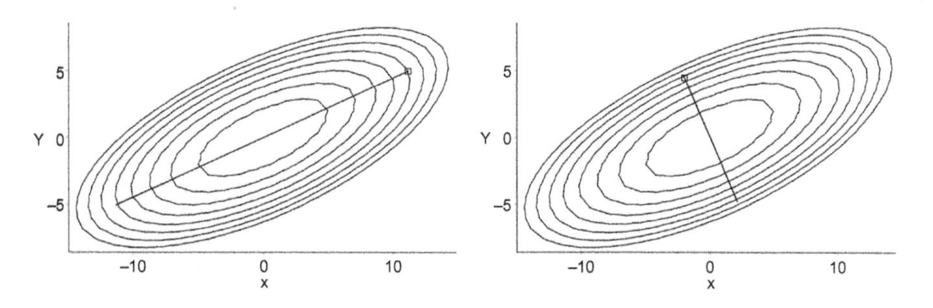

Figure 5.3: Trajectories starting at rest, (left) at a point on the line $x = y\sqrt{5}$; (right) at a point on the line $y = -x\sqrt{5}$.

We thus have two directions in which the force is directed toward the minimum, and they are mutually perpendicular: the first direction has $dy/dx = 1/\sqrt{5}$; for the second, $dy/dx = -\sqrt{5}$.

We can easily check our eigenvectors and eigenvalues. For λ_1 and $\mathbf{r}_1$,

$$\mathsf{H}\mathbf{r}_1 = \begin{pmatrix} -2 & \sqrt{5} \\ \sqrt{5} & -6 \end{pmatrix} \begin{pmatrix} C\sqrt{5} \\ C \end{pmatrix} = C \begin{pmatrix} -\sqrt{5} \\ -1 \end{pmatrix} = (-1) \begin{pmatrix} C\sqrt{5} \\ C \end{pmatrix} = \lambda_1 \mathbf{r}_1 \, .$$

Because our equation system is homogeneous, its solutions are arbitrary as to scale. However, it is often useful to **normalize** eigenvectors, which we can do by choosing the constant (C or C') to make $\mathbf{r}$ of magnitude unity. In the present example,

$$\mathbf{r}_1 = \begin{pmatrix} \sqrt{5/6} \\ \sqrt{1/6} \end{pmatrix} , \qquad \mathbf{r}_2 = \begin{pmatrix} -\sqrt{1/6} \\ \sqrt{5/6} \end{pmatrix} . \tag{5.21}$$

Each of these normalized eigenvectors is still arbitrary as to overall sign.

Before leaving this example, we make three further observations: (1) the number of eigenvalues was equal to the dimension of the matrix H. This is a consequence of the **fundamental theorem of algebra**, namely that an equation of degree n will have n roots; (2) although the secular equation was of degree 2 and quadratic equations can have complex roots, our eigenvalues were real; and (3) our two eigenvectors are orthogonal.

■

Our current example is easily understood physically. The directions in which the displacement and the force are collinear are the symmetry directions of the elliptical potential field, and they are associated with different eigenvalues (the proportionality constant between position and force) because the ellipse has axes of different lengths. We have, as expected, identified the **principal axes** of our basin. With the parameters of Example 5.5.1, the potential could have been written (using the normalized eigenvectors)

$$V = \frac{1}{2}\left(\frac{\sqrt{5}\,x + y}{\sqrt{6}}\right)^2 + \frac{7}{2}\left(\frac{x - \sqrt{5}\,y}{\sqrt{6}}\right)^2 = \tfrac{1}{2}(x')^2 + \tfrac{7}{2}(y')^2 \, ,$$

which shows that V divides into two quadratic terms, with each term involving a new coordinate that is proportional to one of our eigenvectors. The new coordinates are related to the original x, y by a rotation (or rotation/reflection) with orthogonal transformation U:

$$\mathsf{U}\mathbf{r} = \begin{pmatrix} \sqrt{5/6} & \sqrt{1/6} \\ \sqrt{1/6} & -\sqrt{5/6} \end{pmatrix} \begin{pmatrix} x \\ y \end{pmatrix} = \begin{pmatrix} (\sqrt{5}\,x + y)/\sqrt{6} \\ (x - \sqrt{5}\,y)/\sqrt{6} \end{pmatrix} = \begin{pmatrix} x' \\ y' \end{pmatrix} .$$

Finally, we note that when we calculate the force in the primed coordinate system, we get

$$F_{x'} = -x' \, , \qquad F_{y'} = -7y' \, ,$$

corresponding to the eigenvalues we found.

ANOTHER EIGENPROBLEM

Example 5.5.1 is not complicated enough to provide a full illustration of the matrix eigenvalue problem. Consider next the following example:

Example 5.5.2. Block-Diagonal Matrix

Find the eigenvalues and eigenvectors of

$$\mathsf{H} = \begin{pmatrix} 0 & 1 & 0 \\ 1 & 0 & 0 \\ 0 & 0 & 2 \end{pmatrix} . \tag{5.22}$$

Writing the secular equation and expanding in minors using the third row, we have

$$\begin{vmatrix} -\lambda & 1 & 0 \\ 1 & -\lambda & 0 \\ 0 & 0 & 2-\lambda \end{vmatrix} = (2-\lambda) \begin{vmatrix} -\lambda & 1 \\ 1 & -\lambda \end{vmatrix} = (2-\lambda)(\lambda^2 - 1) = 0 . \tag{5.23}$$

We see that the eigenvalues are 2, +1, and −1.

To obtain the eigenvector corresponding to $\lambda = 2$, we examine the equation set $[\mathsf{H} - 2(\mathbf{1})]\,\mathbf{x} = 0$. Letting $\mathbf{x}$ be a vector with components c_1, c_2, c_3, we have

$$\begin{aligned} -2c_1 + c_2 &= 0 , \\ c_1 - 2c_2 &= 0 , \\ 0 &= 0 . \end{aligned}$$

The first two equations of this set lead to $c_1 = c_2 = 0$. The third obviously conveys no information, and we are led to the conclusion that c_3 is arbitrary. Thus, at this point our eigenvalue and eigenvector, which we will designate λ_1 and $\mathbf{x}_1$, are

$$\lambda_1 = 2 , \qquad \mathbf{x}_1 = \begin{pmatrix} 0 \\ 0 \\ C \end{pmatrix} . \tag{5.24}$$

Taking next $\lambda = +1$, the coefficients satisfy

$$\begin{aligned} -c_1 + c_2 &= 0 , \\ c_1 - c_2 &= 0 , \\ c_3 &= 0 . \end{aligned}$$

We clearly have $c_1 = c_2$ and $c_3 = 0$, so our next eigenvalue and eigenvector are

$$\lambda_2 = +1 , \qquad \mathbf{x}_2 = \begin{pmatrix} C \\ C \\ 0 \end{pmatrix} . \tag{5.25}$$

Similar operations for $\lambda = -1$ yield

$$\lambda_3 = -1 , \qquad \mathbf{x}_3 = \begin{pmatrix} C \\ -C \\ 0 \end{pmatrix} . \tag{5.26}$$

Collecting our results, and normalizing the eigenvectors (often useful, but not always necessary), we have

$$\lambda_1 = 2, \quad \mathbf{x}_1 = \begin{pmatrix} 0 \\ 0 \\ 1 \end{pmatrix}, \quad \lambda_2 = 1, \quad \mathbf{x}_2 = \begin{pmatrix} 2^{-1/2} \\ 2^{-1/2} \\ 0 \end{pmatrix}, \quad \lambda_3 = -1, \quad \mathbf{x}_3 = \begin{pmatrix} 2^{-1/2} \\ -2^{-1/2} \\ 0 \end{pmatrix}.$$

Notice that because H was block-diagonal, with an upper-left 2×2 block and a lower-right 1×1 block, the secular equation then separated into a product of the determinants for the two blocks, and its solutions corresponded to those of an individual block, with coefficients of value zero for the other block(s). Thus, $\lambda = 2$ was a solution for the 1×1 block in row/column 3, and its eigenvector involved only the coefficient c_3. The λ values ± 1 came from the 2×2 block in rows/columns 1 and 2, with eigenvectors involving only coefficients c_1 and c_2.

■

In the case of a 1×1 block in row/column i, we saw, for $i = 3$ in Example 5.5.2, that its only element was the eigenvalue, and that the corresponding eigenvector is proportional to $\hat{\mathbf{e}}_i$ (a unit vector whose only nonzero element is $c_i = 1$). A generalization of this observation is that if a matrix H is diagonal, its diagonal elements h_{ii} will be the eigenvalues λ_i, and that the eigenvectors $\mathbf{c}_i$ will be the unit vectors $\hat{\mathbf{e}}_i$.

DEGENERACY

If the secular equation has a multiple root, the eigensystem is said to be **degenerate** or to exhibit **degeneracy**. Here is an example.

Example 5.5.3. Degenerate Eigenproblem

Let's find the eigenvalues and eigenvectors of

$$\mathsf{H} = \begin{pmatrix} 0 & 0 & 1 \\ 0 & 1 & 0 \\ 1 & 0 & 0 \end{pmatrix}. \tag{5.27}$$

The secular equation for this problem is

$$\begin{vmatrix} -\lambda & 0 & 1 \\ 0 & 1-\lambda & 0 \\ 1 & 0 & -\lambda \end{vmatrix} = \lambda^2(1-\lambda) - (1-\lambda) = (\lambda^2 - 1)(1-\lambda) = 0 \tag{5.28}$$

with the three roots $+1$, $+1$, and -1. Let's consider first $\lambda = -1$. Then we have

$$\begin{aligned} c_1 + c_3 &= 0\,, \\ 2c_2 &= 0\,, \\ c_1 + c_3 &= 0\,. \end{aligned}$$

Thus,

$$\lambda_1 = -1, \qquad \mathbf{x}_1 = C \begin{pmatrix} 1 \\ 0 \\ -1 \end{pmatrix}. \tag{5.29}$$

For the double root $\lambda = +1$,

$$\begin{aligned} -c_1 + c_3 &= 0\,, \\ 0 &= 0\,, \\ c_1 - c_3 &= 0\,. \end{aligned}$$

Notice that of the three equations, only one is now linearly independent; the double root signals **two** linear dependencies, and we have solutions for **any** values of c_1 and c_2, with only the condition that $c_3 = c_1$. The eigenvectors for $\lambda = +1$ thus span a two-dimensional **subspace**, in contrast to the trivial one-dimensional subspace characteristic of nondegenerate solutions. The general form for these eigenvectors is

$$\lambda = +1, \qquad \mathbf{x} = \begin{pmatrix} C \\ C' \\ C \end{pmatrix}. \tag{5.30}$$

It is convenient to describe the degenerate eigenspace for $\lambda = 1$ by identifying two mutually orthogonal vectors that span it. We can pick the first vector by choosing arbitrary values of C and C' (an obvious choice is to set one of these, say C', to zero). Then, using the Gram-Schmidt process (or in this case by simple inspection), we find a second eigenvector orthogonal to the first. Here, this leads to

$$\lambda_2 = \lambda_3 = +1, \quad \mathbf{x}_2 = C \begin{pmatrix} 1 \\ 0 \\ 1 \end{pmatrix}, \quad \mathbf{x}_3 = C' \begin{pmatrix} 0 \\ 1 \\ 0 \end{pmatrix}. \tag{5.31}$$

Normalizing, our eigenvalues and eigenvectors become

$$\lambda_1 = -1, \ \mathbf{x}_1 = \begin{pmatrix} 2^{-1/2} \\ 0 \\ -2^{-1/2} \end{pmatrix}, \quad \lambda_2 = \lambda_3 = 1, \ \mathbf{x}_2 = \begin{pmatrix} 2^{-1/2} \\ 0 \\ 2^{-1/2} \end{pmatrix}, \quad \mathbf{x}_3 = \begin{pmatrix} 0 \\ 1 \\ 0 \end{pmatrix}.$$

■

SYMBOLIC COMPUTATION: EIGENVALUE PROBLEMS

In contrast to hand solution of eigenvalue problems, which rapidly becomes laborious as the matrix dimension increases, symbolic computation remains convenient even for matrices of significant size. The commands for generating the eigenvalues and eigenvectors of a matrix A take the following forms:

MAPLE: `> Eigenvalues(A);`	Produces eigenvalues of A as a column vector (does not produce eigenvectors)
`> Eigenvectors(A);`	Produces eigenvalues of A as a column vector, followed by a comma and a matrix whose columns are the eigenvectors of A
`> V,M := Eigenvectors(A);`	Assigns eigenvalues to `V` and eigenvectors to `M`.

Those MAPLE commands need to be preceded by the command `with(LinearAlgebra)` to activate the linear algebra package.

MATHEMATICA: `Eigenvalues[A]`	Produces eigenvalues of A as a vector (does not produce eigenvectors)
`Eigenvectors[A]`	Produces eigenvectors of A as rows of a matrix (does not produce eigenvalues)
`Eigensystem[A]`	Produces {eigenvalues,eigenvectors} of A
`{V,M} = Eigensystem[A]`	Assigns eigenvalues to `V` and eigenvectors to `M`.

If a matrix whose eigenvectors is sought is given in decimal form, both languages produce normalized eigenvectors. Degenerate eigenvectors will be normalized and linearly independent but not necessarily orthogonal to each other.

Example 5.5.4. Symbolic Computation, Eigenvalue Problem

Let's obtain the eigenvalues and eigenvectors of $\mathsf{H} = \begin{pmatrix} 1 & 2 & 0 & 0 \\ 2 & 0 & 1 & 2 \\ 0 & 1 & 0 & 0 \\ 0 & 2 & 0 & -1 \end{pmatrix}$.

Using MAPLE first, we define H and access the `LinearAlgebra` package:

```
> H := Matrix([[1,2,0,0],[2,0,1,2],[0,1,0,0],[0,2,0,-1]]):
> with(LinearAlgebra):
```

Let's first just look at the eigenvalues:

```
> E := Eigenvalues(H);
```

$$E := \begin{bmatrix} -\sqrt{3}-\sqrt{2} \\ \sqrt{3}+\sqrt{2} \\ -\sqrt{3}+\sqrt{2} \\ \sqrt{3}-\sqrt{2} \end{bmatrix}$$

We were fortunate in that the result was not extremely complicated. Under most conditions it is advisable to seek a solution in decimal form. So try

```
E := Eigenvalues(evalf(H)):
```

The result is still complicated, because MAPLE's default procedure is to keep 20 significant figures in the eigenvalue computation and to write all the output in complex form, causing each of the eigenvalues, all of which are real, to include a term "+0.*I*." We can simplify the result greatly by setting `Digits` to 6 and using the command `simplify`. Thus, continue with

```
> Digits := 6:
```

```
> simplify(E);
```

$$\begin{bmatrix} 3.14626 \\ -3.14626 \\ 0.317837 \\ -0.317837 \end{bmatrix}$$

Now that we understand how to get convenient results, we see that a more useful MAPLE session for getting both the eigenvalues and the eigenvectors could be the following (with `Digits` set to 6):

```
> with(LinearAlgebra):
> H := Matrix
      ([[1.,2.,0.,0.],[2.,0.,1.,2.],[0.,1.,0.,0.],[0.,2.,0.,-1.]]):
> E,V := Eigenvectors(H):
> EE := simplify(E);
```

$$EE := \begin{bmatrix} 3.14626 \\ -3.14626 \\ 0.317837 \\ -0.317837 \end{bmatrix}$$

```
> VV := simplify(V);
```

$$VV := \begin{bmatrix} 0.627963 & -0.325058 & -0.627963 & -0.325058 \\ 0.673887 & 0.673887 & 0.214186 & 0.214886 \\ 0.214186 & -0.214886 & 0.673887 & -0.673887 \\ 0.325058 & -0.627963 & 0.325058 & 0.627963 \end{bmatrix}$$

The eigenvalue in the ith row of `EE` corresponds to the eigenvector which is the ith column of `VV`. If we want to work with this eigenvalue and eigenvector, they can be accessed (for $i = 2$) as

```
> EE[2];
```

$$-3.14626$$

```
> Column(VV,2);
```

$$\begin{bmatrix} -0.325058 \\ 0.673887 \\ -0.214186 \\ -0.627963 \end{bmatrix}$$

The command `Column` selects one or more columns of its argument. For a fuller explanation, see Appendix C. We can check that we have actually obtained an eigenvalue and eigenvector by applying `H` to the second eigenvector and verifying that the result is this eigenvector multiplied by the eigenvalue:

```
> simplify(H . Column(VV,2));
```

$$\begin{bmatrix} 1.02272 \\ -2.12023 \\ 0.673887 \\ 1.97574 \end{bmatrix}$$

```
> simplify(EE[2] * Column(VV,2));
```

$$\begin{bmatrix} 1.02272 \\ -2.12023 \\ 0.673887 \\ 1.97574 \end{bmatrix}$$

In MATHEMATICA, learning from our experience with MAPLE that we usually want to do eigenvalue problems with the matrix elements in decimal form, we write

```
H = { {1.,2.,0.,0.}, {2.,0.,1.,2.}, {0.,1.,0.,0.}, {0.,2.,0.,-1} };
{E,V} = Eigensystem[H];
```

Viewing the output in matrix form:

`MatrixForm[E]`
$$\begin{pmatrix} 3.14626 \\ -3.14626 \\ 0.317837 \\ -0.317837 \end{pmatrix}$$

`MatrixForm[V]`
$$\begin{pmatrix} -0.627963 & -0.673887 & -0.214186 & -0.325058 \\ 0.325058 & -0.673887 & 0.214186 & 0.627963 \\ 0.627963 & -0.214186 & -0.673887 & -0.325058 \\ 0.325058 & -0.214186 & 0.673887 & -0.627963 \end{pmatrix}$$

The eigenvalues obtained by MATHEMATICA and MAPLE are clearly in agreement. To compare the eigenvectors, note that a MATHEMATICA eigenvector is a **row** of V. Also, remember that any multiple of an eigenvector is still an eigenvector of the same eigenvalue, and in particular an eigenvector remains valid if it is multiplied by -1 (i.e., if its sign is reversed). With those observations, we can verify that the MATHEMATICA and MAPLE eigenvectors are consistent with each other.

Finally, we may want to work further with the MATHEMATICA eigenvectors and eigenvalues. The ith eigenvalue and eigenvector can be accessed from their arrays (illustrated for $i = 2$) by

```
E[[2]]                -3.14626
V[[2]]                {0.325058, -0.673887, 0.214186, 0.627963}
```

Notice that `V[[2]]` selects the entire second row of `V`. For a fuller explanation, see Appendix C. We now check that these results are correct:

```
H . V[[2]]            {-1.02272, 2.12023, -0.673887, -1.97574}
E[[2]] * V[[2]]       {-1.02272, 2.12023, -0.673887, -1.97574}
```

The reader can verify that in both symbolic systems the eigenvectors are normalized.

■

Exercises

First find the eigenvalues and corresponding normalized eigenvectors of the matrices in these exercises by hand computation. Use the Gram-Schmidt process to convert any degenerate sets of eigenvectors to orthonormal form. Then verify all your eigenvalues and eigenvectors (including their normalization and the orthogonality of degenerate eigenvectors) using symbolic computation.

5.5.1. $\begin{pmatrix} 2.23 & 0 & 0 \\ 0 & 1.23 & 1 \\ 0 & 1 & 1.23 \end{pmatrix}$.

5.5.2. $\begin{pmatrix} 3.14 & 1 & 1 \\ 1 & 3.14 & 1 \\ 1 & 1 & 3.14 \end{pmatrix}$.

5.6 HERMITIAN EIGENVALUE PROBLEMS

We did not specifically call attention to the fact that all the eigenvalue problems we have studied involved real symmetric (and therefore also Hermitian) matrices. That observation is important, because the properties we observed (real eigenvalues, orthogonal eigenvectors) are consequences of the fact that our matrix eigenvalue problems were Hermitian. Recall that a matrix is Hermitian if it is equal to its conjugate-transpose (called its adjoint): $\mathsf{H} = \mathsf{H}^\dagger$.

To develop these properties of Hermitian matrices, consider such a matrix H, with $\mathbf{x}_i$ and $\mathbf{x}_j$ two of its eigenvectors with respective eigenvalues λ_i and λ_j. Then,

$$\mathsf{H}\mathbf{x}_i = \lambda_i \mathbf{x}_i \,, \qquad \mathsf{H}\mathbf{x}_j = \lambda_j \mathbf{x}_j \,. \tag{5.32}$$

Multiplying on the left the first of these by $\mathbf{x}_j^\dagger$ and the second by $\mathbf{x}_i^\dagger$,

$$\mathbf{x}_j^\dagger \mathsf{H}\mathbf{x}_i = \lambda_i \mathbf{x}_j^\dagger \mathbf{x}_i \,, \qquad \mathbf{x}_i^\dagger \mathsf{H}\mathbf{x}_j = \lambda_j \mathbf{x}_i^\dagger \mathbf{x}_j \,. \tag{5.33}$$

We next take the adjoint of the second of these equations, noting that $(\mathbf{x}_i^\dagger \mathbf{x}_j)^\dagger = \mathbf{x}_j^\dagger \mathbf{x}_i$, that the adjoint operation requires that we complex conjugate the occurrence of λ_j, and that (because H is Hermitian)

$$(\mathbf{x}_i^\dagger \mathsf{H}\mathbf{x}_j)^\dagger == \mathbf{x}_j^\dagger \mathsf{H}\mathbf{x}_i \,. \tag{5.34}$$

The adjoint operation therefore converts Eqs. (5.33) into

$$\mathbf{x}_j^\dagger \mathsf{H}\mathbf{x}_i = \lambda_i \mathbf{x}_j^\dagger \mathbf{x}_i \,, \qquad \mathbf{x}_j^\dagger \mathsf{H}\mathbf{x}_i = \lambda_j^* \mathbf{x}_j^\dagger \mathbf{x}_i \,. \tag{5.35}$$

Equations (5.35) permit us to obtain two important results: First, if $i = j$, we note that because $\mathbf{x}_i$ is nonzero (it has been assumed to be an eigenvector) the product $\mathbf{x}_i^\dagger \mathbf{x}_i$ is inherently positive, from which we may conclude that $\lambda_i = \lambda_i^*$. In other words, λ_i must be real. Emphasizing,

The eigenvalues of a Hermitian matrix are real.

This result applies even if H is complex so long as it is Hermitian.

Next, if $i \neq j$, combining the two equations of Eq. (5.35), and remembering that the λ_i are real,

$$(\lambda_i - \lambda_j)\mathbf{x}_j^\dagger \mathbf{x}_i = 0 \,, \tag{5.36}$$

so that either $\lambda_i = \lambda_j$ or $\mathbf{x}_j^\dagger \mathbf{x}_i = 0$. This tells us that

Eigenvectors of a Hermitian matrix corresponding to different eigenvalues are orthogonal, where "orthogonal" is generalized for complex vectors $\mathbf{x}_i$ and $\mathbf{x}_j$ to $\mathbf{x}_i^\dagger \mathbf{x}_j$.

Note, however, that if $\lambda_i = \lambda_j$, which will occur if i and j refer to two degenerate eigenvectors, we know nothing about their orthogonality. In fact, in Example 5.5.3 we examined a pair of degenerate eigenvectors, noting that they spanned a two-dimensional subspace and were not required to be orthogonal. However, we also noted in that context that we could **choose** them to be orthogonal. Since the eigenvectors are solutions to a homogeneous equation system, they can also be normalized. Thus,

Every Hermitian matrix possesses a complete orthonormal set of eigenvectors.

Here the word **complete** means that the number of linearly independent eigenvectors is equal to the dimension of the matrix.

Exercises

5.6.1. Show that the eigenvectors found in Example 5.5.4 are orthogonal as well as normalized.

5.7 MATRIX DIAGONALIZATION

Another approach to the Hermitian matrix eigenvalue problem can be developed if we place the orthonormal eigenvectors of a matrix H as columns of a matrix V, with the ith column of V containing the ith orthonormal eigenvector $\mathbf{x}_i$ of H, whose eigenvalue is λ_i. For simplicity, let's assume H and the $\mathbf{x}_i$ to be real, so V is an orthogonal matrix. If we then form HV, the ith column of this matrix product is $\lambda_i\mathbf{x}_i$. Moreover, if we let Λ be a diagonal matrix whose elements Λ_{ii} are the eigenvalues λ_i, we then see that the matrix product $\mathsf{V}\Lambda$ is a matrix whose columns are also $\lambda_i\mathbf{x}_i$. This situation is illustrated schematically as follows:

$$\begin{pmatrix} \mathsf{H} \end{pmatrix}\begin{pmatrix} \mathbf{x}_1 & \cdots & \mathbf{x}_n \end{pmatrix} = \begin{pmatrix} \lambda_1\mathbf{x}_1 & \cdots & \lambda_n\mathbf{x}_n \end{pmatrix} = \begin{pmatrix} \mathbf{x}_1 & \cdots & \mathbf{x}_n \end{pmatrix}\begin{pmatrix} \lambda_1 & 0 & 0 \\ \vdots & \ddots & \vdots \\ 0 & 0 & \lambda_n \end{pmatrix},$$

corresponding to the equation

$$\mathsf{HV} = \mathsf{V}\Lambda\,. \tag{5.37}$$

We now multiply Eq. (5.37) on the left by V^T, obtaining the matrix equation

$$\mathsf{V}^T\mathsf{HV} = \Lambda\,. \tag{5.38}$$

Equation (5.38) has a nice interpretation. That equation has the form of a orthogonal transformation by the matrix V^T. In other words, V is the inverse (and also the transpose) of the matrix U that rotates H into the diagonal matrix Λ. We therefore have the following important result:

> *A real symmetric matrix* H *can be brought to diagonal form by the transformation* $\mathsf{UHU}^T = \Lambda$, *where* U *is an orthogonal matrix; the diagonal matrix* Λ *has the eigenvalues of* H *as its diagonal elements and the columns of* U^T *are the orthonormal eigenvectors of* H, *in the same order as the corresponding eigenvalues in* Λ.

More casually, one says that a real symmetric matrix can be diagonalized by an orthogonal transformation.

The fact that the eigenvectors and eigenvalues of a real symmetric matrix can be found by diagonalizing it suggests that a route to the solution of eigenvalue problems might be to search for (and hopefully find) a diagonalizing orthogonal transformation. We cannot expect to find an explicit and direct matrix diagonalization method, because that would be equivalent to finding an explicit method for solving algebraic equations of arbitrary order, and it is known that no explicit solution exists for such equations of degree larger than 4. However, numerical methods have been developed for approaching diagonalization via successive approximations, and the insights of this section have contributed to those developments. Matrix diagonalization has been one of the most studied problems of applied numerical mathematics, and methods of high efficiency are now widely available for both numerical and symbolic computation. Already as long ago as 1990 researchers had published communications[1] that report

[1] See, for example, J. Olsen, P. Jørgensen, and J. Simons, Passing the one-billion limit in full configuration-interaction calculations, *Chem. Phys. Lett.* **169**: 463 (1990).

the finding of some eigenvalues and eigenvectors of matrices of dimension larger than 10^9. Extrapolating the increase in computer power to the date of publication of this text, an estimate of the largest matrix that could be handled in 2012 would be of a dimension somewhat larger than 10^{10}.

SIMULTANEOUS EIGENFUNCTIONS

Because a quantum-mechanical system in a state which is an eigenvector of some Hermitian matrix A is postulated to have the corresponding eigenvalue as the unique definite value of the physical quantity associated with A, it is of great interest to know when it will also always be possible to observe at the same time a unique definite value of another quantity that is associated with a Hermitian matrix B. From a mathematical point of view, the question we are asking deals with the possibility that A and B have a complete common set of eigenvectors.

The key result here is simple:

Hermitian matrices have a complete set of simultaneous eigenvectors if and only if they commute.

For proof the reader is referred to Arfken et al in the Additional Readings.

It may happen that we have three matrices A, B, and C, and that $[A, B] = 0$ and $[A, C] = 0$, but $[B, C] \neq 0$. In that case, which is actually quite common in atomic physics, we have a choice. We can insist upon a set of vectors that are simultaneous eigenvectors of A and B, in which case not all of them can be eigenvectors of C, or we can have simultaneous eigenvectors of A and C, but not B. In atomic physics, those choices typically correspond to descriptions in which different angular momenta are required to have definite values.

Example 5.7.1. Simultaneous Eigenvectors

Consider the three matrices

$$\mathsf{A} = \begin{pmatrix} 1 & -1 & 0 & 0 \\ -1 & 1 & 0 & 0 \\ 0 & 0 & 2 & 0 \\ 0 & 0 & 0 & 2 \end{pmatrix}, \qquad \mathsf{B} = \begin{pmatrix} 0 & 0 & 0 & 0 \\ 0 & 0 & 0 & 0 \\ 0 & 0 & 0 & -i \\ 0 & 0 & i & 0 \end{pmatrix},$$

$$\mathsf{C} = \begin{pmatrix} 0 & 0 & -i/\sqrt{2} & 0 \\ 0 & 0 & i/\sqrt{2} & 0 \\ i/\sqrt{2} & -i/\sqrt{2} & 0 & 0 \\ 0 & 0 & 0 & 0 \end{pmatrix}.$$

The reader can verify that these matrices are such that $[\mathsf{A}, \mathsf{B}] = [\mathsf{A}, \mathsf{C}] = 0$, but $[\mathsf{B}, \mathsf{C}] \neq 0$, i.e., $\mathsf{B\,C} \neq \mathsf{C\,B}$. An orthogonal matrix U that diagonalizes A is

$$\mathsf{U} = \begin{pmatrix} 1/\sqrt{2} & 1/\sqrt{2} & 0 & 0 \\ 1/\sqrt{2} & -1/\sqrt{2} & 0 & 0 \\ 0 & 0 & 1 & 0 \\ 0 & 0 & 0 & 1 \end{pmatrix};$$

when U is applied to A, B, and C, we get

$$\mathsf{UAU}^T = \begin{pmatrix} 0 & 0 & 0 & 0 \\ 0 & 2 & 0 & 0 \\ 0 & 0 & 2 & 0 \\ 0 & 0 & 0 & 2 \end{pmatrix}, \quad \mathsf{UBU}^T = \begin{pmatrix} 0 & 0 & 0 & 0 \\ 0 & 0 & 0 & 0 \\ 0 & 0 & 0 & -i \\ 0 & 0 & i & 0 \end{pmatrix},$$

$$\mathsf{UCU}^T = \begin{pmatrix} 0 & 0 & 0 & 0 \\ 0 & 0 & -i & 0 \\ 0 & i & 0 & 0 \\ 0 & 0 & 0 & 0 \end{pmatrix}.$$

At this point, neither UBU^T nor UCU^T is also diagonal, but we can choose to diagonalize one of them (we choose UBU^T) by a further orthogonal transformation that will modify the lower 3×3 block of UBU^T (note that because this block of UAU^T is proportional to a unit matrix the transformation we plan to make will not change it).

An orthogonal matrix V that diagonalizes UBU^T is

$$\mathsf{V} = \begin{pmatrix} 1 & 0 & 0 & 0 \\ 0 & 1 & 0 & 0 \\ 0 & 0 & 1/\sqrt{2} & i/\sqrt{2} \\ 0 & 0 & 1/\sqrt{2} & -i/\sqrt{2} \end{pmatrix};$$

as already stated, further transformation with V leaves UAU^T unchanged, and converts UBU^T and UCU^T to

$$\mathsf{VUBU}^T\mathsf{V}^T = \begin{pmatrix} 0 & 0 & 0 & 0 \\ 0 & 0 & 0 & 0 \\ 0 & 0 & -1 & 0 \\ 0 & 0 & 0 & 1 \end{pmatrix}, \quad \mathsf{VUCU}^T\mathsf{V}^T = \begin{pmatrix} 0 & 0 & 0 & 0 \\ 0 & 0 & -i/\sqrt{2} & -i/\sqrt{2} \\ 0 & i/\sqrt{2} & 0 & 0 \\ 0 & i/\sqrt{2} & 0 & 0 \end{pmatrix}.$$

The columns of $\mathsf{U}^T\mathsf{V}^T = (\mathsf{VU})^T$ are the simultaneous eigenvectors of A and B (but not C). It is not possible to diagonalize simultaneously both B and C, but we could have chosen to diagonalize C rather than B.

■

Exercises

5.7.1. For the matrix

$$\mathsf{A} = \begin{pmatrix} 1.23 & 0.00 & 0.71 & -0.26 \\ 0.00 & 2.34 & 1.46 & -3.01 \\ 0.71 & 1.46 & 3.45 & 0.00 \\ -0.26 & -3.01 & 0.00 & 4.56 \end{pmatrix},$$

working with at least six significant figures of precision,

(a) Find all its eigenvalues, and for each identify the corresponding eigenvector.

(b) From the eigenvectors form a matrix U such that $\mathsf{A}' = \mathsf{UAU}^{-1}$ is diagonal and show that the diagonal elements of A' are the eigenvalues of A.

5.7.2. Given the three matrices

$$\mathsf{A} = \begin{pmatrix} 1 & 0 & 0 \\ 0 & 1 & 0 \\ 0 & 0 & -1 \end{pmatrix}, \quad \mathsf{B} = \begin{pmatrix} 0.64 & 0.48 & 0 \\ 0.48 & 0.36 & 0 \\ 0 & 0 & 1 \end{pmatrix}, \quad \mathsf{C} = \begin{pmatrix} 0.96 & -0.28 & 0 \\ -0.28 & -0.96 & 0 \\ 0 & 0 & 0 \end{pmatrix},$$

(a) Find an orthogonal transformation that will simultaneously diagonalize A and B and identify their simultaneous eigenvectors, for each giving the eigenvalues of A and B.

(b) Find an orthogonal transformation that will simultaneously diagonalize A and C and identify their simultaneous eigenvectors, for each giving the eigenvalues of A and C.

(c) How do you know it will not be possible to find an orthogonal transformation that will simultaneously diagonalize A, B, and C?

5.8 MATRIX INVARIANTS

It is clear from the foregoing section that the eigenvalues of a square matrix A are not changed when it is subjected to a similarity transformation, i.e., a transformation of the form $\mathsf{A} \longrightarrow \mathsf{UAU}^{-1}$. An explicit demonstration to this effect is obtained by looking at its secular equation and applying the determinant product theorem:

$$\det(\mathsf{UAU}^{-1} - \lambda\mathbf{1}) = \det(\mathsf{U}[\mathsf{A} - \lambda\mathbf{1}]\mathsf{U}^{-1}) = \det([\mathsf{A} - \lambda\mathbf{1}]\mathsf{U}^{-1}\mathsf{U}),$$

$$= \det(\mathsf{A} - \lambda\mathbf{1}) = 0\,. \tag{5.39}$$

We see that A and UAU^{-1} satisfy the same secular equation, and therefore must have as solutions the same λ values. We can therefore regard the λ values as **invariants** of A, meaning that they remain unchanged under similarity transformation.

Again applying the determinant product theorem, we note that for a square matrix A,

$$\det(\mathsf{UAU}^{-1}) = \det(\mathsf{AU}^{-1}\mathsf{U}) = \det(\mathsf{A})\,, \tag{5.40}$$

showing that $\det(\mathsf{A})$ is also an invariant. In fact, if we transform A to diagonal form, we see that $\det(\mathsf{A})$ is just the product of its eigenvalues.

Another important invariant of a square matrix A is a quantity known as its **trace**, defined as the sum of its diagonal elements.

$$\mathrm{Tr}(\mathsf{A}) = \sum_i a_{ii}\,. \tag{5.41}$$

To show the invariance, form the trace of UAU^{-1}:

$$\mathrm{Tr}(\mathsf{UAU}^{-1}) = \sum_i (\mathsf{UAU}^{-1})_{ii} = \sum_{ijk} \mathsf{U}_{ij}\mathsf{A}_{jk}\left[\mathsf{U}^{-1}\right]_{ki} = \sum_{jk}\Big(\sum_i \left[\mathsf{U}^{-1}\right]_{ki}\mathsf{U}_{ij}\Big)\mathsf{A}_{jk}$$

$$= \sum_{jk}\delta_{kj}\mathsf{A}_{jk} = \sum_j \mathsf{A}_{jj} = \mathrm{Tr}(\mathsf{A}). \tag{5.42}$$

If we choose U to be a transformation that diagonalizes A, then UAU^{-1} will have the eigenvalues λ_i as its diagonal elements, showing that

$$\mathrm{Tr}(\mathsf{A}) = \sum_i \lambda_i\,, \tag{5.43}$$

obviously an invariant of A.

SYMBOLIC COMPUTING

We have previously identified the symbolic commands that produce the determinant of a matrix A:

MAPLE: `determinant(A)`, MATHEMATICA: `Det[A]`.

The trace is accessed by

MAPLE: `Trace(A)`, MATHEMATICA: `Tr[A]`.

In MAPLE, these commands must be preceded by `with(LinearAlgebra)`. Examples of these commands:

MAPLE:

```
> with(LinearAlgebra):
> M := Matrix([[1,2],[3,4]]):
> Determinant(M);                     -2
> Trace(M);                            5
```

MATHEMATICA:

```
M = { {1,2},{3,4} };
Det[M]                                -2
Tr[M]                                  5
```

Exercises

5.8.1. Use symbolic computing to find the determinant and trace of

$$\begin{pmatrix} 2 & 7 & 11 \\ 3 & 8 & -1 \\ 6 & 9 & 12 \end{pmatrix}.$$

Additional Readings

Arfken, G. B., Weber, H. J., & Harris, F. E. (2013). *Mathematical methods for physicists* (7th ed.). New York: Academic Press.

Bickley, W. G., & Thompson, R. S. H. G. (1964). *Matrices—Their meaning and manipulation.* Princeton, NJ: Van Nostrand (A comprehensive account of matrices in physical problems, their analytic properties, and numerical techniques).

Byron, F. W., Jr. & Fuller, R. W. (1969). *Mathematics of classical and quantum physics.* Reading, MA: Addison-Wesley (Reprinted, Dover (1992)).

Gilbert, J., & Gilbert, L. (1995). *Linear algebra and matrix theory.* San Diego: Academic Press.

Golub, G. H., & Van Loan, C. F. (1996). *Matrix computations* (3rd ed.). Baltimore: JHU Press (Detailed mathematical background and algorithms for the production of numerical software, including methods for parallel computation. A classic computer science text).

Halmos, P. R. (1958). *Finite-dimensional vector spaces* (2nd ed.). Princeton, NJ: Van Nostrand (Reprinted, Springer (1993)).

Heading, J. (1958). *Matrix theory for physicists.* London: Longmans, Green and Co. (A readable introduction to determinants and matrices, with applications to mechanics, electromagnetism, special relativity, and quantum mechanics).

Hirsch, M. (1974). *Differential equations, dynamical systems, and linear algebra.* San Diego: Academic Press.

Jain, M. C. (2007). *Vector spaces and matrices in physics* (2nd ed.). Oxford, UK: Alpha Science International.

Miller, K. S. (1963). *Linear differential equations in the real domain.* New York: Norton.

Parlett, B. N. (1980). *The symmetric eigenvalue problem.* Englewood Cliffs, NJ: Prentice-Hall (Reprinted, SIAM (1998)).

Titchmarsh, E. C. (1962). *Eigenfunction expansions associated with second-order differential equations, Part 1* (2nd ed.). London: Oxford University Press (Part 2 (1958)).

Watkins, D.S. (1991). *Fundamentals of matrix computations.* New York: Wiley.

Wilkinson, J. H. (1965). *The algebraic eigenvalue problem.* London: Oxford University Press (Reprinted (2004). Classic treatise on numerical computation of eigenvalue problems. Perhaps the most widely read book in the field of numerical analysis).

Chapter 6

MULTIDIMENSIONAL PROBLEMS

This chapter deals with various aspects of the application of the methods of calculus to functions of more than one variable. Its topics include partial differentiation, curvilinear coordinate systems, multiple integrals, and other issues that do not arise when dealing with functions of a single variable. For the most part, the material in this chapter does not translate directly into opportunities to solve problems using symbolic computing, but is instead part of the background knowledge that can be used to pose mathematical questions in forms that lead to their solutions.

Also included in this chapter is a discussion of the Dirac delta function, which though not exclusively multidimensional is of great importance in many multidimensional problems.

6.1 PARTIAL DIFFERENTIATION

For a function $f(x)$ of a single variable x, the derivative df/dx, sometimes denoted $f'(x)$, is the limit

$$\frac{df}{dx} = \lim_{dx\to 0} \frac{f(x+dx) - f(x)}{dx} .$$

The derivative is a measure of the amount the function changes per unit change in x, but measured in the limit that the change in x is arbitrarily small. If $f(x)$ is plotted as a curve, df/dx evaluated at any point is the slope of the curve at that point. For most functions of interest in physics, the limit defining the derivative exists except perhaps at isolated points. Using the derivative notation, we can write that the change in f due to a small change in x is given as

$$df = \left(\frac{df}{dx}\right) dx . \tag{6.1}$$

This formula is valid in the limit of small dx; it corresponds to the first two terms in the Taylor series expansion of f about the point x; the omitted terms are proportional to $(dx)^n$ with $n > 1$ and for small enough dx and nonzero df/dx they become negligible compared to the term retained in Eq. (6.1).

Extending these notions to a function $f(x, y)$ of two variables, we define the **partial derivative** of f with respect to x, denoted $\partial f/\partial x$ (in spoken language

Mathematics for Physical Science and Engineering.
http://dx.doi.org/10.1016/B978-0-12-801000-6.00006-7

"partial f dee x"), as the result that we would obtain if y is regarded as fixed, so

$$\frac{\partial f}{\partial x} = \lim_{dx\to 0} \frac{f(x+dx,y)-f(x,y)}{dx} ,$$

with an analogous definition for $\partial f/\partial y$. Extensions to three or more variables, e.g., partial derivatives of $f(x,y,z)$, are defined similarly. If we now plot a function $f(x,y)$ with the points (x,y) in a plane and $f(x,y)$ plotted in the z direction, our graph will be that of a surface rather than a curve, and $\partial f/\partial x$ will be the slope of the surface as measured in the x direction, with $\partial f/\partial y$ its slope in the y direction.

The change in f due to small changes in both x and y can be written

$$\begin{aligned} df &= \left[\frac{f(x+dx,y)-f(x,y)}{dx}\right] dx + \left[\frac{f(x+dx,y+dy)-f(x+dx,y)}{dy}\right] dy \\ &\approx \left(\frac{\partial f}{\partial x}\right) dx + \left(\frac{\partial f}{\partial y}\right) dy , \end{aligned} \tag{6.2}$$

where the second line of Eq. (6.2) is the limit of the first line when dx and dy are small. Although the quantity $\partial f/\partial y$ is actually evaluated for the x value $x + dx$, that change in x alters this derivative only by an amount proportional to dx, so the contribution to df from that source will be proportional to $dxdy$ and has been neglected. The key idea here, which is very useful, is that the first-order changes are (to first order) additive.

Sometimes it is clear from the context that the notation $\partial f/\partial x$ implies that the partial derivative is to be taken keeping some variable y fixed. However, there may be ambiguity as to which variables are to be regarded as independent, and we then need a more detailed notation. For example, a function f might be specified either in terms of its Cartesian coordinates x, y or in terms of the polar coordinates r, θ. Now the partial derivative with respect to r with θ fixed is a different quantity than the partial derivative with respect to r keeping x fixed. This sort of ambiguity can be resolved by listing, as subscripts, the variable(s) that are to be kept fixed when taking the derivative. In this more detailed notation, Eq. (6.2) would be written

$$df = \left(\frac{\partial f}{\partial x}\right)_y dx + \left(\frac{\partial f}{\partial y}\right)_x dy .$$

The subscript notation carries the implication that any other variables that are present in f are to be expressed in terms of those listed before the derivative is to be taken. For example, if f is formally a function of x, y, and z but $z^2 = x^2 + y^2$, we consider z as replaced by its equivalent in terms of x and y when we carry out the differentiation.

The need for a subscript notation is particularly acute in thermodynamics, where systems are studied with the independent variables sometimes taken as the volume V and temperature T, but sometimes as the pressure P and the temperature. If U is the energy of the thermodynamic system, it is important to distinguish $(\partial U/\partial T)_V$ from $(\partial U/\partial T)_P$; they are not the same.

HIGHER DERIVATIVES

Just as for functions of a single variable, we have a need for second and higher derivatives in multidimensional systems. If the function involved has sufficient continuity, **cross derivatives**, meaning those which involve differentiation with respect to more than one variable, yield results that are independent of the order in which the derivatives are taken. It is also traditional to collect multiple derivatives together into a more compact notation. We illustrate both these points:

$$\frac{\partial}{\partial x}\frac{\partial f}{\partial y} \equiv \frac{\partial^2 f}{\partial x \partial y}\,, \qquad \frac{\partial}{\partial y}\frac{\partial f}{\partial x} \equiv \frac{\partial^2 f}{\partial y \partial x}\,; \qquad \text{usually} \;\; \frac{\partial^2 f}{\partial x \partial y} = \frac{\partial^2 f}{\partial y \partial x}\,.$$

When cross derivatives are equal, we can also make simplifications like

$$\frac{\partial}{\partial x}\frac{\partial}{\partial y}\frac{\partial f}{\partial x} = \frac{\partial^3 f}{\partial y \partial x^2}\,.$$

The higher derivatives appear in the higher-dimension analogs of the Maclaurin expansion; for two variables we have

$$f(x,y) = \sum_{p,q=0}^{\infty} a_{pq}\frac{x^p y^q}{p!\,q!}\,, \qquad a_{pq} = \left.\frac{\partial^{p+q} f}{\partial x^p \partial y^q}\right|_{x=y=0}\,. \tag{6.3}$$

Example 6.1.1. Partial Derivatives

If $z = x^y$, then (regarding y as a constant) $\partial z/\partial x = yx^{y-1}$, while (treating x as a constant) $\partial z/\partial y = x^y \ln x$.

Forming the cross derivatives,

$$\frac{\partial}{\partial y}\left(\frac{\partial z}{\partial x}\right) = x^{y-1} + yx^{y-1}\ln x\,, \qquad \frac{\partial}{\partial x}\left(\frac{\partial z}{\partial y}\right) = yx^{y-1}\ln x + x^y\left(\frac{1}{x}\right)\,.$$

The cross derivatives are equal.

■

CHAIN RULE

We frequently encounter situations in which the quantities appearing in a function are themselves functions of another variable. Often in mechanics problems we have expressions like $V(x, y, z)$ where x, y, and z are functions of the time t. One way to see how V changes as we change t is to begin by writing

$$dV = \left(\frac{\partial V}{\partial x}\right)_{yz} dx + \left(\frac{\partial V}{\partial y}\right)_{xz} dy + \left(\frac{\partial V}{\partial z}\right)_{xy} dz\,, \tag{6.4}$$

after which we divide through by dt to obtain the **chain rule**

$$\frac{dV}{dt} = \left(\frac{\partial V}{\partial x}\right)_{yz}\frac{dx}{dt} + \left(\frac{\partial V}{\partial y}\right)_{xz}\frac{dy}{dt} + \left(\frac{\partial V}{\partial z}\right)_{xy}\frac{dz}{dt}\,. \tag{6.5}$$

Most writers do not include the subscripts, but we have written them here to emphasize that the derivatives of V on the right-hand side of Eq. (6.5) are computed considering V to be a function of the three independent variables x, y, z. The t derivative of V on the left-hand side of Eq. (6.5) refers to all its t dependence; from that viewpoint t is the only independent variable so we write here an "ordinary" (i.e., total) derivative.

An interesting and useful application of the chain rule is obtained if, for $z=z(x,y)$, we write

$$dz = \left(\frac{\partial z}{\partial x}\right)_y dx + \left(\frac{\partial z}{\partial y}\right)_x dy \tag{6.6}$$

and then set $dz = 0$. Noting that this means that dx and dy on the right-hand side of Eq. (6.6) are to be taken keeping z constant, we can rearrange Eq. (6.6) to the form

$$\left(\frac{\partial y}{\partial x}\right)_z = -\frac{\left(\frac{\partial z}{\partial x}\right)_y}{\left(\frac{\partial z}{\partial y}\right)_x} . \tag{6.7}$$

Notice that we cannot cancel ∂z in the right-hand side of Eq. (6.7).

Example 6.1.2. Chain Rule and Coordinate Rotation

This example illustrates typical applications of the chain rule when we have a number of variables that are connected by linear relationships.

We consider a situation in which a rotation of the coordinate system causes coordinates $\mathbf{r} = (x, y)$ to be transformed into $\mathbf{r}' = (x', y')$ according to the matrix equation $\mathbf{r}' = \mathsf{U}\mathbf{r}$. We recall that U is orthogonal so the inverse transformation is $\mathbf{r} = U^{\mathrm{T}}\mathbf{r}'$. Note that in this problem, primes indicate transformed coordinates, not derivatives. These transformations correspond to the ordinary equations

$$x' = U_{11}\,x + U_{12}\,y \qquad\qquad x = U_{11}\,x' + U_{21}\,y'$$

$$y' = U_{21}\,x + U_{22}\,y \qquad\qquad y = U_{12}\,x' + U_{22}\,y' .$$

From the equations in the top line we can read out

$$\left(\frac{\partial x'}{\partial x}\right)_y = U_{11} , \qquad \left(\frac{\partial x}{\partial x'}\right)_{y'} = U_{11} ,$$

showing that these derivatives are not mutually reciprocal; in fact, they are equal! They do not have to be reciprocal because they do not involve the same set of independent variables.

Suppose we also needed $(\partial x'/\partial x)_{y'}$. Starting from the x' equation,

$$dx' = U_{11}\,dx + U_{12}\,dy ,$$

we divide through by dx keeping y' constant:

$$\left(\frac{\partial x'}{\partial x}\right)_{y'} = U_{11}\left(\frac{\partial x}{\partial x}\right)_{y'} + U_{12}\left(\frac{\partial y}{\partial x}\right)_{y'} = U_{11} + U_{12}\left(\frac{\partial y}{\partial x}\right)_{y'} . \tag{6.8}$$

Notice that $\partial x/\partial x$ is unity no matter what other variable(s) are held constant.

We now evaluate the last derivative in Eq. (6.8) by applying Eq. (6.7), in the form

$$\left(\frac{\partial y}{\partial x}\right)_{y'} = -\frac{\left(\frac{\partial y'}{\partial x}\right)_y}{\left(\frac{\partial y'}{\partial y}\right)_x} = -\frac{U_{21}}{U_{22}} .$$

We get

$$\left(\frac{\partial x'}{\partial x}\right)_{y'} = \frac{U_{11}U_{22} - U_{12}U_{21}}{U_{22}} = \frac{1}{U_{22}}\,, \tag{6.9}$$

the final simplification occurring because rotational orthogonal transformations have determinants of value unity.

■

IMPLICIT DIFFERENTIATION

If we have an explicit formula for $f(x, y)$ it is straightforward to apply the differentiation rules to f and thereby obtain the partial derivatives of f. However, sometimes we may be presented with the need to evaluate derivatives for a quantity that is only given implicitly, i.e., as an equation in which x and y appear, but not in a solved form. Illustrating for a single independent variable, suppose we have

$$e^y = x(y-1)\,. \tag{6.10}$$

We can differentiate this equation with respect to x even though it is not in solved form:

$$e^y\,y' = (y-1) + xy'\,.$$

This equation can now be solved for y', with the result

$$y' = \frac{y-1}{e^y - x}\,, \quad \text{which can be simplified to } y' = \frac{(y-1)^2\,e^{-y}}{y-2}\,.$$

This result could also have been obtained by solving Eq. (6.10) for x, forming dx/dy, and then noticing that

$$\frac{dx}{dy} = \left(\frac{dy}{dx}\right)^{-1}\,. \tag{6.11}$$

We can apply the same ideas to functions of several variables, but with one important difference. An equation similar to Eq. (6.11) can only be invoked if the same variable(s) are held constant for the derivatives on both sides of the equation. Otherwise there is no reciprocal relationship.

Example 6.1.3. Implicit Differentiation

Let $x = ye^z + z$; we require $(\partial z/\partial x)_y$ and $(\partial z/\partial y)_x$.

Differentiating first with respect to x, keeping y fixed, and then with respect to y, keeping x fixed:

$$1 = ye^z\left(\frac{\partial z}{\partial x}\right)_y + \left(\frac{\partial z}{\partial x}\right)_y\,, \qquad 0 = e^z + ye^z\left(\frac{\partial z}{\partial y}\right)_x + \left(\frac{\partial z}{\partial y}\right)_x\,.$$

Solving for the derivatives, we have

$$\left(\frac{\partial z}{\partial x}\right)_y = \frac{1}{ye^z + 1}\,, \qquad \left(\frac{\partial z}{\partial y}\right)_x = -\frac{e^z}{ye^z+1}\,.$$

It is obvious that obtaining these derivatives would have been far more difficult if we had to start by finding an explicit form for $z(x, y)$.

■

DIFFERENTIATION OF INTEGRALS

Occasionally we encounter a need to differentiate an integral with respect to a variable that occurs parametrically within the integral or in one of its limits. Those situations correspond to the following formulas:

$$\frac{d}{dy}\int_a^b f(x,y)\,dx = \int_a^b \frac{\partial f(x,y)}{\partial y}\,dx\,, \tag{6.12}$$

$$\frac{d}{dy}\int_a^y f(x)\,dx = f(y)\,, \tag{6.13}$$

$$\frac{d}{dy}\int_y^b f(x)\,dx = -f(y)\,. \tag{6.14}$$

If a limit contains a function of the variable to be differentiated, we then use the chain rule to carry out the differentiation. For example,

$$\frac{d}{dy}\int_a^{g(y)} f(x)\,dx = f(g(y))\,\frac{dg}{dy}\,.$$

Exercises

6.1.1. For $z = x/(x^2+y^2)$, find $\partial z/\partial x$ and $\partial z/\partial y$.

6.1.2. For $z = uvw \ln\sqrt{u^2+v^2+w^2}$, find $\partial z/\partial u$, $\partial z/\partial v$, and $\partial z/\partial w$.

6.1.3. For $z = \cos(x^2+y)$, verify that $\dfrac{\partial^2 z}{\partial x \partial y} = \dfrac{\partial^2 z}{\partial y \partial x}$.

6.1.4. For $z = x^2 - y^2$, with $x = r\cos\theta$, $y = r\sin\theta$, find the following partial derivatives:

$$\left(\frac{\partial z}{\partial x}\right)_y,\quad \left(\frac{\partial z}{\partial x}\right)_r,\quad \left(\frac{\partial z}{\partial x}\right)_\theta,\quad \left(\frac{\partial z}{\partial y}\right)_x,\quad \left(\frac{\partial z}{\partial y}\right)_r,\quad \left(\frac{\partial z}{\partial y}\right)_\theta,$$

$$\left(\frac{\partial z}{\partial r}\right)_x,\quad \left(\frac{\partial z}{\partial r}\right)_y,\quad \left(\frac{\partial z}{\partial r}\right)_\theta,\quad \left(\frac{\partial z}{\partial \theta}\right)_x,\quad \left(\frac{\partial z}{\partial \theta}\right)_y,\quad \left(\frac{\partial z}{\partial \theta}\right)_r.$$

6.1.5. If $z^2 = e^{p^2-q^2}$, with $p = e^s$ and $q = e^{2s}$, find dz/ds.

6.1.6. If $u = e^{p^2-q^2}$, with $p = r\cos\theta$ and $q = r\sin\theta$, find $(\partial u/\partial r)_\theta$ and $(\partial u/\partial\theta)_r$.

6.1.7. Given that $ye^{-xy} = \cos x$, find dy/dx.

6.1.8. Let $x^2u + u^2v = 1$ and $uv = x + y$. Find $(\partial x/\partial u)_v$ and $(\partial x/\partial u)_y$.

6.1.9. Evaluate:

(a) $\displaystyle \frac{d}{dx}\int_0^{\sin(x)} e^{-t^2}\,dt,$

(b) $\displaystyle \frac{d}{dx}\int_0^{\sin(x)} e^{-xt^2}\,dt.$

You may leave unevaluated any integrals that cannot be reduced to elementary functions.

6.1.10. Let $I = \int_1^\infty e^{-x^n t}\,dt$. Calculate dI/dx in the two following ways:

(a) Differentiate I with respect to x and **after doing so** evaluate the resulting integral.

(b) Perform the integration in I and **after doing so** differentiate the result, thereby checking the result of part (a).

6.2 EXTREMA AND SADDLE POINTS

A function of a single variable has relative maxima at points where its first derivative vanishes and its second derivative is negative. Relative minima occur when the first derivative vanishes and the second derivative is positive. If both first and second derivatives vanish the function is locally flat but eventually may increase in one direction and decrease in the other; such points are neither maxima nor minima but are called **stationary points** of the function.

The situation is more complicated when there is more than one independent variable. Points where a function has zero first derivatives with respect to all its independent variables are called its **critical points**. If a critical point is a relative maximum (minimum), all straight-line paths through the point will exhibit a maximum (minimum) there. But it is also possible for some paths through a critical point to have a maximum while others have a minimum there. A function with this type of behavior is said to have a **saddle point**; the name arises from the qualitative appearance of the 3-D plot of a function of two independent variables in the neighborhood of the critical point. See Fig. 6.1.

A full discussion of the properties of functions of many variables is beyond the scope of this text. We can, however, identify the conditions that indicate an extremum (maximum or minimum). For a critical point to be a minimum, a function must have

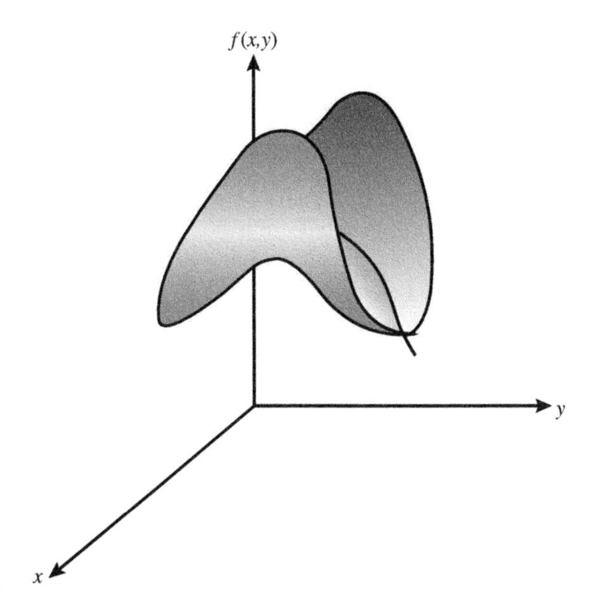

Figure 6.1: A saddle point.

a minimum there for every straight-line path through the critical point; that means that its second derivative along any straight line must be positive at the critical point. This condition can be satisfied only if the matrix of second derivatives of the function (evaluated at the critical point) is positive definite, a condition equivalent to requiring that all its eigenvalues be positive. This matrix of second derivatives, known as the **Hessian**, has for a function $f(x, y)$ of two variables the form

$$\mathsf{H} = \begin{pmatrix} \dfrac{\partial^2 f}{\partial x^2} & \dfrac{\partial^2 f}{\partial x \partial y} \\ \dfrac{\partial^2 f}{\partial y \partial x} & \dfrac{\partial^2 f}{\partial y^2} \end{pmatrix}. \tag{6.15}$$

Conversely, a critical point will be a maximum if the Hessian is negative definite. A saddle point results if the Hessian has both positive and negative eigenvalues, corresponding to the property that $f(x, y)$ will be a maximum on some lines through the critical point and a minimum on others.

Exercises

6.2.1. (a) Verify that $f(x, y) = x^2 + xy - y^2$ has a critical point at $(x, y) = (0, 0)$. Calculate its Hessian at (0,0) and from the eigenvalues of the Hessian determine the nature of the critical point.

(b) Repeat the process of part (a) for $f(x, y) = x^2 - xy + y^2$.

6.2.2. For $f(x, y) = x^3 + y^3 - 6xy$, find all the critical points and classify each as a maximum, minimum, or saddle point.

6.3 CURVILINEAR COORDINATE SYSTEMS

Cartesian coordinate systems have two great virtues: (1) they are uniform, with the local geometry identical at all points, and (2) they are orthogonal; in two dimensions, the lines of constant x are perpendicular to the lines of constant y. In three dimensions, the points of constant x are planes; they intersect the planes of constant y or constant z at right angles.

Despite these obvious advantages, many problems in physics are handled advantageously in other coordinate systems that may reflect some symmetry of the physical system under study. For example, the electric field of a point charge can be expected to take a simple form in a spherical polar coordinate system when the point charge is placed at the coordinate origin. The electrodynamics of a signal in a coaxial cable will be most easily treated in a circular cylindrical coordinate system. Before the widespread deployment of digital computers, still other coordinate systems were frequently introduced for the study of specific problems. Though the proliferation of computers has in recent years reduced the need to use exotic coordinate systems, the use of spherical polar and cylindrical coordinates remains important in physics.

Although the lines of constant coordinate values in curvilinear coordinates may not always be straight lines, it may still be that they intersect at right angles. See, for example, the diagram for 2-D polar coordinates, Fig. 6.2. Here the lines of constant r, though circles, cross the lines of constant θ at right angles, and for this reason the polar coordinates are identified as an **orthogonal** coordinate system. The orthogonality is most important when one discusses the behavior in the neighborhood

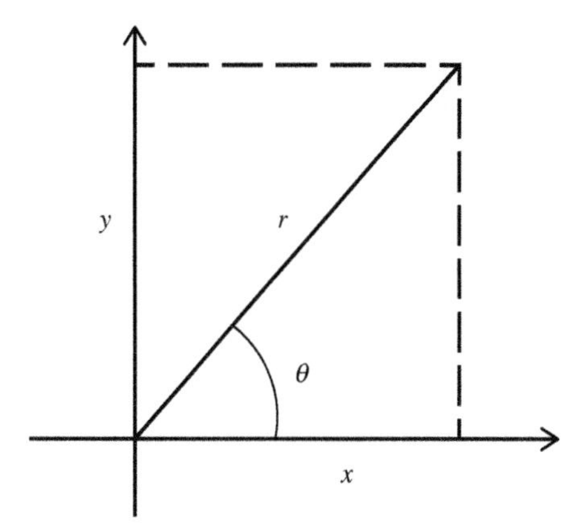

Figure 6.2: Plane polar coordinates r, θ.

of a coordinate point; the infinitesimal displacements that define derivatives in the different coordinate directions are orthogonal (and more independent than if that were not the case). Because of their usefulness and relative simplicity, we restrict further discussion here to the two orthogonal curvilinear systems that are used most frequently in physics and engineering.

SPHERICAL POLAR COORDINATES

In spherical polar coordinates, the coordinates are r, θ, φ, where r is the distance from the origin, θ is the angle from the polar direction (on the Earth, colatitude, which is $90°-$latitude), and φ the azimuthal angle (longitude). It is customary to align the polar direction with the Cartesian coordinate z and to measure φ from a zero (our Greenwich meridian) along the $+x$ direction, with the direction of φ such that the $+y$ direction is at $\varphi = \pi/2$ $(90°)$. Therefore, points with a given value of r lie on a sphere of radius r centered at the origin and points of given θ lie on a cone with vertex at the origin, axis in the z direction and an opening angle of rotation θ. Points of given φ lie on a half-plane which extends from the polar axis to infinity in the direction given by φ. In order for coordinate sets and arbitrary spatial points to be unambiguously related, we need to restrict the range of r to $0 \le r < \infty$, with θ in the range $0 \le \theta \le \pi$ and φ within $0 \le \varphi < 2\pi$. See Fig. 6.3.

The equations connecting the two sets of coordinates are

$$
\begin{aligned}
x &= r\sin\theta\cos\varphi\,, & r &= \sqrt{x^2+y^2+z^2}\,,\\
y &= r\sin\theta\sin\varphi\,, & \cos\theta &= \frac{z}{\sqrt{x^2+y^2+z^2}}\,,\\
z &= r\cos\theta\,, & \tan\varphi &= \frac{y}{x}\,.
\end{aligned}
\tag{6.16}
$$

Notice that Eq. (6.16) gives formulas for $\cos\theta$ and $\tan\varphi$ rather than for θ and φ. When we convert $\cos\theta$ into θ we must use the principal value of the $\cos^{-1}$ function so as to obtain a result within the range $0 \le \theta \le \pi$. For φ we must be even more careful, as the range for the principal value of $\tan^{-1} y/x$ is only of length π, while the range of φ is of length 2π. We must choose the value of φ that is in the azimuthal quadrant consistent with the individual signs of x and y.

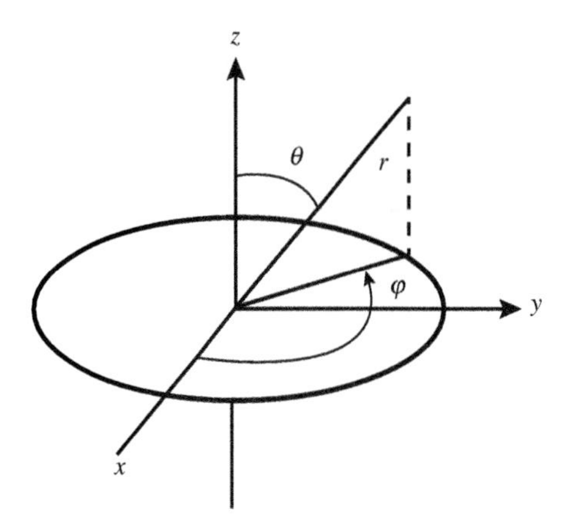

Figure 6.3: Spherical polar coordinates r, θ, φ.

CIRCULAR CYLINDRICAL COORDINATES

Circular cylindrical coordinates use the plane polar coordinates ρ and φ (in place of x and y) and the z Cartesian coordinate. The variable ρ is the distance of a coordinate point from the z Cartesian axis, and φ is its azimuthal angle. The ranges of these coordinates are $0 \le \rho < \infty$, $0 \le \varphi < 2\pi$, and of course $-\infty < z < \infty$. Thus, points of given ρ lie on a cylinder about the z axis of radius ρ, points of given φ lie on a half-plane extending from the entire z axis to infinity in the φ direction, and points of given z lie on the plane with that value of z. See Fig. 6.4.

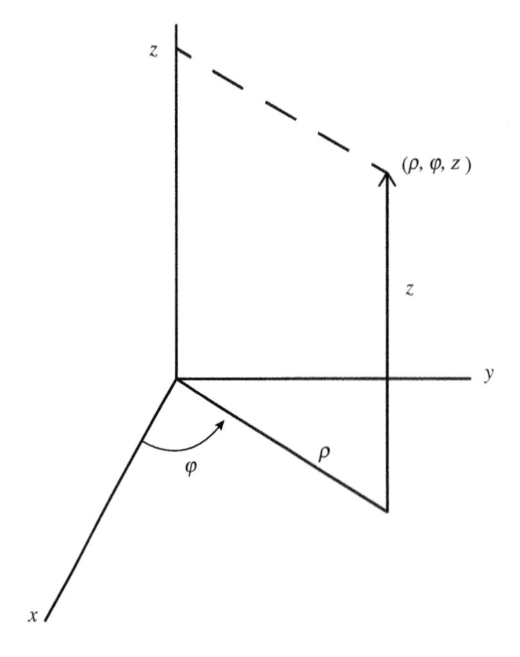

Figure 6.4: Circular cylindrical coordinates ρ, φ, z.

The conversion formulas between circular cylindrical and Cartesian coordinates are

$$\begin{aligned} x &= \rho \cos\varphi\,, & \rho &= \sqrt{x^2+y^2}\,, \\ y &= \rho \sin\varphi\,, & \tan\varphi &= \frac{y}{x}\,, \\ z &= z\,, & z &= z\,. \end{aligned} \tag{6.17}$$

The coordinate φ has the same definition as for spherical polar coordinates, and, just as there, it must be identified as the value of $\tan^{-1} y/x$ consistent with the individual signs of x and y. We have used ρ (not r) as the coordinate denoting distance from the z axis; we are reserving r to always mean distance from the origin and will reserve ρ for the axial distance defined here.

VECTORS IN CURVILINEAR COORDINATES

In Cartesian coordinates, a unit vector $\hat{\mathbf{e}}_x$ is of unit length and in the x direction. That is simple and straightforward because the "x direction" is everywhere the same direction. However, in spherical polar coordinates the "r direction" is surely not the same everywhere and we need to define it unambiguously. Because the direction associated with the change in a coordinate may depend upon the value of it (and the other coordinates), it is most useful to define the "r direction" and other directions as those generated by infinitesimal changes in the coordinate values.

Our definition of the "r direction" is that of a vector from (r, θ, φ) to $(r+dr, \theta, \varphi)$. The unit vector $\hat{\mathbf{r}}$ or $\hat{\mathbf{e}}_r$ is then a vector in the "r direction" and of unit length.

Continuing with spherical polar coordinates, we now wish to consider a unit vector in the θ direction. This direction is that of an infinitesimal vector from (r, θ, φ) to $(r, \theta+d\theta, \varphi)$, and it (and the corresponding unit vector $\hat{\boldsymbol{\theta}}$ or $\hat{\mathbf{e}}_\theta$) will be perpendicular to the unit vector $\hat{\mathbf{r}}$. The third unit vector, $\hat{\boldsymbol{\varphi}}$ or $\hat{\mathbf{e}}_\varphi$, will be perpendicular to $\hat{\mathbf{r}}$ and $\hat{\boldsymbol{\theta}}$, so our spherical polar coordinate system is **orthogonal**.

Observations similar to those of the preceding paragraph indicate that circular cylindrical coordinates also form an orthogonal system. The orthogonality is also apparent from drawings showing the intersections of contours of constant coordinate values.

The unit vectors can be used to decompose vectors into their components in curvilinear systems. However, it is important to notice that vector components cannot be combined (either for addition or for forming dot products) unless the vectors are associated with the same point in space. Violation of this rule would cause the same unit-vector symbol to have different meanings at different occurrences in a single expression, thereby surely causing errors.

If two vectors $\mathbf{A}$ and $\mathbf{B}$ are indeed associated with the same spatial point, then (using spherical polar coordinates as an example), they have respective component decompositions

$$\mathbf{A} = A_r\hat{\mathbf{r}} + A_\theta\hat{\boldsymbol{\theta}} + A_\varphi\hat{\boldsymbol{\varphi}}\,, \qquad \mathbf{B} = B_r\hat{\mathbf{r}} + B_\theta\hat{\boldsymbol{\theta}} + B_\varphi\hat{\boldsymbol{\varphi}}\,,$$

and (adding components)

$$\mathbf{A} + \mathbf{B} = (A_r + B_r)\hat{\mathbf{r}} + (A_\theta + B_\theta)\hat{\boldsymbol{\theta}} + (A_\varphi + B_\varphi)\hat{\boldsymbol{\varphi}}\,.$$

Rewriting the above in a general notation in which A_i refers to the component of $\mathbf{A}$ in the direction of the unit vector $\hat{\mathbf{e}}_i$, we have

$$\mathbf{A} + \mathbf{B} = (A_1 + B_1)\hat{\mathbf{e}}_1 + (A_2 + B_2)\hat{\mathbf{e}}_2 + (A_3 + B_3)\hat{\mathbf{e}}_3 \ . \tag{6.18}$$

Since we have restricted the discussion to orthogonal coordinate systems, we have

$$\hat{\mathbf{e}}_i \cdot \hat{\mathbf{e}}_j = \delta_{ij} \ , \tag{6.19}$$

and it is straightforward to compute the dot product $\mathbf{A} \cdot \mathbf{B}$:

$$\mathbf{A}\cdot\mathbf{B} = (A_1\hat{\mathbf{e}}_1 + A_2\hat{\mathbf{e}}_2 + A_3\hat{\mathbf{e}}_3)\cdot(B_1\hat{\mathbf{e}}_1 + B_2\hat{\mathbf{e}}_2 + B_3\hat{\mathbf{e}}_3) = A_1B_1 + A_2B_2 + A_3B_3 \ . \tag{6.20}$$

This is the same as the formula that applies in Cartesian coordinates.

DISPLACEMENTS IN CURVILINEAR COORDINATES

Here there are significant differences from Cartesian systems. In spherical polar coordinates, a unit change in the coordinate r produces a unit displacement (change in position) of a point, but a unit change in the coordinate θ produces a displacement whose magnitude depends upon the current value of r and (because the displacement is the chord of a circular arc and not the arc-length itself) the displacement is not even linear in the change in θ. Similar remarks apply to the displacement produced by a change in the coordinate φ, which depends on the current values of both r and θ. (This is so because the quantity controlling the relation between displacement and $d\varphi$ is ρ, which is $r\sin\theta$.) We therefore choose to focus attention on the properties of infinitesimal displacements and their relations to the corresponding infinitesimal changes in the coordinate values.

An infinitesimal displacement specified by dx, dy, dz or alternatively given in terms of general curvilinear but orthogonal coordinates q_1, q_2, q_3 can be written

$$\begin{aligned} d\mathbf{r} &= \hat{\mathbf{x}}\,dx + \hat{\mathbf{y}}\,dy + \hat{\mathbf{z}}\,dz \\ &= h_1\hat{\mathbf{q}}_1\,dq_1 + h_2\hat{\mathbf{q}}_2\,dq_2 + h_3\hat{\mathbf{q}}_3\,dq_3 \ . \end{aligned}$$

The quantities h_i are sometimes called **scale factors**, and are needed because a unit change in q_i does not necessarily produce a unit change in displacement.

There are general methods for finding the scale factors from the equations defining the q_i, but in the cases of interest here, the h_i can be obtained easily. Taking cylindrical coordinates as our first example, we note that a change $d\rho$ in ρ causes a displacement $d\rho$, so $h_\rho = 1$. However, if φ is changed an amount $d\varphi$, the result is a displacement (in the $\hat{\boldsymbol{\varphi}}$ direction) of amount $\rho\,d\varphi$, showing that $h_\varphi = \rho$. Finally, we have $h_z = 1$.

In spherical polar coordinates, $h_r = 1$, and h_φ, which has the same meaning as in cylindrical coordinates, has the value $h_\varphi = \rho$; if we express ρ in the spherical coordinates we get $h_\varphi = r\sin\theta$. Finally, we note that $h_\theta = r$.

Summarizing, the above discussion corresponds to

$$d\mathbf{r} = \hat{\mathbf{r}}\,dr + r\,\hat{\boldsymbol{\theta}}\,d\theta + r\sin\theta\,\hat{\boldsymbol{\varphi}}\,d\varphi \ , \qquad \text{spherical polar,} \tag{6.21}$$

$$= \hat{\boldsymbol{\rho}}\,d\rho + \rho\,\hat{\boldsymbol{\varphi}}\,d\varphi + \hat{\mathbf{z}}\,dz \ , \qquad \text{cylindrical,} \tag{6.22}$$

and for future use we list the scale factors explicitly:

$$\text{Spherical: } h_r = 1,\ h_\theta = r,\ h_\varphi = r\sin\theta; \quad \text{Cylindrical: } h_\rho = h_z = 1,\ h_\varphi = \rho. \tag{6.23}$$

Exercises

6.3.1. Express each of these Cartesian coordinate points (x, y, z) in cylindrical and spherical coordinates: $(1,0,0)$, $(1,1,1)$, $(-1,1,0)$, $(0,-1,-1)$, $(-1,-1,1)$.

6.3.2. Convert each of these points expressed in the spherical coordinates (r, θ, φ) into Cartesian coordinates: $(1, \pi/2, 7\pi/4)$, $(2, \pi/3, \pi)$, $(3, 2\pi/3, 3\pi/2)$, $(1,1,1)$, $(1,2,3)$.

6.3.3. Convert each of these points expressed in the cylindrical coordinates (ρ, φ, z) into Cartesian coordinates: $(1, \pi/2, 0)$, $(1, 0, \pi/2)$, $(2, 3\pi/2, 3)$, $(\pi, 1, -2)$, $(2, \pi/3, -1)$.

6.3.4. In Cartesian coordinates two 3-D vectors are $\mathbf{a} = (1,2,3)$ and $\mathbf{b} = (-1,1,2)$, and assume them to be associated with the point $(x, y, z) = (1,1,0)$. Convert $\mathbf{a}$ and $\mathbf{b}$ to cylindrical and to spherical coordinates. Then form the scalar product $\mathbf{a} \cdot \mathbf{b}$ using each of the three coordinate representations.

6.3.5. Repeat the computations of Exercise 6.3.4 assuming the two vectors to be associated with the Cartesian coordinate point (1,1,1).

6.3.6. In spherical polar coordinates (r, θ, φ) two points P and Q are respectively at $(1, \pi/4, \pi/3)$ and $(3, 5\pi/8, 7\pi/6)$. Find the distance between P and Q.

6.4 MULTIPLE INTEGRALS

Multiple integrals often arise when we integrate a physical quantity over a region (or all) of three-dimensional space, or when we integrate such quantities over a surface. Multiple integrals can also occur in mathematical analyses not associated with coordinate systems, as for example when we use an integral to represent some function which is then itself integrated. We therefore consider multiple integrations from a perspective that is not limited to physical-space applications.

Multiple integrals can be written in a variety of notational forms. If the differentials of individual variables are written explicitly (e.g., $dx\,dy$ or $dx\,dy\,dz$) it is usual to write as many integral signs as there are integration variables. However, sometimes the differentials are written collectively, as for example dA or dS to indicate an element of area, or even $d\boldsymbol{\sigma}$ to indicate an element of area together with a unit vector normal to the area. In three dimensions common notations are dV (not used in this book), $d\tau$ (used for any number of dimensions), or (for 3-D) d^3r. Some authors write $d\mathbf{r}$ to mean a 3-D volume element; we do not use that notation because we intend that $d\mathbf{r}$ always represent a differential vector (not a volume). When these collective differentials are used, most often a single integral sign is written. The reader must then determine from the entire expression and its context exactly what integration is implied.

We start this section by looking at integrals of a function $f(r, s)$, over a region of r and s. The contribution to this integral from each infinitesimal rectangular region bounded by r, $r + dr$, s, and $s + ds$ is defined to be $f(r, s)dA$, where dA is the area of the infinitesimal rectangle, or $dA = drds$.

It is often helpful to plot the region of integration on a graph with coordinates labeled r and s. For example, the plot for the region $0 < s < R$, $0 < r < s$ is that shown in Fig. 6.5. From the plot, we see that the region of integration can be covered either by integrating first in strips of fixed r (as illustrated in the left panel of the graph) and then summing the strips by integrating over r. Alternatively, one can first

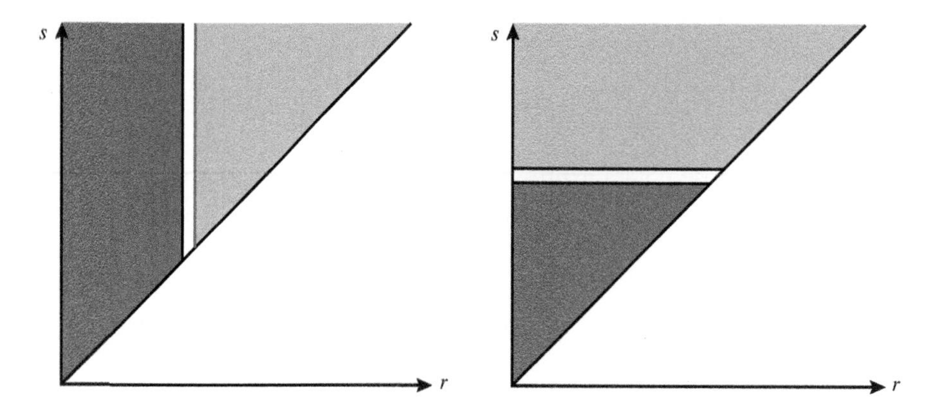

Figure 6.5: Region of integration and integration schemes: left, integrate first over s; right, integrate first over r.

integrate over strips of fixed s (see the right panel) and then sum those strips by integrating over s. One virtue of making a plot is that we can see what the limits of the individual integrations must be. From the left panel, we see that the range of the inner integration (that over s) must be from $s = r$ to $s = R$, while the outer (r) integration is from $r = 0$ to $r = R$. Thus, letting A stand for the region of integration, those observations correspond to

$$I = \int_A f(r,s)\,dA = \int_0^R dr \int_r^R ds f(r,s)\,.$$

On the other hand, integrating first over r we see that its integration range must be from $r = 0$ to $r = s$, with the outer integral (over s) from $s = 0$ to $s = R$, corresponding to the formula

$$I = \int_A f(r,s)\,dA = \int_0^R ds \int_0^s dr f(r,s)\,.$$

Depending upon the actual form of $f(r,s)$, one of these formulas may turn out to be easier than the other to evaluate.

Sometimes the formulation of the integral may be simpler if a particular order of integration is chosen. Suppose we want to integrate $f(r,s)$ over the region shown in Fig. 6.6, where the upper boundary of the region of integration is $\sin r$. If we integrate in strips of fixed r, the s integration will be from 0 to $\sin r$. But if we integrate in strips of fixed s, the r integration will be from the principal value of $\sin^{-1} s$ to $\pi - \sin^{-1} s$. Need we say more?

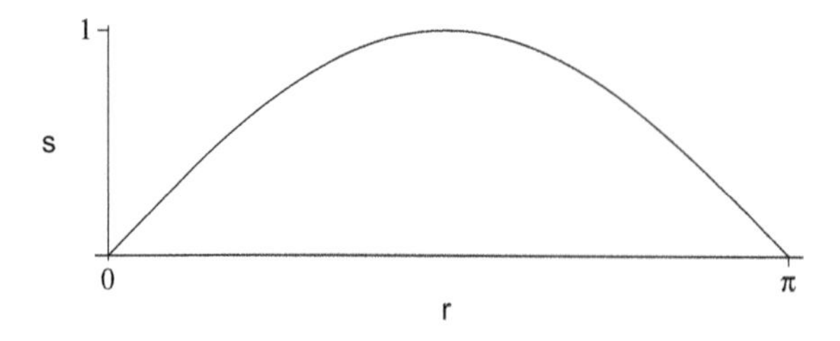

Figure 6.6: Region for double integral. Curve is $s = \sin r$.

INTEGRATION IN DIFFERENT COORDINATE SYSTEMS

Sometimes either the form of a function being integrated or the shape of the region of integration suggests that the integration be carried out in a particular coordinate system. For example, we may require the integral of $f(x,y) = 1/\sqrt{x^2+y^2}$ over a disk of radius R centered at the origin. While one might be able to summon the considerable skill needed to carry out this integration in Cartesian coordinates x, y, it is both easier and wiser to convert this problem to plane polar coordinates r, θ. To make the conversion we need to (1) change the function to its form in the new coordinates, (2) replace $dxdy$ by its equivalent in terms of dr and $d\theta$, and (3) identify the region of integration in terms of the polar coordinates. For the present problem, the first and third of these tasks are simple; in polar coordinates, $f(x,y) \longrightarrow 1/r$; the region of integration is $0 \le r \le R$, $0 \le \theta < 2\pi$. The second task, the conversion of $dxdy$, is the key issue of the present discussion.

What is required is that $dxdy$, which is the area of the region bounded in one direction by x and $x+dx$ and in the other (orthogonal) direction by y and $y+dy$, be replaced by the area of the infinitesimal region defined by dr and $d\theta$. Using the scale factors for dr and $d\theta$ to obtain the corresponding displacements, namely $h_r = 1$ and $h_\theta = r$, we have

$$dxdy \longrightarrow h_r h_\theta dr\, d\theta = r\, dr\, d\theta \,. \tag{6.24}$$

This situation is illustrated in Fig. 6.7.

Returning to our current problem, we now have

$$\int_{\text{disk}} \frac{1}{\sqrt{x^2+y^2}}\, dx\, dy = \int_{\text{disk}} \frac{1}{r}\, r\, dr\, d\theta = \int_0^R dr \int_0^{2\pi} d\theta = 2\pi R \,.$$

The other conversion we encounter frequently is from Cartesian to spherical polar coordinates. The volume associated with infinitesimal changes in r, θ, φ is, referring to Eq. (6.23),

$$dxdydz \longrightarrow h_r h_\theta h_\varphi\, dr\, d\theta\, d\varphi = r^2 \sin\theta\, dr\, d\theta\, d\varphi \,. \tag{6.25}$$

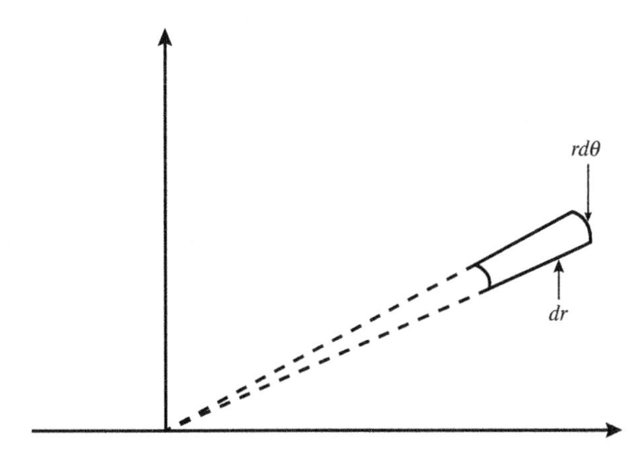

Figure 6.7: Element of area in plane polar coordinates.

JACOBIANS

Conversions of integrations between orthogonal coordinate systems are, from a practical viewpoint, completely covered by the preceding subsection. However, we also encounter changes of variables in multiple integrals that are not of that type. For example, consider the integral

$$I = \int_0^\infty dx \int_0^\infty dy \, \frac{x\, e^{-(xy+x/y)}}{y} \,. \tag{6.26}$$

It appears that we might simplify this integral by making the substitutions $s = xy$ and $t = x/y$, but to do so we would need to know the relationship between $dx\, dy$ and $ds\, dt$. The conversion factor between these quantities is known as the **Jacobian** (denoted J) of the coordinate transformation, with definition and notation

$$J = \frac{dx\, dy}{ds\, dt} \equiv \frac{\partial(x,y)}{\partial(s,t)} = \begin{vmatrix} \left(\dfrac{\partial x}{\partial s}\right)_t & \left(\dfrac{\partial x}{\partial t}\right)_s \\ \left(\dfrac{\partial y}{\partial s}\right)_t & \left(\dfrac{\partial y}{\partial t}\right)_s \end{vmatrix} \,. \tag{6.27}$$

Thus, the Jacobian is a determinant involving partial derivatives connecting the two sets of coordinates. In larger numbers of dimensions the Jacobian formula generalizes to similar determinants. It can also be shown (as suggested by the notation) that interchanging the roles of s, t and x, y will convert J into $1/J$; one of these may be easier than the other to evaluate. A fuller discussion and derivation of the Jacobian formula can be found in Arfken et al. in the Additional Readings.

The sign of J depends on the order chosen for writing x, y and s, t, and may need to be reversed if necessary to make the element of area have a sign consistent with the integral being evaluated.

Example 6.4.1. Use of Jacobian

Let's now return to the integral that motivated this discussion. We rewrite it in terms of s and t, noting that the ranges of s and t will each be $(0, \infty)$ and that we will need the absolute value of the Jacobian of the coordinate transformation. Thus, with $s = xy$ and $t = x/y$,

$$I = \int_0^\infty ds \int_0^\infty dt \left| \frac{\partial(x,y)}{\partial(s,t)} \right| s\, e^{-s-t} \,. \tag{6.28}$$

At this point we could solve for x and y in terms of s and t and proceed to evaluate the Jacobian, but we already have s and t in terms of x and y, so it is simpler to compute

$$\frac{\partial(s,t)}{\partial(x,y)} = \begin{vmatrix} \left(\dfrac{\partial s}{\partial x}\right)_y & \left(\dfrac{\partial s}{\partial y}\right)_x \\ \left(\dfrac{\partial t}{\partial x}\right)_y & \left(\dfrac{\partial t}{\partial y}\right)_x \end{vmatrix} = \begin{vmatrix} y & x \\ 1/y & -x/y^2 \end{vmatrix} = -\frac{2x}{y} = -2s \,.$$

From the above, we have $\dfrac{\partial(x,y)}{\partial(s,t)} = -\dfrac{1}{2s}$.

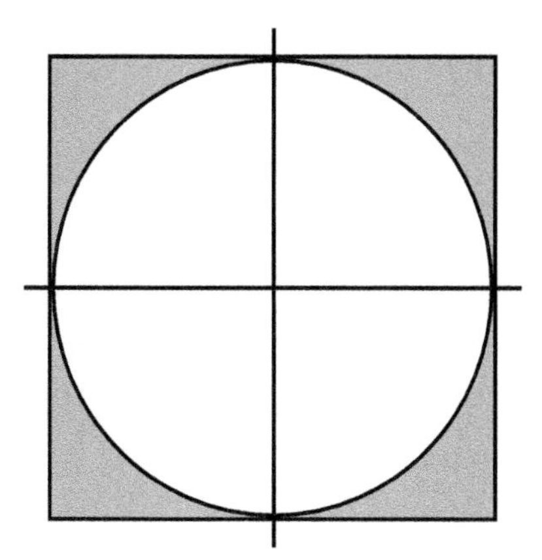

Figure 6.8: Comparison of integral in Cartesian and polar coordinates.

Taking the absolute value, and noting with satisfaction that the factors s cancel, our integral becomes

$$I = \frac{1}{2}\int_0^\infty ds \int_0^\infty dt\, e^{-s-t}\,,$$

which is easily integrated in both s and t to give the final result $I = 1/2$.

■

A common misconception (found in some other texts and even in research articles) may arise when, for example, one converts an integral expressed in Cartesian coordinates into the equivalent integration in terms of polar coordinates. Then (see Fig. 6.8) a reader may wonder if the residual regions in the corners of the square in the figure are somehow being neglected (therefore suggesting the need that they be shown to be negligible). To resolve this question one simply needs to observe that when the polar-coordinate integration is extended to large values of r, any point in these corners will eventually be included in the integration. Thus, if the integral converges, there is no problem.

Exercises

6.4.1. Evaluate by changing the order of integration. Check your work using symbolic computation.

$$\int_0^\infty dx \int_x^\infty dy\, \frac{y-x}{y}\, e^{-y}\,.$$

6.4.2. Evaluate $\displaystyle\int_0^\infty dx \int_x^\infty dy\, y^2\, e^{-y^2} \sin xy\,.$

Check this result using symbolic computation.

6.4.3. Evaluate $\displaystyle\int_A \left(c + \sqrt{x^2+y^2}\right) dA\,,$

where A is a circle of radius c centered at the origin.

6.4.4. Find the center of mass of a uniform solid hemisphere H of radius a.

6.4.5. Evaluate $\iiint x^2 y^2 z^2 \, dx\, dy\, dz$

for a cylinder of height h and radius a with its center at the origin and its axis in the x-direction.

6.4.6. Write the Jacobian matrices $\mathsf{J} = \partial(x,y)/\partial(s,t)$ and $\mathsf{K} = \partial(s,t)/\partial(x,y)$ in forms such that the matrix product $\mathsf{J}\,\mathsf{K} = \mathbf{1}$, showing that $\mathsf{K} = \mathsf{J}^{-1}$.

6.4.7. Introduce a change of variables that makes this double integral elementary, and then evaluate it. Use symbolic computation to check your work.

$$\int_0^\infty dy \int_y^\infty dx \, \frac{y}{x} \, e^{-(x+y)} \, .$$

6.5 LINE AND SURFACE INTEGRALS

When a space has more than one dimension, we have the possibility of specifying integrations that are to be evaluated along a path more general than along a coordinate axis. For example, we may want to find the length of an arbitrary trajectory between two points in 2-D or 3-D space, or possibly the average or total of some quantity along such a path. In 3-D, we may also want to find the area of a surface or the total amount of some quantity on the surface. Problems of these sorts can be formulated using the standard concepts of integral calculus, but attention must be paid to aspects influenced by the geometry.

LINE INTEGRALS

In elementary physics you have probably learned that moving a particle subject to a force $\mathbf{F}$ an infinitesimal amount $d\mathbf{s}$ involves an amount of mechanical work $dw = \mathbf{F}\cdot d\mathbf{s}$. The total amount of work from a finite displacement can be represented by an integral of the form $\int \mathbf{F}\cdot d\mathbf{s}$, where the contributions to the integral are those corresponding to the path of the displacement. An integral of this sort is called a **line integral**. Through the vector nature of $d\mathbf{s}$ it includes a specification of the direction of the traversal along the path. Another example of a line integral, this time not involving vectors or direction of traversal, is the computation of the length of a curve that connects two spatial points. This integral has the deceptively simple form $\int ds$, where ds is an element of length along the curve. Note that a full definition of either of those preliminary examples must include a specification of the path over which it is to be evaluated.

Example 6.5.1. Work Integral

We consider here (in arbitrary units) a line integral representing mechanical work,

$$W = \int \mathbf{F}\cdot d\mathbf{s}, \qquad \text{with} \qquad \mathbf{F} = y\,\hat{\mathbf{e}}_x - (x+y)\hat{\mathbf{e}}_y \, ,$$

which we evaluate for several different paths between the points $P = (0,0,0)$ and $Q = (1,1,0)$. Referring to Fig. 6.9, we consider first the path from P to Q consisting of the segments labeled A and B, giving the line integral for this path the name W_1. Our plan is to evaluate the line integrals of the individual segments and then add them to obtain a final result.

For Segment A, $y = 0$, $d\mathbf{s} = \hat{\mathbf{e}}_x\,dx$, $\mathbf{F} = -x\hat{\mathbf{e}}_y$, so $\mathbf{F}\cdot d\mathbf{s} = 0$, and the contribution to W from this segment is zero. For Segment B, $x = 1$, $d\mathbf{s} = \hat{\mathbf{e}}_y\,dy$, $\mathbf{F} = y\mathbf{e}_x - (1+y)\hat{\mathbf{e}}_y$,

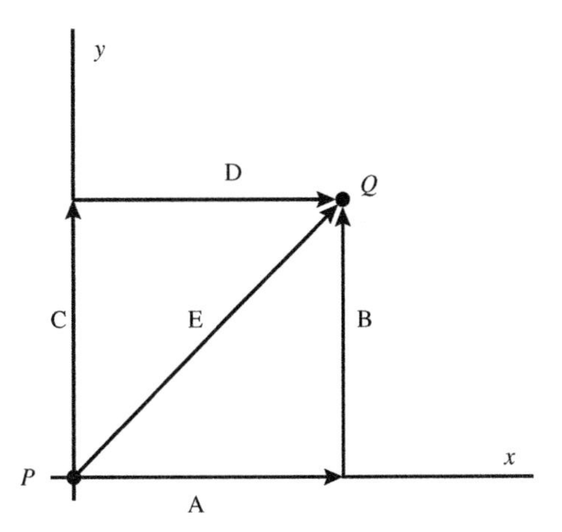

Figure 6.9: Paths for a line integral.

so $\mathbf{F} \cdot d\mathbf{s} = -(1+y)\,dy$, and we must integrate this expression from $y = 0$ to $y = 1$:

$$w_B = -\int_0^1 (1+y)\,dy = -\frac{3}{2}\,.$$

We therefore get $W_1 = 0 + w_B = -3/2$.

A second possibility is to integrate over Segments C and D, naming the line integral for this path W_2. For Segment C, $x = 0$, $d\mathbf{s} = \hat{\mathbf{y}}\,dy$, $\mathbf{F} = y\hat{\mathbf{e}}_x - y\hat{\mathbf{e}}_y$, so $\mathbf{F} \cdot d\mathbf{s} = -y\,dy$. Integration from $y = 0$ to $y = 1$ yields

$$w_C = -\int_0^1 ydy = -\frac{1}{2}\,.$$

For Segment D, $y = 1$, $d\mathbf{s} = \hat{\mathbf{e}}_x\,dx$, $\mathbf{F} = \hat{\mathbf{e}}_x - (x+1)\hat{\mathbf{e}}_y$, so $\mathbf{F} \cdot d\mathbf{s} = dx$, which we integrate from $x = 0$ to $x = 1$:

$$w_D = \int_0^1 dx = 1\,.$$

We therefore get $W_2 = w_C + w_D = 1/2$.

The line integrals for both the above paths are fairly simple because the path segments are in the coordinate directions. Let's now take the path consisting solely of Segment E. For this segment, x and y are both varying, but note that only one of them is independent; the requirement that we be on the indicated line corresponds to the condition $x = y$. On this path, $d\mathbf{s} = \hat{\mathbf{e}}_x\,dx + \hat{\mathbf{e}}_y\,dy$, and therefore

$$\mathbf{F} \cdot d\mathbf{s} = \Big(y\,\hat{\mathbf{e}}_x - (x+y)\hat{\mathbf{e}}_y\Big) \cdot \Big(\hat{\mathbf{e}}_x\,dx + \hat{\mathbf{e}}_y\,dy\Big) = y\,dx - (x+y)\,dy\,.$$

We may now integrate $\mathbf{F} \cdot d\mathbf{s}$ using either x or y as the independent variable; we don't even need to make the same choice for both of the terms. Integrating the first term with x as the independent variable, we write y in terms of x (easy because $y = x$) and identify the integration limits as the start and end points of x. We then take y as the independent variable for the second term. The result is

$$W_3 = \int \mathbf{F} \cdot d\mathbf{s} = \int_0^1 x\,dx - \int_0^1 2y\,dy = 0\,.$$

Note that $W_1 \neq W_2 \neq W_3$, showing that the amount of work (for this force) depends upon the path between the two endpoints.

We continue with this example by considering what happens if we take the reverse of one of the paths, namely Segment B (backwards) followed by Segment A (backwards). The issues involved are fully illustrated if we limit further discussion to Segment B.

Reverse traversal of Segment B involves the same value of $\mathbf{F}$ as for a forward traverse, but now $d\mathbf{s} = -\hat{\mathbf{e}}_y\,dy$, so $\mathbf{F}\cdot d\mathbf{s} = +(1+y)\,dy$. The reader might be tempted to write the integral for "reverse-B" as

$$w_{\text{revB}} = \int_{y=1}^{0} \mathbf{F}\cdot d\mathbf{s} = \int_{1}^{0} (1+y)\,dy = -3/2 \qquad\qquad \text{WRONG!}$$

That is incorrect because the process represented by the integration is to add up the contributions from the elements of the line, and the inversion of the integration limits causes their entry with the wrong sign. The direction of travel on the path has already been fully accounted for by the value assigned to $d\mathbf{s}$.

A correct computation,

$$w_{\text{revB}} = \int_{y=0}^{1} \mathbf{F}\cdot d\mathbf{s}\,,$$

gives the expected and reasonable result, namely that work for a path and its reverse are equal in magnitude and opposite in sign.

■

Let's now turn to the other problem mentioned in the introduction to this Section, namely that of computing the length of a specified curve connecting two points, P and Q. For simplicity we work in 2-D space. If the curve is described by the equation $y = y(x)$, a change dx in x will cause a point on the curve to move a distance

$$ds = \sqrt{dx^2 + dy^2} = \sqrt{1 + \left(\frac{dy(x)}{dx}\right)^2}\;|dx|\,. \tag{6.29}$$

The present situation differs from that treated in Example 6.5.1 in that the contributions to the line integral do not depend upon the direction of travel along the curve. The concept of "direction of travel" is not relevant here, and that fact is indicated by the occurrence of the absolute value sign of $|dx|$. The absolute value sign may seem inconvenient, but it can be removed if we restrict consideration to curves that are single-valued in x and we choose the endpoints so that $x_P < x_Q$. Then the curve length can be written as the following integral in the independent variable x:

$$L = \int_{x_P}^{x_Q} \sqrt{1 + \left(\frac{dy(x)}{dx}\right)^2}\;dx\,. \tag{6.30}$$

What if somewhere the curve becomes vertical? Then dy/dx becomes infinite, but also $\sqrt{dx^2 + dy^2} = |dy|$, and it becomes more convenient to take y as the independent variable, with $x = x(y)$. If we have a curve that is not single-valued in the independent variable (this includes all closed curves) it is necessary to break the line integral into pieces that satisfy the single-valued requirement.

Generalizing now to the integral of some function $f(x, y)$ on the curve, we have (subject to the requirements discussed for L)

$$I = \int_P^Q f(x,y)\,ds = \int_{x_P}^{x_Q} f(x, y(x))\sqrt{1 + \left(\frac{dy(x)}{dx}\right)^2}\;dx\,. \tag{6.31}$$

It should be obvious that the values of L or I may depend upon the path of integration.

The formulas in Eqs. (6.30) and (6.31) can get relatively complicated for general paths, but in many physics applications the path is simple (e.g., one or more segments parallel to a Cartesian axis, or a circular arc). For paths along a circle of radius r_0, great simplification can often be achieved by using polar coordinates:

$$I = \int_{\theta_P}^{\theta_Q} f(r_0, \theta)\, r_0\, d\theta\,. \tag{6.32}$$

Note that $r_0\, d\theta$ is the path length corresponding to $d\theta$.

Another possibility is that the integration path may be expressed parametrically, as $x = x(t)$, $y = y(t)$. We may then integrate in t, noting that

$$\frac{ds}{dt} = \sqrt{\left(\frac{dx(t)}{dt}\right)^2 + \left(\frac{dy(t)}{dt}\right)^2}\, dt\,,$$

so

$$I = \int_{t_P}^{t_Q} f(x(t), y(t)) \sqrt{\left(\frac{dx(t)}{dt}\right)^2 + \left(\frac{dy(t)}{dt}\right)^2}\, dt\,. \tag{6.33}$$

Example 6.5.2. Average on Circular Arc

Let's find, working in 2-D, the average value of xy, which may be written $\langle xy \rangle$, for the portion of the unit circle in the first quadrant. In terms of line integrals, this problem can be written

$$M = \int xy\, ds, \qquad L = \int ds, \qquad \langle xy \rangle = \frac{M}{L}\,,$$

where the integrals are over the arc in the first quadrant.

It is natural to consider this problem in plane polar coordinates, where $x = \cos\theta$, $y = \sin\theta$, $ds = d\theta$, and the line integrals are from $\theta = 0$ to $\theta = \pi/2$. The resultant value of L is $\pi/2$ (we really knew this without computation), and

$$M = \int_0^{\pi/2} \sin\theta \cos\theta\, d\theta = 1.$$

Thus, $\langle xy \rangle = 2/\pi$.

■

SURFACE INTEGRALS

Surface integrals in 3-D space are the limits of summations in which the integrand is multiplied by the element of area on the surface, and can take the form

$$I = \int f(x, y, z)\, dA\,. \tag{6.34}$$

Other possibilities for surface integrals involve vector expressions in which the element of area is written in the form $d\boldsymbol{\sigma} = \hat{\mathbf{n}}\, dA$, where $\hat{\mathbf{n}}$ is a unit vector normal to the surface. The choice between the two normal directions is a matter of convention;

for closed surfaces it is usually taken to be in the outward direction. A typical form for a vector surface integral is

$$K = \int \mathbf{F} \cdot d\boldsymbol{\sigma} = \int \mathbf{F} \cdot \hat{\mathbf{n}}\, dA\,. \tag{6.35}$$

Formally, this second type of integral reduces to the first if we identify F_n as the component of $\mathbf{F}$ normal to the surface.

We continue with a surface integral of the form given in Eq. (6.34), and consider a surface in 3-D space that is defined (for a range of x and y) by the equation $z = g(x, y)$. An obvious approach will be to integrate over x and y, but if we do so we will need to express dA in the form $h(x, y)\, dx\, dy$. If our surface is parallel to the xy-plane (i.e., if g is a constant), then the element of area is simply $dx\, dy$ and we have an ordinary double integral. In more general cases, the expression of dA in terms of $dx\, dy$ depends upon the angle dA makes with the xy-plane. It can be shown that the element of surface area then takes the form

$$dA = \sqrt{\left(\frac{\partial g}{\partial x}\right)^2 + \left(\frac{\partial g}{\partial y}\right)^2 + 1}\;\; dx\, dy\,, \tag{6.36}$$

the 3-D generalization of Eq. (6.29). If the surface is vertical (causing one or both the partial derivatives of g to become infinite), we will need to make z one of the independent variables and express the element of area in terms of $dx\, dz$ or $dy\, dz$. Closed surfaces can be treated by integrating their upper and lower parts as separate integrals which can then be added to give a final result.

Surface integrals that arise in physics often involve spherical or cylindrical surfaces and may be simpler to evaluate in a corresponding coordinate system. For a surface at $r = r_0$ of a spherical coordinate system we have

$$dA = h_\theta h_\varphi\, d\theta\, d\varphi = r_0^2 \sin\theta\, d\theta\, d\varphi\,. \tag{6.37}$$

Here h_θ and h_φ are the scale factors given in Eq. (6.23).

For the curved surface at $\rho = \rho_0$ of a cylindrical coordinate system, the corresponding formula is

$$dA = h_\varphi h_z\, d\varphi\, dz = \rho_0\, d\varphi\, dz\,. \tag{6.38}$$

In vector analysis, surface integrals are often assigned a sign dependent upon conventions and the geometry. Those sign assignments are reviewed where they become relevant.

Example 6.5.3. Inclined Surface

A tetrahedron has vertices at the points $(0, 0, 0)$, $(1, 0, 0)$, $(0, 1, 0)$, and $(0, 0, 1)$ of a Cartesian coordinate system. We wish to calculate $\int \mathbf{B} \cdot d\boldsymbol{\sigma}$ over the area A of the tetrahedron's oblique face, given that $\mathbf{B} = B\hat{\mathbf{e}}_z$.

The equation for the plane containing the oblique face is $x + y + z = 1$, which can be written $z = g(x, y) = 1 - x - y$. We take $d\boldsymbol{\sigma}$ to be the normal to this plane that points out of the tetrahedron, defined by the unit vector $\hat{\mathbf{n}} = (1, 1, 1)/\sqrt{3}$. Using the above information,

$$K = \int_A \mathbf{B} \cdot d\boldsymbol{\sigma} = \int B_z \hat{\mathbf{e}}_z \cdot \hat{\mathbf{n}}\, dA = \int_{A_0} \left(B_z/\sqrt{3}\right) \sqrt{\left(\frac{\partial g}{\partial x}\right)^2 + \left(\frac{\partial g}{\partial y}\right)^2 + 1} \;\; dx\, dy$$

$$= \int_{A_0} \left(B_z/\sqrt{3}\right) \sqrt{3} \;\; dx\, dy\,,$$

where A_0 is the region of x and y corresponding to A. The integrand has the constant value B_z and A_0 has area 1/2, so we get $K = B_z/2$.

■

Example 6.5.4. Centroid of Hemispherical Shell

Let's find the centroid of a unit hemispherical shell whose center of curvature is at the origin and with an orientation such that $z \geq 0$. This problem considers only the curved surface.

In general, the centroid of a shell will be at a point (X, Y, Z) such that X is the mass-weighted average of x for all elements of area belonging to the shell; Y and Z are defined analogously. Here we have two important simplifying features: (1) the shell is uniform in density, and (2) it is easily described in spherical polar coordinates. Moreover, the symmetry of the problem indicates that the centroid is on the polar axis and we only need to compute an average value of z for the elements of surface area:

$$M = \int z\, dA, \qquad A = \int dA, \qquad Z = \frac{M}{A}\,,$$

where both integrals are over the hemispherical surface.

Using spherical coordinates (r, θ, φ), we have $r = 1$ for the entire shell, and we also know from elementary geometry that $A = 2\pi r^2 = 2\pi$. It remains to compute M. We therefore identify $z = \cos\theta$ and $dA = h_\theta h_\varphi\, d\theta\, d\varphi$, where $h_\theta = r$ and $h_\varphi = r\sin\theta$ are coordinate scale factors. Thus, for the present problem, $dA = \sin\theta\, d\theta\, d\varphi$, and the integration range is $0 \leq \varphi < 2\pi$, $0 \leq \theta \leq \pi/2$.

The above data lead us to

$$M = \int_0^{2\pi} d\varphi \int_0^{\pi/2} d\theta\, \cos\theta \sin\theta = 2\pi \int_0^{\pi/2} \cos\theta \sin\theta\, d\theta = \pi\,.$$

The centroid is at $Z = \pi/2\pi = 1/2$.

■

GREEN'S THEOREM IN THE PLANE

In one dimension, a function and its derivative are related by the formula

$$f(b) - f(a) = \int_a^b \frac{df(s)}{ds}\, ds\,. \tag{6.39}$$

That idea generalizes, in a way, to larger numbers of dimensions. In 2-D, the values of a function on a closed curve can be related to an integral (containing derivatives of the function) over the area bounded by the curve. A precise statement of this relationship is known as Green's theorem in the plane.

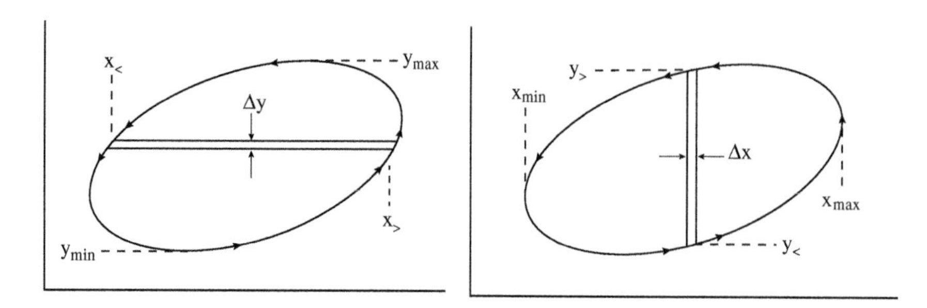

Figure 6.10: Areas and bounding curves.

Our starting point for the development of Green's theorem is to consider the following line integral in the xy-plane,

$$L_1 = \oint P(x, y)\, dy\,. \tag{6.40}$$

The integration is to be in the **counterclockwise** direction around a closed curve that bounds an area, where $P(x, y)$ is a function that is differentiable on the curve and within the bounded area but is otherwise arbitrary. For simplicity we assume the curve to be everywhere convex, implying that no line parallel to either coordinate axis intersects the curve more than twice. A curve and enclosed area meeting these specifications is illustrated in Fig. 6.10.

We now focus attention on the horizontal strip of width Δy shown in the left panel of Fig. 6.10. The strip intersects the curve at two x values of which the larger and smaller are respectively $x_>(y)$ and $x_<(y)$; those intersections are at points that depend on the value of y at which the strip is placed. There are two contributions to the line integral L_1 within the strip. One is at $x_>$, where $P(x, y) = P(x_>, y)$ and (because the integration is counterclockwise) $dy = +\Delta y$. The other contribution is as $x_<$, with $P(x, y) = P(x_<, y)$ and $dy = -\Delta y$. Adding those together and rewriting the result using Eq. (6.39), we have

$$\Big[P(x_>, y) - P(x_<, y)\Big]\, \Delta y = \Delta y \int_{x_<}^{x_>} \frac{\partial P}{\partial x}\, dx\,. \tag{6.41}$$

The sum of the contributions to L_1 of all the horizontal strips (and therefore the value of L_1), is the integral of the expression in Eq. (6.41) over the entire range of y for the boundary curve. That integral,

$$L_1 = \int_{y_{\min}}^{y_{\max}} dy \int_{x_<(y)}^{x_>(y)} \frac{\partial P}{\partial x}\, dx\,,$$

can be recognized as an integration over the area enclosed by the boundary curve, and (inserting the definition of L_1) we reach

$$\oint_{\partial A} P(x, y)\, dy = \int_A \frac{\partial P(x, y)}{\partial x}\, dA\,. \tag{6.42}$$

Equation (6.42) illustrates two conventions that it is useful to remember: first, the notation ∂A denotes the closed curve that bounds A; in a similar vein, we write ∂V to indicate the closed surface that bounds a volume V. Second, line integrals around closed curves are, unless specifically indicated otherwise, in the "mathematically positive," i.e., counterclockwise, direction. The corresponding 3-D convention is that elements of area for closed surfaces have normal vectors that point out of the enclosed volume.

We now repeat the analysis that led from Eq. (6.40) to Eq. (6.42) for another line integral:

$$L_2 = \oint Q(x,y)\,dx\,. \tag{6.43}$$

For that line integral we use the vertical strips shown in the right panel of Fig. 6.10. Note that there is one significant difference relative to our previous discussion: for the intersection of the curve and strip at larger y the value of dx is $-\Delta x$, which is reversed in sign from what we had before. A similar sign reversal occurs for the intersection at smaller y; there $dx = +\Delta x$. As a result, the contribution to L_2 from a vertical strip at x is

$$\Big[Q(x,y_<) - Q(x,y_>)\Big]\,\Delta x = -\Delta x \int_{y_<}^{y_>} \frac{\partial Q}{\partial y}\,dy\,. \tag{6.44}$$

Integrating over all vertical strips, we obtain a formula similar to Eq. (6.42), but reversed in sign:

$$\oint_{\partial A} Q(x,y)\,dx = -\int_A \frac{\partial Q(x,y)}{\partial y}\,dA\,. \tag{6.45}$$

The sign reversal is understandable; it is related to the fact that the y-direction is reached from the x-direction by a positive 90° rotation; a rotation of −90° is needed to go from the y-direction to the x-direction.

Each of Eqs. (6.42) and (6.45) can be regarded as an instance of Green's theorem in the plane; a single equation that captures all the information from both (and which can be regarded as a standard statement of the theorem) is

$$\oint_{\partial A} \Big[P(x,y)\,dy + Q(x,y)\,dx\Big] = \int_A \left[\frac{\partial P}{\partial x} - \frac{\partial Q}{\partial y}\right] dA\,. \tag{6.46}$$

An important feature of this theorem is that it permits us to evaluate a line integral as a double integral over an area or the reverse, so we may choose the formulation that is easier to use. Green's theorem in the plane also has important uses in vector analysis and in complex variable theory. You will encounter it again!

Exercises

6.5.1. Evaluate the line integral $\int (2xy - y^2)\,dx$, where the integration is on the curve $y = x^{3/2}$ from $x = 0$ to $x = 4$.

6.5.2. Evaluate the line integral $\int \mathbf{B}\cdot d\mathbf{r}$, where $\mathbf{B} = x\,\hat{\mathbf{e}}_x - xy\,\hat{\mathbf{e}}_y$, for a straight-line path from $(x,y) = (0,0)$ to $(x,y) = (2,1)$.

6.5.3. A wire of uniform density lies along the portion of the parabola $y = x^2/2$ that extends from $(x, y) = (0, 0)$ to $(x, y) = (2, 2)$. Find the position of its centroid.

Hint. Use symbolic computation to evaluate the necessary integrals.

6.5.4. For the curved surface of a cylinder of radius r_0 with axis on the z axis and extending from $z = -h$ to $z = +h$, evaluate

$$\int (x^2 + y^2 - z^2)\, dA \,.$$

6.5.5. For the curved cylindrical surface of Exercise 6.5.4, evaluate

(a) $\displaystyle\int (x^2 - y^2) z^2 \, dA \,,$

(b) $\displaystyle\int (x^2 - xy) z^2 \, dA \,.$

6.5.6. For the curved surface of the hemisphere of radius r_0 and $z \geq 0$, evaluate

$$\int (x^2 + y^2) z^2 \, dA \,.$$

6.5.7. What does Green's theorem tell us (for an arbitrary area A) if we choose $P(x, y) = x$ and $Q(x, y) = 0$? What is the general result if we apply the theorem to $P(x, y) = 0$, $Q(x, y) = y$?

6.5.8. Verify a particular case of Green's theorem by evaluation of the integrals on both sides of Eq. (6.46) when we choose ∂A to be the unit circle, $P(x, y) = 0$, and $Q(x, y) = x^3 - y^3$.

6.6 REARRANGEMENT OF DOUBLE SERIES

Just as multiple integrals can be evaluated in ways that differ in the coordinate variables and the integration order, so also can we rearrange the summation order in multiple sums. An absolutely convergent double series (one whose terms are identified by two summation indices) presents interesting rearrangement opportunities. Consider

$$S = \sum_{m=0}^{\infty} \sum_{n=0}^{\infty} a_{n,m} \,. \tag{6.47}$$

In addition to the obvious possibility of reversing the order of summation (i.e., doing the m sum first), we can make rearrangements that are more innovative. One reason for doing so is that we may be able to reduce the double sum to a single summation, or even evaluate the entire double sum in closed form.

As an example, suppose we make the following index substitutions in our double series: $m = q$, $n = p - q$. Then we cover all $n \geq 0$, $m \geq 0$ by assigning p the range $(0, \infty)$, and q the range $(0, p)$, so our double series can be written

$$S = \sum_{p=0}^{\infty} \sum_{q=0}^{p} a_{p-q,q} \,. \tag{6.48}$$

In the nm-plane our region of summation is the entire quadrant $m \geq 0$, $n \geq 0$; in the pq-plane our summation is over the triangular region sketched in Fig. 6.11. This same pq region can be covered when the summations are carried out in the reverse

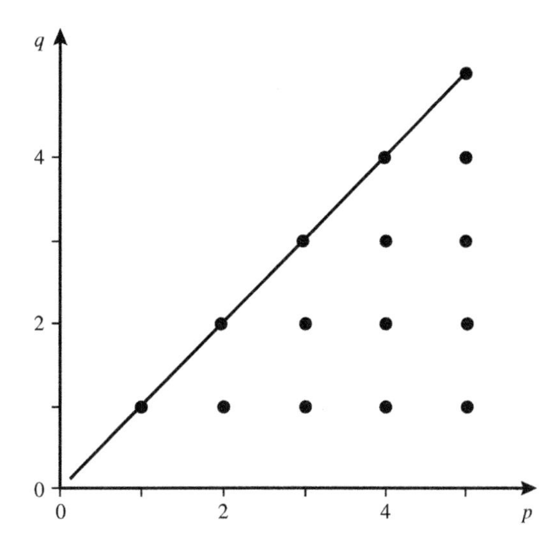

Figure 6.11: The pq index space.

order, but with limits

$$S = \sum_{q=0}^{\infty} \sum_{p=q}^{\infty} a_{p-q,q} \,.$$

The important thing to notice here is that all the above schemes have in common the feature that, by allowing the indices to run over their designated ranges, every $a_{n,m}$ is eventually encountered, and is encountered exactly once.

Another possible index substitution is to set $n = s$, $m = r - 2s$. If we sum over s first, its range must be $(0, [r/2])$, where $[r/2]$ is the integer part of $r/2$, i.e., $[r/2] = r/2$ for r even and $(r-1)/2$ for r odd. The range of r is $[0, \infty)$. That situation corresponds to

$$S = \sum_{r=0}^{\infty} \sum_{s=0}^{[r/2]} a_{s,r-2s} \,. \tag{6.49}$$

The sketches in Figs. 6.12–6.14 show the order in which the $a_{n,m}$ are summed when using the forms given in Eqs. (6.47)–(6.49).

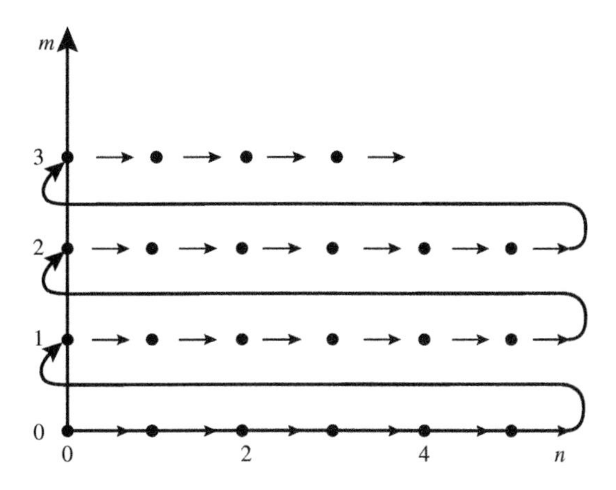

Figure 6.12: Order in which terms are summed with m, n index set, Eq. (6.47).

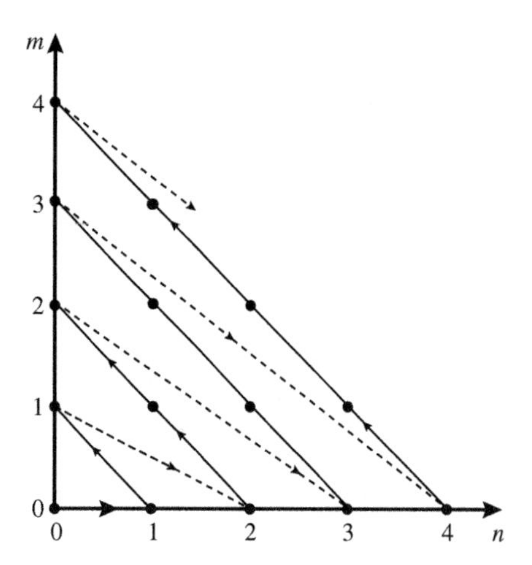

Figure 6.13: Order in which terms are summed with p, q index set, Eq. (6.48).

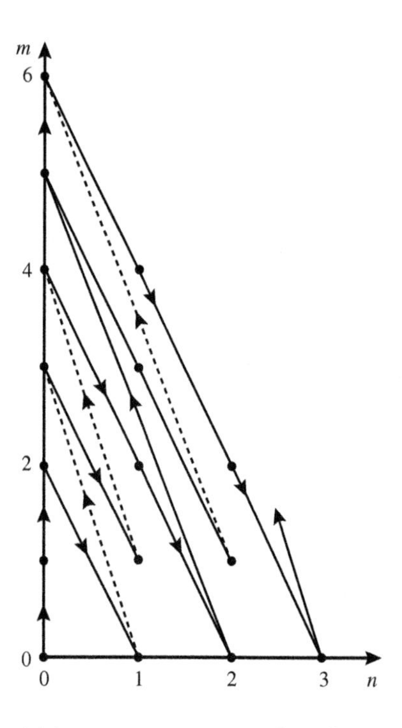

Figure 6.14: Order in which terms are summed with r, s index set, Eq. (6.49).

Because we have assumed the double series introduced originally as Eq. (6.47) to be absolutely convergent, all those rearrangements give the same ultimate result.

Example 6.6.1. Two-Variable Maclaurin Series

Let's compute the two-variable Maclaurin series for e^{x+y}. For this series we need the derivatives identified as a_{pq} in Eq. (6.3). Taking first the qth derivative of e^{x+y} with respect to y, we differentiate (regarding x as a constant), obtaining

$$\frac{d^q e^{x+y}}{dy^q} = e^{x+y} .$$

Next, differentiating p times with respect to x (now regarding y as a constant), we reach e^{x+y}. After taking the derivatives we set x and y to their assigned values $x = y = 0$, reaching the result that for all p and q, $a_{pq} = 1$. Our Maclaurin series is therefore

$$e^{x+y} = \sum_{p,q=0}^{\infty} \frac{x^p y^q}{p!\,q!} .$$

Now let's compare that result with what we would get if we expanded the exponential in powers of $x+y$ and then further expanded those powers using the binomial theorem:

$$e^{x+y} = \sum_{n=0}^{\infty} \frac{(x+y)^n}{n!} = \sum_{n=0}^{\infty}\sum_{p=0}^{n} \binom{n}{p} \frac{x^p\, y^{n-p}}{n!} = \sum_{n=0}^{\infty}\sum_{p=0}^{n} \frac{x^p\, y^{n-p}}{p!\,(n-p)!} .$$

That double sum, which is of the form of Eq. (6.48), is equivalent to one of the type given in Eq. (6.47) when $n-p$ is replaced by q and the summation limits are adjusted properly.

■

Exercises

6.6.1. Verify that the sums claimed to be equivalent in Example 6.6.1,

$$\sum_{p,q=0}^{\infty} \frac{x^p y^q}{p!\,q!} \qquad \text{and} \qquad \sum_{n=0}^{\infty}\sum_{p=0}^{n} \frac{x^p\, y^{n-p}}{p!\,(n-p)!} ,$$

actually give identical results for $x = 0.1$, $y = -0.3$.

6.6.2. Verify the equivalence of the following two double summations by comparing the coefficients of $x^p\, t^q$ for all p and q when both are no larger than 6.

$$\sum_{n=0}^{\infty}\sum_{k=0}^{n} \frac{(-1)^k\,(2n)!}{2^{2n} k!\, n!\,(n-k)!} (2x)^{n-k} t^{n+k} ,$$

$$\sum_{n=0}^{\infty}\sum_{k=0}^{[n/2]} \frac{(-1)^k\,(2n-2k)!}{2^{2n-2k} k!\,(n-k)!\,(n-2k)!} (2x)^{n-2k} t^{n} .$$

The notation $[n/2]$ denotes the largest integer less than or equal to $n/2$.

6.7 DIRAC DELTA FUNCTION

In mechanics problems one may encounter a force whose duration in time is very short but which produces a significant (but finite) amount of momentum transfer. An example is when a speeding automobile is stopped by hitting a massive concrete barrier. Under such circumstances a detailed description of the time dependence of the force may become irrelevant; what certainly matters is the **impulse**, i.e., the total momentum transfer, which is the integral of the force over the time during which it acts. For many purposes the impulse can be treated as though it all occurred at a single instant of time. In that limit, the force is infinite, but its integral (the impulse) has a finite value.

A similar situation arises if we have a continuous distribution of charge, described by a charge density $\rho(\mathbf{r})$, and consider a hypothetical process in which the charge is squeezed until it is concentrated at a single point. At the concentration point the charge density is now infinite, but its integral over any spatial region containing that point continues to have the finite value corresponding to the original charge distribution.

The above-described types of problems arise frequently, and their common element is a need for a quantity which is zero everywhere except at a single point, but whose integral about any region containing that point has the same finite value. Accordingly, we **define** a quantity $\delta(x)$, called the **Dirac delta function**, with the properties

$$\delta(x) = 0, \quad x \neq 0, \qquad \int_a^b f(x)\delta(x)\,dx = f(0)\,. \tag{6.50}$$

Here a and b are arbitrary (except that their interval must contain the origin) and f is an arbitrary well-behaved function. An obvious corollary of Eq. (6.50) is that

$$\int_{-\infty}^{\infty} \delta(x)\,dx = 1\,. \tag{6.51}$$

From a mathematical viewpoint there exists no function $\delta(x)$ with the properties demanded in Eq. (6.50), but those properties can be realized as a limit reached by using a sequence of functions, a procedure that mathematicians view as defining a **generalized function**. A rigorous development of this idea, referred to as a theory of distributions, is beyond the scope of the present text, but distribution theory provides justification for what we shall do. An essential point is that the properties of the Dirac delta function are independent of the function sequence by which it may be implemented, so our focus can be entirely upon the properties presented in Eq. (6.50).

It may be instructive to look briefly at some possible δ-sequences. Two of the infinite number of possibilities are the following, in which the sequence members are characterized by increasing values of n:

$$\delta_n(x) = \begin{cases} n, & |x| < 1/2n\,, \\ 0, & \text{otherwise,} \end{cases} \tag{6.52}$$

$$\delta_n(x) = \frac{n}{\sqrt{\pi}}\, e^{-n^2x^2}\,. \tag{6.53}$$

A member of each of these sequences is shown in Fig. 6.15. Their common feature is that as n increases, the sequence members become increasingly localized near $x = 0$ and that for all members of the sequence the curves enclose unit area.

PROPERTIES

The most important properties of the Dirac delta function are those that define it, Eq. (6.50). Additional properties that are relevant include the following:

$$\int_{-\infty}^{\infty} f(x)\delta(x - x_0)\,dx = f(x_0)\,, \tag{6.54}$$

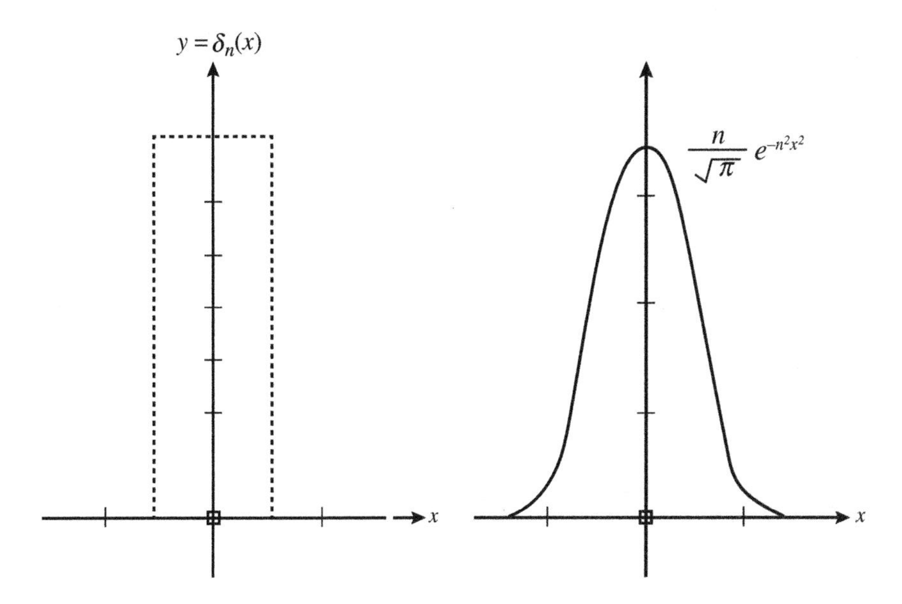

Figure 6.15: Two sequences that can represent $\delta(x)$. Left, Eq. (6.52); right, Eq. (6.53).

$$\delta(-x) = \delta(x) \quad \text{and therefore} \quad \delta(x_0 - x) = \delta(x - x_0)\,, \tag{6.55}$$

$$\delta(ax) = \frac{1}{|a|}\,\delta(x)\,, \tag{6.56}$$

$$\delta\big(f(x)\big) = \sum_i \frac{\delta(x - x_i)}{|f'(x_i)|}, \quad \text{for } x_i \text{ such that } f(x_i) = 0 \text{ and } f'(x_i) \neq 0\,, \tag{6.57}$$

$$\delta[(x - x_1)(x - x_2)] = \frac{1}{|x_1 - x_2|}\,[\delta(x - x_1) + \delta(x - x_2)]\,, \quad x_1 \neq x_2\,, \tag{6.58}$$

$$\int_{-\infty}^{\infty} f(x)\delta'(x - x_0)\,dx = -f'(x_0)\,, \tag{6.59}$$

$$\int_{-\infty}^{\infty} f(x)\delta^{(n)}(x - x_0)\,dx = (-1)^n f^{(n)}(x_0)\,. \tag{6.60}$$

These results can be obtained by applying integrations by parts to the basic definition of $\delta(x)$ or by changing the integration variable. Some of the formulas are topics of the exercises for this section.

UNIT STEP FUNCTION

In some contexts, particularly in discussions of Laplace transforms, one encounters another generalized function, the **Heaviside function**, also more descriptively called the **unit step function**. The Heaviside function $u(x)$ is, like the Dirac delta function, a generalized function that has a clear meaning when it occurs within an integral of the type shown here. Its defining property is that

$$\int_{-\infty}^{\infty} f(x)u(x - x_0)\,dx = \int_{x_0}^{\infty} f(x)\,dx\,, \tag{6.61}$$

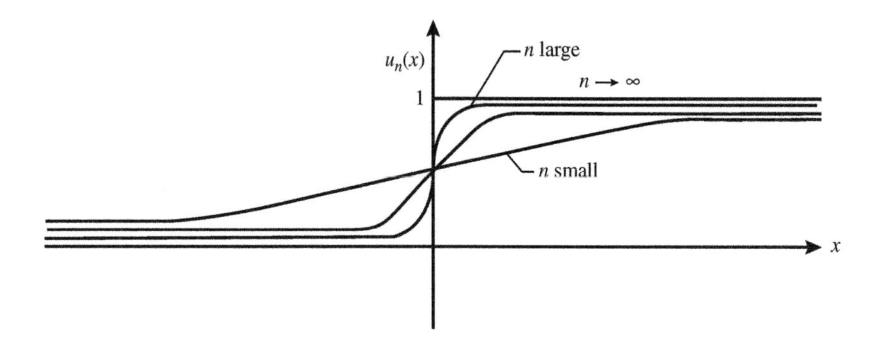

Figure 6.16: Sequence for unit step function.

applicable when the integral on the right-hand side converges. We see that the unit step function has the basic property

$$u(x - x_0) = \begin{cases} 1, & x > x_0, \\ 0, & x < x_0, \end{cases} \tag{6.62}$$

and can be specified more formally as the limit of a sequence of functions as shown in Fig. 6.16. We now observe that $u(x - x_0)$ is related to the Dirac delta function according to

$$u'(x - x_0) = \delta(x - x_0). \tag{6.63}$$

Equations like Eq. (6.63) can be checked by verifying that the value of an appropriate integral (for an arbitrary well-behaved function) is not changed when one side of this equation is replaced by the other. The verification is the topic of Exercise 6.7.5.

DELTA FUNCTIONS IN THREE DIMENSIONS

In Cartesian coordinates, the obvious 3-D analog of the 1-D Dirac delta function is simply $\delta(\mathbf{r}-\mathbf{r}_0) = \delta(x-x_0)\delta(y-y_0)\delta(z-z_0)$, which leads to the expected relationship

$$\int_V \delta(\mathbf{r} - \mathbf{r}_0) f(x, y, z)\, d^3r = \int \delta(x - x_0)\, dx \int \delta(y - y_0)\, dy \int \delta(z - z_0)\, dz f(x, y, z)$$

$$= \begin{cases} f(x_0, y_0, z_0), & V \text{ includes } \mathbf{r}_0, \\ 0, & V \text{ does not include } \mathbf{r}_0. \end{cases} \tag{6.64}$$

In curvilinear coordinate systems the formula for $\delta(\mathbf{r} - \mathbf{r}_0)$ must take account of the Jacobian of the coordinate transformation. We therefore have the following:

$$\delta(r-r_0, \theta-\theta_0, \varphi-\varphi_0) = \frac{\delta(r-r_0)\delta(\theta-\theta_0)\delta(\varphi-\varphi_0)}{r^2 \sin\theta} \quad \text{(spherical coords.)}, \tag{6.65}$$

$$\delta(\rho-\rho_0, \varphi-\varphi_0, z-z_0) = \frac{\delta(\rho-\rho_0)\delta(\varphi-\varphi_0)\delta(z-z_0)}{r} \quad \text{(cylindrical coords.)}. \tag{6.66}$$

Exercises

6.7.1. Show that $\delta(ax) = \delta(x)/|a|$. Make a detailed argument justifying the absolute value signs that enclose a. Your argument should not depend upon the specific representation describing $\delta(x)$.

6.7.2. Apply an integration by parts to establish Eq. (6.59).

6.7.3. By considering the expansion of $f(x)$ about a point x_i such that $f(x_i) = 0$, justify Eq. (6.57).

6.7.4. Justify Eq. (6.58).

6.7.5. Show that the Heaviside and Dirac delta functions are related by $u'(x-x_0) = \delta(x - x_0)$ by verifying the identity

$$\int_a^b u'(x - x_0)f(x)\,dx = \int_a^b \delta(x - x_0)f(x)\,dx$$

for a general "well-behaved" $f(x)$ and arbitrary values of a and b.

Additional Readings

Arfken, G. B., Weber, H. J., & Harris, F. E. (2013). *Mathematical methods for physicists* (7th ed.). New York: Academic Press.

Buck, R. C. (2003). *Advanced calculus* (3rd ed.). Long Grove, IL: Waveland Press (A popular current text).

Edwards, C. H. Jr. (1973). *Advanced calculus of several variables.* Dover (Reprinted (1995). Clear and detailed).

Widder, D. V. (1947). *Advanced calculus* (2nd ed.). Dover (Reprinted (1989). Clear, readable, and still relevant).

Chapter 7

VECTOR ANALYSIS

In Chapter 4 we introduced vectors in a way that was not limited to physical two- or three-dimensional (2-D or 3-D) space and identified their universal basic properties (addition by components, multiplication by a scalar, and the dot product). Then, in Chapter 5, we examined the ways vectors transform when the coordinate system is rotated or reflected, with some emphasis on their behavior in 2-D and 3-D systems. Although these basic properties are important and are relevant to many branches of physics and engineering, vector systems in 3-D space have additional features that are extremely useful in describing the mechanical and electromagnetic properties of matter.

The present chapter deals specifically with vector systems in 3-D space, and includes the study of vectors that are functions of the coordinates in that space. Such collections of vectors, referred to as **vector fields**, can be characterized not only by the values of the vectors at individual points, but also by the rate at which they change as we move the point with which they are associated. This functional dependence can be analyzed by defining and using derivatives of the vector components with respect to position, thereby indicating that our study of 3-D vectors must include the concepts of differential calculus. Many physical quantities are also related to integrals of vector quantities along paths, on surfaces, and over volumes; we therefore also study integrations involving vectors. This whole subject is sometimes referred to as **vector analysis**; the calculus-related aspects of it can also be called **vector calculus**.

The objective of this chapter is to present the basic ideas of vector analysis, identifying typical application areas in mechanics and electromagnetic theory.

7.1 VECTOR ALGEBRA

We start our development of vector analysis using a right-handed Cartesian coordinate system in the 3-D physical space. If the thumb of the right hand is pointed in the $+x$-direction, and the right index finger (at right angles to the thumb) is pointed in the $+y$-direction, then the right middle finger (when bent perpendicular to the thumb and index finger) defines the positive $+z$-direction of a right-handed system. While right-handed systems are used in a large majority of current writings, some of the older literature (particularly from Europe) used left-handed coordinates. Either system, if used consistently, can predict the same physics, but formulas in systems of different handedness will have systematic sign differences in some formulas.

Mathematics for Physical Science and Engineering.
http://dx.doi.org/10.1016/B978-0-12-801000-6.00007-9

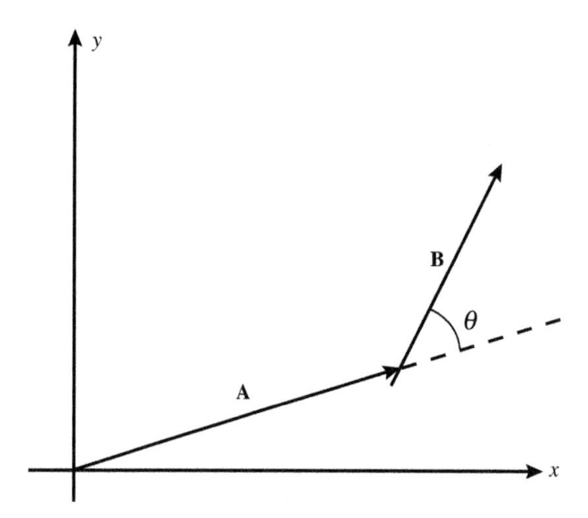

Figure 7.1: Angle for dot product of **A** and **B**.

A 3-D vector **A** in a Cartesian system is characterized by three components A_x, A_y, A_z and using unit vectors can be written

$$\mathbf{A} = A_x \hat{\mathbf{e}}_x + A_y \hat{\mathbf{e}}_y + A_z \hat{\mathbf{e}}_z \,. \tag{7.1}$$

Two such vectors, **A** and **B**, can be added componentwise, multiplied by scalars, and it was shown in Chapter 4 that they have the dot product

$$\mathbf{A} \cdot \mathbf{B} = AB \cos\theta = A_x B_x + A_y B_y + A_z B_z \,, \tag{7.2}$$

where θ is the angle between the directions of **A** and **B** (shown in Fig. 7.1). As pointed out previously, the dot product depends only on the magnitudes and relative orientation of its operands, and is therefore a scalar (independent of the orientation of the coordinate system).

A simple application of the dot product is to the computation of mechanical work, which (for a force that is constant in direction and magnitude and a displacement in a constant direction) is equal to the (component of the force in the direction of motion) multiplied by (the distance moved), or, equivalently, the force multiplied by the component of displacement in the direction of the force. These statements are equivalent to the formula

$$\text{work} = (\text{force})(\text{displacement}) \cos\theta = Fd\cos\theta = \mathbf{F} \cdot \mathbf{d} \,,$$

where θ is the angle between **F** and **d**.

Example 7.1.1. Inclined Plane

A block of mass M is pushed a distance d straight uphill on a frictionless inclined plane whose angle relative to the horizontal is θ, as shown in Fig. 7.2. Let's compute the work w done by the gravitational force. From the figure, we determine that the angle between the gravitational force and the displacement is $\theta + \pi/2$, and the formula for the work w yields

$$w = (Mg)(d)\cos(\theta + \pi/2) = -Mgd\sin\theta \,.$$

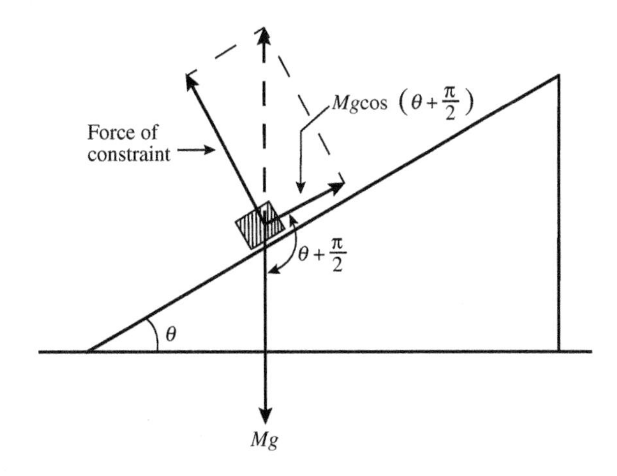

Figure 7.2: Inclined plane.

Alternatively, we can decompose the gravitational force into components perpendicular and parallel to the inclined plane. The perpendicular component is offset by an equal and opposite "force of constraint" imposed by the rigid planar surface, and the component parallel to the inclined plane must be offset by an applied force if the block is to be held stationary or moved (slowly) uphill. This parallel component is $Mg\sin\theta$, which acts (in its direction) for the distance d. This corresponds to work **against** gravity (and therefore with a negative sign) in the amount $-Mgd\sin\theta$, in agreement with our earlier analysis. These results both agree with the work $Mg\Delta h$ needed to lift the block the vertical distance between points separated on the plane by a distance d, namely $\Delta h = d\sin\theta$.

■

The use of vectors and their dot products can simplify many geometric problems. Here is an illustration.

Example 7.1.2. Distance Between a Point and a Plane

The shortest path between a point P and a plane will be along the normal to the plane that passes through P. If the equation of the plane is $a_1x_1 + a_2x_2 + a_3x_3 = c$, which we can write $\mathbf{a}\cdot\mathbf{x} = c$, then any two points in the plane, $\mathbf{x}$ and $\mathbf{y}$, will have the property that $\mathbf{a}\cdot\mathbf{x} = c$, $\mathbf{a}\cdot\mathbf{y} = c$, and therefore $\mathbf{a}\cdot(\mathbf{x}-\mathbf{y}) = 0$, showing that the vector $\mathbf{a}$ is perpendicular to all vectors in the plane. We conclude that $\mathbf{a}$ is in the direction of the normal to the plane. A unit vector in the normal direction is therefore $\mathbf{a}/a$.

The distance between P and the plane can now be computed as the magnitude of the projection onto $\mathbf{a}$ of the vector from P to **any** point Q in the plane. For example, if $P = (2, 1, -1)$ and the equation of the plane is $x - 2y + 2z = 4$, the normal is in the direction $(1, -2, 2)$, and a unit vector in that direction is $\mathbf{n} = (1/3, -2/3, 2/3)$. It is easy to see that $Q = (4, 0, 0)$ is in the plane, so

$$\mathbf{Q} - \mathbf{P} = (2, -1, 1)\,,$$

and the projection of $\mathbf{Q} - \mathbf{P}$ normal to the plane is given by

$$[\mathbf{n}\cdot(\mathbf{Q}-\mathbf{P})]\mathbf{n} = \left[\left(\frac{1}{3}, -\frac{2}{3}, \frac{2}{3}\right)\cdot(2, -1, 1)\right]\mathbf{n} = 2\mathbf{n}\,.$$

If we want the point on the plane that is closest to P, we can find it by computing $\mathbf{P} + 2\mathbf{n} = (8/3, -1/3, 1/3)$.

■

Exercises

7.1.1. A particle is moved, subject to a force $\mathbf{F} = 2\hat{\mathbf{e}}_x + \hat{\mathbf{e}}_y - \hat{\mathbf{e}}_z$, on a path from $\mathbf{A} = (1, 1, 0)$ to $\mathbf{B} = (2, 0, 1)$. Compute the work required for this process:

(a) If the path is a straight line from $\mathbf{A}$ to $\mathbf{B}$, and

(b) If the path consists of the two straight-line segments $\mathbf{A}$–$\mathbf{O}$ and $\mathbf{O}$–$\mathbf{B}$, where $\mathbf{O} = (0, 0, 0)$.

7.1.2. Find the angle between the normals to the two planes $2x + 5y - z = 10$ and $x - y + z = 6$.

7.1.3. Find the distance between the point $(1, 2, -1)$ and the plane $x - y + z = 5$.

7.1.4. Compute the distance of closest approach of the circle $x^2 + y^2 = 1$ (in the xy-plane) to the plane $2x + 2y + z = 5$, and identify the points of closest approach: (x_0, y_0) on the circle and (x_1, y_1, z_1) in the plane.

CROSS PRODUCT

Vectors in 3-D have another product, called the **cross product**, or sometimes the **vector product**, which produces as its result another 3-D vector. The notation for the cross product is

$$\mathbf{C} = \mathbf{A} \times \mathbf{B} . \tag{7.3}$$

The product vector $\mathbf{C}$ *has magnitude* $AB \sin\theta$, *with* θ *the angle* ($\le 180°$) *between* $\mathbf{A}$ *and* $\mathbf{B}$. *The direction of* $\mathbf{C}$ *is perpendicular to both* $\mathbf{A}$ *and* $\mathbf{B}$, *and a choice between the two perpendicular directions can be made by pointing the fingers of the right hand in the direction of* $\mathbf{A}$ *in such a way that the fingers can be bent to point in the direction of* $\mathbf{B}$. *The thumb will then point in the direction of* $\mathbf{C}$.

Note that the cross product is defined in a way that makes $\mathbf{C}$ vanish if $\mathbf{A}$ and $\mathbf{B}$ are either parallel or antiparallel, and that $\mathbf{C}$ will have maximum magnitude (equal to AB) when $\mathbf{A}$ and $\mathbf{B}$ are perpendicular. Moreover, the definition is such that

$$\mathbf{B} \times \mathbf{A} = -\mathbf{A} \times \mathbf{B} . \tag{7.4}$$

Geometrically, $\mathbf{A} \times \mathbf{B}$ is a vector whose magnitude is equal to the area of the parallelogram with sides $\mathbf{A}$ and $\mathbf{B}$ and whose direction is normal to that parallelogram. See Fig. 7.3. This situation illustrates a general principle worthy of notice. Vectors that represent areas (or elements of area) are oriented normal to the area they represent. The choice between the two (opposite) normal orientations is a matter of convention.

It is useful to obtain an algebraic formula for $\mathbf{A} \times \mathbf{B}$ in terms of the components of $\mathbf{A}$ and $\mathbf{B}$. A first step in this direction is to evaluate all the cross products of unit

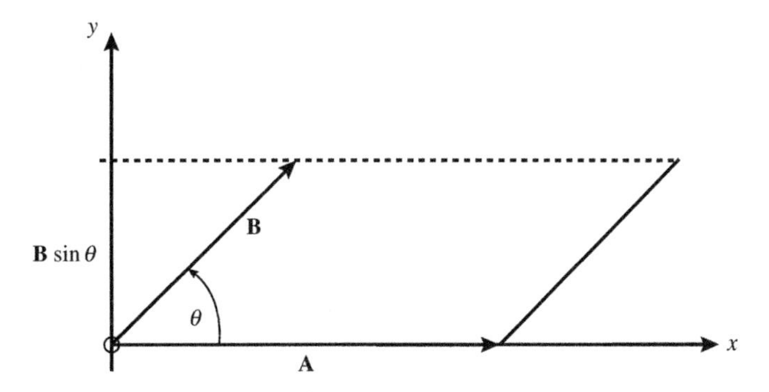

Figure 7.3: Parallelogram of $\mathbf{A} \times \mathbf{B}$.

vectors in the coordinate directions. Applying the rule given after Eq. (7.3),

$$\begin{gathered}
\hat{\mathbf{e}}_x \times \hat{\mathbf{e}}_x = \hat{\mathbf{e}}_y \times \hat{\mathbf{e}}_y = \hat{\mathbf{e}}_z \times \hat{\mathbf{e}}_z = 0, \\
\hat{\mathbf{e}}_x \times \hat{\mathbf{e}}_y = \hat{\mathbf{e}}_z, \qquad \hat{\mathbf{e}}_y \times \hat{\mathbf{e}}_z = \hat{\mathbf{e}}_x, \qquad \hat{\mathbf{e}}_z \times \hat{\mathbf{e}}_x = \hat{\mathbf{e}}_y, \\
\hat{\mathbf{e}}_y \times \hat{\mathbf{e}}_x = -\hat{\mathbf{e}}_z, \qquad \hat{\mathbf{e}}_z \times \hat{\mathbf{e}}_y = -\hat{\mathbf{e}}_x, \qquad \hat{\mathbf{e}}_x \times \hat{\mathbf{e}}_z = -\hat{\mathbf{e}}_y.
\end{gathered} \tag{7.5}$$

One way to remember the above formulas is to note that these vector products are invariant under any cyclic permutation of x, y, z but change sign if the order of the factors is reversed. Thus, once for example we have established $\hat{\mathbf{e}}_x \times \hat{\mathbf{e}}_y = \hat{\mathbf{e}}_z$, the remaining formulas follow directly. Another way of systematizing these formulas is to write them in terms of the Levi-Civita symbol, Eq. 4.62. Equations (7.5) then become

$$\hat{\mathbf{e}}_i \times \hat{\mathbf{e}}_j = \sum_k \varepsilon_{ijk} \hat{\mathbf{e}}_k \,. \tag{7.6}$$

We next need to use the fact that for any vectors $\mathbf{A}$, $\mathbf{B}$, $\mathbf{C}$,

$$\mathbf{A} \times (\mathbf{B} + \mathbf{C}) = \mathbf{A} \times \mathbf{B} + \mathbf{A} \times \mathbf{C} \,. \tag{7.7}$$

In other words, the cross product is distributive relative to addition. Applying Eqs. (7.5) and (7.7), we expand and simplify $\mathbf{A} \times \mathbf{B}$, getting

$$\begin{aligned}
\mathbf{A} \times \mathbf{B} &= (A_x \hat{\mathbf{e}}_x + A_y \hat{\mathbf{e}}_y + A_z \hat{\mathbf{e}}_z) \times (B_x \hat{\mathbf{e}}_x + B_y \hat{\mathbf{e}}_y + B_z \hat{\mathbf{e}}_z) \\
&= (A_y B_z - A_z B_y)\, \hat{\mathbf{e}}_x + (A_z B_x - A_x B_z)\, \hat{\mathbf{e}}_y + (A_x B_y - A_y B_x)\, \hat{\mathbf{e}}_z \,.
\end{aligned} \tag{7.8}$$

Since the signs in this equation are those given by the Levi-Civita symbol, it is not surprising that the right-hand side of Eq. (7.8) can be identified as equivalent to a determinant:

$$\mathbf{A} \times \mathbf{B} = \begin{vmatrix} \hat{\mathbf{e}}_x & \hat{\mathbf{e}}_y & \hat{\mathbf{e}}_z \\ A_x & A_y & A_z \\ B_x & B_y & B_z \end{vmatrix} \,. \tag{7.9}$$

Since a determinant changes sign on row interchange, Eq. (7.9) makes it obvious that $\mathbf{B} \times \mathbf{A} = -\mathbf{A} \times \mathbf{B}$ and that for any $\mathbf{A}$, we have $\mathbf{A} \times \mathbf{A} = 0$.

Example 7.1.3. Perpendicular Vectors

Given any two vectors $\mathbf{A}$ and $\mathbf{B}$ that are not collinear, we wish to find a third that is perpendicular to both. We can use the cross product for this purpose, as any vector proportional to $\mathbf{A} \times \mathbf{B}$ will meet this specification.

A useful generalization of the above observation is to compute the distance of closest approach of two skew lines (linear trajectories in 3-D space). **Skew lines** are lines that are not parallel but do not intersect. Our two lines can be represented parametrically as $\mathbf{P} + s\mathbf{a}$ and $\mathbf{Q} + t\mathbf{b}$. Here the vectors $\mathbf{a}$ and $\mathbf{b}$ define the directions of the lines, the scalars s and t are parameters, and $\mathbf{P}$ and $\mathbf{Q}$ are arbitrary points on their respective lines.

Our key observation is that the line segment of closest approach will be perpendicular to both the skew lines and therefore in the direction of the cross product of $\mathbf{a}$ and $\mathbf{b}$; a unit vector in this direction will be $\mathbf{u} = \mathbf{a} \times \mathbf{b}/|\mathbf{a} \times \mathbf{b}|$. The distance we seek will be the magnitude of the projection of $\mathbf{Q} - \mathbf{P}$ onto $\mathbf{u}$.

Working the specific case of the two lines $(0,1,2) + s(1,-2,0)$ and $(1,0,-1) + t(0,2,1)$, we have

$$\mathbf{a} \times \mathbf{b} = (1,-2,0) \times (0,2,1) = (-2,-1,2)\,, \qquad |\mathbf{a} \times \mathbf{b}|^2 = 9,$$

$$\mathbf{u} = \frac{\mathbf{a} \times \mathbf{b}}{|\mathbf{a} \times \mathbf{b}|} = \left(-\frac{2}{3}, -\frac{1}{3}, \frac{2}{3}\right).$$

The distance between the two lines is therefore given by

$$\Big|(\mathbf{Q} - \mathbf{P}) \cdot \mathbf{u}\Big| = \left|(1,-1,-3) \cdot \left(-\frac{2}{3}, -\frac{1}{3}, \frac{2}{3}\right)\right| = \frac{7}{3}.$$

■

Cross products frequently occur in physics in contexts involving rotational motion. It is convenient and useful to describe a rotational displacement by a vector that is oriented along the axis of rotation, with a magnitude that describes the amount of the displacement, measured in radians, and with its choice of the two axial directions determined by the right-hand rule (the direction of the right thumb when the curved fingers are in the direction of the displacement). Angular velocity, conventionally identified with the symbol $\boldsymbol{\omega}$, is defined as the displacement (in radians) per unit time, and is also represented by a vector along the rotational axis in the direction given by the right-hand rule.

An important concept involving angular motion is that of **angular momentum**. A particle moving with (linear) momentum $\mathbf{p}$ at a point $\mathbf{r}$ (measured from a reference point $\mathbf{r} = 0$) is defined to have angular momentum about $\mathbf{r} = 0$ given by

$$\mathbf{L} = \mathbf{r} \times \mathbf{p}. \tag{7.10}$$

When Newton's law is applied to angular motion, the mechanical quantity that causes a change in angular momentum about $\mathbf{r} = 0$ is called **torque** (more precisely, the torque about $\mathbf{r} = 0$) and is usually denoted $\boldsymbol{\tau}$. The rotational analog of $\mathbf{F} = d\mathbf{p}/dt$ is $\boldsymbol{\tau} = d\mathbf{L}/dt$, with

$$\boldsymbol{\tau} = \mathbf{r} \times \mathbf{F}. \tag{7.11}$$

Here $\mathbf{r}$ is the point of application of the force $\mathbf{F}$ giving rise to the torque, and if several forces are applied to an extended object (such as the lever that is the topic of the

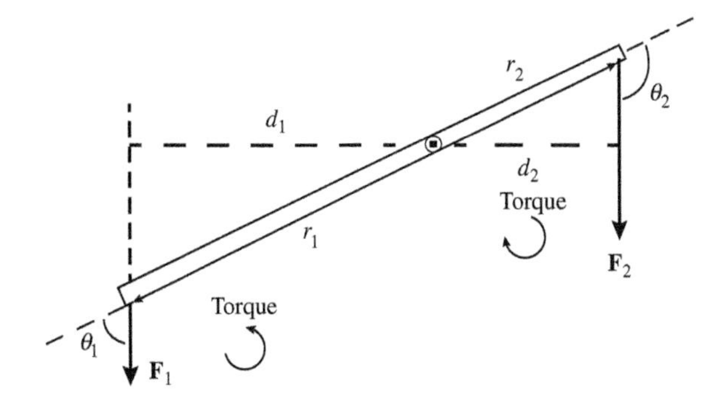

Figure 7.4: Lever.

next example), the net torque on the object will be the vector sum of the individual torques (all measured about the same reference point). If an object is to remain in rotational equilibrium about a reference point, the torques about that point must sum to zero.

Example 7.1.4. Lever

Consider a lever that can rotate in the xy-plane about the origin (its **fulcrum**), as shown in Fig. 7.4. The lever is initially stationary, but also subject to vertically downward forces $\mathbf{F}_1$ and $\mathbf{F}_2$, applied to points on the lever at respective distances r_1 and r_2 from the fulcrum. The lever will remain stationary if the two forces exert equal and opposite torques about the fulcrum. In elementary physics courses this lever problem is usually solved by computing the perpendicular distances between the forces and the fulcrum (marked d_1 and d_2 in the figure), then identifying the condition for equilibrium as $d_1F_1 = d_2F_2$.

Now approaching this problem using vector notation, we note that the torque about the fulcrum due to $\mathbf{F}_1$ is $\boldsymbol{\tau}_1 = \mathbf{r}_1 \times \mathbf{F}_1$. Here $\mathbf{r}_1$ is a vector **from** the fulcrum **to** the point of application of $\mathbf{F}_1$, and

$$\boldsymbol{\tau}_1 = \mathbf{r}_1 \times \mathbf{F}_1 = r_1F_1 \sin\theta_1 \,\hat{\mathbf{n}} = d_1F_1 \,\hat{\mathbf{n}}\,,$$

where $\hat{\mathbf{n}}$ is a unit vector normal to, and **out of** the plane of the figure. A similar analysis of the torque $\boldsymbol{\tau}_2$ reveals that it is

$$\boldsymbol{\tau}_2 = \mathbf{r}_2 \times \mathbf{F}_2 = -r_2F_2 \sin\theta_2 \,\hat{\mathbf{n}} = -d_2F_2 \,\hat{\mathbf{n}}\,,$$

the minus sign occurring because $\hat{\mathbf{n}}$ has the same meaning as before but this torque is directed **into** the plane of the figure. We now observe that $\theta_2 = \pi - \theta_1$, and therefore $\sin\theta_2 = \sin\theta_1$. At equilibrium the sum of these torques is zero.

■

Example 7.1.5. Angular Momentum

1. Consider a particle of mass M on a frictionless track that consists of a circle of radius a centered at the origin and in the xy-plane of a right-handed coordinate system. The particle is moving at speed v in the direction indicated in the left panel of Fig. 7.5. Find its angular velocity and angular momentum about the origin.

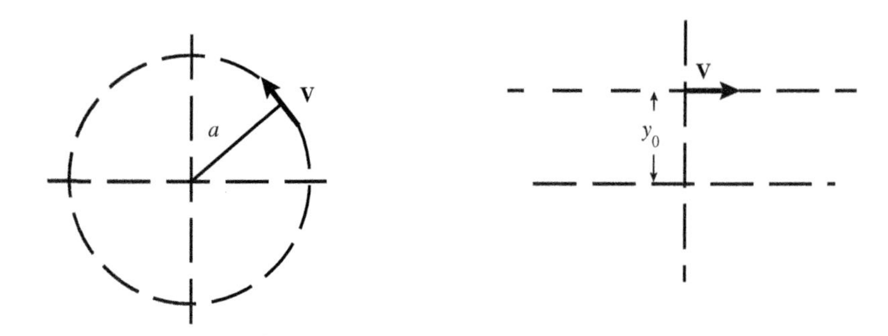

Figure 7.5: Particle motion for Example 7.1.5. Left: Circular motion. Right: Straight line displaced from reference point.

The angular velocity has the constant magnitude v/a and, applying the right-hand rule, is in the $+z$ direction. To compute the angular momentum, note that $\mathbf{r}$ and $\mathbf{v}$ are at all times mutually perpendicular, so $\mathbf{r} \times \mathbf{p} = \mathbf{r} \times (M\mathbf{v})$ will have the constant magnitude $Mvr = Mva$; since $\mathbf{r}$ and $\mathbf{p}$ are always in the same plane, their cross product will always be in the same direction, normal to that plane, and in the same direction as the angular velocity.

2. Consider a particle of mass M moving on a trajectory defined by the equation $\mathbf{r}(t) = (0, y_0, 0) + (v, 0, 0)t$. This trajectory describes motion at constant speed v in the $+x$ direction. See the right panel of Fig. 7.5. Find the angular momentum of the particle about $\mathbf{r} = 0$ as a function of t.

From the given data, $\mathbf{p} = (Mv, 0, 0)$ and

$$\mathbf{L} = \mathbf{r} \times \mathbf{p} = \Big[(0, y_0, 0) + (v, 0, 0)t\Big] \times (Mv, 0, 0) = (0, y_0, 0) \times (Mv, 0, 0) = (0, 0, My_0v)\,.$$

When the first factor of the cross product is expanded, the term involving $(v, 0, 0)t$ drops out because it is in the same direction as $(Mv, 0, 0)$. We thus see that the angular momentum is constant but nonzero, and that it is in a direction normal to the plane defined by the origin and the trajectory line, i.e., in the $\pm z$-direction (with a sign that depends on the value of y_0). The nonzero value can be understood by noting that the trajectory causes a continual change in the direction of the particle as viewed from the origin.

■

Exercises

7.1.5. Find the equation of a plane through $\mathbf{A} = (0, 1, 2)$, $\mathbf{B} = (1, -1, 1)$, and $\mathbf{C} = (2, 0, 0)$. Is this plane unique?

Hint. Start by finding a direction perpendicular to $\mathbf{A} - \mathbf{B}$ and $\mathbf{A} - \mathbf{C}$.

7.1.6. Find the distance from the point $(1, -2, 1)$ to the line passing through the points $(0, 0, 0)$ and $(2, 1, 2)$.

7.1.7. Find a vector that lies in the intersection of the two planes $x + y - z = 3$ and $2x - y + 3z = 4$.

7.1.8. A particle of mass m undergoes rotation at 8 revolutions per second around a circle of radius 1 m in the xy-plane and centered at the origin, with the

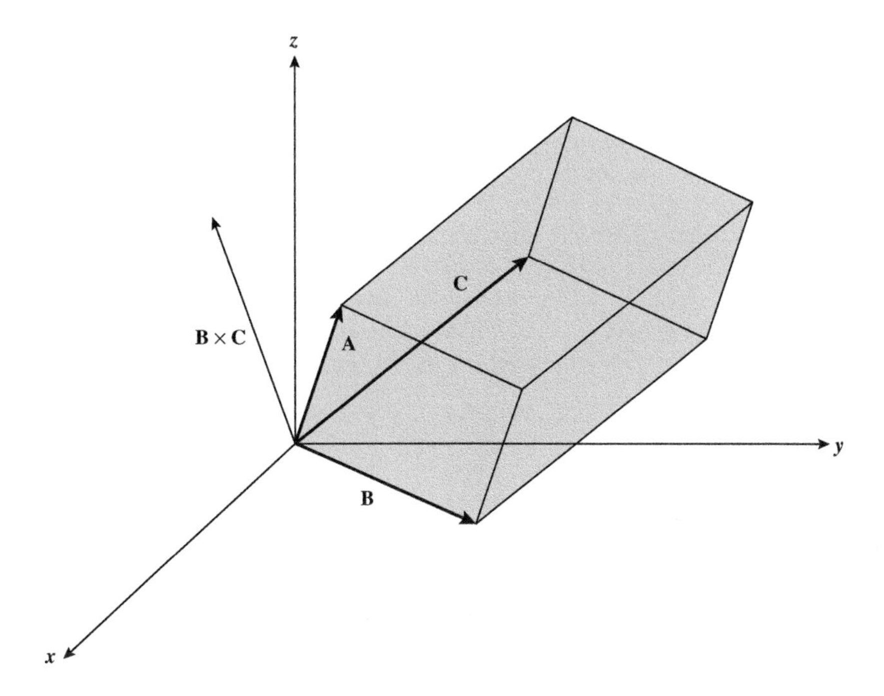

Figure 7.6: Parallelepiped for scalar triple product.

travel in the clockwise direction as viewed from positive z. Find the vector describing the angular momentum of the particle about the origin.

7.1.9. Compute the torque about the origin if a force $\mathbf{F} = 2\hat{\mathbf{e}}_x + 3\hat{\mathbf{e}}_y$ is applied at the point $\mathbf{r} = \hat{\mathbf{e_x}} - \hat{\mathbf{e}}_z$.

7.1.10. A lever of length 3 m pivots about a fulcrum 1 m from its end **A** (and therefore 2 m from its end **B**). The lever can rotate in the xy-plane. If a force $\hat{\mathbf{e}}_x + \hat{\mathbf{e}}_y$ is applied at **A**, what force at **B** must be applied to maintain the lever at equilibrium irrespective of the lever's orientation?

MULTIPLE VECTOR PRODUCTS

Now that we have two different ways of multiplying 3-D vectors, we can form products of more than two vectors in various ways. With three vectors, there are two possibilities: the first, yielding a result which is a scalar, is called the **scalar triple product**; the second, whose result is a vector, is called the **vector triple product**.

SCALAR TRIPLE PRODUCT

The scalar triple product has the form $\mathbf{A} \cdot (\mathbf{B} \times \mathbf{C})$. If we construct a parallelepiped (Fig. 7.6) with edges defined by **A**, **B**, and **C** and with base BC, its base area will be the magnitude of $\mathbf{B} \times \mathbf{C}$, its height h will be the component of **A** perpendicular to the base (i.e., the projection of **A** in the direction of $\mathbf{B} \times \mathbf{C}$), and the volume of the parallelepiped will be h times the area of the base. We may compute h as

$$h = \frac{|\mathbf{A} \cdot (\mathbf{B} \times \mathbf{C})|}{|\mathbf{B} \times \mathbf{C}|},$$

taking the absolute value because the sign of the dot product may come out to be negative. The volume of the parallelepiped is then identified as the absolute value of the scalar triple product $\mathbf{A}\cdot(\mathbf{B}\times\mathbf{C})$.

Notice that we could have taken any side of our parallelepiped as its base and that the cross-product factor of the triple product could have been written with either operand first. We conclude that all of the twelve triple products $\mathbf{A}\cdot(\mathbf{B}\times\mathbf{C})$, $\mathbf{B}\cdot(\mathbf{A}\times\mathbf{C})$, $\mathbf{B}\cdot(\mathbf{C}\times\mathbf{A})$, etc. must have the same magnitude; they can only differ in sign.

Let's now work out the value of the triple product in terms of the components of $\mathbf{A}$, $\mathbf{B}$, and $\mathbf{C}$. Starting with $\mathbf{B}\times\mathbf{C}$ written as a determinant, see Eq. (7.9), its dot product with $\mathbf{A} = A_x\hat{\mathbf{e}}_x + A_y\hat{\mathbf{e}}_y + A_z\hat{\mathbf{e}}_z$ will simply cause the unit vectors in the determinant to be replaced by the components of $\mathbf{A}$, so that

$$\mathbf{A}\cdot(\mathbf{B}\times\mathbf{C}) = \mathbf{A}\cdot\begin{vmatrix} \hat{\mathbf{e}}_x & \hat{\mathbf{e}}_y & \hat{\mathbf{e}}_z \\ B_x & B_y & B_z \\ C_x & C_y & C_z \end{vmatrix} = \begin{vmatrix} A_x & A_y & A_z \\ B_x & B_y & B_z \\ C_x & C_y & C_z \end{vmatrix}. \tag{7.12}$$

This form makes clear the symmetry of the triple scalar product. Since a determinant changes sign when any two of its rows are interchanged, the triple scalar product will change sign if any two of its symbols are interchanged. We also note a complete symmetry in the roles of $\mathbf{A}$, $\mathbf{B}$, and $\mathbf{C}$ despite the fact that they appear in different contexts in the product. From the rules for determinants we can now verify that $\mathbf{A}\cdot(\mathbf{B}\times\mathbf{C})$ will be unchanged by any even permutation of its factors (these are the cyclic permutations), and will change sign under odd permutations of the factors.

VECTOR TRIPLE PRODUCT

The vector triple product has the form $\mathbf{A}\times(\mathbf{B}\times\mathbf{C})$. The parentheses are necessary, because the cross product is not associative, meaning that $\mathbf{A}\times(\mathbf{B}\times\mathbf{C})$ is not necessarily equal to $(\mathbf{A}\times\mathbf{B})\times\mathbf{C}$. If $\mathbf{B}$ and $\mathbf{C}$ are proportional, making them collinear, the vector triple product is zero and we need not discuss it further. But if $\mathbf{B}$ and $\mathbf{C}$ are not collinear, they determine a plane, and $\mathbf{B}\times\mathbf{C}$ will be a nonzero vector in the direction perpendicular (normal) to the BC-plane. That, in turn, means that $\mathbf{A}\times(\mathbf{B}\times\mathbf{C})$ will be perpendicular to the BC-normal and will therefore lie in the BC-plane. Since the vectors $\mathbf{B}$ and $\mathbf{C}$ form a basis that spans the BC-plane, we reach the preliminary conclusion that

$$\mathbf{A}\times(\mathbf{B}\times\mathbf{C}) = c_1\mathbf{B} + c_2\mathbf{C},$$

where c_1 and c_2 are scalars whose values we have not yet determined. By writing out components (possibly using symbolic computation to help), or by other more formal means, it can be established that the vector triple product satisfies the vector identity

$$\mathbf{A}\times(\mathbf{B}\times\mathbf{C}) = (\mathbf{A}\cdot\mathbf{C})\mathbf{B} - (\mathbf{A}\cdot\mathbf{B})\mathbf{C}. \tag{7.13}$$

Some authors call this formula the **BAC rule**, based on writing the first term of the product as $\mathbf{B}(\mathbf{A}\cdot\mathbf{C})$. A proof of this formula using the Levi-Civita symbol is given in Arfken et al., Supplementary Readings.

It is now obvious that if we had used parentheses around the first two members of the triple product we would have

$$(\mathbf{A} \times \mathbf{B}) \times \mathbf{C} = -\mathbf{C} \times (\mathbf{A} \times \mathbf{B}),$$

which has as its result a vector in the AB-plane, clearly not the same as Eq. (7.13).

Vector triple products arise in the discussion of rotational motion. For example, a particle in circular motion at angular velocity $\boldsymbol{\omega}$ and at a position $\mathbf{r}$ measured from a point on the rotation axis has centripetal acceleration $\mathbf{a} = \boldsymbol{\omega} \times (\boldsymbol{\omega} \times \mathbf{r})$.

SYMBOLIC COMPUTATION

In contrast to the situation for linear algebra, where matrix operations, eigenvalue problems, and determinant evaluation in practical problems are almost impossible without symbolic or digital computation, the role of symbolic computation for vector analysis is far less significant, and cannot be a substitute for an adequate background in vector manipulations. However, symbolic methods do provide useful checks on work in vector analysis and it seems justified to explain their use.

The discussion of this subsection assumes the use of Cartesian coordinates. Both MAPLE and MATHEMATICA also support a variety of curvilinear coordinate systems; symbolic methods for vector algebra in curvilinear coordinates are treated later in this text.

In MAPLE, vector analysis is supported by the `VectorCalculus` package, so the operations discussed here must be preceded by the command `with(VectorCalculus)`. Vectors may be of arbitrary dimensionality, but some operations are only defined for 3-D vectors. When the `VectorCalculus` package is active, vectors are by default written in what MAPLE calls "basis format" instead of the column matrix form. This behavior can be controlled by a command `BasisFormat(...)` with argument `true` or `false`.

In MATHEMATICA, vector algebra (not involving differential operators) is part of the basic language. Note, however, that a supplementary package must be invoked when differential vector operators are involved.

Example 7.1.6. Symbolic Computation, Vector Algebra

Formulas are for Cartesian coordinates only.

MAPLE	MATHEMATICA
`> with(VectorCalculus);`	(no command needed)

Define some vectors:

MAPLE	MATHEMATICA
`> A := Vector([Ax,Ay,Az]);`	`A = {Ax,Ay,Az}`
$(Ax)\,\mathbf{e}_x + (Ay)\,\mathbf{e}_y + (Az)\,\mathbf{e}_z$	`{Ax, Ay, Az}`
`> B := Vector([Bx,By,Bz]):`	`B = {Bx,By,Bz};`
`> C := Vector([Cx,Cy,Cz]):`	`G = {Gx,Gy,Gz};`

In MATHEMATICA the symbol `C` has a fixed meaning and cannot be used to designate a vector, so we use `G` instead. Notice also that in MAPLE the output (shown above for `A`) is in basis format.

```
> A . B;                                A . B
```

$$A_xB_x + A_yB_y + A_zB_z \text{ (in both languages)}$$

In MAPLE only; the cross-product command is &x:

```
> A &x B;
```

$$(AyBz - AzBy)\mathbf{e}_x + (AzBx - AxBz)\mathbf{e}_y + (AxBy - AyBx)\mathbf{e}_z$$

```
> BasisFormat(false):
```

```
> C;
```

$$\begin{bmatrix} Cx \\ Cy \\ Cz \end{bmatrix}$$

```
> BasisFormat(true):
> 0 * C;
```

$$0\,\mathbf{e}_x$$

MAPLE gives the first component of the result even if all are zero (to show that it is a vector).

For MATHEMATICA:

```
Cross[A,B]
{-AzBy+AyBz, AzBx-AxBz, -AyBx+AxBy}
```

Triple products:

```
> A . B &x C;                           A . Cross[B,G]
```

(MAPLE output) $Ax(ByCz - BzCy) + Ay(BzCx - BxCz) + Az(BxCy - ByCx)$

```
> A &x (B &x C);                        Cross[A,Cross[B,G]]
```

(MATHEMATICA output)
```
{-AyByGx-AzBzGx+AyBxGy+AzBxGz,
 AxByGx-AxBxGy-AzBzGy+AzByGz, AxBzGx+AyBzGy-AxBxGz-AyByGz}
```

While the output of the scalar triple product might be recognizable as a determinant, that of the vector triple product does not reveal the structure of what is known as the BAC rule. This illustrates our earlier statement that symbolic programming of vector identities is not an easy substitute for appropriate background knowledge. ■

Exercises

7.1.11. Show that a necessary and sufficient condition that three nonvanishing vectors **A**, **B**, and **C** be coplanar is that $\mathbf{A} \cdot (\mathbf{B} \times \mathbf{C}) = 0$.

7.1.12. Given that $\mathbf{A} = 2\hat{\mathbf{e}}_x + \hat{\mathbf{e}}_y - 3\hat{\mathbf{e}}_z$, $\mathbf{B} = \hat{\mathbf{e}}_x - 2\hat{\mathbf{e}}_y + 5\hat{\mathbf{e}}_z$, and $\mathbf{C} = -3\hat{\mathbf{e}}_x + 2\hat{\mathbf{e}}_y - \hat{\mathbf{e}}_z$, evaluate (both by hand and using your symbolic computing system)

(a) $(\mathbf{A} \cdot \mathbf{C})\,\mathbf{B}$, (b) $(\mathbf{A} \times \mathbf{B}) \cdot \mathbf{C}$, (c) $(\mathbf{A} \times \mathbf{B}) \times \mathbf{C}$.

7.1.13. Letting **A**, **B**, and **C** be the displacements forming a triangle (in directions such that $\mathbf{A} + \mathbf{B} + \mathbf{C} = 0$), use the properties of the vector cross product to derive the law of sines, i.e.,

$$\frac{\sin\alpha}{A} = \frac{\sin\beta}{B} = \frac{\sin\gamma}{C},$$

where α, β, γ are respectively the angles of the triangle opposite the sides **A**, **B**, **C**.

7.1.14. The vector triple product $\mathbf{A} \times (\mathbf{B} \times \mathbf{C})$ is found by symbolic computation as

$$[-A_y B_y C_x - A_z B_z C_x + A_y B_x C_y + A_z B_x C_z]\,\hat{\mathbf{e}}_x + [A_x B_y C_x - A_x B_x C_y - A_z B_z C_y + A_z B_y C_z]\,\hat{\mathbf{e}}_y + [A_x B_z C_x + A_y B_z C_y - A_x B_x C_z - A_y B_y C_z]\,\hat{\mathbf{e}}_z\,.$$

Verify that this expression is equivalent to Eq. (7.13).

7.1.15. Use your symbolic computing system to prove the **Lagrange identity**,

$$(\mathbf{A} \times \mathbf{B}) \cdot (\mathbf{U} \times \mathbf{V}) = (\mathbf{A} \cdot \mathbf{U})(\mathbf{B} \cdot \mathbf{V}) - (\mathbf{A} \cdot \mathbf{V})(\mathbf{B} \cdot \mathbf{U})\,.$$

7.1.16. Use your symbolic computing system to prove the **Jacobi identity**,

$$\mathbf{A} \times (\mathbf{B} \times \mathbf{G}) + \mathbf{B} \times (\mathbf{G} \times \mathbf{A}) + \mathbf{G} \times (\mathbf{A} \times \mathbf{B}) = 0\,.$$

7.1.17. A force $\mathbf{F} = \hat{\mathbf{e}}_x + 2\hat{\mathbf{e}}_y - 3\hat{\mathbf{e}}_z$ acts at the point $(1, 4, -2)$. Find the torque of **F**

(a) about the origin;

(b) about the z-axis;

(c) about the axis (through the origin) $\mathbf{p} = 2\hat{\mathbf{e}}_x - 3\hat{\mathbf{e}}_y + \hat{\mathbf{e}}_z$.

7.1.18. Let a rigid body undergo rotation described by the angular velocity $\boldsymbol{\omega}$ ($\boldsymbol{\omega}$ is in the direction of the rotational axis, with a magnitude equal to that of the angular velocity, in radians per unit time). A point **P** of the rigid body is reached by a displacement **r** from an arbitrary point **O** on the rotational axis.

(a) Show that the linear velocity of **P** is $\mathbf{v} = \boldsymbol{\omega} \times \mathbf{r}$, and that this result is independent of the location of **O** on the rotational axis.

(b) The angular momentum about a point **O** of a mass m at a point **P** moving at velocity **v** is $\mathbf{L} = \mathbf{r} \times (m\mathbf{v})$, where **r** is the displacement from **O** to **P**. Assuming **P** to be undergoing the rotational motion described in part (a), find **L** in terms of $\boldsymbol{\omega}$ and **r**.

(c) Assume that the rigid body of part (a) is moving at angular velocity 8 radians per second about an axis through the point $(2, -1, 2)$, with the axial direction $\hat{\mathbf{e}}_x - \hat{\mathbf{e}}_y$, and that the point **P** is $(1, -5, 2)$ (with all coordinates given in meters). Find the angular momentum about the rotational axis (in MKS units) of a 2.75 kg mass at **P**.

7.2 VECTOR DIFFERENTIAL OPERATORS

If we have a physical system containing a distribution of electric charges, we can (at least in principle) place a test charge at any point and observe the direction and magnitude of the electrical force the distribution exerts upon the test charge. In this way we can associate every point in space with a vector indicating the force (per unit strength of the test charge), thereby defining a vector field which is called an **electric field**. In most situations of interest, the force on a test charge will vary from point to point, usually in a continuous fashion. For example, the electric field of a positive point charge will weaken (following an inverse square law) as the distance from the charge is increased, and the direction of the field will be radially outward from the point charge (and therefore not always in the same direction). The magnitude and direction will vary continuously with position except at the position of the point charge, where these quantities become singular. These features of the electric field provide an opportunity to develop insights based on the way in which a vector field depends upon the spatial position, i.e., on the derivatives of the field with respect to the coordinates.

The notion of a field is actually one with which most readers already have experience. For example, an ordinary function of the spatial variables can also be called a field, but we would identify it as a **scalar field**. New and important here are the special relationships exhibited by the spatial derivatives of vector fields in three dimensions.

DERIVATIVES

The usual way of specifying a vector field is to write it as a vector whose components are functions of the spatial variables. In some cases the components may also be functions of additional variable(s), such as time. For example, an electric field could be a function

$$\mathbf{E}(x,y,z,t) = E_x(x,y,z,t)\,\hat{\mathbf{e}}_x + E_y(x,y,z,t)\,\hat{\mathbf{e}}_y + E_z(x,y,z,t)\,\hat{\mathbf{e}}_z\,.$$

The level of detail shown here is often unnecessary, and we frequently simplify the notation by writing $\mathbf{r} = (x, y, z)$, leaving

$$\mathbf{E}(\mathbf{r},t) = E_x(\mathbf{r},t)\,\hat{\mathbf{e}}_x + E_y(\mathbf{r},t)\,\hat{\mathbf{e}}_y + E_z(\mathbf{r},t)\,\hat{\mathbf{e}}_z\,.$$

We may even suppress all the arguments when their presence is not needed to convey understanding. An important thing to keep in mind is that each of the components E_x, E_y, E_z may in general depend upon all the spatial variables.

The rules for differentiating vector components are the same as for all other quantities; derivatives with respect to nonspatial variables (e.g., t) raise no special issues and do not need extensive discussion:

$$\frac{\partial \mathbf{E}}{\partial t} = \frac{\partial E_x}{\partial t}\,\hat{\mathbf{e}}_x + \frac{\partial E_y}{\partial t}\,\hat{\mathbf{e}}_y + \frac{\partial E_z}{\partial t}\,\hat{\mathbf{e}}_z\,. \tag{7.14}$$

If we take the time derivatives of expressions involving vectors or vector fields, an examination of their component expansions confirms the following, in which ψ is a scalar function of t and $\mathbf{r}$ while $\mathbf{A}$ and $\mathbf{B}$ are vector fields dependent on the same variables:

$$\begin{aligned}
\frac{\partial}{\partial t}(\psi\mathbf{A}) &= \frac{\partial \psi}{\partial t}\,\mathbf{A} + \psi\,\frac{\partial \mathbf{A}}{\partial t}\,,\\
\frac{\partial}{\partial t}(\mathbf{A}\cdot\mathbf{B}) &= \mathbf{A}\cdot\frac{\partial \mathbf{B}}{\partial t} + \frac{\partial \mathbf{A}}{\partial t}\cdot\mathbf{B}\,,\\
\frac{\partial}{\partial t}(\mathbf{A}\times\mathbf{B}) &= \mathbf{A}\times\frac{\partial \mathbf{B}}{\partial t} + \frac{\partial \mathbf{A}}{\partial t}\times\mathbf{B}\,.
\end{aligned} \tag{7.15}$$

Remember that because the cross product is a noncommutative operation you must retain the order of the factors in the derivative of $\mathbf{A}\times\mathbf{B}$ or introduce compensating minus sign(s).

GRADIENT

Suppose we have a scalar field whose value varies with position. A simple 2-D example is the gravitational potential V in hilly terrain on the Earth's surface, which has the value $V = g\,h(x,y)$, where g is the acceleration of gravity and $h(x,y)$ is the elevation above sea level at the coordinate point (x,y). If we are at some point on a hillside, we can, by going in different directions, take a path that increases our elevation, take one that decreases it, or (by going sideways to the slope) take a path of constant elevation. The rates of change in V for these paths are called **directional derivatives**; their values obviously depend upon the paths we choose. For a 2-D infinitesimal displacement, let $\mathbf{ds} = \hat{\mathbf{e}}_x\,dx + \hat{\mathbf{e}}_y\,dy$, with magnitude $ds = \sqrt{dx^2 + dy^2}$. Then the directional derivative of V in the direction $\mathbf{ds}$ is

$$\frac{dV}{ds} = \left(\frac{\partial V}{\partial x}\right)_y \left(\frac{dx}{ds}\right) + \left(\frac{\partial V}{\partial y}\right)_x \left(\frac{dy}{ds}\right). \tag{7.16}$$

One problem of interest here is to identify the direction of maximum slope (up in one direction; down in the opposite direction). If we lost our footing on the slope, the downward direction of maximum slope (i.e., maximum steepness) will be the direction in which we would slide. Taking a more formal viewpoint, and switching to an example that does not involve forces of constraint (the force exerted by the ground on the sliding student), let's continue with a charge in 3-D space that is subject to an electrostatic potential $V(x,y,z)$. The direction of maximum downward slope in V will be the direction of the electrostatic force, and this is surely of importance in physics. It is conventional to define a vector field whose value at each coordinate point gives the direction and magnitude of the maximum upward slope at that point; this field is called the **gradient** of V. Thus, we can say that the electrical force (on a unit charge) has at each point a magnitude and direction given by **minus** the gradient of V.

Our next step is to learn how to obtain the gradient from the functional form of $V(x,y,z)$. If we write the 3-D equivalent of Eq. (7.16) in differential form,

$$dV = \left(\frac{\partial V}{\partial x}\right) dx + \left(\frac{\partial V}{\partial y}\right) dy + \left(\frac{\partial V}{\partial z}\right) dz\,, \tag{7.17}$$

one approach might be a direct search for the values of dx, dy, and dz that maximize dV subject to a requirement that $dx^2 + dy^2 + dz^2$ be fixed at the constant value ds^2. This is a known type of extremum problem; it can be handled by what is known as Lagrange's method of undetermined multipliers. That method is the topic of Appendix G. Another approach, which we will use here, is to **define** a quantity we denote $\boldsymbol{\nabla} V$ (and often call "del V"),

$$\boldsymbol{\nabla} V \equiv \left(\frac{\partial V}{\partial x}\right) \hat{\mathbf{e}}_x + \left(\frac{\partial V}{\partial y}\right) \hat{\mathbf{e}}_y + \left(\frac{\partial V}{\partial z}\right) \hat{\mathbf{e}}_z\,. \tag{7.18}$$

Then, writing also $\mathbf{ds} = \hat{\mathbf{e}}_x\,dx + \hat{\mathbf{e}}_y\,dy + \hat{\mathbf{e}}_z\,dz$, Eq. (7.17) takes the form of a dot product

$$dV = (\boldsymbol{\nabla} V) \cdot \mathbf{ds} \quad \text{or} \quad \frac{dV}{ds} = (\boldsymbol{\nabla} V) \cdot \hat{\mathbf{s}}\,. \tag{7.19}$$

The last member of Eq. (7.19) follows because $\mathbf{ds}/ds$ is a unit vector in the $\mathbf{s}$ direction.

Looking further at Eq. (7.19), it is clear that dV/ds is maximized when $\hat{\mathbf{s}}$ is in the direction of $\boldsymbol{\nabla} V$ and that when it is in that direction the magnitude of dV/ds will be

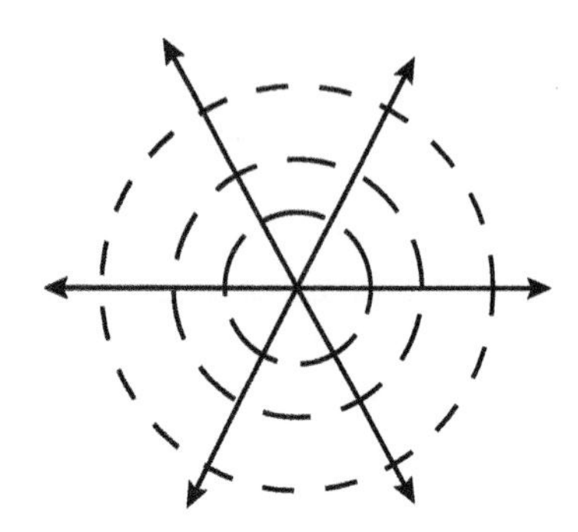

Figure 7.7: Lines of force (solid lines) and equipotentials (dashed lines) of an isolated point charge.

that of ∇V. In other words, ∇V is a vector field giving for each coordinate point the direction and magnitude of the maximum rate of increase in V. We therefore identify ∇V as the gradient of V. Another notation for ∇V is **grad** V.

It is useful to regard ∇ (often called "del") as a **vector operator**:

$$\nabla = \hat{\mathbf{e}}_x \frac{\partial}{\partial x} + \hat{\mathbf{e}}_y \frac{\partial}{\partial y} + \hat{\mathbf{e}}_z \frac{\partial}{\partial z}, \tag{7.20}$$

so that Eq. (7.18), instead of being just a definition, becomes a statement of the result of applying the operator ∇ to the function V.

Equation (7.19) leads us to other interesting observations. If we choose $\mathbf{s}$ to be perpendicular to ∇V, the directional derivative vanishes, and we have identified a direction in which V does not change. Thus, at any coordinate point ∇V is perpendicular (normal) to the equipotential surface of V that passes through that point. If we draw lines that at each point are in the direction of ∇V, we obtain a diagram that shows the pattern of the force associated with the potential field V. In physics applications these lines are known as **lines of force**, and they are everywhere perpendicular to the equipotential surfaces. The lines of force and equipotentials of an isolated point charge are shown in Fig. 7.7.

Example 7.2.1. Gradient of Function of x, y, and z

Sometimes the determination of the gradient is straightforward. Given

$$f(x,y,z) = \frac{xy}{z},$$

we easily find

$$\frac{\partial f}{\partial x} = \frac{y}{z}, \qquad \frac{\partial f}{\partial y} = \frac{x}{z}, \qquad \frac{\partial f}{\partial z} = -\frac{xy}{z^2},$$

leading to $\nabla f = \dfrac{y}{z}\,\hat{\mathbf{e}}_x + \dfrac{x}{z}\,\hat{\mathbf{e}}_y - \dfrac{xy}{z^2}\,\hat{\mathbf{e}}_z$.

■

Example 7.2.2. Gradient of Distance from Origin

Now consider ∇r. Since we only know how to proceed in Cartesian coordinates, write $r = \sqrt{x^2+y^2+z^2}$ before applying the gradient operator. We then get

$$\frac{\partial r}{\partial x} = \frac{x}{\sqrt{x^2+y^2+z^2}} = \frac{x}{r}, \tag{7.21}$$

with corresponding results $\partial r/\partial y = y/r$ and $\partial r/\partial z = z/r$. Combining these formulas,

$$\nabla r = \frac{x}{r}\,\hat{\mathbf{e}}_x + \frac{y}{r}\,\hat{\mathbf{e}}_y + \frac{z}{r}\,\hat{\mathbf{e}}_z = \frac{\mathbf{r}}{r} = \hat{\mathbf{r}}\,. \tag{7.22}$$

■

Example 7.2.3. Gradient of a Function of r

Many important problems involve the gradient of a spherically symmetric function, i.e., a function that depends only on r, the distance from the coordinate origin. Evaluation proceeds as follows:

$$\frac{\partial f(r)}{\partial x} = f'(r)\,\frac{\partial r}{\partial x} = \frac{x}{r}\,f'(r)\,,$$

where we have used Eq. (7.21) to evaluate $\partial r/\partial x$. After similar processing of the y and z derivatives, we reach

$$\nabla f(r) = f'(r)\left[\frac{x}{r}\,\hat{\mathbf{e}}_x + \frac{y}{r}\,\hat{\mathbf{e}}_y + \frac{z}{r}\,\hat{\mathbf{e}}_z\right] = f'(r)\,\hat{\mathbf{r}}\,, \tag{7.23}$$

a result that might have been expected. The quantity $f(r)$ changes value only when there is a change in r.

■

Example 7.2.4. Potential of Point Charge

The electrostatic potential V of a point charge q at the origin of a coordinate system is

$$V = \frac{1}{4\pi\varepsilon_0}\,\frac{q}{r}\,,$$

where $r = \sqrt{x^2+y^2+z^2}$. The electrostatic force on a unit charge (i.e., the **electric field E**), is given by minus the gradient of the potential.

We could treat this problem as a special case of Eq. (7.23), but let's do it independently. To compute ∇V we need

$$\frac{\partial(1/r)}{\partial x} = -\frac{1}{2}\,\frac{2x}{(x^2+y^2+z^2)^{3/2}} = -\frac{x}{r^3}\,; \quad \frac{\partial(1/r)}{\partial y} = -\frac{y}{r^3}\,; \quad \frac{\partial(1/r)}{\partial z} = -\frac{z}{r^3}\,,$$

and we therefore get the expected result

$$\mathbf{E} = -\nabla V = \left(\frac{q}{4\pi\varepsilon_0}\right)\frac{x\hat{\mathbf{e}}_x + y\hat{\mathbf{e}}_y + z\hat{\mathbf{e}}_z}{r^3} = \left(\frac{q}{4\pi\varepsilon_0}\right)\frac{\hat{\mathbf{r}}}{r^2}\,.$$

■

Exercises

7.2.1. Show that $\nabla(uv) = v\nabla u + u\nabla v$.

7.2.2. Evaluate ∇x^2yz^3 at $(2, 5, -1)$.

7.2.3. Assuming $\mathbf{A}$ and $\mathbf{B}$ to be constant vectors, show that $\nabla[\mathbf{A}\cdot(\mathbf{B}\times\mathbf{r})] = \mathbf{A}\times\mathbf{B}$.

7.2.4. Given the electrostatic potential $\psi = \dfrac{\mathbf{p}\cdot\mathbf{r}}{4\pi\varepsilon_0 r^3}$, with $\mathbf{p}$ a constant, find $\mathbf{E} = -\nabla\psi$.

7.2.5. Find the derivative $\dfrac{d(x^2y + yz)}{ds}$ at $(1, -1, 3)$ in the direction of $\hat{\mathbf{e}}_x + 2\hat{\mathbf{e}}_y - \hat{\mathbf{e}}_z$.

7.2.6. Find the derivative $\dfrac{dr^3}{ds}$ at $(1, 1, 1)$ in each of the six directions

$$\mathbf{s}_1 = \hat{\mathbf{e}}_x + \hat{\mathbf{e}}_y + \hat{\mathbf{e}}_z, \qquad \mathbf{s}_2 = -\hat{\mathbf{e}}_x - \hat{\mathbf{e}}_y - \hat{\mathbf{e}}_z, \qquad \mathbf{s}_3 = \hat{\mathbf{e}}_x - \hat{\mathbf{e}}_y,$$

$$\mathbf{s}_4 = \hat{\mathbf{e}}_x - \hat{\mathbf{e}}_z, \qquad \mathbf{s}_5 = \hat{\mathbf{e}}_y - \hat{\mathbf{e}}_z, \qquad \mathbf{s}_6 = \hat{\mathbf{e}}_x - 2\hat{\mathbf{e}}_y + \hat{\mathbf{e}}_z.$$

Compare these directional derivatives with ∇r^3 at $(1, 1, 1)$ and give a qualitative explanation of the directional derivatives.

DIVERGENCE

If, instead of applying the vector operator ∇ to a scalar, we take its dot product with a vector $\mathbf{V}$, we obtain a quantity called the **divergence** of $\mathbf{V}$. Since we have taken a dot product, the divergence will be a scalar. Sometimes we find **div** written to indicate the divergence. Thus,

$$\operatorname{div}\mathbf{V} \equiv \nabla\cdot\mathbf{V} = \left(\frac{\partial}{\partial x}\hat{\mathbf{e}}_x + \frac{\partial}{\partial y}\hat{\mathbf{e}}_y + \frac{\partial}{\partial z}\hat{\mathbf{e}}_z\right)\cdot(V_x\hat{\mathbf{e}}_x + V_y\hat{\mathbf{e}}_y + V_z\hat{\mathbf{e}}_z)$$

$$\equiv \frac{\partial V_x}{\partial x} + \frac{\partial V_y}{\partial y} + \frac{\partial V_z}{\partial z}. \tag{7.24}$$

The divergence is useful in analyzing situations in which a vector field is used to describe the flow of a fluid (or of a theoretical quantity that is thought of as having fluidic properties). Consider a vector field $\mathbf{F}$ that represents the direction and magnitude of the flow (which may vary from point to point). We interpret $\mathbf{F}$ as indicating a flow rate $F\cdot\hat{\mathbf{n}}\,dA$ through an element of area dA, where $\hat{\mathbf{n}}$ is a unit vector normal to the element of area. The amount of the flow is called the **flux** of $\mathbf{F}$ through dA. This situation can be pictured by drawing lines that at each point are in the direction of $\mathbf{F}$. These lines can be identified as the **stream lines** of the flow, and those that pass through dA contribute to the flux of $\mathbf{F}$ through dA. See Fig. 7.8.

We now show that the divergence of a vector field (at each coordinate point) gives its net flux (per unit volume) out of a volume element at that coordinate point. To confirm this claim, consider a vector field $\mathbf{F}$ that describes a flow in the neighborhood of a volume element that has dimensions dx, dy, dz and is centered at a point (x_0, y_0, z_0).

This volume element (see Fig. 7.9) will have faces of area $dy\,dz$ at $x_0 - \frac{1}{2}\,dx$ and $x_0 + \frac{1}{2}\,dx$ (there are four other faces that we will talk about later). The flow **out** of the volume element through the face at $x_0 + \frac{1}{2}\,dx$ will be $F_x\,dy\,dz$, evaluated at $x_0 + \frac{1}{2}\,dx$

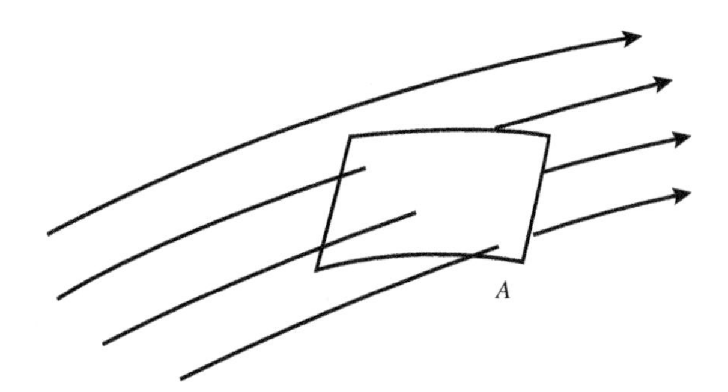

Figure 7.8: Stream lines of a vector field **F** and the flux through the area A.

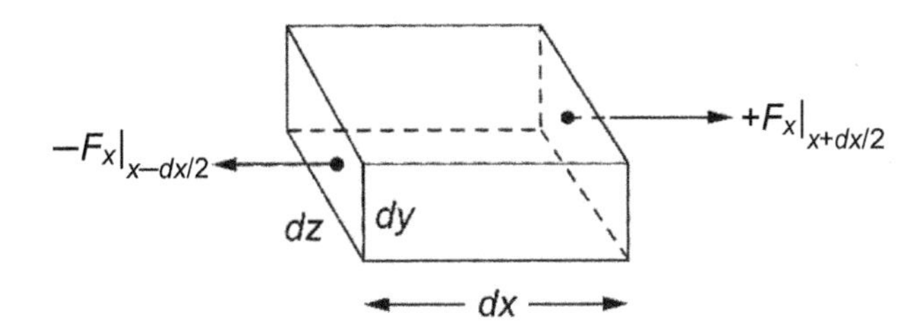

Figure 7.9: Outward flow of **F** from a volume element in the x-direction.

and averaged in y and z over the face, or (to first order) $F_x(x_0 + \frac{1}{2}\,dx, y_0, z_0)\,dy\,dz$. If F_x happened to be negative, this would be a flow into our volume element, i.e., a negative outward flow. Note that the y and z components of **F** cannot cause flow through this face, as they are parallel to it (and thereby perpendicular to the normal to the face).

Next we compute the outward flow through the face at $x_0 - \frac{1}{2}\,dx$. Since outward flow through this face is toward decreasing x, its amount is given by the value of $-F_x(x_0 - \frac{1}{2}\,dx, y_0, z_0)\,dy\,dz$. Thus, the net outward flow from the two faces now under consideration is

$$\Big[F_x(x_0 + \tfrac{1}{2}\,dx, y_0, z_0) - F_x(x_0 - \tfrac{1}{2}\,dx, y_0, z_0)\Big]\,dy\,dz \approx \frac{\partial F_x}{\partial x}\,dx\,dy\,dz\,, \qquad (7.25)$$

evaluated at (x_0, y_0, z_0). We have approximated the difference of the two F_x values by dx times $\partial F_x/\partial x$.

The net flow through the faces at $y_0 \pm \frac{1}{2}\,dy$ and $z_0 \pm \frac{1}{2}\,dz$ give results analogous to those already obtained, so the net outflow through all faces of our volume element is

$$\left[\frac{\partial F_x}{\partial x} + \frac{\partial F_y}{\partial y} + \frac{\partial F_z}{\partial z}\right] dx\,dy\,dz = \boldsymbol{\nabla}\cdot\mathbf{F}\;d^3r\,.$$

If a quantity with flow described by a vector field is only added or removed at points where there are sources or sinks (as for example water flow in a reservoir that has input and output pipes and loses no water by evaporation), the divergence of the vector field must vanish everywhere except at a source or sink.

Example 7.2.5. A Simple Divergence

Let $\mathbf{F} = xy\,\hat{\mathbf{e}}_x + xy\,\hat{\mathbf{e}}_y + \dfrac{xy}{z}\,\hat{\mathbf{e}}_z$.

Compute $\boldsymbol{\nabla}\cdot\mathbf{F}$ as follows:

$$\frac{\partial F_x}{\partial x} = y\,, \qquad \frac{\partial F_y}{\partial y} = x\,, \qquad \frac{\partial F_z}{\partial z} = -\frac{xy}{z^2}\,,$$

so $\boldsymbol{\nabla}\cdot\mathbf{F} = y + x - \dfrac{xy}{z^2}\,.$

■

Example 7.2.6. Divergence of Radius Vector

Given that $\mathbf{F} = \mathbf{r} = x\,\hat{\mathbf{e}}_x + y\,\hat{\mathbf{e}}_y + z\,\hat{\mathbf{e}}_z$,

we have

$$\frac{\partial F_x}{\partial x} = \frac{\partial x}{\partial x} = 1\,, \qquad \frac{\partial F_y}{\partial y} = 1\,, \qquad \frac{\partial F_z}{\partial z} = 1\,,$$

and therefore

$$\boldsymbol{\nabla}\cdot\mathbf{r} = 3\,. \tag{7.26}$$

■

Example 7.2.7. Divergence of Spherically Symmetric Vector Field

Consider $\boldsymbol{\nabla}\cdot[f(r)\mathbf{r}]$. Noting that the x-component of $f(r)\mathbf{r}$ is $xf(r)$, we see that we must compute

$$\frac{\partial}{\partial x}\left[xf(r)\right] = f(r) + xf'(r)\,\frac{\partial r}{\partial x} = f(r) + \frac{x^2}{r}\,f'(r)\,.$$

Combining the above with similar expressions for the y and z derivatives, we find

$$\boldsymbol{\nabla}\cdot[f(r)\mathbf{r}] = 3f(r) + f'(r)\left[\frac{x^2+y^2+z^2}{r}\right] = 3f(r) + rf'(r)\,. \tag{7.27}$$

If we set $f(r) = r^n$, we get

$$\boldsymbol{\nabla}\cdot r^n\,\mathbf{r} = (n+3)r^n\,. \tag{7.28}$$

In the special case $n = 0$, i.e., $f(r) = 1$, Eq. (7.28) reduces to $\boldsymbol{\nabla}\cdot\mathbf{r} = 3$, the result obtained in Example 7.2.6.

■

Example 7.2.8. Electric Field of Point Charge

The electric field of a point charge has the property that it can be thought of as describing a divergenceless flow. To check the validity of this interpretation, let's compute the divergence of the electric field $\mathbf{E}$ found in Example 7.2.4. We could use the result of Example 7.2.7, but let's proceed independently.

Dropping the factor $q/4\pi\varepsilon_0$, and taking

$$E_x = \frac{x}{(x^2+y^2+z^2)^{3/2}},$$

we get

$$\frac{\partial E_x}{\partial x} = \frac{1}{r^3} - \frac{3}{2}\frac{2x^2}{r^5} = \frac{r^2-3x^2}{r^5}.$$

Adding to this the analogous expressions for $\partial E_y/\partial y$ and $\partial E_z/\partial z$, we get

$$\boldsymbol{\nabla}\cdot\mathbf{E} = \frac{r^2-3x^2}{r^5} + \frac{r^2-3y^2}{r^5} + \frac{r^2-3z^2}{r^5} = 0.$$

Note that this analysis does not apply at $\mathbf{r} = 0$, as the derivatives entering the divergence formula do not exist at that point. The behavior at $\mathbf{r} = 0$ (where the point charge is considered to be a source) is important and will be discussed in detail later. ■

Exercises

7.2.7. Prove that $\dfrac{d}{dt}(\mathbf{A}\cdot\mathbf{B}) = \dfrac{d\mathbf{A}}{dt}\cdot\mathbf{B} + \dfrac{d\mathbf{B}}{dt}\cdot\mathbf{A}$.

7.2.8. Evaluate $\boldsymbol{\nabla}\cdot(z\hat{\mathbf{e}}_x + y\hat{\mathbf{e}}_y + x\hat{\mathbf{e}}_z)$.

7.2.9. Evaluate $\boldsymbol{\nabla}\cdot\left(e^{-r^2}\hat{\mathbf{r}}\right)$.

7.2.10. Show that $\boldsymbol{\nabla}\cdot[f(r)\mathbf{A}] = f(r)\boldsymbol{\nabla}\cdot\mathbf{A} + [\boldsymbol{\nabla}f(r)]\cdot\mathbf{A}$.

CURL

Another way in which the operator $\boldsymbol{\nabla}$ can enter vector analysis is as its cross product with a vector. Letting $\mathbf{B}$ be a 3-D vector, we write this expression $\boldsymbol{\nabla}\times\mathbf{B}$. The quantity $\boldsymbol{\nabla}\times\mathbf{B}$ is called the **curl** of $\mathbf{B}$ and an alternate notation for it is **curl B**. In the evaluation of $\boldsymbol{\nabla}\times\mathbf{B}$, the components of $\boldsymbol{\nabla}$ and $\mathbf{B}$ combine in accord with the formula for the cross product. However, since $\boldsymbol{\nabla}$ is both a differential operator and a vector, the product must be written in a way that shows the differentiation to be applied to the components of $\mathbf{B}$. Based on these considerations, we have (for pedagogic purposes writing $\boldsymbol{\nabla}$ initially as $\nabla_x\hat{\mathbf{e}}_x + \nabla_y\hat{\mathbf{e}}_y + \nabla_z\hat{\mathbf{e}}_z$)

$$\begin{aligned}\text{curl}\,\mathbf{B} = \boldsymbol{\nabla}\times\mathbf{B} &= (\nabla_y B_z - \nabla_z B_y)\hat{\mathbf{e}}_x + (\nabla_z B_x - \nabla_x B_z)\hat{\mathbf{e}}_y + (\nabla_x B_y - \nabla_y B_x)\hat{\mathbf{e}}_z \\ &= \left(\frac{\partial B_z}{\partial y} - \frac{\partial B_y}{\partial z}\right)\hat{\mathbf{e}}_x + \left(\frac{\partial B_x}{\partial z} - \frac{\partial B_z}{\partial x}\right)\hat{\mathbf{e}}_y + \left(\frac{\partial B_y}{\partial x} - \frac{\partial B_x}{\partial y}\right)\hat{\mathbf{e}}_z\,. \end{aligned} \tag{7.29}$$

We can write Eq. (7.29) as the determinant

$$\boldsymbol{\nabla}\times\mathbf{B} = \begin{vmatrix} \hat{\mathbf{e}}_x & \hat{\mathbf{e}}_y & \hat{\mathbf{e}}_z \\ \partial/\partial x & \partial/\partial y & \partial/\partial z \\ B_x & B_y & B_z \end{vmatrix}, \tag{7.30}$$

but with the special condition that the determinant must be evaluated in a way that causes the derivatives in its second row to be applied to the components of $\mathbf{B}$ in the third row. A simple recipe that accomplishes this is to require each term of the expansion of the determinant to have its factors written in the order of the rows from which they came; one way of phrasing this requirement is to state that the determinant is to be evaluated "from the top down."

The name **curl** is appropriate because it can be shown that if $\nabla \times \mathbf{B}$ is nonzero at a point, the stream lines of $\mathbf{B}$ in the vicinity of that point will correspond to a flow that has a nonzero rotational component. In fluids this phenomenon is called vorticity; in physics contexts it is sometimes called **circulation**. Circulation occurs in a variety of physical phenomena. For example, the magnetic field in the neighborhood of a current-carrying wire has stream lines that form closed loops surrounding the wire. Vorticity is also frequently encountered in fluid flow (e.g., eddies in flowing water, hurricanes and tornadoes in the atmosphere).

Our starting point for understanding the **curl** is to note that if a flow described by a vector field $\mathbf{B}$ has circulation in a given region, the average of the component of $\mathbf{B}$ along closed loops in some orientations must be nonzero (otherwise the overall flow could not have a rotational component). Since we want to identify circulation with points, we consider loops of infinitesimal size and, for simplicity, planar, with the coordinate point under study in the region enclosed by the loop. We then define the circulation associated with such a loop as having a direction $\hat{\mathbf{n}}$ normal to the loop and given as the small-area limit of

$$\text{Circulation} = \frac{\hat{\mathbf{n}}}{A} \oint \mathbf{B} \cdot d\mathbf{r}\,, \tag{7.31}$$

where A is the area of the loop. The circle on the integral sign is a reminder that this line integral is over a closed loop. The normal $\hat{\mathbf{n}}$ has the sign given by the right-hand rule: when the curved fingers of the right hand point in the direction of the line integration, the thumb indicates the positive direction of $\hat{\mathbf{n}}$.

It suffices to consider a counterclockwise loop in the xy-plane. For such a loop, the normal $\hat{\mathbf{n}}$ is in the positive z-direction, while $d\mathbf{r} = \hat{\mathbf{e}}_x\, dx + \hat{\mathbf{e}}_y\, dy$, and Eq. (7.31) then takes the form

$$\text{Circulation} = \frac{\hat{\mathbf{e}}_z}{A} \oint\limits_{\partial A} \mathbf{B} \cdot (\hat{\mathbf{e}}_x\, dx + \hat{\mathbf{e}}_y\, dy) = \frac{\hat{\mathbf{e}}_z}{A} \oint\limits_{\partial A} (B_x\, dx + B_y\, dy)\ . \tag{7.32}$$

The line integral in Eq. (7.32), in which the notation ∂A indicates that the integration is around the boundary of A, is exactly of the form corresponding to Green's theorem in the plane, Eq. (6.46), with $P = B_y$ and $Q = B_x$, so it can be written as an integral over the area of the loop:

$$\text{Circulation} = \frac{\hat{\mathbf{e}}_z}{A} \int\limits_{A} \left[\frac{\partial B_y}{\partial x} - \frac{\partial B_x}{\partial y} \right] dA\,. \tag{7.33}$$

As the loop shrinks toward a point $\mathbf{r}$, the integral over the area will approach A times the value of the integrand at $\mathbf{r}$, and in that limit we get

$$\text{Circulation} = \hat{\mathbf{e}}_z \left[\frac{\partial B_y}{\partial x} - \frac{\partial B_x}{\partial y} \right]\,. \tag{7.34}$$

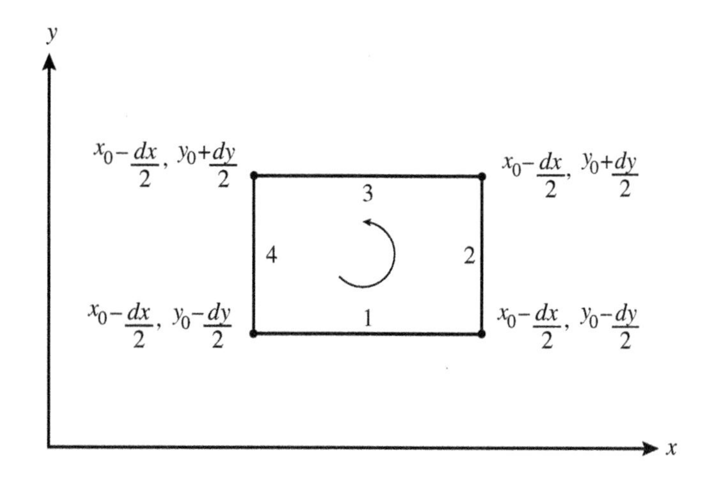

Figure 7.10: Rectangular loop for computation of curl **B**.

The right-hand side of Eq. (7.34) is the contribution to $\boldsymbol{\nabla}\times\mathbf{B}$ that is in the z-direction, i.e., the direction perpendicular to the loop.

Because the above analysis could have been carried out for a planar path in any orientation, our overall conclusion is that at any point (x, y, z) the circulation for a vector field $\mathbf{B}$ is given by $\boldsymbol{\nabla}\times\mathbf{B}$.

An alternate derivation of Eq. (7.34) can be developed by analyzing the line integral of $\mathbf{B}$ on a small rectangular loop in the xy-plane. Referring to Fig. 7.10, we identify the four segments of the integral $\oint \mathbf{B}\cdot d\mathbf{r}$ on this loop as shown in the figure, where beside each segment appears its contribution to the line integral. The positive direction for the normal to this loop is toward $+z$, and the line integral has the approximate value

$$\Big[B_y(x+\tfrac{1}{2}\,dx) - B_y(x-\tfrac{1}{2}\,dx)\Big]\,dy - \Big[B_x(y+\tfrac{1}{2}\,dy) - B_x(y-\tfrac{1}{2}\,dy)\Big]\,dx$$

$$\approx \left(\frac{\partial B_y}{\partial x} - \frac{\partial B_x}{\partial y}\right) dA = (\boldsymbol{\nabla}\times\mathbf{B})\cdot\hat{\mathbf{e}}_z\,dA\,,$$

the approximation becoming exact as the loop area $dx\,dy$ approaches zero.

Example 7.2.9. A Simple Curl

Let's evaluate $\boldsymbol{\nabla}\times\mathbf{B}$, where $\mathbf{B} = xy\,\hat{\mathbf{e}}_x + xy\,\hat{\mathbf{e}}_y + \dfrac{xy}{z}\,\hat{\mathbf{e}}_z$.

The nonzero derivatives of components of $\mathbf{B}$ that we need are

$$\frac{\partial B_x}{\partial y} = x\,, \qquad \frac{\partial B_y}{\partial x} = y\,, \qquad \frac{\partial B_z}{\partial x} = \frac{y}{z}\,, \qquad \frac{\partial B_z}{\partial y} = \frac{x}{z}\,,$$

and by direct substitution into Eq. (7.29) we get

$$\boldsymbol{\nabla}\times\mathbf{B} = \frac{x}{z}\,\hat{\mathbf{e}}_x - \frac{y}{z}\,\hat{\mathbf{e}}_y + (y-x)\,\hat{\mathbf{e}}_z\,.$$

■

Example 7.2.10. Curl of Radius Vector

We now evaluate $\nabla \times \mathbf{r}$. Writing $\mathbf{r} = x\,\hat{\mathbf{e}}_x + y\,\hat{\mathbf{e}}_y + z\,\hat{\mathbf{e}}_z$, we note that

$$\frac{\partial r_x}{\partial y} = \frac{\partial x}{\partial y} = 0\,,$$

and all the other derivatives needed for the curl also vanish, so we get

$$\nabla \times \mathbf{r} = 0\,. \tag{7.35}$$

This result could have been expected because the stream lines of $\mathbf{r}$ radiate out from the origin on straight-line trajectories. There isn't even a hint of possible circulation.

■

Example 7.2.11. Curl of Spherically Symmetric Vector Field

Consider $\nabla \times \mathbf{F}$, where $\mathbf{F} = f(r)\,\mathbf{r}$.

Noting that $F_x = xf(r)$ and that $F_y = yf(r)$, we look at

$$\frac{\partial F_y}{\partial x} = yf'(r)\,\frac{\partial r}{\partial x} = yf'(r)\,\frac{x}{r} = \frac{xy}{r}\,f'(r)\,,$$

$$\frac{\partial F_x}{\partial y} = xf'(r)\,\frac{\partial r}{\partial y} = xf'(r)\,\frac{y}{r} = \frac{xy}{r}\,f'(r)\,,$$

and therefore find

$$(\nabla \times \mathbf{F})_z = \frac{\partial F_y}{\partial x} - \frac{\partial F_x}{\partial y} = \frac{xy}{r}\,f'(r) - \frac{xy}{r}\,f'(r) = 0\,.$$

Similar evaluations can be obtained for the other components of $\nabla \times \mathbf{F}$, so our overall computation yields

$$\nabla \times [f(r)\,\mathbf{r}] = 0\,, \tag{7.36}$$

irrespective of the specific functional form of $f(r)$. Again, this is an expected result.

■

Example 7.2.12. Some Other Curls

To illustrate the fact that a lack of symmetry in a vector field $\mathbf{B}$ is not by itself an indicator that $\nabla \times \mathbf{B}$ will not vanish, consider $\nabla \times \mathbf{F}$ and $\nabla \times \mathbf{G}$, where $\mathbf{F} = y\,\hat{\mathbf{e}}_x + x\,\hat{\mathbf{e}}_y$ and $\mathbf{G} = y\,\hat{\mathbf{e}}_x - x\,\hat{\mathbf{e}}_y$. These curls evaluate to

$$\nabla \times \mathbf{F} = 0\,, \qquad \nabla \times \mathbf{G} = -2\,\hat{\mathbf{e}}_z\,.$$

Verification of these formulas is the topic of Exercise 7.2.11.

■

Example 7.2.13. Circulation: Field of a Point Charge

We have already found, in Example 7.2.8, that the electric field of a point charge has zero divergence everywhere except at the charge itself (where the derivatives appearing in the divergence are not defined). That fact makes it natural to identify the electric field $\mathbf{E}$ with a "flow" in which (except at the charge) there are no sources or sinks. Since the stream lines of $\mathbf{E}$ are its lines of force, we do not expect them to form closed loops; if they did, particles in the field of a point charge could move in continuously accelerating circular motion, violating both common sense and the law of conservation of energy. We therefore expect that if $\mathbf{E}$ is the electric field of a point charge,

$$\mathbf{E} = \left(\frac{q}{4\pi\varepsilon_0}\right)\left[\frac{x}{r^3}\,\hat{\mathbf{e}}_x + \frac{y}{r^3}\,\hat{\mathbf{e}}_y + \frac{z}{r^3}\,\hat{\mathbf{e}}_z\right],$$

then $\boldsymbol{\nabla}\times\mathbf{E} = 0$.

We could identify this result as a special case of Example 7.2.11, but let's work it out explicitly. Checking the x component of $\boldsymbol{\nabla}\times\mathbf{E}$:

$$\begin{aligned}(\boldsymbol{\nabla}\times\mathbf{E})_x &= \left(\frac{q}{4\pi\varepsilon_0}\right)\left[\frac{\partial E_z}{\partial y} - \frac{\partial E_y}{\partial z}\right] = \left(\frac{q}{4\pi\varepsilon_0}\right)\left[z\,\frac{\partial r^{-3}}{\partial y} - y\,\frac{\partial r^{-3}}{\partial z}\right] \\ &= \left(\frac{q}{4\pi\varepsilon_0}\right)\left[z\left(\frac{-3y}{r^5}\right) - y\left(\frac{-3z}{r^5}\right)\right] = 0\,.\end{aligned}$$

By symmetry, the y and z components of the curl must also vanish.

■

Exercises

7.2.11. Verify the results of Example 7.2.12, namely that for $\mathbf{F} = y\,\hat{\mathbf{e}}_x + x\,\hat{\mathbf{e}}_y$ and $\mathbf{G} = y\,\hat{\mathbf{e}}_x - x\,\hat{\mathbf{e}}_y$, we have $\boldsymbol{\nabla}\times\mathbf{F} = 0$ and $\boldsymbol{\nabla}\times\mathbf{G} = -2\,\hat{\mathbf{e}}_z$.

7.2.12. Evaluate $\boldsymbol{\nabla}\times(z\hat{\mathbf{e}}_x + y\hat{\mathbf{e}}_y + x\hat{\mathbf{e}}_z)$.

7.2.13. Show that for $\mathbf{m}$ a constant vector, and for $\mathbf{B}$ such that $\boldsymbol{\nabla}\times\mathbf{B} = \boldsymbol{\nabla}\cdot\mathbf{B} = 0$,

$$\boldsymbol{\nabla}\times(\mathbf{B}\times\mathbf{m}) = \boldsymbol{\nabla}(\mathbf{m}\cdot\mathbf{B})\,.$$

7.2.14. Show that $\mathbf{A}\times(\boldsymbol{\nabla}\times\mathbf{A}) = \frac{1}{2}\,\boldsymbol{\nabla}(\mathbf{A}^2) - (\mathbf{A}\cdot\boldsymbol{\nabla})\mathbf{A}$.

SYMBOLIC COMPUTATION

There are major differences in the way MAPLE and MATHEMATICA treat vector differential operators. Strictly speaking, these operators apply to **vector fields**. A detailed discussion of the symbolic treatment of vector fields can be found in Appendix H. Much of the material in that Appendix becomes significant only when vectors are written in curvilinear coordinate systems and is not needed here because we are presently assuming that all vectors and vector operators are in Cartesian coordinates.

In MAPLE, it is necessary to identify vectors as vector fields and to specify the names of the coordinates prior to the application of differential operators. One way to do so is (in this order) to (1) specify the names of the coordinates, using the command `SetCoordinates`, (2) define a vector, with components that will typically be functions of the chosen set of coordinates, and (3) designate the vector as a vector field. After these steps have been taken, then MAPLE will recognize and know how to process the basic vector differential operators `Gradient` (also called `Del`), `Divergence`, and `Curl`.

In MATHEMATICA, commands involving differential vector operators are supported by a package that is accessed by the command `<<VectorAnalysis‘`. The symbols `<<` and the grave accent ‘ are essential parts of the designation. The package only supports 3-D vectors. In MATHEMATICA it is not necessary to identify vectors as vector fields, nor is it required to specify the coordinate names if one accepts the default coordinate system (Cartesian) and the default coordinate names (which are `Xx, Yy, Zz`). However, the user is free to choose other coordinate names. The basic vector operators are designated `Grad, Div`, and `Curl`.

Example 7.2.14. Symbolic Computing, Vector Operators

Formulas are for Cartesian coordinates only.

In MAPLE:

```
> with(VectorCalculus);
> SetCoordinates(cartesian[x,y,z]);
```

The above statement is required and applies to all subsequent statements.

```
> A := Vector([x^2,x-y,x*z]):
> A := VectorField(A);
```
$$A = x^2\overline{\mathbf{e}}_x + (x-y)\overline{\mathbf{e}}_y + xz\overline{\mathbf{e}}_z$$
```
> B := VectorField([0,0,z]);
```
$$z\overline{\mathbf{e}}_z$$

The overbars tell the user that `A` and `B` are vector fields.

```
> G := x^2*exp(z);
```
$$G := x^2\,e^z$$
```
> Gradient(G);
```
$$2xe^z\overline{\mathbf{e}}_x + x^2e^z\overline{\mathbf{e}}_z$$
```
> Del(G);
```
$$2xe^z\overline{\mathbf{e}}_x + x^2e^z\overline{\mathbf{e}}_z$$

The above shows that `Del` and `Gradient` are interchangeable here.

```
> Divergence(A);
```
$$-1 + 3x$$
```
> Curl(A);
```
$$-z\overline{\mathbf{e}}_y + \overline{\mathbf{e}}_z$$

If a vector is zero MAPLE reports its zero first component (thereby showing the vector-field overbar).

```
> Curl(B);
```
$$0\overline{\mathbf{e}}_x$$

One can also use `Del` as an operator before the dot and cross-product symbols:

```
> Del . A;
```
$$-1 + 3x$$
```
> Del &x A;
```
$$-z\overline{\mathbf{e}}_y + \overline{\mathbf{e}}_z$$

In MATHEMATICA:

```
<<VectorAnalysis‘
```

The default coordinates `Xx, Yy, Zz` are awkward; change to `x,y,z`:

```
SetCoordinates[Cartesian[x,y,z]]
```

Those coordinates remain applicable unless or until changed.

```
A = {x^2, x-y, x*z}                    {x^2, x-y, x*z}
G = x^2*E^z                            e^z x^2
```

No further steps are needed prior to applying the vector differential operators. The results are in MATHEMATICA's standard vector format.

```
Grad[G]                                {2e^z x, 0, e^z x^2}
Div[A]                                 -1 + 3x
Curl[A]                                {0, -z, 1}
```

■

Exercises

Use your symbolic computation system to evaluate the following expressions, and then check your work with hand computation. Here $f = xyz$ and $\mathbf{B} = xy\hat{\mathbf{e}}_x - yz\hat{\mathbf{e}}_y + xz\hat{\mathbf{e}}_z$.

7.2.15. (a) $\boldsymbol{\nabla} f$, (b) $(\boldsymbol{\nabla} f) \cdot (\boldsymbol{\nabla} f)$, (c) $(\boldsymbol{\nabla} f) \cdot \mathbf{B}$.

7.2.16. (a) $\boldsymbol{\nabla} \cdot \mathbf{B}$, (b) $\boldsymbol{\nabla} \times \mathbf{B}$, (c) $(\boldsymbol{\nabla} \times \mathbf{B}) \cdot (\boldsymbol{\nabla} \times \mathbf{B})$.

7.3 VECTOR DIFFERENTIAL OPERATORS: FURTHER PROPERTIES

A number of useful expressions result if we successively apply two vector operators in various ways.

LAPLACIAN

An important example of successive vector operations is the divergence of the gradient of a function, i.e.,

$$\text{div}\,(\text{grad}\,\psi(\mathbf{r})) = \boldsymbol{\nabla} \cdot \boldsymbol{\nabla}\psi(\mathbf{r})\,. \tag{7.37}$$

This formula shows that $\boldsymbol{\nabla} \cdot \boldsymbol{\nabla}$ is a vector operator that can be applied to a scalar function to produce a result that is also a scalar; it is called the **Laplacian** and is often denoted $\boldsymbol{\nabla}^2$. Inserting the explicit form of $\boldsymbol{\nabla}$, we find that the Laplacian can be written as a sum of second derivatives:

$$\begin{aligned}\boldsymbol{\nabla}^2 = \boldsymbol{\nabla} \cdot \boldsymbol{\nabla} &= \left[\frac{\partial}{\partial x}\hat{\mathbf{e}}_x + \frac{\partial}{\partial y}\hat{\mathbf{e}}_y + \frac{\partial}{\partial z}\hat{\mathbf{e}}_z\right] \cdot \left[\frac{\partial}{\partial x}\hat{\mathbf{e}}_x + \frac{\partial}{\partial y}\hat{\mathbf{e}}_y + \frac{\partial}{\partial z}\hat{\mathbf{e}}_z\right] \\ &= \frac{\partial^2}{\partial x^2} + \frac{\partial^2}{\partial y^2} + \frac{\partial^2}{\partial z^2}\,. \end{aligned} \tag{7.38}$$

The Laplacian appears in many of the fundamental partial differential equations of physics, such as

$$\nabla^2\psi = 0 \qquad \text{Laplace equation,}$$

$$\nabla^2\psi = -\frac{\rho}{\varepsilon_0} \qquad \text{Poisson equation,}$$

$$\nabla^2\psi = -k^2\psi \qquad \text{Helmholtz equation,}$$

$$\nabla^2\psi = \frac{1}{a^2}\frac{\partial^2\psi}{\partial t^2} \qquad \text{Wave equation,}$$

$$\nabla^2\psi = \frac{1}{a}\frac{\partial\psi}{\partial t} \qquad \text{Diffusion equation, Schrödinger equation.}$$

Methods for solving these partial differential equations are taken up in Chapter 15.

Another product involving two vector operations is $\nabla \times (\nabla \times \mathbf{V})$, where $\mathbf{V}$ is a vector field. If we evaluate this product using Eq. (7.13), being careful to write the result in a form that shows both vector operators to be applied to $\mathbf{V}$, we obtain

$$\nabla \times (\nabla \times \mathbf{V}) = \nabla(\nabla \cdot \mathbf{V}) - (\nabla \cdot \nabla)\mathbf{V}. \qquad (7.39)$$

The term $(\nabla \cdot \nabla)\mathbf{V}$, often written $\nabla^2\mathbf{V}$, is called the **vector Laplacian** of $\mathbf{V}$. In Cartesian coordinates it is a vector whose components are $\nabla^2 V_x$, $\nabla^2 V_y$, $\nabla^2 V_z$. However, in curvilinear coordinates we do not get this component separation and it avoids potential errors to identify the vector Laplacian with the result obtained by rearranging Eq. (7.39):

$$\nabla^2\mathbf{V} \equiv \nabla(\nabla \cdot \mathbf{V}) - \nabla \times (\nabla \times \mathbf{V}). \qquad (7.40)$$

Example 7.3.1. A Laplacian

The Laplacian of $g = x^3 - y^3$ is

$$\nabla^2 g = \left[\frac{\partial^2}{\partial x^2} + \frac{\partial^2}{\partial y^2} + \frac{\partial^2}{\partial z^2}\right](x^3 - y^3) = 6x - 6y.$$

■

Example 7.3.2. Laplacian of r

We start the evaluation with

$$\frac{dr}{dx} = \frac{x}{(x^2+y^2+z^2)^{1/2}}, \qquad \frac{d^2r}{dx^2} = \frac{1}{(x^2+y^2+z^2)^{1/2}} - \frac{x^2}{(x^2+y^2+z^2)^{3/2}}.$$

Combining the above with corresponding expressions for the y and z derivatives and writing $(x^2+y^2+z^2)^{1/2} = r$, we get

$$\nabla^2 r = \left(\frac{1}{r} - \frac{x^2}{r^3}\right) + \left(\frac{1}{r} - \frac{y^2}{r^3}\right) + \left(\frac{1}{r} - \frac{z^2}{r^3}\right) = \frac{3}{r} - \frac{x^2+y^2+z^2}{r^3} = \frac{2}{r}. \qquad (7.41)$$

■

Example 7.3.3. Laplacian of $1/r$

An important equation results if we apply the Laplacian operator to the potential of a point charge. Disregarding the constant factor $q/4\pi\varepsilon_0$, we consider $\nabla^2(1/r)$. Writing $r = \sqrt{x^2+y^2+z^2}$, the evaluation seems straightforward. Applying first $\partial^2/\partial x^2$, we have

$$\frac{\partial^2}{\partial x^2}\left(\frac{1}{r}\right) = \frac{\partial}{\partial x}\left(-\frac{x}{r^3}\right) = -\frac{1}{r^3} + \frac{3x^2}{r^5}\,.$$

Combining that result with the corresponding expressions for the y and z derivatives, we get

$$\nabla^2\left(\frac{1}{r}\right) = \left[-\frac{1}{r^3} + \frac{3x^2}{r^5}\right] + \left[-\frac{1}{r^3} + \frac{3y^2}{r^5}\right] + \left[-\frac{1}{r^3} + \frac{3z^2}{r^5}\right]$$

$$= -\frac{3}{r^3} + \frac{3x^2+3y^2+3z^2}{r^5} = 0\,. \tag{7.42}$$

What may be less obvious is that this result is only valid for $r \neq 0$, as the derivatives we need to use do not exist there. We revisit this problem later and consider its behavior at $r = 0$.

■

IRROTATIONAL AND SOLENOIDAL VECTOR FIELDS

Other interesting examples of successive vector operations are the expressions $\nabla\times\nabla\psi$ and $\nabla\cdot(\nabla\times\mathbf{V})$. Using the determinantal forms of the cross product and the scalar triple product, these expressions can be evaluated as follows:

$$\nabla\times\nabla\psi = \begin{vmatrix} \hat{\mathbf{e}}_x & \hat{\mathbf{e}}_y & \hat{\mathbf{e}}_z \\ \partial/\partial x & \partial/\partial y & \partial/\partial z \\ \partial\psi/\partial x & \partial\psi/\partial y & \partial\psi/\partial z \end{vmatrix} = \begin{vmatrix} \hat{\mathbf{e}}_x & \hat{\mathbf{e}}_y & \hat{\mathbf{e}}_z \\ \partial/\partial x & \partial/\partial y & \partial/\partial z \\ \partial/\partial x & \partial/\partial y & \partial/\partial z \end{vmatrix} \psi = 0\,, \tag{7.43}$$

$$\nabla\cdot(\nabla\times\mathbf{V}) = \begin{vmatrix} \partial/\partial x & \partial/\partial y & \partial/\partial z \\ \partial/\partial x & \partial/\partial y & \partial/\partial z \\ V_x & V_y & V_z \end{vmatrix} = 0\,. \tag{7.44}$$

Those expressions vanish because the determinants contain two identical rows.

Equation (7.43) makes an important general statement, namely that any vector field that is a gradient has a zero curl (and therefore no circulation). We already noted this behavior in Example 7.2.13 for the field of a point charge, but we now see that it is general, applying to any field that is the gradient of a potential. Fields having a zero curl are called **irrotational**.

Equation (7.44) is in a way complementary to Eq. (7.43); it states that any field that is a curl must have zero divergence, i.e., no sources or sinks. Fields without sources or sinks are termed **solenoidal**, because their stream lines cannot start or end anywhere, and therefore can only be closed loops. The prime physical example of a solenoidal field is the magnetic field; the nonexistence of magnetic "charges" (monopoles) is a physical property showing that a magnetic induction field **B** can have no sources or sinks. The fact that the magnetic field is solenoidal permits it to be represented as the curl of some vector field $\mathbf{A}$; a vector field whose curl is $\mathbf{B}$ is called a **vector potential.**

The conditions that a vector field be irrotational or solenoidal are of sufficient importance that we set them forth specifically here:

$$\begin{aligned}&\text{If } \mathbf{E} = \nabla\psi, \text{ then } \nabla\times\mathbf{E} = \nabla\times\nabla\psi = 0, \text{ and } \mathbf{E} \text{ is } \textbf{irrotational}.\\ &\text{If } \mathbf{B} = \nabla\times\mathbf{A}, \text{ then } \nabla\cdot\mathbf{B} = \nabla\cdot(\nabla\times\mathbf{A}) = 0, \text{ and } \mathbf{B} \text{ is } \textbf{solenoidal}.\end{aligned} \tag{7.45}$$

SYMBOLIC COMPUTATION

Many expressions with multiple instances of ∇ can be constructed using the techniques in earlier sections of this chapter. Both MAPLE and MATHEMATICA contain explicit definitions of the Laplacian operator, so that it is not necessary to construct it as a symbolic rendition of $\nabla\cdot\nabla$. In both languages, the same symbol is used for both scalar and vector arguments. Thus, for computation in Cartesian coordinates,

MAPLE

```
>with(VectorCalculus);
> G := x^2 * exp(z):
> A := Vector([x^2,x-y,x*z]):
> Laplacian(G);
```

$$2e^z + x^2e^x$$

```
> Laplacian(VectorField(A));
```

$$2\overline{\mathrm{e}}_x$$

MATHEMATICA

```
<<VectorAnalysis`
G = x^2 * E^z;
A = {x^2, x-y, x*z};
Laplacian[G]
```

$$2e^z + e^zx^2$$

```
Laplacian[A]
```

$$\{2, 0, 0\}$$

VECTOR IDENTITIES

The vector operators that we have introduced can be applied in ways that lead to a large number of identities, and use of these identities can resolve or simplify many problems in vector analysis. Table 7.1 gives a list of useful identities. These identities can be verified by expanding all the operators and vector fields into components, but such a process may be quite cumbersome and perhaps not very enlightening. Most of the identities can also be checked using symbolic computation. An approach that may provide more insight is to work by hand, using the known properties of the operators in appropriate ways. Here are some techniques that are useful in proving vector identities.

Example 7.3.4. Vector Identity #4

We wish to prove the identity $\nabla\times(\varphi\mathbf{A}) = \varphi(\nabla\times\mathbf{A}) - \mathbf{A}\times\nabla\varphi$.

Expand the x component of the left-hand side and organize the result appropriately:

$$\nabla\times(\varphi\mathbf{A})\Big|_x = \frac{\partial}{\partial y}(\varphi A_z) - \frac{\partial}{\partial z}(\varphi A_y) = \frac{\partial\varphi}{\partial y}A_z + \varphi\frac{\partial A_z}{\partial y} - \frac{\partial\varphi}{\partial z}A_y - \varphi\frac{\partial A_y}{\partial z}$$

$$= \varphi\left[\frac{\partial A_z}{\partial y} - \frac{\partial A_y}{\partial z}\right] + \left[\frac{\partial\varphi}{\partial y}A_z - \frac{\partial\varphi}{\partial z}A_y\right] = \varphi\nabla\times\mathbf{A}\Big|_x + (\nabla\varphi)\times\mathbf{A}\Big|_x.$$

Table 7.1: Vector Identities.

1.	$\mathbf{A} \times (\mathbf{B} \times \mathbf{C}) = \mathbf{B}(\mathbf{A} \cdot \mathbf{C}) - \mathbf{C}(\mathbf{A} \cdot \mathbf{B})$
2.	$\boldsymbol{\nabla}(\mathbf{A} \cdot \mathbf{B}) = \mathbf{A} \times (\boldsymbol{\nabla} \times \mathbf{B}) + \mathbf{B} \times (\boldsymbol{\nabla} \times \mathbf{A}) + (\mathbf{A} \cdot \boldsymbol{\nabla})\mathbf{B} + (\mathbf{B} \cdot \boldsymbol{\nabla})\mathbf{A}$
3.	$\boldsymbol{\nabla} \cdot (\varphi \mathbf{A}) = \varphi (\boldsymbol{\nabla} \cdot \mathbf{A}) + \mathbf{A} \cdot \boldsymbol{\nabla} \varphi$
4.	$\boldsymbol{\nabla} \times (\varphi \mathbf{A}) = \varphi (\boldsymbol{\nabla} \times \mathbf{A}) - \mathbf{A} \times \boldsymbol{\nabla} \varphi$
5.	$\boldsymbol{\nabla} \cdot (\mathbf{A} \times \mathbf{B}) = \mathbf{B}(\boldsymbol{\nabla} \times \mathbf{A}) - \mathbf{A}(\boldsymbol{\nabla} \times \mathbf{B})$
6.	$\boldsymbol{\nabla} \times (\mathbf{A} \times \mathbf{B}) = \mathbf{A}(\boldsymbol{\nabla} \cdot \mathbf{B}) - \mathbf{B}(\boldsymbol{\nabla} \cdot \mathbf{A}) - (\mathbf{A} \cdot \boldsymbol{\nabla})\mathbf{B} + (\mathbf{B} \cdot \boldsymbol{\nabla})\mathbf{A}$
7.	$\boldsymbol{\nabla} \cdot (\boldsymbol{\nabla} \times \mathbf{A}) = 0$
8.	$\boldsymbol{\nabla} \times \boldsymbol{\nabla} \varphi = 0$
9.	$\boldsymbol{\nabla} \cdot (\boldsymbol{\nabla} \varphi \times \boldsymbol{\nabla} \psi) = 0$
10.	$\boldsymbol{\nabla} \times (\boldsymbol{\nabla} \times \mathbf{A}) = \boldsymbol{\nabla}(\boldsymbol{\nabla} \cdot \mathbf{A}) - \boldsymbol{\nabla}^2 \mathbf{A}$
11.	$\boldsymbol{\nabla} r^n = n r^{n-1} \hat{\mathbf{r}}$
12.	$\boldsymbol{\nabla} f(r) = \hat{\mathbf{r}} f'(r)$
13.	$\boldsymbol{\nabla} \cdot (\hat{\mathbf{r}} r^n) = (n+2) r^{n-1}$
14.	$\boldsymbol{\nabla} \cdot \Big(\hat{\mathbf{r}} f(r)\Big) = 2r^{-1} f(r) + f'(r)$
15.	$\boldsymbol{\nabla} \times \Big(\hat{\mathbf{r}} f(r)\Big) = 0$
16.	$\boldsymbol{\nabla}^2 \left(\frac{1}{r}\right) = -4\pi \delta(\mathbf{r})$

A, **B**, **C** are vectors; φ, ψ are scalars; $r = \sqrt{x^2 + y^2 + z^2}$.

This result is correct if its last term is interpreted as meaning that the $\boldsymbol{\nabla}$ in $(\boldsymbol{\nabla}\varphi)$ only applies to the material within the parentheses, i.e., only to φ. That interpretation is inherently ambiguous, and the need for its use can be avoided by changing the order of the factors in that term. This interchange introduces a sign change, so our final formula for the x component of the formula becomes

$$\boldsymbol{\nabla} \times (\varphi \mathbf{A})\Big|_x = \varphi\, \boldsymbol{\nabla} \times \mathbf{A}\Big|_x - \mathbf{A} \times (\boldsymbol{\nabla}\varphi)\Big|_x .$$

Extending this formula to all components, we confirm Vector Identity #4.

Note that, to avoid ambiguity, all the identities in our table have been written in forms such that the differential operators act on all quantities to their right, irrespective of the presence of parentheses.

■

Example 7.3.5. Vector Identity #6

We now wish to prove the identity

$$\nabla \times (\mathbf{A} \times \mathbf{B}) = \mathbf{A}(\nabla \cdot \mathbf{B}) - \mathbf{B}(\nabla \cdot \mathbf{A}) - (\mathbf{A} \cdot \nabla)\mathbf{B} + (\mathbf{B} \cdot \nabla)\mathbf{A}\,.$$

When a differentiation is applied to a product containing two factors, the result is the sum of the application to the first factor and the application to the second factor. In the present example it is useful to make this feature explicit by introducing the temporary notations ∇_A and ∇_B to indicate which factor is to be operated on by ∇. We can then arrange our formula to cause all quantities being differentiated to be written to the right of ∇ while those not being differentiated are placed to the left (with whatever sign changes may be needed to make the rearrangement legitimate). The way it all works will become clearer as we work this example.

We therefore write $\nabla \times (\mathbf{A} \times \mathbf{B}) = \nabla_A \times (\mathbf{A} \times \mathbf{B}) + \nabla_B \times (\mathbf{A} \times \mathbf{B})$. Without paying attention to the order of the factors, we expand the ∇_A term according to Vector Identity #1, getting

$$\nabla_A \times (\mathbf{A} \times \mathbf{B}) \longrightarrow \mathbf{A}(\nabla_A \cdot \mathbf{B}) - \mathbf{B}(\nabla_A \cdot \mathbf{A})\,,$$

which we make legitimate and correct by rearranging to a form in which ∇_A always occurs after $\mathbf{B}$ and before $\mathbf{A}$:

$$\nabla_A \times (\mathbf{A} \times \mathbf{B}) = (\mathbf{B} \cdot \nabla_A)\mathbf{A} - \mathbf{B}(\nabla_A \cdot \mathbf{A})\,. \tag{7.46}$$

Now that the factors have been brought to an order such that ∇_A operates on everything to its right and on nothing to its left, the subscript A is no longer necessary and need not be retained when Eq. (7.46) is used.

A similar procedure for the ∇_B term yields

$$\nabla_B \times (\mathbf{A} \times \mathbf{B}) = -(\mathbf{A} \cdot \nabla_B)\mathbf{B} + \mathbf{A}(\nabla_B \cdot \mathbf{B})\,. \tag{7.47}$$

The subscripts B in this equation have also become redundant and can be dropped.

Finally, adding together Eqs. (7.46) and (7.47), we confirm Vector Identity #6.

■

Exercises

7.3.1. Confirm Vector Identity #3 by hand computation in which you apply the differentiations to the product $\varphi\mathbf{A}$. Check your work using symbolic computation.

7.3.2. Confirm Vector Identity #5 by hand computation in which you expand $\mathbf{A}\times\mathbf{B}$ and differentiate. Check your work using symbolic computation.

7.3.3. Verify that the symbolic computation command `Laplacian` (operating on a scalar f) produces the same result as the more explicit expressions `Del.Del(f)` or `Div[Grad[f]]`.

7.3.4. Use symbolic computation to confirm that all curls are solenoidal and that all gradients are irrotational (Vector Identities #7 and #8).

7.3.5. Confirm Vector Identity #9.

7.3.6. Evaluate (for $r \neq 0$) $\boldsymbol{\nabla}^2 \left(\dfrac{xe^r}{r}\right)$, where $r^2 = x^2 + y^2 + z^2$. Check your work using symbolic computation.

7.3.7. Use symbolic computation to confirm Vector Identity #10.

7.4 INTEGRAL THEOREMS

The properties of the divergence and the curl lead to important theorems involving integrals containing vector differential operators. That this is the case is not really a surprise from a mathematically sophisticated viewpoint, as we have already learned (from study of Green's theorem in the plane, Section 6.5) that in two dimensions we have equations connecting a line integral of functions P and Q and a surface integral containing their derivatives. Some sort of extension to three dimensions might be expected. The key theorems we shall develop are (1) the **divergence theorem** (sometimes also called Gauss' theorem) and (2) **Stokes' theorem**, a fairly direct 3-D extension of the 2-D Green's theorem.

DIVERGENCE THEOREM

The basic content of the divergence theorem is the following: given that the divergence is a measure of the net outflow of flux from a volume element, the sum of the net outflows from all volume elements of a 3-D region (as calculated from the divergence) must be equal to the total outflow from the region (as calculated from the flux through the closed surface bounding the region). A more formal statement of the theorem takes the form

$$\int_V \boldsymbol{\nabla} \cdot \mathbf{B}\, d\tau = \oint_{\partial V} \mathbf{B} \cdot \hat{\mathbf{n}}\, dA\,. \tag{7.48}$$

Here V is a volume and ∂V denotes its bounding surface. Since the surface integral is to evaluate outflow, the unit vector $\hat{\mathbf{n}}$ must be the normal to ∂V in the direction pointing **out** of V.

The divergence theorem continues to be valid even if ∂V is not a single surface. For example, V may be the region between two concentric spheres. Then ∂V consists of both spherical surfaces, but because $\hat{\mathbf{n}}$ must point out of V it will point toward larger $\mathbf{r}$ on the outer sphere but toward smaller $\mathbf{r}$ on the inner sphere.

Example 7.4.1. Divergence Theorem: Simple Example

A unit cube is placed with corners at the points (0,0,0), (1,0,0), (0,1,0), and (0,0,1) of a Cartesian coordinate system; let V denote the volume enclosed by the cube. Check the divergence theorem by comparing the integrals $\int_{\partial V} \mathbf{F} \cdot d\boldsymbol{\sigma}$ and $\int_V \boldsymbol{\nabla} \cdot \mathbf{F}\, d\tau$, where $\mathbf{F} = (xyz)(\hat{\mathbf{e}}_x + \hat{\mathbf{e}}_y + \hat{\mathbf{e}}_z)$.

Start by computing

$$\nabla \cdot \mathbf{F} = \frac{\partial(xyz)}{\partial x} + \frac{\partial(xyz)}{\partial y} + \frac{\partial(xyz)}{\partial z} = yz + xz + xy\,.$$

Then integrate

$$\int_0^1 dx \int_0^1 dy \int_0^1 dz\,(yz + xz + xy) = \frac{1}{4} + \frac{1}{4} + \frac{1}{4} = \frac{3}{4}.$$

The surface integral over the cube face $x = 0$ vanishes because $xyz = 0$ there; that over the face $x = 1$ has outward normal $\hat{\mathbf{e}}_x$, and $\mathbf{F} \cdot \hat{\mathbf{e}}_x = xyz|_{x=1} = yz$, so the integral over this face is

$$\int_0^1 dy \int_0^1 dz\; yz = \frac{1}{4}\,.$$

Similar analyses indicate that the surface integrals over the faces $y = 0$ and $z = 0$ vanish while the integrals for the faces $y = 1$ and $z = 1$ are each $1/4$. The sum of all the surface integrals is $3/4$, in agreement with the volume integral of $\nabla \cdot \mathbf{F}$.

■

Example 7.4.2. Divergence Theorem: Radius Vector

A unit cube is placed with corners at $(\pm\frac{1}{2}, \pm\frac{1}{2}, \pm\frac{1}{2})$; let V denote the volume within the cube. With $\mathbf{F}(\mathbf{r}) = \mathbf{r}$, compare $\int_{\partial V} \mathbf{F} \cdot d\boldsymbol{\sigma}$ and $\int_V \nabla \cdot \mathbf{F}\, d\tau$.

From Eq. (7.26) we have $\nabla \cdot \mathbf{r} = 3$. Our cube has unit volume, so

$$\int_V \nabla \cdot \mathbf{F}\, d\tau = 3\,.$$

We next evaluate the surface integral over the cube face $x = +\frac{1}{2}$. The outward direction for this face is toward positive x, so

$$\int_{x=1/2} \mathbf{F} \cdot d\boldsymbol{\sigma} = \int_{x=1/2} \mathbf{r} \cdot d\boldsymbol{\sigma} = \int_{x=1/2} \mathbf{r} \cdot \hat{\mathbf{e}}_x\, dA = \int_{x-1/2} x\, dA = \frac{1}{2}\, dA = \frac{1}{2}\,.$$

Continuing with the cube face $x = -\frac{1}{2}$, we note that this face has outward direction toward $-x$, so

$$\int_{x=-1/2} \mathbf{F} \cdot d\boldsymbol{\sigma} = \int_{x=-1/2} \mathbf{r} \cdot d\boldsymbol{\sigma} = \int_{x=-1/2} \mathbf{r} \cdot (-\hat{\mathbf{e}}_x)\, dA$$

$$= \int_{x=-1/2} (-x)\, dA = \int_{x=-1/2} \frac{1}{2}\, dA = \frac{1}{2}\,.$$

The same result is obtained for each of the other four cube faces, so the surface integrals sum to $6 \cdot (1/2) = 3$. Again the divergence theorem is confirmed.

■

Example 7.4.3. Function that Vanishes on Boundary

The divergence theorem is often used in situations where a function vanishes on the boundary of the region involved. Here we apply the theorem to $\mathbf{F} = \exp(-r^2)\mathbf{r}$ over the entire 3-D space to obtain a formula connecting two transcendental integrals.

We start by computing $\boldsymbol{\nabla} \cdot \mathbf{F}$, noting that

$$\frac{\partial F_x}{\partial x} = \frac{\partial}{\partial x}\left(xe^{-x^2+y^2+z^2}\right) = (1-2x^2)\,e^{-x^2-y^2-z^2} = (1-2x^2)\,e^{-r^2}.$$

Combining with similar results for the y and z derivatives, we get

$$\boldsymbol{\nabla} \cdot \mathbf{F} = (3-2r^2)\,e^{-r^2}.$$

Because $\mathbf{F}$ vanishes exponentially at infinity, the surface integral $\int \mathbf{F} \cdot d\boldsymbol{\sigma}$ is zero, so the divergence theorem tells us that

$$\int \boldsymbol{\nabla} \cdot \mathbf{F}\, d\tau = \int (3-2r^2)\,e^{-r^2}\, d\tau = 0, \qquad \text{implying} \qquad \int r^2 e^{-r^2}\, d\tau = \frac{3}{2}\int e^{-r^2}\, d\tau\,,$$

where the integrals are over all space.

One way to check the above result is to write the integrals in spherical coordinates, where the above formula becomes

$$4\pi \int_0^\infty r^4 e^{-r^2}\, dr = \left(\frac{3}{2}\right) 4\pi \int_0^\infty r^2 e^{-r^2}\, dr\,.$$

We can complete a verification by evaluating the resulting one-dimensional integrals, using symbolic computing if helpful.

■

GREEN'S THEOREM

The divergence theorem can be applied to obtain a useful result known as Green's theorem. If u and v are scalar functions, they satisfy the identities

$$\boldsymbol{\nabla} \cdot (u\boldsymbol{\nabla} v) = u\boldsymbol{\nabla}^2 v + (\boldsymbol{\nabla} u) \cdot (\boldsymbol{\nabla} v)\,, \tag{7.49}$$

$$\boldsymbol{\nabla} \cdot (v\boldsymbol{\nabla} u) = v\boldsymbol{\nabla}^2 u + (\boldsymbol{\nabla} u) \cdot (\boldsymbol{\nabla} v)\,. \tag{7.50}$$

Integrating Eqs. (7.49) and (7.50) over a volume V and applying the divergence theorem to the left-hand sides of the resulting equations,

$$\int_{\partial V} u\boldsymbol{\nabla} v \cdot \hat{\mathbf{n}}\, dA = \int_V u\boldsymbol{\nabla}^2 v\, d\tau + \int_V (\boldsymbol{\nabla} u) \cdot (\boldsymbol{\nabla} v)\, d\tau\,, \tag{7.51}$$

$$\int_{\partial V} v\boldsymbol{\nabla} u \cdot \hat{\mathbf{n}}\, dA = \int_V v\boldsymbol{\nabla}^2 u\, d\tau + \int_V (\boldsymbol{\nabla} u) \cdot (\boldsymbol{\nabla} v)\, d\tau\,. \tag{7.52}$$

Both these equations can be regarded as forms of Green's theorem, but its most widely recognized presentation is the difference of the two equations,

$$\int_{\partial V} (u\boldsymbol{\nabla} v - v\boldsymbol{\nabla} u) \cdot \hat{\mathbf{n}}\, dA = \int_V \left(u\boldsymbol{\nabla}^2 v - v\boldsymbol{\nabla}^2 u\right) d\tau\,. \tag{7.53}$$

In quantum mechanics, Eq. (7.53) finds frequent use when V is the entire 3-D space (sometimes denoted $\Re^3$) and the functions u and v are negligible at infinity. Then the surface integral vanishes, leaving the useful result

$$\int_{\Re^3} u\nabla^2 v\, d\tau = \int_{\Re^3} v\nabla^2 u\, d\tau\,. \tag{7.54}$$

Example 7.4.4. Green's Theorem

The average value $\langle T\rangle$ of the kinetic energy T of an electron in a quantum state described by a wave function $\psi(\mathbf{r})$ is given in hartree units ($m = e = h/2\pi = 1$) by the integral

$$\langle T\rangle = -\frac{1}{2}\int_V \psi^*(\mathbf{r})\nabla^2\psi(\mathbf{r})\, d\tau\,,$$

with the integration over the region V where the electron has a nonzero probability of being found. The wave function ψ is required to be continuous and therefore must vanish on the boundary ∂V.

We can use Green's theorem to confirm that $\langle T\rangle$ is positive, in accord with our beliefs about the nature of the universe. From Eq. (7.51) with $v = \psi$ and $u = \psi^*$, we get

$$\int_V \psi^*(\mathbf{r})\nabla^2\psi(\mathbf{r})\, d\tau = \oint_{\partial V} \psi^*\nabla\psi\cdot\hat{\mathbf{n}}\, dA - \int_V (\nabla\psi)^*\cdot(\nabla\psi)\, d\tau\,.$$

The surface integral vanishes, and the integral for $\langle T\rangle$ becomes

$$\langle T\rangle = +\frac{1}{2}\int_V (\nabla\psi)^*\cdot(\nabla\psi)\, d\tau = +\frac{1}{2}\int_V |\nabla\psi|^2\, d\tau\,.$$

Because ψ cannot be identically zero and still describe a probability distribution, that integral must evaluate to a positive result.

■

STOKES' THEOREM

Stokes' theorem is the result obtained when we extend the analysis of the curl that led to Eq. (7.34) so that it applies to surfaces of finite size and more general shape. Therefore, consider a vector field $\mathbf{B}$ in a region within which we have defined a surface S that is closed except for a single opening (a hole), with the opening specified by the curve which is its boundary. Our objective is to relate the rotational flow of $\mathbf{B}$ around the boundary with an integral over the surface it bounds. The situation is illustrated in Fig. 7.11.

We proceed by dividing the surface into elements of area; for convenience we have drawn them as approximately rectangular. From our earlier discussion of the curl, we know that the line integral $\oint \mathbf{B}\cdot d\mathbf{r}$ around any element of area is equal to $(\nabla\times\mathbf{B})\cdot\hat{\mathbf{n}}\, dA$ for that element. Focusing for now on the shaded element of area in the figure, we identify $\hat{\mathbf{n}}$ as pointing **out** of the surface when the boundary of the element is traversed in the direction indicated by the arrows inside its boundary.

Now let's combine the line integrals for all elements of area. Because the contributions to the same line segment from adjacent elements cancel, we can see that the only surviving contributions are those at the boundary of the surface, and these

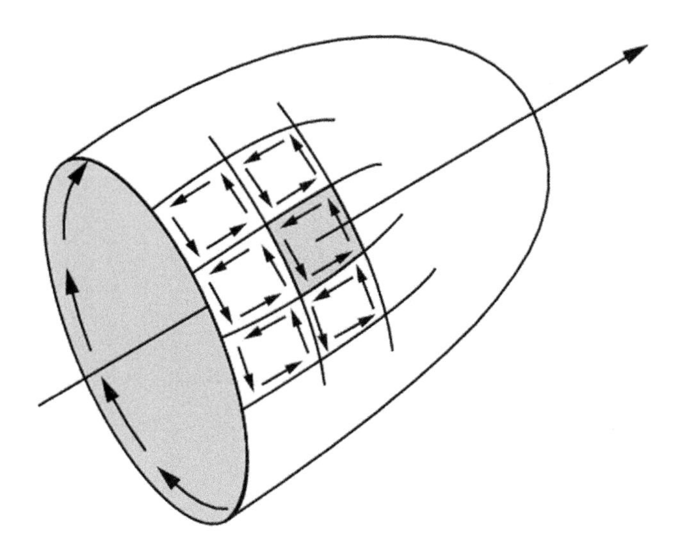

Figure 7.11: Relation between line integrals around elements of a surface and that around its boundary.

contributions are represented by the line integral $\int_{\partial S} \mathbf{B} \cdot d\mathbf{r}$, with the path traversed in the direction shown.

Alternatively, we can combine the surface integrals over the elements of area; we then get $\int_S (\boldsymbol{\nabla} \times \mathbf{B}) \cdot \hat{\mathbf{n}}\, dA$. Equating these two results, we get Stokes' theorem,

$$\int_{\partial S} \mathbf{B} \cdot d\mathbf{r} = \int_S (\boldsymbol{\nabla} \times \mathbf{B}) \cdot \hat{\mathbf{n}}\, dA \,. \tag{7.55}$$

The above equation, Stokes' theorem, applies when the direction of travel for the line integral over ∂S and the direction of the normal to S are related by the right-hand rule: the path direction is that indicated by the curved fingers of the right hand when the thumb points in a direction of flow that will pass through S in the outward direction.

Stokes' theorem is valid when the circulation can be calculated for all paths on S; this condition is equivalent to a requirement that $\boldsymbol{\nabla} \times \mathbf{B}$ exist for every point of S.

It may seem paradoxical that the curl of $\mathbf{B}$ can be integrated on **any** surface that is bounded by a specific curve, and yet always give the same result. This is true because, like all curls, $\boldsymbol{\nabla} \times \mathbf{B}$ is solenoidal. The flux of $\boldsymbol{\nabla} \times \mathbf{B}$ through the bounding loop will be the same as that flowing through any surface bounded by the loop (a solendoidal vector field has no sources or sinks).

We make one final observation. Suppose our surface is closed (it has no hole and completely surrounds a volume V). This situation corresponds to a hole that is collapsed to a single point, and line integrals of finite quantities for such a path (of zero length) will vanish. In turn, we conclude that

$$\int_{\partial V} (\boldsymbol{\nabla} \times \mathbf{B}) \cdot \hat{\mathbf{n}}\, dA = 0 \,, \tag{7.56}$$

where $\mathbf{B}$ and the volume V are arbitrary. Note that ∂V is the closed surface enclosing V.

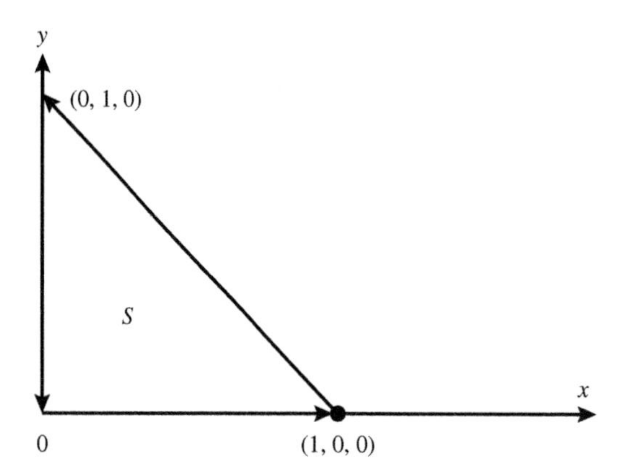

Figure 7.12: Triangular loop.

Example 7.4.5. Triangular Loop

Let's check Stokes' theorem for a vector $\mathbf{B} = y^2\,\hat{\mathbf{e}}_x - x^2\,\hat{\mathbf{e}}_y$ and a triangular loop in the xy-plane that passes (in the order given) through the points $(0,0,0)$, $(1,0,0)$, and $(0,1,0)$ and thereby enclosing an area S. See Fig. 7.12.

We first compute the line integral of $\mathbf{B}$, breaking it into three pieces corresponding to the sides of our triangle.

$$\int_{\partial S} \mathbf{B}\cdot\mathbf{r} = \int_0^1 B_x\,dx + \int_{y=0}^1 (B_x\,dx + B_y\,dy) + \int_{y=1}^0 B_y\,dy\,.$$

For the first of the above integrals $B_x = y^2$ is evaluated for $y = 0$ and therefore vanishes; for the third integral $B_y = -x^2$ is zero and that integral also vanishes. We can rewrite the second integral entirely in terms of y if we note that $dx = -dy$ and that $B_y = -x^2 = -(1-y)^2$. The line integral then assumes the form

$$\int_{\partial S} \mathbf{B}\cdot\mathbf{r} = \int_0^1 \left[-y^2 - (1-y)^2\right]\,dy = -\frac{2}{3}\,.$$

Next we compute

$$\boldsymbol{\nabla}\times\mathbf{B}\Big|_z = \frac{\partial B_y}{\partial x} - \frac{\partial B_x}{\partial y} = -2x - 2y\,.$$

The other components of the curl vanish, so $\boldsymbol{\nabla}\times\mathbf{B} = -2(x+y)\,\hat{\mathbf{e}}_z$.

Finally, we evaluate the surface integral over S: $\displaystyle\int_S \boldsymbol{\nabla}\times\mathbf{B}\cdot d\boldsymbol{\sigma}\,.$

Because of the direction in which we evaluated the line integral, the normal to S is in the $+z$ direction, so the surface integral can be written

$$\int_S (\boldsymbol{\nabla}\times\mathbf{B})\cdot d\boldsymbol{\sigma} = -2\int_0^1 dx \int_0^{1-x} dy(x+y) = -\frac{2}{3}\,.$$

The agreement of the line and surface integrals confirms this case of Stokes' theorem.

■

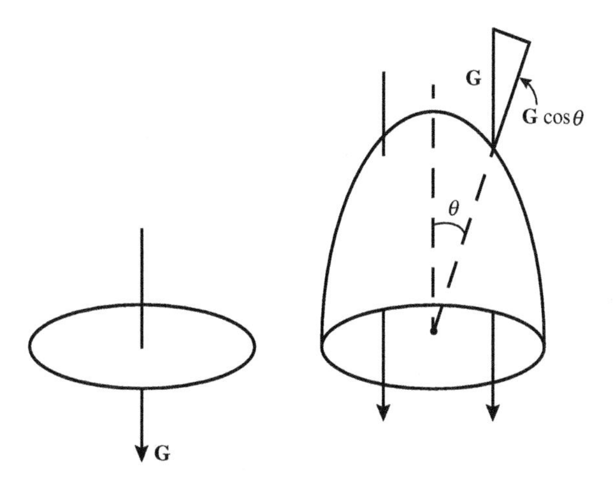

Figure 7.13: Left: Disk for Stokes' theorem. Right: Hemispherical surface with same boundary curve. $\mathbf{G} = \boldsymbol{\nabla} \times \mathbf{B}$.

Example 7.4.6. Different Surfaces

In this example our boundary ∂S is a unit circle in the xy-plane and we take a vector $B = y\,\hat{\mathbf{e}}_x - x\,\hat{\mathbf{e}}_y$, so $\nabla \times \mathbf{B} = -2\,\hat{\mathbf{e}}_z$. Now let's apply Stokes' theorem in the two cases (1) that the area S with boundary ∂S is a unit disk in the xy-plane and (2) that S is a unit hemisphere whose surface extends toward positive z. See Fig. 7.13.

Stokes' theorem states that so long as we consider surfaces with the same boundary we should get the same value of $\int_S \boldsymbol{\nabla} \times \mathbf{B} \cdot d\boldsymbol{\sigma}$, because in all cases that integral must be equal to the line integral $\int_{\partial S} \mathbf{B} \cdot d\mathbf{r}$. Let's check.

For reference we first evaluate the line integral:

$$\int_{\partial S} \mathbf{B} \cdot d\mathbf{r} = \int_{\partial S} (y\,\hat{\mathbf{e}}_x - x\,\hat{\mathbf{e}}_y) \cdot (\hat{\mathbf{e}}_x\,dx + \hat{\mathbf{e}}_y\,dy) = \int_{\partial S} (y\,dx - x\,dy)\,.$$

We take the line integral in the mathematically positive direction (counterclockwise as viewed from positive z), noting for later use that the normal to the enclosed area is $+\hat{\mathbf{e}}_z$.

We now use Green's theorem in the plane to evaluate the line integral. Applying Eq. (6.46) with $P = -x$ and $Q = y$, the integral reduces to $-2\int dA = -2\pi$.

Returning next to the surface integrals of $\boldsymbol{\nabla} \times \mathbf{B}$, the integral for the disk, Case (1), is

$$\int_{\text{disk}} (\boldsymbol{\nabla} \times \mathbf{B}) \cdot \hat{\mathbf{e}}_z\,dA = -2\int_{\text{disk}} dA = -2\pi\,,$$

as expected.

Case (2) is more complicated, and if this were not a demonstration of Stokes' theorem we would use that theorem to evaluate its surface integral by recognizing that it has the same value as for Case (1). Pressing on, we identify position on the hemisphere by its spherical coordinates and note that the normal component of $\boldsymbol{\nabla} \times \mathbf{B}$ at polar angle θ has the value $-2\cos\theta$. See Fig. 7.13. Thus, the contribution to $\boldsymbol{\nabla} \times \mathbf{B} \cdot d\boldsymbol{\sigma}$ for the element of surface area dA is $-2\cos\theta\,dA$. Writing $dA = \sin\theta\,d\theta\,d\varphi$, the integral we must evaluate is

$$\int_{\text{hemisphere}} \nabla \times \mathbf{B} \cdot d\boldsymbol{\sigma} = \int_0^{\pi/2} \sin\theta \, d\theta \int_0^{2\pi} d\varphi \, (-2\cos\theta) = -2\pi \, .$$

This case of Stokes' theorem is confirmed.

■

Example 7.4.7. Ampère's Law

Stokes' theorem plays an important role in the computation of the magnetic fields generated by currents. Consider a straight wire that carries a steady current I. Experiments (in the 19th Century) revealed that the current flow generates a magnetic field that is perpendicular to the direction of the wire and that the lines of magnetic force are circular, with the line integral $\oint \mathbf{B} \cdot d\mathbf{r}$ around a closed circular stream line proportional to the strength of the current. The generalization of this observation to general current densities and geometries leads to Ørsted's law (valid for situations without time dependence):

$$\nabla \times \mathbf{B} = \mu_0 \mathbf{J} \, . \tag{7.57}$$

Here $\mathbf{B}$ is the magnetic induction field, μ_0 is the permeability, and J is the current density.

The situation described above is ideal for the application of Stokes' theorem, because that theorem states that a line integral of $\mathbf{B}$ is unambiguously related to the total current encircled by the integration path. In situations of high symmetry, Eq. (7.57) provides enough information to solve for $\mathbf{B}$, leading to a formula known as Ampère's law.

Thus, consider a wire placed along the z-axis of a Cartesian system, with a total current I flowing through it in the $+z$ direction. The current could be thought of as a current density $\mathbf{J}$ distributed in some circularly symmetrical fashion through the cross section of the wire; the details of the distribution turn out not to be important because we will integrate $\mathbf{J}$ over that cross section and thereby recover I.

Next draw a circle of arbitrary radius a in the xy-plane with the wire at its axis of symmetry. The symmetry of the problem dictates that such a circle must be a stream line of $\mathbf{B}$. The stream lines cannot spiral either in or out (or have z-components) because such features would require the existence of magnetic poles to provide a source or sink for $\mathbf{B}$. The only features of $\mathbf{B}$ not determined by symmetry are its sense (clockwise vs. counterclockwise) and magnitude.

We are now ready to apply Stokes' theorem to our circle ∂S that is a stream line and the circular disk S bounded by the circle. We equate the line integral of $\mathbf{B}$ around ∂S and the integral of $\nabla \times \mathbf{B}$ on S:

$$\int_{\partial S} \mathbf{B} \cdot d\mathbf{r} = \int_S (\nabla \times \mathbf{B}) \cdot d\boldsymbol{\sigma} = \int_S \mu_0 \mathbf{J} \cdot d\boldsymbol{\sigma} = \mu_0 I \, . \tag{7.58}$$

Since I is in the $+z$ direction, application of the right-hand rule tells us that the positive direction for $\mathbf{B}$ will be in the mathematically positive direction ($+x \to +y \to -x \to -y \cdots$).

To complete the computation, we note that because $\mathbf{B}$ is along the circle (of radius a), the line integral will have the value $2\pi aB$, and Eq. (7.58) yields

$$2\pi aB = \mu_0 I\,, \qquad \text{or} \qquad B = \frac{\mu_0 I}{2\pi a}\,.$$

■

Exercises

Use the divergence theorem or Stokes' theorem wherever they can make these exercises easier.

7.4.1. Given $\mathbf{V} = (x^2 + y^2)\hat{\mathbf{e}}_x + 3y\hat{\mathbf{e}}_y - 2xz\hat{\mathbf{e}}_z$, for the surface of a unit cube in the first octant ($0 \le x, y, z \le 1$), compute $\displaystyle\int \mathbf{V}\cdot d\boldsymbol{\sigma}$.

7.4.2. For the vector in the xy-plane $\mathbf{A} = (2x^2 - y^2)\hat{\mathbf{e}}_x + xy\hat{\mathbf{e}}_y$,

(a) Compute $\boldsymbol{\nabla}\times\mathbf{A}$.

(b) For a rectangle defined by $0 \le x \le a$, $0 \le y \le b$, compute $\displaystyle\int (\boldsymbol{\nabla}\times\mathbf{A})\cdot d\boldsymbol{\sigma}$.

(c) Compute $\int \mathbf{A}\cdot d\mathbf{r}$ for the perimeter of the rectangle of part (b) and thereby confirm Stokes' theorem for these integrals.

7.4.3. Evaluate $\displaystyle\int_S (\boldsymbol{\nabla}\times\mathbf{V})\cdot d\boldsymbol{\sigma}$, where $\mathbf{V} = 2xy\hat{\mathbf{e}}_x + (x^2 + 2xy^2)\hat{\mathbf{e}}_y - x^2z^2\hat{\mathbf{e}}_z$,

and S is the portion of the surface defined by $z = 4 - x^2 - y^2$ for which $z \ge 0$. Identify the direction of $d\boldsymbol{\sigma}$ for which your answer is correct.

7.4.4. Use Stokes' theorem to show that $\displaystyle\oint u\boldsymbol{\nabla}v\cdot d\mathbf{r} = \int_S (\boldsymbol{\nabla}u\times\boldsymbol{\nabla}v)\cdot d\boldsymbol{\sigma}$.

How is the region S defined?

Note. The symbol $\oint$ denotes a line integral around a closed loop.

7.4.5. Using the result of Exercise 7.4.4, show that $\displaystyle\oint u\boldsymbol{\nabla}v\cdot d\mathbf{r} = -\oint v\boldsymbol{\nabla}u\cdot d\mathbf{r}$.

7.4.6. If $\boldsymbol{\nabla}^2\psi = 0$ at all points of a volume V, show that

$$\int_{\partial V} \boldsymbol{\nabla}\psi\cdot d\boldsymbol{\sigma} = 0\,,$$

where ∂V is the closed surface bounding V.

7.4.7. Evaluate $\displaystyle\int r^2\boldsymbol{\nabla}^2 e^{-r^2}\,d^3r$, where the integration is over the full 3-D space.

Hint. After using an appropriate integral theorem, work in Cartesian coordinates.

7.5 POTENTIAL THEORY

Vector fields are extremely useful in the description of physical phenomena, particularly in mechanics and electromagnetism. However, many problems can be simplified and their understanding enhanced if we can introduce the notion of a scalar potential from which the properties of the vector field can be determined. The essential features of this idea are fairly easy to examine in the case where the vector field is the gravitational force on an object (assumed constant), and we are interested in the work involved in moving the object from place to place. Keep in mind that the essence of our discussion applies also to fields (here forces) that vary with position, so our results have greater generality than our initial example.

CONSERVATIVE FIELDS

Most students are familiar, from a course in elementary physics or by experience, with the fact that if friction is disregarded the mechanical work of moving an object of mass M in a gravitational field between two nearby points at respective heights h_1 and h_2 is $Mg(h_2 - h_1)$, irrespective of the path by which the motion takes place. Here g is a constant ("the acceleration of gravity") characterizing the gravitational field. This statement corresponds to the following equation containing a line integral along the chosen path (designated C):

$$w_{1\to 2} = \int_C \mathbf{F} \cdot d\mathbf{r} = Mg(h_2 - h_1)\,. \tag{7.59}$$

Here $\mathbf{F}$ is the force needed to overcome gravity, which in a coordinate system with $+x$ in the upward direction is $Mg\,\hat{\mathbf{e}}_x$ and $d\mathbf{r} = \hat{\mathbf{e}}_x\,dx + \hat{\mathbf{e}}_y\,dy + \hat{\mathbf{e}}_z\,dz$. Our equation then has the more detailed evaluation

$$w_{1\to 2} = \int_{h_1}^{h_2} Mg\,\hat{\mathbf{e}}_x \cdot (\hat{\mathbf{e}}_x\,dx + \hat{\mathbf{e}}_y\,dy + \hat{\mathbf{e}}_z\,dz) = Mg\int_{h_1}^{h_2} dx = Mg(h_2 - h_1)\,, \tag{7.60}$$

confirming that $w_{1\to 2}$ depends on C only through the values of h at its endpoints.

When a vector field has an integral of the type in Eq. (7.59) that depends only on the endpoints of the path, it is said to be a **conservative field** or (if the field represents a force) we may call the force **conservative**. Note that if we travel a path C in reverse, $\mathbf{F}$ stays the same as before but $d\mathbf{r}$ changes sign, so $w_{2\to 1} = -w_{1\to 2}$. Since $w_{1\to 2}$ is independent of the path, we will still have $w_{2\to 1} = -w_{1\to 2}$ even if the forward and reverse journeys are on different paths, with the result that any closed path (one that starts and ends at the same point) will have $w = 0$. That is the essence of the meaning of the term **conservative**.

Not all forces are conservative; in our current example we could have included friction. Because friction is always in a direction that opposes the motion, to overcome it will always require (positive) work, and travel along a closed path (even if level) will have a value of w that depends upon the path.

Once we have identified our gravitational force as conservative, we can associate with every point the amount of work needed to get there from some reference point. This process will define a **scalar** field ψ, which we call the **scalar potential** (or just **potential**) of our (vector) gravitational field. In the present problem we have $\psi = Mgh$, where the reference point is assigned the value $h = 0$. It is important to notice that the zero of ψ depends on our choice of reference point, and that computations of work depend upon differences in ψ and not on their individual values. The potential has a clear physical significance; its value at $x = h$ represents the amount of energy (positive or negative) that can be recovered by moving back to a point where $x = 0$.

It is of extreme importance to note that a vector field can be identified with a scalar potential only if the field is conservative; a nonconservative force will not have fixed amounts of energy associated with specific points. An important example of a nonconservative field is the magnetic field B produced by an electric current. The integral $\oint \mathbf{B} \cdot d\mathbf{r}$ for a loop around a current-carrying wire (see Example 7.4.7) is nonzero. (Energy conservation in physics is saved because there exist no magnetic monopoles that can capture additional energy each time they go around the loop.)

An objective of the present discussion is to identify and understand the key properties of conservative fields. Let's start by assuming a vector field $\mathbf{F}$ has components that are continuous and have continuous partial derivatives, and that $\mathbf{F}$ is conservative in a simply connected region of space.[1] The connectedness requirement is necessary to make some of the following statements true.

The fact that $\mathbf{F}$ has been assumed conservative has already led us to conclude that

$$\int_P^Q \mathbf{F} \cdot d\mathbf{r} = \psi(Q) - \psi(P) \tag{7.61}$$

for some single-valued function ψ. Equation (7.61) is equivalent to the statement that $\mathbf{F} \cdot d\mathbf{r}$ is the differential of ψ, i.e., $d\psi = \mathbf{F} \cdot d\mathbf{r}$.

Next, note that

$$\frac{\partial \psi}{\partial x} = \mathbf{F} \cdot \left(\frac{\partial \mathbf{r}}{\partial x}\right) = F_x\,, \tag{7.62}$$

and that $\partial\psi/\partial y = F_y$ and $\partial\psi/\partial z = F_z$. In other words, $\mathbf{F} = \nabla\psi$, showing that if $\mathbf{F}$ is conservative, it must be the gradient of some ψ.

Since $\mathbf{F}$ is a gradient,

$$\nabla \times \mathbf{F} = \nabla \times \nabla\psi = 0\,, \tag{7.63}$$

where we have used Eq. (7.43). This equation confirms that a conservative force has a vanishing curl.

Finally, we can use Stokes' theorem and the vanishing of curl $\mathbf{F}$ to conclude that

$$\oint \mathbf{F} \cdot d\mathbf{r} = 0 \tag{7.64}$$

for every simple closed curve in the region, thereby recovering our original assumption that $\mathbf{F}$ is conservative. The overall result of this circular discussion is that any of the conditions we have derived implies all the others.

We close this subsection with two further remarks. First, the condition curl $\mathbf{F} = 0$ is equivalent to the statement that $\mathbf{F}$ is irrotational, and we now have the deeper understanding that this property prevents us from following stream lines of $\mathbf{F}$ to reach the same point with multiple values of ψ. Second, we must stress the importance of the connectedness requirement. If our region were multiply connected (an example is a torus—the mathematical equivalent of a doughnut or life preserver), some of the above equivalences could not be established.

[1]A region is called **simply connected** if every simple closed curve within the region can be shrunk (staying within the region) to a single point within the region; loosely speaking, a simple closed curve is a closed loop that contains no permanent knots.

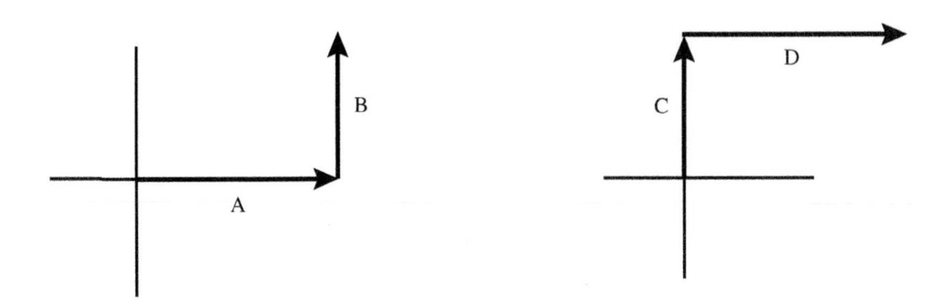

Figure 7.14: Two paths for the integrations in Example 7.5.1.

Example 7.5.1. Conservative and Nonconservative Forces

Consider the two force fields studied in Example 7.2.12,

$$\mathbf{F} = y\hat{\mathbf{e}}_x + x\hat{\mathbf{e}}_y \qquad \text{and} \qquad \mathbf{G} = y\hat{\mathbf{e}}_x - x\hat{\mathbf{e}}_y \,.$$

In that Example, we found that $\boldsymbol{\nabla} \times \mathbf{F} = 0$, while $\boldsymbol{\nabla} \times \mathbf{G} = -2\hat{\mathbf{e}}_z$. These relationships indicate that $\mathbf{F}$ is a conservative force and that $\mathbf{G}$ is nonconservative.

Let's check by computing $\int \mathbf{F} \cdot d\mathbf{r}$ and $\int \mathbf{G} \cdot d\mathbf{r}$ for the two paths between $(0, 0, 0)$ and $(x_0, y_0, 0)$ shown in Fig. 7.14. Identifying the individual line segments of the paths by their labels in the figure, we evaluate for $\mathbf{F}$

$$\int_A \mathbf{F} \cdot d\mathbf{r} = \int_0^{x_0} y(0)\, dx = 0 \,, \qquad \int_B \mathbf{F} \cdot d\mathbf{r} = \int_0^{y_0} x(x_0)\, dy = x_0 y_0 \,,$$

$$\int_C \mathbf{F} \cdot d\mathbf{r} = \int_0^{y_0} x(0)\, dy = 0 \,, \qquad \int_D \mathbf{F} \cdot d\mathbf{r} = \int_0^{x_0} y(y_0)\, dx = x_0 y_0 \,.$$

We have written $x(0)$, $x(x_0)$, $y(0)$, and $y(y_0)$ instead of their respective values (0, x_0, 0, y_0) to make these evaluations clearer. Combining now the integrals over segments A and B, and those over segments C and D, we find that

$$\int_{A+B} \mathbf{F} \cdot d\mathbf{r} = \int_{C+D} \mathbf{F} \cdot d\mathbf{r} = x_0 y_0 \,,$$

consistent with the claim that the force $\mathbf{F}$ is conservative. Our calculation also reveals that a potential from which we can recover $\mathbf{F}$ is $\psi(x, y, z) = xy$. We can verify that ψ is a valid scalar potential for $\mathbf{F}$ by computing

$$\boldsymbol{\nabla}\psi = y\hat{\mathbf{e}}_x + x\hat{\mathbf{e}}_y \,.$$

Continuing now with the force $\mathbf{G}$, we have

$$\int_A \mathbf{G} \cdot d\mathbf{r} = \int_0^{x_0} y(0)\, dx = 0 \,, \qquad \int_B \mathbf{G} \cdot d\mathbf{r} = \int_0^{y_0} [-x(x_0)]\, dy = -x_0 y_0 \,,$$

$$\int_C \mathbf{G} \cdot d\mathbf{r} = \int_0^{y_0} x(0)\, dy = 0 \,, \qquad \int_D \mathbf{G} \cdot d\mathbf{r} = \int_0^{x_0} y(y_0)\, dx = x_0 y_0 \,.$$

Combining now the integrals over segments A and B, and those over segments C and D, we get

$$\int_{A+B} \mathbf{F}\cdot d\mathbf{r} = -x_0 y_0, \qquad \int_{C+D} \mathbf{F}\cdot d\mathbf{r} = x_0 y_0\,,$$

showing that the integral of $\mathbf{G}$ from $(0,0,0)$ to $(x_0, y_0, 0)$ has a value that depends upon the path. This confirms that $\mathbf{G}$ is nonconservative and that there is no scalar potential for $\mathbf{G}$.

■

VECTOR POTENTIAL

Instead of having a vector field that is irrotational (and therefore related to a scalar potential) we now consider a vector field that is solenoidal, and it becomes natural to seek to describe it in terms of a potential of some sort. It turns out to be convenient to represent a solenoidal vector field $\mathbf{B}$ as $\mathbf{B} = \boldsymbol{\nabla}\times\mathbf{A}$, where $\mathbf{A}$ is called a **vector potential**. Remembering that **solenoidal** means divergenceless, we see that the form chosen for $\mathbf{B}$ will automatically be solenoidal because

$$\boldsymbol{\nabla}\cdot\mathbf{B} = \boldsymbol{\nabla}\cdot\boldsymbol{\nabla}\times\mathbf{A} = 0\,. \tag{7.65}$$

Applying Eq. (7.44), we get zero for arbitrary $\mathbf{A}$.

In the previous subsection we showed that any irrotational vector field could be represented in terms of a scalar potential; the question at issue here is whether an arbitrary solenoidal vector field can be written in terms of a vector potential. One way to answer that question is, given an arbitrary solenoidal $\mathbf{B}$, to construct a vector field $\mathbf{A}$ such that $\mathbf{B} = \boldsymbol{\nabla}\times\mathbf{A}$. There are an infinite number of ways to do this (and we will shortly have a better understanding as to why that is so). Right now we simply write down a form for $\mathbf{A}$ and show that its curl is equal to $\mathbf{B}$. Our expression is

$$\mathbf{A} = \hat{\mathbf{e}}_y \int_{x_0}^{x} B_z(x,y,z)\,dx + \hat{\mathbf{e}}_z\left[\int_{y_0}^{y} B_x(x_0,y,z)\,dy - \int_{x_0}^{x} B_y(x,y,z)\,dx\right]. \tag{7.66}$$

Here x_0 and y_0 can have any desired values. Because we have chosen an expression for which $A_x = 0$, we get

$$(\boldsymbol{\nabla}\times\mathbf{A})_y = -\frac{\partial A_z}{\partial x} = \frac{\partial}{\partial x}\int_{x_0}^{x} B_y(x,y,z)\,dx = B_y\,,$$

$$(\boldsymbol{\nabla}\times\mathbf{A})_z = \frac{\partial A_y}{\partial x} = \frac{\partial}{\partial x}\int_{x_0}^{x} B_z(x,y,z)\,dx = B_z\,,$$

$$\begin{aligned}
(\boldsymbol{\nabla}\times\mathbf{A})_x &= \frac{\partial A_z}{\partial y} - \frac{\partial A_y}{\partial z} \\
&= \frac{\partial}{\partial y}\left[\int_{y_0}^{y} B_x(x_0,y,z)\,dy - \int_{x_0}^{x} B_y(x,y,z)\,dx\right] - \frac{\partial}{\partial z}\int_{x_0}^{x} B_z(x,y,z)\,dx \\
&= B_x(x_0,y,z) - \int_{x_0}^{x}\left[\frac{\partial B_y(x,y,z)}{\partial y} + \frac{\partial B_z(x,y,z)}{\partial z}\right] dx\,.
\end{aligned}$$

To complete the evaluation of $(\nabla \times \mathbf{A})_x$, we use the fact that $\nabla \cdot \mathbf{B} = 0$ to replace the two terms within square brackets by $-\partial B_x(x, y, z)/\partial x$, after which we get

$$(\nabla \times \mathbf{A})_x = B_x(x_0, y, z) + \int_{x_0}^{x} \frac{\partial B_x(x, y, z)}{\partial x}\, dx = B_x(x, y, z)\,,$$

completing the verification of our form for $\mathbf{A}$.

Once we have a particular $\mathbf{A}$ whose curl is $\mathbf{B}$, we can add to $\mathbf{A}$ any gradient and still get the same $\mathbf{B}$. This is so because

$$\nabla \times (\mathbf{A} + \nabla\varphi) = \nabla \times \mathbf{A} + \nabla \times \nabla\varphi = \nabla \times \mathbf{A}.$$

The term containing φ vanishes because it is a case of Eq. (7.43).

Example 7.5.2. Vector Potential of Constant Field

Let's first choose the constant field to be in the z direction with magnitude B_z, so

$$\mathbf{B} = B_z\, \hat{\mathbf{e}}_z\,.$$

If we now construct a vector potential $\mathbf{A}$ using Eq. (7.66) with $x_0 = 0$, we get the relatively simple result

$$\mathbf{A} = \hat{\mathbf{e}}_y \int_0^x B_z\, dx = x B_z \hat{\mathbf{e}}_y\,.$$

A somewhat more devious way to proceed is to choose the constant field to be in the y direction; with $x_0 = 0$, Eq. (7.66) now gives

$$\mathbf{A}_1 = -x B_y \hat{\mathbf{e}}_z\,.$$

If we then rotate this formula by making the permutation $(xyz) \to (yzx)$ (because cross products are involved the coordinate system must remain right-handed), our constant field will again be in the z direction, but our formula for $\mathbf{A}$ now becomes

$$\mathbf{A}' = -y B_z \hat{\mathbf{e}}_x\,,$$

clearly different from what we had before.

If we try to resolve this dilemma by looking in various elementary texts, we most often find neither of the above formulas, but instead discover

$$\mathbf{A}'' = \frac{1}{2}(\mathbf{B} \times \mathbf{r}) = \frac{B_z}{2}(x\hat{\mathbf{e}}_y - y\hat{\mathbf{e}}_x)\,.$$

All three formulas for $\mathbf{A}$ are correct, and their difference is indicative of the large amount of freedom in the choice of $\mathbf{A}$. We can convert each of the formulas into the other two by adding to $\mathbf{A}$ an expression of the form $C\,\nabla\psi$, which (as we have already observed) does not change the value of $\mathbf{B} = \nabla \times \mathbf{A}$.

In the present situation, we take $\psi = xy$, from which we get

$$\nabla\psi = y\hat{\mathbf{e}}_x + x\hat{\mathbf{e}}_y\,, \qquad \text{showing that} \qquad \nabla \times (y\hat{\mathbf{e}}_x + x\hat{\mathbf{e}}_y) = 0\,.$$

All three vector potentials yield the same $\mathbf{B}$ because

$$\mathbf{A} - \frac{B_z}{2}(y\hat{\mathbf{e}}_x + x\hat{\mathbf{e}}_y) = \mathbf{A}' + \frac{B_z}{2}(y\hat{\mathbf{e}}_x + x\hat{\mathbf{e}}_y) = \mathbf{A}''\,.$$

■

SOURCES AND SINKS

In our earlier discussion of the divergence, we constructed the stream lines associated with the electric field of a point charge at the coordinate origin and in Example 7.2.8 showed that (except possibly at $\mathbf{r} = 0$) the flow was divergenceless. However, this cannot be the whole story because there is clearly an outward flow of electric flux from the origin in all directions. We can resolve this dilemma by determining the total amount of this outward flux and identifying it as a point contribution to the divergence of $\mathbf{E}$.

We proceed by placing a small sphere (of radius a) about a point charge of magnitude q and making a direct computation (in the limit of small a) of the net outflow for an electric field $\mathbf{E} = (q/4\pi\varepsilon_0)\hat{\mathbf{r}}/r^2$. The symmetry of the problem indicates that the electric field will be normal to the spherical surface, so the total outward flux will be the area of the sphere ($4\pi a^2$) times the magnitude of the electric field ($q/4\pi\varepsilon_0 a^2$). Combining these factors, we note that the radius a cancels (as does the factor 4π), leaving a net outflow equal to q/ε_0.

Our interpretation is that there is a point outflow at the origin that is infinite in outflow density but localized in such a way that its integral evaluates to q/ε_0. This is the type of situation for which the Dirac delta function was invented (see Section 6.7), so a more complete description of the divergence of E for the point charge is

$$\nabla \cdot \mathbf{E} = \frac{q}{\varepsilon_0}\,\delta(\mathbf{r})\,. \tag{7.67}$$

Note that Eq. (7.67) says both that div $\mathbf{E}$ is zero everywhere except at $\mathbf{r} = 0$ and that at $\mathbf{r} = 0$ it is infinite in such a way that its integrated contribution is q/ε_0.

The result in Eq. (7.67) has broader implications than its application to electrostatics. If we recall from Example 7.2.4 that the electric field of a point charge q can be derived from a potential $\varphi = q/4\pi\varepsilon_0 r$, we can derive the formula

$$\nabla \cdot \mathbf{E} = \nabla \cdot (-\nabla\varphi) = -\frac{q}{4\pi\varepsilon_0}\nabla^2\left(\frac{1}{r}\right) = \frac{q}{\varepsilon_0}\,\delta(\mathbf{r})\,,$$

which simplifies to the important and more general result

$$\nabla^2\left(\frac{1}{r}\right) = -4\pi\,\delta(\mathbf{r})\,. \tag{7.68}$$

GAUSS' LAW

Consider now a point charge q at the coordinate origin, surrounded by a small sphere of radius a, but surrounded also by a closed but otherwise arbitrary "larger" surface that is completely outside the small sphere. See Fig. 7.15. Between the two surfaces $\mathbf{r}$ is everywhere nonzero and div $\mathbf{E} = 0$, so the net outflow from the region between the two surfaces must add to zero. But according to Eq. (7.67) we have an outflow from the small sphere (and therefore flow into our intersurface region) in the amount q/ε_0, so we must have an outflow in this same amount through the larger surface. That result can be restated in the following simpler form:

> *If a charge q is surrounded by a closed surface of any shape or dimensions, the net outward flux of the electric field of the charge through the surface will be q/ε_0.*

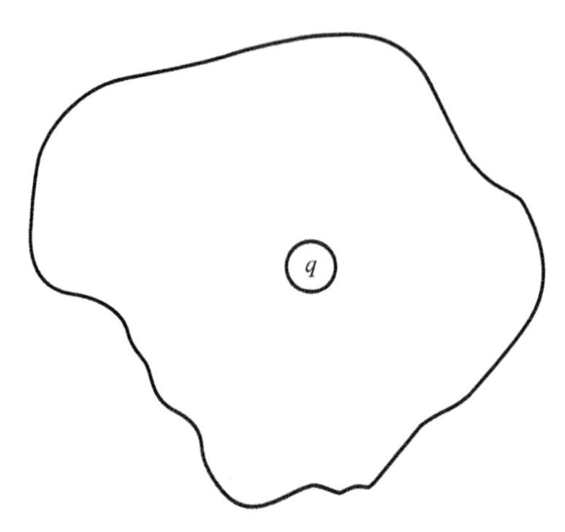

Figure 7.15: Surfaces for Gauss' law.

Because electromagnetic theory is linear, we can generalize the above statement to the form usually called Gauss' law:

The net outward flux through an arbitrary closed surface enclosing one or more charges or a continuous charge distribution will be Q/ε_0, where Q is the total amount of charge enclosed.

We can summarize Gauss' law by the following equation:

$$\int_{\partial V} \mathbf{E} \cdot d\boldsymbol{\sigma} = \frac{Q}{\varepsilon_0}, \qquad (7.69)$$

where Q is the total charge within V.

Note also that because of the linearity of electromagnetic theory we can conclude that div $\mathbf{E} = 0$ everywhere except where there is charge, with div $\mathbf{E} = \rho(\mathbf{r})/\varepsilon_0$ in regions where there is a continuous distribution of charge with density $\rho(\mathbf{r})$, and with delta-function contributions to div $\mathbf{E}$ at the positions of point charges.

Example 7.5.3. Gauss' Law

Gauss' law permits the direct solution of various high-symmetry problems in electrostatics. Consider a spherical conductor of radius h which has been given a total charge Q. Let's now use Gauss' law to analyze the flux through an imaginary sphere of radius r, with $r > h$. The total flux through this sphere will be Q/ε_0, and because of the symmetry it will be directed radially outward and be uniform in magnitude. Since the area of our imaginary sphere is $4\pi r^2$, Eq. (7.69) yields

$$4\pi r^2 |\mathbf{E}| = \frac{Q}{\varepsilon_0}, \qquad \text{leading to} \quad |\mathbf{E}| = \frac{1}{4\pi\varepsilon_0}\frac{Q}{r^2}. \qquad (7.70)$$

We have obtained the well-known result that for all points external to a spherically symmetric charge distribution, the electric field has the value corresponding to placing all the charge at the center of the sphere.

What can we say about the electric field for $r < h$? First, it does not matter whether the conductor is a thin spherical shell or a solid ball; the repulsion between the elements of its charge will cause them to migrate to $r = h$. Then within the region

$r < h$ **E** will be divergenceless everywhere (including $r = 0$, as there is no charge there). If we consider an imaginary sphere of any radius $r < h$, the only way we can have zero outflow and spherical symmetry is for **E** to vanish. We thus conclude that the interior of the conductor will be field-free.

■

MAXWELL'S EQUATIONS

The equations governing electromagnetic theory provide an excellent example of the use of differential vector operators, and we are now ready to appreciate their content. We first write Maxwell's equations, in a form appropriate for charges and currents in otherwise empty space, and then supply relevant comments.

$$\nabla \cdot \mathbf{B} = 0\,, \tag{7.71}$$

$$\nabla \cdot \mathbf{E} = \frac{\rho}{\varepsilon_0}\,, \tag{7.72}$$

$$\nabla \times \mathbf{B} = \varepsilon_0 \mu_0 \frac{\partial \mathbf{E}}{\partial t} + \mu_0 \mathbf{J}\,, \tag{7.73}$$

$$\nabla \times \mathbf{E} = -\frac{\partial \mathbf{B}}{\partial t}\,. \tag{7.74}$$

Here **E** is the electric field, **B** is the magnetic induction field, ρ is the charge density, **J** is the current density, ε_0 is the electric permittivity, and μ_0 is the magnetic permeability. It can be shown that the product $\varepsilon_0\mu_0$ has the value $1/c^2$, where c is the velocity of light.

Equation (7.71) states that **B** is divergenceless everywhere, reflecting the nonexistence of magnetic monopoles whose presence would create sources and sinks for **B**. This equation forces **B** to be solenoidal. Equation (7.72), discussed in previous subsections, identifies the charge density as the source and sinks of the electric field, and causes **E** to be divergenceless in places where there is no charge density. When Eq. (7.72) is integrated over a region of volume, it yields Gauss' law, and it can be said that this equation is the differential form corresponding to Gauss' law.

In the absence of time dependence, Eq. (7.73) corresponds to Ørsted's law for the magnetic field of a steady current; if integrated over an open surface S and Stokes' theorem is used to convert the surface integral to a line integral around ∂S, we obtain Ampére's law of magnetism, as shown in Example 7.4.7. The $\partial\mathbf{E}/\partial t$ term of Eq. (7.73) is what is sometimes called the **displacement current**; when current flow is not continuous but instead causes charge accumulation or depletion (e.g., by charging or discharging a capacitor), the lack of continuity is accompanied by changes that then occur in the electric field.

Finally, Eq. (7.74) corresponds to Faraday's law of electromagnetic induction: a changing magnetic field generates an electric field.

A set of four coupled partial differential equations is capable of describing a wide variety of phenomena. Two possibilities not previously discussed in this text are indicated here.

Example 7.5.4. Current Loop

In our earlier study of Stokes' theorem we saw that a steady current is surrounded by a magnetic field that encircles it in a direction dictated by the right-hand rule.

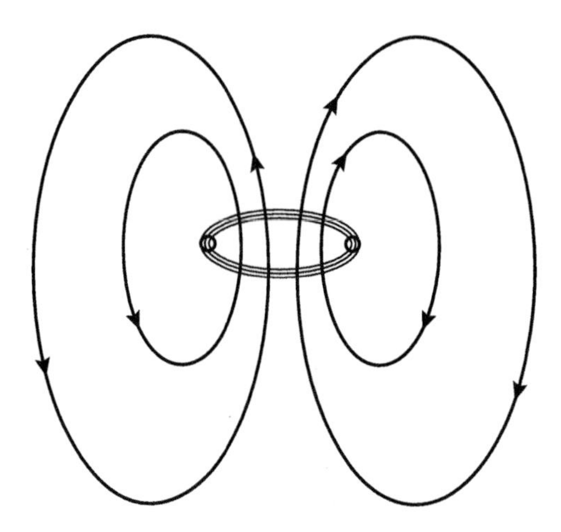

Figure 7.16: Stream lines of **B** through current loop.

If the current is formed into a loop, there will be a flux of **B** through the loop in the right-hand-rule direction. In Fig. 7.16 we sketch some stream lines of **B**. If the loop is small, the stream lines are (when not too near the loop) similar to those produced by an electric dipole (a closely spaced pair of charges of equal magnitude but opposite signs). This analogy can be pursued quantitatively, and small current loops are thereby identified as magnetic dipoles. Such loops at an atomic scale (describing electron currents) are the sources of magnetic phenomena in nature. This topic is treated in more detail in the Additional Readings.

■

Example 7.5.5. Electromagnetic Waves

In empty space (no charges or currents) the two Maxwell's curl equations can be combined in an interesting and important way. We start by taking the curl of Eq. (7.74), reaching

$$\nabla \times (\nabla \times \mathbf{E}) = -\frac{\partial}{\partial t}(\nabla \times \mathbf{B})\,.$$

We are assuming that **E** and **B** are continuous and differentiable, so that the order of the space and time derivatives can be interchanged. We next replace $\nabla \times \mathbf{B}$ by its value as given by Eq. (7.73), thereby obtaining a partial differential equation in **E** alone:

$$\nabla \times (\nabla \times \mathbf{E}) = -\varepsilon_0\mu_0 \frac{\partial^2 \mathbf{E}}{\partial t^2}\,.$$

We now use Eq. (7.39), in which we drop the term containing $\nabla \cdot \mathbf{E}$ because it is zero in empty space, to rearrange our partial differential equation to the form

$$\nabla^2 \mathbf{E} = \varepsilon_0\mu_0 \frac{\partial^2 \mathbf{E}}{\partial t^2}\,.$$

This is a wave equation, with solutions that have been identified as electromagnetic waves with a velocity c such that $c^2 = 1/\varepsilon_0\mu_0$. Because our original equations coupled **E** and **B**, an electromagnetic wave describing **E** must be accompanied by a wave for **B** that satisfies a similar differential equation. A detailed discussion of electromagnetic waves is complicated and largely outside the scope of this text.

■

The introduction of potentials can make some problems in electromagnetic theory easier to solve. If we restrict consideration to situations without time dependence (fixed charges, steady currents), and let φ and $\mathbf{A}$ be respectively the scalar and vector potentials, we have

$$\mathbf{B} = \boldsymbol{\nabla} \times \mathbf{A}, \qquad \mathbf{E} = -\boldsymbol{\nabla}\varphi, \tag{7.75}$$

and it can be shown that these potentials are related to the sources (charge and current) by the separate differential equations

$$-\boldsymbol{\nabla}^2\varphi = \frac{\rho}{\varepsilon_0}, \tag{7.76}$$

$$-\boldsymbol{\nabla}^2\mathbf{A} = \mu_0\mathbf{J}. \tag{7.77}$$

Equation (7.76) is known as the Poisson equation; we have already encountered it when discussing the potential of a point charge.

SYMBOLIC COMPUTING

MAPLE has some features relevant to potential theory. The MAPLE command

```
V := ScalarPotential(E)
```

attempts to compute a scalar potential V such that $\boldsymbol{\nabla}V = \mathbf{E}$. Note that this MAPLE equation does **not** have the minus sign that is customary in physics, so V would produce a force in the direction of **increasing**, rather than decreasing potential. We therefore need to change the sign of the computer output to obtain agreement with the formulas in this text. If no scalar potential exists (because $\mathbf{E}$ is not irrotational), no output is returned. The command

```
A := VectorPotential(B)
```

attempts to compute a vector potential $\mathbf{A}$ such that $\boldsymbol{\nabla} \times \mathbf{A} = \mathbf{B}$. Here the sign of the MAPLE formula agrees with customary usage (and this text). The computation fails and no output is returned if $\mathbf{B}$ is not solenoidal. Thus,

```
with(VectorCalculus):
SetCoordinates(cartesian[x,y,z]):
E := VectorField([x,y,z]);
B := VectorField([y,x,0]);
```

$$E := x\,\overline{e}_x + y\,\overline{e}_y + z\,\overline{e}_z$$

$$B := y\,\overline{e}_x - x\,\overline{e}_y$$

We can check that $\mathbf{E}$ is irrotational but not solenoidal:

```
Curl(E);              0
Divergence(E);        3
```

and that $\mathbf{B}$ is both irrotational and solenoidal:

```
Curl(B);              0
Divergence(B);        0
```

We now try to obtain scalar and vector potentials:

`V := ScalarPotential(E);`	$V := \frac{1}{2}x^2 + \frac{1}{2}y^2 + \frac{1}{2}z^2$
`A := VectorPotential(E);`	(no output)
`V := ScalarPotential(B);`	$V := yx$
`A := VectorPotential(B);`	$A := (xz)\overline{e}_x - yz\overline{e}_y$

In MATHEMATICA, it appears that there are no single commands that yield scalar or vector potentials of a vector field $\mathbf{F}$, but one can use less direct ways to obtain those potentials. The most obvious way to proceed is in two steps, of which the first would be to determine whether a scalar or vector potential exists, by computing $\boldsymbol{\nabla} \times \mathbf{F}$ or $\boldsymbol{\nabla} \cdot \mathbf{F}$. If a scalar potential exists, its value at points $\mathbf{r}'$ can then be obtained as an integral $\int^{\mathbf{r}=\mathbf{r}'} \mathbf{F} \cdot d\mathbf{r}$; if a vector potential exists, its value can be obtained by the evaluation of Eq. (7.66).

Neither MAPLE nor MATHEMATICA reports the delta function as the output of $\boldsymbol{\nabla}^2(1/r)$ (or as a part of the output in related operator expressions). Both return expressions which are valid only when the operand is nonsingular.

Exercises

7.5.1. For each of the following force fields, determine whether it is conservative, and if it is, find the scalar potential $\varphi(x, y, z)$ such that $\mathbf{F} = -\boldsymbol{\nabla}\varphi$.

(a) $\mathbf{F} - x^2\,\hat{\mathbf{e}}_x + y^2\,\hat{\mathbf{e}}_y + z^2\,\hat{\mathbf{e}}_z\,,$

(b) $\mathbf{F} = y\,\hat{\mathbf{e}}_x + z\,\hat{\mathbf{e}}_y + y\,\hat{\mathbf{e}}_z\,,$

(c) $\mathbf{F} = y^2 z \sinh(2xz)\,\hat{\mathbf{e}}_x + 2y\cosh^2(xz)\,\hat{\mathbf{e}}_y + xy^2\sinh(2xz)\,\hat{\mathbf{e}}_z\,.$

7.5.2. For each of the following force fields, determine whether it is solenoidal, and if it is, find a vector potential $\mathbf{A}(x, y, z)$ such that $\mathbf{F} = \boldsymbol{\nabla} \times \mathbf{A}$.

(a) $\mathbf{F} = y\,\hat{\mathbf{e}}_x + z\,\hat{\mathbf{e}}_y + x\,\mathbf{e}_z\,,$

(b) $\mathbf{F} = y\,\hat{\mathbf{e}}_x - x\,\hat{\mathbf{e}}_y + z\,\mathbf{e}_z\,,$

(c) $\mathbf{F} = x^2 y\,\hat{\mathbf{e}}_x - xy^2\,\hat{\mathbf{e}}_y + xy\,\mathbf{e}_z\,.$

7.5.3. A sphere of radius a contains charge of uniform density ρ_0 (SI units). Find the magnitude of the electric field:

(a) At all points $r < a$ within the sphere.

(b) At all points $r > a$ outside the sphere.

7.5.4. Use your symbolic computing system to compute $\boldsymbol{\nabla}^2(1/r)$ and $\boldsymbol{\nabla}^2(e^{-r}/r)$, where $r = \sqrt{x^2 + y^2 + z^2}$.

For $r = 0$ compare the computer output with hand computations.

7.5.5. In the presence of a time-dependent magnetic field the electric field $\mathbf{E}$ is not irrotational, because one of the Maxwell equations is

$$\boldsymbol{\nabla} \times \mathbf{E} = -\frac{\partial \mathbf{B}}{\partial t}\,.$$

Write $\mathbf{B}$ in terms of a vector potential and explain why it is legitimate to write

$$\mathbf{E} = -\boldsymbol{\nabla}\varphi - \frac{\partial \mathbf{A}}{\partial t},$$

where φ is a scalar potential that reduces to the electrostatic potential when the magnetic field is time-independent.

7.6 VECTORS IN CURVILINEAR COORDINATES

In Chapter 6 we have already seen that algebraic operations involving vectors have no meaning unless the vectors are associated with the same spatial point, and that in that case we can combine vectors expressed in orthogonal curvilinear coordinates using the same formulas as for Cartesian systems. For a general orthogonal system with coordinates q_i and corresponding unit vectors $\hat{\mathbf{e}}_i$, and letting vectors $\mathbf{A}$ and $\mathbf{B}$ have component expansions $\mathbf{A} = A_1\hat{\mathbf{e}}_1 + A_2\hat{\mathbf{e}}_2 + A_3\hat{\mathbf{e}}_3$ and $\mathbf{B} = B_1\hat{\mathbf{e}}_1 + B_2\hat{\mathbf{e}}_2 + B_3\hat{\mathbf{e}}_3$, then

$$a\mathbf{A} + b\mathbf{B} = (aA_1 + bB_1)\hat{\mathbf{e}}_1 + (aA_2 + bB_2)\hat{\mathbf{e}}_2 + (aA_3 + bB_3)\hat{\mathbf{e}}_3,$$

$$\mathbf{A}\cdot\mathbf{B} = \sum_i A_i B_i, \quad \mathbf{A}\times\mathbf{B} = \begin{vmatrix} \hat{\mathbf{e}}_1 & \hat{\mathbf{e}}_2 & \hat{\mathbf{e}}_3 \\ A_1 & A_2 & A_3 \\ B_1 & B_2 & B_3 \end{vmatrix}. \tag{7.78}$$

The scalar and vector triple products therefore must also have the forms given in Section 7.1.

In Chapter 6 we also saw that infinitesimal displacements of vectors do **not** obey the same formulas in curvilinear and Cartesian coordinates, and this means that formulas involving differential operators (e.g., the gradient, divergence, and curl) must (for orthogonal coordinate systems) contain the scale factors h_i that relate changes dq_i in the coordinates to the displacements $ds_i = h_i\,dq_i$ thereby produced. Nonorthogonal systems present additional complications and we do not discuss them further in the present context.

VECTOR DIFFERENTIAL OPERATORS

Restricting discussion to orthogonal curvilinear systems, we start by noting that $\boldsymbol{\nabla}\psi$ is a vector whose components are the directional derivatives $d\psi/ds_i$ in the three coordinate directions. This observation leads directly to the formula

$$\boldsymbol{\nabla}\psi = \frac{\hat{\mathbf{e}}_1}{h_1}\frac{\partial\psi}{\partial q_1} + \frac{\hat{\mathbf{e}}_2}{h_2}\frac{\partial\psi}{\partial q_2} + \frac{\hat{\mathbf{e}}_3}{h_3}\frac{\partial\psi}{\partial q_3}, \tag{7.79}$$

and indicates that in the curvilinear system we have

$$\boldsymbol{\nabla} = \frac{\hat{\mathbf{e}}_1}{h_1}\frac{\partial}{\partial q_1} + \frac{\hat{\mathbf{e}}_2}{h_2}\frac{\partial}{\partial q_2} + \frac{\hat{\mathbf{e}}_3}{h_3}\frac{\partial}{\partial q_3}. \tag{7.80}$$

When we wrote the corresponding formula for Cartesian coordinates, Eq. (7.20), it did not matter whether we put the unit vectors $\hat{\mathbf{e}}_x$, $\hat{\mathbf{e}}_y$, $\hat{\mathbf{e}}_z$ before or after the derivative

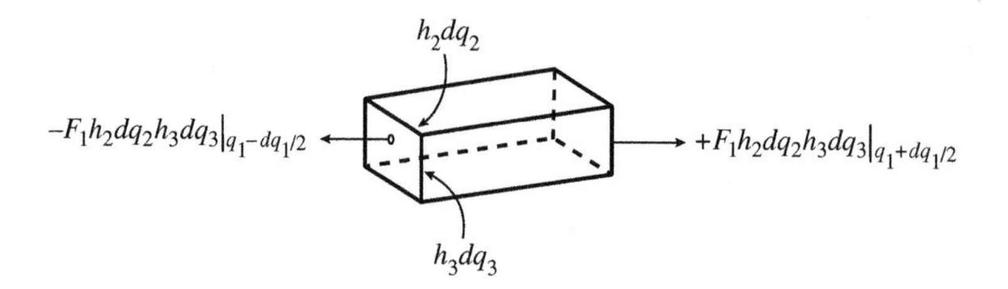

Figure 7.17: Outward flow for divergence (curvilinear coordinates).

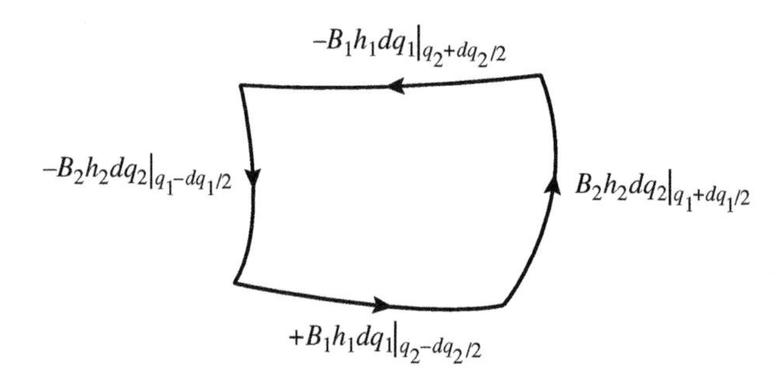

Figure 7.18: Line integral for curl (curvilinear coordinates).

(because they are constants), but here the $\hat{\mathbf{e}}_i$ may be variables and must be placed where they will not be differentiated.

Continuing to $\boldsymbol{\nabla}\cdot\mathbf{F}$, we make a computation similar to that which led to Eq. (7.25), but now the areas of the faces through which the flux flows outward include scale factors. See Fig. 7.17. The result is that the net outflow from the pictured volume element in the q_1 direction has the value

$$\left[(h_2h_3F_1)(q_1+\tfrac{1}{2}\,dq_1,q_2,q_3)-(h_2h_3F_1)(q_1-\tfrac{1}{2}\,dq_1,q_2,q_3)\right]dq_2dq_3\,, \tag{7.81}$$

where the notation $(h_2h_3F_1)(q_1,q_2,q_3)$ indicates that the entire product $h_2h_3F_1$ is to be evaluated at the indicated coordinate point. We are therefore taking into account that not only F_1, but also h_2 and h_3 may vary as q_1 is changed.

We now approximate the outflow given in Eq. (7.81) as

$$\begin{aligned}\frac{\partial}{\partial q_1}(h_2h_3F_1)dq_1\,dq_2\,dq_3 &= \frac{1}{h_1h_2h_3}\frac{\partial}{\partial q_1}(h_2h_3F_1)h_1\,dq_1h_2\,dq_2h_3\,dq_3\\ &= \frac{1}{h_1h_2h_3}\frac{\partial}{\partial q_1}(h_2h_3F_1)d^3r\,.\end{aligned} \tag{7.82}$$

We may form div $\mathbf{F}$ by combining Eq. (7.82) with similar expressions for outflows through the other faces of our volume element.

To compute the curl in curvilinear coordinates, we evaluate the line integral of $\mathbf{B}\cdot d\mathbf{r}$ much as in the analysis for Cartesian coordinates that used Fig. 7.10. However, the presence of scale factors changes the details. Referring to Fig. 7.18, the formula for circulation about a small area in the q_1q_2 plane is

$$\text{Circulation} = \frac{\hat{\mathbf{e}}_3}{A}\oint_{\partial A}(B_1h_1\,dq_1+B_2h_2\,dq_2)\,. \tag{7.83}$$

Here we have used the fact that $ds_i = h_i\,dq_i$. If we now insert $dA = h_1h_2\,dq_1\,dq_2$ and apply Green's theorem in the plane, Eq. (6.46), we can reduce Eq. (7.83) to

$$(\boldsymbol{\nabla}\times\mathbf{B})_3 = \frac{1}{h_1h_2}\left[-\frac{\partial}{\partial q_2}(h_1B_1) + \frac{\partial}{\partial q_1}(h_2B_2)\right]. \tag{7.84}$$

Equation (7.84) can be generalized to obtain the remaining components of curl $\mathbf{B}$ in an arbitrary orthogonal coordinate system.

We are now ready to collect the above analyses into general formulas for the gradient, divergence, and curl. In addition, we combine the formulas for the gradient and divergence to obtain a general formula for the (scalar) Laplacian operator. In this way we reach

$$\boldsymbol{\nabla}\psi(q_1,q_2,q_3) = \hat{\mathbf{e}}_1\frac{1}{h_1}\frac{\partial\psi}{\partial q_1} + \hat{\mathbf{e}}_2\frac{1}{h_2}\frac{\partial\psi}{\partial q_2} + \hat{\mathbf{e}}_3\frac{1}{h_3}\frac{\partial\psi}{\partial q_3}, \tag{7.85}$$

$$\boldsymbol{\nabla}\cdot\mathbf{F}(q_1,q_2,q_3) = \frac{1}{h_1h_2h_3}\left[\frac{\partial}{\partial q_1}(F_1h_2h_3) + \frac{\partial}{\partial q_2}(F_2h_3h_1) + \frac{\partial}{\partial q_3}(F_3h_1h_2)\right], \tag{7.86}$$

$$\boldsymbol{\nabla}^2\psi(q_1,q_2,q_3) =$$

$$\frac{1}{h_1h_2h_3}\left[\frac{\partial}{\partial q_1}\left(\frac{h_2h_3}{h_1}\frac{\partial\psi}{\partial q_1}\right) + \frac{\partial}{\partial q_2}\left(\frac{h_3h_1}{h_2}\frac{\partial\psi}{\partial q_2}\right) + \frac{\partial}{\partial q_3}\left(\frac{h_1h_2}{h_3}\frac{\partial\psi}{\partial q_3}\right)\right], \tag{7.87}$$

$$\boldsymbol{\nabla}\times\mathbf{B}(q_1,q_2,q_3) = \frac{1}{h_1h_2h_3}\begin{vmatrix} \hat{\mathbf{e}}_1h_1 & \hat{\mathbf{e}}_2h_2 & \hat{\mathbf{e}}_3h_3 \\ \partial/\partial q_1 & \partial/\partial q_2 & \partial/\partial q_3 \\ h_1B_1 & h_2B_2 & h_3B_3 \end{vmatrix}. \tag{7.88}$$

CIRCULAR CYLINDRICAL COORDINATES

We now specialize the formulas in Eqs. (7.85)–(7.88) to cylindrical coordinates (ρ,φ,z), for which $h_\rho = 1$, $h_\varphi = \rho$, and $h_z = 1$. The results are

$$\boldsymbol{\nabla}\psi(\rho,\varphi,z) = \hat{\mathbf{e}}_\rho\frac{\partial\psi}{\partial\rho} + \hat{\mathbf{e}}_\varphi\frac{1}{\rho}\frac{\partial\psi}{\partial\varphi} + \hat{\mathbf{e}}_z\frac{\partial\psi}{\partial z}, \tag{7.89}$$

$$\boldsymbol{\nabla}\cdot\mathbf{F}(\rho,\varphi,z) = \frac{1}{\rho}\frac{\partial}{\partial\rho}(\rho F_\rho) + \frac{1}{\rho}\frac{\partial F_\varphi}{\partial\varphi} + \frac{\partial F_z}{\partial z}, \tag{7.90}$$

$$\boldsymbol{\nabla}^2\psi(\rho,\varphi,z) = \frac{1}{\rho}\frac{\partial}{\partial\rho}\left(\rho\frac{\partial\psi}{\partial\rho}\right) + \frac{1}{\rho^2}\frac{\partial^2\psi}{\partial\varphi^2} + \frac{\partial^2\psi}{\partial z^2}, \tag{7.91}$$

$$\boldsymbol{\nabla}\times\mathbf{B}(\rho,\varphi,z) = \frac{1}{\rho}\begin{vmatrix} \hat{\mathbf{e}}_\rho & \rho\hat{\mathbf{e}}_\varphi & \hat{\mathbf{e}}_z \\ \partial/\partial\rho & \partial/\partial\varphi & \partial/\partial z \\ B_\rho & \rho B_\varphi & B_z \end{vmatrix}. \tag{7.92}$$

The use of curvilinear coordinates has made the expressions for the various operators a bit more complicated than in Cartesian coordinates. However, note that the Laplacian, though no longer a straightforward sum of second derivatives, still has the feature that it contains no mixed second derivatives; each term of the Laplacian contains derivatives with respect to only one variable. This separation of the derivatives provides important opportunities for simplification when we try to solve partial differential equations that contain the Laplacian operator, and is a consequence of the fact that our coordinate system is orthogonal.

In cylindrical coordinates the vector Laplacian does not separate into Laplacians of the individual components of its argument. For the form of that operator the reader is referred to Arfken et al. in the Additional Readings.

Example 7.6.1. Axially Symmetric Functions

If $\mathbf{F}$ and ψ depend only on the cylindrical coordinate ρ, Eqs. (7.90) and (7.91) take the forms

$$\boldsymbol{\nabla}\cdot\mathbf{F}(\rho) = \frac{dF_\rho}{d\rho} + \frac{F_\rho}{\rho}, \tag{7.93}$$

$$\boldsymbol{\nabla}^2\psi(\rho) = \frac{d^2\psi}{d\rho^2} + \frac{1}{\rho}\frac{d\psi}{d\rho}. \tag{7.94}$$

The results differ from those in Cartesian coordinates.

■

SPHERICAL POLAR COORDINATES

The other coordinate system of sufficient importance to be examined here uses the coordinates (r, θ, φ), where θ is the polar angle and φ is sometimes called an azimuthal angle. In this coordinate system we have $h_r = 1$, $h_\theta = r$, and $h_\varphi = r\sin\theta$. Our differential-operator formulas reduce to

$$\boldsymbol{\nabla}\psi(r,\theta,\phi) = \hat{\mathbf{e}}_r\frac{\partial\psi}{\partial r} + \hat{\mathbf{e}}_\theta\frac{1}{r}\frac{\partial\psi}{\partial\theta} + \hat{\mathbf{e}}_\varphi\frac{1}{r\sin\theta}\frac{\partial\psi}{\partial\varphi}, \tag{7.95}$$

$$\boldsymbol{\nabla}\cdot\mathbf{F}(r,\theta,\phi) = \frac{1}{r^2\sin\theta}\left[\sin\theta\frac{\partial}{\partial r}(r^2F_r) + r\frac{\partial}{\partial\theta}(\sin\theta F_\theta) + r\frac{\partial F_\varphi}{\partial\varphi}\right], \tag{7.96}$$

$$\boldsymbol{\nabla}^2\psi(r,\theta,\phi) = \frac{1}{r^2\sin\theta}\left[\sin\theta\frac{\partial}{\partial r}\left(r^2\frac{\partial\psi}{\partial r}\right) + \frac{\partial}{\partial\theta}\left(\sin\theta\frac{\partial\psi}{\partial\theta}\right) + \frac{1}{\sin\theta}\frac{\partial^2\psi}{\partial\varphi^2}\right], \tag{7.97}$$

$$\boldsymbol{\nabla}\times\mathbf{B}(r,\theta,\phi) = \frac{1}{r^2\sin\theta}\begin{vmatrix} \hat{\mathbf{e}}_r & r\,\hat{\mathbf{e}}_\theta & r\sin\theta\,\hat{\mathbf{e}}_\varphi \\ \partial/\partial r & \partial/\partial\theta & \partial/\partial\varphi \\ B_r & rB_\theta & r\sin\theta B_\varphi \end{vmatrix}. \tag{7.98}$$

Here also the Laplacian contains no mixed second derivatives because the spherical polar coordinates are orthogonal. The vector Laplacian does not separate into Laplacians of the individual components of its argument. For that operator see Arfken et al. in the Additional Readings.

Example 7.6.2. Laplacian of Spherically Symmetric Function

If ψ is a function only of r, i.e., $\psi(r)$, the formula for the Laplacian in spherical coordinates simplifies greatly. Directly from Eq. (7.97) we get

$$\nabla^2\psi(r) = \frac{d^2\psi}{dr^2} + \frac{2}{r}\frac{d\psi}{dr}. \tag{7.99}$$

An alternative form that is mathematically equivalent is

$$\nabla^2\psi(r) = \frac{1}{r}\frac{d^2}{dr^2}(r\psi). \tag{7.100}$$

■

Example 7.6.3. Hydrogen Ground State

In Hartree atomic units ($m = e = h/2\pi = 1$) the spherically symmetric states of the hydrogen atom satisfy the following eigenvalue equation (the Schrödinger equation):

$$-\frac{1}{2}\nabla^2\psi - \frac{1}{r}\,\psi = E\,\psi.$$

The eigenvalues E are eigenstate energies. Let's verify that the hydrogenic ground state, with wave function $\psi(r) = e^{-r}$, has $E = -1/2$ hartree.

Using Eq. (7.99) to evaluate $\nabla^2\psi$, we get

$$-\frac{1}{2}\left[e^{-r} - \frac{2}{r}\,e^{-r}\right] - \frac{1}{r}\,e^{-r} = Ee^{-r},$$

which is satisfied only if $E = -1/2$.

■

Example 7.6.4. Comparison with Cartesian Formula

In earlier sections we computed the gradient and the divergence for quantities with spherical symmetry (i.e., quantities that are functions only of r). Let's compare with the spherical-coordinate formulas.

Taking first the gradient, Eq. (7.23) is seen to be in exact agreement with Eq. (7.95). Proceeding next to the divergence, note that Eq. (7.27) is for a vector $f(r)\,\mathbf{r}$ whose r component is $rf(r)$, so we must insert $F_r = rf(r)$ into Eq. (7.96) to make a proper comparison. Then we get

$$\nabla\cdot\mathbf{F} = \frac{1}{r^2}\frac{d}{dr}\left[r^3 f(r)\right] = 3f(r) + r\,f'(r),$$

in agreement with Eq. (7.27).

To check the Laplacian, we can use Eq. (7.27) with f replaced by $\psi'(r)/r$. The result is $3\psi'(r)/r + r[\psi'(r)/r]'$, which expands to $\psi'' + 2\psi'/r$, as required.

It is important to observe that even though the r coordinate has the same scaling as a Cartesian coordinate ($h = 1$), the curvature of the coordinate system affects the differential operators even when they are applied to spherically symmetric functions.

■

SYMBOLIC COMPUTING

If we have a problem whose formulation is in a curvilinear coordinate system, we may want to input vector fields as their curvilinear components and use the vector differential operators in the curvilinear system. This approach requires that we activate the relevant coordinate system and define the names of its coordinates. In MAPLE, the available coordinate systems have the names `cartesian`, `cylindrical`, `spherical`, plus others that we will not discuss here. In MATHEMATICA, the corresponding names are `Cartesian`, `Cylindrical`, and `Spherical`. In MAPLE, the coordinates do not have default names, but in MATHEMATICA, they do have defaults. The Cartesian defaults (given previously) are `Xx`, `Yy`, `Zz`; the Spherical defaults are `Rr`, `Ttheta`, `Pphi`, and the Cylindrical defaults are `Rr`, `Ttheta`, `Zz`.

In MAPLE, a vector entry in curvilinear coordinates that has not been designated as a vector field consists of the values of the three coordinates of its head when its tail is placed at the origin, with the entries permanently identified as being in the coordinate system that was active when the vector was defined. However, for differential-operator purposes the coordinate values are not the most useful quantities. What we really need is a vector field, represented by the values of the coefficients of the three unit vectors $\overline{e}_i$ when the vector field is evaluated at the point with coordinates q_i. This point and its consequences are the main topic of Appendix H, which provides detail that may be helpful in working with vector operators in curvilinear coordinates.

Example 7.6.5. Curvilinear Coordinates in Maple

We start by creating some vectors in spherical coordinates:

```
> with(VectorCalculus):
> SetCoordinates(spherical[r,theta,phi]);
```

$$spherical_{r,\theta,\phi}$$

```
> A := Vector([1,Pi/2,0]);
```

$$A := e_r + \tfrac{1}{2}\,\pi\, e_\theta$$

```
> B := Vector([1,Pi/2,Pi/2]);
```

$$B := e_r + \tfrac{1}{2}\,\pi\, e_\theta + \tfrac{1}{2}\,\pi\, e_\phi$$

```
> C := Vector([1,0,0]);
```

$$C := e_r$$

Here MAPLE's output notation is a bit deficient, as the coefficients preceding e_r, e_θ, and e_ϕ are not really the vector components they appear to be; they are actually the values of r, θ, and φ for the point of the vector under study when its tail is at the origin. Because spherical coordinates were active when A, B, and C were defined, MAPLE knows that the coefficients belong to the spherical-coordinate representations of those vectors, and the mathematically inexact notation should be regarded simply as a reminder that the vectors are written in spherical coordinates.

For application of vector operators in MAPLE it is necessary to have scalar and vector fields. We therefore define

```
> B := VectorField([r, 0, r*cos(phi)]);
```

$$B := r\,\overline{e}_r + (r\,\cos(\phi))\,\overline{e}_\phi$$

```
> f := r*cos(theta);
```

$$f := r\,\cos(\theta)$$

The output for B is a legitimate mathematical form; by assigning the vector field attribute we are causing the quantities preceding the unit vectors to be identified as the possibly position-dependent components of B. Because MAPLE knows both the coordinate system and the names of the coordinates, it has the information needed to apply vector differential operators. Therefore,

```
> Gradient(f);
```

$$\cos(\theta)\,\overline{e}_r - \sin(\theta)\,\overline{e}_\theta$$

```
> Divergence(B);
```

$$\frac{3r^2\sin(\theta) - r^2\sin(\phi)}{r^2\sin(\theta)}$$

```
> simplify(%);
```

$$\frac{3\sin(\theta) - \sin(\phi)}{\sin(\theta)}$$

```
> simplify(Laplacian(f^2));
```

$$2$$

```
> Curl(B);
```

$$\left(\frac{\cos(\theta)\cos(\phi)}{\sin(\theta)}\right)\overline{e}_r - 2\cos(\phi)\overline{e}_\theta$$

The vector operations often produce results that can be simplified. Before simplification the Laplacian formula was rather complicated.

■

In MATHEMATICA, the lists representing vectors (or vector fields) have meanings that depend upon the coordinate system that is active when the list is used, rather than when it is created. That point becomes clearer as we examine some illustrative computations.

Example 7.6.6. Curvilinear Coordinates in Mathematica

First create some vectors intended for use in spherical coordinates:

```
<<VectorAnalysis`
A = {1, Pi/2, 0}                    {1, π/2, 0}
B = {1, Pi/2, Pi/2}                 {1, π/2, π/2}
C = {1, 0, 0}                       {1, 0, 0}
```

The above has simply defined three ordered lists, each containing three elements. If interpreted as the coordinates of the points of vectors with tails at the origin and spherical coordinates are used, they are unit vectors, respectively in the x-, y-, and z-directions. In order for MATHEMATICA to make this interpretation we must activate the spherical coordinate system, after which we can easily make the partial check that A, B, and C are mutually orthogonal and of unit length:

```
SetCoordinates[Spherical[r,theta,phi]]      Spherical[r,theta,phi]
A . A                                       1
A . B                                       0
```

The above results are consistent with the notion that **A** and **B** are orthogonal unit vectors.

To show that the vectors do not carry coordinate information, look at

```
SetCoordinates[Cartesian]                   Cartesian[Xx, Yy, Zz]
```

`A . A` $\quad 1+\frac{\pi^2}{4}$

`A . B` $\quad 1+\frac{\pi^2}{4}$

A and **B** are now interpreted as vectors in Cartesian coordinates. With this interpretation they are neither orthogonal nor unit vectors.

We now define two quantities that we intend to interpret as vector and scalar fields in spherical coordinates:

```
B = {r, 0, r*Cos[phi]}                    {r, 0, r*Cos[phi]}
f = r * Cos[theta]                        r * Cos[theta]
```

Because the Cartesian coordinates are still active (with coordinates named Xx, Yy, Zz), neither B nor f has any coordinate dependence, so we get

```
Grad[f]                                   0
Div[B]                                    0
```

Now switch to spherical coordinates:

```
SetCoordinates[Spherical]                 Spherical[Rr,Ttheta,Pphi]
```

The vector operators will still not work as expected because in defining f and B we did not use the somewhat awkward default coordinate names. We need to enter

```
SetCoordinates[Spherical[r,theta,phi]]    Spherical[r,theta,phi]
```

and now we get

```
Grad[f]                      {Cos[theta], -Sin[theta], 0}
Div[B]                       Csc[theta](- r^2 Sin[phi] + 3r^2 Sin[theta]) / r^2
Simplify[%]                  3 - Csc[theta]Sin[phi]
Simplify[Laplacian[f^2]]     2
Curl[B]                      {Cos[phi]Cot[theta], -2Cos[phi], 0}
```

The vector operations often produce results that can be simplified. Before simplification the Laplacian formula was rather complicated.

■

Exercises

Where it is convenient to do so, check your answers using symbolic computation.

7.6.1. Evaluate the following expressions in circular cylindrical coordinates:

$$\text{(a)}\ \nabla\cdot\hat{\mathbf{e}}_\rho\,,\qquad \text{(b)}\ \nabla\times\hat{\mathbf{e}}_\rho\,,\qquad \text{(c)}\ \nabla\cdot\hat{\mathbf{e}}_\varphi\,,\qquad \text{(d)}\ \nabla\times\hat{\mathbf{e}}_\varphi\,.$$

7.6.2. Evaluate the following expressions in spherical polar coordinates:

$$\text{(a)}\ \nabla\cdot\hat{\mathbf{e}}_r\,,\qquad \text{(b)}\ \nabla\times\hat{\mathbf{e}}_r\,,\qquad \text{(c)}\ \nabla\cdot\hat{\mathbf{e}}_\theta\,,$$
$$\text{(d)}\ \nabla\times\hat{\mathbf{e}}_\theta\,,\qquad \text{(e)}\ \nabla\cdot\hat{\mathbf{e}}_\varphi\,,\qquad \text{(f)}\ \nabla\times\hat{\mathbf{e}}_\varphi\,.$$

7.6.3. Evaluate, in circular cylindrical coordinates, with $f(\rho)$ an arbitrary differentiable function of ρ only:

$$\text{(a)}\ \nabla \cdot \cos\rho\, \hat{\mathbf{e}}_\rho\,, \qquad \text{(b)}\ \nabla \times [f(\rho)\hat{\mathbf{e}}_\rho]\,, \qquad \text{(c)}\ \nabla^2 \left[\rho^2 e^{-\rho}\right].$$

7.6.4. Evaluate, in spherical polar coordinates, with $f(r)$ an arbitrary differentiable function of r only:

$$\text{(a)}\ \nabla \cdot \left[e^{-r}\hat{\mathbf{e}}_r\right], \qquad \text{(b)}\ \nabla \times [f(r)\,\hat{\mathbf{e}}_r]\,, \qquad \text{(c)}\ \nabla^2 \left[r\, e^{-r^2}\right].$$

For the next four exercises it may be helpful to draw a sketch identifying the directions of the various unit vectors when at a specific coordinate point.

7.6.5. Establish the following formulas giving the Cartesian unit vectors in terms of their circular cylindrical components.

$$\hat{\mathbf{e}}_x = \cos\varphi\,\hat{\mathbf{e}}_\rho - \sin\varphi\,\hat{\mathbf{e}}_\varphi\,, \qquad \hat{\mathbf{e}}_y = \sin\varphi\,\hat{\mathbf{e}}_\rho + \cos\varphi\,\hat{\mathbf{e}}_\varphi\,, \qquad \hat{\mathbf{e}}_z = \hat{\mathbf{e}}_z\,.$$

7.6.6. Establish the following formulas giving the Cartesian components of the circular cylindrical unit vectors.

$$\hat{\mathbf{e}}_\rho = \cos\varphi\,\hat{\mathbf{e}}_x + \sin\varphi\,\hat{\mathbf{e}}_y\,, \qquad \hat{\mathbf{e}}_\varphi = -\sin\varphi\,\hat{\mathbf{e}}_x + \cos\varphi\,\hat{\mathbf{e}}_y\,, \qquad \hat{\mathbf{e}}_z = \hat{\mathbf{e}}_z\,.$$

7.6.7. Establish the following formulas giving the Cartesian unit vectors in terms of their spherical polar components.

$$\begin{aligned}
\hat{\mathbf{e}}_x &= \sin\theta\cos\varphi\,\hat{\mathbf{e}}_r + \cos\theta\cos\varphi\,\hat{\mathbf{e}}_\theta - \sin\varphi\,\hat{\mathbf{e}}_\varphi\,,\\
\hat{\mathbf{e}}_y &= \sin\theta\sin\varphi\,\hat{\mathbf{e}}_r + \cos\theta\sin\varphi\,\hat{\mathbf{e}}_\theta + \cos\varphi\,\hat{\mathbf{e}}_\varphi\,,\\
\hat{\mathbf{e}}_z &= \cos\theta\,\hat{\mathbf{e}}_r - \sin\theta\,\hat{\mathbf{e}}_\theta\,.
\end{aligned}$$

7.6.8. Establish the following formulas giving the Cartesian components of the spherical polar unit vectors.

$$\begin{aligned}
\hat{\mathbf{e}}_r &= \sin\theta\cos\varphi\,\hat{\mathbf{e}}_x + \sin\theta\sin\varphi\,\hat{\mathbf{e}}_y + \cos\theta\,\hat{\mathbf{e}}_z\,,\\
\hat{\mathbf{e}}_\theta &= \cos\theta\cos\varphi\,\hat{\mathbf{e}}_x + \cos\theta\sin\varphi\,\hat{\mathbf{e}}_y - \sin\theta\,\hat{\mathbf{e}}_z\,,\\
\hat{\mathbf{e}}_\varphi &= -\sin\varphi\,\hat{\mathbf{e}}_x + \cos\varphi\,\hat{\mathbf{e}}_y\,.
\end{aligned}$$

7.6.9. (a) If a particle described by circular cylindrical coordinates (ρ, φ, z) is in motion, with the time derivatives of its coordinates written as $(\dot\rho, \dot\varphi, \dot z)$, show that the unit vectors at its position in the coordinate directions have time derivatives given as

$$\frac{d\hat{\mathbf{e}}_\rho}{dt} = \dot\varphi\,\hat{\mathbf{e}}_\varphi\,, \qquad \frac{d\hat{\mathbf{e}}_\varphi}{dt} = -\dot\varphi\,\hat{\mathbf{e}}_\rho\,, \qquad \frac{d\hat{\mathbf{e}}_z}{dt} = 0\,.$$

Hint. It may be helpful to start by computing the time derivatives from the results in Exercise 7.6.6.

(b) Noting now that the displacement of the particle from the origin has the form $\mathbf{s} = \rho\hat{\mathbf{e}}_\rho + z\hat{\mathbf{e}}_z$, show that its velocity has the component expansion

$$\frac{d\mathbf{s}}{dt} = \dot\rho\,\hat{\mathbf{e}}_\rho + \rho\dot\varphi\,\hat{\mathbf{e}}_\varphi + \dot z\,\hat{\mathbf{e}}_z\,.$$

(c) Find the component expansion of the acceleration, $d^2\mathbf{s}/dt^2$. (Write second derivatives with respect to time in the dot notation illustrated by $\ddot\rho$.)

7.6.10. (a) Following the method illustrated in Exercise 7.6.9, show that the unit vectors at the position of a moving particle in spherical polar coordinates (r,θ,φ) have time derivatives

$$\frac{d\hat{\mathbf{e}}_r}{dt} = \dot\theta\,\hat{\mathbf{e}}_\theta + \sin\theta\,\dot\varphi\,\hat{\mathbf{e}}_\varphi\,, \qquad \frac{d\hat{\mathbf{e}}_\theta}{dt} = -\dot\theta\,\hat{\mathbf{e}}_r + \cos\theta\,\dot\varphi\,\hat{\mathbf{e}}_\varphi\,,$$

$$\frac{d\hat{\mathbf{e}}_\varphi}{dt} = -\sin\theta\,\dot\varphi\,\hat{\mathbf{e}}_r - \cos\theta\,\dot\varphi\,\hat{\mathbf{e}}_\theta\,.$$

(b) Writing the position of the particle as $\mathbf{s} = r\,\hat{\mathbf{e}}_r$, find the component expansion of the velocity of the particle in spherical polar coordinates.

(c) Find the spherical component expansion of the acceleration, $d^2\mathbf{s}/dt^2$.

7.6.11. Show that the following expressions for $\nabla^2\psi(\rho)$, written in cylindrical coordinates, are equivalent:

$$\frac{d^2\psi}{d\rho^2} + \frac{1}{\rho}\frac{d\psi}{d\rho} \qquad \text{and} \qquad \frac{1}{\rho}\frac{d}{d\rho}\left(\rho\frac{d\psi}{d\rho}\right).$$

7.6.12. Show that the following expressions for $\nabla^2\psi(r)$, written in spherical polar coordinates, are equivalent:

$$\text{(a)}\ \frac{d^2\psi}{dr^2} + \frac{2}{r}\frac{d\psi}{dr}\,, \qquad \text{(b)}\ \frac{1}{r}\frac{d^2}{dr^2}[r\psi]\,, \qquad \text{(c)}\ \frac{1}{r^2}\frac{d}{dr}\left[r^2\frac{d\psi}{dr}\right].$$

7.6.13. The angular momentum operator $\mathbf{L}$ has in quantum mechanics the form (in dimensionless units) $\mathbf{L} = -i(\mathbf{r}\times\nabla)$. Verify the following operator identities:

$$\text{(a)}\ \nabla = \hat{\mathbf{e}}_r\frac{\partial}{\partial r} - i\frac{\mathbf{r}\times\mathbf{L}}{r^2}\,, \qquad \text{(b)}\ \mathbf{r}\nabla^2 - \nabla\left(1 + r\frac{\partial}{\partial r}\right) = i(\nabla\times\mathbf{L})\,.$$

7.6.14. Write $\partial/\partial x$, $\partial/\partial y$, and $\partial/\partial z$ in spherical polar coordinates. Your answers should be entirely in terms of the spherical coordinates.

Hint. Start by using the chain rule to form equations such as

$$\frac{\partial}{\partial x} = \left(\frac{\partial r}{\partial x}\right)_{yz}\frac{\partial}{\partial r} + \left(\frac{\partial\theta}{\partial x}\right)_{yz}\frac{\partial}{\partial\theta} + \left(\frac{\partial\varphi}{\partial x}\right)_{yz}\frac{\partial}{\partial\varphi}\,.$$

7.6.15. (a) Use the results from Exercise 7.6.14 to show that

$$L_z = -i\left(x\frac{\partial}{\partial y} - y\frac{\partial}{\partial x}\right) = -i\frac{\partial}{\partial\varphi}\,.$$

(b) Show that L_z is the z-component of the quantum angular-momentum operator $-i(\mathbf{r}\times\nabla)$.

7.6.16. This exercise illustrates the fact that Stokes' theorem is applicable only if the curl appearing in the surface integral is defined for all points of the integration region. Consider

$$\mathbf{B} = -\frac{y}{x^2+y^2}\,\hat{\mathbf{e}}_x + \frac{x}{x^2+y^2}\,\hat{\mathbf{e}}_y\,.$$

(a) Convert $\mathbf{B}$ entirely into cylindrical coordinates. The result of Exercise 7.6.6 will be helpful.

(b) Compare the line integral of $\mathbf{B}$ around the unit circle with the integral $\int(\boldsymbol{\nabla}\times\mathbf{B})\cdot\hat{\mathbf{n}}\,dA$ for the area within the unit circle.

(c) Identify the reason for the failure of Stokes' theorem.

7.6.17. Find the electric field of a dipole of moment $\mathbf{p}$ at the origin by evaluating the expression $\mathbf{E} = -\boldsymbol{\nabla}\psi$, where

$$\psi = \frac{1}{4\pi\varepsilon_0}\frac{\mathbf{p}\cdot\mathbf{r}}{r^3}\,.$$

7.6.18. A force field $\mathbf{F}$ is given in spherical coordinates by

$$\mathbf{F} = \frac{2\cos\theta}{r^3}\,\hat{\mathbf{e}}_r + \frac{\sin\theta}{r^3}\,\hat{\mathbf{e}}_\theta\,.$$

(a) Calculate $\boldsymbol{\nabla}\times\mathbf{F}$ and verify that a scalar potential ψ exists, with $\mathbf{F} = -\boldsymbol{\nabla}\psi$.

(b) Determine ψ.

7.6.19. A long straight wire carrying a steady current I generates a magnetic field

$$\mathbf{B} = \frac{\mu_0 I}{2\pi}\left[-\frac{y}{x^2+y^2}\,\hat{\mathbf{e}}_x + \frac{x}{x^2+y^2}\,\hat{\mathbf{e}}_y\right].$$

The objective of this exercise is to find a vector potential $\mathbf{A}$ such that $\boldsymbol{\nabla}\times\mathbf{A} = \mathbf{B}$.

(a) Write $\mathbf{B}$ entirely in cylindrical coordinates. (This is the same as part (a) of Exercise 7.6.16).

(b) The symmetry of the problem indicates the possibility to make $\mathbf{A}$ independent of z. In that case, show that in order to recover the correct value of $\mathbf{B}$, A_z must be independent of φ, and therefore $A_z = A_z(\rho)$.

(c) Now show that setting $A_\rho = A_\varphi = 0$ is consistent with obtaining the correct value of $\mathbf{B}$.

(d) Complete the exercise by finding a vector potential $\mathbf{A}$ that satisfies the conditions identified in parts (b) and (c).

Additional Readings

Arfken, G. B., Weber, H. J., & Harris, F. E. (2013). *Mathematical methods for physicists* (7th ed.). New York: Academic Press.

Borisenko, A. I., & Tarpov, I. E. (1968). *Vector and tensor analysis with applications.* Englewood Cliffs, NJ: Prentice-Hall (Reprinted, Dover (1980)).

Davis, H. F., & Snider, A. D. (1995). *Introduction to vector analysis* (7th ed.). Boston: Allyn & Bacon.

Kellogg, O. D. (1929). *Foundations of potential theory.* New York: Springer (Reprinted, Dover (2010). The classic text on potential theory).

Lewis, P. E., & Ward, J. P. (1989). *Vector analysis for engineers and scientists.* Reading, MA: Addison-Wesley.

Mathews, P. C. (2000). *Vector calculus.* New York: Springer (Thorough and detailed, not too advanced).

Spiegel, M. R. (1989). *Vector analysis.* New York: McGraw-Hill.

Wrede, R. C. (1963). *Introduction to vector and tensor analysis.* New York: Wiley (Reprinted, New York: Dover (1972). Fine historical introduction. Excellent discussion of differentiation of vectors and applications to mechanics).

Chapter 8

TENSOR ANALYSIS

Tensors are generalizations of the multicomponent quantities we have already encountered, namely vectors and matrices. In three-dimensional space, we define a **tensor** of **rank** n as an object with 3^n components, with properties we shortly discuss in some detail. Tensors are defined in such a way that those of rank 0, 1, and 2 can be respectively identified as scalars, vectors, and matrices. The essence of this identification is that tensors must have properties that are independent of the coordinate system used for their description and, in particular, independent of the orientation of the coordinate axes.

We have already encountered in this text a need to use tensors of ranks 0, 1, and 2 to describe physical phenomena. Tensors of rank 0 (scalars) are independent of the orientation of the coordinates and are therefore appropriate for the description of quantities such as temperature or the dot products of vectors. Tensors of rank 1 (vectors) are needed to describe forces, displacements, momenta, and many other physical variables, while those of rank 2 (matrices) occur in contexts where a vector that is the source of a phenomenon produces a result that is a vector in another direction: examples are the relation between an electric field and the polarization it induces in an anisotropic material, or the relation between angular velocity and angular momentum in an asymmetric rigid body.

Physics and engineering also involve phenomena whose description requires the use of tensors of rank greater than 2. As an example, consider what happens when a **stress** is applied to a material object (this may include shear as well as compressive forces); stress can be described by a $3^2 = 9$-component tensor of rank 2. The response to the stress, called **strain**, may involve both compression (or dilation) and distortion, and can also be described by a tensor of rank 2. In the most general case, involving anisotropic materials, the relation between the stress and strain involves a $9^2 = 81$-component tensor of rank 4. Tensors of ranks 3 and 4 also occur in the description of electro-optic and piezoelectric phenomena. Tensors of these ranks are also encountered in relativistic physics.

Handling these objects with large numbers of components can become complicated and messy. It is therefore important to approach tensor analysis in ways that promote clarity and that focus on the underlying generalities.

Mathematics for Physical Science and Engineering.
http://dx.doi.org/10.1016/B978-0-12-801000-6.00008-0

8.1 CARTESIAN TENSORS

The emphasis in this chapter will be on **Cartesian tensors**, which have components that are related to the three mutually perpendicular directions of a Cartesian coordinate system, and which are defined to have appropriate properties when the coordinate system is rotated.

Our starting point will be tensors of rank 1, which (calling them vectors) we have already studied. To qualify as a 3-D vector, we required a quantity $\mathbf{B}$ to have three components, which when written in some Cartesian coordinate system we identify as B_1, B_2, and B_3. It is notationally convenient to label the coordinates x_1, x_2, x_3 instead of x, y, z, so B_i is the vector component in the x_i-direction. We further required that when the coordinate system is subjected to a rotation, our vector in the rotated coordinates would have components

$$B'_\alpha = \sum_{i=1}^{3} U_{\alpha i} B_i\,, \qquad (\alpha = 1,\, 2,\, 3), \tag{8.1}$$

where the coefficients $U_{\alpha i}$ are the elements of an orthogonal transformation matrix. Equation (8.1) is to be regarded as the defining property of a Cartesian tensor B of rank 1; this tensor has an infinite number of possible representations (such as B') because it can be described using coordinates in any one of an infinite number of orientations.

We now define a tensor T of rank 2 as a quantity with components T_{ij} satisfying an obvious generalization of Eq. (8.1):

$$T'_{\alpha\beta} = \sum_{i=1}^{3}\sum_{j=1}^{3} U_{\alpha i} U_{\beta j} T_{ij}\,, \qquad (\alpha,\, \beta = 1,\, 2,\, 3)\,. \tag{8.2}$$

This definition identifies the values of the subscripts of T (i.e., i, j =1, 2, or 3) with the original Cartesian axes, with the values of α and β associated with the rotated axes. Note also that because the quantities being summed are individual elements, they can be written in any order. However, the order of the subscripts attached to T, T', or either U is significant, as different subscript orderings point to different positions in the array being referenced. Moreover, the definition in Eq. (8.2) creates a connection between corresponding subscripts in T and T'. Note also that because a rotational transformation in Cartesian coordinates is described by a matrix U that is independent of the variables x_i (it depends only on the angles that specify the rotation), all the instances of U in Eq. (8.2) refer to the same matrix.

This is also a good time to make observations about the symbols used as subscripts. In Eq. (8.2), i and j are called **dummy indices**; their function is only to indicate elements to be summed, and the names given them do not affect the value of the sum, assuming that one has avoided index assignments that create ambiguity. What is important is that indices that are supposed to be the same are given the same symbol. Indices that are not summed are called **free indices**; they will have individual values in ranges suitable to the problem under study.

We have previously suggested that matrices are tensors of rank 2. In that connection we now note that the relation between T and T' corresponds to what we called in Chapter 5 a matrix similarity transformation; if we use the orthogonality property of U to rewrite $U_{\beta j}$ as $(\mathsf{U}^{-1})_{j\beta}$, Eq. (8.2) can be rewritten

$$T'_{\alpha\beta} = \sum_{i=1}^{3}\sum_{j=1}^{3} U_{\alpha i} T_{ij} (\mathsf{U}^{-1})_{j\beta}\,,$$

which shows that T' is the matrix product $\mathsf{T}' = \mathsf{UTU}^{-1}$. This observation confirms that our Cartesian tensor of rank 2 transforms like a matrix.

Finally, we continue by defining Cartesian tensors of rank n to be n-index quantities whose rotational transformation is given by

$$T'_{\alpha\beta\gamma\cdots} = \sum_{ijk\cdots} U_{\alpha i} U_{\beta j} U_{\gamma k} \cdots T_{ijk\cdots}\,, \tag{8.3}$$

where the index sets $\alpha\beta\gamma\cdots$ and $ijk\cdots$ each have n members, and the equation contains n factors U. The generalization embodied in Eq. (8.3) includes the scalar case ($n = 0$), for which $T' = T$. Note that in Eq. (8.3) we did not specify the summation range or that the equation is valid for all $\alpha, \beta, \ldots$ Here and in subsequent equations these features are assumed to be known.

Formulas of the type represented by Eq. (8.3) can be constructed for arbitrary arrays T. However, such formulas are useful in science or engineering only to the extent that T describes something that actually transforms as a tensor.

OPERATIONS ON TENSORS

Like scalars, vectors, and matrices, tensors of the same rank can be added together, and a tensor of any rank can be multiplied by a scalar. Each of these operations produces a new tensor whose rank is the same as that of the operand(s), and the operations are element-by-element (just as for vectors and matrices). For example, if the following tensors are all of rank 3,

$$\mathsf{T} = \mathsf{P} - 3\mathsf{Q} \quad \text{implies} \quad T_{ijk} = P_{ijk} - 3Q_{ijk} \quad \text{for all } i,\ j,\ k. \tag{8.4}$$

Direct Product. Another tensor operation is the formation of a product in which each component of one tensor is multiplied by a component of another tensor, with the products having components identified by all the indices of both tensors. Examples of the components of such products are $T_{ij} = P_i Q_j$ and $W_{ijk} = T_{ij} P_k$. An important feature of these **direct products**, sometimes also called **outer products** or **Kronecker products**, is that if their factors are components of tensors, the direct products are tensors too. Illustrating for T_{ij}, given that

$$P'_\alpha = \sum_i U_{\alpha i} P_i \qquad \text{and} \qquad Q'_\beta = \sum_j U_{\beta j} Q_j\,,$$

we have

$$T'_{\alpha\beta} = P'_\alpha Q'_\beta = \sum_{ij} U_{\alpha i} U_{\beta j} P_i Q_j = \sum_{ij} U_{\alpha i} U_{\beta j} T_{ij}\,.$$

The result of the direct-product operation is a tensor whose rank is the sum of the ranks of the factors. It is customary to assign indices to the direct-product tensor in the order in which they occur in the factors; manual or computer operations that form a direct product must label or store the product components in a way that is consistent with this index ordering.

Example 8.1.1. Direct Product

Given two vectors $\mathbf{A} = (1, 3, 2)$ and $\mathbf{B} = (2, 0, -1)$, their direct product C is a rank-2 tensor with elements $C_{ij} = A_i B_j$. In matrix form,

$$\mathsf{C} = \begin{pmatrix} 2 & 0 & -1 \\ 6 & 0 & -3 \\ 4 & 0 & -2 \end{pmatrix}.$$

Let's transform A, B, and C using the orthogonal matrix

$$\mathsf{U} = \begin{pmatrix} 0.8 & 0.6 & 0.0 \\ -0.6 & 0.8 & 0.0 \\ 0.0 & 0.0 & 1.0 \end{pmatrix}.$$

We get

$$A'_i = \sum_j U_{ij} A_j = \begin{pmatrix} 2.6 \\ 1.8 \\ 2.0 \end{pmatrix}, \qquad B'_i = \sum_j U_{ij} B_j = \begin{pmatrix} 1.6 \\ -1.2 \\ -1.0 \end{pmatrix},$$

$$C'_{ij} = \sum_{mn} U_{im} U_{jn} C_{mn} = \begin{pmatrix} 4.16 & -3.12 & -2.60 \\ 2.88 & -2.16 & -1.80 \\ 3.20 & -2.40 & -2.00 \end{pmatrix}.$$

We can verify the tensor nature of C by checking that $C'_{ij} = A'_i B'_j$: we have $C'_{11} = (2.6)(1.6) = 4.16$, $C'_{12} = (1.8)(-1.2) = -3.12$, etc.

■

Contraction. This is an operation that is a generalization of ordinary matrix multiplication, and is carried out by assigning the same symbol to two indices (any two) and summing over the duplicated index. This operation can combine two tensors, as in these examples:

$$W_{ikm} = \sum_j T_{ijk} P_{jm}, \tag{8.5}$$

$$W_{ijm} = \sum_k T_{ijk} P_{mk}, \tag{8.6}$$

or it can be used within a single tensor, as in

$$W_j = \sum_i T_{iji}. \tag{8.7}$$

Again it may be useful to remind the reader that tensor indices are just labels and may be given any convenient names. What is important is the identification of indices that are identical and those that are to be summed. For example, the equation we just wrote is entirely equivalent to

$$W_\alpha = \sum_\mu T_{\mu\alpha\mu}.$$

Cases of contraction that we have encountered earlier in this text include vector dot products and various matrix-matrix and matrix-vector products, e.g.,

$$W = \mathbf{A} \cdot \mathbf{B} = \sum_i A_i B_i \,,$$

$$C_{ik} = \sum_j A_{ij} B_{jk} \,,$$

$$y_i = \sum_j A_{ij} x_j \,.$$

A contraction has the general property that it reduces the overall rank of a tensor expression by 2, as can be seen by looking at the above examples.

At this point one might ask whether a tensor contraction necessarily produces a tensor as output. To see that it does, let's transform T and P in Eq. (8.5) and examine the transformation rule thereby obtained for W. We have

$$\begin{aligned}
W'_{\alpha\beta\gamma} &= \sum_{\mu=1}^{3} T'_{\alpha\mu\beta} P'_{\mu\gamma} = \sum_{\mu=1}^{3} \left(\sum_{i,j,k=1}^{3} U_{\alpha i} U_{\mu j} U_{\beta k} T_{ijk} \right) \left(\sum_{n,m=1}^{3} U_{\mu n} U_{\gamma m} P_{nm} \right) \\
&= \sum_{i,k,m=1}^{3} U_{\alpha i} U_{\beta k} U_{\gamma m} \sum_{j,n=1}^{3} \left(\sum_{\mu=1}^{3} U_{\mu j} U_{\mu n} \right) T_{ijk} P_{nm} \\
&= \sum_{i,k,m=1}^{3} U_{\alpha i} U_{\beta k} U_{\gamma m} \sum_{j,n=1}^{3} \delta_{jn} T_{ijk} P_{nm} \\
&= \sum_{i,k,m=1}^{3} U_{\alpha i} U_{\beta k} U_{\gamma m} \sum_{j=1}^{3} T_{ijk} P_{jm} = \sum_{i,k,m=1}^{3} U_{\alpha i} U_{\beta k} U_{\gamma m} W_{ikm} \,. \qquad (8.8)
\end{aligned}$$

We went from the second to the third line of the above equation sequence by using the fact that the columns of an orthogonal matrix are orthonormal vectors; see Eq. (5.7). The final result is that W transforms as a tensor.

NOTATION

It is convenient to identify tensors by exhibiting a typical element. Thus one often uses a notation such as W_{ijk} to refer to the entire tensor W, making it unnecessary also to state in addition that it is of rank 3. Literally, a statement such as "W_{ijk} is a Cartesian tensor..." makes no sense (since W_{ijk} is only one element), but it is ordinarily clear from the context whether the writing refers to that specific element or to the tensor as a whole. This notational convention becomes particularly useful when we later introduce general tensors, for which it is customary to use both subscripts and superscripts.

Another notational convention that is widely used is to omit the explicit writing of summation signs, of which many tend to occur in tensor equations. Instead, it is assumed that any index that occurs exactly twice in a tensor or tensor product (thereby producing a contraction) is to be summed. In its original form, called the **Einstein summation convention**, it was assumed that summation takes place only when an index occurs (once each) as a subscript and a superscript in a general tensor expression. However, this summation convention is also used (and we use it) for Cartesian tensors when the identical indices are both subscripts. Thus, for example,

$$\begin{aligned}
T_{iji} &\longrightarrow T_{1j1} + T_{2j2} + T_{3j3}\,, \\
a_{\alpha\alpha} &\longrightarrow a_{11} + a_{22} + a_{33}\,, \\
T_{ijk}a_{km} &\longrightarrow T_{ij1}a_{1m} + T_{ij2}a_{2m} + T_{ij3}a_{3m}\,, \\
T_{\beta\beta\alpha}a_{\alpha\gamma} &\longrightarrow (T_{111} + T_{221} + T_{331})a_{1\gamma} + (T_{112} + T_{222} + T_{332})a_{2\gamma} \\
&\qquad + (T_{113} + T_{223} + T_{333})a_{3\gamma}\,.
\end{aligned}$$

It is obvious that if an expression is to involve multiple contractions, a different duplicated index must be used for each.

From now on, we will ordinarily assume use of the summation convention when writing tensor expressions, usually without making comments to warn the reader.

QUOTIENT RULE

We have already established that a product of two Cartesian tensors (either a direct product or one involving contraction) yields as its result a tensor. We now want to turn that observation around to make the observation that if an indexed quantity S satisfies an equation of the type $\mathsf{ST} = \mathsf{V}$, where T and V are tensors of some specific ranks and T is arbitrary, then S must also be a tensor. Note that we have not defined division for tensors, so we cannot just state that S must be a tensor because "it is V/T." A theorem to the effect that S is a tensor (and not just a quantity that happens to have some indices) is known as the **quotient rule**.

One important use of the quotient rule is to establish the tensor nature of a physical quantity. For example, the electric polarization $\mathbf{P}$ of an anisotropic material placed in an electric field $\mathbf{E}$ has components that satisfy the relation $P_i = X_{ij}E_j$ (using the summation convention). An equation of this type holds for all $\mathbf{E}$, and must hold irrespective of the orientation of the coordinate system (because $\mathbf{E}$ and $\mathbf{P}$ are vectors, i.e., rank-1 tensors). Then the quotient rule permits us to conclude that the coefficient array X_{ij} must also be a tensor. Because X has two indices, we identify it as a tensor of rank 2.

Typical situations for which the quotient rule establishes the tensor nature of S are equations such as the following (all written using the summation convention):

$$\begin{aligned}
S_{ij}T_j &= V_i\,, \\
S_iT_i &= V\,, \\
S_iT_{jk} &= V_{ijk}\,, \\
S_{ijk}T_{mk} &= V_{ijm}\,.
\end{aligned}$$

The validity of the quotient rule for $\mathsf{ST} = \mathsf{V}$ depends upon the fact that an equation of that type is assumed to exist for arbitrary T. We indicate now how the rule may be proved by demonstrating its proof for the first case in the above list: $S_{ij}T_j = V_i$.

Let's begin by writing transformation formulas for $\mathbf{T}$. This will also provide practice in using the summation convention. Letting U be a rotation matrix and primes represent quantities after rotation,

$$T'_\alpha = U_{\alpha j} T_j \quad \text{and} \quad T_j = (\mathsf{U}^{-1}\mathsf{T})_j = (\mathsf{U}^{-1})_{j\alpha} T'_\alpha = U_{\alpha j} T'_\alpha \,. \tag{8.9}$$

The second of these equations follows from the first because U is an orthogonal matrix.

Now, start from $S_{ij}T_j = V_i$ and (1) multiply each side of the equation on the left by $U_{\alpha i}$ (note that this implies a summation over i), and (2) replace T_j on the left-hand side by $U_{\beta j}T'_\beta$. We use β for the dummy index in step 2 to avoid a conflict with α, which is already in use. The first of these two steps converts the right-hand side of the equation to V'_α, and we get

$$U_{\alpha i} S_{ij} U_{\beta j} T'_\beta = V'_\alpha \,. \tag{8.10}$$

Also, rotating the original equation (and calling the indices α and β), we have

$$S'_{\alpha\beta} T'_\beta = V'_\alpha \,. \tag{8.11}$$

Subtracting Eq. (8.10) from Eq. (8.11),

$$(S'_{\alpha\beta} - U_{\alpha i} U_{\beta j} S_{ij}) T'_\beta = 0 \,. \tag{8.12}$$

Because we require Eq. (8.12) to be satisfied for all $\mathbf{T}'$, the parenthesized quantity multiplying T'_β must vanish, so

$$S'_{\alpha\beta} = U_{\alpha i} U_{\beta j} S_{ij} \,, \tag{8.13}$$

showing that S has the transformation properties of a tensor.

Incidentally, this method of proof that involves the introduction of an arbitrary quantity is a rather common device in mathematical analysis; other examples following a similar line of reasoning appear elsewhere in this text.

OTHER PROPERTIES

If a tensor remains the same when the order of its indices is changed, it is termed **symmetric**; if all tensor components change sign when any two indices are interchanged, it is called **antisymmetric**. An arbitrary tensor of rank 2 can be decomposed into symmetric and antisymmetric parts:

$$T_{ij} = \frac{1}{2}(T_{ij} + T_{ji}) + \frac{1}{2}(T_{ij} - T_{ji}) \,. \tag{8.14}$$

While it is possible to extract symmetric and completely antisymmetric parts from tensors of rank greater than 2, arbitrary tensors of rank 3 or higher will have parts that are neither fully symmetric nor antisymmetric.

Some tensors are **isotropic**, meaning that they have the same value (i.e., the same components) irrespective of the orientation of the coordinate system. An example of an isotropic tensor of rank 2 is the Kronecker delta δ_{ij}. Perhaps that is obvious, as when written in matrix form, we have

$$\delta_{ij} = \begin{pmatrix} 1 & 0 & 0 \\ 0 & 1 & 0 \\ 0 & 0 & 1 \end{pmatrix} , \tag{8.15}$$

and we expect a unit matrix to be unaffected by similarity transformations. Algebraically, we confirm that δ_{ij} is isotropic by writing

$$\delta'_{\alpha\beta} = U_{\alpha i}U_{\beta j}\delta_{ij} = U_{\alpha i}U_{\beta i} = \begin{cases} 1, & \alpha = \beta\,, \\ 0, & \alpha \neq \beta\,. \end{cases} \tag{8.16}$$

The last step of Eq. (8.16) follows because U is orthogonal.

Levi-Civita symbol. This symbol, originally introduced in this text in connection with determinants, has in 3-D space the definition

$$\varepsilon_{ijk} = \begin{cases} 1, & i,\ j,\ k \text{ an even permutation of } 1,2,3, \\ -1, & i,\ j,\ k \text{ an odd permutation of } 1,2,3, \\ 0, & i,\ j,\ k \text{ not a permutation of } 1,2,3. \end{cases} \tag{8.17}$$

The ε_{ijk} are the components of an isotropic tensor of rank 3. To show this, recall from our discussion of determinants in Chapter 4 that for any matrix U of dimension 3,

$$\det \mathsf{U} = U_{1i}U_{2j}U_{3k}\varepsilon_{ijk}\,, \tag{8.18}$$

and that if we permute the rows of this determinant from $(1, 2, 3)$ to α, β, γ we get

$$\varepsilon_{\alpha\beta\gamma} \det \mathsf{U} = U_{\alpha i}U_{\beta j}U_{\gamma k}\varepsilon_{ijk}\,. \tag{8.19}$$

If we now observe that the matrix U for a rotation is orthogonal, with $\det \mathsf{U} = 1$, we can reduce Eq. (8.19) to

$$\varepsilon_{\alpha\beta\gamma} = U_{\alpha i}U_{\beta j}U_{\gamma k}\varepsilon_{ijk}\,, \tag{8.20}$$

which is a statement that ε_{ijk} transforms into itself and therefore is isotropic.

There is no isotropic tensor of rank 1 (no nonzero vector exists that remains unchanged under rotation). However, by explicit construction we have identified isotropic tensors of ranks 2 and 3, and isotropic tensors of all higher ranks can be formed by direct products (possibly with contractions) of the isotropic tensors we have already found.

Example 8.1.2. Tensor Transformation

Consider the tensor equation $T_{ij} = M_{ijk}A_k$, where A is a vector of components $(3, 1, 0)$ and M is a rank-3 tensor with components (in each expression, k runs from 1 to 3)

$$\begin{array}{lll} M_{11k} = (2,0,1), & M_{12k} = (1,2,3), & M_{13k} = (0,0,-1), \\ M_{21k} = (-1,0,0), & M_{22k} = (0,1,0), & M_{23k} = (2,1,1), \\ M_{31k} = (0,0,2), & M_{32k} = (0,-1,0), & M_{33k} = (1,1,1). \end{array}$$

Inserting the components of M and A and carrying out the summation over the index k, we find

$$\mathsf{T} = \begin{pmatrix} 6 & 5 & 0 \\ -3 & 1 & 7 \\ 0 & -1 & 4 \end{pmatrix}.$$

To confirm the tensor transformation properties, let's introduce the orthogonal transformation

$$\mathsf{U} = \begin{pmatrix} 1 & 0 & 0 \\ 0 & 0.8 & -0.6 \\ 0 & 0.6 & 0.8 \end{pmatrix}$$

and form the transformed quantities T', M', A'.

With some labor, we carry out the transformations as specified in Eq. (8.3), finding

$$\mathsf{A}' = (0.3,\ 0.8,\ 0.6), \qquad \mathsf{T}' = \begin{pmatrix} 6.0 & 4.0 & 3.0 \\ -2.4 & -0.8 & 3.4 \\ -1.8 & -4.6 & 5.8 \end{pmatrix},$$

and

$$\begin{aligned}
&M'_{11k} = (2.0, -0.6, 0.8), \quad M'_{12k} = (0.80, -0.52, 3.36), \quad M'_{13k} = (0.60, 0.36, 1.52),\\
&M'_{21k} = (-0.80, 0.72, -0.96), \qquad M'_{22k} = (-0.60, 0.872, 0.504),\\
&M'_{23k} = (0.800, 0.704, 0.728), \qquad M'_{31k} = (-0.60, -0.96, 1.28),\\
&M'_{32k} = (-1.200, -0.296, -1.272), \quad M'_{33k} = (1.600, 0.128, 1.496).
\end{aligned}$$

The reader will be more easily able to confirm these results using symbolic computing after it is introduced later in this chapter.

We can now check that our tensor equation transforms properly by comparing T'_{ij} with $M'_{ijk}A'_k$. That task is left to the reader.

■

Example 8.1.3. Constructing a Tensor

Sometimes a physical problem provides the information needed to construct a tensor.

The angular momentum $\mathbf{L}$ of a particle of mass m at a point $\mathbf{r}$ moving at angular velocity $\boldsymbol{\omega}$, where $\mathbf{L}$, $\mathbf{r}$, and $\boldsymbol{\omega}$ are all relative to the same origin, is given by the formula

$$\mathbf{L} = m[\mathbf{r} \times (\boldsymbol{\omega} \times \mathbf{r})]. \tag{8.21}$$

From this equation we see that the components of $\mathbf{L}$ will be linear combinations of the components of $\boldsymbol{\omega}$ and will therefore satisfy an equation of the form

$$L_i = I_{ij}\,\omega_j\,. \tag{8.22}$$

Since $\mathbf{L}$ and $\boldsymbol{\omega}$ are vectors, $\mathbf{I}$ must by the quotient rule be a tensor. It is called the **inertia tensor**.

To find $\mathbf{I}$, we write out a typical component of $\mathbf{L}$ as given by Eq. (8.21), finding

$$\begin{aligned}
L_x &= m[\mathbf{r}\cdot\mathbf{r}\,\omega_x - (\mathbf{r}\cdot\boldsymbol{\omega})\,x]\\
&= m\big[(x^2+y^2+z^2)\omega_x - (x^2\,\omega_x + xy\,\omega_y + xz\,\omega_z)\big]\\
&= m\big[(y^2+z^2)\,\omega_x - xy\,\omega_y - xz\,\omega_z\big]\,.
\end{aligned}$$

Comparing with Eq. (8.22), we read out

$$I_{xx} = m(y^2+z^2), \qquad I_{xy} = -mxy, \qquad I_{xz} = -mxz\,. \tag{8.23}$$

By symmetry, the complete inertia tensor therefore must have the form

$$\mathbf{I} = \begin{pmatrix} m(y^2+z^2) & -mxy & -mxz \\ -mxy & m(x^2+z^2) & -myz \\ -mxz & -myz & m(x^2+y^2) \end{pmatrix}. \tag{8.24}$$

■

Exercises

These exercises are expressed using the Einstein summation convention.

8.1.1. Confirm the details of the work in Example 8.1.1 by doing the following:

(a) Verify that with $\mathbf{A}$, $\mathbf{B}$, and C as given in the Example, C is the direct product of $\mathbf{A}$ and $\mathbf{B}$; this relationship is often written $\mathsf{C} = \mathbf{A} \otimes \mathbf{B}$.

(b) Verify that, using the given U, the transformed tensors $\mathbf{A}'$, $\mathbf{B}'$, and C' are as given in the Example.

(c) Verify that $\mathsf{C}' = \mathbf{A}' \otimes \mathbf{B}'$.

8.1.2. Confirm the quotient rule for the tensor equation $S_i T_{jk} = V_{ijk}$, by showing that if T and V are tensors, then $\mathbf{S}$ must be a tensor of rank 1 (i.e., a vector).

8.1.3. The inertia tensor of a collection of particles is given by a generalization of Eq. (8.24) in which each element of $\mathbf{I}$ is replaced by a sum over all particles, e.g.,

$$-mxy \longrightarrow -\sum_i m_i x_i y_i \quad \text{or} \quad m(y^2+z^2) \longrightarrow \sum_i m_i(y_i^2+z_i^2).$$

For a continuous mass distribution (i.e., a rigid body), the elements of $\mathbf{I}$ become integrals. For a mass distribution $\rho(x,y,z)$,

$$-mxy \longrightarrow -\int \rho(x,y,z)xy\,d\tau, \quad \text{etc.}$$

Find the inertia tensors of the following objects, assumed centered at the origin of a Cartesian coordinate system and of uniform mass density:

(a) A spherical ball of total mass M and radius a.

(b) A thin circular disk of total mass M and radius a located on the xy-plane (assume its thickness to be negligible).

(c) A solid cube of side a.

8.1.4. Four unit masses are located (in the xy-plane) at the points $(2,2)$, $(-2,-2)$, $(1,-1)$, and $(-1,1)$.

(a) Construct the inertia tensor $\mathbf{I}$ of this system (relative to the origin).

(b) Explain how the form of $\mathbf{I}$ shows that for rotation about the z-axis the angular momentum will also be about the z-axis, but that the angular velocity and the angular momentum will not be collinear for a rotation about either the x-axis or the y-axis. (This will cause a rotation initially about the z axis to be stable, while a rotation initially about the x or y axis will not remain in that orientation.)

(c) The condition that $\mathbf{L}$ and $\boldsymbol{\omega}$ be collinear is given by the matrix equation $\mathbf{I}\boldsymbol{\omega} = \lambda\boldsymbol{\omega}$, where λ is a constant. Solve this equation (which is an eigenvalue problem) and find the directions in which $\mathbf{L}$ and $\boldsymbol{\omega}$ are collinear and the form of $\mathbf{I}$ when using a coordinate system aligned in those directions. These directions are known as the **principal axes of inertia**, and the eigenvalues are called the **principal moments of inertia**.

(d) Relate the principal axes of inertia to the symmetry of the four-mass system and explain qualitatively the relative magnitudes of the principal moments of inertia.

8.1.5. Given $\mathbf{A} = \begin{pmatrix} 1 \\ -3 \\ 2 \end{pmatrix}$, $\quad \mathsf{B} = \begin{pmatrix} 2 & 1 & -1 \\ 0 & -2 & 2 \\ 1 & 2 & 2 \end{pmatrix}$, $\quad \mathbf{F} = \begin{pmatrix} -4 \\ 2 \\ 1 \end{pmatrix}$,

compute C, where $C_{ik} = A_i B_{jk} F_j$.

8.1.6. Simplify:

(a) $\delta_{ij}\varepsilon_{ijk}$, (b) $\varepsilon_{irs}\varepsilon_{jrs}$, (c) $\varepsilon_{ijk}\varepsilon_{ijk}$.

8.1.7. Verify that $\varepsilon_{ijk}\varepsilon_{pqk} = \delta_{ip}\,\delta_{jq} - \delta_{iq}\,\delta_{jp}$.

8.1.8. (a) Verify that $\mathbf{A} \times \mathbf{B} = \varepsilon_{ijk} A_j B_k \,\hat{\mathbf{e}}_i$.

(b) Using the result of part (a), show that

$$\mathbf{A} \times (\mathbf{B} \times \mathbf{C}) = (\mathbf{A} \cdot \mathbf{C})\,\mathbf{B} - (\mathbf{A} \cdot \mathbf{B})\,\mathbf{B}.$$

8.1.9. Using the Levi-Civita symbol, show that

$$\boldsymbol{\nabla} \times (\boldsymbol{\nabla} \times \mathbf{A}) = \boldsymbol{\nabla}(\boldsymbol{\nabla} \cdot \mathbf{A}) - \boldsymbol{\nabla}^2 \mathbf{A}.$$

8.2 PSEUDOTENSORS AND DUAL TENSORS

When we originally introduced orthogonal transformations in Chapter 5, we noted that matrices U describing rotations had determinant $\det \mathsf{U} = 1$. In addition, we observed that reflections could also be described using orthogonal matrices, but that reflection transformations were characterized by matrices with $\det \mathsf{U} = -1$. We now proceed to a more detailed discussion of the properties of reflection transformations, with emphasis on their behavior in three-dimensional space.

PSEUDOTENSORS

When we perform a transformation on the coordinate system, the components we use for the description of a fixed vector will change. If the coordinate system is changed by a reflection through the xy-plane (to which the normal is in the z-direction), the z-component of any vector will change in sign. This behavior is part of what is included when we identify a quantity as a vector. Suppose now we consider the cross product of two vectors $\mathbf{A}$ and $\mathbf{B}$, for simplicity assuming both vectors to be in the xy-plane.

Then their cross product $\mathbf{C} = \mathbf{A} \times \mathbf{B}$ will be in the z-direction, with z component $C_z = A_x B_y - A_y B_x$. The reflection does not change the x and y components of $\mathbf{A}$ and $\mathbf{B}$, so this equation indicates that the sign of C_z should not change. That is not consistent with the application of a vector transformation rule to $\mathbf{C}$.

The resolution to the dilemma posed by the preceding paragraph involves a recognition that $\mathbf{C}$ is a different type of quantity than $\mathbf{A}$ and $\mathbf{B}$, and that in order to keep the physics invariant $\mathbf{C}$ must transform differently than $\mathbf{A}$ and $\mathbf{B}$ under reflection transformations. This distinction has been known for a long time; quantities like $\mathbf{A}$ and $\mathbf{B}$ have been called **polar vectors**, while those like $\mathbf{C}$ were called **axial vectors**. The name "axial" refers to the fact that vectors that transform like $\mathbf{C}$ also arise when used to identify an axis of rotation. If a rotational motion (and the axis defining it) are viewed through a mirror whose reflection plane is parallel to the rotational axis, the handedness of the trajectory will be reversed, but the axial direction (the axial vector) will remain the same.

Because axial vectors do not satisfy the vector transformation equations for reflections, it has now become customary to call them **pseudovectors**, reserving the name **vector** for quantities that satisfy orthogonal transformation equations for both rotations and reflections of the coordinates. Since the failure of pseudovectors to be described by orthogonal transformations under reflection is associated with a change in the handedness of the coordinate system, a convenient way to detect a pseudovector is to note that a right-hand rule is involved in its specification.

It is useful to have a formal rule for carrying out transformations on pseudovectors. Since the transformation is for rotations the same as for vectors while for reflections there is a sign reversal, a suitable transformation formula is

$$C'_\alpha = (\det \mathsf{U})\, U_{\alpha i} C_i\,, \qquad (\mathbf{C} \text{ a pseudovector}). \tag{8.25}$$

If we now extend this transformation rule to direct products or contractions involving pseudovectors, we see that its generalization to quantities of higher rank will be

$$T'_{\alpha\beta\cdots} = (\det \mathsf{U})\, U_{\alpha i} U_{\beta j} \cdots T_{ij\cdots}\,. \tag{8.26}$$

Analogous to the terminology for pseudovectors, we call quantities that transform according to Eq. (8.26) **pseudotensors**. A pseudotensor of rank 0 is referred to as a **pseudoscalar**. It will be a single-component quantity that changes sign under reflection transformations of the coordinate system.

Applying Eq. (8.26) to the factors in a direct or contracted product, we see that we will get one factor $\det \mathsf{U}$ for each pseudo quantity in the product. Since $\det \mathsf{U} = \pm 1$, an even number of such factors reduces to unity, while an odd number leads to the result $\det \mathsf{U}$. We conclude that pseudo status can be determined by counting the number of pseudo factors in a tensor product.

We have already encountered several pseudotensors, though we did not previously identify them as such.

Example 8.2.1. Levi-Civita Symbol

The Levi-Civita symbol is a pseudotensor. Revisiting Eq. (8.19) and moving $\det \mathsf{U}$ to the right-hand side (permitted because it is ± 1), we have

$$\varepsilon_{\alpha\beta\gamma} = (\det \mathsf{U}) U_{\alpha i} U_{\beta j} U_{\gamma k}\, \varepsilon_{ijk}\,, \tag{8.27}$$

the transformation formula for a rank-3 pseudotensor.

■

Example 8.2.2. Expressions Containing Cross Products

The pseudo status of expressions containing cross products can be analyzed using the fact that they can be written using the Levi-Civita symbol. We return to Eq. (7.6), written here using the summation convention:

$$\hat{\mathbf{e}}_i \times \hat{\mathbf{e}}_j = \varepsilon_{ijk}\hat{\mathbf{e}}_k \,. \tag{8.28}$$

Let's now take $\mathbf{A} = A_i\hat{\mathbf{e}}_i$ and $\mathbf{B} = B_j\hat{\mathbf{e}}_j$, and form

$$\mathbf{G} = G_k\hat{\mathbf{e}}_k = \mathbf{A} \times \mathbf{B} = A_iB_j \; \hat{\mathbf{e}}_i \times \hat{\mathbf{e}}_j = A_iB_j \, \varepsilon_{ijk} \; \hat{\mathbf{e}}_k \,, \tag{8.29}$$

which translates into the component formula

$$G_k = A_iB_j \; \varepsilon_{ijk} \,. \tag{8.30}$$

Now, assuming that $\mathbf{A}$ and $\mathbf{B}$ are vectors, we have verified that $\mathbf{G}$ is a pseudovector by virtue of the fact that its representation includes one pseudotensor.

The vector triple product $\mathbf{A} \times (\mathbf{B} \times \mathbf{C})$ has components that can be written as a tensor contraction:

$$[\mathbf{A} \times (\mathbf{B} \times \mathbf{C})]_k = \varepsilon_{kij}A_i(\mathbf{B} \times \mathbf{C})_j = \varepsilon_{kij}A_i\varepsilon_{jmn}B_mC_n \,.$$

From the presence of the two pseudotensors ε_{kij} and ε_{jmn} we conclude that the vector triple product is a vector, not a pseudovector.

Finally, form a scalar triple product by taking the dot product of $\mathbf{G}$ as given in Eq. (8.29) with a vector $\mathbf{C}$. We get

$$\mathbf{C} \cdot (\mathbf{A} \times \mathbf{B}) = C_kA_iB_j \; \varepsilon_{ijk} \,. \tag{8.31}$$

Equation (8.31) could alternatively have been obtained as the determinant formula for the scalar triple product. The scalar triple product is a pseudoscalar, containing one pseudo factor, namely the Levi-Civita tensor. We now better understand why the result ($\pm$ the volume enclosed by the parallelepiped defined by $\mathbf{A}$, $\mathbf{B}$, and $\mathbf{C}$) can have either sign, depending on the order of the factors.

■

DUAL TENSORS

Let's look now at the most general antisymmetric tensor T_{ij} of rank 2 in 3-D space. Invoking the antisymmetry, we see that it can be written

$$\mathsf{T} = \begin{pmatrix} 0 & T_{12} & -T_{31} \\ -T_{12} & 0 & T_{23} \\ T_{31} & -T_{23} & 0 \end{pmatrix}, \tag{8.32}$$

showing that T has only three independent components. If we form a contraction between T and the Levi-Civita symbol ε_{ijk}, we get

$$V_i = \varepsilon_{ijk}T_{jk}. \tag{8.33}$$

Evaluating Eq. (8.33), we get

$$V_1 = 2T_{23}, \quad V_2 = 2T_{31}, \quad V_3 = 2T_{12} \,. \tag{8.34}$$

Our conclusion from Eqs. (8.33) and (8.34) is that, invoking the quotient rule, $\mathbf{V}$ is a vector, or more precisely, a pseudovector, with components that are (except for a factor 2) the same as the components of our original antisymmetric tensor.

We express this correspondence between a pseudovector and an antisymmetric tensor by identifying each as a **dual** of the other. These quantities can be viewed as alternative representations of the same mathematical object. An example of dual tensors is provided by the vector cross product, which we have already identified as a pseudovector. If we form the antisymmetric tensor $T_{ij} = (A_i B_j - A_j B_i)/2$, then Eq. (8.33) is equivalent to the vector equation $\mathbf{V} = \mathbf{A} \times \mathbf{B}$.

The reader should take note that the specific duality we have just described is unique to three-dimensional space; in four dimensions (appropriate for relativity) an antisymmetric rank-2 tensor has six independent components and cannot be expected to provide an alternate representation of a four-vector.

Exercises

8.2.1. Show that the two-index Levi-Civita symbol ε_{ij} is (in 2-D space) a pseudotensor.

8.2.2. Identify each quantity in the following equation as a tensor or a pseudotensor: $\mathbf{v} = \boldsymbol{\omega} \times \mathbf{r}$.

8.2.3. Determine whether $\varepsilon_{ijk}\varepsilon_{rst}$ is a tensor or a pseudotensor.

8.2.4. Let $\mathbf{C} = (C_1, C_2, C_3)$ be an arbitrary pseudovector. Then form the array

$$\mathsf{G} = \begin{pmatrix} G_{11} & G_{12} & G_{13} \\ G_{21} & G_{22} & G_{23} \\ G_{31} & G_{32} & G_{33} \end{pmatrix} = \begin{pmatrix} 1 & 2C_3 & 0 \\ 0 & 2 & 2C_1 \\ 2C_2 & 0 & 3 \end{pmatrix}.$$

(a) Verify that the equation

$$C_i = \frac{1}{2}\,\varepsilon_{ijk} G_{jk}$$

is satisfied and that G is a tensor. This shows that G is dual to $\mathbf{C}$.

(b) Show that G can be decomposed into symmetric and antisymmetric parts, and that the antisymmetric part of G corresponds to the duality illustrated by Eqs. (8.32) through (8.34).

8.3 NONCARTESIAN TENSORS

When a Cartesian coordinate system is subjected to an orthogonal transformation (a rotation or reflection), the old and new coordinates are related by a linear transformation, causing the coefficients in the tensor transformation formulas to be constants. However, there are circumstances in which we need to use transformations that are not linear; two such situations of relevance in physics are the use of curvilinear coordinates and the introduction of nonEuclidean coordinates in general relativity. The nonlinearity prevents us from carrying out some of the operations that were discussed earlier in this chapter, and an alternate method of approach is needed.

METRIC TENSOR

In previous chapters we have introduced unit vectors in curvilinear coordinate systems. For tensor analysis it is instead sometimes useful to employ basis vectors $\mathbf{a}_i$ that are not required to be of unit length. Accordingly, we define

$$\mathbf{a}_i = \frac{\partial x}{\partial q_i}\,\hat{\mathbf{e}}_x + \frac{\partial y}{\partial q_i}\,\hat{\mathbf{e}}_y + \frac{\partial z}{\partial q_i}\,\hat{\mathbf{e}}_z\,. \tag{8.35}$$

These basis vectors $\mathbf{a}_i$ are associated with the changes in the Cartesian coordinates per unit change in the curvilinear coordinates, which we designate q_i. In terms of the $\mathbf{a}_i$, a differential displacement $d\mathbf{s}$ takes the form

$$d\mathbf{s} = \mathbf{a}_i\,dq_i\,. \tag{8.36}$$

Note that here and in the formulas that follow we are using the summation convention. From Eq. (8.36) we get

$$d\mathbf{s}^2 = (\mathbf{a}_i\,dq_i)\cdot(\mathbf{a}_j\,dq_j) = dq_i\,g_{ij}\,dq_j\,, \tag{8.37}$$

with

$$g_{ij} = \mathbf{a}_i\cdot\mathbf{a}_j\,. \tag{8.38}$$

The quantities g_{ij} form a tensor of rank 2, called the **metric tensor**. From Eq. (8.38) we can see that the metric tensor is symmetric and that if the curvilinear coordinates are orthogonal, its matrix representation will be diagonal. For orthogonal coordinates, the diagonal elements are the squares of the scale factors h_i introduced in Section 6.3: $g_{ii} = h_i^2$. In the special case of Cartesian coordinates, the $\mathbf{a}_i$ will be unit vectors, making the g_{ij} a unit matrix.

Another way to form basis vectors that are related to changes in the curvilinear coordinates is to define them to be the gradients of the individual q_i: accordingly, we write $\mathbf{b}_i = \boldsymbol{\nabla} q_i$. This definition corresponds to the equation

$$\mathbf{b}_i = \boldsymbol{\nabla} q_i = \frac{\partial q_i}{\partial x}\,\hat{\mathbf{e}}_x + \frac{\partial q_i}{\partial y}\,\hat{\mathbf{e}}_y + \frac{\partial q_i}{\partial z}\,\hat{\mathbf{e}}_z\,, \tag{8.39}$$

which is not the same as Eq. (8.35). However, in the special case that the q_i are Cartesian coordinates (at any orientation), the derivatives in these two formulas become equal, as in that case

$$\frac{\partial x}{\partial q_i} = \frac{\partial q_i}{\partial x} = \cos\alpha_i\,, \tag{8.40}$$

where α_i is the angle between the x- and q_i-directions. Similar remarks apply to derivatives involving the coordinates y and z. Although these derivatives seem (superficially) to be mutually reciprocal, remember that they are to be computed with different other variables held constant, explaining why they can nevertheless be equal.

The two bases $\mathbf{a}_i$ and $\mathbf{b}_i$ are not completely independent; they are related by the formula

$$\mathbf{a}_i = g_{ij}\,\mathbf{b}_j\,. \tag{8.41}$$

Equation (8.41) can be verified by inserting the explicit forms of all the quantities and simplifying by use of the chain rule.

In contrast to the situation for Cartesian coordinates, the basis vectors $\mathbf{a}_i$ or $\mathbf{b}_i$ have directions that are functions of the coordinates. For example, the vector $\mathbf{a}_r$

in spherical coordinates (a displacement associated with a change in r) will have a direction that depends on the values of the angles θ and φ. In general coordinate systems, displacements may not be linear functions of the coordinates. For these reasons the basis vectors are determined by differentials dq_i at their point of action rather than from the values of the q_i themselves.

Example 8.3.1. A Metric Tensor

Consider a spherical polar coordinate system, for which $(q_1, q_2, q_3) = (r, \theta, \varphi)$. Starting from the defining equations,

$$x = r\sin\theta\cos\varphi\,, \qquad y = r\sin\theta\sin\varphi\,, \qquad z = r\cos\theta\,,$$

and using Eq. (8.35), we find the basis vectors

$$\begin{aligned}
\mathbf{a}_1 &= \sin\theta\cos\varphi\,\hat{\mathbf{e}}_x + \sin\theta\sin\varphi\,\hat{\mathbf{e}}_y + \cos\theta\,\hat{\mathbf{e}}_z\,,\\
\mathbf{a}_2 &= r\cos\theta\cos\varphi\,\hat{\mathbf{e}}_x + r\cos\theta\sin\varphi\,\hat{\mathbf{e}}_y - r\sin\theta\,\hat{\mathbf{e}}_z\,,\\
\mathbf{a}_3 &= -r\sin\theta\sin\varphi\,\hat{\mathbf{e}}_x + r\sin\theta\cos\varphi\,\hat{\mathbf{e}}_y\,.
\end{aligned} \tag{8.42}$$

We now form $g_{ij} = \mathbf{a}_i \cdot \mathbf{a}_j$. Because the $\mathbf{a}_i$ are orthogonal, individual elements g_{ij} will vanish unless $i = j$. Evaluating the g_{ii} and simplifying by applying trigonometric identities, we get

$$g_{ij} = \begin{pmatrix} 1 & 0 & 0 \\ 0 & r^2 & 0 \\ 0 & 0 & r^2\sin^2\theta \end{pmatrix}. \tag{8.43}$$

The diagonal elements of this matrix are the squares of the scale factors h_i. In agreement with our earlier findings, we have $h_1 = h_r = 1$, $h_2 = h_\theta = r$, $h_3 = h_\varphi = r\sin\theta$.

■

NOTATION

Before we go further it is advisable to introduce some definitions and notation that will clarify and simplify subsequent discussion. Let's start by observing that the basis vectors $\mathbf{a}_i$ that we introduced are customarily referred to as **covariant** vectors, and (for reasons that will become more apparent shortly) the vectors $\mathbf{b}_i$ are called **contravariant**. In general (nonCartesian) tensor analysis, a contravariant vector is usually written with its index as a superscript (**upper index**), while covariant vectors are written with a subscript (i.e., a **lower index**). It is important for the reader to remember that these superscripts are not exponents; it is usually obvious from the context whether a superscript is an index or an exponent.

When we examine contractions in general coordinates we will find that they are defined as summations in which one of the duplicated indices is upper and the other lower. We therefore modify the summation convention so that it assumes contraction when an upper and a lower index in an expression are assigned the same symbol. In that connection we consider an upper index in a denominator as a lower index and vice versa, while coordinate indices are assumed to be contravariant (upper). Thus, for example, we would now rewrite $\partial x/\partial q_i$ as $\partial x/\partial q^i$ and identify i as a lower index for contraction purposes.

Finally, to agree with the bulk of the literature, instead of using the symbols $\mathbf{b}_i$ for the contravariant basis vectors we will write them $\mathbf{a}^i$, thereby conveying also that they are related to the $\mathbf{a}_i$ by a transformation involving the metric tensor.

CONTRAVARIANT AND COVARIANT VECTORS

With our new notational conventions, we can now write a vector V in terms of either of the bases we have introduced:

$$\mathsf{V} = V^i\,\mathbf{a}_i \qquad \text{or} \qquad \mathsf{V} = V_i\,\mathbf{a}^i\,. \tag{8.44}$$

Equation (8.44) not only implies (via the summation convention) that each representation of V is a sum over the basis, but also indicates that the V^i can be identified as the **contravariant** components of V, with the V_i called its **covariant** components. Frequently used are the less precise (but clearly understandable) statements that V^i is a contravariant vector, V_i is a covariant vector, and that these two vectors are **associated**, meaning that they actually represent the same thing.

From the requirement that both forms of Eq. (8.44) represent the same V, we can relate V^i and V_i. Writing

$$V^i\mathbf{a}_i = V^i\,g_{ij}\,\mathbf{a}^j = \left(g_{ji}V^i\right)\mathbf{a}^j\,,$$

we readily see that

$$V_i = g_{ij}V^j\,. \tag{8.45}$$

Central to the definition of a vector are its transformational properties. Given a set of basis vectors $\mathbf{a}_i$ in the q_i coordinate system, this basis in a q_i' coordinate system becomes

$$\mathbf{a}_i' = \left(\frac{\partial q^j}{\partial (q')^i}\right)\mathbf{a}_j\,, \tag{8.46}$$

a result that can be verified by inserting the explicit form for $\mathbf{a}_j$ and invoking the chain rule. If we also write

$$(V')^i = \left(\frac{\partial (q')^i}{\partial q^k}\right)V^k, \tag{8.47}$$

we find

$$(V')^i\,\mathbf{a}_i' = \left(\frac{\partial q^j}{\partial (q')^i}\right)\left(\frac{\partial (q')^i}{\partial q^k}\right)V^k\,\mathbf{a}_j = V^k\,\mathbf{a}_k\,. \tag{8.48}$$

Here the sum over i in the central member of this equation is an instance of the chain rule that evaluates to $\partial q^j/\partial q^k$ and is therefore unity when $j = k$ and zero otherwise.

Our conclusion is that V will be invariant with respect to a coordinate transformation if the V^i transform according to Eq. (8.47). Summarizing, we identify Eq. (8.47) as the transformation rule for a general **contravariant vector**.

A similar analysis of the form $V_i\,\mathbf{a}^i$ leads to the conclusion that a **covariant vector** transforms according to

$$V_i' = \left(\frac{\partial q^k}{\partial (q')^i}\right)V_k\,. \tag{8.49}$$

We now have more insight relative to the meanings of the terms **covariant** and **contravariant**; they describe complementary variations under coordinate transformation such that the contraction of a covariant and a contravariant quantity remains unchanged: when one of these quantities changes in some way, the other quantity undergoes a compensating change. In the present case one of the quantities is the set of components of a vector, while the other quantity is the basis in which the vector is described.

CONTRAVARIANT AND COVARIANT TENSORS

With covariant and contravariant vectors defined, we are now ready to extend our analysis to tensors of arbitrary rank. We do so by generalizing the Cartesian-tensor transformation rule, Eq. (8.3). We need to replace the matrix elements U_{ij} in that equation by partial derivatives of the kinds occurring in Eqs. (8.47) or (8.49). Since this gives us two choices for each transformation coefficient, we can define tensors in which all the indices transform covariantly (a **covariant tensor**), or in which all indices transform contravariantly (a **contravariant tensor**), or in which some indices transform covariantly while others transform contravariantly (a **mixed tensor**). Here are some examples:

$$(T')^{\alpha\beta} = \left(\frac{\partial(q')^\alpha}{\partial q^i}\right)\left(\frac{\partial(q')^\beta}{\partial q^j}\right) T^{ij} ,$$

$$T'_{\alpha\beta\gamma} = \left(\frac{\partial q^i}{\partial(q')^\alpha}\right)\left(\frac{\partial q^j}{\partial(q')^\beta}\right)\left(\frac{\partial q^k}{\partial(q')^\gamma}\right) T_{ijk} ,$$

$$(T')^\alpha_\beta = \left(\frac{\partial(q')^\alpha}{\partial q^i}\right)\left(\frac{\partial q^j}{\partial(q')^\beta}\right) T^i_j .$$

Note that in the above formulas, the scope of each partial differentiation is limited to within its enclosing parentheses. We also observe that the rules for index placement cause there to be consistency between the locations of the free indices on the two sides of these equations and that the dummy indices occur in a way that implies summations.

Example 8.3.2. Kronecker Delta

The Kronecker delta is a mixed rank-2 tensor. To verify both this statement and our expectation that this tensor is isotropic, let's look at its transformation formula:

$$(\delta')^\alpha_\beta = \left(\frac{\partial(q')^\alpha}{\partial q^i}\right)\left(\frac{\partial q^j}{\partial(q')^\beta}\right)\delta^i_j = \left(\frac{\partial(q')^\alpha}{\partial q^i}\right)\left(\frac{\partial q^i}{\partial(q')^\beta}\right) = \frac{\partial(q')^\alpha}{\partial(q')^\beta} = \delta^\alpha_\beta .$$

We get the expected transformation formula. We would fail if we attempted to carry out this analysis assuming the Kronecker delta to be a tensor that was either completely covariant or completely contravariant.

■

OTHER OPERATIONS

Contraction. For Cartesian tensors we used the fact that the transformation coefficients were elements of orthogonal matrices to show that the result of a contraction was a tensor expression whose rank had been decreased by 2. For our present more general tensors we can still prove that the result of a contraction is a tensor, but the key to the proof is the use of the chain rule with one covariant and one contravariant factor. To see this, examine

$$(T')^{\cdots m \cdots}_{\cdots n \cdots} = \cdots \left(\frac{\partial(q')^m}{\partial q^i}\right)\cdots\left(\frac{\partial q^j}{\partial(q')^n}\right)\cdots T^{\cdots i \cdots}_{\cdots j \cdots} \tag{8.50}$$

with n set to m and therefore summed. The quantities directly participating in the summation are

$$\left(\frac{\partial q^j}{\partial(q')^m}\right)\left(\frac{\partial(q')^m}{\partial q^i}\right) = \delta^j_i\,, \tag{8.51}$$

so the summed index drops from the right-hand side of Eq. (8.50), leaving a standard tensor transformation with two less indices than before. It is important to note that the simplification following from the use of Eq. (8.51) would not have occurred if both indices had been either covariant or contravariant. Thus the upper/lower index duplication is an essential feature of the specification of a contraction.

Quotient rule. We do not prove it here, but an argument along the lines used for establishing the quotient rule for Cartesian tensors can be extended to general curvilinear tensors, so this rule can be viewed as having general validity.

SUMMARY

We summarize and to some extent here extend without proof the key formulas of the present section.

1. Tensors in curvilinear coordinates q^i are conveniently described using either a **covariant** basis $\mathbf{a}_i$ or a **contravariant** basis $\mathbf{a}^i$:

$$\mathbf{a}_i = \frac{\partial x}{\partial q^i}\,\hat{\mathbf{e}}_x + \frac{\partial y}{\partial q^i}\,\hat{\mathbf{e}}_y + \frac{\partial z}{\partial q^i}\,\hat{\mathbf{e}}_z\,, \tag{8.52}$$

$$\mathbf{a}^i = \frac{\partial q^i}{\partial x}\,\hat{\mathbf{e}}_x + \frac{\partial q^i}{\partial y}\,\hat{\mathbf{e}}_y + \frac{\partial q^i}{\partial z}\,\hat{\mathbf{e}}_z\,. \tag{8.53}$$

   These bases are not necessarily unit vectors or orthogonal; their relationship and scaling is indicated by the **metric tensor**, which (as a whole) we denote (g_{ij}), with elements

$$g_{ij} = \mathbf{a}_i \cdot \mathbf{a}_j\,. \tag{8.54}$$

   In general, covariant quantities have subscript indices; contravariant quantities have superscript indices.

2. The tensor (g_{ij}) is symmetric; if the coordinates form an orthogonal system it is also diagonal, with diagonal elements $g_{ii} = h_i^2$, where h_i is the scale factor for q^i.

   The inverse of (g_{ij}) is denoted (g^{ij}). It can be shown that

$$\mathbf{a}_i = g_{ij}\,\mathbf{a}^j\,, \quad \text{so} \quad \mathbf{a}^i = \sum_j (g^{-1})_{ij}\,\mathbf{a}_j = g^{ij}\,\mathbf{a}_j\,, \quad \text{with} \quad g_{ik}\,g^{kj} = \delta^j_i\,. \tag{8.55}$$

   The placement of indices for (g_{ij}) and (g^{ij}) is consistent with the upper-index/lower-index requirement of the summation convention, the use of which is assumed throughout this summary.

3. The transformation rules for the bases $\mathbf{a}_i$ and $\mathbf{a}^i$ are respectively

$$\mathbf{a}'_i = \left(\frac{\partial q^j}{\partial(q')^i}\right)\mathbf{a}_j \quad \text{and} \quad (\mathbf{a}')^i = \left(\frac{\partial(q')^i}{\partial q^j}\right)\mathbf{a}^j\,. \tag{8.56}$$

4. The same tensor T can be represented either by its contravariant or its covariant components:
$$\mathsf{T} = T^{ij\cdots}\mathbf{a}_i\,\mathbf{a}_j\cdots = T_{ij\ldots}\mathbf{a}^i\,\mathbf{a}^j\cdots, \tag{8.57}$$
or by a mixed component set, such as
$$\mathsf{T} = T^{i\cdots}_{j\ldots}\mathbf{a}_i\,\mathbf{a}^j\cdots. \tag{8.58}$$
These component sets are often referred to as contravariant, covariant, or mixed tensors, and the various versions of the same tensor are called **associated**. By virtue of the relationship between their bases, we have (for each index that is lowered or raised)
$$\begin{aligned} T^{\cdots}_{\alpha\cdots} &= g_{\alpha i}T^{i\cdots}_{\cdots}\,, \\ T^{\alpha\cdots}_{\cdots} &= g^{\alpha i}T^{\cdots}_{i\cdots}\,. \end{aligned} \tag{8.59}$$

5. By virtue of the foregoing definitions and properties, contravariant, covariant, and mixed tensors have the following transformation rules:
$$\begin{aligned} (T')^{\alpha\beta\cdots} &= \left(\frac{\partial(q')^\alpha}{\partial q^i}\right)\left(\frac{\partial(q')^\beta}{\partial q^j}\right)\cdots T^{ij\cdots}\,, \\ T'_{\alpha\beta\cdots} &= \left(\frac{\partial q^i}{\partial(q')^\alpha}\right)\left(\frac{\partial q^j}{\partial(q')^\beta}\right)\cdots T_{ij\cdots}\,, \\ (T')^{\alpha\cdots}_{\beta\ldots} &= \left(\frac{\partial(q')^\alpha}{\partial q^i}\right)\left(\frac{\partial q^j}{\partial(q')^\beta}\right)\cdots T^{i\cdots}_{j\ldots}\,. \end{aligned} \tag{8.60}$$

6.
   - Direct products of general tensors are tensors of a rank equal to the sum of the ranks of the factors.
   - The result of a contraction is a tensor whose rank is 2 less than that of the tensor expression before contraction.
   - Because the coefficients in the tensor transformation formulas are not necessarily constants, derivatives of tensors with respect to the spatial coordinates may not be tensors. Methods for dealing with this complication are outside the scope of this text; they are covered in several of the Additional Readings.

Example 8.3.3. Cylindrical Coordinates

The cylindrical coordinates ρ, φ, z and the Cartesian coordinates x, y, z are related by the nonlinear equations
$$\begin{aligned} x &= \rho\cos\varphi\,, \\ y &= \rho\sin\varphi\,, \\ z &= z\,. \end{aligned}$$
The covariant and contravariant bases, obtained by the use of Eqs. (8.52) and (8.53), are
$$\mathbf{a}_1 = \cos\varphi\,\hat{\mathbf{e}}_x + \sin\varphi\,\hat{\mathbf{e}}_y\,, \quad \mathbf{a}_2 = -\rho\sin\varphi\,\hat{\mathbf{e}}_x + \rho\cos\varphi\,\hat{\mathbf{e}}_y\,, \quad \mathbf{a}_3 = \hat{\mathbf{e}}_z\,,$$
$$\mathbf{a}^1 = \cos\varphi\,\hat{\mathbf{e}}_x + \sin\varphi\,\hat{\mathbf{e}}_y\,, \quad \mathbf{a}^2 = -(\sin\varphi/\rho)\,\hat{\mathbf{e}}_x + (\cos\varphi/\rho)\,\hat{\mathbf{e}}_y\,, \quad \mathbf{a}^3 = \hat{\mathbf{e}}_z\,.$$

Incidentally, note that the coefficient of $\hat{\mathbf{e}}_x$ in $\mathbf{a}_2$ is $\partial x/\partial\varphi$, while the coefficient of $\hat{\mathbf{e}}_x$ in $\mathbf{a}^2$ is $\partial\varphi/\partial x$; in the present curvilinear coordinate system these are neither equal nor reciprocal.

The metric tensor can now be formed from $g_{ij} = \mathbf{a}_i \cdot \mathbf{a}_j$. In matrix form, it is

$$g_{ij} = \begin{pmatrix} 1 & 0 & 0 \\ 0 & \rho^2 & 0 \\ 0 & 0 & 1 \end{pmatrix},$$

and its inverse is

$$g^{ij} = \begin{pmatrix} 1 & 0 & 0 \\ 0 & 1/\rho^2 & 0 \\ 0 & 0 & 1 \end{pmatrix}.$$

From g_{ij}, which is diagonal, we can obtain the scale factors as the square roots of the diagonal elements: $h_e = h_z = 1$, $h_\varphi = \rho$.

In Cartesian coordinates, now consider a rank-2 tensor T whose only nonzero element is $T^{11} = A$, a constant. More precisely, T is a tensor field which has the same value at all spatial points; it can be represented as $A\hat{\mathbf{e}}_x\,\hat{\mathbf{e}}_x$. We have designated T as contravariant; in the present Cartesian coordinates the associated covariant tensor would also have $T_{11} = A$ as its only nonzero element.

Let's now use the tensor transformation formula to find the contravariant T^{ij} in cylindrical coordinates. Writing the first formula of Eq. (8.60) with $(q^1, q^2, q^3) = (x, y, z)$ and $((q')^1, (q')^2, (q')^3) = (\rho, \varphi, z)$, and remembering that the only nonzero element of T is T^{11}, we have

$$(T')^{\alpha\beta} = A\left(\frac{\partial(q')^\alpha}{\partial x}\right)\left(\frac{\partial(q')^\beta}{\partial x}\right).$$

Evaluating the partial derivatives, the nonzero elements of T' are found to be

$$(T')^{11} = A\cos^2\varphi\,, \quad (T')^{12} = (T')^{21} = -A\rho\sin\varphi\cos\varphi\,, \quad (T')^{22} = A\rho^2\sin^2\varphi\,.$$

We see that in cylindrical coordinates this constant tensor field has components that are position dependent. The explanation is that the basis vectors are functions of position, and the components that multiply them must take this into account. We have here a generalization of the corresponding behavior for vector fields that is discussed in Appendix H.

We can check the T' elements by forming $(T')^{ij}\mathbf{a}_i\mathbf{a}_j$:

$$\begin{aligned}(T')^{ij}\mathbf{a}_i\mathbf{a}_j &= A\cos^2\varphi[\cos\varphi\,\hat{\mathbf{e}}_x + \sin\varphi\,\hat{\mathbf{e}}_y][\cos\varphi\,\hat{\mathbf{e}}_x + \sin\varphi\,\hat{\mathbf{e}}_y] \\ &\quad - 2A\rho\sin\varphi\cos\varphi[\cos\varphi\,\hat{\mathbf{e}}_x + \sin\varphi\,\hat{\mathbf{e}}_y][-(\sin\varphi/\rho)\,\hat{\mathbf{e}}_x + (\cos\varphi/\rho)\,\hat{\mathbf{e}}_y] \\ &\quad + A\rho^2[-(\sin\varphi/\rho)\,\hat{\mathbf{e}}_x + (\cos\varphi/\rho)\,\hat{\mathbf{e}}_y][-(\sin\varphi/\rho)\,\hat{\mathbf{e}}_x + (\cos\varphi/\rho)\,\hat{\mathbf{e}}_y] \\ &= A\hat{\mathbf{e}}_x\hat{\mathbf{e}}_x\,.\end{aligned}$$

We have recovered our constant tensor.

■

Exercises

8.3.1. Check the details of Example 8.3.3 by doing the following:

(a) Using Eq. (8.52), verify that the covariant basis vectors $\mathbf{a}_i$ are given by the formulas in the Example.

(b) Obtain ρ and φ in terms of x, y, and z and use Eq. (8.53) to confirm the expressions given for the contravariant basis vectors $\mathbf{a}^i$.

(c) Verify by use of the defining equations that $g_{22} = \rho^2$ and that $g^{22} = \rho^{-2}$.

(d) Confirm that (g_{ij}) and (g^{ij}) are mutually inverse.

8.3.2. Taking the q_i to be the spherical polar coordinates (r, θ, φ), repeat the entire analysis of Exercise 8.3.1.

8.3.3. A contravariant tensor field $\mathbf{F}$, written in terms of cylindrical coordinates $q_1 = \rho$, $q_2 = \varphi$, $q_3 = z$, has the matrix form

$$(F^{ij}) = \begin{pmatrix} \rho^2 \cos\varphi & \rho \sin\varphi & 0 \\ -\rho \sin\varphi & \rho^2 & 0 \\ 0 & 0 & \rho^2 + z^2 \end{pmatrix}.$$

(a) Find the corresponding covariant field (F_{ij}).

(b) Form the contracted quantities $G^i_k = F^{ij} F_{kj}$ and $H^i_k = F^{ij} F_{jk}$.

8.4 SYMBOLIC COMPUTATION

Both MAPLE and MATHEMATICA support operations involving tensors, but do so in rather different ways. MAPLE's approach is largely designed to facilitate computations involving the mathematics of relativity, and therefore stresses capabilities needed for nonEuclidean geometry. On the other hand, MATHEMATICA provides relatively convenient support for Euclidean geometry in Cartesian coordinates, while avoiding issues such as the distinction between covariant and contravariant quantities. As a result, MATHEMATICA provides a simpler route to the capabilities of most interest here, while MAPLE has features that may be attractive in more advanced applications.

This survey of symbolic computation methods for tensors is restricted to Cartesian systems (but does include both rotation and reflection coordinate transformations).

MATHEMATICA

We discuss MATHEMATICA first. A tensor of rank n in 3-D space corresponds to an n-dimensional table of element values, with the index values for each dimension ranging from 1 to 3. If the tensor elements are the values of a function $f(i_1, i_2, \ldots)$ of the indices i_n, we can enter all the elements by a command of the type illustrated here for a tensor of rank 4 with $f(i_1, i_2, i_3, i_4) = 1000i_1 + 100i_2 + 10i_3 + i_4$:

```
t4 = Table[1000*i1+100*i2+10*i3+i4, {i1,3}, {i2,3}, {i3,3}, {i4,3}]

        {{{{1111,1112,1113},{1121,1122,1123},{1131,1132,1133}},
        {{1211,1212,1213},{1221,1222,1223},{1231,1232,1233}},
        {{1311,1312,1313},{1321,1322,1323},{1331,1332,1333}}},
```

```
{{{2111,2112,1113},{2121,2122,2123},{2131,2132,2133}},
{{2211,2212,2213},{2221,2222,2223},{2231,2232,2233}},
{{2311,2312,2313},{2321,2322,2323},{2331,2332,2333}}},
{{{3111,3112,3113},{3121,3122,3123},{3131,3132,3133}},
{{3211,3212,3213},{3221,3222,3223},{3231,3232,3233}},
{{3311,3312,3313},{3321,3322,3323},{3331,3332,3333}}}}
```

Here `{i1,3}, {i2,3}, {i3,3}, {i4,3}` indicates that the range of each i_n is from 1 to 3. We see that the output is a nested list, with the nesting depth equal to the rank of the tensor. A tensor with completely arbitrary elements can be entered using the standard MATHEMATICA list construction:

```
A = {{{a111,a112,a113},{a121,a122,a123},{a131,a132,a133}},
        {{a211,a212,a213},{a221,a222,a223},{a231,a232,a233}},
        {{a311,a312,a313},(a321,a322,a323},{a331,a332,a333}}};
```

MATHEMATICA forces the input into a two-dimensional format if it is examined using **MatrixForm**. These are the formats used for tensors of ranks 3 and 4:

```
MatrixForm[A]
```

$$\begin{pmatrix} \begin{pmatrix} a111 \\ a112 \\ a113 \end{pmatrix} & \begin{pmatrix} a121 \\ a122 \\ a123 \end{pmatrix} & \begin{pmatrix} a131 \\ a132 \\ a133 \end{pmatrix} \\ \begin{pmatrix} a211 \\ a212 \\ a213 \end{pmatrix} & \begin{pmatrix} a221 \\ a222 \\ a223 \end{pmatrix} & \begin{pmatrix} a231 \\ a232 \\ a233 \end{pmatrix} \\ \begin{pmatrix} a311 \\ a312 \\ a313 \end{pmatrix} & \begin{pmatrix} a321 \\ a322 \\ a323 \end{pmatrix} & \begin{pmatrix} a331 \\ a332 \\ a333 \end{pmatrix} \end{pmatrix}$$

```
MatrixForm[t4]
```

$$\begin{pmatrix} \begin{pmatrix} 1111 & 1112 & 1113 \\ 1121 & 1122 & 1123 \\ 1131 & 1132 & 1133 \end{pmatrix} & \begin{pmatrix} 1211 & 1212 & 1213 \\ 1221 & 1222 & 1223 \\ 1231 & 1232 & 1233 \end{pmatrix} & \begin{pmatrix} 1311 & 1312 & 1313 \\ 1321 & 1322 & 1323 \\ 1331 & 1332 & 1333 \end{pmatrix} \\ \begin{pmatrix} 2111 & 2112 & 2113 \\ 2121 & 2122 & 2123 \\ 2131 & 2132 & 2133 \end{pmatrix} & \begin{pmatrix} 2211 & 2212 & 2213 \\ 2221 & 2222 & 2223 \\ 2231 & 2232 & 2233 \end{pmatrix} & \begin{pmatrix} 2311 & 2312 & 2313 \\ 2321 & 2322 & 2323 \\ 2331 & 2332 & 2333 \end{pmatrix} \\ \begin{pmatrix} 3111 & 3112 & 3113 \\ 3121 & 3122 & 3123 \\ 3131 & 3132 & 3133 \end{pmatrix} & \begin{pmatrix} 3211 & 3212 & 3213 \\ 3221 & 3222 & 3223 \\ 3231 & 3232 & 3233 \end{pmatrix} & \begin{pmatrix} 3311 & 3312 & 3313 \\ 3321 & 3322 & 3323 \\ 3331 & 3332 & 3333 \end{pmatrix} \end{pmatrix}$$

Tensors of ranks 1 and 2 are displayed as vectors and matrices.

MATHEMATICA supports several operations for combining or manipulating tensors. First, there is an inner product; the inner product of two tensors A and B is a contraction using the last index of A and the first index of B. It is therefore a tensor analog of matrix multiplication, and is written in MATHEMATICA using the dot operator. Also supported is an outer product `Outer`, also called a **direct product** or a **Kronecker product**, which is a product (without contraction) containing all the indices of both factors. Contraction within a tensor is supported by the operation `Tr` (for **trace**), which can be used to form a contraction on the first two indices of a tensor (it is thus related in a way to the matrix trace operation).

The `dot` and `Tr` operations can be used to produce contractions of arbitrarily located tensor indices by making use of the operation `Transpose`, which supports arbitrary index permutations. We can permute the indices to be contracted to the necessary special positions, perform the contractions, and then carry out any further index permutation that may be needed to reach the proper index order.

Many tensor operations involve the Levi-Civita symbol (called `Signature` in MATHEMATICA) or the Kronecker delta. These quantities are supported as functions, with the following syntax:

```
KroneckerDelta[2,2]         1
KroneckerDelta[1,2]         0
Signature[{1,2,3}]          1
Signature[{1,2,2}]          0
Signature[{1,3,2}]         -1
```

Use of the MATHEMATICA commands referred to above is illustrated in the following comprehensive example.

Example 8.4.1. Tensor Operations in Mathematica

Let's introduce two tensors for illustrative purposes:

```
B = {{{b111,b112,b113},{b121,b122,b123},{b131,b132,b133}},
   {{b211,b212,b213},{b221,b222,b223},{b231,b232,b233}},
   {{b311,b312,b313},{b321,b322,b323},{b331,b332,b333}}};

G = {{g11,g12,g13},{g21,g22,g23},{g31,g32,g33}};
```

Let's also form the rank-3 Levi-Civita tensor:

```
Levi = Table[Signature[{i1,i2,i3}], {i1,3}, {i2,3}, {i3,3}]

                    {{{0, 0, 0}, {0, 0, 1}, {0, -1, 0}},
                     {{0, 0, -1}, {0, 0, 0}, {1, 0, 0}},
                     {{0, 1, 0}, {-1, 0, 0}, {0, 0, 0}}}
```

We can verify that `Levi` is antisymmetric by examining the result of permuting its indices:

```
L2 = Transpose[Levi, {2, 1, 3}]

                    {{{0, 0, 0}, {0, 0, -1}, {0, 1, 0}},
                     {{0, 0, 1}, {0, 0, 0}, {-1, 0, 0}},
                     {{0, -1, 0}, {1, 0, 0}, {0, 0, 0}}}
```

The second argument of **Transpose** gives the order to which the indices of the first argument are permuted; for **Levi**, the permutation we used is odd, so L2 = − **Levi**. The reader can verify that even permutations leave **Levi** unchanged.

Next let's make a tensor which is the contraction of BG corresponding to the index assignment $H_\mu = B_{\alpha\beta\alpha} G_{\beta\mu}$.[1] One way to make this computation is to transpose B_{ijk} to obtain $\overline{B}_{ikj}$, then form the inner product $\overline{H}_{ik\mu} = \overline{B}_{ik\beta} G_{\beta\mu}$, and finally take the trace of the first two indices of $\overline{H}$, reaching $H_\mu = \overline{H}_{\alpha\alpha\mu}$. These steps correspond to the MATHEMATICA statements

```
B1 = Transpose[B, {1, 3, 2}];
H1 = B1 . G;
H = Tr[H1, Plus, 2]

    {b111 g11 + b212 g11 + b313 g11 + b121 g21 + b222 g21
              + b323 g21 + b131 g31 + b232 g31 + b333 g31,
     b111 g12 + b212 g12 + b313 g12 + b121 g22 + b222 g22
              + b323 g22 + b131 g32 + b232 g32 + b333 g32,
     b111 g13 + b212 g13 + b313 g13 + b121 g23 + b222 g23
              + b323 g23 + b131 g33 + b232 g33 + b333 g33}
```

The command Tr forms a generalized trace of H1; the generalization is that the third argument of Tr tells how many initial indices are to be set equal while the second argument indicates that the elements with different values of the duplicated index are to be summed (the arguments **Plus** and "2" therefore cause this command to be a contraction on the first two indices).

In principle one could carry out hand computations to verify the formula for H. Instead, let's generate H in a different way and compare the results: form the outer product $M_{ijkmn} = B_{ijk} G_{mn}$, then transpose to $\overline{M}_{ikjmn}$, next apply Tr to contract to $\overline{H}_{jmn} = \overline{M}_{\alpha\alpha jmn}$ and then apply Tr a second time to further contract to $H_\mu = \overline{H}_{\beta\beta\mu}$. Commands that accomplish this are

```
M = Outer[Times, B, G];
M1 = Transpose[M, {1, 3, 2, 4, 5}];
H1 = Tr[M1, Plus, 2];
H2 = Tr[H1, Plus, 2];
```

We previously encountered **Outer** when we studied matrices; recall that the argument **Times** signals that the elements of B and G are to be combined multiplicatively. The present example shows that **Outer** can be used to form an outer product for tensors of arbitrary rank.

The tensors H and H2 should be equal. We can check:

```
H - H2

                    {0, 0, 0}
```

Since tensors are standard MATHEMATICA arrays, their elements can be accessed or changed in the usual way:

[1]Here and below we are using the Einstein summation convention: α and β are to be summed.

`B[[1,2,3]]`	b123
`B[[1,2,3]] = 17`	17
`B[[1,2,3]]`	17

■

MAPLE

MAPLE provides support for tensor analysis in a package called **tensor**, but for elementary operations involving Cartesian tensors in Euclidean space our approach will be to use standard MAPLE procedures.

Tensors of rank 1 or 2 can be introduced as vectors or matrices. Those of higher rank can be defined using the **Array** command. The arguments of this command give the range for each dimension of the array (for tensors in 3-D space, each dimension will be the sequence `1..3`), followed by a nested list containing the array elements.

```
> A := Array(1..3,1..3,1..3,[[[1,2,3],[4,5,6],[7,8,9]],
              [[10,11,12],[13,14,15],[16,17,18]],
              [[19,20,21],[22,23,24],[25,26,27]]]);
```

(output is information about A)

For tensors of rank higher than 2 the elements themselves are not displayed, and there is no command that produces a user-friendly display. One can, however, get an element list:

```
> ArrayElems(%);
```

$$\{(1,1,1)=1,\ (1,1,2)=2,\ (1,1,3)=3,\ (1,2,1)=4,\ (1,2,2)=5,\ (1,2,3)=6,$$
$$(1,3,1)=7,\ (1,3,2)=8,\ (1,3,3)=9,\ (2,1,1)=10,\ (2,1,2)=11,\ (2,1,3)=12,$$
$$(2,2,1)=13,\ (2,2,2)=14,\ (2,2,3)=15,\ (2,3,1)=16,\ (2,3,2)=17,\ (2,3,3)=18,$$
$$(3,1,1)=19,\ (3,1,2)=20,\ (3,1,3)=21,\ (3,2,1)=22,\ (3,2,2)=23,\ (3,2,3)=24,$$
$$(3,3,1)=25,\ (3,3,2)=26,\ (3,3,3)=27\}$$

A tensor whose elements are given as values of a function f can be built using MAPLE coding such as

```
> B := Array(1..3, 1..3, 1..3):
> for i from 1 to 3 do for j from 1 to 3 do for k from 1 to 3 do
>    B[i,j,k] := f(i,j,k)
> end do end do end do:
```

Tensor products can be formed in MAPLE by statements that define the product array, explicitly writing out the operations that carry out any contractions that may be involved. More general manipulations (e.g., index permutations) can also be generated explicitly. Warning: the dot operator does not generate an inner product when applied to tensors of rank greater than 2; its use in that context should probably be avoided.

Tensor operations frequently make use of the Kronecker delta or the Levi-Civita symbol. These quantities are defined in various MAPLE packages but are not present in the basic language. We assume them to be available with the respective names KD and Eps. MAPLE code for these functions is given in Appendix I; this or equivalent code should be executed in a MAPLE session prior to their use. Because these quantities are functions (not arrays), they are accessed with the following syntax:

```
> KD(1,2);            0
> KD(2,2);            1
> Eps(1,2,3);         1
> Eps(2,1,3);        -1
> Eps(1,2,2);         0
```

Example 8.4.2. Tensor Operations in Maple

As an initial effort, let's make a Levi-Civita tensor, assuming that the procedure Eps in Appendix I has already been executed in the current MAPLE session:

```
> Levi := Array(1..3, 1..3, 1..3):
> for i from 1 to 3 do for j from 1 to 3 do for k from 1 to 3 do
>    Levi[i,j,k] := Eps(i,j,k)
> end do end do end do:
```

Note that we defined Levi to be an array of the expected dimensions before we started to use it. Some of the operations illustrated here do not work properly if an array is undefined when its elements are given values. Note that Levi is an array and that Eps is a function.

Let's now check that Levi is antisymmetric. We can do so by subjecting it to index permutations. The permutation interchanging its first two indices is carried out as follows:

```
> L2 := Array(1..3, 1..3, 1..3):
> for i from 1 to 3 do for j from 1 to 3 do for k from 1 to 3 do
>    L2[i,j,k] := Levi[j,i,k]
> end do end do end do:
```

This permutation should cause a sign change. Checking,

```
> Levi + L2;                 (output is information only)
```

To see what happened, look at the tensor elements:

```
> ArrayElems(%);                 { }
```

The reader now needs to know that ArrayElems only lists nonzero elements; the empty set { } shows that every element of Levi + L2 is zero, consistent with L2 = −Levi.

We go on now to the formation of tensor products. If B is a rank-3 tensor with elements B_{ijk} and G is a tensor of rank 2 with elements G_{mn}, their direct product M, of rank 5 and with elements $\mathsf{M}_{ijkmn} = \mathsf{B}_{ijk}\,\mathsf{G}_{mn}$, can be constructed by the following five-fold loop:

```
> M := Array(1..3, 1..3, 1..3, 1..3, 1..3):
> for i1 from 1 to 3 do for i2 from 1 to 3 do for i3 from 1 to 3 do
```

```
> for i4 from 1 to 3 do for i5 from 1 to 3 do
>    M[i1,i2,i3,i4,i5] := B[i1,i2,i3] * G[i4,i5]
> end do end do end do end do end do:
```

Contractions can in a similar way be written as sums, with no particular complications that depend upon the locations of the contracted indices. Suppose we want to contract the first and the third indices of a rank-3 tensor B to make a vector $\overline{\mathsf{B}}$. The coding is as follows:

```
> B1 := Vector(1..3):
> for i from 1 to 3 do
>    B1[i] := add(B[alpha, i, alpha], alpha = 1 .. 3)
> end do:
```

A contraction that corresponds (except for index placement) with an inner product is that represented by $\overline{\mathsf{H}} = \mathsf{B}_{i\beta j}\mathsf{G}_{\beta\mu}$. Coding for this product is

```
> H1 := Array(1..3, 1..3, 1..3):
> for i1 from 1 to 3 do for i2 from 1 to 3 do for i3 from 1 to 3 do
>    H1[i1,i2,i3] := add(B[i1, beta, i2] * G[beta, i3], beta = 1 .. 3)
> end do end do end do:
```

Using the expressions we have already generated, we can combine them further to make $\mathsf{H}_\mu = \mathsf{B}_{\alpha\beta\alpha}G_{\beta\mu}$, either as $\overline{B}_\beta\mathsf{G}_{\beta\mu}$ or as $\overline{\mathsf{H}}_{\alpha\alpha\mu}$. Coding for these alternatives is

```
> H := Vector(3):
> H2 := Vector(3):
> for i from 1 to 3 do
>    H[i] := add(B1[beta] * G[beta, i], beta = 1 .. 3):
>    H2[i] := add(H1[alpha, alpha, i], alpha = 1 .. 3):
> end do:
```

The vectors `H` and `H2` should be identical. Checking,

```
> simplify(H - H2);
```

$$\begin{bmatrix} 0 \\ 0 \\ 0 \end{bmatrix}$$

Finally, we remind the reader that individual elements of all tensor quantities are accessible by standard MAPLE methods:

```
> B[1, 2, 3];
```

$$f(1,2,3)$$

```
> B[1, 2, 3] := 17;
```

$$B[1,2,3] := 17$$

```
> B[1, 2, 3];
```

$$17$$

■

TENSOR TRANSFORMATIONS

A defining property of tensors is that they must transform properly under rotations or reflections of the coordinate system. In the discussion that follows, we continue with the restriction to Cartesian coordinates that was used in developing the tensor formulas found earlier in this Section. Since our emphasis will be on problems in the 3-D physical space, we further assume that all coordinate transformations can be described by real orthogonal 3×3 matrices, with determinant $+1$ for pure rotations and determinant -1 for transformations that include a reflection.

Referring to the general transformation formulas in Section 8.3, and letting U be an orthogonal transformation matrix and B be a tensor to be transformed (for illustrative purposes of rank 3), we need to evaluate expressions (written using the Einstein summation convention) such as

$$\overline{B}_{ijk} = U_{ir}U_{js}U_{kt}B_{rst}\,. \tag{8.61}$$

In MATHEMATICA, one way of evaluating Eq. (8.61) is by the following steps:

1. Form W_{ist} as the inner product $U_{ir}B_{rst}$.
2. Transpose W, forming $\overline{W}_{sit} = W_{ist}$.
3. Form Y_{jit} as the inner product $U_{js}\overline{W}_{sit}$.
4. Form U^T, the transpose of U.
5. Form Z_{jik} as the inner product $Y_{jit}U^T_{tk}$.
6. Transpose Z, forming $\overline{B}_{ijk} = Z_{jik}$.

MATHEMATICA code carrying out the above prescription is

```
W = U . B;
W1 = Transpose[W,{2,1,3}];
Y = U . W1;
UT = Transpose[U,{2,1}];
Z = Y . UT;
B1 = Transpose[Z,{2,1,3}];
```

In MAPLE, we do not need to transpose to reach specific index positions, but instead we have explicit index loops. MAPLE code for the evaluation of Eq. (8.61) is

```
> for i from 1 to 3 do for j from 1 to 3 do for k from 1 to 3 do
>   B1[i,j,k] := add(add(add(U[i,r]*U[j,s]*U[k,t]*B[r,s,t], r=1..3),
>                    s=1..3), t=1..3)
> end do end do end do:
```

Example 8.4.3. Levi-Civita Symbol

Let's use symbolic computing to verify that the Levi-Civita symbol is an isotropic pseudotensor. We assume that the rank-3 Levi-Civita tensor, named `Levi`, has been formed by the procedure described in Example 8.4.1 (MATHEMATICA) or

Example 8.4.2 (MAPLE). Our plan is to transform `Levi` using the two matrices

$$\mathsf{U} = \begin{pmatrix} 0.8 & 0.6 & 0 \\ -0.6 & 0.8 & 0 \\ 0 & 0 & 1 \end{pmatrix} \quad \text{and} \quad \mathsf{V} = \begin{pmatrix} 0.8 & 0.6 & 0 \\ -0.6 & 0.8 & 0 \\ 0 & 0 & -1 \end{pmatrix}.$$

Both U and V are orthogonal, but $\det(\mathsf{U}) = +1$ and $\det(\mathsf{V}) = -1$, showing that U is a rotation while V includes a reflection.

Using the coding that precedes this Example, we find with MATHEMATICA that the tensors obtained using transformations U and V have the following `MatrixForm` displays:

$$\begin{pmatrix} \begin{pmatrix} 0. \\ 0. \\ 0. \end{pmatrix} & \begin{pmatrix} 0. \\ 0. \\ 1. \end{pmatrix} & \begin{pmatrix} 0. \\ -1. \\ 0. \end{pmatrix} \\ \begin{pmatrix} 0. \\ 0. \\ -1. \end{pmatrix} & \begin{pmatrix} 0. \\ 0. \\ 0. \end{pmatrix} & \begin{pmatrix} 1. \\ 0. \\ 0. \end{pmatrix} \\ \begin{pmatrix} 0. \\ 1. \\ 0. \end{pmatrix} & \begin{pmatrix} -1. \\ 0. \\ 0. \end{pmatrix} & \begin{pmatrix} 0. \\ 0. \\ 0. \end{pmatrix} \end{pmatrix}, \quad \begin{pmatrix} \begin{pmatrix} 0. \\ 0. \\ 0. \end{pmatrix} & \begin{pmatrix} 0. \\ 0. \\ -1. \end{pmatrix} & \begin{pmatrix} 0. \\ 1. \\ 0. \end{pmatrix} \\ \begin{pmatrix} 0. \\ 0. \\ 1. \end{pmatrix} & \begin{pmatrix} 0. \\ 0. \\ 0. \end{pmatrix} & \begin{pmatrix} -1. \\ 0. \\ 0. \end{pmatrix} \\ \begin{pmatrix} 0. \\ -1. \\ 0. \end{pmatrix} & \begin{pmatrix} 1. \\ 0. \\ 0. \end{pmatrix} & \begin{pmatrix} 0. \\ 0. \\ 0. \end{pmatrix} \end{pmatrix}.$$

(with U) (with V)

Except for the decimal points, the transformation with U is the same as `Levi`; that with V is also the same except for a sign reversal that shows `Levi` actually to be a pseudotensor.

If these transformations are processed with MAPLE, we may see what happened by applying `ArrayElems` to the results. For the transformation with U, we get

```
{(1,2,3) = 1.00, (1,3,2) = -1.00, (2,1,3) = -1.00, (2,3,1) = 1.00,
(3,1,2) = 1.00, (3,2,1) = -1.00}
```

For the transformation with V, we get

```
{(1,2,3) = -1.00, (1,3,2) = 1.00, (2,1,3) = 1.00, (2,3,1) = -1.00,
(3,1,2) = -1.00, (3,2,1) = 1.00}
```

Remembering that all elements that are not listed are zero, we see that the MAPLE and MATHEMATICA results agree.

■

Exercises

8.4.1. Using the symbolic system of your choice, reproduce the computations of Example 8.4.3.

Additional Readings

Arfken, G. B., Weber, H. J., & Harris, F. E. (2013). *Mathematical methods for physicists* (7th ed.). New York: Academic Press.

Hassani, S. (1991). *Foundations of mathematical physics.* Boston, MA: Allyn and Bacon.

Jeffreys, H. (1952). *Cartesian tensors.* Cambridge: Cambridge University Press (This is an excellent discussion of Cartesian tensors and their application to a wide variety of fields of classical physics).

Margenau, H., & Murphy, G. M. (1956). *The mathematics of physics and chemistry* (2nd ed.). Princeton, NJ: Van Nostrand (Chapter 5 covers curvilinear coordinates and 13 specific coordinate systems).

Sokolnikoff, I. S. (1964). *Tensor analysis—theory and applications* (2nd ed.). New York: Wiley (Particularly useful for its extension of tensor analysis to non-Euclidean geometries).

Young, E. C. (1993). *Vector and tensor analysis* (2nd ed.). New York: Dekker.

Chapter 9

GAMMA FUNCTION

In this chapter we study the gamma function (a generalization to nonintegers of the factorial function $n!$), together with some related additional functions. The gamma function is not itself the solution of any of the differential equations that are important in the physical sciences; its usual role is as a factor in expansions, in formulas for integrals, in probability and statistics, and in other expressions that describe solutions to physical problems. In that role, the gamma function appears with a frequency rivaling that of the exponential, the logarithm, and the trigonometric functions. It is clearly useful to acquire skill in dealing with this ubiquitous function.

While symbolic computer systems are well equipped to produce numerical evaluations of gamma functions, scientists and engineers still need to be able to manipulate them and to obtain analytical formulas that can provide both accuracy and insight. Our study of gamma functions also serves as a model illustrating the use of special functions; our aim is not to make the student an expert in all functions (an impossible task), but to permit him/her to gain the confidence and experience that will facilitate the handling of less familiar functions whose properties are described in reference works or the literature.

In addition to a discussion of the gamma function, this chapter introduces two related special functions, namely the error function (important not only for probability theory but also in more general contexts), and the exponential integral (which occurs in a variety of areas, including astrophysics.

9.1 DEFINITION AND PROPERTIES

We start by evaluating the definite integral

$$I(n) = \int_0^\infty t^n e^{-t}\, dt \tag{9.1}$$

for the special case $n = 0$. For this case the integrand has indefinite integral $-e^{-t}$ and we get $I(0) = 1$. We can proceed to $n = 1, 2, \cdots$ by repeated integrations by parts. In particular,

$$I_n = \int_0^\infty t^n e^{-t}\, dt = \big[-t^n\, e^{-t} \big]_0^\infty + n \int_0^\infty t^{n-1} e^{-t}\, dt\,.$$

For positive n the integrated terms vanish and, using Eq. (9.1) for the integral on the right-hand side, we get (for positive integer n)

$$I(n) = n\, I(n-1)\,. \tag{9.2}$$

Mathematics for Physical Science and Engineering.
http://dx.doi.org/10.1016/B978-0-12-801000-6.00009-2

Equation (9.2) is a recurrence relation that enables us to compute $I(n)$ from $I(n-1)$; since we already have a value for $I(0)$, we easily find that, for all nonnegative integers n,

$$I(n) = n!\,. \tag{9.3}$$

Equation (9.3) provides additional justification for our earlier decision to assign 0! the value unity.

We cannot use Eq. (9.2) to obtain $I(-1)$ from I_0 because that equation then becomes $I_0 = 0 \cdot I(-1)$, and solution for $I(-1)$ would involve a division by zero. This is not surprising, since the integral $I(-1)$ is divergent.

Both the integral of Eq. (9.1) and the analysis leading to Eq. (9.2) remain valid if n is permitted to become nonintegral. It is thus natural to extend the factorial function to noninteger values (which we designate s), providing that s is restricted to a range such that the integrals involved converge.

EULER INTEGRAL

With the observations of the above paragraph as motivation, we define the **gamma function**, designated $\Gamma(s)$, as the value of the integral

$$\Gamma(s) = \int_0^\infty t^{s-1} e^{-t}\, dt\,. \tag{9.4}$$

This definition is assumed to apply whenever the integral over t converges; if s is restricted to real values, the range of convergence is $s > 0$. Note that the definition in Eq. (9.4) took the power of t to be $s - 1$, while in Eq. (9.1) the power of t was n. This shift in the definition of $\Gamma(s)$ has the effect that

$$\Gamma(n) = (n-1)! \quad \text{and therefore also} \quad n! = \Gamma(n+1)\,. \tag{9.5}$$

The definition in Eq. (9.4) is known as the **Euler integral** for the gamma function. Some users of the gamma function regard the shifted correspondence with the factorial as an inconvenience; others note that there may be some virtue to having the convergence limit of the integral at $s = 0$. What remains important is that the definition of $\Gamma(s)$ is universally accepted and we must work with it as it is.

FACTORIAL NOTATION

Some authors choose to use the factorial notation even when its argument is not an integer, e.g., writing $(-1/2)!$ in place of $\Gamma(1/2)$. While that notation is not ambiguous, it is not widely used in serious work. We do not write factorials with noninteger arguments in this book.

RECURRENCE FORMULA

One of the important results from complex variable theory (discussed in Chapter 17) is that if two formulas describe the same function of s everywhere on a line segment of finite length in the complex plane, either formula is a valid representation of that function for all complex s for which it converges (this notion is the basic principle behind what is called **analytic continuation**). This principle is relevant here because we have the recurrence formula, Eq. (9.2). In our present notation, it is

$$\Gamma(s+1) = s\,\Gamma(s)\,. \tag{9.6}$$

Even though we used the Euler integral to derive the recurrence formula, the analytic continuation principle tells us that it must apply for all complex values of s for which the formula is well-defined, including values of s for which the Euler integral diverges.

Focusing on real s, and noting that the Euler integral gives positive values of $\Gamma(s)$ for all $s > 0$, we see that application of the recurrence formula to $0 < s < 1$ in the form

$$\Gamma(s-1) = \frac{\Gamma(s)}{s-1}, \tag{9.7}$$

will give us values of $\Gamma(s)$ for $-1 < s < 0$, and these values will be negative. The nonexistence of $\Gamma(0)$ will prevent us from obtaining $\Gamma(-1)$. In a similar way, we can obtain values of $\Gamma(s)$ for all negative s except at the negative integers; a graph of $\Gamma(s)$ for both positive and negative s is shown in Fig. 9.1. Note that recurrence based on the negative values of $\Gamma(s)$ for $-1 < s < 0$ causes $\Gamma(s)$ to be positive for $-2 < s < -1$; for negative s the sign of $\Gamma(s)$ alternates in successive intervals of unit length.

SPECIAL VALUES

From the recurrence formula we have already seen that $\Gamma(s)$ becomes infinite for $s = 0$ and for all negative integer s. From the relation to the factorial we also have $\Gamma(1) = \Gamma(2) = 1$, and (as shown in Fig. 9.1) $\Gamma(s)$ is positive for all positive s, with a minimum between $s = 1$ and $s = 2$. $\Gamma(s)$ increases rapidly with increasing positive s; since $e^{x+1} = e\,(e^x)$ while $\Gamma(s+1) = s\,\Gamma(s)$, the increase is faster than that of the exponential.

Closed forms are available for $\Gamma(s)$ of half-integer s. To obtain the value of $\Gamma(1/2)$, first write the Euler integral with the integration variable replaced by u^2:

$$\Gamma(1/2) = \int_0^\infty t^{-1/2}\,e^{-t}\,dt = \int_0^\infty u^{-1}\,e^{-u^2}(2u\,du) = 2\int_0^\infty e^{-u^2}\,du\,. \tag{9.8}$$

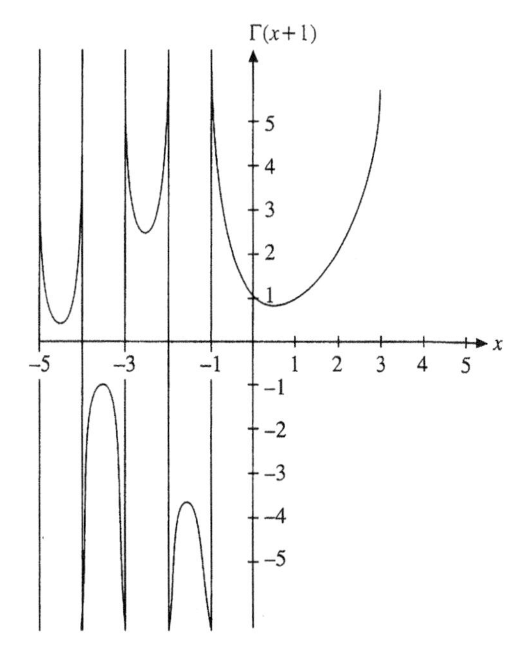

Figure 9.1: Gamma function $\Gamma(x+1)$ for real x.

This definite integral (including the premultiplier, 2) has the value $\sqrt{\pi}$. To prove this, form $[\Gamma(1/2)]^2$, taking u and v as the integration variables of the two instances of $\Gamma(1/2)$. Then rewrite the resulting double integral in polar coordinates r, θ, replacing $du\,dv$ by $r\,dr\,d\theta$. The region $u > 0$, $v > 0$ corresponds to the quadrant $0 \le \theta \le \pi/2$, $0 \le r < \infty$ in the polar coordinates, so

$$[\Gamma(1/2)]^2 = 4\int_0^\infty e^{-u^2}\,du \int_0^\infty e^{-v^2}\,dv = 4\int_0^\infty du \int_0^\infty dv\, e^{-u^2-v^2}$$

$$= 4\int_0^\infty r\,e^{-r^2}\,dr \int_0^{\pi/2} d\theta = \pi\,. \tag{9.9}$$

The double integral in the second line of Eq. (9.9) is elementary and evaluates, as indicated, to π. Taking the square root of both sides of Eq. (9.9), we reach the final result,

$$\Gamma(1/2) = \sqrt{\pi}\,. \tag{9.10}$$

Starting from Eq. (9.10), we can apply the recurrence formula to evaluate $\Gamma(n + \frac{1}{2})$ for any positive or negative integer n.

REFLECTION FORMULA

Equations connecting different values of the same function are called **functional relations**; an example we have already encountered for the gamma function is the recurrence formula, Eq. (9.6). Another useful functional relation for the gamma function is a formula connecting its values for two s that are related by reflection about $s = 1/2$:

$$\Gamma(s)\Gamma(1-s) = \frac{\pi}{\sin s\pi}\,. \tag{9.11}$$

We derive Eq. (9.11) in Chapter 17. However, we can at this time check its consistency with Eq. (9.10). Setting $s = 1/2$, Eq. (9.11) yields $[\Gamma(1/2)]^2 = \pi$.

DUPLICATION FORMULA

Another useful functional relation for the gamma function is the **Legendre duplication formula**,

$$\Gamma(s+1)\Gamma(s+\tfrac{1}{2}) = 2^{-2s}\sqrt{\pi}\,\Gamma(2s+1)\,. \tag{9.12}$$

Equation (9.12) can be proved for general s, but we confirm it here only for $s = n$, an integer. Then $\Gamma(s+1) = \Gamma(n+1) = n!$, $\Gamma(2s+1) = \Gamma(2n+1) = (2n)!$, and

$$\Gamma(s+\tfrac{1}{2}) = \Gamma(n+\tfrac{1}{2}) = \sqrt{\pi}\left[\frac{1}{2}\cdot\frac{3}{2}\cdots\frac{2n-1}{2}\right] = \frac{\sqrt{\pi}}{2^n}(2n-1)!! = \frac{\sqrt{\pi}}{2^n}\frac{(2n)!}{2^n\,n!}$$

$$= \frac{\sqrt{\pi}}{2^{2n}}\frac{\Gamma(2n+1)}{\Gamma(n+1)}\,. \tag{9.13}$$

Equation (9.13) rearranges to a form consistent with Eq. (9.12).

RELATION TO FACTORIALS

The confirmation of the duplication formula involved an equation we will encounter again. It is repeated here along with some other potentially useful formulas:

$$(2n)!! = 2^n \Gamma(n+1), \tag{9.14}$$

$$(2n-1)!! = \frac{\Gamma(2n+1)}{2^n \Gamma(n+1)}, \tag{9.15}$$

$$\Gamma(n+\tfrac{1}{2}) = \frac{\sqrt{\pi}\,(2n-1)!!}{2^n}, \tag{9.16}$$

$$\binom{n+\frac{1}{2}}{j} = \frac{\Gamma(n+\frac{3}{2})}{j!\,\Gamma(n+\frac{3}{2}-j)}. \tag{9.17}$$

SYMBOLIC COMPUTING

The gamma function is known to both MAPLE and MATHEMATICA. In MAPLE, it is `GAMMA`; by writing entirely in uppercase `Gamma` remains available as the name of a variable. Note: the MAPLE name `gamma` is **not** an available variable name; it is reserved for the Euler-Mascheroni constant.

In MATHEMATICA, the gamma function is `Gamma`.

Example 9.1.1. Symbolic Computing, Gamma Function

In MAPLE, these function values are computed exactly:

```
> GAMMA(5);
```
$$24$$
```
> GAMMA(5/2);
```
$$\frac{3}{4}\sqrt{\pi}$$

The following two cases do not have exact closed evaluations; MAPLE returns the first unevaluated, but transforms the second to what it considers a computationally simpler form (using the reflection transformation).

```
> GAMMA(3/4);
```
$$\Gamma\left(\frac{3}{4}\right)$$
```
> GAMMA(1/3);
```
$$\frac{2}{3}\frac{\pi\sqrt{3}}{\Gamma\left(\frac{2}{3}\right)}$$

A gamma function with any decimal argument is evaluated as a decimal quantity (using methods discussed later in this chapter):

```
> GAMMA(2.5);          1.329340388
> GAMMA(5.0);          24.00000000
> GAMMA(0.75);         1.225416702
```

Checking the Euler integral,

```
> evalf(int(t^0.713*exp(-t), t=0 .. infinity));     0.9111663772
> GAMMA(1.713);                                      0.9111663772
```

In MATHEMATICA, these function values are computed exactly:

`Gamma[5];`	24
`Gamma[5/2];`	$\frac{3\sqrt{\pi}}{4}$

No closed evaluation; MATHEMATICA returns unevaluated:

`Gamma[1/3]`	Gamma$\left[\frac{1}{3}\right]$

A gamma function with any decimal argument is evaluated as a decimal quantity (using methods discussed later in this chapter):

`Gamma[2.5]`	1.32934
`Gamma[5.0]`	24.
`Gamma[0.75]`	1.22542

Checking the Euler integral,

`Integrate[t^0.713*E^(-t), {t,0,Infinity}]`	0.911166
`Gamma[1.713]`	0.911166

■

Exercises

First use formal methods to reduce these exercises to expressions involving gamma functions, and then evaluate these expressions using symbolic computing where it is helpful. Compare your numerical answers for integrals with the results of numerical quadratures produced by symbolic computation.

9.1.1. Simplify: (a) $\frac{\Gamma(8)}{\Gamma(6)}$, (b) $\frac{\Gamma(2/5)}{\Gamma(12/5)}$.

9.1.2. Evaluate: (a) $\int_0^\infty x^{3/4}\, e^{-x}\, dx$,

(b) $\int_0^\infty x^{1/2}\, e^{-2x}\, dx$,

(c) $\int_0^\infty x^3\, e^{-x^2}\, dx$,

(d) $\int_0^1 (\ln 1/x)^{1/3}\, dx$,

(e) $\int_0^\infty e^{-x^4}\, dx$.

9.1.3. Verify the formulas of Eqs. (9.14)—(9.17).

9.1.4. Show that $\Gamma(s)$ (for $s > 0$) may be written

$$\Gamma(s) = 2\int_0^\infty e^{-t^2} t^{2s-1}\,dt\,,$$

$$\Gamma(s) = \int_0^1 \left[\ln\left(\frac{1}{t}\right)\right]^{s-1} dt\,.$$

9.1.5. Show that, for $n > -1$,

$$\int_0^1 x^n \ln x\,dx = -\frac{1}{(n+1)^2}\,.$$

9.1.6. Show that, for integer n,

$$\Gamma(\tfrac{1}{2}-n)\,\Gamma(\tfrac{1}{2}+n) = (-1)^n\,\pi\,.$$

9.1.7. Show that

$$\frac{d\Gamma(s)}{ds} = \int_0^\infty x^{s-1}e^{-x}\ln x\,dx\,.$$

9.2 DIGAMMA AND POLYGAMMA FUNCTIONS

Information about the derivatives of the gamma function is useful for further development of its properties. In this connection it turns out that it is more convenient to work with derivatives of ln gamma than with gamma itself. Accordingly, we define a function

$$\psi(s) = \frac{d}{ds}\ln\Gamma(s) = \frac{d\Gamma(s)/ds}{\Gamma(s)}\,, \tag{9.18}$$

calling $\psi(s)$ the **digamma function**, or sometimes just **Psi**.

Euler has shown that $\psi(s+1)$ can be written as the series expansion

$$\psi(s+1) = -\gamma + \sum_{m=1}^\infty \frac{s}{m(m+s)}\,, \tag{9.19}$$

where γ is the Euler-Mascheroni constant, first introduced in this book at Eq. (2.13). For reference, we repeat here the definition:

$$\gamma = \lim_{n\to\infty}\left(\sum_{m=1}^{n} m^{-1} - \ln n\right) = 0.577\,215\,664\,901\,\cdots\,. \tag{9.20}$$

A derivation of Eq. (9.19) is given in Arfken et al., Additional Readings.

If $s = 0$, Eq. (9.19) reduces to the simple result $\psi(1) = -\gamma$, while if s is a positive integer n, the summation in that equation can be written

$$\sum_{m=1}^\infty \frac{n}{m(m+n)} = \sum_{m=1}^\infty\left(\frac{1}{m} - \frac{1}{n+m}\right) = \sum_{m=1}^{n}\frac{1}{m}\,,$$

so we get

$$\psi(n+1) = -\gamma + \sum_{m=1}^{n} \frac{1}{m}. \tag{9.21}$$

More explicitly,

$$\psi(1) = -\gamma,$$

$$\psi(2) = -\gamma + 1,$$

$$\psi(3) = -\gamma + 1 + \frac{1}{2},$$

$$\psi(4) = -\gamma + 1 + \frac{1}{2} + \frac{1}{3}, \quad \text{etc.}$$

Figure 9.2 compares the behavior of $\Gamma(x+1)$ with that of $\psi(x+1)$; because of its logarithmic form $\psi(x)$ increases far less rapidly than $\Gamma(x)$ as x is increased.

Higher logarithmic derivatives of the gamma function are also useful; they are termed **polygamma functions** and have the forms

$$\psi^{(n)}(s+1) = \frac{d^{n+1}}{ds^{n+1}} \ln\Gamma(s+1) = \frac{d^n}{ds^n}\left[-\gamma + \sum_{m=1}^{\infty}\left(\frac{1}{m} - \frac{1}{s+m}\right)\right]. \tag{9.22}$$

Note that the polygamma function identified with (n) is the $(n+1)$th derivative of $\ln\Gamma(s+1)$.

Of particular interest are the polygamma functions $\psi^{(n)}(1)$. If we carry out the indicated differentiations in Eq. (9.22) and then set $s = 0$, we get (for $n > 0$)

$$\psi^{(n)}(1) = (-1)^{n+1} n! \sum_{m=1}^{\infty} \frac{1}{m^{n+1}} = (-1)^{n+1} n!\, \zeta(n+1), \tag{9.23}$$

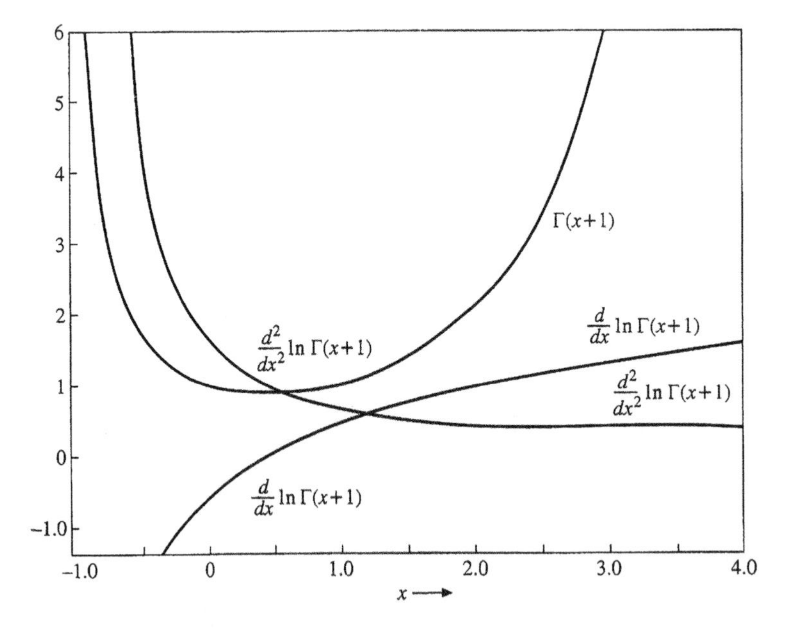

Figure 9.2: Gamma function and its first two logarithmic derivatives.

where we have recognized the m summation as a Riemann zeta function, defined at Eq. (2.12).

The $\psi^{(n)}(1)$ can be used to form a power series expansion of $\psi(s+1)$. Here is the Taylor series:

$$\psi(s+1) = \psi(1) + \frac{1}{1!}\,\psi^{(1)}(1)\,s - \frac{1}{2!}\,\psi^{(2)}(1)\,s^2 + \cdots$$

$$= -\gamma + \frac{1!\,\zeta(2)}{1!}\,s - \frac{2!\,\zeta(3)}{2!}\,s^2 + \cdots .$$

Integrating both sides of this equation, we reach an expansion for $\ln\Gamma(s+1)$:

$$\ln\Gamma(s+1) = -\gamma s + \sum_{n=2}^{\infty}(-1)^n\,\zeta(n)\,\frac{s^n}{n}\,. \tag{9.24}$$

Noting that $\zeta(n)$ approaches unity for large n, we see that this series converges for $-1 < s \le 1$, thereby providing a way to compute $\ln\Gamma(s+1)$, and therefore also $\Gamma(s+1)$, for a range of s of at least unit length. Then, using the recurrence formula, we can extend the computation to all $\Gamma(s)$ except the integers for which $\Gamma(s)$ is singular.

It should be mentioned that for $s \gg 1$ it is less efficient to use Eq. (9.24) and recursion than to compute $\Gamma(s)$ using Stirling's formula (discussed in Section 9.3). But Stirling's formula is an asymptotic expansion, and it is desirable to have at least one method for computing $\Gamma(s)$ that has no theoretical accuracy limitation.

SYMBOLIC COMPUTING

In MAPLE, the digamma function $\psi(s)$ is named `Psi(s)` and the polygamma function $\psi^{(n)}(s)$ is accessed as `Psi(n,s)`. In MATHEMATICA these functions are respectively designated `PolyGamma[s]` and `PolyGamma[n,s]`. The Euler-Mascheroni constant is named `gamma` in MAPLE and `EulerGamma` in MATHEMATICA. The Riemann zeta function $\zeta(s)$, which occurs in Eq. (9.24), is called `Zeta(s)` (MAPLE) and `Zeta[s]` (MATHEMATICA).

Example 9.2.1. Finding Specific Values

Suppose we want to find a value of s for which $d\Gamma(s)/ds = 2$. Writing

$$\frac{d\Gamma(s)}{ds} = \psi(s)\Gamma(s)\,,$$

we can use symbolic computing to solve the transcendental equation $\psi(s)\Gamma(s) = 2$. In MAPLE,

```
> solve(Psi(s)*GAMMA(s)=2,s);
```

$$RootOf(\Psi(\_Z)\,\Gamma(\_Z) - 2)$$

We can force numerical evaluation using `allvalues`:

```
> allvalues(%);
```

$$RootOf(\Psi(\_Z)\,\Gamma(\_Z) - 2,\ 3.059424875)$$

From the above we read out $s = 3.059424875$.

In MATHEMATICA, `Solve` is less helpful here than `FindRoot`:

```
FindRoot[PolyGamma[s]*Gamma[s]==2, {s, 1}]
```

$\{s \to 3.05942\}$

In `FindRoot`, the second argument lists the independent variable (s) and the starting point for the root search.

■

Example 9.2.2. Convergence Rate of Taylor Series for ln Γ

Let's see how fast the series in Eq. (9.23) converges when used to evaluate $\Gamma(1/2)$ (corresponding to $s = -1/2$), which has the known value $\pi^{1/2} = 1.772\ 453\ 851 \cdots$. Here is a procedure, with argument n, that evaluates the series through n and reports the error.

First using MAPLE,

```
> S := proc(n) local A,j;
>   A := gamma * 0.5;
>   for j from 2 to n do A := A + Zeta(j) * (0.5)^j/j  end do;
>   printf(" Series for Gamma(1/2):%10.6f    Error:%10.6f\n",
>               exp(A), exp(A)-GAMMA(0.5));
> end proc;
```

A similar procedure in MATHEMATICA can take the form

```
s[n_] := Module[{a, j},
      a = EulerGamma*0.5;
      Do[a = a + Zeta[j] * (0.5)^j/j, {j, 2, n}];
      Print[" Series for Gamma(1/2):", PaddedForm[N[E^a,10],{10, 6}],
        "   Error:", PaddedForm[N[E^a-Gamma[0.5],10],{10,6}]]]
```

Trying this out:

```
> S(4);  S(8);

      Series for Gamma(1/2): 1.752812   Error: -0.019641
      Series for Gamma(1/2): 1.771749   Error: -0.000705

s[12]

      Series for Gamma(1/2): 1.772423   Error: -0.000031
```

■

Exercises

First use formal methods to reduce these exercises to expressions involving gamma-related functions, and then evaluate these expressions using symbolic computing where it is helpful. Compare your numerical answers for integrals with the results of numerical quadratures produced by symbolic computation.

9.2.1. Using di- and polygamma functions, sum the series

$$(a)\quad \sum_{n=1}^{\infty} \frac{1}{n(n+1)}\,, \qquad (b)\quad \sum_{n=2}^{\infty} \frac{1}{n^2-1}\,.$$

9.2.2. An expansion of $\ln \Gamma(s+1)$ that apparently differs from Eq. (9.24) is

$$\ln \Gamma(s+1) = -\ln(1+s) + s(1-\gamma) + \sum_{n=2}^{\infty} (-1)^n \Big[\zeta(n) - 1\Big] \frac{s^n}{n}\,.$$

(a) Show that this expansion agrees with Eq. (9.24) for $|s| < 1$.

(b) What is the range of convergence of this new expansion?

9.2.3. Show that

$$\sum_{n=1}^{\infty} \frac{1}{(n+a)(n+b)} = \frac{1}{(b-a)} \Big[\psi(1+b) - \psi(1+a)\Big],$$

where $a \neq b$, and neither a nor b is a negative integer. It is of some interest to compare this summation with the corresponding integral,

$$\int_1^{\infty} \frac{dx}{(x+a)(x+b)} = \frac{1}{b-a} \Big[\ln(1+b) - \ln(1+a)\Big].$$

9.2.4. Derive the difference relation for the polygamma function

$$\psi^{(m)}(s+2) = \psi^{(m)}(s+1) + (-1)^m \frac{m!}{(s+1)^{m+1}}\,, \qquad m = 0, 1, 2, \ldots$$

9.2.5. Verify

$$(a)\quad \int_0^{\infty} e^{-r} \ln r \, dr = -\gamma\,,$$

$$(b)\quad \int_0^{\infty} r e^{-r} \ln r \, dr = 1 - \gamma\,,$$

$$(c)\quad \int_0^{\infty} r^n e^{-r} \ln r \, dr = (n-1)! + n \int_0^{\infty} r^{n-1} e^{-r} \ln r \, dr, \qquad n = 1, 2, 3, \ldots$$

Hint. These may be verified by integration by parts, or by differentiating the Euler integral formula for $\Gamma(n+1)$ with respect to n.

9.2.6. Find the value of s for which $\Gamma(s)$ is a minimum.

Hint. You can do this graphically, or by the method illustrated in Example 9.2.1.

9.3 STIRLING'S FORMULA

Stirling's formula is an asymptotic expansion of $\ln\Gamma(s)$ that can be derived using the Euler-Maclaurin summation formula. While this expansion shares the limitations of all asymptotic formulas, its rate of approach to exactness as s increases is remarkably rapid, and it is often the best way to evaluate gamma functions. For applications in statistical mechanics, which involve factorials of numbers representing numbers of atoms or molecules in a macroscopic sample, Stirling's formula is virtually completely exact.

While your symbolic computing system knows about Stirling's formula and uses it automatically whenever appropriate, you will still need to know something about it yourself when you study statistical physics or even relatively elementary topics in probability.

Here is Stirling's formula:

$$\ln\Gamma(s+1) \sim \frac{1}{2}\ln 2\pi + \left(s+\frac{1}{2}\right)\ln s - s + \sum_{n=1}^{N}\frac{B_{2n}}{2n(2n-1)s^{2n-1}}\,. \tag{9.25}$$

The B_{2n} are the Bernoulli numbers (with the definitions used in this book); see Eq. (2.73) and Table 2.2. For a derivation of Eq. (9.25), the reader is referred to Arfken et al. in the Additional Readings. Substituting for the first few Bernoulli numbers, Eq. (9.25) is equivalent to

$$\ln\Gamma(s+1) \sim \frac{1}{2}\ln 2\pi + \left(s+\frac{1}{2}\right)\ln s - s + \frac{1}{12s} - \frac{1}{360s^3} + \frac{1}{1260s^3} - \cdots\,. \tag{9.26}$$

Taking the exponential of Eq. (9.26) and collecting similar terms, we get an expansion for the gamma function:

$$\Gamma(s+1) \sim \sqrt{2\pi}\,s^{s+1/2}e^{-s}\left[1+\frac{1}{12s}+\frac{1}{288s^2}-\frac{139}{51840s^3}+\cdots\right]. \tag{9.27}$$

Stirling's series is formally divergent, but its error when truncated at any finite N will be less than the first omitted term. We provide below some examples illustrating the accuracy obtainable from the formula.

When used in statistical mechanics, s is ordinarily large enough compared to unity that it is only useful to retain terms that increase at least as fast as s, so we have the extremely simple approximation

$$\ln\Gamma(s) \approx s\ln s - s\,. \tag{9.28}$$

An illustration of the use of this formula is given in Appendix G; see Example G.2.

For s as small as 10, a series truncated at $N=1$ is often satisfactory, and truncation at $N=2$ gives three significant figures for s as small as 2, i.e., already for $\Gamma(3)$. See Table 9.1 and Fig. 9.3.

Exercises

9.3.1. Use Stirling's formula to estimate 52!, the number of possible rearrangements of cards in a standard deck of playing cards. What can you say about the error if you use Eq. (9.28) for your estimate?

9.3.2. Show that $\lim\limits_{x\to\infty} x^{b-a}\dfrac{\Gamma(x+a+1)}{\Gamma(x+b+1)} = 1\,.$

Table 9.1: Ratios of one- and two-term Stirling series to exact values of $\Gamma(s+1)$.

s	$\dfrac{1}{\Gamma(s+1)}\sqrt{2\pi}s^{s+1/2}e^{-s}$	$\dfrac{1}{\Gamma(s+1)}\sqrt{2\pi}s^{s+1/2}e^{-s}\left(1+\dfrac{1}{12s}\right)$
1	0.92213	0.99898
2	0.95950	0.99949
3	0.97270	0.99972
4	0.97942	0.99983
5	0.98349	0.99988
6	0.98621	0.99992
7	0.98817	0.99994
8	0.98964	0.99995
9	0.99078	0.99996
10	0.99170	0.99998

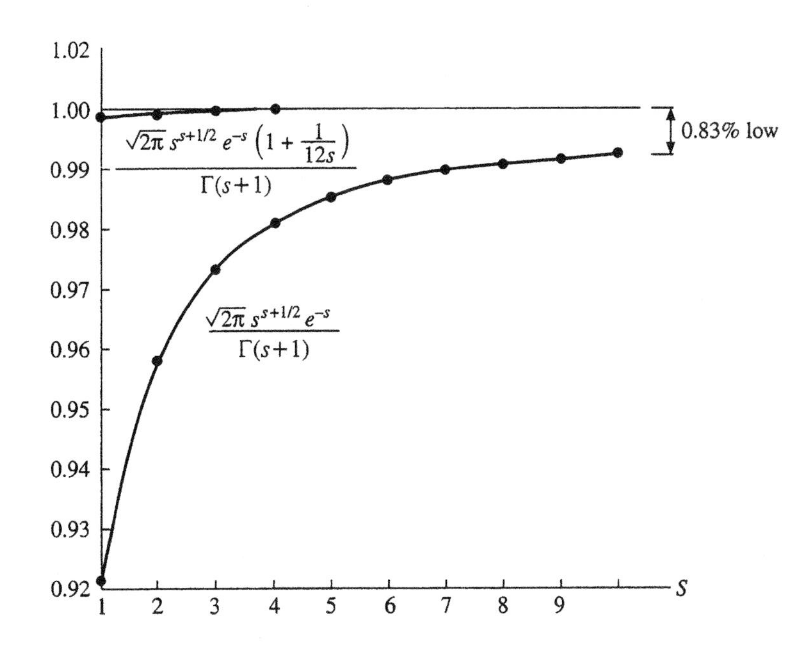

Figure 9.3: Accuracy of Stirling's formula.

9.3.3. Show that $\lim\limits_{n\to\infty}\dfrac{(2n-1)!!}{(2n)!!}\,n^{1/2}=\pi^{-1/2}$.

9.3.4. Find $\lim\limits_{n\to\infty}\dfrac{\Gamma(n+\frac{3}{2})}{\sqrt{n}\;\Gamma(n+1)}$.

9.4 BETA FUNCTION

The **beta function**, defined as

$$B(p,q)=\frac{\Gamma(p)\,\Gamma(q)}{\Gamma(p+q)}, \tag{9.29}$$

is useful, in large part because many important definite integrals can be written in terms of beta functions. This function is called "beta p, q" and the "B" is an uppercase beta. Note that it is obvious from the definition that $B(q,p) = B(p,q)$; this function is symmetric in its two arguments.

TRIGONOMETRIC INTEGRALS

We start our analysis of the beta function by writing $\Gamma(p)\,\Gamma(q)$ as a product of two integrals, which we then manipulate to reduce to a single integration. Our first step is to rewrite Eq. (9.4), changing the integration variable from t to x^2:

$$\Gamma(p) = \int_0^\infty t^{p-1} e^{-t}\, dt = \int_0^\infty x^{2p-2} e^{-x^2} (2x\, dx) = 2\int_0^\infty x^{2p-1} e^{-x^2}\, dx\,. \tag{9.30}$$

Writing a similar formula for $\Gamma(q)$, but now using y as the dummy variable of integration, we reach

$$\Gamma(p)\,\Gamma(q) = 4\int_0^\infty dx \int_0^\infty dy\, x^{2p-1} y^{2q-1} e^{-x^2-y^2}\,. \tag{9.31}$$

We now rewrite Eq. (9.31) in polar coordinates (r,θ) with $x = r\cos\theta$, $y = r\sin\theta$, so $x^2+y^2 = r^2$ and $dx\, dy = r\, dr\, d\theta$. In the polar coordinates the integration is over the entire first quadrant, i.e., $0 \le r < \infty,\ \ 0 \le \theta \le \pi/2$. We get

$$\begin{aligned}\Gamma(p)\,\Gamma(q) &= 4\int_0^{\pi/2} d\theta \int_0^\infty r\, dr\, (r\cos\theta)^{2p-1} (r\sin\theta)^{2q-1}\, e^{-r^2} \\ &= 4\left[\int_0^\infty r^{2p-2q-1} e^{-r^2}\, dr\right] \int_0^{\pi/2} \cos^{2p-1}\theta\, \sin^{2q-1}\theta\, d\theta\,.\end{aligned} \tag{9.32}$$

Making the substitution $r^2 = t$ in the r integration, we have

$$\int_0^\infty r^{2p+2q-1}\, e^{-r^2}\, dr = \frac{1}{2}\int_0^\infty t^{p+q-1}\, e^{-t}\, dt = \frac{1}{2}\Gamma(p+q)\,. \tag{9.33}$$

Substituting Eq. (9.33) into Eq. (9.32) and rearranging, we reach the first of a set of integral formulas for which the beta function is useful:

$$B(p,q) = \frac{\Gamma(p)\,\Gamma(q)}{\Gamma(p+q)} = 2\int_0^{\pi/2} \cos^{2p-1}\theta\, \sin^{2q-1}\theta\, d\theta\,. \tag{9.34}$$

Equation (9.34) is often needed for integer values of p and q; in that case, with $p \to p+1$ and $q \to q+1$ it can be written

$$B(p+1,q+1) = \frac{p!\, q!}{(p+q+1)!} = 2\int_0^{\pi/2} \cos^{2p+1}\theta\, \sin^{2q+1}\theta\, d\theta\,, \tag{9.35}$$

thereby producing trigonometric definite integrals with odd powers of the sine and cosine.

If in Eq. (9.34) we make the shift $p \to p+1/2$ and $q \to q+1$ we produce an integral in which $\cos\theta$ is raised to an even integer power; because gamma functions of half-integer argument have convenient closed expressions, the resulting formulas are useful. Thus,

$$B(p+1/2,q+1) = \frac{\Gamma(p+1/2)\, q!}{\Gamma(p+q+3/2)} = 2\int_0^{\pi/2} \cos^{2p}\theta\, \sin^{2q+1}\theta\, d\theta\,. \tag{9.36}$$

Similar formulas can be written for the odd/even- and even/even-power trigonometric integrals.

ADDITIONAL DEFINITE INTEGRALS

By making changes of the integration variable, Eq. (9.36) can be converted into other useful integrals. With $\cos^2\theta = t$, we get

$$B(p+1, q+1) = \int_0^1 t^p (1-t)^q \, dt \,. \tag{9.37}$$

From that equation, the further substitution $t = x^2$ yields

$$B(p+1, q+1) = 2\int_0^1 x^{2p+1}(1-x^2)^q \, dx \,, \tag{9.38}$$

while the substitution $t = u/(1+u)$ into Eq. (9.37) leads to

$$B(p, q) = \int_0^\infty \frac{u^{p-1}}{(1+u)^{p+q}} \, du \,. \tag{9.39}$$

SYMBOLIC COMPUTING

Both MAPLE and MATHEMATICA have a knowledge of the beta function built into their integral evaluators, so it will ordinarily not be necessary to identify an integral as a beta function before attempting its symbolic evaluation. However, there may be instances in which a user may wish to introduce a beta function explicitly. Commands that do this are `Beta(p,q)` (MAPLE) and `Beta[p,q]` (MATHEMATICA).

Example 9.4.1. Beta Function

Let's evaluate

$$I = \int_0^\infty \frac{x^{3/2}}{(1+x)^5} \, dx \,.$$

From Eq. (9.39), we see that $I = B(5/2,\, 5/2)$. By hand computation we find

$$B(5/2,\, 5/2) = \frac{\Gamma(5/2)\,\Gamma(5/2)}{\Gamma(5)}, \quad \text{and} \quad \Gamma(5/2) = \frac{1}{2} \cdot \frac{3}{2}\,\Gamma(1/2) = \frac{3\sqrt{\pi}}{4} \,.$$

Using also $\Gamma(5) = 4! = 24$, we reach $I = 3\pi/128$.

Checking the above:

```
> int(x^(3/2)/(1+x)^5, x=0 .. infinity);
```

$$\frac{3}{128}\,\pi$$

```
Beta[5/2,5/2]
```

$$\frac{3\pi}{128}$$

■

Exercises

9.4.1. Verify the following beta function identities:

(a) $B(a, b) = B(a+1, b) + B(a, b+1)$,

(b) $B(a,b) = \frac{a+b}{b} B(a, b+1)$,

(c) $B(a,b) = \frac{b-1}{a} B(a+1, b-1)$,

(d) $B(a,b)B(a+b,c) = B(b,c)B(a,b+c)$.

9.4.2. Show that $\int_0^1 (1-x^4)^{-1/2}\,dx = \frac{[\Gamma(5/4)]^2 \cdot 4}{(2\pi)^{1/2}} = 1.311028777$.

9.4.3. Evaluate the following integrals:

$$(a)\ \int_0^1 \frac{dx}{\sqrt{1-x^2}}, \qquad (b)\ \int_0^\infty \frac{y^2\,dy}{(1+y)^6},$$

$$(c)\ \int_0^{\pi/2} \frac{d\theta}{\sqrt{\sin\theta}}, \qquad (d)\ \int_0^{\pi/2} (\cos x)^{5/2}\,dx.$$

9.4.4. Evaluate $\int_0^5 x^{-1/3}(5-x)^{10/3}\,dx$.

9.4.5. (a) Show that

$$\int_{-1}^1 (1-x^2)^{1/2} x^{2n}\,dx = \begin{cases} \pi/2, & n=0, \\ \pi\dfrac{(2n-1)!!}{(2n+2)!!}, & n=1,2,3,\ldots \end{cases}$$

(b) Show that

$$\int_{-1}^1 (1-x^2)^{-1/2} x^{2n}\,dx = \begin{cases} \pi, & n=0, \\ \pi\dfrac{(2n-1)!!}{(2n)!!}, & n=1,2,3,\ldots \end{cases}$$

9.4.6. Show that, for integer p and q,

(a) $\displaystyle\int_0^1 x^{2p+1}(1-x^2)^{-1/2}\,dx = \frac{(2p)!!}{(2p+1)!!}$,

(b) $\displaystyle\int_0^1 x^{2p}\,(1-x^2)^q\,dx = \frac{(2p-1)!!\,(2q)!!}{(2p+2q+1)!!}$.

9.4.7. Show that

$$\int_0^\infty \frac{\sinh^\alpha x}{\cosh^\beta x}\,dx = \frac{1}{2}\,B\left(\frac{\alpha+1}{2}, \frac{\beta-\alpha}{2}\right), \qquad -1 < \alpha < \beta.$$

Hint. Let $\sinh^2 x = u$.

9.4.8. From

$$\lim_{n\to\infty} \frac{\displaystyle\int_0^{\pi/2} \sin^{2n}\theta\,d\theta}{\displaystyle\int_0^{\pi/2} \sin^{2n+1}\theta\,d\theta} = 1,$$

derive the Wallis formula for π:

$$\frac{\pi}{2} = \frac{2 \cdot 2}{1 \cdot 3} \cdot \frac{4 \cdot 4}{3 \cdot 5} \cdot \frac{6 \cdot 6}{5 \cdot 7} \cdots .$$

9.4.9. Show that $\lim_{n\to\infty} \left[n^x B(x,n) \right] = \Gamma(x)$.

9.4.10. Show, by means of the beta function, that

$$\int_t^z \frac{dx}{(z-x)^{1-\alpha}(x-t)^\alpha} = \frac{\pi}{\sin \pi\alpha}, \qquad 0 < \alpha < 1.$$

9.5 ERROR FUNCTION

The gamma function can be generalized by replacing one of the limits of the Euler integral by a variable, thereby defining a function of two arguments. The resulting functions are called **incomplete gamma functions**; the two most obvious ways of defining such functions are

$$\gamma(s,x) = \int_0^x t^{s-1} e^{-t}\, dt\,, \tag{9.40}$$

$$\Gamma(s,x) = \int_x^\infty t^{s-1} e^{-t}\, dt\,. \tag{9.41}$$

We do not undertake a detailed analysis of this wider class of functions. However, special cases of the incomplete gamma function are important in physical science; this and the next section reviews two such functions.

The function of interest in the present section is called the **error function** and is designated **erf** (and pronounced the way it is spelled). Its formal definition is as an integral:

$$\text{erf}\, x = \frac{2}{\sqrt{\pi}} \int_0^x e^{-t^2}\, dt\,. \tag{9.42}$$

The factor premultiplying the integral has the effect that in the limit $x \to \infty$, we have erf $x \to 1$. This scaling is convenient when erf x appears in computations of probabilities, where an infinite upper integration limit may correspond to "certainty," i.e., unit probability. If (for positive x) we change the dummy variable in Eq. (9.42) from t to $u^{1/2}$, we get (for $x > 0$)

$$\text{erf}\, x = \frac{2}{\sqrt{\pi}} \int_0^{x^2} e^{-u}(u^{-1/2}du/2) = \pi^{-1/2}\gamma(1/2, x^2)\,, \tag{9.43}$$

showing that erf is an incomplete gamma function.

Returning to Eq. (9.42), we note that the integrand of the erf integral is a bell-shaped curve centered at zero and known as the Gauss normal probability distribution; because of this symmetry, erf is an odd function, with erf(0) = 0, erf($+\infty$) = 1, and erf($-\infty$) = -1. See Fig. 9.4.

Related to erf is the **complementary error function**, defined as

$$\text{erfc}\, x = 1 - \text{erf}\, x\,, \tag{9.44}$$

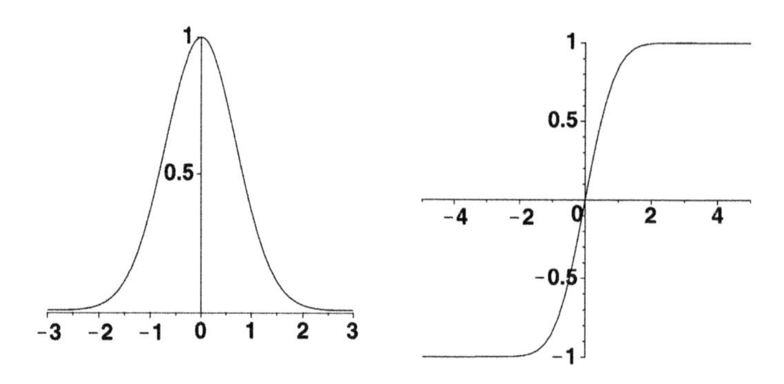

Figure 9.4: Left panel, e^{-x^2}; right panel, error function, erf x.

which has values $\mathrm{erfc}(+\infty) = 0$, $\mathrm{erfc}(0) = 1$, and $\mathrm{erfc}(-\infty) = 2$. Another related function is the **Gauss standard normal cumulative distribution** $\Phi(x)$, with definition

$$\Phi(x) = \frac{1}{\sqrt{2\pi}} \int_{-\infty}^{x} e^{-t^2/2}\, dt\,. \tag{9.45}$$

It is straightforward to show that

$$\Phi(x) = \frac{1}{2}\left[1 + \mathrm{erf}\left(\frac{x}{\sqrt{2}}\right)\right]. \tag{9.46}$$

$\Phi(x)$ is sometimes also called an error function, so it behooves the reader to make sure that the definitions entering an analysis are entirely consistent. However, we note that except in applications of probability, most workers use erf or erfc, not Φ.

COMPUTATION OF ERROR FUNCTIONS

A power series expansion of erf x can be obtained simply by expanding the exponential in Eq. (9.42) and integrating term-by term. The result is

$$\mathrm{erf}\; x = \frac{2}{\sqrt{\pi}} \sum_{n=0}^{\infty} \frac{(-1)^n\, x^{2n+1}}{(2n+1)\, n!}\,. \tag{9.47}$$

This series converges for all x, but the convergence becomes extremely slow if x significantly exceeds unity. A good computational strategy for large x is to use an asymptotic expansion for erfc x, then computing erf x (if that is what is actually desired) as $1 - \mathrm{erfc}\; x$.

The asymptotic expansion for erfc x was the topic of Exercise 2.10.1. It can be obtained by repeated integrations by parts on the integral

$$\mathrm{erfc}\; x = \frac{2}{\sqrt{\pi}} \int_{x}^{\infty} e^{-x^2}\, dx, \tag{9.48}$$

taking each time the factor to be integrated as $2xe^{-x^2}\, dx$, with a differentiation applied to the remainder of the integrand. The result (truncated as shown) is

$$\mathrm{erf}\; x = 1 - \frac{e^{-x^2}}{\sqrt{\pi}\, x}\left(1 - \frac{1}{2x^2} + \frac{1\cdot 3}{2^2\, x^4} - \frac{1\cdot 3\cdot 5}{2^3\, x^6} + \cdots + (-1)^n\, \frac{(2n-1)!!}{2^n\, x^{2n}}\right). \tag{9.49}$$

SYMBOLIC COMPUTING

In MAPLE, error functions are accessed as `erf(x)` and `erfc(x)`; in MATHEMATICA, they are denoted `Erf[x]` and `Erfc[x]`. The incomplete gamma function is also directly available. The function $\Gamma(s,x)$ of Eq. (9.41) is called `GAMMA(s,x)` in MAPLE and `Gamma[s,x]` in MATHEMATICA.

Example 9.5.1. Symbolic Computation, Error Functions

When a closed-form exact result is available, both symbolic systems give it. When erf or erfc is called with an exact (nondecimal) argument and there is no closed exact result, the expression is returned unevaluated. Function calls with decimal arguments are returned as numeric values. However, note that MAPLE does not reduce expressions with symbolic constants (e.g., π, γ) to numeric form unless forced by the use of `evalf`.

In MAPLE,

```
> erf(0), erf(infinity), erfc(-infinity),erfc(0);
```

$$0,\ 1,\ 2,\ 1.$$

```
> erf(0.5*gamma), erf(1.23), erfc(Pi/2);
```

$$\mathrm{erf}(0.5\,\gamma),\ 0.9180501041,\ \mathrm{erfc}\left(\frac{1}{2}\pi\right)$$

```
> evalf(erf(0.5*Pi));
```

$$0.9736789251$$

In MATHEMATICA,

`Erfc[0]`	1
`Erf[Pi/2]`	$\mathrm{Erf}\left[\frac{\pi}{2}\right]$
`Erf[0.5*Pi]`	0.973679
`Erf[1.23]`	0.91805

Check of Eq. (9.43):

```
Pi^(-1/2)*(Gamma[1/2] - Gamma[1/2, 1.23^2])
```

0.91805

■

Example 9.5.2. Gaussian Orbitals

An electron is said to be in a Gaussian-type orbital centered at the origin if its wave function $\psi(\mathbf{r})$ is of the form

$$\psi(\mathbf{r}) = Ne^{-\alpha r^2}, \quad \text{with} \quad N = \left(\frac{2\alpha}{\pi}\right)^{3/4}.$$

Here N is a **normalization constant**, chosen so that

$$\int |\psi(\mathbf{r})|^2\, d^3r = 1\,.$$

Before proceeding further let's verify that ψ is normalized. Writing the integral in spherical coordinates:

$$\int |\psi(\mathbf{r})|^2\, d^3r = N^2 \int_0^{2\pi} d\varphi \int_0^{\pi} \sin\theta\, d\theta \int_0^{\infty} e^{-2\alpha r^2}\, r^2\, dr$$

$$= 4\pi \left(\frac{2\alpha}{\pi}\right)^{3/2} \int_0^{\infty} e^{-2\alpha r^2}\, r^2\, dr\,.$$

Using symbolic computing, we evaluate the r integral:

$$\int_0^{\infty} e^{-2\alpha r^2}\, r^2\, dr = \frac{\sqrt{2\pi}}{16\alpha^{3/2}}\,.$$

Using this result the normalization is confirmed.

Now let's compute the probability that the electron is within one distance unit of the origin. That probability is given by the integral

$$P = 4\pi N^2 \int_0^{1} e^{-2\alpha r^2}\, r^2\, dr\,.$$

Using symbolic computing, we find

$$P = \left(\frac{128\alpha^3}{\pi}\right)^{1/2} \left[-\frac{e^{-2\alpha}}{4\alpha} + \frac{\sqrt{2\pi}\,\mathrm{erf}(\sqrt{2\alpha})}{16\alpha^{3/2}}\right].$$

Both symbolic systems regard reduction to an expression involving erf as an evaluation of the integral. If we have a numerical value for α this probability can be reduced to a numerical value. Taking $\alpha = 1$ and forcing complete numerical evaluation with **evalf** or `N`, we get $P = 0.739$.

■

Exercises

9.5.1. Using symbolic computation, compute erf(1) and $\pi^{-1/2}\gamma(1/2, 1)$ and show that they are equal.

9.5.2. Integrals of the form $G_p(\alpha) = \displaystyle\int_0^{\infty} t^p\, e^{-\alpha t^2}\, dt$ arise in many different contexts.

(a) Verify (and then check by symbolic computation) the formulas

$$G_0(\alpha) = \frac{1}{2}\sqrt{\frac{\pi}{\alpha}}\,, \qquad G_1(\alpha) = \frac{1}{2\alpha}\,.$$

(b) By differentiating the formulas of part (a) with respect to α, obtain the more general formulas

$$G_{2p}(\alpha) = \frac{(2p-1)!!\,\sqrt{\pi}}{2^{p+1}\,\alpha^{p+1/2}}\,, \qquad G_{2p+1}(\alpha) = \frac{p!}{2\alpha^{p+1}}\,.$$

(c) Using your symbolic computing system, find (for general p) a formula for $G_p(\alpha)$. Show that the general formula reduces to the results of part (b) when p is a nonnegative integer.

Comment. Does your symbolic system take note of the fact that $G_p(\alpha)$ is only defined for specific ranges of p and α?

9.5.3. As a function of α, find the values of x for which $e^{-\alpha x^2}$ has half of its maximum value (the maximum is reached at $x = 0$).

9.6 EXPONENTIAL INTEGRAL

Another incomplete gamma function of some interest is the **exponential integral** $\mathrm{Ei}(x)$, customarily defined (for historical reasons) as

$$-\mathrm{Ei}(-x) = \int_x^\infty \frac{e^{-t}}{t}\,dt \equiv E_1(x)\,. \tag{9.50}$$

This is an awkward definition, with ambiguities for negative x due to the fact that the defining integral is divergent at $x = 0$, but most difficulties can be avoided by using E_1 rather than Ei and restricting evaluation to the range $x > 0$. The relation between E_1 and an incomplete gamma function is

$$E_1(x) = \Gamma(0, x)\,.$$

This equation shows that $E_1(0)$ is equivalent to $\Gamma(0)$, which is undefined. A graph of $E_1(x)$ is presented in Fig. 9.5.

An alternate form for $E_1(x)$, obtained by a change of scale of the integration variable, is

$$E_1(x) = \int_1^\infty \frac{e^{-xt}}{t}\,dt\,. \tag{9.51}$$

Equation (9.51) can be generalized to the set of related functions

$$E_n(x) = \int_1^\infty \frac{e^{-xt}}{t^n}\,dt\,. \tag{9.52}$$

COMPUTATION OF $E_1(x)$

The integral defining $E_1(x)$ diverges at $x = 0$, so there is not a pure power series expansion about that point. However, if we recognize that the divergence is logarithmic, we can isolate the divergent behavior and supplement it by a convergent

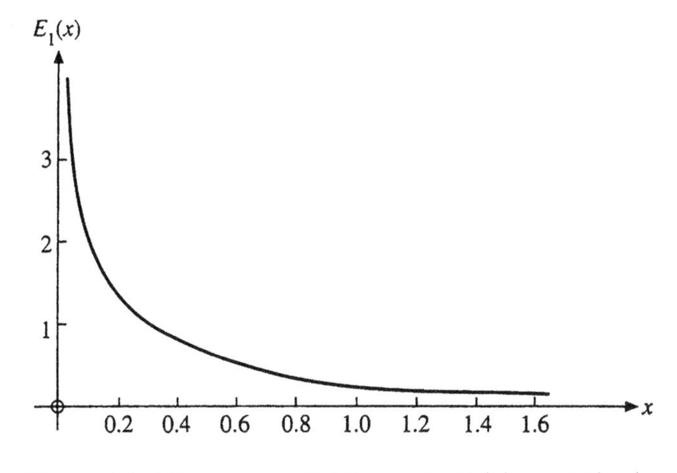

Figure 9.5: The exponential integral, $E_1(x) = -\mathrm{Ei}(-x)$.

expansion of the remainder. The details of a method for doing this can be found in Arfken et al. in the Additional Readings. The resulting formula is

$$E_1(x) = -\gamma - \ln x - \sum_{n=1}^{\infty} \frac{(-1)^n x^n}{n \cdot n!} . \tag{9.53}$$

Here γ is the Euler-Mascheroni constant. The series in Eq. (9.53) is convergent for all $x > 0$, but the convergence rate is too slow to be useful when $x \gg 1$.

For large x the most practical computational approach is via an asymptotic expansion. Such an expansion for $E_1(x)$ was used as an example in the general discussion of asymptotic expansions in Section 2.10. It was there shown, in Eq. (2.83), that repeated integrations by parts on the integral of Eq. (9.51) yields the formula

$$E_1(x) \sim e^{-x} \left[\frac{1}{x} - \frac{1!}{x^2} + \frac{2!}{x^3} - \frac{3!}{x^3} + \cdots \right] . \tag{9.54}$$

SYMBOLIC COMPUTING

It is recommended that symbolic computations of exponential integrals be carried out using the function defined as $E_1(x)$. In both MAPLE and MATHEMATICA a command to evaluate $E_n(x)$, Eq. (9.52), is available. In MAPLE, it is `Ei(n,x)`; in MATHEMATICA it is `ExpIntegralE[n,x]`. The function E_1 can therefore be obtained from these commands by setting $n = 1$. Both languages also contain quantities corresponding to Ei(x), namely `Ei(x)` and `ExpIntegralEi[x]`. These are not synonymous with $E_1(x)$ and cannot be used when E_1 is intended.

Example 9.6.1. Symbolic Computing, Exponential Integral

In MAPLE,

`> Digits:=6: x := 1.75:`		
`> Ei(1,x);`	0.0694887	This is $E_1(x)$
`> Ei(x);`	4.08365	This isn't!
`> -Ei(-x);`	0.0694887	E_1 again.

The function you get depends on whether `Ei` is called with one or with two arguments.

In MATHEMATICA,

`x = 1.75;`		
`ExpIntegralE[1,x]`	0.0694887	This is $E_1(x)$
`ExpIntegralEi[x]`	4.08365	This isn't!
`-ExpIntegralEi[-x]`	0.0694887	E_1 again.

Here `ExpIntegralE` and `ExpIntegralEi` must be called with the correct numbers of arguments.

■

Exercises

9.6.1. Show that $E_1(z)$ may be written as

$$E_1(z) = e^{-z} \int_0^\infty \frac{e^{-zt}}{1+t}\, dt.$$

9.6.2. A generalized exponential integral $E_n(x)$ was defined in Eq. (9.53). Show that $E_n(x)$ satisfies the recurrence relation

$$E_{n+1}(x) = \frac{1}{n}\, e^{-x} - \frac{x}{n}\, E_n(x), \quad n = 1, 2, 3, \cdots.$$

9.6.3. With $E_n(x)$ as defined in Eq. (9.53), show that for $n > 1$, $E_n(0) = 1/(n-1)$.

9.6.4. Verify by symbolic computing that the exponential integral has the expansion

$$\int_x^\infty \frac{e^{-t}}{t}\, dt = -\gamma - \ln x - \sum_{n=1}^{\infty} \frac{(-1)^n x^n}{n \cdot n!},$$

where γ is the Euler-Mascheroni constant (`gamma` in MAPLE, `EulerGamma` in MATHEMATICA).

Proceed by comparing the above formula with `Ei(1,x)` or `ExpIntegralE[1,x]` for $x = 0.1$, 0.01, and 0.001.

Additional Readings

Abramowitz, M., & Stegun, I. A. (Eds.) (1972). *Handbook of mathematical functions with formulas, graphs, and mathematical tables (AMS-55).* Washington, DC: National Bureau of Standards (Reprinted, Dover (1974). Contains a wealth of information about gamma functions, incomplete gamma functions, exponential integrals, error functions, and related functions in Chapters 4–6).

Arfken, G. B., Weber, H. J., & Harris, F. E. (2013). *Mathematical methods for physicists* (7th ed.). New York: Academic Press.

Davis, H. T. (1933). *Tables of the higher mathematical functions.* Bloomington, IN: Principia Press (Volume I contains extensive information on the gamma function and the polygamma functions).

Gradshteyn, I. S., & Ryzhik, I. M. (2007). In A. Jeffrey, & D. Zwillinger (Eds.), *Table of integrals, series, and products* (7th ed.). New York: Academic Press.

Luke, Y. L. (1975). *Mathematical functions and their approximations.* New York: Academic Press (This is an updated supplement to *Handbook of mathematical functions with formulas, graphs, and mathematical tables (AMS-55).* Chapter 1 deals with the gamma function. Chapter 4 treats the incomplete gamma function and a host of related functions).

Olver, F. W. J., Lozier, D. W., Boisvert, R. F., & Clark, C. W. (Eds.) (2010). *NIST handbook of mathematical functions.* Cambridge: Cambridge University Press (Update of *AMS-55* (Abramowitz and Stegun, 1972), but links to computer programs are provided instead of tables of data).

Chapter 10

ORDINARY DIFFERENTIAL EQUATIONS

10.1 INTRODUCTION

Many problems in physics and engineering involve differential equations. This is not surprising, as many physical laws connect the rate at which some physical quantity changes (in space or in time) with other physical quantities. For example, Newton's second law for a particle on a trajectory $x(t)$ and subject to an applied force $F(x) = \sin x$ corresponds to the equation

$$m\frac{d^2x}{dt^2} = \sin x\,. \tag{10.1}$$

If the force had been $F(t) = \sin t$, the differential equation would be

$$m\frac{d^2x}{dt^2} = \sin t\,. \tag{10.2}$$

An electric inductance L has a voltage across it which is proportional to the rate at which the current I through it is changing with time. The voltage across a capacitor C is proportional to its charge q, while the voltage across a resistor R is proportional to the current. In the circuit shown in Fig. 10.1, $I = dq/dt$, and the sum of the voltages across L, C, and R must at all times add to V, the applied voltage. We may therefore write

$$L\frac{dI}{dt} + \frac{q}{C} + RI = V\,.$$

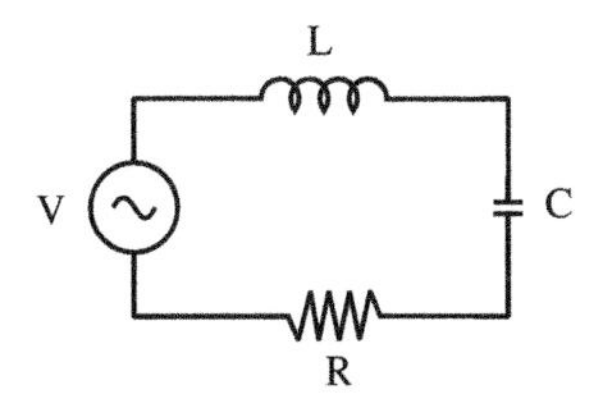

Figure 10.1: Electronic circuit.

Mathematics for Physical Science and Engineering.
http://dx.doi.org/10.1016/B978-0-12-801000-6.00010-9

Since q and I are related: $I = dq/dt$, we differentiate this equation with respect to t, obtaining

$$L\frac{d^2I}{dt^2} + \frac{I}{C} + R\frac{dI}{dt} = \frac{dV}{dt}\,. \tag{10.3}$$

The rate at which the temperature distribution $T(x)$ of a rod (initially not at a uniform temperature) equilibrates is a function of both position on the rod (x) and the time t. The differential equation governing this process can be shown to have the form

$$\frac{\partial T}{\partial t} = C\frac{\partial^2 T}{\partial x^2}\,. \tag{10.4}$$

The rate at which the concentration y of a reactant in a process in which the reactant is destroyed by binary collisions satisfies an equation of the type

$$\frac{dy}{dt} = -k\,y^2\,. \tag{10.5}$$

Equations (10.1)–(10.5) are all differential equations. However, these equations vary markedly in their properties and in the methods by which they can solved, and we classify them accordingly.

A differential equation has as its **independent variables** the quantities with respect to which derivatives are taken; in Eq. (10.1) the independent variable is t; in Eq. (10.4) the independent variables are t and x. The quantity being differentiated is called a **dependent variable**. When there is only one independent variable, we have an **ordinary differential equation** (abbreviated **ODE**); when there is more than one independent variable (and the derivatives are therefore partial derivatives), we have a **partial differential equation**, or **PDE**. Thus, Eqs. (10.1)–(10.3) and (10.5) are ODEs; Eq. (10.4) is a PDE.

The **order** of a differential equation is the order of the highest derivative occurring in it; Eqs. (10.4) and (10.5) are first-order; Eqs. (10.1)–(10.3) are second-order.

If each term of a differential equation is independent of, or linear in the **dependent** variable, a differential equation is said to be **linear**. Note that since the differentiation operator is linear, y' and y'' are regarded as linear in y. Thus, Eq. (10.3) is a linear ODE and Eq. (10.4) is a linear PDE. Equation (10.2) is also linear; note that it is linearity with respect to the **dependent variable** that determines linearity. Equations (10.1) and (10.5) are nonlinear.

A differential equation is **homogeneous** if all its terms contain the dependent variable to the same aggregate power (with y' and y'' regarded as of power 1). A differential equation that is not homogeneous is called **inhomogeneous**.

A **solution** to a differential equation (in the variables x and y) is a relation between x and y (and not involving any derivatives) that yields an identity when substituted into the differential equation. Such a relation is regarded as a solution even if it is not in the explicit form $y = f(x)$. Expressions such as $y = \int f(x)\,dx$, where $f(x)$ is a known function, are also regarded as solutions even when one cannot perform the integration analytically. In such cases it is sometimes said that the differential equation has been reduced to a **quadrature**. The essential point is that an unambiguous recipe for the solution has been supplied.

The present chapter deals only with ODEs. However, PDEs, which present additional and more complicated issues (but are of great importance to physics, which lives in three-dimensional space) are treated in a later chapter.

This chapter does not cover several elementary topics that for pedagogical reasons are discussed later in this book. We therefore call attention that integral-transform methods for ODEs, particularly the use of Fourier and Laplace transforms, are discussed in Chapter 13, while the series-expansion method for second-order linear ODEs (the method of Frobenius) and some of its important applications are discussed in Chapter 14. Finally, we point out that our discussion is not as comprehensive as in the now-standard treatises on the subject, and readers who need more material than is treated here are referred to the Additional Readings, and in particular to the classic work by Ince.

Before proceeding to detailed discussions of specific methods for solving ODEs, we examine here some general properties of linear ODEs.

HOMOGENEOUS LINEAR ODEs

If $y(x)$ is a solution to a homogeneous ODE, then $Cy(x)$ is also a solution for any nonzero value of the constant C. This observation is valid because the substitution of Cy for y will cause each term in the ODE to be multiplied by the same power of C. If the ODE is also linear, and it has two solutions $y_1(x)$ and $y_2(x)$, then any linear combination of y_1 and y_2 is also a solution. Let's check this, using as an example the following second-order ODE:

$$y'' + p(x)y' + q(x)y = 0\,.$$

We write

$$(C_1y_1 + C_2y_2)'' + p(x)(C_1y_1 + C_2y_2)' + q(x)(C_1y_1 + C_2y_2) =$$

$$C_1\Big[y_1'' + p(x)y_1' + q(x)y_1\Big] + C_2\Big[y_2'' + p(x)y_2' + q(x)y_2\Big] = 0 + 0 = 0\,.$$

The above considerations indicate that the most general solution to a homogeneous linear ODE (usually just called its **general solution**) can be written

$$y(x) = \sum_{i=1}^{N} C_i y_i(x)\,, \tag{10.6}$$

where the y_i can be required to be linearly independent. Linear dependence of functions was discussed in Section 4.8, where it was also shown how to use the Wronskian of the function set to determine whether or not the functions are linearly dependent. The linear independence asserted for Eq. (10.6) means that any y_i that is linearly dependent upon others has been removed.

We next ask how many linearly independent solutions a homogeneous linear ODE can possess. Since the removal of each derivative in the ODE can correspond to an integration, one might expect that the general solution to an nth-order ODE would involve n independent constants of integration. For homogeneous linear equations these occur as coefficients of the individual solutions, so we expect an nth-order homogeneous linear ODE to have as its general solution the form given in Eq. (10.6) containing n terms. A more general ODE of order n will also have a solution containing n independent constants, but the constants will not appear as the coefficients in a linear form corresponding to Eq. (10.6).

To determine the number of solutions for a first-order ODE, of the form $y' + q(x)y = 0$, let's rearrange it to the form

$$\frac{y'}{y} = -q(x)\,,$$

and we can then conclude that any two of its solutions, y_1 and y_2, must satisfy

$$\frac{y_1'}{y_1} = \frac{y_2'}{y_2} \tag{10.7}$$

for every value of x, with the result that y_1 and y_2 must be (for all x) proportional, i.e., linearly dependent. We thus have the expected result:

A first-order homogeneous linear ODE has a general solution of the form of Eq. (10.6) *with* $N = 1$, *i.e., it is a single function with an undetermined coefficient.*

Returning to Eq. (10.7), we can multiply that equation by $y_1 y_2$ and rearrange the result to obtain

$$y_1 y_2' - y_2 y_1' = W(y_1, y_2) = 0\,. \tag{10.8}$$

We have identified the expression obtained from Eq. (10.7) as the Wronskian W of y_1 and y_2 (see Section 4.8). The result $W = 0$ is also a signal of the linear dependence of y_1 and y_2.

Using Wronskians of the solutions of ODEs of orders higher than $n = 1$, one can show that the general solution to a homogeneous linear ODE of order n consists of n linearly independent solutions with arbitrary coefficients. Details are in Appendix J.

INHOMOGENEOUS LINEAR ODEs

We now consider inhomogeneous linear ODEs. Our first observation is that if we have any solution y_i to the corresponding homogeneous ODE (that resulting if we set the inhomogeneous term to zero), it can be added to a solution y_p of the inhomogeneous equation, producing a function that is still a solution to the inhomogeneous ODE. To see how this comes about, write the inhomogeneous ODE as $\mathcal{L}y_p = f(x)$, and its homogeneous counterpart as $\mathcal{L}y_i = 0$. Then

$$\mathcal{L}(y_p + y_i) = \mathcal{L}y_p + \mathcal{L}y_i = f(x) + 0 = f(x)\,.$$

It can be proved that an inhomogeneous linear ODE of order n has a general solution containing n independent constants, so we may now conclude that

The general solution to an inhomogeneous linear ODE of order n consists of any solution to the inhomogeneous ODE (often called a **particular solution***), plus the general solution to the corresponding homogeneous ODE.*

Inhomogeneous linear ODEs are therefore often solved by taking the following steps:

(1) Find the general solution to the corresponding homogeneous ODE;

(2) Find **any** solution to the inhomogeneous ODE;

(3) Add the above to obtain a general solution to the inhomogeneous ODE.

INITIAL AND BOUNDARY CONDITIONS

The constants in the general solution to an ODE can often be used to obtain a solution that meets additional conditions. The types of additional conditions usually encountered for a linear ODE of order n are either **initial conditions** (e.g., values of $y(x)$ and its first $n - 1$ derivatives at some initial value of x) or **boundary conditions** (values of $y(x)$ and/or derivatives at both ends of an interval in x). A detailed discussion of the conditions under which initial or boundary conditions can be satisfied is outside the scope of this text, but under a wide variety of circumstances an ODE of order n can be solved subject to n additional conditions.

An important exception to this common situation is encountered when we look at a homogeneous linear second-order ODE with boundary conditions of the type $y(a) = y(b) = 0$. To satisfy the boundary conditions, we need to use a linear combination of the linearly independent solutions y_1 and y_2 to the ODE, i.e., a solution with C_1 and C_2 such that

$$\begin{cases} y(a) = y_1(a)\,C_1 + y_2(a)\,C_2 = 0\,, \\ y(b) = y_1(b)\,C_1 + y_2(b)\,C_2 = 0\,. \end{cases}$$

But these equations have only the solution $C_1 = C_2 = 0$ unless their left-hand sides are linearly dependent, an seemingly improbable occurrence. However, if the ODE contains an adjustable parameter, there may be values of that parameter that produce linear dependence and therewith a nonzero solution to the ODE that satisfies the boundary conditions. Situations of this sort are the **eigenvalue problems** discussed in Chapter 11.

Example 10.1.1. Initial Conditions

The general solution of the ODE

$$y''(t) + \omega^2 y(t) = 0$$

is $y(t) = C_1 \cos\omega t + C_2 \sin\omega t$. Find a solution such that $y(0) = a$ and $y'(0) = b$.

We proceed by first differentiating the general solution $y(t)$:

$$y'(t) = -\omega C_1 \sin\omega t + \omega C_2 \cos\omega t$$

and we therefore find $y(0) = C_1 = a$, $y'(0) = \omega C_2 = b$, so the required specific solution is

$$y(t) = a\cos\omega t + \frac{b}{\omega}\sin\omega t\,.$$

Note that if our initial conditions were homogeneous ($a = b = 0$), we would get $y(t) = 0$, which we do not regard as a solution.

■

Example 10.1.2. Boundary Conditions

Consider again the ODE of Example 10.1.1, but now with the boundary conditions $y(0) = a$, $y(1) = b$. From the ODE we get

$$y(0) = C_1 = a\,, \qquad y(1) = C_1 \cos\omega + C_2 \sin\omega = b\,.$$

We already have a value for C_1. Solving for C_2, we get

$$C_2 = \frac{b - a\cos\omega}{\sin\omega}\,.$$

If a and b are not both zero (and $\sin\omega \neq 0$) we get a unique solution to our ODE and its boundary conditions. But if $a = b = 0$ we get no solution at all unless $\sin\omega = 0$.

■

CHECKING ODE SOLUTIONS

It is a good idea to check the ODE solutions you find, to verify that they actually solve the ODE. Problems are most likely to arise for nonlinear ODEs, but checking your solutions will also detect errors that you might have made in the solution process.

Example 10.1.3. An Interesting Nonlinear ODE

Here is a nonlinear ODE of a type known as a **Clairaut equation**:

$$\mathcal{L}(x) \equiv {y'}^2 + xy' - y = 0 .$$

A special technique for solving this ODE is to start by differentiating it. The result simplifies:

$$\frac{d\mathcal{L}(x)}{dx} = 2y'y'' + y' + xy'' - y' = y''(2y' + x) = 0 .$$

We therefore can apparently have solutions if either $y'' = 0$ or if $2y' + x = 0$. The general solution for $y'' = 0$ is $y = C_1x + C_2$; the general solution for $y' = -x/2$ is $y = -x^2/4 + C_3$.

However, if we substitute $y = C_1x + C_2$ into the original ODE, we find that it is a solution only if $C_2 = C_1^2$; we also find that $y = -x^2/4 + C_3$ is a solution only if $C_3 = 0$. This is a case where checking an alleged solution produces nontrivial results. ■

Exercises

10.1.1. The ODE $y'' - y = 0$ has the general solution $y(t) = Ae^t + Be^{-t}$. Find a solution such that $y(0) = 0$ and $y(\ln 2) = 3/2$.

10.1.2. An object falling under the influence of gravity (and no air resistance or other frictional force) moves subject to the ODE $y''(t) = -g$, where g is the (constant) acceleration of gravity, and $y(t)$ is the vertical position of the object at time t. The general solution to this ODE is $y = C_1 + C_2t - gt^2/2$. Find solutions giving $y(t)$ if

(a) At $t = 0$ the object is at rest at $y = 0$.

(b) At $t = t_0$ the object is at $y = y_0$ moving upward at velocity $y' = v_0$.

10.1.3. The ODE $y'' + y = 0$ has general solution $y(t) = A\sin t + B\cos t$. Find a formula for $y(t)$ if $y(0) = 0$ and $y'(0) = v_0$.

10.1.4. A water droplet (assumed spherical) evaporates at a rate proportional to its surface area. Show that the rate at which its radius shrinks is constant.

10.1.5. The general solution of the ODE $x^2y''(x) + xy'(x) - y(x) = x^2$ is

$$y(x) = \frac{x^2}{3} + C_1x + \frac{C_2}{x} .$$

Find a solution such that $y(0) = y(1) = 0$.

10.1.6. Check the "solutions" found for the ODE of Example 10.1.3 and verify that valid solutions exist only if $C_2 = C_1^2$ and $C_3 = 0$.

10.2 SYMBOLIC COMPUTING

The use of symbolic computing to solve differential equations appears early in this chapter so that symbolic methods can be used to check ODE solutions as they are encountered. Symbolic computing should not be used as a excuse to avoid becoming familiar with basic methods for dealing with ODEs.

Both MAPLE and MATHEMATICA can solve differential equations. The coding to accomplish this and handle the results differs in the two languages, so we discuss them separately.

MAPLE

An ODE to be solved can be entered explicitly into the solving command, but it is often convenient to define a variable and assign to it the entire ODE. If, for example, our ODE is $y'(x) + 2y(x) = x^2$, we might define

```
> ODE := diff(y(x),x) + 2*y(x) = x^2:
```

When the above command is executed it is necessary that both x and y be undefined. Note also that the operator `:=` causes everything to its right to be the object assigned to `ODE` and that the equals sign of the ODE is represented by `=` without a preceding colon. It is important to notice that y always appears in the form $y(x)$; omission of the argument is regarded by MAPLE as an error, but MAPLE's error message may not make it obvious what has gone wrong.

We can now obtain the solution to our ODE by invoking the command `dsolve`:

```
> dsolve(ODE);
```

$$y(x) = \frac{1}{4} - \frac{x}{2} + \frac{x^2}{2} + e^{(-2x)}\_C1$$

Here `dsolve` identifies the dependent and independent variables (y and x) as those occurring in a derivative; in situations where application of this rule leads to ambiguity, an additional argument $y(x)$ must be included, as in `dsolve(ODE, y(x))`. Note also that `dsolve` gave us the general solution to our ODE; it contains one undetermined constant, which MAPLE gave the name $\_C1$.

The solution is written as an equation (not an assignment), which is not a particularly useful form if we plan to use the solution in further computations. To retrieve the solution as a named expression, we next write

```
> Y := op(2,%);
```

$$Y := \frac{1}{4} - \frac{x}{2} + \frac{x^2}{2} + e^{(-2x)}\_C1$$

Our solution has now been assigned to Y. (What has happened here is that the equation $y(x) = \cdots$ is a MAPLE expression with two operands, of which the second is its right-hand side, and `op(2,%)` is a command whose output is the second operand of %, the most recent output.)

Note that the above coding does not establish Y as a function. It is, at this point, simply a variable with the content shown above. If we want Y to be a function, we must define a procedure making it one. For example, we could write

```
> y := proc(t);   eval(Y, x=t);   end proc:
```

The `eval` statement is necessary because we must force the explicit evaluation of Y rather than just referring to it. We need the argument $x = t$ because the evaluation does not really take place unless Y is changed (the evaluation replaces x by t).

Moreover, we could not have made y into a function just by writing `Y(x) := ...` when defining `Y`.

If desired, we can specialize Y, or our function $y(x)$, by assigning a value to $\_C1$. Alternatively, we may solve our ODE with an "initial condition" whose satisfaction determines the value of $\_C1$. For example,

```
> icond := y(1) = 0:
```
Define an initial condition

```
> dsolve({ODE, icond});
```
$$y(x) = \frac{1}{4} - \frac{x}{2} + \frac{x^2}{2} - \frac{1}{4}\frac{e^{(-2x)}}{e^{(-2)}}$$

```
> op(2,%);
```
$$\frac{1}{4} - \frac{x}{2} + \frac{x^2}{2} - \frac{1}{4}\frac{e^{(-2x)}}{e^{(-2)}}$$

Note that the ODE and its initial condition are within braces; they are members of a set of equations, all of which are to be solved together. The initial conditions, like the ODE, are written as equations with equality indicated by `=`, not `:=`.

Possible initial conditions for an ODE in y can include setting the value of y at a specific point to infinity. The number of initial conditions should not exceed the order of the ODE; if this condition is violated, MAPLE may generate meaningless results.

For ODEs of second or higher order, initial conditions can include specifications of the values of derivatives. MAPLE insists that its operator `D` be used to set such initial conditions. `D` is defined such that `D(y)(a)` denotes the derivative of y with respect to its independent variable, evaluated at the point a. Thus, one might write

```
> icond := y(1)=4, D(y)(1)=3;
```
For $y(1)$ to be 4 and $y'(1)$ to be 3

```
> icond :- D(D(y))(2)=1;
```
For $y''(2)$ to be 1

When there is more than one initial condition, note that they are separated by commas.

Example 10.2.1. Solving an ODE with MAPLE

Let's solve the ODE

$$y'' + 5y' + 4y = e^{-2x}\,,$$

subject to the initial conditions $y(0) = 0$, $y'(0) = 1$. After we get the solution let's verify that the initial conditions are satisfied, plot the solution, and characterize the extremum shown on the plot. MAPLE code for these operations can be as follows:

```
> Digits := 6:
```
To avoid carrying too many digits

Set up the ODE and its initial conditions:

```
> ODE := diff(y(x),x,x) + 5*diff(y(x),x) + 4*y(x) = exp(-2*x):
> icond := y(0)=0, D(y)(0)=1:
```

Now solve the ODE. The result is an equation.

```
> dsolve({ODE,icond});
```
$$y(x) = -\frac{1}{6}e^{(-4x)} + \frac{2}{3}e^{(-x)} - \frac{1}{2}e^{(-2x)}$$

We need to save the solution for further use.

```
> Y := op(2,%};
```
$$Y := -\frac{1}{6}e^{(-4x)} + \frac{2}{3}e^{(-x)} - \frac{1}{2}e^{(-2x)}$$

Next generate dY/dx and place the result in YY:

```
> YY := diff(Y,x);
```
$$YY := \frac{2}{3}\,e^{(-4x)} - \frac{2}{3}\,e^{(-x)} + e^{(-2x)}$$

Now evaluate Y and YY at $x = 0$ to check the initial conditions:

```
> x := 0: [Y, YY];
```
$$[0, 1]$$

```
> x := 'x':
```
Undefine x

We need x to be undefined for the next operation; even if that were not the case, potential errors later on may be avoided by undefining quantities as soon as their definitions are no longer needed.

Assuming that we are interested in $y(x)$ for $x > 0$ we plot it for $0 < x < 6$:

```
> plot(Y, x=0 .. 6);
```

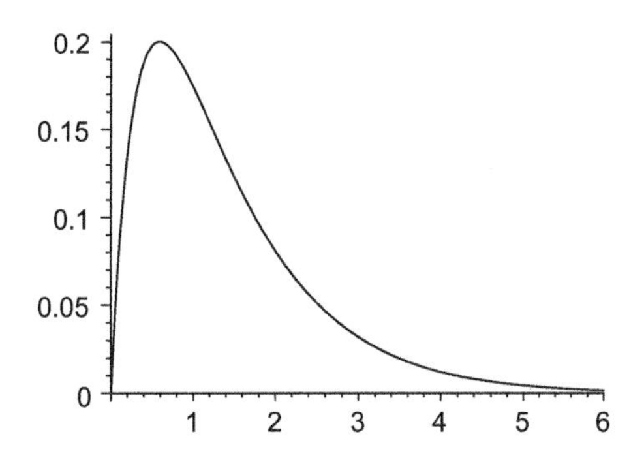

We characterize the maximum in $y(x)$ by finding the x value at which $dy/dx = 0$ (this is a zero of YY). Thus,

```
> [solve(YY = 0., x)];
```
$$[0.591360, \; -0.295680 - 1.77822\,I, \; -0.2956580 + 1.77822\,I]$$

The coding of this MAPLE statement deserves several comments. First, because we are solving an algebraic (and not a differential) equation, the command is `solve`, not `dsolve`. Second, because we want a decimal result and not a complicated analytical expression, we set YY equal to a decimal zero so that the computation will be reduced to decimal numbers. Third, we placed the command within square brackets so that the result would be in the form of a list, thereby making it easy to select the relevant solution (the positive real result). We now set x equal to that result and evaluate Y:

```
> x := %[1];
```
$$x := 0.591360$$

```
> Y;
```
$$0.200176$$

This output shows that the maximum is at $x = 0.591$, with $y(x) = 0.200$.

Sometimes we will want our ODE solution established as a function, thereby permitting us to invoke it with an arbitrary argument. We proceed as follows:

```
> x:='x': y:='y':
```
Undefine, in case it is necessary

```
> y = proc(t); eval(Y, x=t); end proc:
```
y is now a function

To check what we have done,

```
> y(z);
```

$$-\frac{1}{6}e^{(-4z)}+\frac{2}{3}e^{(-z)}-\frac{1}{2}e^{(-2z)}$$

■

MATHEMATICA

It may be convenient to define a variable and assign to it the entire ODE to be solved. If, for example, our ODE is $y'(x) + 2y(x) = x^2$, we might define

```
ode = y'[x] + 2*y[x] == x^2
```

Derivatives are indicated by primes (right single quotes); the second derivative of y is `y''`, written using two single quotes, not a double quote. Derivatives can also (with identical results) be written using MATHEMATICA's `D` operator. When the above command is executed it is necessary that both x and y be undefined. Note also that the operator `=` causes everything to its right to be the object assigned to `ode` and that the equals sign of the ODE is represented by `==`.

The command for solving an ODE is `DSolve`, which must contain as arguments the ODE to be solved, the dependent variable, and the independent variable. Thus, we might write

```
DSolve[ode,y[x],x]
```

$$\left\{\left\{y[x] \to \frac{1}{4}(1-2x+2x^2)+e^{-2x}C[1]\right\}\right\}$$

Note that both in `ode` and the `DSolve` command, y is always written with its argument: `y[x]`. Failure to include arguments in a consistent manner is regarded by MATHEMATICA as an error. MATHEMATICA does permit the argument to be omitted from `y[x]` and its derivatives (both in `ode` and `DSolve`), but doing so produces different output than we are showing here.

The output identifies the general solution to the ODE, containing one undetermined constant, given the name `C[1]`. If all we wanted was the solution we are done. But if we want to do further work with the solution, we need to retrieve it in a more usable form. Since each set of braces defines an ordered list, our solution is the first (and only) element of the first (and only) element of the output nested lists, and it therefore can be obtained as

```
%[[1]][[1]]
```

$$y[x] \to \frac{1}{4}(1-2x+2x^2)+e^{-2x}C[1]$$

To recover the solution from this MATHEMATICA construct we need to know that it is represented as a two-element list, so our solution can be isolated as

```
Y = %[[2]]
```

$$\frac{1}{4}(1-2x+2x^2)+e^{-2x}C[1]$$

This expression, which we named `Y`, can be processed by the usual MATHEMATICA commands (including those for differentiation and integration), but it has not been defined as a function and notations like `Y'` have no meaning. We can create a function $y(x)$ from `Y` by writing

```
y[x_] := Evaluate[Y]
```

The command `Evaluate` is needed because MATHEMATICA does not substitute a value for `Y` before analyzing its dependence on x.

The solution, in either of the forms Y or $y(x)$, can be specialized to meet an "initial condition" by making an appropriate choice of $C[1]$. Alternatively, MATHEMATICA can find the $C[i]$ if the initial condition(s) are included in `DSolve`. To do so, we replace `ode` as the first argument of `DSolve` by a list containing the ODE and its initial conditions. To solve our present ODE subject to the condition $y(1) = 0$, we write

```
DSolve[ {ode, y[1]==0}, y[x], x];
```

$$\left\{\left\{y[x] \to \frac{1}{4}\, e^{-2x}\left(-e^2 + e^{2x} - 2e^{2x}x + 2e^{2x}x^2\right)\right\}\right\}$$

Possible initial conditions for an ODE in y can include setting the value of y at a specific point to infinity. The number of initial conditions should not exceed the order of the ODE; if this condition is violated, MATHEMATICA may generate meaningless results.

For ODEs of second or higher order, initial conditions can include specifications of the values of derivatives. When more than one initial condition is to be specified, they can be included as additional elements in the list that contains `ode`. For example,

```
DSolve[{ode, y[1]==4, y'[1]==0}, y[x], x];
```

Input of the following kind is also accepted:

```
DSolve[{ode, y[0]==y'[0]==0}, y[x], x];
```

Example 10.2.2. Solving an ODE with MATHEMATICA

Let's solve the ODE

$$y'' + 5y' + 4y = e^{-2x}\,,$$

subject to the initial conditions $y(0) = 0$, $y'(0) = 1$. After we get the solution let's verify that the initial conditions are satisfied, plot the solution, and characterize the extremum shown on the plot. MATHEMATICA code for these operations can be as follows:

Set up the ODE:

```
ode = y''[x] + 5*y'[x] + 4*y[x] = E^(-2*x);
```

(y" would be wrong!)

Now insert the initial conditions and solve the ODE:

```
DSolve[ {ode, y[0]==0, y'[0]==1}, y[x], x]
```

$$\left\{\left\{y[x] \to \frac{1}{6}\, e^{-4x}(-1 - 3\,e^{2x} + 4\,e^{3x})\right\}\right\}$$

We need to save the solution for further use.

```
Y = %[[1]][[1]][[2]]
```

$$\frac{1}{6}\, e^{-4x}(-1 - 3\,e^{2x} + 4\,e^{3x})$$

Here the `[[1]][[1]]` removes the two sets of braces, and the `[[2]]` selects the material to the right of the symbol $\to$.

We continue by computing dY/dx, placing the result in `YY`:

```
YY = Expand[D[Y,x]]
```

$$\frac{2\,e^{-4x}}{3} + e^{-2x} - \frac{2\,e^{-x}}{3}$$

The command `Expand` is needed to force the collection of similar terms.

To check the initial conditions we now evaluate `Y` and `YY` at $x = 0$:

```
x = 0; {Y, YY}                                  {0,1}

Clear[x]                                        Undefine x
```

The next command will work even if `x` remains defined. However, potential errors later on may be avoided by undefining quantities as soon as their definitions are no longer needed.

Assuming that we are interested in $y(x)$ for $x > 0$ we plot it for $0 < x < 6$:

```
Plot[Y, {x,0,6}]
```

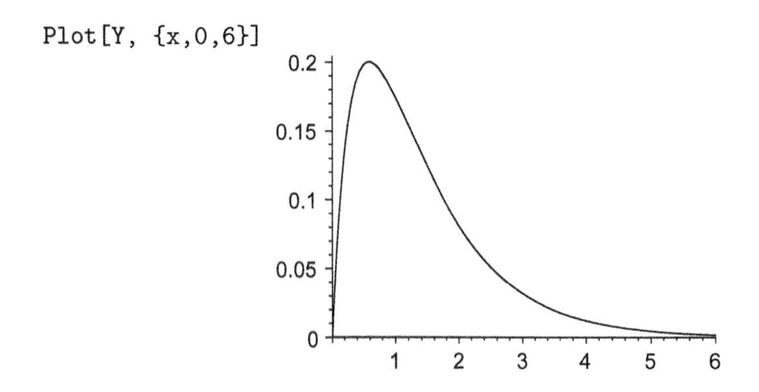

We characterize the maximum in $y(x)$ by finding the x value at which $dy/dx = 0$ (this is a zero of `YY`). Thus,

```
Solve[YY == 0., x]
```

$$\{\{x \to -0.29568 - 1.77822\,i\}, \{x \to -0.29568 + 1.77822\,i\}, \{x \to 0.59136\}\}$$

The coding of this MATHEMATICA statement deserves two comments. First, because we are solving an algebraic (and not a differential) equation, the command is `Solve`, not `DSolve`. Second, because we want a decimal result and not a complicated analytical expression, we set `YY` equal to a decimal zero so that the computation will be reduced to decimal numbers.

The positive real root is the one we want. To isolate it we need to select the third item of the outer list, the first (and only item) of the list that is that third item, and then take the second argument of the construct $x \to 0.59136$. This can be done all at once by the command

```
x = %[[3]][[1]][[2]]                            0.59136
```

The final step is to evaluate $y(x)$ for this `x` value:

```
Y                                               0.200176
```

We have therefore established that the maximum is at $x = 0.591$, with $y(x) = 0.200$.

Sometimes we will want our ODE solution established as a function, thereby permitting us to invoke it with an arbitrary argument. We proceed as follows:

```
Clear[x]; Clear[y];                  Undefine, in case it is necessary

y[x_] := Evaluate[Y];                y is now a function
```

To check what we have done,

```
y[t]
```

$$\frac{1}{6}\,e^{-4t}(-1-3\,e^{2t}+4\,e^{3t})$$

■

Exercises

10.2.1. Consider the ODE

$$x^2y''(x)+2xy'(x)-6y(x)=x^2\,.$$

Using symbolic computing for all steps of this exercise,

(a) Find the general solution to the above ODE.

(b) Find a solution such that $y(1)=0$ and $y'(1)=1/5$.

(c) Define a function $y(x)$ corresponding to the solution found in part (b).

(d) Differentiate $y(t)$ from part (c) and form the expression $t^2y''+2ty'-6y$. Show that this expression is equal to t^2.

The next two exercises show that the commands for symbolic solution of ODEs are able to generate solutions that involve special functions. Solve each ODE symbolically, and then show (by hand computation) that the solutions are correct. The special functions involved are defined and discussed in Sections 9.5 and 9.6 of this text.

10.2.2. $y''+2xy'=x\,.$

10.2.3. $xy''+(x+1)y'=x\,.$

10.3 FIRST-ORDER EQUATIONS

The most easily treated class of first-order ODEs deals with those that are termed **separable**, meaning that the equation can be written with only the dependent variable occurring on one side and only the independent variable on the other. An ODE in this form can be integrated directly.

Example 10.3.1. A Separable ODE

A radioactive isotope decays at a rate that is proportional to the number n of its nuclei that are still present. The ODE describing the decay is

$$\frac{dn}{dt}=-kn\,, \tag{10.9}$$

where the constant k determines the decay rate. Because we are not required to keep the fraction describing the derivative in its original form, Eq. (10.9) can be rearranged to

$$\frac{dn}{n}=-k\,dt\,, \tag{10.10}$$

and we may now integrate each side of this equation because it contains only a single variable. If we integrate in t from t_0 to t, the n integration must be from $n_0=n(t_0)$ to $n=n(t)$, and we get

$$\ln n-\ln n_0=-k(t-t_0)\,,\quad\text{equivalent to}\quad \ln n=-kt+C\,, \tag{10.11}$$

where $C = \ln n_0 + kt_0$, a constant whose value depends upon the state of the system at an initial time t_0. Because the situation illustrated by the last formula of Eq. (10.11) is of general occurrence, it is useful to note that it corresponds to evaluating indefinite integrals for the two sides of Eq. (10.10) and adding a constant of integration.

Continuing from Eq. (10.11) and regarding C as an as-yet unknown constant, we exponentiate both sides and give e^C the new name C', reaching

$$n(t) = C' e^{-kt} \,. \tag{10.12}$$

We see that this linear first-order ODE has a general solution containing one constant, consistent with the general analysis of the preceding section. The presence of this constant means that we have found a **family** of solutions whose members have different values of C'. If we set $t = 0$, we see that different members of the family correspond to different values of $n(0)$, i.e., to different initial conditions. Thus, a **particular** solution to this ODE with $n(0)$ having the specific value n_0 is achieved by setting $C' = n_0$.

A symbolic computing check for this exercise is

MAPLE
```
> dsolve({diff(n(t),t)=-k*n(t),n(0)=n0});
```

$$n(t) = n0\, e^{(-kt)}$$

MATHEMATICA
```
DSolve[{n'[t]==-k*n[t], n[0]==n0},n[t],t]
```

$$\{\{n[t] \to e^{-kt}\, n0\}\}$$

■

Separable ODEs need not be linear. All that is required is that they be of a general form

$$y'(x) = F(x)G(y)\,, \quad \text{equivalent to} \quad \frac{dy}{G(y)} = F(x)\,dx\,. \tag{10.13}$$

From the viewpoint of differential-equation theory, the form

$$\int^y \frac{dy}{G(y)} = \int^x F(x)\,dx + C$$

is a solution to Eq. (10.13). The usefulness of the solution obviously depends upon the specific forms of $F(x)$ and $G(y)$.

EXACT ODEs

If a first-order ODE can be brought to the form

$$d\,F(x,y) = h(x)\,dx, \tag{10.14}$$

it then has the simple solution $F(x,y) = \int h(x)\,dx$, and the ODE is called **exact**, referring to the fact that its left-hand side is an exact differential (the differential of some function). Note that dF refers to the differential of F associated with changes dx in x and dy in y; to be more explicit, it is

$$dF = \left(\frac{\partial F}{\partial x}\right) dx + \left(\frac{\partial F}{\partial y}\right) dy\,. \tag{10.15}$$

Of course, ODEs do not usually come to us in forms that tell us they are exact. One way to investigate this issue is to start by writing our ODE as

$$P(x,y)\,dx + Q(x,y)\,dy = h(x)\,dx\,. \tag{10.16}$$

If the left-hand side of Eq. (10.16) is the differential of some function $F(x,y)$, then, comparing with Eq. (10.15), we identify

$$\frac{\partial F}{\partial x} = P(x,y)\,, \qquad \frac{\partial F}{\partial y} = Q(x,y)\,. \tag{10.17}$$

We now form the mixed second derivatives of F,

$$\frac{\partial^2 F}{\partial y \partial x} = \frac{\partial P}{\partial y}\,, \qquad \frac{\partial^2 F}{\partial x \partial y} = \frac{\partial Q}{\partial x}\,.$$

These derivatives must be equal, so we find

$$\frac{\partial P}{\partial y} = \frac{\partial Q}{\partial x}\,. \tag{10.18}$$

Equation (10.18) is therefore a test for exactness; if exactness is found, we may rewrite Eq. (10.16) as

$$dF = h(x)\,dx\,,$$

which can be integrated to reach

$$F(x,y) = \int^x h(x)\,dx + C\,. \tag{10.19}$$

To complete the solution process we now find F by integrating Eqs. (10.17). To see how that is done, look at the following Example.

Example 10.3.2. Simple Exact and Inexact ODEs

Consider first the ODE $y' + x^{-1}y = x$. Expanding, we have

$$x\,dy + y\,dx = x^2\,dx\,, \qquad \text{so} \quad P = y,\ Q = x, \text{ and } \frac{\partial P}{\partial y} = \frac{\partial Q}{\partial x} = 1\,.$$

The above shows that $dF = x\,dy + y\,dx$ is an exact differential. In the present case it is rather obvious that $F = xy$, but it is instructive to see how we can find F by means other than clairvoyant inspection.

We start by integrating the first of Eqs. (10.17). Because we have a partial derivative, the integration is with respect to x treating y as a constant. Thus,

$$\frac{\partial F}{\partial x} = P \quad \longrightarrow \quad F = \int P\,dx = \int y\,dx = xy + f(y)\,. \tag{10.20}$$

We wrote F as an indefinite integral. Because the constant of integration may contain y, we call it $f(y)$. At this point $f(y)$ is entirely arbitrary.

We next process the second of Eqs. (10.17). This time we integrate with respect to y treating x as a constant. We get

$$F = \int Q\,dy = \int x\,dy = xy + g(x)\,, \tag{10.21}$$

Here $g(x)$ is arbitrary. We now need to reconcile the two forms given for F in Eqs. (10.20) and (10.21); in the present problem this is an easy task because consistency is achieved if $f(y) = g(x) = 0$. Our final result is that $F = xy$. Our ODE now has the form

$$dF = x^2\,dx\,,$$

which we can integrate to reach $F(x, y) = x^3/3 + C$, or, more explicitly,

$$xy = \frac{1}{3}x^3 + C\,. \tag{10.22}$$

Suppose we had been somewhat idiosyncratic and had written our expanded ODE as

$$x^2\,dy + xy\,dx = x^3\,dx\,.$$

This is really the same ODE but the check for exactness now fails:

$$P = xy, \quad Q = x^2, \quad \frac{\partial P}{\partial y} = x, \quad \frac{\partial Q}{\partial x} = 2x\,.$$

We must conclude that exactness is not an intrinsic property of an ODE but that it depends upon the way in which the ODE is presented.

■

Example 10.3.3. Exact vs. Separable

Here is an example showing that exactness and separability are not synonymous. An ODE in which the variables have been separated is exact (that is the topic of Exercise 10.3.1), but not all exact ODEs are separable. Consider

$$\cos(x - y)\sin y\,dx + \sin(x - 2y)\,dy = 0\,. \tag{10.23}$$

This ODE is not separable. However, it is exact:

$$\frac{\partial}{\partial y}\Big[\cos(x - y)\sin y\Big] = \sin(x - y)\sin y + \cos(x - y)\cos y = \cos(x - 2y)\,,$$

$$\frac{\partial}{\partial x}\sin(x - 2y) = \cos(x - 2y)\,.$$

To solve the ODE, we compute F by integrating each of Eqs. (10.17), requiring the two resulting expressions to be compatible:

$$F_1 = \int \cos(x - y)\sin y\,dx = \sin(x - y)\sin y + f(y)\,, \tag{10.24}$$

$$F_2 = \int \sin(x - 2y)\,dy = \frac{1}{2}\cos(x - 2y) + g(x)\,. \tag{10.25}$$

Knowing it must be possible to choose $g(x)$ and $f(y)$ in a way such that $F_1 = F_2$, we expand all the trigonometric functions with compound arguments, reaching

$$F_1 = \sin x \sin y \cos y - \cos x \sin^2 y + f(y)\,,$$

$$F_2 = \frac{1}{2}\Big[\cos x \cos^2 y - \cos x \sin^2 y + 2\sin x \sin y \cos y\Big] + g(x)\,.$$

If in F_2 we make the substitution $\cos^2 y = 1 - \sin^2 y$, we find that F_1 and F_2 become equal if we set $f(y) = 0$ and $g(x) = -(1/2)\cos x$. Thus, returning to the more compact form for F_1 given in Eq. (10.24), we see that the ODE has a solution given by the equation $\sin(x - y)\sin y = C$.

■

MAKING A LINEAR ODE EXACT

In Example 10.3.2 we saw that the ODE under discussion could be written in two forms, only one of which was exact. This observation suggests the possibility of taking an arbitrary ODE and making it exact by multiplying the entire equation by a suitable function of the independent variable. The quantity by which the ODE is multiplied is called an **integrating factor**. It can be proved that any first-order ODE can be made exact by applying an integrating factor, but a systematic way of finding an integrating factor is only known if the ODE is linear.

Given a linear first-order ODE of the generic form

$$y'(x) + p(x)y(x) = f(x)\,, \tag{10.26}$$

let's multiply it by a presently unknown function $w(x)$, reaching

$$w(x)y'(x) + w(x)p(x)y(x) = w(x)f(x)\,.$$

We now note that

$$[w(x)y]' = w(x)y' + w'(x)y\,,$$

and if $w(x)$ has been chosen in a way such that $w'(x) = w(x)p(x)$, our ODE would take the form

$$[w(x)y]' = w(x)f(x)\,. \tag{10.27}$$

This equation is exact, as each term is an exact differential—the second term is the differential of its integral! We therefore can integrate Eq. (10.27) to obtain

$$y = \frac{1}{w(x)}\int w(x)f(x)\,dx\,. \tag{10.28}$$

Remembering that Eq. (10.28) contains an indefinite integral, a more explicit way of indicating the general solution of the ODE is to write

$$y = \frac{1}{w(x)}\left[\int^x w(x)f(x)\,dx + C\right]. \tag{10.29}$$

Note that C occurs where it is multiplied by $1/w(x)$.

Arrival at Eq. (10.29) depends upon $w(x)$ satisfying the condition $w' = w\,p(x)$. This is a separable ODE that is easy to solve. We have

$$w' = w\,p(x) \longrightarrow \frac{dw}{w} = p(x)\,dx \longrightarrow \ln w = \int p(x)\,dx\,,$$

which corresponds to the useful formula

$$w = \exp\left(\int p(x)\,dx\right). \tag{10.30}$$

We don't have to worry here about a constant of integration, as it would only cause w to be multiplied by a constant factor, with no ultimate effect on the analysis.

The above development contains a trap for the unwary; our ODE was specifically defined to have coefficient unity for its y' term. If given an ODE in which y' has a coefficient other than unity we must divide the equation by the value of the coefficient to bring the ODE to the standard form assumed here.

Example 10.3.4. Integrating Factor

Let's use an integrating factor to solve $2y' + 4xy = 3x$.

We start by bringing the ODE to our standard form

$$y' + 2xy = \frac{3x}{2}. \tag{10.31}$$

Using Eq. (10.30), we find

$$w = \exp\left(\int (2x)\,dx\right) = e^{x^2},$$

so Eq. (10.31) becomes

$$e^{x^2} y' + 2xe^{x^2} y = \frac{3x}{2}\, e^{x^2}. \tag{10.32}$$

The left-hand side of this equation can be written $\left[e^{x^2} y\right]'$, a result that is expected because we used an appropriate procedure for finding the integrating factor. We can therefore integrate both sides of Eq. (10.32), reaching

$$e^{x^2} y = \int \frac{3x}{2}\, e^{x^2}\, dx = \frac{3}{4}\, e^{x^2} + C. \tag{10.33}$$

Solving for y, we get

$$y = \frac{3}{4} + C\, e^{-x^2}. \tag{10.34}$$

This solution contains one constant and is therefore the general solution to the ODE. It is a good idea to check our solution. It is legitimate to make the check using the original form of the ODE. A check of Eq. (10.34) is the topic of Exercise 10.3.2.

■

In the case of a linear first-order ODE, our procedure for making it exact leads directly to a solution even if the ODE is inhomogeneous, and we do not really need to use the general procedure outlined at the end of Section 10.1. However, that procedure will be valuable for treating ODEs of higher order; later in this chapter we discuss applications to second-order ODEs.

OTHER METHODS FOR FIRST-ORDER ODEs

The procedure involving construction of an integrating factor pretty much provides a complete procedure for the solution of linear first-order ODEs. There are no totally systematic methods for nonlinear first-order ODEs, but there are some techniques we mention here.

• If an ODE for $y(x)$, expressed in differential form, has equal combined powers of x and y in all its terms, it is sometimes referred to as **homogeneous in x and y**, and it can be reduced to separable form by the substitution $y = vx$, leading to an ODE in which v is the dependent variable.

• Sometimes an ODE can be converted into a more easily soluble form by interchanging the roles of the dependent and independent variables. For a first-order equation written in the form $y' = F(x, y)$, we simply change it to $x' = 1/F(x, y)$. If the transformed equation is linear in x, it can be solved by the methods previously discussed.

Example 10.3.5. ODE Homogeneous in x and y

Consider the ODE

$$(x + y)y' - 2y = 0\,. \tag{10.35}$$

This equation is homogeneous in x and y. Expanding and writing $y = vx$, we get

$$(x + y)dy - 2ydx = 0 \quad \longrightarrow \quad (1 + v)(v\,dx + x\,dv) - 2v\,dx = 0\,.$$

Collecting terms, we reach

$$(v^2 - v)\,dx + (1 + v)x\,dv = 0\,,$$

which is separable. Separating and integrating, we have

$$\frac{dx}{x} + \frac{(1 + v)\,dv}{v^2 - v} = 0 \quad \longrightarrow \quad \ln x + 2\ln(v - 1) - \ln v = \ln C\,,$$

which, exponentiating, leads to

$$x = \frac{Cv}{(v - 1)^2} \quad \longrightarrow \quad 1 = \frac{Cy}{(y - x)^2}\,. \tag{10.36}$$

Solving Eq. (10.36) for x (and redefining C),

$$x = y + Cy^{1/2}\,. \tag{10.37}$$

■

Example 10.3.6. Changing the Independent Variable

Here is an alternative method for solving the ODE of Eq. (10.35). If we view y as the independent variable we can rearrange that equation as follows:

$$y' = \frac{2y}{x + y} \quad \longrightarrow \quad x' = \frac{x + y}{2y} \quad \longrightarrow \quad x' - \frac{x}{2y} = \frac{1}{2}\,, \tag{10.38}$$

thereby obtaining a linear ODE. An integrating factor for Eq. (10.38) is $y^{-1/2}$, and that equation becomes

$$\left[y^{-1/2}x\right]' = \frac{y^{-1/2}}{2} \quad \longrightarrow \quad y^{-1/2}x = \frac{1}{2}\int y^{-1/2}\,dx = y^{1/2} + C\,. \tag{10.39}$$

Equation (10.39) is equivalent to the result obtained previously, Eq. (10.37).

Verification of some of the steps taken in this Example is the topic of Exercise 10.3.3.

■

• Sometimes a change of variable can be useful, especially for treating nonlinear ODEs. We give one example.

Example 10.3.7. Bernoulli ODE

A nonlinear ODE of the form

$$y' + P(x)y = Q(x)y^n \tag{10.40}$$

is known as the Bernoulli equation. If we multiply Eq. (10.40) by y^{-n}, we get

$$y^{-n}y' + P(x)y^{1-n} = Q(x)\,,$$

suggesting the substitution $u = y^{1-n}$. Noting that

$$u' = (1-n)\,y^{-n}y'\,,$$

we rewrite the Bernoulli equation as

$$\frac{u'}{1-n} + P(x)\,u = Q(x)\,,$$

which in our standard form becomes

$$u' + (1-n)\,P(x)\,u = (1-n)\,Q(x)\,. \tag{10.41}$$

Equation (10.41) is a linear ODE which can be solved by the methods outlined earlier in this section.

■

Exercises

10.3.1. Show that an ODE that is separable is also exact when written in separated form.

Hint. You need to identify each term as a perfect differential.

10.3.2. Verify that the general solution of the ODE studied in Example 10.3.4 is correctly given by Eq. (10.34).

10.3.3. Confirm the details of the analysis presented in Example 10.3.5 as follows:

(a) Verify that the substitution $y = vx$ leads to the ODE

$$(v^2 - v)\,dx + (1+v)x\,dv = 0\,.$$

(b) Verify that the general solution to this ODE can be brought to the form given in Eq. (10.37).

Solve the following ODEs in $y(x)$ subject to the indicated condition (or if no condition is indicated, obtain the general solution), both by hand computation and by symbolic computing. If the hand and symbolic solutions are different in form, do whatever is necessary to make them equivalent. In some cases this may involve solving an algebraic equation of the form $y = f(x)$ to bring the solution to the form $x = F(y)$. If you are asked to obtain a solution subject to a condition, plot your solution for a range of x and y that includes the point for which $y(x)$ was specified.

10.3.4. $xy' - y = 0\,,$ with $y(1) = 4$.

10.3.5. $y' - y = e^x\,.$

10.3.6. $(1+y)y' = y\,,$ with $y(1) = 1$.

10.3.7. $2xy' + y = 2x^{3/2}\,.$

10.3.8. $xy' = y^2\,,$ with $y(1) = 2$.

10.3.9. $(x+y^2)y' - y = 0\,.$

10.3.10. $y' + 2xy^2 = 0\,,$ with $y(2) = 1$.

10.3.11. $x \ln x\, y' - 2y = \ln x\,.$

10.3.12. $x(1+y)y' = -y\,,$ with $y(1/2) = 2$.

10.3.13. $(x^2y + x^3)y' + y^3 + xy^2 = 0\,.$

10.3.14. $y' \cos x + y = \cos^2 x\,.$

10.3.15. $y^2y' = x\,,$ with $y(2) = 2$.

10.3.16. $y' + x^2y^2 - y = 0\,.$

10.3.17. $2xy' + y = -3x\,.$

10.3.18. $\tan^{-1}x\, y' - \dfrac{y}{1+x^2} = 0\,.$

10.3.19. An electric circuit containing a capacitor C, a resistor R, and a time-dependent applied voltage $V(t) = V_0 \sin \omega t$ is connected as shown in Fig. 10.2. At time $t = 0$ the switch is closed; prior to that time there is no charge q on the capacitor. The behavior of the circuit is described by the ODE

$$R\frac{dq}{dt} + \frac{q}{C} = V(t)\,.$$

(a) Find the charge q on the capacitor as a function of t.

(b) The exponentially decaying portion of $q(t)$ is referred to as its **transient** behavior. Identify (in terms of R and C) the time required for the transient to decay to $1/e$ times its maximum value.

10.3.20. The forward acceleration of a rocket is produced by the rearward discharge at high velocity of the mass of its fuel combustion products. When a mass $-dm$ is discharged at a rearward velocity of magnitude u (relative to the rocket), the velocity of the rocket changes from v to $v + dv$ and its mass

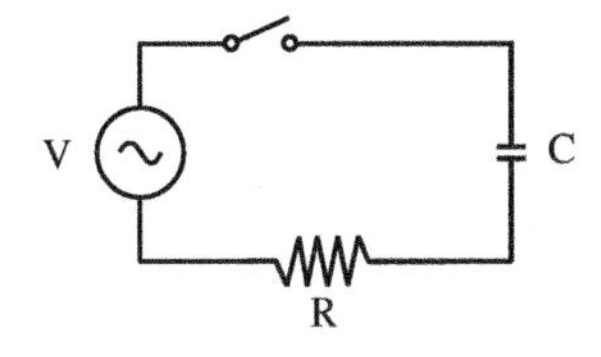

Figure 10.2: Electronic circuit for Exercise 10.3.19.

m changes by an amount dm (obviously dm will be negative). Conservation of momentum (in a space-fixed coordinate system) requires that the momentum discharged, $-(v-u)\,dm$, plus the change in the momentum of the remaining mass of the rocket, $d(mv) = m\,dv + v\,dm$, add to zero. Therefore,

$$-(v-u)\,dm + m\,dv + v\,dm = 0, \qquad \text{or} \qquad m\,dv = -u\,dm\,.$$

Calculate the final velocity of a rocket that is initially stationary, assuming that 90% of its original mass is expelled, and that all the discharged mass is expelled at the same velocity, $u = 300$ m/s. Neglect the effects of gravity and air resistance.

10.4 ODEs WITH CONSTANT COEFFICIENTS

We discuss first homogeneous ODEs with constant coefficients.

HOMOGENEOUS EQUATIONS

Homogeneous ODEs of arbitrary order but with constant coefficients can be solved in a straightforward manner. The key to the process is to recognize that these ODEs have solutions that are exponentials. Consider the ODE

$$y^{(n)}(x) + a_{n-1}y^{(n-1)}(x) + \cdots + a_1 y'(x) + a_0 y(x) = 0\,, \tag{10.42}$$

where the a_i are constants. Assuming a solution of the form $y(x) = e^{mx}$, our ODE becomes

$$m^n e^{mx} + a_{n-1}m^{n-1}e^{mx} + \cdots + a_1 m e^{mx} + a_0 e^{mx} = 0\,,$$

which reduces to the algebraic equation (called the **auxiliary equation** of the ODE)

$$m^n + a_{n-1}m^{n-1} + \cdots + a_1 m + a_0 = 0\,. \tag{10.43}$$

This equation, according to the fundamental theorem of algebra, has n roots, which we designate m_i, $i = 1, \ldots, n$. If all the roots are distinct, our general solution will be

$$y(x) = \sum_{i=1}^{n} C_i e^{m_i x}\,, \tag{10.44}$$

which we know is the general solution because it contains n arbitrary constants.

Equation(10.42) can be looked at in another way that we can discuss by introducing a symbol D to represent differentiation with respect to the independent variable. In the present context, this means

$$D \equiv \frac{d}{dx} \quad \text{and} \quad \Big(D^n + a_{n-1}D^{n-1} + \cdots + a_1 D + a_0\Big)\,y(x) = 0\,, \tag{10.45}$$

and we can write this operator equation in the factored form

$$(D - m_1)(D - m_2)\cdots(D - m_n)\,y(x) = 0\,. \tag{10.46}$$

The roots m_i are the same as those of Eq. (10.43). Because all the quantities $(D - m_i)$ commute with each other, they can be written in Eq. (10.46) in any order.

Next, note that Eq. (10.46) will surely be satisfied if

$$(D - m_n)y(x) = 0\,, \quad \text{or, in a more familiar notation,} \quad y' - m_n y = 0\,, \tag{10.47}$$

which has the solution $y = C\,e^{m_n x}$. Since any factor from the operator product in Eq. (10.46) can be moved so that it occurs last, we have an alternate way of establishing Eq. (10.44).

We are now ready to consider the possibility that Eq. (10.43) may have multiple roots, and in that case the functions $e^{m_i x}$ will include duplicates and therefore be insufficient in number to provide a general solution to the ODE. Our strategy for finding additional roots is to seek solutions (other than e^{mx}) for the equation

$$(D - m)^p\, y = 0\,, \qquad \text{for } p > 1.$$

A key step toward finding additional solutions is the observation that

$$(D - m)\,(x^n\, e^{mx}) = nx^{n-1}\, e^{mx}\,, \tag{10.48}$$

which in particular means that

$$(D - m)(D - m)\;(xe^{mx}) = (D - m)e^{mx} = 0\,. \tag{10.49}$$

Equation (10.49) shows that if the factored form of our ODE shows a double root m_i, then the ODE has, in addition to $e^{m_i x}$, the solution $x\,e^{m_i x}$.

Generalizing, we note that for any positive integers $q < p$,

$$(D - m)^p\;(x^q e^{mx}) = 0\,, \qquad q < p, \tag{10.50}$$

producing additional ODE solutions for larger values of p. Thus,

If m_i is a p-fold degenerate multiple root of our ODE, meaning that the factor $(D - m_i)$ occurs p times in the factored expression given in Eq. (10.46), then the solutions $y(x)$ corresponding to this m_i take the form

$$y(x) = \Big(C_0 + C_1 x + \cdots + C_{p-1} x^{p-1}\Big)\, e^{m_i x}\,. \tag{10.51}$$

Another possibility is that Eq. (10.43) may have a pair of roots that are complex conjugates: $m = \alpha + i\beta$ and $m = \alpha - i\beta$. Then, because $e^{\pm i\beta x} = \cos\beta x \pm i\sin\beta x$, these roots, which contribute to the general solution to the ODE as

$$y = Ae^{(\alpha+i\beta)x} + Be^{(\alpha-i\beta)x} = e^{\alpha x}\big(Ae^{i\beta x} + Be^{-i\beta x}\big)\,, \tag{10.52}$$

can also be described as

$$y = e^{\alpha x}(C\cos\beta x + C'\sin\beta x)\,. \tag{10.53}$$

Example 10.4.1. Harmonic Oscillator

A spring attached to a ceiling supports a weight that can oscillate up and down. Letting the rest position of the weight correspond to $y = 0$ (with positive y at higher elevation), the combined force of gravity and that due to stretching or compression of the spring is approximately $-ky$, where k (assumed positive) is called the **spring**

constant. According to Newton's Second Law, the weight will experience **acceleration** (defined as d^2y/dt^2), with t the time, in an amount proportional to the net force $\mathbf{F}$ on it. For a weight of mass m, Newton's Law corresponds to the equation

$$m\frac{d^2y}{dt^2} = -ky, \quad \text{often written} \quad y'' + \left(\frac{k}{m}\right) y = 0\,. \tag{10.54}$$

Equation (10.54) is of the form

$$(D^2 + a^2)y = (D + ia)(D - ia) = 0\,, \qquad \text{with } a = \sqrt{\frac{k}{m}}, \tag{10.55}$$

and with general solution

$$y = C_1 e^{iat} + C_2 e^{-iat} \quad \text{or} \quad y = A \sin at + B \cos at\,.$$

This general solution describes sinusoidal oscillation at arbitrary amplitude and phase, at angular frequency $\omega = a$. In the approximation with which we are currently treating this problem, the oscillation would continue indefinitely at the same amplitude.

If we were to change the sign of the term ky, our ODE would then have the generic form

$$(D^2 - a^2)y = (D + a)(D - a)y = 0\,, \quad \text{with solution} \quad y = C_1 e^{-at} + C_2 e^{at}\,.$$

The motion is now unstable. Unless the initial conditions are actually such that $C_2 = 0$, y will grow exponentially with increasing t. What is really going on here is that with $F = -ky$, the force is in the direction opposite to the displacement from equilibrium, tending to restore the equilibrium position. The change to $F = +ky$ causes the force to be toward greater displacement, with an exponential decay toward the equilibrium position only possible if the initial conditions are that the weight is moving toward $y = 0$ and the force is just sufficient to stop the weight as it approaches that point.

Before leaving this example we note that the same ODE applies to an electric circuit containing an inductor (L) and a capacitor (C), but no resistance, often called an LC circuit. See Fig. 10.3. An LC circuit (with no external driving force) satisfies Eq. (10.55) with $a = 1/\sqrt{LC}$, and therefore supports oscillation of arbitrary amplitude and phase at the angular frequency $\omega = a$.

■

Example 10.4.2. Damped Oscillator

The harmonic oscillator of Example 10.4.1 is unrealistic to the extent that it takes no account of frictional or resistive forces that will ultimately damp out an initially

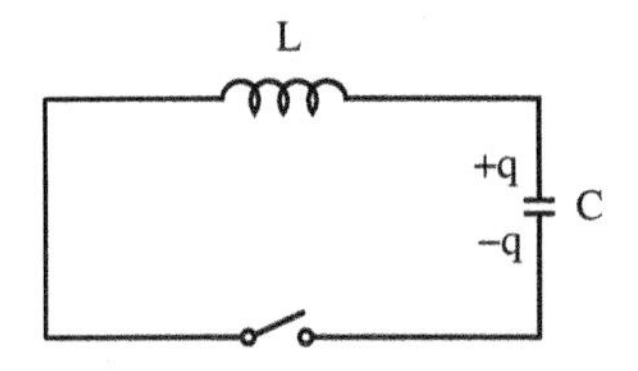

Figure 10.3: LC circuit.

induced oscillation. The simplest reasonable model for a frictional force is to make it proportional to the velocity (i.e., y') and opposite in direction (thereby tending to reduce the magnitude of the velocity), thus of the form $-ly'$, with $l > 0$. When this force is added to the right-hand side of Eq. (10.54), the resulting ODE has the form

$$my'' = -ky - ly' ,$$

which for later convenience we write in the form

$$y'' + 2by' + \omega^2 y = 0 , \quad \text{with } \omega^2 = \frac{k}{m} \text{ and } b = \frac{l}{2m} . \tag{10.56}$$

The ODE of Eq. (10.56) can be treated by the methods of the current discussion. We write it in the form

$$(D - \alpha_+)(D - \alpha_-) = 0, \quad \text{with} \quad \alpha_\pm = -b \pm \sqrt{b^2 - \omega^2} . \tag{10.57}$$

Looking at our ODE as written in Eq. (10.57), we see that we can expect its behavior to depend qualitatively on the quantity $p = \sqrt{b^2 - \omega^2}$. If $b > \omega$, p will be positive, lying in the range $0 < p < b$. In that case, both α_+ and α_- will be negative, and our general solution will be a linear combination of two negative exponentials:

$$y = C_1 e^{-(b-p)t} + C_2 e^{-(b+p)t} , \qquad b > \omega, \; 0 < p < b . \tag{10.58}$$

This behavior, called **overdamped**, corresponds to a unidirectional decay toward $y = 0$; there are no oscillations.

If $b = \omega$, the system is referred to as **critically damped**. In this case, $p = 0$, the auxiliary equation has two equal roots (both $-b$), and our ODE therefore has the general solution

$$y = (C_1 + C_2 t)e^{-bt} , \qquad b = \omega . \tag{10.59}$$

We get either zero or one oscillation, depending upon the initial conditions.

The final, and most interesting possibility, is that $b < \omega$, causing p to be imaginary; let's designate it

$$p = i\omega', \qquad \text{with} \quad \omega' = \sqrt{\omega^2 - b^2} . \tag{10.60}$$

This situation is referred to as **underdamped** or **oscillatory**. The general solution to our ODE can then be written

$$y = C_1 e^{(-b+i\omega')t} + C_2 e^{(-b-i\omega')t} = (A \sin \omega' t + B \cos \omega' t)e^{-bt} , \qquad b < \omega . \tag{10.61}$$

The solution is oscillatory, but with an amplitude that decays exponentially with time. Note that the oscillation frequency is decreased from the value it would have for a frictionless oscillator. This is the behavior we get when the frictional forces are not too large.

An electric-circuit problem of the type shown in Fig. 10.1 satisfies the same ODE as a damped mechanical oscillator, with the inductance L in the role of m, the resistance R in place of the coefficient of friction, and the inverse of the capacitance $1/C$ in place of the force constant. The dependent variable (in the mechanical problem, the displacement from equilibrium) is now the charge q on the capacitor. We thus have an ODE of the form

$$Lq'' + Rq' + \frac{q}{C} = 0 ,$$

with solutions entirely analogous to those of the mechanical problem.

■

INHOMOGENEOUS EQUATIONS

An approach toward solving inhomogeneous ODEs with constant coefficients can consist of the above analysis to obtain the general solution to the corresponding homogeneous equation, followed by whatever further work is needed to obtain a particular solution to the full inhomogeneous equation. The methods of Section 10.6 provide a straightforward route to the particular solution if our equation is second-order. Extensions of those methods can in principle be used for equations of higher order, but the procedure rapidly becomes tedious.

If the inhomogeneous term has only a small number of linearly independent derivatives, one can make the assumption that there exists a particular solution that is a linear combination of the inhomogeneous term and its derivatives. It may then be possible to find the coefficients of the terms in our assumed solution. This approach is likely to work when the inhomogeneous term is an exponential, a sine or cosine, a low-degree polynomial, or a combination of those elements. In the following examples we look at some typical situations involving an ODE of the general form

$$\mathcal{L}y(x) = f(x)\,, \quad \text{where}$$
$$\mathcal{L} \equiv y^{(n)}(x) + a_{n-1}y^{(n-1)}(x) + \cdots + a_1 y'(x) + a_0 y(x)\,. \tag{10.62}$$

Example 10.4.3. Exponential Inhomogeneous Term

Consider the ODE of Eq. (10.62), with $f(x) = a\,e^{\alpha x}$. This ODE is easy to deal with, particularly if α is not equal to any of the roots m_i of the corresponding homogeneous equation.

We try to find a particular solution of the form $y_p = A\,e^{\alpha x}$. Each term of $\mathcal{L}$ simply multiplies y_p by a power of α, so we get

$$\mathcal{L}y_p = \Big(\alpha^n + a_{n-1}\alpha^{n-1} + \cdots + a_0\Big)\,A\,e^{\alpha x}\,,$$

and we need only choose the constant A at the value that makes the overall coefficient of $e^{\alpha x}$ equal to a.

A complication arises, however, if α is equal to one of the roots m_i of the auxiliary equation for the homogeneous ODE. Then $\mathcal{L}\,(Ae^{\alpha x}) = 0$, and we get no y_p. The solution to this dilemma becomes obvious if we write our inhomogeneous equation in the form

$$(D - m_1)\cdots(D - \alpha)y_p = a\,e^{\alpha x}\,. \tag{10.63}$$

We try a y_p of the modified form $Ax\,e^{\alpha x}$, so that the factor $(D - \alpha)$ in $\mathcal{L}$ will give us $e^{\alpha x}$ instead of zero. Actual computation to determine the coefficient A in y_p is most easily done from the expanded form of $\mathcal{L}$, using Leibniz' formula to differentiate the product $x\,e^{\alpha x}$:

$$\frac{d^n}{dx^n}\,x\,e^{\alpha x} = \left(\alpha^n x + n\alpha^{n-1}\right)e^{\alpha x}\,.$$

The details are left to the Exercises.

■

Example 10.4.4. More than One Inhomogeneous Term

Suppose we have an ODE with two inhomogeneous terms: $f(x) + g(x)$. We can deal with these individually. If y_f is a particular solution to $\mathcal{L}y = f$ and y_g is a particular solution to $\mathcal{L}y = g$, then a particular solution to $\mathcal{L}y = f + g$ is $y_f + y_g$.

■

Example 10.4.5. Trigonometric Inhomogeneous Term

Suppose now that we have the ODE of Eq. (10.62), with $f(x) = a \sin \alpha x$. This problem is a case of Example 10.4.4 in which the inhomogeneous term is a linear combination of complex exponentials:

$$f(x) = \frac{a}{2i}\left[e^{i\alpha x} - e^{-i\alpha x}\right].$$

We can either use the above form for $f(x)$ or take $f(x) = ae^{i\alpha x}$ and keep the imaginary part of our solution to the ODE. In that case the real part of the ODE solution will correspond to an inhomogeneous term of the form $a \cos \alpha x$.

■

Example 10.4.6. x^p Inhomogeneous Term

Consider next the ODE of Eq. (10.62), with $f(x) = ax^p$. Take for y_p the trial form

$$y_p = b_p x^p + \cdots + b_1 x + b_0 .$$

The application of $\mathcal{L}$ to this y_p will produce a polynomial in x of degree p, with coefficients c_i that are functions of the b_i. Setting $c_p = a$ and $c_0, \ldots, c_{p-1}$ to zero, we can solve for the b_i. The details are deferred to the Exercises.

■

We continue now with two examples in which inhomogeneous ODEs describe physical problems.

Example 10.4.7. Undamped Driven Oscillator

Consider an undamped oscillator with a sinusoidal inhomogeneous term:

$$y''(t) + \omega^2 y(t) = F \sin \alpha t . \tag{10.64}$$

The associated homogeneous equation (i.e., that with $F = 0$) has general solution $C_1 e^{i\omega t} + C_2 e^{-i\omega t}$, more conveniently written as $A \sin \omega t + B \cos \omega t$. To complete a general solution we need a particular solution to the complete ODE.

Using the fact that $F \sin \alpha t$ is the imaginary part of $Fe^{i\alpha t}$, we seek a particular solution Y_p to

$$Y''(t) + \omega^2 Y(t) = Fe^{i\alpha t} ; \tag{10.65}$$

the imaginary part of Y_p will be a solution y_p to our original ODE. We assume that Y_p will have the form $Y_p = Ce^{i\alpha t}$; this form will yield a solution if $\alpha \neq \omega$. If $\alpha = \omega$, we would instead need to use $Y_p = Cte^{i\omega t}$.

Continuing for $\alpha \neq \omega$, substitution of the assumed form for Y_p leads to

$$C\left[(i\alpha)^2 + \omega^2\right] e^{i\alpha t} = F e^{i\alpha t},$$

from which we extract

$$C\left[\omega^2 - \alpha^2\right] = F, \quad \text{leading to} \quad Y_p(t) = \frac{F}{\omega^2 - \alpha^2}\, e^{i\alpha t}.$$

Adding the imaginary part of Y_p to the general solution of the homogeneous equation, we get (for $\alpha \neq \omega$)

$$y(t) = A \sin \omega t + B \cos \omega t + \frac{F \sin \alpha t}{\omega^2 - \alpha^2}. \tag{10.66}$$

A particular solution for the initial conditions $y(0) = y'(0) = 0$ describes an oscillator that was inactive for all $t < 0$, but starting at $t = 0$ is driven by the sinusoidal force term on the right-hand side of Eq. (10.64). To apply the initial condition $y(0) = 0$ we must set to zero the constant B in Eq. (10.66). For $y'(0) = 0$ we need to compute y' from that equation:

$$y'(t) = A\omega \cos \omega t + \frac{\alpha F \cos \omega t}{\omega^2 - \alpha^2}.$$

Putting $t = 0$ in this expression and setting the result to zero, we find

$$A\omega + \frac{\alpha F}{\omega^2 - \alpha^2} = 0 \quad \longrightarrow \quad A = -\frac{\alpha F}{\omega(\omega^2 - \alpha^2)}.$$

Our final result (for $\alpha \neq \omega$) is

$$y(t) = F\left[\frac{\omega \sin \alpha t - \alpha \sin \omega t}{\omega(\omega^2 - \alpha^2)}\right]. \tag{10.67}$$

To treat the case $\alpha = \omega$, we need to repeat the above analysis using $Y_p = Cte^{i\omega t}$. We find

$$Y_p''(t) + \omega^2 Y_p(t) = 2i\omega C e^{i\omega t} = F e^{i\omega t}, \quad \text{which yields} \quad C = \frac{F}{2i\omega}.$$

Inserting this result into the expression for Y_p, we get

$$Y_p = \frac{Fte^{i\omega t}}{2i\omega} = Ft\left[\frac{-i\cos \omega t}{2\omega} + \frac{\sin \omega t}{2\omega}\right].$$

Adding the imaginary part of this expression to the general solution of the homogeneous equation, we get (for $\alpha = \omega$)

$$y(t) = A \sin \omega t + B \cos \omega t - \frac{Ft \cos \omega t}{2\omega}. \tag{10.68}$$

Again considering the initial conditions $y(0) = y'(0) = 0$, we find from $y(0) = 0$ that $B = 0$. We next compute

$$y'(t) = \omega A \cos \omega t - \frac{F \cos \omega t}{2\omega} + \frac{Ft \sin \omega t}{2}$$

and obtain therefrom

$$y'(0) = \omega A - \frac{F}{2\omega} = 0 \quad \longrightarrow \quad A = \frac{F}{2\omega^2}.$$

The final result (for $\alpha = \omega$) is

$$y(t) = F\left[\frac{\sin \omega t}{2\omega^2} - \frac{t \cos \omega t}{2\omega}\right]. \tag{10.69}$$

Equation (10.69) indicates that $y(t)$, the response of the oscillator to a driving force at its natural angular frequency ω, increases in amplitude without limit as t increases. One way to understand what is going on is to note that the driving force is (for large t) in phase with y', thereby continually reinforcing the oscillation.

While the unlimited increase in the amplitude of $y(t)$ is not physically realistic (there will be damping from frictional or resistive forces), the large amplitude of $y(t)$ is qualitatively reasonable, as it results from a force that repeats periodically at the natural oscillation frequency. This phenomenon is known as **resonance**; it is important (and desirable) in electric circuits (e.g., in radios), because it can cause selective amplification of a signal at a desired frequency. It may, however, be less desirable in mechanical structures (e.g., bridges), where excessive oscillation can damage or even destroy the structure.

■

Example 10.4.8. Damped Driven Oscillator

Here we have a more realistic treatment of a driven oscillator that includes a frictional or resistive term in the governing ODE. Our ODE is an inhomogeneous version of that discussed in Example 10.4.2, with a sinusoidal driving force:

$$y''(t) + 2by'(t) + \omega^2 y(t) = F \sin \alpha t\,. \tag{10.70}$$

As we saw in Example 10.4.2, the damping force causes the motion of an undriven oscillator to decay exponentially, with or without oscillations of decreasing amplitude, depending upon the strength of the damping. For a driven oscillator, we are most often only interested in that part of the motion that does not decay with time; we find that an oscillator with a periodic driving force exhibits long-term oscillation only at the frequency with which it is driven. Relevant questions include the amplitude of the oscillation that a given driving force can produce, and the identification of frequencies of maximum response (i.e., **resonance**).

Proceeding as for the undamped oscillator in Example 10.4.7, we seek a particular solution $Y_p(t)$ to

$$Y''(t) + 2bY'(t) + \omega^2 Y(t) = Fe^{i\alpha t}\,,$$

and identify the imaginary part of Y_p as a solution to Eq. (10.70).

Our trial solution will be $Y_p = Ce^{i\alpha t}$; insertion of this form into the equation for Y yields

$$(-\alpha^2 + 2ib\alpha + \omega^2)Ce^{i\alpha t} = Fe^{i\alpha t}\,,$$

which we can solve for C:

$$C = \frac{F}{(\omega^2 - \alpha^2) + 2ib\alpha}\,.$$

For what we plan to do next, it is useful to rewrite C in polar form. Letting $\delta = \arctan[2b\alpha/(\omega^2 - \alpha^2)]$, we get

$$C = |C|\, e^{-i\delta}\,, \quad \text{with } |C| = \frac{F}{\sqrt{(\omega^2 - \alpha^2)^2 + 4b^2\alpha^2}}\,, \quad \text{and } Y_p = |C|\, e^{i(\alpha t - \delta)}\,.$$

Then

$$y_p = \frac{F}{\sqrt{(\omega^2 - \alpha^2)^2 + 4b^2\alpha^2}} \sin(\alpha t - \delta)\,. \tag{10.71}$$

The particular solution y_p is the only part of the general solution to Eq. (10.70) that does not decay exponentially; it is often called the steady-state solution (with the decaying terms called **transients**). From the formula for y_p, we see that if we have a physical situation in which we can vary ω, y_p will have its maximum amplitude when $\omega = \alpha$, and that value of ω is identified as the resonant frequency.

■

Exercises

10.4.1. (a) Starting from Eq. (10.71), show that if a mechanical or electrical system is adjusted by changing ω (at fixed α), the oscillatory displacement (or charge) y_p has maximum amplitude when $\omega = \alpha$; this value of ω is usually called the **resonance** angular frequency. (The resonance frequency is $\omega/2\pi$.) The changing of ω is the process involved when a radio is tuned to the frequency of an incoming signal.

(b) Find the value of α that maximizes the displacement (or charge) for a fixed value of ω; it is NOT ω! This datum is significant if we need to minimize the maximum magnitude of oscillation of a mechanical structure, such as a bridge, and could be regarded as the resonant angular frequency of the structure.

(c) Show that if α is varied, as in part (b), the value of α that maximizes the oscillation amplitude of dy_p/dt (which for an electric circuit is the current) IS at $\alpha = \omega$. This observation rationalizes the identification of ω as the resonant angular frequency of an electric circuit, even if it is tuned by changing α.

10.4.2. Radioactive nuclei decay with a probability k that is independent of time, and (in the absence of any process that replenishes their number), the number of nuclei present at the time t, $N(t)$, is therefore described by the ODE

$$\frac{dN}{dt} = -kN\,.$$

(a) Solve this ODE subject to the initial condition $N(0) = N_0$.

(b) Show that the time τ required for half the N_0 nuclei to decay (called the **half-life** of the isotope) is independent of N_0 and find the relation between τ and the decay constant k.

10.4.3. In naturally occurring radioactivity (and often also for artificially produced nuclear isotopes), there may be a series of decay reactions. For example, Radium 226 decays to produce Radon 222, which decays into Polonium 218, which decays further (by many steps), finally reaching a stable isotope of lead. Radium 222 is itself a decay product (in a chain originating with Uranium 238), but it is produced so slowly that we can ignore here its replenishment. Letting N_{Ra} and N_{Rn} respectively denote the numbers of radium and radon nuclei,

(a) Write ODEs for $N_{\rm Ra}$ and $N_{\rm Rn}$, taking into account that the radium nuclei decay (at half-life $\tau_{\rm Rn}$) but are not replaced, while the radon nuclei decay (at half-life $\tau_{\rm Rn}$) but that the radon nuclei are also replenished by the decay of the radium nuclei.

(b) Solve the ODEs of part (a) subject to the initial conditions $N_{\rm Ra}(0) = N_0$ and $N_{\rm Rn}(0) = 0$: solve the $N_{\rm Ra}$ equation first and then substitute its solution into the $N_{\rm Rn}$ equation.

(c) Given that the half-lives of Ra and Rn are respectively 1600 years and 38 days, compute the time T required for $N_{\rm Rn}$ to rise from zero to its maximum value.

10.4.4. Find the general solution of $y''(x) + y'(x) - 2y(x) = x\,.$

10.4.5. Find the general solution of $2y''(x) - 9y'(x) + 4y(x) = e^{2x}\,.$

10.4.6. Find the general solution of $2y''(x) - 9y'(x) + 4y(x) = e^{4x} + 2x^2\,.$

Exercises 10.4.7–10.4.9 deal with electric circuits of the types shown in Fig. 10.4. The switch is assumed to be open prior to time $t = 0$, when it is closed. For numerical computations, note that the ODEs for these circuits correspond to the use of the SI units of L, R, C, V, I, q, and t, which are respectively henry, ohm, farad, volt, ampere, coulomb, and second.

10.4.7. This exercise deals with the LR circuit with $L = 10$ H, $R = 1000\ \Omega$, and $V_0 = 100$ V. The ODE for this circuit is

$$LI'(t) + RI(t) = V_{\rm app}(t) = V_0 \sin \alpha t\,.$$

Make computations for the two angular frequencies $\alpha_1 = 120\pi$ (60 Hz) and $\alpha_2 = 2000\pi$ (1 kHz).

(a) Solve the ODE subject to the initial condition $I(0) = 0$. Identify the transient and steady-state currents.

(b) Plot the transient current for each α value. Suggested range of t (both frequencies), 0 to 0.1 sec. Explain how the condition $I(0) = 0$ causes there to be a nonzero transient, and discuss the similarities and differences in the transient terms for the two α values. Is the transient term oscillatory? Explain.

(c) Plot the total current $I(t)$ for each α value. Suggested ranges: for α_1, 0 to 0.1 sec and 10 to 10.1 sec; for α_2, 0 to 0.01 sec and 10 to 10.01 sec. Explain the difference between the earlier and later-time plots.

(d) Compare the amplitudes of the steady-state oscillations at the two α values. Give a qualitative explanation of the differences.

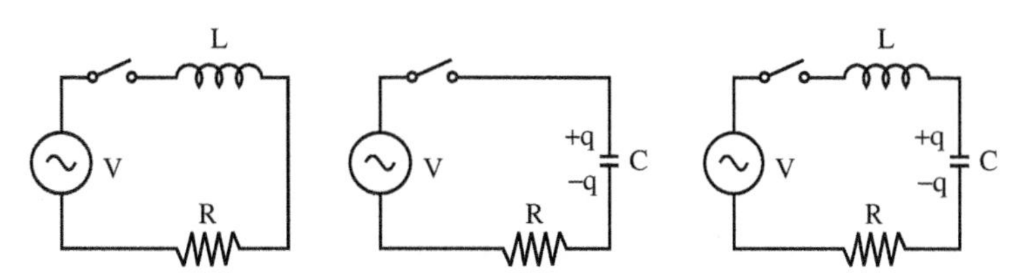

Figure 10.4: Some electric circuits.

(e) Compare the phases of $V_{\rm app}(t)$ and $I(t)$ in the large-t limit (obtain this from the steady-state term). Explain the phase shift in terms of the properties of the circuit elements.

10.4.8. This exercise deals with the RC circuit in Fig. 10.4, with $R = 1000\ \Omega$, $C = 10 \times 10^{-6}$ F (10 microfarad, written 10 μF), $V_0 = 100$ V. The ODE for this circuit is

$$Rq'(t) + \frac{1}{C}\, q(t) = V_{\rm app}(t) = V_0 \sin \alpha t\,,$$

where $q(t)$ is the charge on the plates of the capacitor at time t. Make computations for the two angular frequencies $\alpha_1 = 120\pi$ (60 Hz) and $\alpha_2 = 2000\pi$ (1 kHz).

(a) Solve the ODE subject to the initial condition $q(0) = 0$. Identify the transient and steady-state charges.

(b) Differentiate $q(t)$ to obtain the current $I(t)$. Identify its transient and steady-state components. Explain why $I(0)$ is zero.

(c) Plot the transient charge and current for each α value. Suggested range of t (both frequencies), 0 to 0.1 sec. Explain how, even though $q(0) = 0$, there are nonzero transients, and discuss the similarities and differences in the transient terms for the two α values. Is the transient term oscillatory? Explain.

(d) Plot the total charge $q(t)$ and current $I(t)$ for each α value. Suggested ranges: for α_1, 0 to 0.1 sec and 10 to 10.1 sec; for α_2, 0 to 0.01 sec and 10 to 10.01 sec. Explain the difference between the earlier and later-time plots.

(e) Compare the amplitudes of the steady-state oscillations at the two α values. Give a qualitative explanation of the differences.

(f) Compare the phases of $V_{\rm app}(t)$ and $I(t)$ in the large-t limit (obtain this from the steady-state term). Explain the phase shift in terms of the properties of the circuit elements.

10.4.9. This exercise deals with the LRC circuit in Fig. 10.4, with $L = 10^4$ H, $R = 100{,}000\ \Omega$, $C = 10^{-4}$ F (100 μF). $V_0 = 1$ V. The ODE for this circuit is

$$Lq''(t) + Rq'(t) + \frac{1}{C}\, q(t) = V_{\rm app}(t) = V_0 \sin \alpha t\,,$$

where $q(t)$ is the charge on the plates of the capacitor at time t. Make computations for the two angular frequencies $\alpha_1 = 120\pi$ (60 Hz) and $\alpha_2 = 600\pi$ (300 Hz).

(a) Solve the ODE subject to the initial conditions $q(0) = 0$, $q'(0) = I(0) = 0$. Identify the transient and steady-state charges.

(b) Differentiate $q(t)$ to obtain the current $I(t)$. Identify its transient and steady-state components.

(c) Plot the transient charge and current for each α value. Suggested range of t (both frequencies), 0 to 0.5 sec. Explain how, despite the initial conditions, there are nonzero transients, and discuss the similarities and differences in the transient terms for the two α values. Is the transient term oscillatory? Explain.

(d) Plot the total current $I(t)$ for each α value. Suggested ranges: for α_1, 0 to 0.2 sec and 10 to 10.2 sec; for α_2, 0 to 0.06 sec and 100 to 10.06 sec. Explain the difference between the earlier and later-time plots.

(e) Compare the amplitudes of the steady-state oscillations at the two α values. Give a qualitative explanation of the differences.

(f) Compare the phases of $V_{\rm app}(t)$ and $I(t)$ in the large-t limit (obtain this from the steady-state term). Explain the phase shift in terms of the properties of the circuit elements.

10.4.10. A parachutist falls subject to the gravitational force and to the air resistance provided by the parachute, assumed in this exercise to be proportional to the speed of descent. If the distance the parachutist falls is denoted by y, the ODE governing her motion is

$$my'' = mg - ky' ,$$

where m is the mass of the parachutist, g is the acceleration of gravity, and k is the coefficient of the air-resistance force.

(a) Assuming the parachutist's descent starts at $t = 0$, and that $y(0) = y'(0) = 0$, obtain formulas for $y(t)$ and $y'(t)$ (the velocity of descent).

(b) Show that the velocity asymptotically approaches a constant value, called the **terminal velocity**, and identify its value in two different ways: (i) from the solution of the ODE, and (ii) from the fact that the velocity will be constant when the net force on the parachutist is zero.

10.4.11. A chain of length L hangs from a frictionless pulley (of negligible diameter relative to L). Nothing is attached to the ends of the chain, and at $t = 0$ it is stationary, but with its two ends differing in vertical position by an amount $2h_0$.

(a) Construct the ODE for the motion of the chain, i.e., for $h(t)$, recognizing that the net force on the chain arises from the different amounts of chain on the two sides of the pulley, but the force must result in motion of the entire chain.

(b) Obtain a solution to the ODE subject to the given initial conditions, with your solution valid for $h \le L/2$.

(c) Calculate the kinetic energy of the chain (assuming its total mass to be M) at the time that one end of the chain reaches the pulley. Write your answer in a form that does not involve any transcendental functions.

10.5 MORE GENERAL SECOND-ORDER EQUATIONS

The bulk of our analysis of second-order ODEs is focused upon linear equations, which are encountered more frequently in science and engineering than their nonlinear counterparts. Even for linear second-order ODEs, there is no universal procedure that leads to closed-form solutions. In fact, many rather simple-looking ODEs have solutions that are only expressible as infinite series or as the "special functions" such series represent. For nonlinear ODEs the situation is bleaker; only a few nonlinear ODEs with special characteristics have formal solutions that are expressible in terms of well-known functions. When practical solutions are needed for nonlinear ODEs, most often they are obtained numerically, typically with massive amounts of computational effort by techniques that are outside the scope of the present text.

Linear second-order ODEs are of great importance in physics and engineering, in part because conventional electric circuit analysis involves linear second-order ODEs, in part because Maxwell's equations of electromagnetic theory (at least in vacuum) are linear, and in part because the basic equation of quantum mechanics (**Schrödinger's equation**) is linear and second order. Many important applications of ODEs have been covered due to our study (in the preceding section) of linear equations with constant coefficients, and a number of ODEs of great importance in the physical sciences are examined in detail in Chapter 14. The analysis of that chapter is based on power-series expansion methods that will be developed there.

The present section surveys various methods for solving second-order ODEs or for reducing them to first-order problems.

DEPENDENT VARIABLE MISSING

A second-order ODE with independent variable x and dependent variable y may be such that y only appears as the derivatives y' and y''. The ODE can then be converted into a first-order equation by the substitution

$$y' = u, \qquad y'' = u' . \tag{10.72}$$

If our ODE was nonlinear, the equation (now first-order, and with dependent variable u) may be recognizable as one of the few nonlinear first-order ODEs for which a solution is known. But if our original ODE was linear, we now have a linear first-order equation which we know how to solve (for $u(x)$, its dependent variable). We complete the solution process by integrating u to obtain $y(x)$.

Example 10.5.1. ODE with y Missing

Consider the ODE

$$2xy'' - y' = 3x .$$

Letting $u = y'$, this equation becomes

$$2xu' - u = 3x \quad \longrightarrow \quad u' - \left(\frac{1}{2x}\right) u = \frac{3}{2}, \tag{10.73}$$

which is a linear first-order equation in u, written to the right of the arrow in standard form (with unit coefficient for u'). From the standard form of this ODE, we identify p, the coefficient of u, as $p = -1/2x$, so an integrating factor for Eq. (10.73) (in standard form) is

$$w = \exp\left[\int^x \left(-\frac{1}{2x}\right) dx\right] = e^{-(1/2)\ln x} = x^{-1/2} .$$

Applying this integrating factor to the standard form of the ODE, we reach

$$[x^{-1/2}u]' = \frac{3}{2}\, x^{-1/2} \quad \longrightarrow \quad x^{-1/2}u = \frac{3}{2}\int^x x^{-1/2}\, dx + C = 3x^{1/2} + C ,$$

equivalent to

$$u = 3x + Cx^{1/2}\,.$$

To complete the solution of our original ODE, we now write $u = y'$:

$$u = y' \quad \longrightarrow \quad (3x + Cx^{1/2})\,dx = dy\,,$$

which we can integrate, getting (after redefining C)

$$y(x) = \frac{3x^2}{2} + Cx^{3/2} + C'\,.$$

This final result should be checked.

■

INDEPENDENT VARIABLE MISSING

If our ODE, for $y(x)$, is linear and does not contain x, it is homogeneous and with constant coefficients, so we need do nothing other than apply the solution method of Section 10.4. If the ODE is nonlinear, we can make the substitutions

$$y' = u\,, \qquad y'' = \frac{du}{dx} = \frac{du}{dy}\frac{dy}{dx} = u\,\frac{du}{dy}\,, \tag{10.74}$$

and thereby reduce our problem to a nonlinear first-order ODE, but with u and y now respectively the dependent and independent variables. If we are able to solve this ODE to get $u = u(y, C)$, we may then solve the separable ODE

$$y' = u \quad \longrightarrow \quad \frac{dy}{dx} = u(y, C) \quad \longrightarrow \int^y \frac{dy}{u(y, C)} = \int dx = x + C'\,. \tag{10.75}$$

Example 10.5.2. ODE with x Missing

Because our objective is only to show how this method of solution works, let's apply it to the linear equation

$$y'' + \omega^2 y = 0\,. \tag{10.76}$$

Substituting for y'' as given in Eq. (10.74), the ODE in Eq. (10.76) becomes

$$u\,\frac{du}{dy} + \omega^2 y = 0\,.$$

This equation is nonlinear, but separable; we have (calling the constant of integration $C^2/2$)

$$u\,du + \omega^2 y\,dy = 0 \quad \longrightarrow \quad \frac{u^2}{2} + \frac{\omega^2 y^2}{2} = \frac{C^2}{2}\,. \tag{10.77}$$

Solving for u, we get $u = \pm\sqrt{C^2 - \omega^2 y^2}$.

To find y, we insert this value of u into Eq. (10.75), obtaining

$$\int^y \frac{dy}{\pm\sqrt{C^2 - \omega^2 y^2}} = \pm\frac{1}{\omega}\arcsin\left(\frac{\omega y}{C}\right) = x + C'\,.$$

Writing this equation as

$$\arcsin\left(\frac{\omega y}{C}\right) = \pm\omega(x + C'),$$

and taking the sine of both sides, we reach

$$\frac{\omega y}{C} = \pm \sin[\omega(x + C')],$$

which by redefining the constants we can write in the less cluttered form

$$y = A\sin(\omega x + \delta). \tag{10.78}$$

We recognize this as the general solution to Eq. (10.76), confirming our analysis. ■

EULER EQUATION

An ODE of the general form

$$x^2 y'' + a_1 x\, y' + a_0\, y = f(x) \tag{10.79}$$

is called an Euler ODE. If we note that each derivative (which would decrease the power of x in x^p) is preceded by the multiplication of a power of x that will restore the original power, we can guess that a possible form for the solution of the homogeneous Euler ODE (i.e., with $f(x)$ set to zero) might be simply $y(x) = x^p$. Substituting this trial solution into Eq. (10.79), we get

$$p(p-1)x^p + a_1 p x^p + a_0 x^p = 0 \quad \longrightarrow \quad p^2 + (a_1 - 1)p + a_0 = 0, \tag{10.80}$$

indicating that we have a solution x^p for those values of p that solve the algebraic equation given in Eq. (10.80).

If the equation for p yields a double root, the two solutions to the homogeneous Euler ODE are x^p and $x^p \ln x$. Proof of this statement is the topic of Exercise 10.5.6.

A particular solution to the inhomogeneous Euler ODE can then be found either by the techniques illustrated in Section 10.4 or by the general methods developed in Section 10.6.

Example 10.5.3. Euler ODEs

1. Let's first obtain the general solution of

$$\mathcal{L}y \equiv x^2 y'' + 2xy' - 2y = x^3. \tag{10.81}$$

Examining first the associated homogeneous equation, $y = x^p$ is found to be a solution if

$$p^2 + p - 2 = 0, \quad \text{which has roots } p = 1 \text{ and } p = -2.$$

To obtain a particular solution y_p to the inhomogeneous equation, we again use the fact that application of $\mathcal{L}$ will not change the power of x in any function x^q, so it is natural to take $y_p = Ax^3$. Inserting this form into the ODE, we get

$$A\left[6x^3 + 6x^3 - 2x^3\right] = x^3,$$

which we solve for A, finding $A = 1/10$. Thus our ODE has the general solution

$$y(x) = C_1 x + \frac{C_2}{x^2} + \frac{x^3}{10}\,.$$

2. Next, consider

$$\mathcal{L}y(x) \equiv x^2 y'' - 3y' + 4y = 2x\,. \tag{10.82}$$

Inserting $y = x^p$ into the associated homogeneous equation, we find it to be a solution if

$$p^2 - 4p + 4 = 0, \quad \text{which has a double root at } p = 2.$$

We therefore conclude that the homogeneous equation has the independent solutions x^2 and $x^2 \ln x$. The double root does not interfere with our search for a particular solution to the inhomogeneous equation, for which we take the trial solution $y_p = Ax$. Applying $\mathcal{L}$,

$$\mathcal{L}y_p = A[0 - 3x + 4x] = 2x\,,$$

which is a valid equation if $A = 2$. Thus, our ODE has the general solution

$$y(x) = C_1 x^2 + C_2 x^2 \ln x + 2x\,.$$

3. Finally, let's obtain the general solution of

$$\mathcal{L}y \equiv x^2 y'' + 2xy' - 2y = 3x\,. \tag{10.83}$$

This ODE has the same differential operator as that in Eq. (10.81), so its associated homogeneous equation has the same solutions, $y = x$ and $y = x^{-2}$. But if we attempt to find a particular solution to the inhomogeneous equation, we cannot use a trial solution of the form $y_p = Ax$ because $\mathcal{L}(Ax) = 0$. Noting that when p was a double root, a second solution of the homogeneous equation had the form $x^p \ln x$, we see whether y_p can have the form $y_p = Ax \ln x$. We therefore look at

$$\mathcal{L}(Ax \ln x) = A\left[x^2\left(\frac{1}{x}\right) + 2x(\ln x + 1) - 2(x \ln x)\right] = 3Ax\,.$$

Comparing with Eq. (10.83), we see that we have a particular solution if $A = 1$, so our ODE has the general solution

$$y(x) = C_1 x + \frac{C_2}{x^2} + x \ln x\,.$$

■

ANOTHER EQUATION

An ODE that arises frequently enough that we should discuss it here is

$$y'' + f(y) = 0\,. \tag{10.84}$$

This equation can be solved after multiplying it by y'. Performing the multiplication, writing y'' in the first term as dy'/dx and writing y' in the second term as dy/dx, the quantities dx cancel and we are left with

$$y'y'' + f(y)y' = 0 \quad \longrightarrow \quad y'\,dy' + f(y)\,dy = 0\,,$$

which is separable in y and y'. Integrating, we reach

$$\frac{1}{2}(y')^2 + \int^y f(y)\,dy = C\,. \tag{10.85}$$

Equation (10.85) is also separable (after rearrangement):

$$y' = \pm\left[C - 2\int^y f(y)\,dy\right]^{1/2} \quad\longrightarrow\quad \pm\left[C - 2\int^y f(y)\,dy\right]^{-1/2} dy = dx\,.$$

The final result, which is often easier to evaluate than the relatively complicated formula might suggest, is

$$\pm\int^y \left[C - 2\int^y f(z)\,dz\right]^{-1/2} dy = x + C'\,. \tag{10.86}$$

Example 10.5.4. Check of Eqs. (10.85) and (10.86)

Our check ODE will be Eq. (10.76), for which $f(y) = \omega^2 y$. Applying Eq. (10.85) we get

$$\frac{1}{2}(y')^2 + \int^y \omega^2 y\,dy = \frac{1}{2}(y')^2 + \frac{1}{2}\omega^2 y^2 = \frac{C^2}{2}\,.$$

Solving for y', we have

$$y' = \pm\sqrt{C^2 - \omega^2 y^2} \quad\longrightarrow\quad \frac{dy}{\pm\sqrt{C^2 - \omega^2 y^2}} = dx\,.$$

This is the same result as was found in Example 10.5.2 and after integration we can reach the form $y = A\sin(\omega x + \delta)$.

■

Example 10.5.5. A First Integral

Part of the value of the analysis of Eq. (10.84) is in mechanics, where Newton's law for a particle of mass m subject to a force $F(y)$ takes the form

$$m\,\frac{d^2y}{dt^2} = F(y)\,. \tag{10.87}$$

Multiplying Eq. (10.87) by $y' = dy/dt$ and identifying this derivative as the velocity v, Eq. (10.85) for the current problem becomes

$$\frac{1}{2}mv^2 = \int^y F(y)\,dy + C\,.$$

This is the law of conservation of mechanical energy; mathematicians call it a **first integral** of the ODE describing the motion. Note that the first integral was obtained in a way that did not require us to ever complete the solution of Eq. (10.87).

■

Exercises

10.5.1. For the ODE $y'' + yy' = 0$, find a general solution containing two independent constants and, in addition, a singular solution that contains only one constant. Does your symbolic computing system find the singular solution?

Hint. Consider what happens to your solution procedure if one of the constants of integration is zero.

Find general solutions for the following ODEs.

10.5.2. $2yy'' - y'^2 = 0\,.$

10.5.3. $xy'' - y' - y'^2 = 0\,.$ In addition, this ODE has the singular solution $y = C$.

10.5.4. $xy'' + y' = 0\,.$

10.5.5. $x^2y'' - xy' - 3y = x^4\,.$

10.5.6. $x^2y'' - 3xy' + 4y = x\,.$

10.5.7. $x^2y'' - 3xy' + 4y = x^2\,.$

Hint. Try solutions of the form $x^p \ln^q x$.

10.5.8. The ODE satisfied by a quartic oscillator of mass m is of the form

$$m\,\frac{d^2x}{dt^2} = -kx^3\,.$$

Find a first integral of this ODE that can be interpreted as the law of conservation of energy for the oscillator.

10.5.9. Consider the surface of the Earth to be a sphere of radius R. An object of mass m, located at a distance $r > R$ from the Earth's center will experience a gravitational force due to the Earth in amount

$$F = \frac{mgR^2}{r^2}\,.$$

Neglecting air resistance and other extraterrestrial masses, find the minimum upward velocity (**escape velocity**) an object must have when launched from the Earth's surface in order for it to escape from the earth's gravitational field.

10.5.10. A particle of mass m initially at rest at $x = 0$ is subject, starting at time $t = 0$ to a force

$$F = (a - x)\left[1 + 2\ln\left(\frac{a - x}{a}\right)\right].$$

By a method that does not require a complete solution of the equation of motion $mx'' = F$ show that the particle will move only from $x = 0$ to $x = a$.

10.6 GENERAL PROCESSES FOR LINEAR EQUATIONS

This section introduces some techniques that are often helpful in obtaining the general solution to a linear second-order ODE after we have found a single solution to the associated homogeneous ODE.

FINDING A SECOND SOLUTION

It is often easy to find one solution to a homogeneous linear second-order ODE, but intuition or power-series expansion fails to reveal a second solution. In that event, it turns out that we can use properties of the Wronskian W of the two independent solutions to obtain a second solution.

Let

$$y''(x) + p(x)y'(x) + q(x)y(x) = 0 \tag{10.88}$$

be our ODE, and let $y_1(x)$ and $y_2(x)$ be linearly independent solutions of Eq. (10.88). We assume y_1 to be known; our task is to find y_2. As shown in Appendix J, an explicit formula for $y_2(x)$ can be written in terms of $y_1(x)$ and the Wronskian W of the two solutions, and that the Wronskian in turn can be computed from $p(x)$, the coefficient of y' in the ODE. The key result is the following:

$$y_2(x) = y_1(x) \int^x \frac{W(t)}{y_1(t)^2}\, dt = y_1(x) \int^x \frac{\exp(-\int^t p(u)\, du)}{y_1(t)^2}\, dt\,. \tag{10.89}$$

Equation (10.89) is an important result because it enables construction of y_2 using only y_1 and the coefficient $p(x)$ from the original ODE. Despite its seemingly complicated form, Eq. (10.89) reduces in many practical cases to easily managed final expressions.

Example 10.6.1. A Second Solution

The ODE

$$(1 - x^2)y'' - 2xy' + 2y = 0$$

has the solution $y_1(x) = x$. Our task is to find a second solution to this ODE. Unless you already know the form of the second solution, it seems unlikely that you would be able to guess it.

To use the formula for y_2, Eq. (10.89), we need to write our ODE in the form

$$y'' - \frac{2x}{1 - x^2}\, y' + \frac{2}{1 - x^2}\, y = 0$$

because that is the form in which $p(x)$ is the coefficient of y'. Thus, we identify

$$p(x) = \frac{-2x}{1 - x^2}\,.$$

Next we evaluate

$$W(t) = \exp\left(- \int^t p(u)\, du \right) = \exp\left(- \int^t \frac{-2u}{1 - u^2}\, du \right)$$

$$= \exp\Big(- \ln(1 - t^2) \Big) = \frac{1}{1 - t^2}\,.$$

Finally, using Eq. (10.89) and remembering that $y_1(x) = x$, we form

$$y_2(x) = y_1(x)\int^x \frac{W(t)}{y_1(t)^2}\,dt = x\int^x \frac{1}{t^2(1-t^2)}\,dt\,.$$

The integration over t can be carried out making a decomposition into partial fractions:

$$\begin{aligned} y_2(x) &= x\int^x \left[\frac{1}{t^2} + \frac{1}{1-t^2}\right] dt = x\int^x \left[\frac{1}{t^2} + \frac{1}{2}\left(\frac{1}{1-t} + \frac{1}{1+t}\right)\right] dt \\ &= x\left[-\frac{1}{x} + \frac{1}{2}\ln\left(\frac{1+x}{1-x}\right)\right] \\ &= \frac{x}{2}\ln\left(\frac{1+x}{1-x}\right) - 1\,. \end{aligned}$$

Prudence dictates that one should check to make sure that $y_2(x)$ is a solution to our ODE. This task is left as an Exercise.

■

SOLUTIONS TO INHOMOGENEOUS EQUATIONS

Assuming now that we have two independent solutions to a homogeneous linear second-order ODE, we consider how we might obtain in a systematic way a particular solution to a corresponding inhomogeneous ODE. Thus, let

$$y'' + p(x)y' + q(x)y = f(x)\,, \tag{10.90}$$

and we assume that we already know that $y_1(x)$ and $y_2(x)$ are linearly independent solutions of the corresponding homogeneous ODE (that with $f(x)$ replaced by zero). Based on our knowledge of y_1 and y_2, we seek a solution y_p of the inhomogeneous ODE (which will enable the construction of the general solution to our ODE).

We proceed by a method sometimes given the name **variation of constants**, or perhaps more accurately, **variation of parameters**. The method starts by assuming y_p to have the form

$$y_p(x) = u_1(x)y_1(x) + u_2(x)y_2(x)\,, \tag{10.91}$$

the name arising from the fact that Eq. (10.91) is like the form of the general solution to the homogeneous equation, but with the constants in that general solution replaced by unknown functions u_i of the independent variable x. Differentiating Eq. (10.91), we find

$$y_p' = (u_1'y_1 + u_2'y_2) + u_1y_1' + u_2y_2'\,. \tag{10.92}$$

An ingenious feature of the method is that the form chosen for y_p contains more flexibility than necessary, and we can, without creating inconsistency, require that the parenthesized terms in Eq. (10.92) add to zero. Thus, we have created the following condition on u_1 and u_2:

$$u_1'y_1 + u_2'y_2 = 0\,, \tag{10.93}$$

and our formula for y_p' reduces to

$$y_p' = u_1y_1' + u_2y_2'\,. \tag{10.94}$$

Now taking the second derivative, we reach

$$y_p'' = u_1 y_1'' + u_1' y_1' + u_2 y_2'' + u_2' y_2' \,, \tag{10.95}$$

and insertion of Eqs. (10.91), (10.94), and (10.95) into the inhomogeneous ODE for y_p leads to

$$(u_1 y_1'' + u_1' y_1' + u_2 y_2'' + u_2' y_2') + p(u_1 y_1' + u_2 y_2') + q(u_1 y_1 + u_2 y_2) = f(x) \,. \tag{10.96}$$

The three terms containing u_1 vanish because y_1 satisfies the homogeneous ODE, and the three terms containing u_2 vanish because y_2 also satisfies the homogeneous ODE. All that remains in Eq. (10.96) reduces to

$$u_1' y_1' + u_2' y_2' = f(x) \,,$$

and this equation, together with Eq. (10.93), provide a pair of simultaneous equations whose solutions will yield u_1' and u_2'. Together, these equations are

$$\begin{aligned} y_1 u_1' + y_2 u_2' &= 0 \,, \\ y_1' u_1' + y_2' u_2' &= f(x) \,. \end{aligned} \tag{10.97}$$

The final steps in our odyssey are to solve Eqs. (10.97) for u_1' and u_2' and to integrate these quantities to obtain u_1 and u_2. Inserting u_1 and u_2 into Eq. (10.91) then gives us y_p. The above scenario needs to be illustrated by an example.

Example 10.6.2. Particular Solution, Inhomogeneous ODE

We desire the general solution to the inhomogeneous ODE

$$x^2 y'' - 2xy' + 2y = x^3 e^x \,. \tag{10.98}$$

We recognize the homogeneous ODE obtained by setting the right-hand side of Eq. (10.98) to zero as an Euler equation with solutions x^p, and we easily find that its solutions correspond to $p = 1$ and $p = 2$. We therefore designate $y_1 = x$ and $y_2 = x^2$, and search for a solution y_p to the inhomogeneous equation of the form

$$y_p = y_1 \, u_1(x) + y_2 \, u_2(x) = x \, u_1(x) + x^2 \, u_2(x) \,. \tag{10.99}$$

We intend to find u_1 and u_2 by using Eqs. (10.97), for which we need not only y_1 and y_2, but also $y_1' = 1$ and $y_2' = 2x$. In addition, we need $f(x)$, which is NOT $x^3 e^x$ because Eq. (10.98) is not in the standard form with the coefficient of y'' equal to 1. Dividing Eq. (10.98) through by x^2, we find $f(x) = x\, e^x$. Failure to define $f(x)$ properly is a common error that one should be careful to avoid.

With the above identifications, Eqs. (10.97) for the current problem are found to be

$$\begin{aligned} x\, u_1' + x^2 \, u_2' &= 0 \,, \\ u_1' + 2x \, u_2' &= x\, e^x \,. \end{aligned} \tag{10.100}$$

Solving these simultaneous algebraic equations for u_1' and u_2', we get the results shown below, which we then integrate to obtain values for u_1 and u_2:

$$\left\{ \begin{array}{l} u_1' = -x\, e^x \\ u_2' = e^x \end{array} \right\} \longrightarrow \left\{ \begin{array}{l} u_1 = e^x(1-x) \\ u_2 = e^x \end{array} \right\} . \tag{10.101}$$

Inserting these expressions into Eq. (10.99), some simplification results, and we get

$$y_p = x\,e^x\,.$$

This result should be checked. We calculate $y_p' = (x+1)\,e^x$, $y_p'' = (x+2)\,e^x$, and verify that Eq. (10.98) is satisfied.

Our ODE therefore has general solution

$$y(x) = C_1 x + C_2 x^2 + x\,e^x\,. \tag{10.102}$$

■

Exercises

10.6.1. Verify in detail the steps taken to find $y_2(x)$ in Example 10.6.1.

10.6.2. Verify in detail the steps taken in Example 10.6.2 leading to Eqs. (10.100)–(10.102).

10.6.3. The function $y_1(x) = x^i$ is a solution of the ODE

$$x^2 y'' + xy' + y = 0\,.$$

From the fact that all the coefficients of the ODE are real, we can conclude that another solution of the ODE is $y_2(x) = x^{-i}$. Confirm that $y_2(x)$ is the result of applying Eq. (10.89) for this ODE.

10.6.4. Consider the inhomogeneous ODE

$$x^2 y'' + xy' + y = x^2\,.$$

The associated homogeneous ODE was treated in Exercise 10.6.3.

(a) Use the method of variation of parameters to find the general solution of the inhomogeneous ODE.

(b) Confirm your result using symbolic computing. Your computer will probably produce a result that looks different than your hand computations. Resolve any apparent discrepancy.

10.7 GREEN'S FUNCTIONS

In working with an inhomogeneous linear ODE, we have already seen that if we have a sum of two or more inhomogeneous terms, a particular solution can be obtained by adding solutions found for the individual inhomogeneous terms. A useful generalization of this idea is to construct solutions for which the inhomogeneity is localized to a contribution from a single value of the independent variable (call it x'), and to then construct the solution for a more general inhomogeneity by summing contributions associated with the inhomogeneities at different values of x'. More formally, we are suggesting the following procedure, applicable to linear second-order ODEs:

Consider a linear inhomogeneous ODE of the form $\mathcal{L}y(x) = f(x)$, where $\mathcal{L}$ is a differential operator with x as its independent variable (we remind the reader of that fact by writing $\mathcal{L}_x$) and $\mathcal{L}$ contains $y''(x)$ with coefficient unity. If $\mathcal{L}$ as originally given did not have this property, it should be divided by the necessary factor, with the definition of $f(x)$ modified accordingly. Then,

1. Obtain (for general values of x') a solution $G(x,x')$ to the following related ODE, where x is the independent variable and x' is a parameter indicating the location of the inhomogeneity:

$$\mathcal{L}_x G(x,x') = \delta(x-x')\,.$$

The delta function corresponds to making the imhomogeneity at x' of unit strength. We call $G(x,x')$ a **Green's function**, and note that it has the property of giving the contribution to $y(x)$, for all x, from a unit "impulse" (i.e., inhomogeneous term) at $x = x'$.

2. Write the solution to our original ODE as

$$y(x) = \int G(x,x')f(x')\,dx'\,, \tag{10.103}$$

where the interval of integration must include all x' for which $f(x')$ is nonzero. This integral sums the contributions to $y(x)$ from all values of the parameter x'.

This recipe works because Eq. (10.103) adds, for each x', the contribution to $y(x)$ corresponding to a localized inhomogeneous term concentrated at x' and of strength $f(x')$. A proof that $y(x)$ solves the ODE is easily developed:

$$\mathcal{L}_x y(x) = \mathcal{L}_x\left[\int G(x,x')f(x')\,dx\right] = \int\left[\mathcal{L}_x G(x,x')\right]f(x')\,dx'$$

$$= \int \delta(x-x')f(x')\,dx' = f(x)\,.$$

The reader has doubtless encountered the present situation before: the potential from a distribution of charge (it satisfies a linear differential equation with inhomogeneities at the locations of charges) is the sum of the contributions from individual elements of charge, whose potentials can be computed separately.

Green's functions are most useful when we seek a solution $y(x)$ to an inhomogeneous differential equation subject to initial or boundary conditions all of which are of the forms $y(x_i) = 0$ or $y'(x_i) = 0$. Then we can determine $G(x,x')$ subject (for general x') to these conditions, and when $y(x)$ is computed as in Eq. (10.103) the result will, for any $f(x')$, satisfy the given conditions. Thus, a Green's function can be used to construct directly an ODE solution that satisfies common kinds of initial/boundary conditions.

We are now ready to consider how we might find a Green's function. We start by noting that a Green's function corresponding to the inhomogeneous term $\delta(x-x')$ must be a solution to the related homogeneous equation everywhere except at $x = x'$ and must have properties at $x = x'$ that generate the singular inhomogeneous term. A singularity in a function is worsened by taking a derivative, suggesting that we set up a singularity in $y(x)$ at $x = x'$ that will be a delta function in y'', and therefore a finite discontinuity in y' (but with no discontinuity in y). In order for $y''(x)$ (with its unit coefficient) to be a properly scaled delta function, the discontinuity in y' must be a unit step at $x = x'$, so that

$$\lim_{\varepsilon=0+}\int_{x'-\varepsilon}^{x'+\varepsilon} y''(x)\,dx = \lim_{\varepsilon=0+}\Big[y'(x'+\varepsilon) - y'(x'-\varepsilon)\Big] = 1\,.$$

Our recipe for $G(x,x')$ is now formally complete: for $x < x'$, $G(x,x')$ must be a suitable solution to the homogeneous ODE, satisfying whatever conditions we impose in that region of x. (In an initial-conditions problem, we may choose $G(x,x')$ to be identically zero for $x < x'$.) And for $x > x'$, $G(x,x')$ must solve the homogeneous ODE under conditions relevant to that region of x. The solutions for these two regions must connect in such a way that

$$\lim_{\varepsilon\to 0+}\Big[G(x'+\varepsilon,x') - G(x'-\varepsilon,x')\Big] = 0\,, \tag{10.104}$$

$$\lim_{\varepsilon\to 0+}\left[\frac{d}{dx}G(x,x')\bigg|_{x=x'+\varepsilon} - \frac{d}{dx}G(x,x')\bigg|_{x=x'-\varepsilon}\right] = 1\,. \tag{10.105}$$

The significance of the above equations and statements will become clearer as we examine some typical examples.

Example 10.7.1. Green's Function: Initial Conditions

Consider the second-order ODE

$$y''(t) + \omega^2 y(t) = f(t)\,, \qquad \text{with } y(0) = y'(0) = 0\,.$$

The conditions $y(0) = y'(0) = 0$ cause $y(t)$ to remain zero prior to the action of the forcing function $f(t)$, which can be viewed as a time-dependent source term. We desire to introduce a Green's function that can help us to find the function $y(t)$ produced by an arbitrary $f(t)$. Such a Green's function $G(t,t')$ should describe the response to a unit impulse at $t = t'$, with no response prior to the impulse. Thus, our Green's function must be the solution to

$$\frac{\partial^2}{\partial t^2}G(t,t') + \omega^2 G(t,t') = \delta(t-t')\,, \qquad \text{with } G(t,t') = \frac{\partial}{\partial t}G(t,t') = 0 \text{ for } t < t'\,. \tag{10.106}$$

The general solution to Eq. (10.106) for $t > t'$ is simply the solution of the associated homogeneous equation, and is therefore

$$G(t,t') = C_1 \sin[\omega(t-t')] + C_2 \cos[\omega(t-t')]\,.$$

To satisfy the continuity condition of Eq. (10.104) we set $C_2 = 0$, as we must connect to $G(t,t') = 0$ for $t < t'$. To obtain the unit discontinuity in slope required by Eq. (10.105), we require

$$\frac{d}{dt}C_1 \sin[\omega(t-t')]\bigg|_{t=t'} = C_1\omega\cos[\omega(t-t')]\bigg|_{t=t'} = C_1\omega = 1\,,$$

showing that $C_1 = 1/\omega$. Combining these observations, we have

$$G(t,t') = \begin{cases} 0\,, & t < t', \\ \dfrac{1}{\omega}\sin[\omega(t-t')]\,, & t > t'. \end{cases} \tag{10.107}$$

Let's check that this Green's function gives the expected result. Take first $f(t) = \sin\alpha t$, with $\alpha \neq \omega$. The solution predicted by the Green's function is

$$y(t) = \int_0^t \frac{1}{\omega}\sin[\omega(t-t')]\,\sin\alpha t'\,dt' = \frac{\sin\alpha t}{\omega^2-\alpha^2} - \frac{\alpha\sin\omega t}{\omega(\omega^2-\alpha^2)}\,. \tag{10.108}$$

This result agrees with that found in our earlier investigation of this problem in Example 10.4.7. Note that it contains the amount of the general solution that is needed to cause $y(0) = y'(0) = 0$.

The Green's-function method continues to give correct results even if the inhomogeneous term is taken to be $\sin\omega t$. However, the integral for $y(t)$ then evaluates to a different result than was obtained in Eq. (10.108). We now get

$$y(t) = \int_0^t \frac{1}{\omega}\sin[\omega(t-t')]\,\sin\omega t'\,dt' = \frac{\sin\omega t}{2\omega^2} - \frac{t\cos\omega t}{2\omega}\,, \tag{10.109}$$

also in agreement with Example 10.4.7.

■

Green's functions are of great value when it is desired to solve an ODE subject to boundary conditions.

Example 10.7.2. Green's Function: Boundary Conditions

Consider the ODE

$$x^2 y''(x) + x y'(x) - y(x) = g(x)\,, \tag{10.110}$$

subject to the boundary conditions $y(0) = y(1) = 0$. We want to proceed by finding a Green's function that is consistent with these boundary conditions, so that we can solve Eq. (10.110) for a variety of inhomogeneous terms $g(x)$.

Our first step is to write Eq. (10.110) in a standard form in which y'' appears with a unit coefficient; we thus continue from

$$y''(x) + x^{-1}y'(x) - x^{-2}y(x) = f(x)\,, \qquad \text{where } f(x) = g(x)/x^2. \tag{10.111}$$

The important consequence of the passage to Eq. (10.111) is that we must use $f(x)$, not $g(x)$, when we construct a solution to our ODE using its Green's function.

We next obtain solutions to the homogeneous ODE associated with Eq. (10.110). We recognize the equation as an Euler ODE, and use the technique for that equation to find the two linearly independent solutions $y = x$ and $y = x^{-1}$. For the region $x < x'$, we need a solution $y_1(x)$ with the property $y_1(0) = 0$; from our two solutions we easily choose $y_1(x) = x$. For the region $x > x'$, we need a solution $y_2(x)$ such that $y_2(1) = 0$; neither of the solutions as originally found meets this boundary condition, but it can be met by taking $y_2(x) = x^{-1} - x$. We now need to multiply these solutions by respective coefficients C_1 and C_2 in a way that satisfies the continuity/discontinuity conditions given in Eqs. (10.104) and (10.105). We can satisfy Eq. (10.104) by taking $C_1 = Cy_2(x')$ and $C_2 = Cy_1(x')$ (both with the same value of C, which may depend upon x'), thereby reaching

$$G(x,x') = \begin{cases} Cy_2(x')y_1(x) = \dfrac{Cx}{x'}\left(1 - x'^2\right), & a \le x < x', \\[2ex] Cy_1(x')y_2(x) = \dfrac{Cx'}{x}\left(1 - x^2\right), & x' < x \le b. \end{cases}$$

It is apparent that these forms for $G(x,x')$ exhibit continuity at $x = x'$.

Our final step in finding the Green's function is to determine the value of C that will cause a unit step in $\partial G/\partial x$ at $x = x'$, as required by Eq. (10.105). For this purpose we compute

$$\frac{\partial G(x,x')}{\partial x} = \begin{cases} Cy_2(x')y_1'(x) = C\left(\frac{1}{x'} - x'\right), & a \le x < x', \\ Cy_1(x')y_2'(x) = Cx'\left(-\frac{1}{x^2} - 1\right), & x' < x \le b. \end{cases}$$

Evaluating these derivatives at $x = x'$ and forming their difference, which must evaluate to unity, we get

$$\frac{\partial G(x'_+,x')}{\partial x} - \frac{\partial G(x'_-,x')}{\partial x} = C\left[x'\left(-\frac{1}{x'^2} - 1\right) - \left(\frac{1}{x'} - x'\right)\right] = -\frac{2C}{x'} = 1\,,$$

showing that Eq. (10.105) will be satisfied if we take $C = -x'/2$. Our final formula for the Green's function is therefore

$$G(x,x') = \begin{cases} \left(\frac{x'^2 - 1}{2}\right)x\,, & a \le x < x', \\ \frac{x'^2}{2}\left(\frac{x^2 - 1}{x}\right), & x' < x \le b. \end{cases} \tag{10.112}$$

Let's check that this Green's function produces the correct result for the case $g(x) = x^2$. As indicated by Eq. (10.111), this g corresponds to $f(x) = 1$, and the formula for a solution to Eq. (10.110) that satisfies the boundary conditions should be given by

$$y(x) = \int_0^1 G(x,x')f(x')\,dx' = \int_0^x G(x,x')\,dx' + \int_x^1 G(x,x')\,dx'\,.$$

We have written this integral in a peculiar and apparently redundant form because the formula to be inserted for $G(x,x')$ is different for the two partial ranges of x'. In the first integral on the right-hand side, we insert the form of $G(x,x')$ appropriate to $x > x'$, while for the last integral we use the form of G for $x < x'$. These choices lead to

$$y(x) = \left(\frac{x^2 - 1}{x}\right)\int_0^x \frac{x'^2}{2}\,dx' + x\int_x^1 \left(\frac{x'^2 - 1}{2}\right)dx' = \frac{x(x-1)}{3}\,. \tag{10.113}$$

This form for $y(x)$ clearly satisfies the boundary conditions at $x = 0$ and $x = 1$. By substitution into Eq. (10.110) we can also confirm that it is a solution to our ODE with $g(x) = x^2$.

■

Exercises

10.7.1. Verify that the integrals for $y(t)$ given in Eqs. (10.108) and (10.109) give the results there shown.

10.7.2. Perform the integrations appearing in Eq. (10.113) and thereby confirm the formula there given for $y(x)$.

10.7.3. (a) Find the Green's function $G(t,t')$ for the ODE

$$y''(t) + 2y'(t) + y(t) = f(t)\,,$$

subject to the initial conditions $G(t,t') = \partial G(t,t')/\partial t = 0$ for $t < t'$.

(b) Use the Green's function to find a solution to the ODE with the above initial conditions for $f(t) = t^2$ for $t \geq 0$, $f(t) = 0$ otherwise.

(c) Check your solution, including its initial conditions.

10.7.4. For the ODE $x^2y'' - xy' - 3y = f(x)$,

(a) Find its Green's function for the boundary conditions $y(0) = y(1) = 0$.

(b) Use the Green's function to find the solution to this ODE and its boundary conditions for $f(x) = x^2$.

(c) Check your solution, including verification of the boundary conditions.

10.7.5. (a) Construct a Green's function for the ODE

$$x^2y'' + xy' + y = f(x)\,,$$

subject to the boundary conditions $y(1) = y(e^{\pi/2}) = 0$.

(b) For $f(x) = x^2$, use your Green's function to solve the ODE. It may be helpful to note that this ODE was the topic of Exercises 10.6.3 and 10.6.4.

(c) Verify that your answer to part (b) is a solution to the ODE that also satisfies the boundary conditions identified in part (a).

Additional Readings

Bronson, R., & Costa, G. (2009). *Schaum's outline of theory and problems of differential equations* (3rd ed.). New York: McGraw-Hill.

Cohen, H. (1992). *Mathematics for scientists and engineers.* Englewood Cliffs, NJ: Prentice-Hall.

Golomb, M., & Shanks, M. (1965). *Elements of ordinary differential equations.* New York: McGraw-Hill.

Hubbard, J., & West, B. H. (1995). *Differential equations.* Berlin: Springer.

Ince, E. L. (1956). *Ordinary differential equations.* New York: Dover (The classic work in the theory of ordinary differential equations).

Miller, R. K., & Michel, A. N. (1982). *Ordinary differential equations.* New York: Academic Press.

Ritger, P. D., & Rose, N. J. (1986). *Differential equations with applications.* New York: McGraw-Hill.

Chapter 11

GENERAL VECTOR SPACES

In Chapter 2 we discussed in detail the representation of functions by series, with emphasis on power-series expansions, either about the origin of our coordinate system (Maclaurin series) or about more general points (Taylor series). While the possibility of expanding in terms of other sets of functions was mentioned, we did not really explore that topic. Now, using the background provided by Chapters 4 and 5, we are ready to discuss some of the properties of expansions in more general function sets.

This topic is of importance because more or less arbitrary functions can be expanded in terms of known functions, enabling us to work with the coefficients in the expansions rather than directly with the perhaps unknown functions the expansions represent. These expansions have played an important role in the solutions of a wide variety of physics and engineering problems in areas that include quantum mechanics, classical systems involving oscillatory motion, transport of material or energy, even fundamental particle theory.

It turns out that the ideas associated with the description of vectors in terms of their components and with the way in which the components change under rotational transformations can be generalized to apply to expansions in terms of general sets of functions. To pursue this line of investigation, we need to introduce some new concepts.

11.1 VECTORS IN FUNCTION SPACES

Suppose that we have a two-dimensional (**2-D**) space in which the two coordinates, which are real (or in the most general case, complex) numbers that we will call a_1 and a_2, are respectively associated with the two functions $\varphi_1(s)$ and $\varphi_2(s)$. It is important at the outset to understand that our new 2-D space has nothing whatsoever to do with the physical xy space. It is a space in which the coordinate point (a_1, a_2) corresponds to the function

$$f(s) = a_1\varphi_1(s) + a_2\varphi_2(s)\,. \tag{11.1}$$

The analogy with a physical 2-D vector space with vectors $\mathbf{A} = A_1\hat{\mathbf{e}}_1 + A_2\hat{\mathbf{e}}_2$ is that $\varphi_i(s)$ corresponds to $\hat{\mathbf{e}}_i$, while $a_i \longleftrightarrow A_i$, and $f(s) \longleftrightarrow \mathbf{A}$. In other words, the coordinate values in our new space are the **coefficients** of the $\varphi_i(s)$, so each point in the space identifies a different function $f(s)$. Both f and φ are shown above as

Mathematics for Physical Science and Engineering.
http://dx.doi.org/10.1016/B978-0-12-801000-6.00011-0

dependent upon an independent variable we call s. We choose the name s to emphasize the fact that the formulation is not restricted to the spatial variables x, y, z, but can be whatever variable, or set of variables, is needed for the problem at hand. Note further that the variable s is not a continuous analog of the discrete variables x_i of an ordinary vector space. It is a parameter reminding the reader that the φ_i that correspond to the "unit vectors" of our vector space are usually functions of one or more variables. The variable(s) denoted by s may sometimes correspond to physical displacements, but that is not always the case. What should be clear is that s has nothing to do with the coordinates in our vector space; that is the role of the a_i.

Equation (11.1) defines a set of functions (a **function space**) that can be built from the **basis** φ_1, φ_2; we call this space a **linear vector space** because its members are linear combinations of the basis functions and the addition of its members corresponds to component (coefficient) addition. If $f(s)$ is given by Eq. (11.1) and $g(s)$ is given by another linear combination of **the same** basis functions,

$$g(s) = b_1\varphi_1(s) + b_2\varphi_2(s) ,$$

with b_1 and b_2 the coefficients defining $g(s)$, then

$$h(s) = f(s) + g(s) = (a_1 + b_1)\varphi_1(s) + (a_2 + b_2)\varphi_2(s) \tag{11.2}$$

defines $h(s)$, the member of our space (i.e., the function) which is the sum of the members $f(s)$ and $g(s)$. In order for our vector space to be useful, we consider only spaces in which the sum of any two members of the space is also a member. We sometimes (particularly in quantum mechanics) deal with functions that may be complex. For that reason the development of this chapter assumes that the basis functions φ_i and the coefficients may be complex, and from here on we make definitions that are appropriate for complex quantities.

Our space is said to be **linear** because its members are linear combinations of the basis functions. In addition, the notion of linearity includes the requirement that if $f(s)$ is a member of our vector space, then $u(s) = k\,f(s)$, where k is a real or complex number, is also a member, and we can write

$$u(s) = k\,f(s) = ka_1\varphi_1(s) + ka_2\varphi_2(s) . \tag{11.3}$$

Vector spaces for which addition of two members or multiplication of a member by a scalar always produces a result that is also a member are termed **closed** under these operations.

The essence of our findings up to this point is that the **coefficients** of the expansions representing members of our function space obey the rules that apply to the components of vectors. What is more general is that the members of the **basis** for our space are not physical-space unit vectors; they are functions.

The functions that form the basis of our vector space can be ordinary functions, and may be as simple as powers of s, or more complicated, as for example $\varphi_1 = (1 + 3s + 3s^2)e^s$, $\varphi_2 = (1 - 3s + 3s^2)e^{-s}$, or even completely abstract quantities that are defined only by certain properties they may possess. The number of basis functions (i.e., the **dimension** of our basis) may be a small number such as 2 or 3, a larger but finite integer, or even denumerably infinite (as would arise in an untruncated power series). The main universal restriction on the form of a basis is that the basis members be linearly independent, so that any function (member) of our vector space will be described by a unique linear combination of the basis functions. We illustrate the possibilities with two simple examples.

Example 11.1.1. Two Vector Spaces

1. We consider first a vector space of dimension 3, which is **spanned by** (meaning that it has a basis that consists of) the three functions $P_0(s) = 1$, $P_1(s) = s$, $P_2(s) = \frac{3}{2}s^2 - \frac{1}{2}$. Some members of this vector space include the functions

$$s + 3 = 3P_0(s) + P_1(s)\,, \quad s^2 = \frac{1}{3}\,P_0(s) + \frac{2}{3}\,P_2(s)\,, \quad 4 - 3s = 4P_0(s) - 3P_1(s)\,.$$

In fact, because we can write 1, s, and s^2 in terms of our basis, we can see that **any** quadratic form in s will be a member of our vector space, and that our space includes only functions of s that can be written in the form $c_0 + c_1 s + c_2 s^2$.

To illustrate our vector-space operations, we can form

$$s^2 - 2(s+3) = \left[\frac{1}{3}\,P_0(s) + \frac{2}{3}\,P_2(s)\right] - 2\Big[3P_0(s) + P_1(s)\Big] \tag{11.4}$$

$$= \left(\frac{1}{3} - 6\right) P_0(s) - 2P_1(s) + \frac{2}{3}\,P_2(s) \tag{11.5}$$

$$= -\frac{17}{3}\,P_0(s) - 2P_1(s) + \frac{2}{3}\,P_2(s)\,. \tag{11.6}$$

This calculation involves only operations on the coefficients; we do not need to refer to the definitions of the P_n to carry it out.

Notice that we are free to define our basis any way we want, so long as its members are linearly independent. We could have chosen as our basis for this same vector space $\varphi_0 = 1$, $\varphi_1 = s$, $\varphi_2 = s^2$, but we chose not to do so.

2. The set of functions $\varphi_n(s) = s^n$ $(n = 0, 1, 2, \ldots)$ is a basis for a vector space whose members consist of functions that can be represented by a Maclaurin series. To avoid difficulties with this infinite-dimensional basis, we will usually need to restrict consideration to functions and ranges of s for which the Maclaurin series converges. Convergence and related issues are of great interest in pure mathematics; in physics problems we usually proceed in ways such that convergence is assured.

The members of our vector space will have representations

$$f(s) = a_0 + a_1\,s + a_2\,s^2 + \cdots = \sum_{n=0}^{\infty} a_n\,s^n\,,$$

and we can (at least in principle) use the rules for making power-series expansions to find the coefficients that correspond to a given $f(s)$.

■

SCALAR PRODUCT

To make the vector-space concept useful and parallel to that of vector algebra in ordinary space, we need to introduce the concept of a scalar product in our function space.

We write the scalar product of two members of our vector space, f and g, as $\langle f|g\rangle$. This is the notation that is almost universally used in physics; various other notations can be found in the mathematics literature; examples include $[f,g]$ and (f,g).

The scalar product has several features, the full meaning of which may only become clear as we proceed. Defined in a way that is appropriate when the members of our function space can be complex, these features include:

(1) The scalar product of a member with itself, e.g., $\langle f|f\rangle$, must evaluate to a nonnegative numerical value (not a function) that plays the role of the square of the magnitude of that member, corresponding to the dot product of an ordinary vector with itself,

(2) The scalar product $\langle f|g\rangle$ must have the following linearity properties:

$$\langle f|a_1g_1+a_2g_2\rangle = a_1\langle f|g_1\rangle + a_2\langle f|g_2\rangle, \qquad \langle a_1f_1+a_2f_2|g\rangle = a_1^*\langle f_1|g\rangle + a_2^*\langle f_2|g\rangle.$$

(3) Behavior under complex conjugation: $\langle f|g\rangle^* = \langle g|f\rangle$.

There exists an extremely wide range of possibilities for defining scalar products that meet these criteria. The situation that arises most often in physics is that the scalar product of the two members $f(s)$ and $g(s)$ is computed as an integral over a real range $(a \dots b)$ of the type

$$\langle f|g\rangle = \int_a^b f^*(s)g(s)\,w(s)\,ds\,, \tag{11.7}$$

with the choice of a, b, and $w(s)$ dependent upon the particular definition we wish to adopt for our scalar product. In the special case $\langle f|f\rangle$, the scalar product is to be interpreted as the square of a "length," and this scalar product must therefore be positive for any f that is not itself identically zero. Since the integrand in the scalar product is then $f^*(s)f(s)w(s)$ and $f^*(s)f(s) \geq 0$ for all s (even if $f(s)$ is complex), we can see that $w(s)$ must be positive over the entire range $[a,b]$ except possibly for zeros at isolated points.

Let's dispose right now of a commonly held misconception relative to Eq. (11.7). It is **not** appropriate to interpret that equation as a continuum analog of the ordinary dot product, with the variable s thought of as the continuum limit of an index labeling vector components. The integral actually arises pursuant to a decision to compute a "squared length" as a possibly weighted average over the range of values of the parameter s.

Vector spaces that are closed under addition and multiplication by a scalar and which have a scalar product that exists for all pairs of its members are termed **Hilbert spaces**; these are the vector spaces of primary importance in science and engineering.

ORTHOGONALITY AND NORMALIZATION

With now a well-behaved scalar product in hand, we can make the definition that two functions f and g are **orthogonal** if $\langle f|g\rangle = 0$, which means that $\langle g|f\rangle$ will also vanish.

Example 11.1.2. Some Orthogonal Functions

Let's define the scalar product

$$\langle f|g\rangle = \int_{-1}^{1} f(s)g(s)\,ds\,, \tag{11.8}$$

corresponding to taking Eq. (11.7) with $a = -1$, $b = 1$ and $w(s) = 1$. **With this definition of the scalar product**, the basis functions $P_0(s) = 1$, $P_1(s) = s$, and $P_2(s) = \frac{3}{2}s^2 - \frac{1}{2}$ from Example 11.1.1 are orthogonal:

$$\langle P_0|P_1\rangle = \int_{-1}^{1} (1)(s)\, ds = 0\,, \qquad \langle P_0|P_2\rangle = \int_{-1}^{1} (1)\left(\frac{3}{2}\,s^2 - \frac{1}{2}\right) ds = 0\,,$$

$$\langle P_1|P_2\rangle = \int_{-1}^{1} (s)\left(\frac{3}{2}\,s^2 - \frac{1}{2}\right) ds = 0\,.$$

The integrals representing $\langle P_0|P_1\rangle$ and $\langle P_1|P_2\rangle$ vanish due to symmetry in the range of s. However, the integral for $\langle P_0|P_2\rangle$ only vanishes because the coefficients of s^2 and s^0 in $P_2(s)$ have been chosen appropriately.

■

We further define a function f as **normalized** if the scalar product $\langle f|f\rangle = 1$; this is the function-space equivalent of a unit vector. We will find that great convenience results if the basis functions for our function space are normalized and mutually orthogonal, corresponding to the description of a 2-D or 3-D physical vector space based on orthogonal unit vectors. A set of functions that is both normalized and mutually orthogonal is called an **orthonormal** set. If a member f of an orthogonal set is not normalized, it can be made so without disturbing the orthogonality: we simply rescale it to $\overline{f} = f/\langle f|f\rangle^{1/2}$, so any orthogonal set can easily be made orthonormal if desired.

Example 11.1.3. Normalization

Let's normalize the functions P_n of Example 11.1.2. First, compute

$$\langle P_0|P_0\rangle = \int_{-1}^{1} (1)(1)\, ds = 2\,, \qquad \langle P_1|P_1\rangle = \int_{-1}^{1} (s)(s)\, ds = \frac{2}{3}\,,$$
$$\langle P_2|P_2\rangle = \int_{-1}^{1} \left(\frac{3}{2}\,s^2 - \frac{1}{2}\right)^2 ds = \int_{-1}^{1} \left(\frac{9}{4}\,s^4 - \frac{3}{2}\,s^2 + \frac{1}{4}\right) ds = \frac{2}{5}\,. \tag{11.9}$$

We then rescale the P_n to functions we call φ_n:

$$\varphi_0(s) = \sqrt{\frac{1}{2}}\, P_0(s) = \sqrt{\frac{1}{2}}\,, \qquad \varphi_1(s) = \sqrt{\frac{3}{2}}\, P_1(s) = \sqrt{\frac{3}{2}}\; s\,,$$

$$\varphi_2(s) = \sqrt{\frac{5}{2}}\, P_2(s) = \sqrt{\frac{5}{2}}\left(\frac{3}{2}\,s^2 - \frac{1}{2}\right).$$

It is now straightforward to verify that the three functions φ_n satisfy

$$\langle \varphi_m|\varphi_n\rangle = \int_{-1}^{1} \varphi_m^*(s)\varphi_n(s)\, ds = \delta_{mn}.$$

Here δ_{mn} is the **Kronecker delta**; we remind the reader that $\delta_{mn} = 1$ if $m = n$ and is zero otherwise.

■

ORTHOGONAL EXPANSIONS

We now consider how to expand a member of a function space in terms of an orthonormal basis we have chosen for the space. This process is the vector-space analog of finding the components of a physical vector in an arbitrarily oriented unit-vector basis, and it has, in essence, the same solution. Note the following:

- The components of a physical-space vector are its dot products with the unit vectors of an orthogonal basis;
- We therefore expect the coefficients in the expansion of a function to be its scalar products with the orthonormal basis functions.

Let's see how this works by returning to our 2-D example, with the assumption that the φ_i are orthonormal, and consider the result of taking the scalar product of $f(s)$, as given by Eq. (11.1), with $\varphi_1(s)$:

$$\langle\varphi_1|f\rangle = \langle\varphi_1|a_1\varphi_1 + a_2\varphi_2\rangle = a_1\langle\varphi_1|\varphi_1\rangle + a_2\langle\varphi_1|\varphi_2\rangle \,. \tag{11.10}$$

The orthonormality of the φ now comes into play; the scalar product multiplying a_1 is unity, while that multiplying a_2 is zero, so we have the simple and useful result

$$\langle\varphi_1|f\rangle = a_1 \,. \tag{11.11}$$

As claimed earlier, this result is the function-space analog of finding the projection of a vector in a given direction. Here we are projecting f in the φ_1 direction, obtaining the coefficient of φ_1 in its orthonormal expansion.

The above analysis gives us a rather mechanical means of identifying the components in the expansion of f. The general result corresponding to Eq. (11.10) is:

$$\text{If } \langle\varphi_i|\varphi_j\rangle = \delta_{ij} \quad \text{and} \quad f = \sum_{i=1}^{n} a_i\varphi_i \,, \quad \text{then} \quad a_i = \langle\varphi_i|f\rangle \,. \tag{11.12}$$

Equation (11.12) is an extremely important result. We will use it repeatedly; the reader should have a full understanding of this equation and its implications.

Looking once again at Eq. (11.10), we consider what happens if the φ_i are orthogonal but not normalized. Then instead of Eq. (11.12) we would have:

$$\text{If the } \varphi_i \text{ are orthogonal and } f = \sum_{i=1}^{n} a_i\varphi_i \,, \quad \text{then} \quad a_i = \frac{\langle\varphi_i|f\rangle}{\langle\varphi_i|\varphi_i\rangle} \,. \tag{11.13}$$

This form of the expansion will be convenient when normalization of the basis introduces unpleasant factors.

Example 11.1.4. A First Expansion

Let's use Eq. (11.13) to verify the expansion of $f(s) = s^2 - 2s - 6$ that we found in Eq. (11.6). For this purpose we need to know, from Example 11.1.2, the definitions of the $P_n(s)$ and the fact that they are orthogonal; we also need to know that their normalization is that given in Eq. (11.9). In preparation for the application of

Eq. (11.13), we compute

$$\langle P_0|f\rangle = \int_{-1}^{1} (1)(s^2 - 2s - 6)\, ds = -\frac{34}{3}, \quad \langle P_1|f\rangle = \int_{-1}^{1} (s)(s^2 - 2s - 6)\, ds = -\frac{4}{3},$$

$$\langle P_2|f\rangle = \int_{-1}^{1} \left(\frac{3}{2}s^2 - \frac{1}{2}\right)(s^2 - 2s - 6)\, ds = \frac{4}{15}.$$

Then we form

$$a_0 = \frac{\langle P_0|f\rangle}{\langle P_0|P_0\rangle} = \frac{-34/3}{2} = -\frac{17}{3}, \qquad a_1 = \frac{\langle P_1|f\rangle}{\langle P_1|P_1\rangle} = \frac{-4/3}{2/3} = -2,$$

$$a_2 = \frac{\langle P_2|f\rangle}{\langle P_2|P_2\rangle} = \frac{4/15}{2/5} = \frac{2}{3}.$$

The above results correspond to

$$f(s) = -\frac{17}{3} P_0(s) - 2P_1(s) + \frac{2}{3} P_2(s),$$

in agreement with Eq. (11.6).

■

In order for functions to be orthogonal to each other, they need to be of opposite sign for part, but not all of the interval on which the scalar product is defined. For the P_n of the above example, these sign differences occur because each P_n has n nodes. See Fig. 11.1. It is a general feature of orthogonal sets of functions that their members have different amounts of oscillation.

Example 11.1.5. Infinite Series of Orthonormal Functions

Consider the set of functions $\chi_n(x) = \sin nx$, for $n = 1, 2, \ldots$, to be used for x in the interval $0 \le x \le \pi$ with scalar product

$$\langle f|g\rangle = \int_0^{\pi} f^*(x)g(x)\, dx\,. \tag{11.14}$$

We wish to use these functions for the expansion of the function $f(x) = x^2(\pi - x)$.

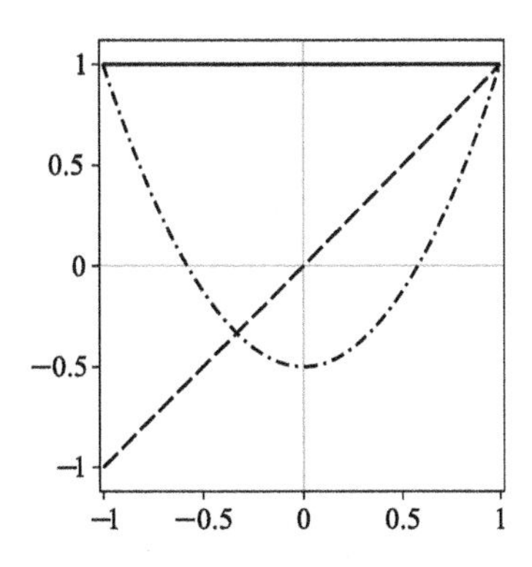

Figure 11.1: P_0 (solid), P_1 (dash), and P_2 (dash-dot) of Example 11.1.4.

First, we check that the $\chi_n(x)$ are orthogonal:

$$S_{nm} = \int_0^\pi \chi_n^*(x)\chi_m(x)\,dx = \int_0^\pi \sin nx \sin mx\,dx\,.$$

For $n \neq m$ this integral vanishes. If this claim is not obvious to the reader, he/she should verify it using symbolic computation. To determine normalization, we need S_{nn}. From symmetry considerations the integrand, $\sin^2 nx = \frac{1}{2}(1 - \cos 2nx)$, can be seen to have average value 1/2 over the range $(0, \pi)$, leading to $S_{nn} = \pi/2$ for all integer n. This means the χ_n are not normalized, but can be made so if we multiply by $\sqrt{2/\pi}$. So our orthonormal basis will be

$$\varphi_n(x) = \left(\frac{2}{\pi}\right)^{1/2} \sin nx\,, \qquad n = 1, 2, 3, \ldots \tag{11.15}$$

To expand $x^2(\pi - x)$, we apply Eq. (11.2), which requires the evaluation of

$$a_n = \langle \varphi_n | x^2(\pi - x) \rangle = \left(\frac{2}{\pi}\right)^{1/2} \int_0^\pi (\sin nx)\, x^2(\pi - x)\,dx\,, \tag{11.16}$$

for use in the expansion

$$x^2(\pi - x) = \left(\frac{2}{\pi}\right)^{1/2} \sum_{n=0}^{\infty} a_n \sin nx\,. \tag{11.17}$$

Evaluating cases of Eq. (11.16) by hand or using a computer for symbolic computation, we have for the first few a_n: $a_1 = 5.0132$, $a_2 = -1.8300$, $a_3 = 0.1857$, $a_4 = -0.2350$.

■

The above example illustrates the possibility of an infinite-series expansion of $f(x)$ that is fundamentally different from a power series. A power series for $f(x)$ is built from information about $f(z)$ and its derivatives at a single point (the expansion point), while our present orthogonal expansion uses information based on weighted averages (integrals) of $f(x)$ over the entire range for which the orthogonal expansion is to apply. This difference causes the convergence properties of the two expansions also to differ.

Example 11.1.6. Comparison of Expansions

Let's compare $f(x) = e^{-x}$ with two approximations to it: Its Maclaurin series $f_1(x)$, and its expansion $f_2(x)$ in the functions $\cos nx$, which are orthogonal with $w(x) = 1$ on the interval $0 \le x \le \pi$. Each expansion has been truncated after its first four terms.

The left panel of Fig. 11.2, for which the range of x is that appearing in the definition of the scalar product, shows that the truncated Maclaurin series is completely accurate at $x = 0$ but becomes less rapidly convergent as x increases, with qualitatively poor results beyond about $x = \pi/4$. This behavior is consistent with the notion that the Maclaurin series is an expansion about the point $x = 0$.

On the other hand, we see from the right panel of Fig. 11.2 that the orthogonal expansion, though not extremely accurate over any significant range of x, is of more or

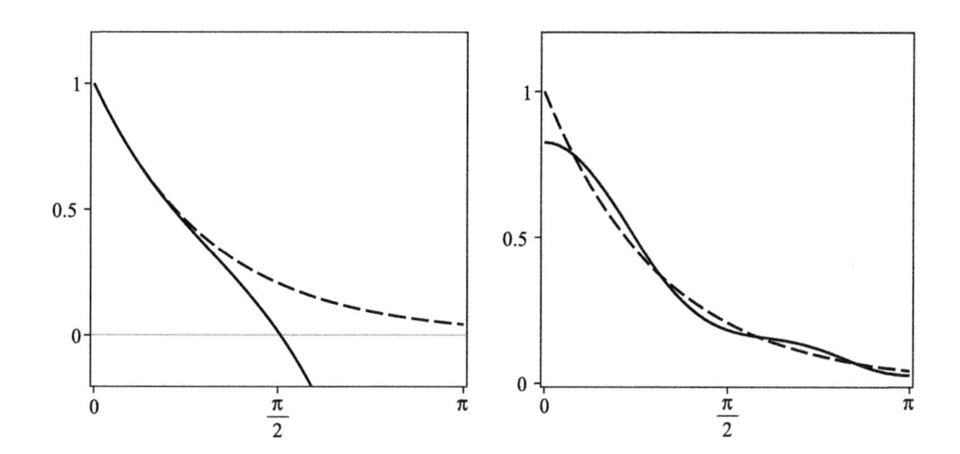

Figure 11.2: Dashed line, e^{-x}; solid line, its four-term expansions: Left, Maclaurin series; right, orthogonal expansion.

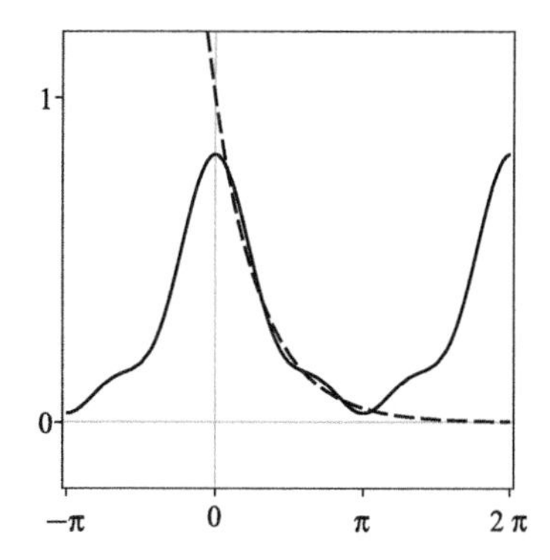

Figure 11.3: Dashed line, e^{-x}; solid line, its four-term orthogonal expansion.

less equivalent quality throughout almost all of the range $0 \le x \le \pi$. This illustrates one of the virtues of orthogonal expansions; they are in some sense a fit over a **region** rather than an expansion about a **point**.

Finally, note what happens for x values outside the range $[0, \pi]$ used in the scalar product. From Fig. 11.3 we see that the quality of the expansion deteriorates markedly as we go beyond that range. This behavior is easily understood; we have only used data for $0 \le x \le \pi$ to determine the coefficients in our orthogonal expansion.

■

EXPANSIONS AND SCALAR PRODUCTS

If we have found the expansions of two functions,

$$f = \sum_\mu a_\mu \varphi_\mu \quad \text{and} \quad g = \sum_\nu b_\nu \varphi_\nu ,$$

then their scalar product can be written

$$\langle f|g\rangle = \sum_{\mu\nu} a_\mu^* b_\nu \langle \varphi_\mu|\varphi_\nu\rangle .$$

If the φ set is orthonormal, the above reduces to

$$\langle f|g\rangle = \sum_\mu a_\mu^* \, b_\mu \,. \tag{11.18}$$

In the special case $g = f$, this reduces to

$$\langle f|f\rangle = \sum_\mu |a_\mu|^2 \,, \tag{11.19}$$

consistent with the requirement that $\langle f|f\rangle \geq 0$, with equality if $f = 0$.

If we regard the set of expansion coefficients a_μ as the elements of a column vector $\mathbf{a}$ representing f, with column vector $\mathbf{b}$ similarly representing g, Eqs. (11.18) and (11.19) correspond to the matrix equations

$$\langle f|g\rangle = \mathbf{a}^\dagger \mathbf{b} \,, \qquad \langle f|f\rangle = \mathbf{a}^\dagger \mathbf{a} \,. \tag{11.20}$$

Notice that by taking the adjoint of $\mathbf{a}$, we both complex conjugate it and convert it into a row vector, so that the matrix products in Eq. (11.20) collapse to scalars, as required. Note also that $\mathbf{a}^\dagger\mathbf{b}$ can be regarded as a complex version of the dot product in which the left member has been complex conjugated. This feature is necessary to make $\langle f|f\rangle$ real.

DIRAC NOTATION

Much of what we have discussed can be brought to a form that promotes clarity and suggests possibilities for additional analysis by using a notational device invented by P. A. M. Dirac. Dirac suggested that instead of just writing a function f, it be written enclosed in the right half of an angle-bracket pair, which he named a **ket**. Thus $f \to |f\rangle$, $\varphi_i \to |\varphi_i\rangle$, etc. Then he suggested that the complex conjugates of functions be enclosed in left half-brackets, which he named **bras**. An example of a bra is $\varphi_i^* \to \langle\varphi_i|$. Finally, he suggested that when the sequence (bra followed by ket = bra ket $\sim$ bracket) is encountered, the pair should be interpreted as a scalar product (with the dropping of one of the two adjacent vertical lines). As an initial example of the use of this notation, take Eq. (11.12), inserting $\langle\varphi_j|f\rangle$ for a_j and writing the equation as

$$|f\rangle = \sum_j a_j|\varphi_j\rangle = \sum_j |\varphi_j\rangle\langle\varphi_j|f\rangle = \left(\sum_j |\varphi_j\rangle\langle\varphi_j|\right)|f\rangle \,. \tag{11.21}$$

This notational rearrangement shows that we can view the expansion in the φ basis as the insertion of a set of basis members in a way which, in sum, has no effect. If the sum is over a complete set of φ_j, the ket-bra sum in Eq. (11.21) will have no net effect when inserted before any ket in the space, and therefore we can view the sum as a **resolution of the identity**. To emphasize this, we write

$$\mathbf{1} = \sum_j |\varphi_j\rangle\langle\varphi_j| \,. \tag{11.22}$$

Many expressions involving expansions in orthonormal sets can be derived by the insertion of resolutions of the identity.

Dirac notation can also be applied to expressions involving vectors and matrices, where it illuminates the parallelism between physical vector spaces and the function spaces here under study. If $\mathbf{a}$ and $\mathbf{b}$ are column vectors and M is a matrix, then we can write $|\mathbf{b}\rangle$ as a synonym for $\mathbf{b}$, we can write $\langle\mathbf{a}|$ to mean $\mathbf{a}^\dagger$, and then $\langle\mathbf{a}|\mathbf{b}\rangle$ is interpreted as equivalent to $\mathbf{a}^\dagger\mathbf{b}$, which is matrix notation for a complex version of the dot product $\mathbf{a}\cdot\mathbf{b}$. Other examples are expressions such as

$$\mathbf{a} = \mathsf{M}\mathbf{b} \qquad \longleftrightarrow \qquad |\mathbf{a}\rangle = |\mathsf{M}\mathbf{b}\rangle = \mathsf{M}|\mathbf{b}\rangle\,,$$

$$\mathbf{a}^\dagger\mathsf{M}\mathbf{b} = (\mathsf{M}^\dagger\mathbf{a})^\dagger\mathbf{b} \qquad \longleftrightarrow \qquad \langle\mathbf{a}|\mathsf{M}\mathbf{b}\rangle = \langle\mathsf{M}^\dagger\mathbf{a}|\mathbf{b}\rangle\,.$$

SUMMARY: VECTORS IN FUNCTION SPACES

- The functions in a Hilbert space can be represented by column vectors whose elements are the coefficients of the basis functions being used to define the space.
- The column vectors representing functions can be added or multiplied by scalars using the rules for vector operations.
- If the basis of the Hilbert space is orthonormal, the scalar product of any two functions can be computed as the (complex) dot product of the vectors representing the functions. If $\mathbf{a}$ and $\mathbf{b}$ are the column vectors respectively representing φ and χ,

$$\langle\varphi|\chi\rangle = \mathbf{a}^\dagger\mathbf{b}\,.$$

- If the φ_i are an orthonormal basis and ψ is a function in the Hilbert space, it has (in Dirac notation) the expansion

$$|\psi\rangle = \sum_i a_i|\varphi_i\rangle\,, \quad \text{with} \quad a_i = \langle\varphi_i|\psi\rangle\,.$$

- Expressions involving Dirac bras and kets can often be derived by use of the **resolution of the identity**, with the φ_i an orthonormal function set:

$$1 = \sum_i |\varphi_i\rangle\langle\varphi_i|\,.$$

Thus, for example, the formula for the expansion of $|\psi\rangle$ can be written

$$\psi = (1)\psi = \left(\sum_i |\varphi_i\rangle\langle\varphi_i|\right)\psi = \sum_i |\varphi_i\rangle\langle\varphi_i|\psi\rangle = \sum_i |\varphi_i\rangle\, a_i\,,$$

with $a_i = \langle\varphi_i|\psi\rangle$.

SYMBOLIC COMPUTING

Work with orthogonal expansions requires the ability to evaluate scalar products involving the members of the expansion basis and/or the function being expanded. It is useful to define a procedure that gives the scalar product of any two quantities u and v, which are assumed for now to be real expressions containing a variable (chosen in the present discussion to be s). In MAPLE,

```
> ScalProd := proc(u,v);
>   int(u*v*w, s = a .. b);
> end proc;
```

In MATHEMATICA,

```
ScalProd[u_,v_] := Integrate[u*v*w, {s, a, b}]
```

In both symbolic languages, w, a, and b must be replaced by appropriate quantities. If the free variable in u and v is not s or if u and v can be complex, the above coding must be changed accordingly.

As an example of the adjustments that may be needed, the scalar product defined in Eq. (11.14) is

MAPLE:
```
> ScalProd := proc(u,v);
>    int(conjugate(u)*v, x = 0 .. Pi)
> end proc:
```

MATHEMATICA:
```
ScalProd[u_,v_] := Integrate[Conjugate[u]*v, {x,0,Pi}]
```

Exercises

11.1.1. (a) Using your symbolic computing system, define the four expressions

$$Z_0 = 1\,, \quad Z_1 = 2x-1\,, \quad Z_2 = 6x^2-6x+1\,, \quad Z_3 = 20x^3-30x^2+12x-1\,,$$

calling them (in either symbolic language) `z[0]`, `z[1]`, `z[2]`, and `z[3]`.

(b) Write a symbolic procedure `ScalProd(U,V)` or `ScalProd[U,V]` that will produce the value of the scalar product $\langle U|V\rangle$, assuming U and V to be real. Define the scalar product to be

$$\langle U|V\rangle = \int_0^1 U(x)V(x)\,dx\,.$$

(c) Using the scalar product and symbolic procedure defined in part (b), verify that the `z[i]` are orthogonal but not all normalized.

(d) Redefine the `z[i]` in a way that preserves their orthogonality but makes them normalized.

(e) Write a symbolic procedure `xpand3` that depends upon your work in parts (b) through (d) and finds the coefficients a_0 through a_3 in the expansion of a function $f(x)$ according to

$$f(x) \approx \sum_{i=0}^{3} a_i Z_i(x)\,.$$

Check your work by verifying that `f=x^3` yields an expansion that can be verified by inserting the explicit forms for the Z_i.

Hint. A procedure returns a single quantity as output, so a set of coefficients can only be returned if packaged into a suitable compound quantity such as a list. Note also that a compound quantity cannot be an output if its number of elements has not been defined. MAPLE users: do not employ the `sum` command. Write the four-term sum explicitly.

(f) Form the four-term expansions of the three functions x, x^2, and x^3. Combine the coefficients of these expansions to obtain an expansion of $f(x) = 2x - 3x^2 + x^3$, and check your result in two ways:

(i) By using scalar products to obtain the expansion of $f(x)$.

(ii) By inserting the explicit form of the Z_i and simplifying the result.

11.1.2. Repeat Exercise 11.1.1 in its entirety for the four functions

$$Z_0 = 1\,, \quad Z_1 = 2x\,, \quad Z_2 = 4x^2 - 2\,, \quad Z_3 = 8x^3 - 12x\,,$$

with a scalar product of the form

$$\langle U|V\rangle = \int_{-\infty}^{\infty} U(x)V(x)\,e^{-x^2}\,dx\,.$$

Note. This exercise is far less work if done entirely by symbolic computing.

11.1.3. Write symbolic code verifying that (for nonnegative integers n and m)

$$\int_0^{\pi} \sin nx \sin mx\,dx = \frac{\pi}{2}\,\delta_{mn}\,.$$

11.1.4. Evaluate, using symbolic computing (for nonnegative integers n and m)

$$\int_0^{\pi} \cos nx \cos mx\,dx\,, \qquad \int_0^{\pi} \sin nx \cos mx\,dx\,, \qquad \int_0^{2\pi} \sin nx \cos mx\,dx\,.$$

Based on this exercise and Exercise 11.1.3, explain the conditions under which the set of functions $\sin nx$ and $\cos nx$ can form the basis for an expansion in orthonormal functions.

11.1.5. You are given two sets of orthonormal functions, $\{\varphi_i(x)\}$ and $\{\chi_i(x)\}$. If the expansions of a function $f(x)$ in the two sets are

$$f(x) = \sum_i a_i\varphi_i(x)\,, \qquad f(x) = \sum_i b_i\chi_i(x)\,.$$

(a) Find a formula relating the b_i to the a_i.

(b) Possibly using a resolution of the identity, show that your formula for the b_i is consistent with a direct expansion of $f(x)$ in the χ_i set.

11.1.6. Write symbolic code that will, all as a single procedure,

(a) Obtain the coefficients that describe the expansion of an arbitrary explicit expression $f(x)$ on the interval $0 \le x \le 1$ in terms of the four functions Z_i of Exercise 11.1.1, using the scalar product defined in that exercise.

(b) Combine the coefficients and the Z_i to obtain an expression that represents the expansion of $f(x)$.

(c) Plot $f(x)$ and its expansion on the same graph, for the interval $0 \le x \le 1$. The MAPLE command for getting these two plots on the same graph is `plot([f1,f2],x=0..1);` in MATHEMATICA, use `Plot[{f1,f2},{x,0,1}]`.

In addition, provide an alternative version of your procedure that plots the single function $f_1 - f_2$ so that if the error is small you will still be able to see its behavior.

Use your procedures to examine the four-term expansions of e^x and $x/(x+1)$, and comment on the qualitative behavior of the error in the expansion. In any case where the difference between the function and its expansion is not apparent when both are plotted, examine this difference directly before making your comments.

11.2 GRAM-SCHMIDT ORTHOGONALIZATION

Because of the great value and utility of an orthonormal expansion basis, we turn now to methods by which a set of functions whose members are linearly independent but not orthogonal can be used to construct an orthonormal set. A typical example are the simple powers: 1, x, $x^2, \ldots$ These functions are not orthogonal with any reasonable definition of the scalar product, but from them we can build orthonormal function sets in various ways, including the Gram-Schmidt process, which was used to orthogonalize vectors in Section 5.4 and which we will use again here.

Because we can write arbitrary functions in our Hilbert space as the vectors containing their coefficients, the orthogonalization process for functions is logically the same as that already developed for ordinary vectors. However, we develop it briefly here in a notation that refers explicitly to the functions involved and to their scalar products.

Let ψ_i, $i = 1, 2, \ldots$ be a set of functions that is neither orthogonal nor normalized. Let ϕ_i be functions that are orthogonal but not yet normalized, with φ_i the final orthonormal functions.

Our first orthonormal function will simply be a normalized version of ψ_1, formed as

$$\varphi_1 = \frac{\psi_1}{\langle\psi_1|\psi_1\rangle^{1/2}} . \tag{11.23}$$

An unnormalized, but orthogonal second function will be

$$\phi_2 = \psi_2 - \langle\varphi_1|\psi_2\rangle\varphi_1 .$$

The orthogonality is easily checked:

$$\langle\varphi_1|\phi_2\rangle = \langle\varphi_1|\psi_2\rangle - \langle\varphi_1|\psi_2\rangle\langle\varphi_1|\varphi_1\rangle = \langle\varphi_1|\psi_2\rangle - \langle\varphi_1|\psi_2\rangle(1) = 0 .$$

We next normalize ϕ_2 by writing

$$\varphi_2 = \frac{\phi_2}{\langle\phi_2|\phi_2\rangle^{1/2}} .$$

Further orthonormal functions are obtained as

$$\phi_n = \psi_n - \sum_{i=1}^{n-1}\langle\varphi_i|\psi_n\rangle\varphi_i , \qquad \varphi_n = \frac{\phi_n}{\langle\phi_n|\phi_n\rangle^{1/2}} . \tag{11.24}$$

This scheme for generating orthonormal functions produces a result that depends not only on the definition of the scalar product but also on the original set of nonorthogonal functions ψ_i, and even on the order in which the ψ_i are arranged. If we choose ψ_n to be s^n (and start with $n = 0$), our first orthogonal function will be proportional to s^0, the next will be a linear combination of s^0 and s^1, etc, so our orthogonal functions will be polynomials of successive degrees. But if for example we had chosen ψ_0 to be s^6, then all the orthogonal functions could contain s^6, and we would not expect to get orthogonal polynomials of successive degrees; we certainly would get different sets of orthogonal polynomials from different orderings of the same set of functions ψ_i.

SYMBOLIC COMPUTING

We assume that prior to carrying out an orthogonalization process we have defined a scalar product procedure `ScalProd(u,v)` (MAPLE) or `ScalProd[u,v]` (MATHEMATICA) appropriate for the functions to be orthogonalized.

We place the orthonormal functions in an array `Q` with numbering starting from zero, and assume that the original nonorthogonal functions are stored in `u[n]`. Then the first orthonormal function is

MAPLE:
```
> Q[0] := u[0]/sqrt(ScalProd(u[0],u[0]));
```

In MATHEMATICA, if we want the function numbering to start from zero it is convenient to have the u_n (and the orthonormal functions Q_n) stored as the values of functions `u` and `Q` with integer arguments; hence we write

MATHEMATICA:
```
Q[0] = u[0]/Sqrt[ScalProd[u[0],u[0]] ]
```

and additional orthonormal functions Q_1 through $\mathrm{Q}_{\mathrm{nmax}}$ are obtained as follows:

MAPLE:
```
> for n from 1 to nmax do
>   QQ := u[n]-add(ScalProd(u[n],Q[j])*Q[j], j=0 .. n-1);
>   Q[n] := QQ/sqrt(ScalProd(QQ,QQ))
> end do;
```

MATHEMATICA:
```
Do[ QQ = u[n] - Sum[ ScalProd[ u[n], Q[j] ] * Q[j], {j, 0, n-1} ];
    Q[n] = QQ/Sqrt[ ScalProd[QQ,QQ] ],
    {n, 1, nmax} ]
```

If the functions u_n are actually polynomials of successive degrees, the Gram-Schmidt process will convert them into polynomials of successive degrees that are orthogonal using the relevant scalar product.

Example 11.2.1. Legendre Polynomials

A set of polynomials known as the **Legendre polynomials** can be generated by orthogonalizing successive powers of s with the scalar product that was defined in Eq. (11.8). This application of the Gram-Schmidt procedure not only yields the functions P_0, P_1, and P_2 of Example 11.1.2 but can also produce an unlimited number of additional orthogonal polynomials P_n of respective degrees n.

Defining the scalar product procedure

MAPLE
```
> ScalProd := proc(u,v); int(u*v, s=-1 .. 1) end proc;
```

MATHEMATICA
```
ScalProd[u_,v_] := Integrate[u*v, {s, -1, 1}]
```

and setting $u_n = s^n$ by executing one of

MAPLE
```
> for n from 0 to nmax do u[n]=s^n end do;
```

MATHEMATICA
```
Do[ u[n] = s^n, {n, 0, nmax} ]
```

we obtain the orthonormal polynomials Q_n using the code that immediately precedes this Example.

We are not quite done, because the conventional definition of the Legendre polynomials P_n is not that they are normalized, but instead that they are scaled such that for all n, $P_n(1) = 1$. To obtain this scaling, all we have to do is compute $Q_n(1)$ and form $P_n(s) = Q_n(s)/Q_n(1)$. Thus, we execute one of

MAPLE:

```
> for n from 0 to nmax do
>    PP := subs(s=1, Q[n]); P[n] := Q[n]/PP
> end do
```

MATHEMATICA:

```
Do[ PP = Q[n]/. s -> 1; P[n] = Q[n]/PP, {n, 0, nmax} ]
```

In MAPLE, the command **subs** causes `Q[n]` to be evaluated with s replaced by 1 (this does not change the stored quantity `Q[n]`). The same result is obtained in MATHEMATICA by ending the assignment to `PP` with `/.s->1`.

The mathematics of the Gram-Schmidt process is now complete, but it is convenient to have the Legendre polynomials in a simplified form with ascending or descending powers of s. In MAPLE, the $P_n(s)$ are already fully simplified but the terms are not in order. The command `sort(P[n])` places the powers of s in `P[n]` in ascending order. Note that it does so by rearrangement "in place," so (1) the original term ordering is lost, and (2) it is not necessary to assign the sorted polynomial to `P[n]`; it is already there. In MATHEMATICA, the P_n are **not** in simplified form, but when they are simplified, their terms are automatically placed in ascending powers of s. Thus, a final clean-up of the Gram-Schmidt output requires

MAPLE

```
for n from 0 to nmax do sort(P[n]) end do
```

MATHEMATICA

```
Do[ P[n] = Simplify[ P[n] ], {n, 0, nmax} ]
```

The reader should run the symbolic code leading to the $P_n(s)$ and verify the result by comparison with the authentic Legendre polynomials given in Table 14.1. The first three Legendre polynomials are plotted in Fig. 11.1.

■

Example 11.2.2. Hermite Polynomials

The **Hermite polynomials** are obtained by orthogonalizing the successive functions s^n for a scalar product defined for the range $(-\infty < s < \infty)$ and with weight e^{-s^2}:

$$\langle f|g\rangle = \int_{-\infty}^{\infty} f^*(s)g(s)\, e^{-s^2}\, ds\,. \tag{11.25}$$

This is a different definition than was used for the Legendre polynomials, so we can expect that the Gram-Schmidt process will yield a different set of orthogonal polynomials.

We now need the following code to define the scalar product:

MAPLE:

```
> ScalProd := proc(u,v);
>    int(u*v*exp(-s^2), s=-infinity .. infinity) end proc;
```

MATHEMATICA:

```
ScalProd[u_,v_] := Integrate[u*v*E^(-s^2), {s,-Infinity,Infinity}]
```

As in the previous example, we set $u_n = s^n$ by executing one of

MAPLE `> for n from 0 to nmax do u[n]=s^n end do`

MATHEMATICA `Do[ u[n] = s^n, {n, 0, nmax} ]`

We now run the code given earlier in this section to obtain the orthonormal functions `Q[n]`.

The conventional definition of the Hermite polynomials is that they are not normalized, but are instead at a scale such that the coefficient of their leading term is 2^n. To obtain this scaling, we compute $H_n(s) = 2^n Q_n(s)/C_n$, where C_n is the coefficient of s^n in Q_n. Code to do this and sort the result by powers of s is

MAPLE:

```
> for n from 0 to nmax do
>    H[n] := sort( 2^n * Q[n]/lcoeff(Q[n],s) ) end do;
```

MATHEMATICA:

```
H[0] = 1
Do[ H[n] = Expand[Simplify[ 2^n*Q[n]/Coefficient[Q[n],s^n] ]],
    {n, 1, nmax} ]
```

Here `lcoeff(Q,s)` (MAPLE) extracts the coefficient of the leading power of s in a polynomial Q and `Coefficient[Q,s^n]` (MATHEMATICA) extracts the coefficient of s^n from Q. `Coefficient` does not work with argument s^0 so H_0 is processed in MATHEMATICA as a special case.

■

SUMMARY: ORTHOGONAL POLYNOMIALS

Because orthogonal polynomials are so useful for making expansions, it is desirable to have their essential data and features collected for easy reference. This information is presented in Table 11.1 for the polynomials examined in this section and for two additional sets of orthogonal polynomials. The normalizations can be deduced from the values given for the scalar products; the names by which these polynomials can be accessed using symbolic computation are also indicated. A reference is also given to the locations later in the book to tabulations of the first few polynomials of each type.

Exercises

11.2.1. Carry out the symbolic computations of Example 11.2.1 and thereby verify that the procedure of that Example produces the Legendre polynomials at their conventional scaling (see Table 11.1). Check your result against the Legendre polynomials as given in Table 14.1.

11.2.2. Carry out the symbolic computations of Example 11.2.2 and thereby verify that the procedure of that Example produces the Hermite polynomials at their conventional scaling (see Table 11.1). Check your result against the Hermite polynomials as given in Table 14.4.

Table 11.1: Orthogonal polynomials generated by the Gram-Schmidt process from functions u^n, $n = 0, 1, \ldots$

Polynomial	Scalar Product	Symbolic Names	Table
Legendre	$\int_{-1}^{1} P_n(s)P_m(s)\,ds = \frac{2}{2n+1}\,\delta_{nm}$	`LegendreP(n,s)` `LegendreP[n,s]`	Table 14.1
Hermite	$\int_{-\infty}^{\infty} H_n(s)H_m(s)e^{-s^2}\,ds = 2^n\pi^{1/2}n!\,\delta_{nm}$	`HermiteH(n,s)` `HermiteH[n,s]`	Table 14.2
Laguerre	$\int_{0}^{\infty} L_n(s)L_m(s)e^{-s}\,ds = \delta_{nm}$	`LaguerreL(n,s)` `LaguerreL[n,s]`	Table 14.3
Chebyshev I	$\int_{-1}^{1} T_n(s)T_m(s)(1-s^2)^{-1/2}\,ds = \frac{\pi}{2-\delta_{n0}}\,\delta_{nm}$	`ChebyshevT(n,s)` `ChebyshevT[n,s]`	Table 14.4

11.2.3. Carry out symbolic computations in which the function set $u_n = x^n$ (starting with $n = 0$) is orthogonalized using the scalar product defined for the Chebyshev I polynomials and brought to their conventional scaling (see Table 11.1). Check your result against those obtained by symbolic computing.

11.3 OPERATORS

An operator is a mapping between functions in its **domain** (those to which it can be applied) and functions in its **range** (those it can produce). While the domain and the range need not be in the same space, our concern here is with operators whose domain and range are both in all or part of the same Hilbert space. To make this discussion more concrete, here are a few examples of operators:

- Multiplication by 2: Converts f into $2f$;
- For a space containing algebraic functions of a variable x, d/dx: Converts $f(x)$ into df/dx;
- An integral operator A defined by $A\,f(x) = \int G(x,x')f(x')\,dx'$; a special case of this is a projection operator $|\varphi_i\rangle\langle\varphi_i|$ which converts f into $\langle\varphi_i|f\rangle\varphi_i$.

In addition to the above mentioned restriction on domain and range, we also for our present purposes restrict attention to operators that are **linear**, meaning that if A and B are linear operators, f and g functions, and k a constant, then

$$(A+B)f = Af + Bf, \qquad A(f+g) = Af + Ag, \qquad A(kf) = k(Af)\,.$$

We are interested in linear operators at this time because they occur in a wide variety of contexts in electromagnetic theory and quantum mechanics.

Among the important linear operators are **differential operators**, namely those that include differentiation of the functions to which they are applied. These operators arise when linear differential equations are written in operator form; for example, the operator

$$\mathcal{L}(x) = (1 - x^2)\frac{d^2}{dx^2} - 2x\frac{d}{dx} \tag{11.26}$$

enables us to write Legendre's differential equation,

$$(1 - x^2)\frac{d^2y(x)}{dx^2} - 2x\frac{dy(x)}{dx} + \lambda y(x) = 0\,,$$

in the form $\mathcal{L}(x)y(x) = -\lambda y(x)$, which is an operator eigenvalue equation. When no confusion thereby results, this equation can be shortened to $\mathcal{L}y = -\lambda y$.

IDENTITY, INVERSE, ADJOINT

An operator that is generally available is the **identity operator**, namely one that leaves functions unchanged. Depending on the context, this operator will be denoted either I or simply **1**. Some, but not all operators will have an inverse, namely an operator that will "undo" its effect. Letting A^{-1} denote the inverse of A, if A^{-1} exists, it will have the property

$$A^{-1}A = AA^{-1} = I\,. \tag{11.27}$$

Associated with many operators will be another operator, called its **adjoint** and denoted $A^\dagger$, which will be such that for all functions f and g in the Hilbert space,

$$\langle f|Ag\rangle = \langle A^\dagger f|g\rangle\,. \tag{11.28}$$

Thus, we see that $A^\dagger$ is an operator which, applied to the left member of **any** scalar product, produces the same result as is obtained if A is applied to the right member of the same scalar product. Equation (11.28) is the defining equation for $A^\dagger$.

Depending upon the specific operator A, and the definitions in use of the Hilbert space and the scalar product, $A^\dagger$ may or may not be equal to A. If $A = A^\dagger$, we call A **self-adjoint**, or equivalently, **Hermitian**. This definition is worth emphasis:

$$\text{If}\quad H^\dagger = H\,, \qquad H \text{ is Hermitian.} \tag{11.29}$$

Another situation of frequent occurrence is that the adjoint of an operator is equal to its inverse, in which case the operator is called **unitary**. A unitary operator U is therefore defined by the statement

$$\text{If}\quad U^\dagger = U^{-1}\,, \qquad U \text{ is unitary.} \tag{11.30}$$

In the special case that U is both real and unitary, it is called **orthogonal**.

The reader will doubtless note that the nomenclature for operators is similar to that previously introduced for matrices. This is not accidental; operator and matrix expressions have many properties in common.

Example 11.3.1. Finding the Adjoint

The operators of most interest to us have adjoints that can be found by rearranging or manipulating a scalar product formula. For example, if the operator A is x (meaning we just multiply by x), then, for any functions f and g in our vector space we have

$$\int_a^b f^*(x)\big[xg(x)\big]\,w(x)\,dx = \int_a^b \big[xf(x)\big]^* g(x)w(x)\,dx\,,$$

showing that x is the adjoint of x, i.e., that the operator x is Hermitian.

Suppose next that $A = d/dx$, that the scalar product has the form

$$\langle f|g\rangle = \int_{-\infty}^{\infty} f^*(x)g(x)\,dx\,, \tag{11.31}$$

and that our space consists of all functions such that $\langle f|f\rangle$ exists (i.e., that the integral converges). This requirement implies that $f(x)$ must approach zero at $x \to \pm\infty$ faster than $x^{-1/2}$. Let's carry out an integration by parts on $\langle f|Ag\rangle$:

$$\langle f|Ag\rangle = \int_{-\infty}^{\infty} f^*(x)\frac{dg(x)}{dx}\,dx = f^*(x)g(x)\Big|_{-\infty}^{\infty} - \int_{-\infty}^{\infty}\left[\frac{d}{dx}f(x)\right]^* g(x)\,dx = -\langle Af|g\rangle\,.$$

The integrated terms vanish because f and g have been assumed to be members of our space. This equation tells us that $A = d/dx$ is not Hermitian, as $A^\dagger = -A$, but because of the complex conjugation of the left half of the scalar product the operator $i\,d/dx$ will be Hermitian:

$$\begin{aligned}\left\langle f\,\middle|\, i\frac{dg}{dx}\right\rangle &= i\int_{-\infty}^{\infty} f^*(x)\,\frac{dg(x)}{dx}\,dx = -i\int_{-\infty}^{\infty}\left[\frac{d}{dx}f(x)\right]^* g(x)\,dx \\ &= \int_{-\infty}^{\infty}\left[i\frac{d}{dx}f(x)\right]^* g(x)\,dx = \left\langle i\frac{df}{dx}\,\middle|\, g\right\rangle\,.\end{aligned}$$

■

Example 11.3.2. Adjoint Depends on Vector Space

Take now a vector space consisting of all functions possessing the scalar product of Eq. (11.8). Note that this scalar product differs from that of Example 11.3.1 in that the range of the scalar product integral is $(-1, 1)$ rather than (∞, ∞). We again consider the operator $A = i\,d/dx$, which we found to be Hermitian under the conditions of Example 11.3.1. If we manipulate the integral $\langle f|Ag\rangle$ as in that example, we get

$$\begin{aligned}\left\langle f\,\middle|\, i\frac{dg}{dx}\right\rangle &= if^*(x)g(x)\Big|_{-1}^{1} - i\int_{-1}^{1}\left[\frac{d}{dx}f(x)\right]^* g(x)\,dx \\ &= i\Big[f^*(1)g(1) - f^*(-1)g(-1)\Big] + \left\langle i\frac{df}{dx}\,\middle|\, g\right\rangle\,.\end{aligned}$$

The adjoint to A in this vector space can be shown to be

$$A^\dagger = -i\Big[[\delta(x-1) - \delta(x+1)\Big] + i\frac{d}{dx}\,. \tag{11.32}$$

We see that A is not self-adjoint in this space because the space includes functions with arbitrary values at $x = \pm 1$.

However, in a different vector space with the scalar product used here (specifically, one in which all members vanish at $x = \pm 1$), our formula for $A^\dagger$ would become just $A^\dagger = i\,d/dx$, showing that in this new space A is indeed self-adjoint. Our overall conclusion is that the adjoint of an operator is a quantity that may depend not only on the definition of the scalar product but also on other features of the vector space.

Finally, note that if A is self-adjoint, so also is A^2. Because Ag and Af are members of our vector space we can write

$$\langle f|A^2 g\rangle = \langle f|A(Ag)\rangle = \langle Af|Ag\rangle = \langle A(Af)|g\rangle = \langle A^2 f|g\rangle.$$

In the present situation $A^2 = -d^2/dx^2$, so we have found that d^2/dx^2 is also a Hermitian operator for a space defined by the current scalar product and possessing only members that vanish at the endpoints $x = \pm 1$.

■

Exercises

11.3.1. Show that linear operators A and B have the following properties in common with matrices. Assuming the existence of the quantities involved,

$$(a)\ \ (AB)^{-1} = B^{-1}A^{-1}, \qquad\qquad (b)\ \ (AB)^\dagger = B^\dagger A^\dagger .$$

(c) If A and B are Hermitian, identify the conditions under which AB is also Hermitian.

(d) If U and V are both unitary, show that UV is also unitary.

11.3.2. The time-independent Schrödinger equation of an electron in a one-dimensional quantum mechanics problem has (in hartree atomic units) the form $H\psi = E\psi$, where

$$H = -\frac{1}{2}\frac{d^2}{dx^2} + V(x),$$

where V is a real multiplicative operator (i.e., it is a real function of x that contains no derivatives or integrals). Show that the operator H is Hermitian if the Hilbert space is for a finite or infinite interval in x, with its members required to vanish at the ends of the interval, and with a scalar product of definition

$$\langle\chi|\varphi\rangle = \int \chi^* \varphi\, dx\,.$$

The integral in the scalar product is over the interval relevant to the Hilbert space.

11.3.3. Explain why $A^\dagger$ as given by Eq. (11.32) is the adjoint of A for the vector space defined in the first paragraph of Example 11.3.2.

11.4 EIGENVALUE EQUATIONS

An important use of operators is to define equations of the general type

$$A\psi(x) = \lambda\,\psi(x)\,, \tag{11.33}$$

where A is a given operator involving the variable x, and both $\psi(x)$ and a constant parameter λ are to be determined. Here we assume A to be a linear second-order differential operator, so Eq. (11.33) is a linear homogeneous second-order ODE containing a parameter λ of unknown value.

If no additional conditions are imposed on our problem, ODEs of the type illustrated in Eq. (11.33) will have solutions for most if not all values of λ. However, in most situations of physical interest, additional requirements on the solution $\psi(x)$ do exist, with the result that there exist solutions only for certain values of λ. Typical additional requirements are that $\psi(x)$ have no singularities within the region relevant to the problem and that ψ meet certain conditions on the boundary of that region. The result is that we have a set of solutions ψ_n to our problem, each ψ_n associated with a corresponding parameter value λ_n.

An equation such as Eq. (11.33) and its associated additional conditions is referred to as an **eigenvalue problem** or more specifically as a **boundary-value problem**; the functions ψ_n that are its solutions are called **eigenfunctions** and the corresponding λ_n are called **eigenvalues**. The eigenvalue problems to be treated in this book will have **discrete** eigenvalues, meaning that the eigenvalues form a discrete set of points (in contrast to being distributed over a continuous range of values).

BOUNDARY-VALUE PROBLEMS

A typical boundary-value problem involves an ODE of the general form

$$p(x)y'' + q(x)y' + r(x)y = \lambda y\,, \tag{11.34}$$

together with the boundary conditions $y(x_1) = y(x_2) = 0$. For any fixed λ this linear homogeneous ODE has two independent solutions, $y_1(x)$ and $y_2(x)$, and therefore its general solution has the form

$$y(x) = c_1 y_1(x) + c_2 y_2(x) = c_1 \left[y_1(x) + \frac{c_2}{c_1}\, y_2(x) \right] .$$

When written this way we see that the solution can be characterized as having a scale parameter (c_1), which does not affect the locations of the zeros (nodes) of $y(x)$, and only one additional parameter (c_2/c_1) that can alter the form of the solution in a way more general than scale. We see by varying c_1 and c_2 we only have enough freedom to place a zero of y at one specific point; we do not have the flexibility to fix the positions of two zeros.

The dilemma of the preceding paragraph is resolved by noticing that we have one additional way to alter the solution to our ODE, namely by varying the parameter λ. We do not in this way magically find additional solutions to our ODE; we are instead changing it to a different (but related) ODE with different solutions, and if we are lucky (or insightful) we can change to an ODE that has a solution consistent with the boundary conditions. This state of affairs is best understood from an example.

Example 11.4.1. A Boundary-Value Problem

Let a string be fixed at the points $x = 0$ and $x = L$; it can have transverse standing-wave vibrations, of amplitude $y(x)$, that satisfy the ODE

$$\frac{d^2y(x)}{dx^2} = -k^2\, y(x)\,.$$

Before solving this problem we do not know the value of k; we must find its possible values by solving the eigenvalue problem

$$y'' = -k^2 y\,, \qquad y(0) = y(L) = 0\,. \tag{11.35}$$

The general solution to this ODE is $y(x) = A\sin kx + B\cos kx$. To satisfy the boundary condition at $x = 0$ we set $B = 0$, thereby reducing our solution to $y = A\sin kx$. We cannot satisfy the boundary condition at $x = L$ by a suitable choice of A; setting $A = 0$ would leave us with the trivial (and useless) formula $y = 0$. Instead, however, we can choose k such that $\sin kL = 0$. There are infinitely many nonzero k values that satisfy that equation, namely $k = n\pi/L$, where n is any nonzero integer, and we therefore have the eigenfunctions

$$y_n(x) = A\sin\left(\frac{n\pi x}{L}\right), \qquad \text{all nonzero integer } n.$$

We are interested in the linearly independent solutions to our eigenvalue problem, and note that the solutions for negative n are -1 times the solutions for corresponding positive n. Moreover, because different A only correspond to different scales for our eigenfunction, we need not keep it as a factor in the solution. Therefore, a linearly independent set of eigenfunctions for this problem can be written as

$$y_n(x) = \sin\left(\frac{n\pi x}{L}\right), \qquad n = 1,\, 2,\, 3, \ldots \tag{11.36}$$

The eigenvalues that correspond to these y_n are $-k^2 = -n^2\pi^2/L^2$.

We see that the eigenfunctions correspond to sinusoidal oscillations with wavelengths such that we can have nodes at both ends of the string. These wavelengths (which are functions of k) have to form a discrete set, as a small change in k from a wavelength that fits will certainly produce a wavelength that does not have the required nodal locations.

■

An alternative way of describing the problem of the foregoing example is to identify it as an eigenvalue problem (without additional conditions) in a space defined on the interval $[0, L]$ and consisting of functions that vanish at $x = 0$ and $x = L$. If we associate with this space the scalar product

$$\langle f|g\rangle = \int_0^L f^*(x)g(x)\,dx\,,$$

we can (referring to Example 11.3.2) further identify d^2/dx^2 as an operator that is self-adjoint (within the context of the present problem). Many of the eigenvalue problems in which we are interested involve self-adjoint operators. This observation motivates our decision to proceed next to a general discussion of self-adjoint (Hermitian) eigenvalue problems.

Exercises

11.4.1. Plot the eigenfunctions y_1, y_2, and y_3 of Example 11.4.1 at the scale given by Eq. (11.36). From the results, indicate the number of nodes (including those at $x = 0$ and $x = L$) exhibited by the y_n of general (positive integer) n, and write an equation for the locations of all nodes of y_n.

11.5 HERMITIAN EIGENVALUE PROBLEMS

Our objective here is to establish properties of the eigenfunctions of operators that are Hermitian (self-adjoint) in the Hilbert space within which they are defined.

Given a Hermitian operator H, we start by identifying eigenfunctions ψ_i and ψ_j and their respective eigenvalues λ_i and λ_j:

$$H\psi_i = \lambda_i \psi_i \,,$$

$$H\psi_j = \lambda_j \psi_j \,.$$

We now take the scalar product of the first of these equations with ψ_j and that of the second with ψ_i:

$$\langle \psi_j | H\psi_i \rangle = \lambda_i \langle \psi_j | \psi_i \rangle \,, \tag{11.37}$$

$$\langle \psi_i | H\psi_j \rangle = \lambda_j \langle \psi_i | \psi_j \rangle \,. \tag{11.38}$$

Next we take the complex conjugate of Eq. (11.38); the conjugate of $\langle \psi_i | H\psi_j \rangle$ is $\langle H\psi_j | \psi_i \rangle$, and because H is Hermitian this is also equal to $\langle \psi_j | H\psi_i \rangle$. The conjugate of λ_j is simply λ_j^*, and that of $\langle \psi_i | \psi_j \rangle$ is $\langle \psi_j | \psi_i \rangle$, so we get

$$\langle \psi_j | H\psi_i \rangle = \lambda_j^* \langle \psi_j | \psi_i \rangle \,. \tag{11.39}$$

We now subtract Eq. (11.39) from Eq. (11.37), reaching

$$(\lambda_i - \lambda_j^*) \langle \psi_j | \psi_i \rangle = 0 \,. \tag{11.40}$$

Taking first Eq. (11.40) with $i = j$, we conclude that because $\langle \psi_i | \psi_i \rangle$ must be positive we have $\lambda_i = \lambda_i^*$, meaning that λ_i must be real. Then, with $i \neq j$, substitution of λ_j for λ_j^* leads us to conclude that if $\lambda_i \neq \lambda_j$, it is necessary that $\langle \psi_j | \psi_i \rangle = 0$. Summarizing,

If an operator on a Hilbert space is Hermitian (using the scalar product and other properties defined for that space), then its eigenvalues are real and its eigenfunctions of different eigenvalues are orthogonal.

The above result has several important implications:

- A Hermitian operator can be viewed as a source of orthogonal functions. These functions (its eigenfunctions) can be used as a basis for orthogonal expansions on the space they span.
- Because an operator H must be Hermitian for its eigenfunctions to have these properties, the scalar product and the membership of the Hilbert space must be defined appropriately.
- If we have degeneracy ($\lambda_i = \lambda_j$ with $i \neq j$), we can orthogonalize the degenerate eigenfunctions so that our entire eigenfunction set is orthogonal. Since our original operator equation is linear, we can also scale the eigenfunctions to make them an orthonormal set.

The reader may have noticed that the eigenfunctions of Hermitian operators have properties that closely parallel those of the eigenvectors of Hermitian matrices. This is not a coincidence, and the relations between these two superficially dissimilar mathematical constructions can be developed in detail. See, for example, Chapter 5 in Arfken et al., Additional Readings.

COMPLETENESS OF EIGENFUNCTIONS

It can be shown that the set of all eigenfunctions of a Hermitian operator H within a Hilbert space forms a **complete** basis for the Hilbert space. **Completeness** is defined to mean that if $F(x)$ is a member of the Hilbert space and $\varphi_n(x)$ are the eigenfunctions of H in that space, then the expansion

$$\overline{F}(x) = \sum_n a_n \varphi_n(x) \tag{11.41}$$

is an approximation to $F(x)$ such that

$$\Big\langle (F - \overline{F}) \Big| (F - \overline{F}) \Big\rangle = 0 , \tag{11.42}$$

where the scalar product is that defined for the Hilbert space under discussion. This condition, which is called **convergence in the mean**, is less stringent than convergence at all x because it permits F and $\overline{F}$ to differ at isolated points. A practical consequence of the theorem represented by Eq. (11.42) is that functions with discontinuities (which cannot be represented by power-series expansions in regions that include a discontinuity) can be expanded in a complete basis for the Hilbert space in which they reside, with discrepancies between the function and its expansion occurring only at the discontinuities.

Example 11.5.1. An Orthogonal Set

In Example 11.4.1 we found the eigenfunctions of an operator ($H = d^2/dx^2$) on a Hilbert space for which H was Hermitian. The work of the present section now enables us to conclude from the Hermiticity of H that its eigenfunctions, $y_n = \sin(n\pi x/L)$, are orthogonal on the interval $[0, L]$ using the unweighted scalar product defined on that interval. We no longer have to evaluate directly the integrals implied by the scalar products $\langle y_n | y_m \rangle$ to establish that result.

■

Exercises

11.5.1. Given the operator and scalar product

$$H = \frac{d^2}{dx^2} - 2x\frac{d}{dx}, \qquad \langle \varphi | \psi \rangle = \int_{-\infty}^{\infty} \varphi^*(x)\psi(x)\, e^{-x^2}\, dx ,$$

for a Hilbert space containing all $\varphi(x)$ such that $\langle \varphi | \varphi \rangle$ is defined,

(a) Show that the operator H is Hermitian.

(b) Show that the following are eigenfunctions of H, and for each determine its eigenvalue:

$$\varphi_0 = 1, \qquad \varphi_1 = 2x, \qquad \varphi_2 = 4x^2 - 2 .$$

(c) Evaluate the integrals $\langle \varphi_i | \varphi_j \rangle$ for all pairs of the above φ (including both $i = j$ and $i \neq j$). Confirm that in each case, Eq. (11.40) is satisfied.

11.6 STURM-LIOUVILLE THEORY

We now make some additional observations about second-order linear differential operators and the conditions under which they can lead to Hermitian eigenvalue problems. The material of this section is part of what is called **Sturm-Liouville theory**.

We consider eigenvalue problems in which

$$\mathcal{L}\psi = \lambda\psi\,, \tag{11.43}$$

with

$$\mathcal{L} = p(x)\,\frac{d^2}{dx^2} + q(x)\,\frac{d}{dx} + r(x)\,. \tag{11.44}$$

The quantities $p(x)$, $q(x)$, and $r(x)$ are assumed to be real.

SELF-ADJOINT ODEs

In differential equation theory, $\mathcal{L}$ is called **self-adjoint** if

$$p'(x) = q(x)\,. \tag{11.45}$$

When Eq. (11.45) is satisfied, we can rearrange $\mathcal{L}$ to the form

$$\mathcal{L} = \frac{d}{dx}\left[p(x)\,\frac{d}{dx}\right] + r(x)\,, \tag{11.46}$$

and application of $\mathcal{L}$ to a function $g(x)$ can be written

$$\mathcal{L}g = (pg')' + rg\,. \tag{11.47}$$

The significance of Eq. (11.47) is that the scalar product $\langle f|\mathcal{L}g\rangle$, with weight $w(x) = 1$, can then be manipulated by applying two successive integrations by parts to the part of the integrand containing p:

$$\begin{aligned}
\langle f|\mathcal{L}g\rangle &= \int_a^b f^*(pg')'\,dx + \int_a^b f^* rg\,dx \\
&= f^*(pg')\Big|_a^b - \int_a^b (f')^*(pg')\,dx + \int_a^b (rf)^* g\,dx \\
&= f^*(pg')\Big|_a^b - \int_a^b (pf')^* g'\,dx + \int_a^b (rf)^* g\,dx \\
&= f^*(pg')\Big|_a^b - (pf')^* g\Big|_a^b + \int_a^b [(pf')']^* g\,dx + \int_a^b (rf)^* g\,dx\,.
\end{aligned} \tag{11.48}$$

In the above analysis we kept in mind that p and r were assumed to be real.

We now recognize that the two integrals in Eq. (11.48) together contain $(\mathcal{L}f)^*$, so that equation can be rewritten

$$\langle f|\mathcal{L}g\rangle = \Big[f^*(pg') - (pf')^* g\Big]_a^b + \langle \mathcal{L}f|g\rangle\,. \tag{11.49}$$

Equation (11.49) shows that if $\mathcal{L}$ is self-adjoint in the differential-equation sense it will also be a Hermitian Hilbert-space operator if the boundary terms of that equation vanish. We therefore note that being a Hermitian operator is a more restrictive property than ODE self-adjointness, but our operator $\mathcal{L}$ will be Hermitian if we also have

$$\Big[f^*(pg') - (pf')^*g\Big]_a^b = p(x)\left[f^*g' - (f')^*g\right]\Big|_a^b = 0\,. \tag{11.50}$$

In many situations of interest, we can define a Hilbert space in such a way that we have an unweighted scalar product and that Eq. (11.50) is satisfied for some choice of the integration endpoints a and b. An obvious way of satisfying the endpoint condition is to choose a and b such that $p(a) = p(b) = 0$. Other possibilities include limiting the Hilbert space to functions that vanish at $x = a$ and $x = b$.

Example 11.6.1. Legendre ODE

Legendre's ODE is

$$(1 - x^2)y'' - 2xy' + \lambda y = 0\,. \tag{11.51}$$

This equation is self-adjoint, and can be cast in the form

$$\mathcal{L}y = \lambda y\,, \qquad \mathcal{L}g = \Big[-(1 - x^2)g'\Big]'\,. \tag{11.52}$$

We want to identify conditions that will make $\mathcal{L}$ Hermitian, so that its eigenfunctions will form an orthogonal set.

Our analysis starts by noting from Eq. (11.52) that the coefficient $p(x)$, which appears in Eq. (11.50), is

$$p(x) = -(1 - x^2)\,, \tag{11.53}$$

and that the boundary terms in that equation will vanish if $p(a) = p(b) = 0$; this condition is met if the endpoints a and b are chosen to be ± 1.

Our conclusion is that for a Hilbert space with an unweighted scalar product on the range $(-1, +1)$ the Legendre operator $\mathcal{L}$ is Hermitian, so its eigenvalues are real and its eigenfunctions are orthogonal with that scalar product. Note that the general considerations of our present analysis do not show us how to find the eigenvalues or eigenfunctions; all we are establishing here are the properties that these quantities must have.

■

MAKING AN ODE SELF-ADJOINT

If an ODE is not self-adjoint in the differential-equation sense, it can be made self-adjoint by multiplying the entire ODE by a quantity we designate (for reasons that will become apparent) $w(x)$. An ODE that originally had the form $\mathcal{L}\psi(x) = \lambda\psi(x)$ then becomes

$$w(x)\mathcal{L}\psi(x) = w(x)p(x)\psi''(x) + w(x)q(x)\psi'(x) + w(x)r(x)\psi(x) = w(x)\lambda\psi(x)\,. \tag{11.54}$$

In order for $w(x)\mathcal{L}$ to be self-adjoint, it is necessary that $(wp)' = wq$. This is a separable first-order ODE that we can solve for w:

$$w\frac{dp}{dx} + p\frac{dw}{dx} = wq \quad\longrightarrow\quad \frac{dp}{p} + \frac{dw}{w} = \frac{q}{p}\,dx \quad\longrightarrow\quad \ln p + \ln w = \int \frac{q}{p}\,dx\,. \tag{11.55}$$

From Eq. (11.55) we get

$$w(x) = \frac{1}{p(x)}\exp\left(\int \frac{q(x)}{p(x)}\,dx\right)\,. \tag{11.56}$$

Since p and q have been assumed real, w will also be real.

Once $w(x)$ has been found, the application of $w\mathcal{L}$ to a function $g(x)$ can be written

$$w\mathcal{L}g = (wpg')' + wrg\,, \tag{11.57}$$

and we proceed to consider the integral

$$\int_a^b f^* w\mathcal{L}g\,dx = \int_a^b f^*(wpg')'\,dx + \int_a^b f^* wrg\,dx\,. \tag{11.58}$$

We now process the right-hand side of Eq. (11.58) by the same steps followed in Eq. (11.48). The result analogous to the last line of Eq. (11.48) is

$$\int_a^b f^* w\mathcal{L}g\,dx = f^*(wpg')\Big|_a^b - (wpf')^* g\Big|_a^b + \int_a^b [(wpf')']^* g\,dx + \int_a^b (wrf)^* g\,dx\,. \tag{11.59}$$

The two integrals on the right-hand side of Eq. (11.59) can together be identified as containing $(w\mathcal{L})^*$, so we have

$$\int_a^b f^* w\mathcal{L}g\,dx = \Big[f^*(wpg') - (wpf')^* g\Big]_a^b + \int_a^b (w\mathcal{L}f)^* g\,dx\,. \tag{11.60}$$

We now see that if we decide to define the scalar product as

$$\langle f|g\rangle = \int_a^b f^*(x) g(x) w(x)\,dx\,, \tag{11.61}$$

then Eq. (11.60) becomes (remembering that w is real)

$$\langle f|\mathcal{L}g\rangle = \Big[f^*(wpg') - (wpf')^* g\Big]_a^b + \langle \mathcal{L}f|g\rangle\,. \tag{11.62}$$

Equation (11.62) indicates that if we define a Hilbert space with a scalar product given by Eq. (11.61) and with properties such that

$$w(x)p(x)\left[f^* g' - (f')^* g\right]\Big|_a^b = 0\,, \tag{11.63}$$

then the present $\mathcal{L}$ will be a Hermitian operator.

If $\mathcal{L}$ has been made Hermitian, it will have real eigenvalues and orthogonal eigenfunctions, with the orthogonality based on the scalar product in use.

Example 11.6.2. Laguerre ODE

Laguerre's ODE is

$$xy'' + (1-x)y' + \lambda y = 0\,. \tag{11.64}$$

Let's write this equation in the form

$$\mathcal{L}y = -\lambda y, \qquad \mathcal{L}g = xg'' + (1-x)g'\,. \tag{11.65}$$

This ODE is clearly not self-adjoint, as $x' \neq 1 - x$, but nevertheless we want to find conditions under which $\mathcal{L}$ will be Hermitian and therefore have orthogonal eigenfunctions (using a suitable scalar product).

We begin by finding the function $w(x)$ by which $\mathcal{L}$ can be multiplied to make it self-adjoint. Applying Eq. (11.56), with $p(x) = x$ and $q(x) = 1 - x$, we have

$$w(x) = \frac{1}{x} \exp\left(\int \frac{1-x}{x}\, dx\right) = \frac{1}{x} \exp(\ln x - x) = e^{-x}\,. \tag{11.66}$$

When we use this value of $w(x)$, $w(x)p(x)$ becomes xe^{-x}, and we can make the endpoint terms, Eq. (11.63), zero by choosing the endpoints to make xe^{-x} vanish. This can be accomplished by taking $a = 0$ and $b = \infty$. Summarizing, eigenfunctions φ_i and φ_j of the Laguerre ODE with respective eigenvalues λ_i and λ_j satisfy the orthogonality relation

$$\int_0^\infty \varphi_i^*(x)\varphi_j(x)e^{-x}\, dx = 0 \quad \text{if } \lambda_i \neq \lambda_j. \tag{11.67}$$

■

Exercises

11.6.1. The operator H of Exercise 11.5.1,

$$H = \frac{d^2}{dx^2} - 2x\frac{d}{dx}\,,$$

was found to be self-adjoint when the scalar product contained the weight $w(x) = e^{-x^2}$. Using the method of this section, derive this formula for $w(x)$ from the form of the operator.

11.6.2. (a) Find the scalar product weight and a choice of endpoints that make the following differential operator Hermitian:

$$(1 - x^2)\,\frac{d^2}{dx^2} - 2x(m+1)\,\frac{d}{dx}\,.$$

(b) If in addition it is known that this operator has eigenfunctions φ_n that are polynomials of successive degrees, find φ_n for $n = 0$, 1, 2, and 3.

Additional Readings

Arfken, G. B., Weber, H. J. & Harris, F. E. (2013). *Mathematical methods for physicists* (7th ed.). New York: Academic Press.

Brown, W. A. (1991). *Matrices and vector spaces.* New York: M. Dekker.

Byron, F. W., Jr., & Fuller, R. W. (1969). *Mathematics of classical and quantum physics.* Reading, MA: Addison-Wesley (Reprinted, Dover (1992)).

Dennery P., & Krzywicki, A. (1967). *Mathematics for physicists.* New York: Harper & Row (Reprinted, Dover (1996)).

Halmos, P. R. (1958). *Finite-dimensional vector spaces* (2nd ed.). Princeton, NJ: Van Nostrand (Reprinted, Springer (1993)).

Jain, M. C. (2007). *Vector spaces and matrices in physics* (2nd ed.). Oxford, UK: Alpha Science International.

Kreyszig, E. (1988). *Advanced engineering mathematics* (6th ed.). New York: Wiley.

Lang, S. (1987). *Linear algebra.* Berlin: Springer.

Roman, S. (2005). *Advanced linear algebra, graduate texts in mathematics 135* (2nd ed.). Berlin: Springer.

Chapter 12

FOURIER SERIES

Periodic phenomena occur in a wide variety of physical problems ranging from wave propagation to rotating machinery. These problems can often be treated using expansions in trigonometric functions. These expansions, named **Fourier series** after their inventor, are the subject of the present chapter, and the use of these techniques is called **Fourier analysis**.

12.1 PERIODIC FUNCTIONS

We begin our study by looking at **sinusoidal** oscillations; this name includes of course sine functions, also cosines, which are sines displaced in phase: $\cos\varphi = \sin(\varphi + \pi/2)$, and similar functions at arbitrary phase, e.g., $\sin(\varphi + c)$. An oscillatory motion that is described by a sinusoidal function is also called **simple harmonic motion**.

SIMPLE HARMONIC MOTION

A weight that undergoes vertical oscillation when suspended from a Hooke's-law spring is an example of simple harmonic motion. The weight has a position y (relative to its stationary equilibrium point) which can be represented by

$$y(t) = A\cos\omega t\,. \tag{12.1}$$

Equation (12.1) applies if the time t is measured from a zero when the oscillation is at $y = A$, the maximum value of y. The coefficient A is called the **amplitude** of the motion and ω is called its **angular frequency**. Because $y(t)$ repeats itself every time ωt increases by 2π, the **period** τ (the time interval of repetition) satisfies

$$\omega\tau = 2\pi \qquad \text{or} \quad \tau = \frac{2\pi}{\omega}\,. \tag{12.2}$$

The **frequency** ν of the periodic motion (the number of periods of oscillation per unit time) is therefore

$$\nu = \frac{1}{\tau} = \frac{\omega}{2\pi}\,. \tag{12.3}$$

Another example of simple harmonic motion is provided by the x coordinate of a point marked on a horizontal circular disk that is rotating at constant velocity about the vertical (z) axis (see Fig. 12.1).

Mathematics for Physical Science and Engineering.
http://dx.doi.org/10.1016/B978-0-12-801000-6.00012-2

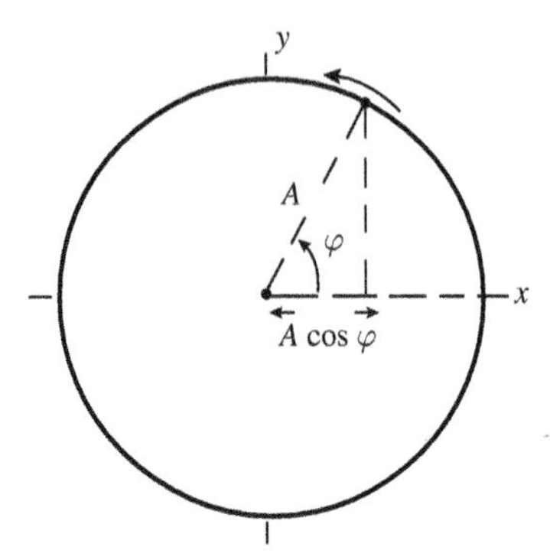

Figure 12.1: Rotating disk.

When the marked point is at an angle φ relative to the $+x$ direction, its x coordinate will be $A\cos\varphi$, where A is the distance between the marked point and the center of the disk. If the angular velocity of the rotation is ω (in radians per unit time), the time dependence of the x coordinate of the marked point will be (if at maximum x at $t = 0$)

$$x(t) = A\cos\omega t\,, \tag{12.4}$$

and this motion will have the same period, frequency, and amplitude as our Hooke's-law example. It is now obvious why ω in the Hooke's-law problem was called an angular frequency.

It is sometimes useful to consider a complex periodic function, such as

$$z(t) = Ae^{i\omega t} = A(\cos\omega t + i\sin\omega t)\,, \tag{12.5}$$

in which case both its real and imaginary parts describe simple harmonic motion (with these two parts differing in phase by 90°).

SUPERPOSITION OF WAVES

We are ordinarily interested in phenomena that are **linear**, with the result that if two oscillations occur simultaneously, the result can be described as a single (possibly more complicated) oscillation that is their sum. This linearity is exact for electromagnetic waves in vacuum, and approaches exactness for small-amplitude oscillations of many types in material media (e.g., sound or water waves). The general case of the addition of waves can yield results that are relatively complicated, but a simple result is obtained when we add waves at the same frequency.

Example 12.1.1. Addition, Waves at Different Phases

Suppose we have two waves of the same (unit) amplitude and frequency but different in phase,

$$x_1(t) = \cos\omega t\,, \qquad x_2(t) = \cos(\omega t + \theta)\,.$$

Then.

$$X(t) = x_1(t) + x_2(t) = \cos\omega t + \cos\theta\cos\omega t - \sin\theta\sin\omega t \tag{12.6}$$

$$= 2\cos(\theta/2)\,\cos(\omega t + \theta/2)\,. \tag{12.7}$$

This result is not difficult to prove; it is the topic of Exercise 12.1.1. We see that $X(t)$ has an amplitude that ranges from 2 (when the two oscillations are in phase: $\theta = 0$) to zero (when the phases are opposite: $\theta = \pi$). We thus get everything from complete reinforcement to complete cancellation, depending on the relative phases of the two oscillations.

■

TRAVELING AND STANDING WAVES

We may have a physical situation in which a periodic wave distribution $f(x)$ is moving at constant velocity v (for now, assume the motion is in the $+x$ direction). A wave distribution moving in this way is called a **traveling wave**. If the traveling wave corresponds to motion of the matter constituting what is called the **medium** supporting the wave motion (as air carrying sound waves or water, for water waves), the velocity of the traveling wave will depend upon the properties of the medium. If the wave, in contrast, describes oscillation of an electromagnetic field in vacuum, its velocity is fixed by the physical constants in the field equations (in that case c, the velocity of light).

This description of a traveling wave means that the entire distribution moves (without change in shape) in such a way that after a time interval δt the distribution is now displaced an amount $v\,\delta t$ toward positive x. Thus, if $f(x, 0)$ describes the distribution at $t = 0$, then the distribution at time t, namely $f(x,t)$, will be

$$f(x,t) = f(x - vt, 0)\,. \tag{12.8}$$

The statement that $f(x)$ is periodic is equivalent to stating that $f(x + \lambda) = f(x)$ for all x; we call λ the **wavelength** of the wave distribution. The frequency ν assigned to our traveling wave is the number of wavelengths that pass a given point per unit time; we clearly have

$$\nu\lambda = v\,, \qquad \text{or} \quad \nu = \frac{v}{\lambda}\,. \tag{12.9}$$

One way to tell whether we are observing a traveling wave is to note the time dependence of the positions where the displacement of the wave is zero. These points are called **nodes**. For our traveling wave, the nodes will be moving toward $+x$ at velocity v.

Sometimes we will have an oscillation in time under conditions such that the nodes are stationary; an example is provided by the vibrations of a string whose ends are fixed, and are therefore stationary nodes. (The vibration may also sometimes have additional stationary nodes at points other than its ends.) These kinds of wave distributions are called **standing waves**, and are described by functions in which the position and time-dependence occur as separate factors:

$$f(x,t) = h(x)\,g(t)\,. \tag{12.10}$$

This type of distribution describes oscillations because $g(t)$ is oscillatory.

One way to look at a standing wave is to think of it as a pair of traveling waves of the same shape and size but of opposite sign and going in opposite directions. If at a particular time and value of x a corresponding node of both waves coincide, that x value will also have zero amplitude at later times because the positive contribution of one of the traveling waves will be exactly compensated by the negative contribution of the other. See Fig. 12.2. The conclusion to be drawn from this analysis is that the pair of traveling waves do indeed create a situation in which the nodes are stationary (and therefore do represent a standing wave).

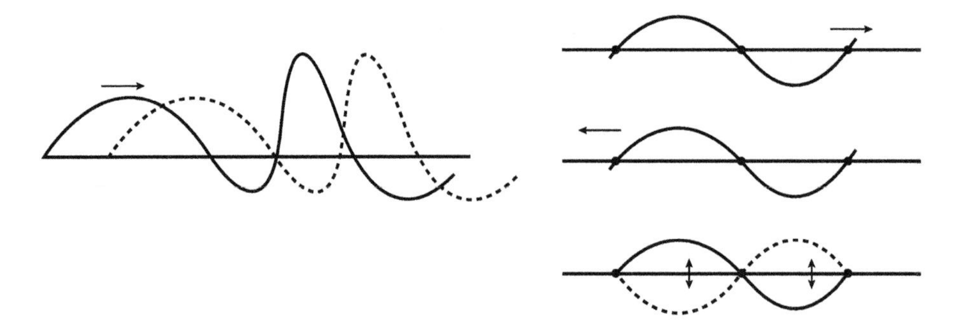

Figure 12.2: Left: traveling wave. Right: standing wave, shown as the sum of two traveling waves moving in opposite directions. Dashed curves are for a time later than that of the corresponding solid curves.

This way of looking at a standing wave identifies its frequency as that of the related traveling waves, because that is the frequency at which corresponding amplitude distributions recur. Summarizing, we have shown that if a standing wave has wavelength λ, its frequency is given by the same formula as for traveling waves, namely, Eq. (12.9).

INTENSITY AND ENERGY

It requires energy to create a wave distribution; this is the case both for oscillations involving matter (e.g., vibrating strings, sound waves, water waves) and for electromagnetic waves. For both these types of situations, the energy in a wave distribution is proportional to the square of the amplitude of the oscillation. This quantity is called the **intensity** of the oscillation. For traveling waves, it is a measure of the energy transfer carried by the waves; for standing waves, it indicates the energy stored in the oscillations.

GENERAL WAVE DISTRIBUTIONS

Although we opened this chapter with a discussion of simple harmonic motion, most of the material of this section applies to waves of a more general form, and the analysis of the later sections of the chapter is designed to include the treatment of these more complicated situations. To illustrate the range of applicability of Fourier analysis, we present in Fig. 12.3 a number of wave forms that arise in practical applications. Some of these wave forms occur in a variety of contexts in electric power engineering; others are important in computer design and construction.

Exercises

12.1.1. Use appropriate trigonometric identities to show that

$$\cos \omega t + \cos(\omega t + \theta) = 2\cos(\theta/2)\cos(\omega t + \theta/2)\,.$$

12.1.2. Plot $\sin(1.5x)$ over a range of x sufficient to identify the smallest interval on which this function is periodic.

12.1.3. (a) Given $f(x) = \sin(\pi x/2)$ defined on the range $(0, 2)$, assume that its extension to x values beyond that range is periodic (with wavelength 2). Sketch or plot $f(x)$ and its periodic extension for the range $(-3, 4)$.

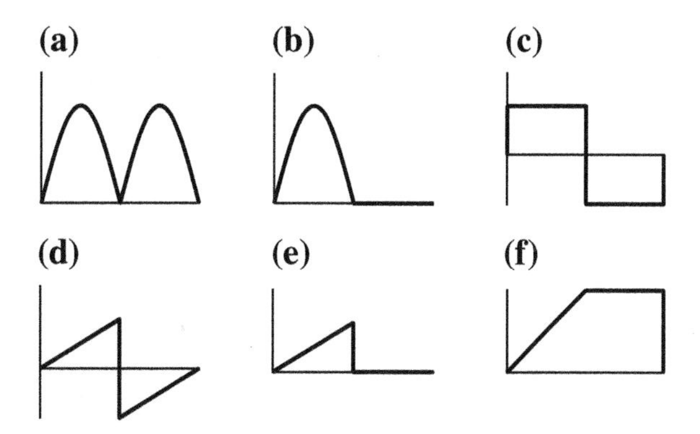

Figure 12.3: (a) Full-wave rectifier, (b) half-wave rectifier, (c) square wave, (d) sawtooth, (e) rectified sawtooth, (f) ramp function.

(b) For the same $f(x)$, assume it to be defined on the range $(-1, 1)$ and sketch or plot it and its periodic extension for the range $(-3, 4)$.

(c) If your answers for parts (a) and (b) differ, explain briefly why.

12.1.4. A wave distribution is described by $f(x, t) = 4\cos(2x+3t-0.4)$. As perceived by an observer at $x = 0.5$,

(a) Determine the velocity with which the distribution passes the observer, and state the direction in which it is moving.

(b) Find the wavelength, frequency, angular frequency, and period of the distribution.

(c) Find the times at which the observer will see (i) a (positive) maximum amplitude, and (ii) a node.

12.1.5. A sinusoidal wave distribution is moving in the direction of the vector $\hat{\mathbf{e}}_x + \hat{\mathbf{e}}_y$ at velocity 250 m/s and with angular frequency 1250 s^{-1}. The maximum amplitude of the wave is 3.0 mm. Determine the wavelength and period of the oscillation, and write an equation for its amplitude as a function of position and time.

12.1.6. The fundamental frequency of a string of length 1 m clamped at both ends is 50 s^{-1}. (This is the standing-wave oscillation of longest wavelength.) On a long string maintained under similar conditions (string density and tension), how long would it take a traveling wave to move 100 m?

12.1.7. A piston is caused to move back and forth horizontally by being attached via a long rod to a point on the circumference of a rotating wheel) see Fig. 12.4. Find (in seconds) the period of oscillation of the piston when the wheel is rotating at 3000 rpm (revolutions per minute).

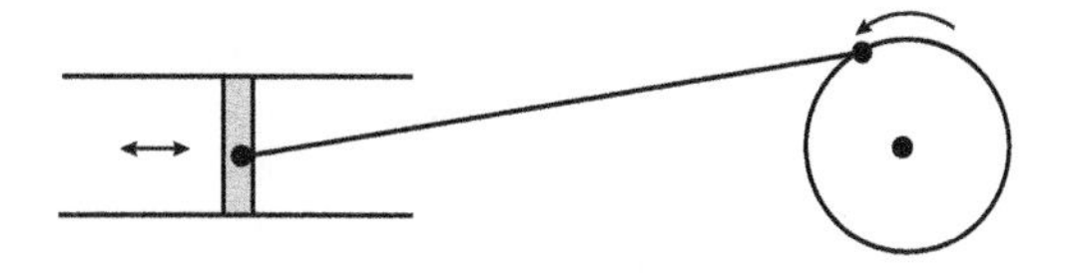

Figure 12.4: Piston for Exercise 12.1.7.

12.1.8. If two waves of the same wavelength and of unit intensity (in arbitrary units) and traveling in the same direction are superposed, determine the intensity of the combined wave distribution:

(a) If the two waves are in phase.

(b) If they differ in phase by 30°.

(c) If they differ in phase by 90°.

(d) If they differ in phase by 180°.

12.1.9. A wave of frequency 100 s^{-1} is multiplied by a wave of frequency 1 MHz ($10^6\ \text{s}^{-1}$). It is sometimes said that the low-frequency signal is used to modulate (more precisely, "amplitude-modulate") the high-frequency signal. What frequencies will be present in the combined frequency distribution?

12.2 FOURIER EXPANSIONS

A periodic wave distribution more complicated than that describing a simple harmonic motion can be represented as a series of terms, provided that our entire series is periodic. One way to accomplish this, for a "standard" periodicity interval of length 2π, is to take the functions $\sin nx$ (for $n = 1, 2, 3, \ldots$) and $\cos nx$ (for $n = 0, 1, 2, 3, \ldots$). The cosine function for $n = 0$ is just a constant; it has (trivially) the required periodicity. All the functions for $n > 1$ are periodic for fractions of our standard interval, but they are also periodic for the entire interval. We are therefore proposing to represent a general function that is periodic (with period 2π) by the **Fourier expansion**

$$f(x) = \frac{a_0}{2} + \sum_{n=1}^{\infty} \Big[a_n \cos nx + b_n \sin nx \Big] . \qquad (12.11)$$

The factor 1/2 multiplying a_0 has been inserted because its presence makes simpler some of the formulas to be derived. This factor is a convenience rather than a necessity.

FOURIER COEFFICIENTS

The Fourier expansion is an expansion in functions that are orthogonal on the interval of periodicity, using the unweighted scalar product

$$\langle f|g\rangle = \int_0^{2\pi} f^*(x)g(x)\,dx\,. \qquad (12.12)$$

One can demonstrate the orthogonality by identifying the functions $\cos nx$ and $\sin nx$ as eigenfunctions of the operator d^2/dx^2 on a Hilbert space with the above scalar product and with members that are periodic on the interval $(0, 2\pi)$. These observations identify the functions $\cos nx$, $\sin nx$ as solutions to a Hermitian Sturm-Liouville problem and therefore assure us that they form a complete basis for our Hilbert space of periodic functions. However, it is also easy to prove the orthogonality directly, as the integrals involved are elementary. If we write

$$\cos nx \cos mx = \frac{1}{2}\Big[\cos(n-m)x + \cos(n+m)x\Big],$$

$$\sin nx \sin mx = \frac{1}{2}\Big[\cos(n-m)x - \cos(n+m)x\Big],$$

$$\sin nx \cos mx = \frac{1}{2}\Big[\sin(n-m)x + \sin(n+m)x\Big],$$

the integrands of all the orthogonality/normalization integrals reduce to terms containing only $\cos Nx$ or $\sin Nx$ over an interval of periodicity, with $N = n \pm m$. Such integrals vanish except when $N = 0$, leading to the following results:

$$\langle \sin nx | \sin mx \rangle = \langle \cos nx | \cos mx \rangle = 0, \qquad (n \neq m), \tag{12.13}$$

$$\langle \sin nx | \cos mx \rangle = 0, \qquad (\text{all } n \text{ and } m), \tag{12.14}$$

$$\langle \sin nx | \sin nx \rangle = \pi, \qquad (n \neq 0), \tag{12.15}$$

$$\langle \cos nx | \cos nx \rangle = (1 + \delta_{n0})\,\pi, \qquad (\text{all } n). \tag{12.16}$$

Using the above integral values and the general formula for coefficients in an orthogonal expansion, Eq. (11.13), we rewrite our Fourier expansion and its coefficients as

$$f(x) = \frac{a_0}{2} + \sum_{n=1}^{\infty} \Big[a_n \cos nx + b_n \sin nx\Big], \qquad \text{with}$$

$$a_n = \frac{1}{\pi}\langle \cos nx | f \rangle = \frac{1}{\pi}\int_0^{2\pi} f(x) \cos nx \, dx \qquad n = 0, 1, 2, \ldots, \tag{12.17}$$

$$b_n = \frac{1}{\pi}\langle \sin nx | f \rangle = \frac{1}{\pi}\int_0^{2\pi} f(x) \sin nx \, dx \qquad n = 1, 2, \ldots$$

The factor $1/2$ attached to a_0 allows the formula for a_n to apply without change for $n = 0$ despite the fact that the normalization integral for $n = 0$ is twice as large as for nonzero n.

Example 12.2.1. Sawtooth Wave

Like orthogonal expansions in general, a Fourier series approximates functions based on their behavior over an interval rather than from data for a single expansion point, and are therefore useful even when power series cannot be used. The present example deals with the sawtooth wave form shown in Fig. 12.3(d). Its functional representation is

$$f(x) = \begin{cases} x, & 0 \le x < \pi, \\ x - 2\pi, & \pi < x \le 2\pi. \end{cases} \tag{12.18}$$

All the a_n vanish due to symmetry, and $b_n = 2(-1)^{n-1}/n$. The Fourier expansion is therefore

$$f(x) = 2\left[\sin x - \frac{\sin 2x}{2} + \frac{\sin 3x}{3} - \cdots + (-1)^{n-1}\frac{\sin nx}{n} + \cdots\right]. \tag{12.19}$$

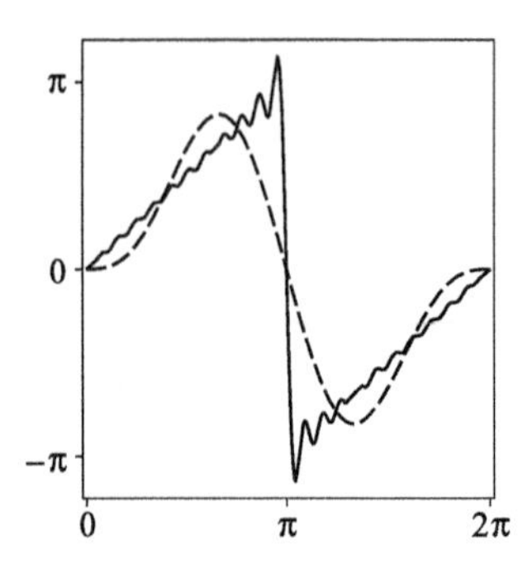

Figure 12.5: Fourier expansion of sawtooth wave. Solid line, truncated after $\sin 20x$; dashed line, truncated after $\sin 2x$.

As the number of terms kept in the expansion is increased, this series rapidly approaches a good approximation to the sawtooth wave. See Fig. 12.5.

■

SHIFT OF INTERVAL

In defining the scalar product for our Fourier series, Eq. (12.12), we chose the integration limits 0 and 2π. But because our Hilbert space was assumed to contain only functions of period 2π, we could, without changing any of the analysis, use any other interval of length 2π. In fact, many workers find it more convenient to use the interval $(-\pi, \pi)$, in part because that choice makes more obvious the even/odd symmetry of the sine and cosine functions. That choice corresponds to the coefficient formulas

$$a_n = \frac{1}{\pi}\langle \cos nx|f\rangle = \frac{1}{\pi}\int_{-\pi}^{\pi} f(x)\cos nx\,dx\,, \qquad n = 0,\,1,\,2,\ldots,$$
$$b_n = \frac{1}{\pi}\langle \sin nx|f\rangle = \frac{1}{\pi}\int_{-\pi}^{\pi} f(x)\sin nx\,dx\,, \qquad n = 1,\,2,\ldots \tag{12.20}$$

However, the implied assumption that the function $f(x)$ is periodic can lead to unexpected results when the form chosen for $f(x)$ does not actually exhibit periodicity.

Example 12.2.2. Nonperiodic Function

Suppose $f(x) = x$ and we expand in Fourier series on the interval $(0, 2\pi)$. The result will be an approximation to the periodic function shown in the left panel of Fig. 12.6. But if we had used the interval $(-\pi, \pi)$ for our expansion, our series would approximate the function shown in the right panel of the figure.

What is obviously going on is that the expansion describes the periodic repetition of the function values that occur within the interval used for the expansion.

■

CHANGE OF SCALE

There is, of course, no reason we must always assume the periodicity to be on an interval of length 2π; we can make a change of scale to make the periodicity for an

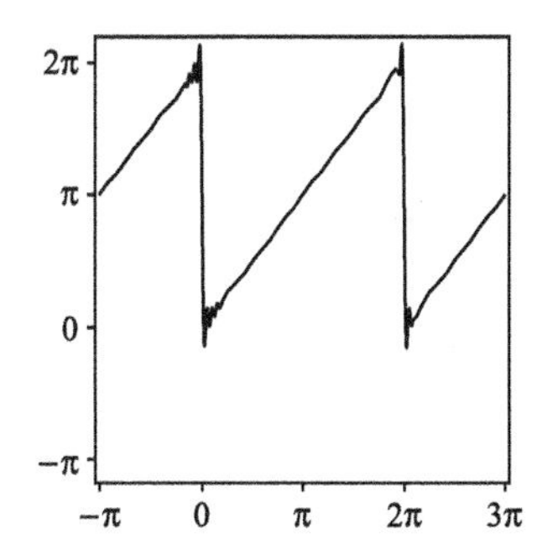

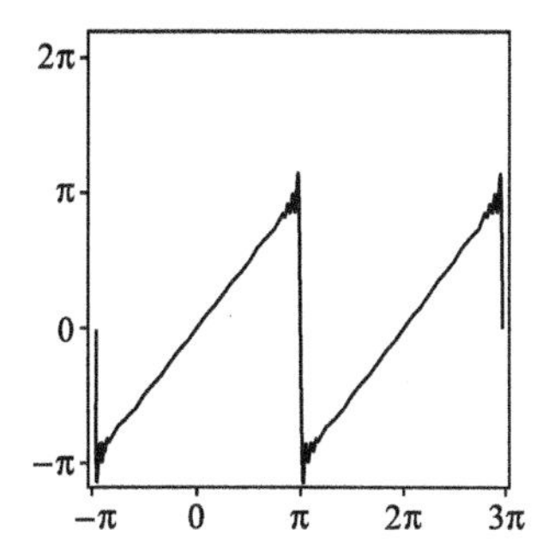

Figure 12.6: Fourier expansions of x: Left, on interval $(0, 2\pi)$; right, on interval $(-\pi, \pi)$. Series truncated after $\sin 40x$.

interval of any length $2L$. Then the formulas of Eq. (12.20) become

$$f(x) = \frac{a_0}{2} + \sum_{n=1}^{\infty} \left[a_n \cos\frac{n\pi x}{L} + b_n \sin\frac{n\pi x}{L} \right], \qquad \text{with}$$

$$a_n = \frac{1}{L} \int_{-L}^{L} f(x) \cos\frac{n\pi x}{L}\, dx\,, \qquad n = 0,\, 1,\, 2, \ldots, \tag{12.21}$$

$$b_n = \frac{1}{L} \int_{-L}^{L} f(x) \sin\frac{n\pi x}{L}\, dx\,, \qquad n = 1,\, 2, \ldots$$

EVEN AND ODD FUNCTIONS

If we make a Fourier expansion of an even function on the interval $(-\pi, \pi)$, we get nonzero coefficients only for the even functions of our expansion set, i.e., only the cosine functions. In that event, we can also use the symmetry to restrict to positive values the range of the integrals defining the coefficients a_n. Moreover, if we have an arbitrary function defined on the range $(0, \pi)$, we can extend it (as an even function) to the additional range $(-\pi, 0)$ and then obtain its expansion as a Fourier series containing only the cosine functions. These ideas make it clear that we can use a **Fourier cosine** series to describe an arbitrary function that is defined on the range $(0, \pi)$; our expansion will be a valid approximation within that range, but (irrespective of the actual behavior of the function being approximated) the expansion will extend, as an even function, to the range $(-\pi, 0)$.

The analysis of the preceding paragraph indicates that an odd function can have a Fourier expansion that includes only the sine functions, and that, by extending an arbitrary function defined for the interval $(0, \pi)$ as an odd function to $(-\pi, 0)$, we can expand it as a **Fourier sine** series that is valid for the range $(0, \pi)$.

Formulas for the Fourier cosine and sine expansions for the interval $(0, L)$ are the following:

$$f(x) = \frac{a_0}{2} + \sum_{n=1}^{\infty} a_n \cos\frac{n\pi x}{L}, \qquad a_n = \frac{2}{L} \int_{0}^{L} f(x) \cos\frac{n\pi x}{L}\, dx\,, \tag{12.22}$$

$$f(x) = \sum_{n=1}^{\infty} b_n \sin\frac{n\pi x}{L}, \qquad b_n = \frac{2}{L} \int_{0}^{L} f(x) \sin\frac{n\pi x}{L}\, dx\,. \tag{12.23}$$

Example 12.2.3. Fourier Cosine and Sine Series

Consider the function

$$f(x) = \begin{cases} 1, & 0 < x < \pi/2, \\ 0, & \pi/2 < x < \pi. \end{cases} \tag{12.24}$$

Expanding $f(x)$ first as a Fourier cosine series, we have

$$a_{2n+1} = \frac{2}{\pi}\int_0^{\pi/2} \cos[(2n+1)x]\,dx = \frac{2(-1)^n}{(2n+1)\pi},$$

$$a_{2n} = \frac{2}{\pi}\int_0^{\pi/2} \cos 2nx\,dx = \delta_{n0}.$$

The cosine expansion is therefore

$$f(x) = \frac{1}{2} + \frac{2}{\pi}\left[\cos x - \frac{\cos 3x}{3} + \frac{\cos 5x}{5} - \cdots\right]. \tag{12.25}$$

Expanding now $f(x)$ as a Fourier sine series, we have

$$b_{4n+1} = \frac{2}{\pi}\int_0^{\pi/2} \sin[(4n+1)x]\,dx = \frac{2}{(4n+1)\pi},$$

$$b_{4n+2} = \frac{2}{\pi}\int_0^{\pi/2} \sin[(4n+2)x]\,dx = \frac{2}{(2n+1)\pi},$$

$$b_{4n+3} = \frac{2}{\pi}\int_0^{\pi/2} \sin[(4n+3)x]\,dx = \frac{2}{(4n+3)\pi},$$

$$b_{4n} = \frac{2}{\pi}\int_0^{\pi/2} \sin 4nx\,dx = 0.$$

Thus, the sine expansion is

$$f(x) = \frac{2}{\pi}\left[\sin x + \sin 2x + \frac{\sin 3x}{3} + \frac{\sin 5x}{5} + \frac{\sin 6x}{3} + \frac{\sin 7x}{7} + \cdots\right]. \tag{12.26}$$

Extending $f(x)$ as functions of appropriate parity, Eqs. (12.25) and (12.26) correspond to the periodic functions sketched in Fig. 12.7.

■

Example 12.2.4. Sine Series of Unity

There are times when we may want to represent a constant by a Fourier sine series. (Representing a constant as a Fourier cosine series is trivial, as that corresponds to the constant term in the cosine expansion.)

Expanding on the interval $(0, L)$ and using Eq. (12.23), we have for $f(x) = 1$

$$b_n = \frac{2}{L}\int_0^L \sin\frac{n\pi x}{L}\,dx = \begin{cases} \dfrac{4}{n\pi}, & n \text{ odd}, \\ 0, & n \text{ even}. \end{cases}$$

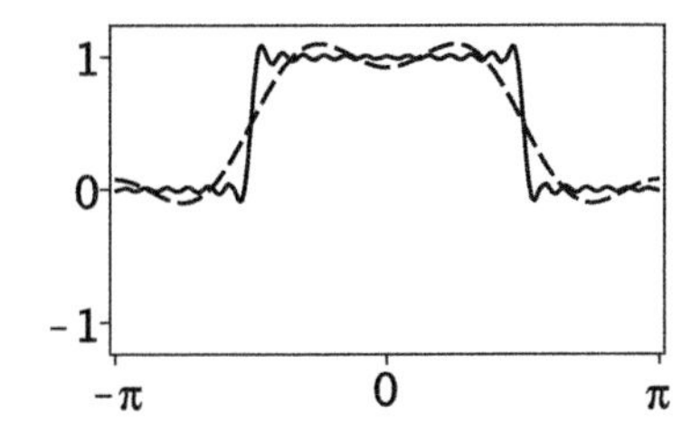

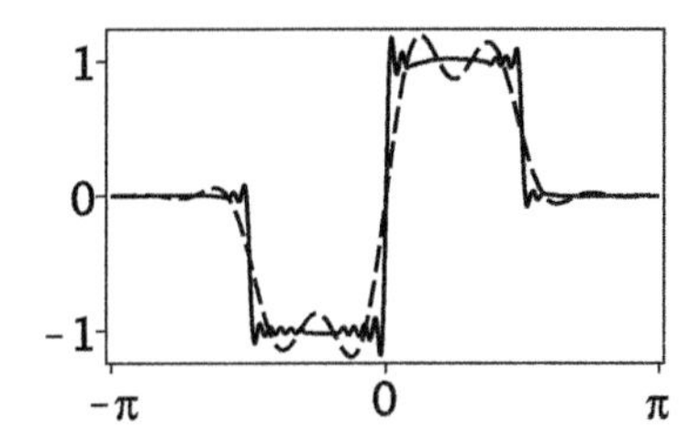

Figure 12.7: Functions represented by the series in Example 12.2.3: Left, Fourier cosine series; right, Fourier sine series. Solid curves are expansions truncated after $\cos 25x$ (left) and $\sin 51x$ (right); dashed curves are for truncation after $\cos 3x$ (left) and $\sin 7x$ (right).

Inserting this result into the expansion of Eq. (12.23),

$$1 = \sum_{n \text{ odd}} \frac{4}{n\pi} \sin \frac{n\pi x}{L} . \tag{12.27}$$

■

COMPLEX FOURIER SERIES

In Chapter 3 we found that $\sin x$ and $\cos x$ could be written as linear combinations of e^{ix} and e^{-ix}, so the Fourier series we are now using can be rewritten as series involving the functions e^{inx}, with n assuming all integer values of both signs. We note also that, for all integer n, e^{inx} has periodicity 2π, and that e^{inx} of different n are orthogonal on any interval of length 2π. If $m \neq n$,

$$\int_0^{2\pi} \left[e^{inx}\right]^* e^{imx}\, dx = \int_0^{2\pi} e^{i(m-n)x}\, dx = \left.\frac{e^{i(m-n)x}}{i(m-n)}\right|_0^{2\pi} = \frac{1-1}{i(m-n)} = 0 .$$

If $m = n$, this integral reduces to $\int_0^{2\pi} dx = 2\pi$, so we have the important general result

$$\int_0^{2\pi} \left[e^{inx}\right]^* e^{imx}\, dx = 2\pi\, \delta_{mn} . \tag{12.28}$$

This result is the same whether or not $n = 0$. From Eq. (12.28) we can determine that the complex Fourier series expansion of a function $f(x)$ takes the form

$$f(x) = \sum_{n=-\infty}^{\infty} c_n\, e^{inx} , \qquad c_n = \frac{1}{2\pi} \int_0^{2\pi} e^{-inx} f(x)\, dx . \tag{12.29}$$

Note that in this form of the expansion we do not have a fraction 1/2 associated with the $n = 0$ term. Also, if $f(x)$ is real, we have $c_{-n} = c_n^*$.

Example 12.2.5. Complex Fourier Series

Let's obtain the complex Fourier expansion of the square wave,

$$f(x) = \begin{cases} 1, & 0 < x < \pi , \\ 0, & -\pi < x < 0 . \end{cases} \tag{12.30}$$

The coefficients are

$$c_{2n} = \frac{1}{2\pi}\int_0^\pi e^{-2inx}\,dx = \frac{1}{2\pi}\left(\frac{e^{-2in\pi}-1}{-2in}\right) = 0, \qquad (n \neq 0),$$

$$c_{2n+1} = \frac{1}{2\pi}\int_0^\pi e^{-i(2n+1)x}\,dx = \frac{1}{2\pi}\left(\frac{-2}{-i(2n+1)}\right) = -\frac{i}{(2n+1)\pi},$$

$$c_0 = \frac{1}{2\pi}\int_0^\pi dx = \frac{1}{2},$$

and the expansion is

$$f(x) = \frac{1}{2}e^0 - \frac{i}{\pi}\left[e^{ix} + \frac{e^{3ix}}{3} + \frac{e^{5ix}}{5} + \cdots\right] + \frac{i}{\pi}\left[e^{-ix} + \frac{e^{-3ix}}{3} + \frac{e^{-5ix}}{5} + \cdots\right]. \tag{12.31}$$

This expansion has the property that $c_{-n} = c_n^*$, consistent with the fact that the function being expanded is real. We can bring the expansion to an explicitly real form by combining the e^{inx} and e^{-inx} terms, obtaining

$$f(x) = \frac{1}{2} + \frac{2}{\pi}\left[\frac{\sin x}{1} + \frac{\sin 3x}{3} + \frac{\sin 5x}{5} + \cdots\right]. \tag{12.32}$$

■

Exercises

12.2.1. Derive the orthogonality/normalization integrals given in Eqs. (12.13)–(12.16).

12.2.2. Verify that the Fourier coefficients for the sawtooth wave of Example 12.2.1 have the values given in that Example.

12.2.3. Verify the orthogonality and determine the normalization of the functions appearing in the expansion of Eq. (12.21).

12.2.4. Verify that the Fourier coefficients appearing in Example 12.2.3 have the values shown.

12.2.5. Verify that the Fourier coefficients appearing in Example 12.2.5 have the values shown.

12.3 SYMBOLIC COMPUTING

Symbolic computing can be helpful in determining Fourier coefficients and in the graphical display of Fourier expansions. Typical coding for computation of the Fourier coefficients of a function $f(x)$ can be as follows, where $f(x)$ is replaced by its explicit coding (not a function call), and `nmax` is set to the truncation point of the expansion:

MAPLE

```
> a0 := 1/Pi*int( f(x) , x = -Pi .. Pi);
> for n from 1 to nmax do
```

```
>    a[n] := 1/Pi*int( f(x) *cos(n*x), x = -Pi .. Pi);
>    b[n] := 1/Pi*int( f(x) *sin(n*x), x = -Pi .. Pi)
> end do;
```

MATHEMATICA

```
a = Table[0,{n,1,nmax}];
b = Table[0,{n,1,nmax}];
a0 = 1/Pi*Integrate[ f(x) , {x, -Pi, Pi} ]
Do[ a[[n]] = 1/Pi*Integrate[ f(x) * Cos[n*x], {x, -Pi, Pi} ];
    b[[n]] = 1/Pi*Integrate[ f(x) * Sin[n*x], {x, -Pi, Pi} ],
    {n, 1, nmax} ]
```

If $f(x)$ is zero or represented by different functional forms in parts of its range, the above coding should be modified accordingly. In addition, if $f(x)$ is such that the explicit forms of the coefficients are unduly complicated or are returned as unevaluated integrals, it may be better to enclose the entire right-hand side of each of the above equations by `evalf(...)` or `N[...]` to produce the coefficients in decimal form.

After the needed Fourier coefficients have been computed and stored, one can define the Fourier series as a function:

MAPLE

```
> ff := proc(x) local n;
>        a0/2 + add( a[n]*cos(n*x) + b[n]*sin(n*x), n = 1 .. nmax);
> end proc;
```

MATHEMATICA

```
ff[x_] := Module[{n},
            a0/2 + Sum[ a[[n]]*Cos[n*x] + b[[n]]*Sin[n*x],
            {n,1,nmax} ] ]
```

If the series is to be plotted, the function representing it must be a function of only the single variable that is associated with the horizontal axis of the plot. We may then display the Fourier expansion by executing

MAPLE: `plot(ff(x), x = -Pi .. Pi);`

MATHEMATICA: `Plot[ff[x], {x, -Pi, Pi}]`

Example 12.3.1. Finding a Fourier Expansion

Let's use symbolic computing to study the Fourier expansion of the square wave

$$f(x) = \begin{cases} 0, & -\pi < x < 0, \\ 1, & 0 < x < \pi. \end{cases}$$

We want to be able to explore expansions of arbitrary truncation, so we write, first using MAPLE, and illustrating for $nmax = 10$,

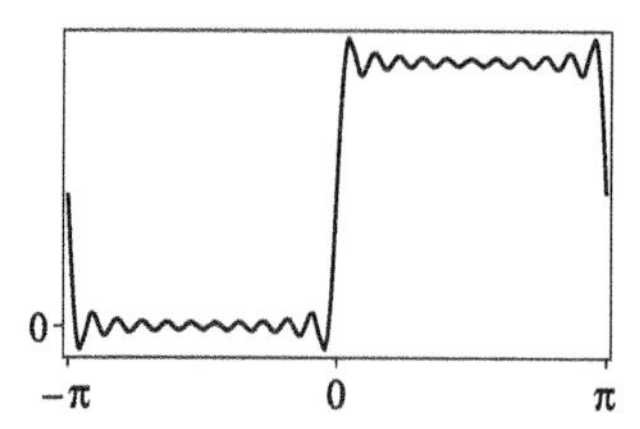

Figure 12.8: Square wave, Fourier series terminated after $\sin 10x$ and $\cos 10x$.

```
> nmax := 10:
> a0 := 1/Pi*int(1, x = 0 .. Pi):
> for n from 1 to nmax do
>   a[n] := 1/Pi*int(cos(n*x), x = 0 .. Pi);
>   b[n] := 1/Pi*int(sin(n*x), x = 0 .. Pi);   end do:
```

To generate and display the Fourier series for the current value of *nmax* we write

```
> ff := proc(x) local n;
>        a0/2 + add( a[n]*cos(n*x) + b[n]*sin(n*x),
>                        n = 1 .. nmax);
> end proc:
> plot(ff(x), x = -Pi .. Pi);
```

This plot is shown as Fig. 12.8, and the individual Fourier coefficients can be accessed by entering a0, or for any appropriate n, a[[n]], and b[[n]].

In MATHEMATICA, the Fourier coefficients for this problem, keeping terms through $nmax = 10$, are obtained from

```
nmax = 10;
a = Table[0,{n,1,nmax}];
b = Table[0,{n,1,nmax}];
a0 = 1/Pi*Integrate[1, {x, 0, Pi} ];
Do[ a[[n]] = 1/Pi*Integrate[Cos[n*x], {x, 0, Pi} ];
    b[[n]] = 1/Pi*Integrate[Sin[n*x], {x, 0, Pi} ],
    {n, 1, nmax} ]
```

To generate and display the Fourier series for the current value of *nmax* we write

```
ff[x_] := Module[{n},
          a0/2 + Sum[ a[[n]]*Cos[n*x] + b[[n]]*Sin[n*x],
          {n,1,nmax} ] ]
Plot[ff[x], {x, -Pi, Pi}]
```

This plot is shown as Fig. 12.8, and the individual Fourier coefficients can be accessed by entering a0, or for any appropriate n, a[[n]], and b[[n]].

■

Exercises

12.3.1. Enter into your computer the code from Example 12.3.1 and reproduce the plot in Fig. 12.8

(a) With $n_{\rm max} = 10$, and

(b) For the additional values $n_{\rm max} = 2$, 6, and 20.

12.3.2. (a) Derive the following formula for the Fourier cosine expansion of x on the interval $(0, \pi)$, and write symbolic code to obtain the coefficients and evaluate the expansion through an arbitrary number of terms.

$$x = \frac{\pi}{2} - \frac{4}{\pi}\sum_{n=0}^{n_{\rm max}} \frac{\cos[(2n+1)x]}{(2n+1)^2}.$$

(b) Plot x and the above expansion on the range $(-2\pi, 2\pi)$ for $n_{\rm max} = 1$, 2, and 4.

(c) To obtain a clearer measure of the errors in your expansions, plot the error in the expansion (your expansion minus x) for the range $(0, \pi)$ for each of the $n_{\rm max}$ values.

12.3.3. (a) Derive formulas for the Fourier series expansion of the half-wave rectifier

$$V(x) = \begin{cases} 0, & -\pi < x < 0, \\ V_0 \sin x & 0 < x < \pi. \end{cases}$$

and write symbolic code to generate the expansion.

(b) Plot the expansion of $V(x)$ (truncated for three different values of $n_{\rm max}$) for the range $(-\pi, \pi)$.

12.3.4. (a) Derive formulas for the Fourier series expansion of the full-wave rectifier,

$$V(x) = V_0|\sin x|, \qquad 0 < x < 2\pi,$$

and write symbolic code to generate the expansion.

(b) Plot $V(x)$ and its expansion (truncated for three different values of $n_{\rm max}$) for the range $(0, 2\pi)$.

12.3.5. Write symbolic code for the coefficients in the Fourier sine series of

$$\frac{x(2-x)}{\sqrt{1+x^2}}\, e^{-x^2}$$

for the range $0 < x < 2$. Do not try to evaluate the necessary integrals analytically; arrange for their numerical evaluation. Then determine the smallest number of terms that must be kept in the sine series to cause the error in the expansion to be no larger than 0.01 at any x value in the interval.

12.4 PROPERTIES OF EXPANSIONS

We now consider issues associated with the convergence of Fourier series and the way in which convergence is affected by operations we can perform on the series.

DIRICHLET'S THEOREM

A useful statement about the convergence of Fourier series is provided by a theorem due to Dirichlet. Dirichlet's theorem states:

If a function $f(x)$ defined on a finite interval (a, b)

(1) has a finite number of maxima and minima,

(2) has at most a finite number of discontinuities, and

(3) has a finite value of $\int_a^b |f(x)|\,dx$,

then its Fourier series for the interval (a, b) converges to $f(x)$ at all points where $f(x)$ is continuous, and converges to the midpoint of the jump in $f(x)$ at each finite discontinuity of $f(x)$ (including any discontinuity that may occur at $x = a$ or $x = b$ if $f(x)$ is extended beyond these points as a periodic function).

For most of the Fourier expansions we will encounter, Dirichlet's theorem not only resolves the question of convergence; it also answers another important question by stating that the convergence is to the value of $f(x)$. In other words, the theorem confirms that the trigonometric functions occurring in a Fourier series form a complete basis in terms of which a large class of periodic functions can be expanded.

Note that Dirichlet's theorem includes a recipe to be followed when a function has a discontinuity at the ends of its interval of definition. Among other things, this means that we cannot remove the effect of a discontinuity by making a redefinition to shift the discontinuity to the endpoints of the expansion interval.

Dirichlet's condition (3) will certainly be satisfied if $f(x)$ is finite throughout its interval of definition. But it will also be satisfied for additional functions that are weakly singular. An important use of Fourier series is in the discussion of wave forms $f(x)$ that are everywhere finite, but with discontinuities, and we see that these functions are included among those to which Dirichlet's theorem applies. We have previously noted that discontinuous functions, which occur in both power-generation and digital-computation problems, cannot be treated using power-series expansion. We see here that Fourier analysis gives us an additional, and powerful tool for the study of these important areas.

PARSEVAL'S THEOREM

Given a function $f(x)$ and its Fourier expansion, written in the standard form, Eq. (12.11), we wish to derive a relation connecting $\langle f|f\rangle$ and the Fourier coefficients of f. Inserting the expansion into $\langle f|f\rangle$, and invoking orthogonality, we get

$$\langle f|f\rangle = \left\langle \frac{a_0}{2} + \sum_{n=1}^{\infty}(a_n \cos nx + b_n \sin nx) \middle| \frac{a_0}{2} + \sum_{m=1}^{\infty}(a_m \cos mx + b_m \sin mx) \right\rangle$$

$$= \frac{a_0^2}{4}\langle 1|1\rangle + \sum_{n=1}^{\infty}\Big(a_n^2\langle \cos nx|\cos nx\rangle + b_n^2\langle \sin nx|\sin nx\rangle\Big). \tag{12.33}$$

Inserting the values of the normalization integrals, Eq. (12.33) reduces to

$$\frac{1}{2\pi}\langle f|f\rangle = \left(\frac{a_0}{2}\right)^2 + \frac{1}{2}\sum_{n=1}^{\infty}(a_n^2 + b_n^2). \tag{12.34}$$

This result is called **Parseval's theorem.**

If one were to carry out the analysis that led to Eq. (12.34) for an interval of periodicity of length $2L$ instead of 2π, the only change to the result would be the replacement of the 2π on the left-hand side by $2L$. One could observe that, irrespective of the value of L, the left-hand side of Eq. (12.34) is the average value of $|f|^2$ over the interval; Parseval's theorem is sometimes stated in that language. Another generalization is to complex Fourier series; then Eq. (12.34) becomes

$$\frac{1}{2\pi}\langle f|f\rangle = \sum_{n=-\infty}^{\infty} |c_n|^2 \,. \tag{12.35}$$

Since every term of Eq. (12.35) is nonnegative, any truncated Fourier expansion must satisfy the inequality

$$\frac{1}{2\pi}\langle f|f\rangle \geq \sum_{n} |c_n|^2 \,. \tag{12.36}$$

This result is called **Bessel's inequality**.

As has been already mentioned several times, Fourier series, like other orthogonal expansions, fit the expansion over an interval rather than at a point. These types of series converge **in the mean** rather than pointwise, meaning (for the Fourier series now under discussion) that

$$\int_{-\pi}^{\pi} \left| f(x) - \sum_{n=-\infty}^{\infty} c_n e^{inx} \right|^2 dx = 0 \,. \tag{12.37}$$

As noted previously, this type of convergence is weaker than pointwise convergence because it permits $f(x)$ and its expansion to differ at individual isolated points that do not contribute to the integral of Eq. (12.37). Evidence of the lack of convergence at individual points is illustrated both by the behavior at discontinuities (see Dirichlet's theorem) and by spikes near the discontinuities that can be seen in Figs. 12.5–12.7. These spikes do not disappear as the extent of the expansion is increased, but converge instead into an overshoot of about 9% on each side of the discontinuity. This behavior is called the **Gibbs phenomenon**, and is of practical importance when one attempts to use Fourier series for highly precise computations near discontinuities.

Parseval's theorem has an interesting interpretation: The individual terms of the right-hand side of Eq. (12.34) or (12.35) are (equally scaled) intensities of the waves that are the individual terms of the expansion. We thus have the nice result that the total intensity of a complicated but periodic wave distribution is the sum of the intensities of the individual sinusoidal waves that form the distribution.

Example 12.4.1. Application of Parseval's theorem

The squares of the coefficients of a Fourier series may correspond to a series we would like to evaluate. Using the following expansion for x^2 (derived in Example 12.4.2 below),

$$x^2 = \frac{\pi^2}{3} - 4\left[\cos x - \frac{\cos 2x}{2^2} + \frac{\cos 3x}{3^2} - \cdots\right],$$

we write the corresponding case of Eq. (12.34):

$$\frac{1}{2\pi}\int_{-\pi}^{\pi} x^4\, dx = \frac{\pi^4}{5} = \left(\frac{\pi^2}{3}\right)^2 + \frac{1}{2}16\left[1 + \frac{1}{2^4} + \frac{1}{3^4} + \cdots\right],$$

which rearranges to

$$1 + \frac{1}{2^4} + \frac{1}{3^4} + \cdots = \zeta(4) = \frac{\pi^4}{90}. \tag{12.38}$$

The zeta function occurring in Eq. (12.38) was defined at Eq. (2.12) and the present evaluation of $\zeta(4)$ can be checked against Eq.(2.79).

■

INTEGRATION AND DIFFERENTIATION OF SERIES

The termwise integration of a Fourier series produces a new series that is more convergent than the original series due to the fact that $\int \cos nx = \sin nx/n$ and $\int \sin nx = -\cos nx/n$. Note, however, that if the original series contained a constant term $a_0/2$, the integrated series from a technical viewpoint is no longer a Fourier series, as it contains a leading term $a_o x/2$.

The situation is quite different when a Fourier series is termwise differentiated. The differentiations make the terms of index n larger by a factor n, and the differentiation yields a valid Fourier series only if the resulting series is convergent. Examples already examined in the text reveal that Fourier series with coefficients a_n or b_n that have n-dependence $1/n$ describe functions with finite discontinuities; when differentiated, the resulting series will not be convergent everywhere.

Extending the discussion of the preceding paragraph, we note that if a function is continuous over the expansion interval, its Fourier coefficients must converge at least proportionally to $1/n^2$, and that a function with m continuous derivatives will have Fourier coefficients that converge at least as rapidly as $1/n^{m+2}$. Thus, the Fourier expansion of a function with m continuous derivatives can be differentiated (to yield a convergent expansion) $m+1$ times. This observation is consistent with the qualitative notion that "smooth" functions (those without discontinuities or rapid oscillations) will have rapidly converging Fourier series. This observation is consistent with the fact that terms $\sin nx$ or $\cos nx$ of large n are needed to describe high-frequency oscillations.

Example 12.4.2. Integrating a Fourier Series

Let's obtain a new Fourier series by integrating the sawtooth expansion of Example 12.2.1. Integrating the expansion of Eq. (12.19) from 0 to x, we get

$$\int_0^x x\,dx = \int_0^x 2\left[\sin x - \frac{\sin 2x}{2} + \frac{\sin 3x}{3} - \cdots\right] \Longrightarrow$$

$$\frac{x^2}{2} = -2\left[\cos x - \frac{\cos 2x}{2^2} + \frac{\cos 3x}{3^2} - \cdots\right] + 2\left[1 - \frac{1}{2^2} + \frac{1}{3^2} - \cdots\right]. \tag{12.39}$$

The quantities in the last set of brackets in this equation sum to $\zeta(2)/2 = \pi^2/12$, so our integrated series can be written

$$x^2 = \frac{\pi^2}{3} - 4\left[\cos x - \frac{\cos 2x}{2^2} + \frac{\cos 3x}{3^2} - \cdots\right]. \tag{12.40}$$

The sawtooth wave has a discontinuity at the ends of the interval of periodicity and its coefficients approach zero for large n as $1/n$. The integrated series is continuous at the endpoints $x = \pm\pi$ (with $\pi^2 = (-\pi)^2$), but these points are characterized by a discontinuity in slope. The series converges as $1/n^2$, consistent with that behavior.

Let's now repeat the above analysis starting from the Fourier cosine series for x, obtained in Exercise 12.3.2. Viewing the cosine series as an even extension of x to negative values, and therefore being an expansion of $|x|$ on $(-\pi, \pi)$, we note that the periodic extension of this function is continuous, but with a discontinuities in its derivative at zero and $\pm\pi$. We have

$$\int_0^x x\,dx = \int_0^x \left[\frac{\pi}{2} - \frac{4}{\pi}\sum_{n=0}^{\infty}\frac{\cos[(2n+1)x]}{(2n+1)^2}\right] dx \Longrightarrow$$

$$\frac{x^2}{2} = \frac{\pi x}{2} - \frac{4}{\pi}\left[\sin x + \frac{\sin 3x}{3^3} + \frac{\sin 5x}{5^3} + \cdots\right], \tag{12.41}$$

valid for $0 \le x \le \pi$. Setting $x = \pi/2$ in Eq. (12.41), we have

$$\frac{\pi^2}{8} = \frac{\pi^2}{4} - \frac{4}{\pi}\left[1 - \frac{1}{3^3} + \frac{1}{5^3} - \cdots\right],$$

which is easily rearranged to obtain the evaluation

$$\left[1 - \frac{1}{3^3} + \frac{1}{5^3} - \cdots\right] = \frac{\pi^3}{32}. \tag{12.42}$$

Note also that Eq. (12.41) is not a Fourier series for $x^2/2$, though it could be interpreted as providing such a series for $(x^2 - \pi x)/2$.

■

Example 12.4.3. Differentiating a Fourier Series

The previous example used two different expansions of x: a Fourier sine series and a Fourier cosine series. If we attempt to differentiate these series (to obtain a Fourier expansion of unity) we find that these series behave differently. Differentiation of the sine series, the integrand in the first line of Eq. (12.39), yields

$$\cos x - \cos 2x + \cos 3x - \cdots,$$

which is not convergent for all x; in particular, for $x = 0$ we get $1 - 1 + 1 - \cdots$. On the other hand, the cosine series, which appears as the integrand in the first line of Eq. (12.41), can be differentiated to give

$$f(x) = 1 = \frac{1}{\pi}\sum_{n=0}^{\infty}\frac{\sin[(2n+1)x]}{2n+1} \tag{12.43}$$

which converges to unity on the interval $(0, \pi)$ except at the discontinuities associated with the odd extension to negative x. This extension corresponds to $f(x) = -1$ for $x < 0$, and the midpoint of the jump at $x = 0$ is at $f(0) = 0$, as required by Dirichlet's theorem.

■

Exercises

12.4.1. (a) Find the Fourier sine series for x on the range $(0, \pi)$.

(b) Use Parseval's theorem and the result of part (a) to show that $\zeta(2) = \pi^2/6$.

12.4.2. Starting from the Fourier expansion of the square wave,

$$f(x) = \begin{cases} 1, & 0 < x < \pi, \\ -1, & -\pi < x < 0, \end{cases}$$

$$= \frac{4}{\pi}\left[\frac{\sin x}{1} + \frac{\sin 3x}{3} + \frac{\sin 5x}{5} + \cdots\right],$$

use Parseval's theorem to show that

$$1 + \frac{1}{3^2} + \frac{1}{5^2} + \frac{1}{7^2} + \cdots = \frac{\pi^2}{8}.$$

12.4.3. (a) Show that $\cos ax = \dfrac{2a \sin a\pi}{\pi}\left[\dfrac{1}{2a^2} - \dfrac{\cos x}{a^2 - 1^2} + \dfrac{\cos 2x}{a^2 - 2^2} - \dfrac{\cos 3x}{a^2 - 3^2} + \cdots\right]$.

(b) Integrate term-by-term the expansion of part (a) to obtain a series for $\sin ax$.

(c) Differentiate term-by-term the expansion of part (a) to obtain another series for $\sin ax$.

(d) Show that the series of parts (b) and (c) are consistent with each other.

Hint. Use the Fourier sine series for x, developed in Exercise 12.4.1.

12.5 APPLICATIONS

Although it is possible to have wave distributions that are inherently nonperiodic (e.g., composed of waves that travel at different speeds), a wide variety of problems of interest are characterized by frequency distributions that are multiples of a lowest (or **fundamental**) frequency, and are therefore conveniently represented by a Fourier series on an interval L corresponding to the wavelength of the fundamental frequency. A typical example is provided by the standing-wave oscillations of a string whose ends are fixed; the possible modes of oscillation correspond to the individual terms of a Fourier sine series. These individual modes of oscillation are often referred to as **harmonics**; the fundamental frequency is also called the **first harmonic**, with the frequency at n times the fundamental called the n**th harmonic**. As we have already observed, the intensities of the terms (harmonics) are additive, and it is well known that the perceived nature of a musical tone depends both on its **pitch** (the frequency of its fundamental) and on the distribution of its higher harmonics.

The distribution of harmonics arising from a vibrating string will depend upon the initial conditions that excite the vibration. For example, a string that is distorted from equilibrium (plucked) by displacement at a point will have a harmonic distribution obtainable from the Fourier expansion of the initial displacement.[1] The identification of the harmonic composition of a wave distribution is often referred to as **harmonic analysis**.

Another application of harmonic analysis arises when a rectifier is used to permit the passage of electric current in only one direction. If the incoming current is

[1] If the string were able to respond equally effectively at all frequencies and there were no frictional forces, the initial triangular configuration would recur periodically; in actuality, the high-frequency part of the oscillation rapidly becomes damped and the initial slope discontinuity does not recur.

sinusoidal and the outgoing current is only the positive lobe of the sinusoidal wave, the harmonic composition of the output can be obtained from its Fourier expansion. This harmonic analysis has the practical feature that it indicates the principal harmonics that are present in the rectifier output; this information may be relevant for the design of electric filters to reduce the variability of the output signal.

Example 12.5.1. Sound Wave

The harmonics generated from a string that is plucked (or stroked with a bow, as is usual for a violin) depend on the location of the displacement, and the tonal quality of a stringed instrument depends upon the mixture of the harmonics. Let's compare the harmonic distribution of a string plucked at its midpoint with that of a string plucked at 1/10th of its length (a typical disturbance point when playing a violin).

Using a Fourier sine series for the interval $(0, \pi)$, a disturbance at the midpoint corresponds to an initial displacement proportional to

$$f(x) = \begin{cases} x, & 0 \le x \le \pi/2, \\ \pi - x, & \pi/2 \le x \le \pi. \end{cases} \tag{12.44}$$

The expansion coefficients are

$$\begin{aligned} b_n &= \frac{2}{\pi}\left[\int_0^{\pi/2} x \sin nx\, dx + \int_{\pi/2}^{\pi} (\pi - x) \sin nx\, dx\right] \\ &= \begin{cases} \dfrac{4(-1)^{(n-1)/2}}{n^2\pi}, & n \text{ odd}, \\ 0, & n \text{ even}. \end{cases} \end{aligned} \tag{12.45}$$

Scaled by dividing by b_1 and squaring (to get intensity), the relative intensities of the first few harmonics are

$$b_1^2 = 1, \quad b_2^2 = 0, \quad b_3^2 = 0.0123, \quad b_4^2 = 0, \quad b_5^2 = 0.0016.$$

The contributions other than the fundamental are quite small.

If the disturbance is at $x = \pi/10$, the initial displacement is proportional to

$$g(x) = \begin{cases} x, & 0 \le x \le \pi/10, \\ \dfrac{\pi - x}{9}, & \pi/10 \le x \le \pi, \end{cases} \tag{12.46}$$

and the expansion coefficients are

$$b_n = \frac{2}{\pi}\left[\int_0^{\pi/10} x \sin nx\, dx + \int_{\pi/10}^{\pi} \left(\frac{\pi - x}{9}\right) \sin nx\, dx\right].$$

This expression is somewhat cumbersome and is most easily handled using symbolic computing. Accordingly, we code (setting **nmax** to a desired value)

MAPLE:
```
> for n from 1 to nmax do
      b[n] := evalf(2/Pi*(int(x * sin(n*x), x = 0 .. Pi/10)
              + int((Pi-x)/9 * sin(n*x), x = Pi/10 .. Pi) ))
  end do;
```

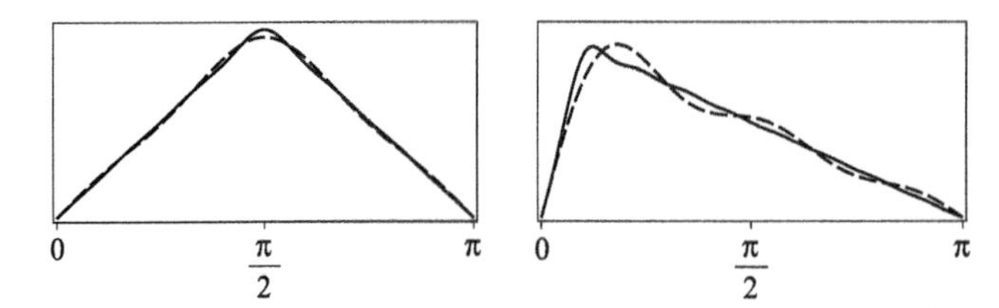

Figure 12.9: Fourier sine series for the initial displacements of a plucked string: Left, plucked at center; right, plucked at 1/10 its length. Solid curves are expansions truncated after $\sin 15x$; dashed curves are for truncation after $\sin 5x$.

MATHEMATICA:

```
b = Table[0,{n,1,nmax}];
Do[ b[[n]] = N[2/Pi*(Integrate[x * Sin[n*x], {x, 0, Pi/10}]
              + Integrate[(Pi - x)/9 * Sin[n*x], {x, Pi/10, Pi}] )],
   {n, 1, nmax}]
```

Scaling to $b_1 = 1$ and squaring to get relative intensities, the above code yields the following output:

$$b_1^2 = 1, \quad b_2^2 = 0.2261, \quad b_3^2 = 0.0846, \quad b_4^2 = 0.0370, \quad b_5^2 = 0.0168.$$

The off-center disturbance causes the generation of substantial intensity in harmonics other than the fundamental, with the largest such contribution from the second harmonic (musically, one octave above the fundamental).

The difference in the rate of convergence can be seen from the plots in Fig. 12.9, in which are plotted the expansions of $f(x)$ and $g(x)$, truncated after $\sin 5x$ and $\sin 15x$. We see that the symmetrical initial displacement, $f(x)$, is nearly converged when truncated at $\sin 5x$; the same cannot be said for the asymmetrical initial displacement $g(x)$.

■

Example 12.5.2. Rectifiers

Some rectifier output waveforms were shown in Fig. 12.3. Consider first a **half-wave** rectifier that accepts an input signal $\sin x$, $-\pi \le x \le \pi$, and produces as output only the positive part of the signal: $V(x) = \sin x$ for $x \ge 0$; $V(x) = 0$ for $x < 0$. The output has Fourier coefficients

$$a_0 = \frac{1}{\pi}\int_0^\pi \sin x\,dx = \frac{2}{\pi},$$

$$a_n = \frac{1}{\pi}\int_0^\pi \cos nx \sin x\,dx = \frac{1}{2\pi}\int_0^\pi \Big(\sin[(n+1)x] - \sin[(n-1)x]\Big)\,dx$$

$$= \begin{cases} \dfrac{1}{\pi}\left(\dfrac{1}{n+1} - \dfrac{1}{n-1}\right), & n > 0 \text{ and even,} \\ 0, & n \text{ odd,} \end{cases}$$

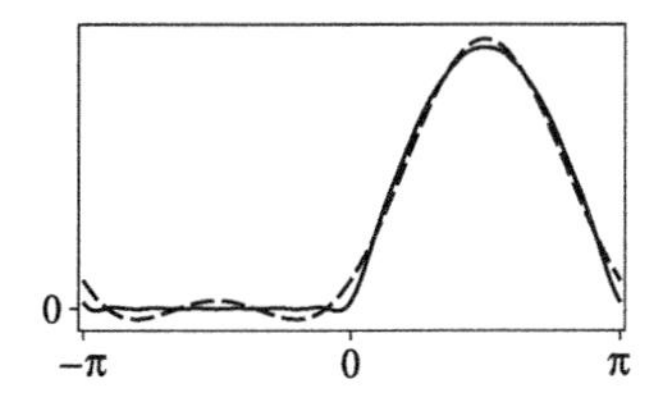

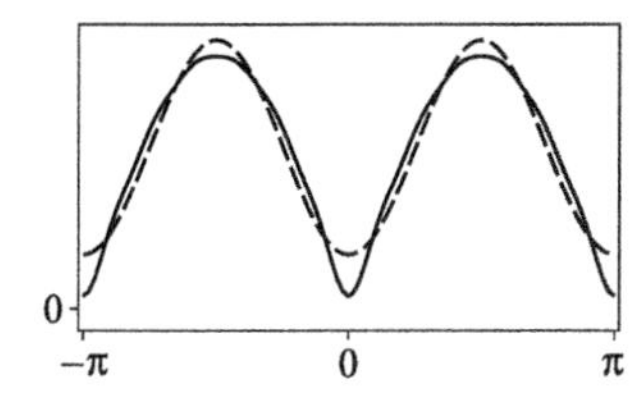

Figure 12.10: Fourier expansions of rectifier waveforms: Left, half-wave; right, full-wave. Dashed line, truncated after $n = 1$ in Eqs. (12.47) and (12.48); solid line, truncated after $n = 6$ in those equations.

$$b_1 = \frac{1}{\pi}\int_0^\pi \sin^2 x\,dx = \frac{1}{2},$$

$$b_n = \frac{1}{\pi}\int_0^\pi \sin nx \sin x\,dx = 0, \qquad n > 1.$$

The output signal therefore has the form

$$V(x) = \frac{1}{\pi} + \frac{1}{2}\sin x - \sum_{n=1}^{\infty} \frac{2\cos 2nx}{\pi(4n^2-1)}. \tag{12.47}$$

The relatively rapid convergence of this series is illustrated in Fig. 12.10.

Referring to Eq.(12.34), we identify the total energy of the incoming signal as proportional to $I = \langle \sin x|\sin x\rangle/2\pi = 1/2$, with the output energies of the Fourier components distributed in the proportions

$$I_0 = \frac{(a_0/2)^2}{I} = 2(a_0/2)^2, \quad I_{n(\cos)} = \frac{a_n^2}{2I} = a_n^2, \quad I_{n(\sin)} = \frac{b_n^2}{2I} = b_n^2.$$

These formulas show that the incoming energy is distributed upon output as follows:

Blocked by the rectifier:		50.00%,
DC output:	$2\left(\frac{1}{\pi}\right)^2$	$= 20.26\%$,
$\sin x$:	$\left(\frac{1}{2}\right)^2$	$= 25.00\%$,
$\cos 2x$:	$\left(\frac{2}{\pi(3)}\right)^2$	$= 4.50\%$,
$\cos 2nx \quad (n > 1)$:		0.24%.

Note that the main AC component is a relatively large unrectified sinusoidal signal, and that the DC output represents only a minor fraction of the input energy.

A **full-wave** rectifier does a far better job of conversion to direct current. To analyze it we need the Fourier series for $|\sin x|$:

$$V(x) = \frac{2}{\pi} - \sum_{n=1}^{\infty} \frac{4\cos 2nx}{\pi(4n^2-1)}. \tag{12.48}$$

This Fourier series is also illustrated in Fig. 12.10. The output energy distribution now becomes

DC output:	81.04%,
$\sin x$:	0.00%,
$\cos 2x$:	18.00%,
Other:	0.96%.

Now the main remaining AC component is the second harmonic. It is sufficiently large that action is often taken to filter it out.

■

Exercises

12.5.1. A square-wave signal, of period τ and amplitude

$$f(t) = \begin{cases} 0, & 0 < t < \tau/2, \\ 1, & \tau/2 < t < \tau, \end{cases}$$

is attenuated (in amplitude) by a factor that depends upon frequency ν: $e^{-k\nu}$.

(a) Generate the Fourier series for the unattenuated signal and (by examining a plot of truncated expansions of $f(t)$ for a time period of length τ) select a truncation of the series that gives a good representation of the square wave.

(b) Set the period to $\tau = 0.001$ s, take $k = 0.0002$ s, and plot the attenuated waveform. Then change k to larger and smaller values and observe the result.

12.5.2. Analyze (as in Example 12.5.2) the energy distribution of the harmonics of a rectified sawtooth wave defined by

$$f(x) = \begin{cases} 0, & -\pi < x < 0, \\ x, & 0 < x < \pi. \end{cases}$$

It suffices to examine the first eight harmonics.

12.5.3. (a) Obtain the Fourier series expansion of $\delta(x - x_0)$, on the range $-\pi < x < \pi$, with x_0 an arbitrary point in the interior of the range.

(b) Show that when this expansion is used in

$$\int_{-\pi}^{\pi} f(x)\, \delta(x - x_0)\, dx\,,$$

you get the expected result, a Fourier series for $f(x_0)$.

12.5.4. A musical instrument creates a sound nominally at middle C (frequency $\nu_0 = 262\ \text{s}^{-1}$). The actual amplitude profile of the sound wave is, in arbitrary units (written here for one period)

$$f(t) = \begin{cases} 1.00, & -\dfrac{1}{2\nu_0} < t < -\dfrac{1}{4\nu_0}, \\ -0.75, & -\dfrac{1}{4\nu_0} < t < 0, \\ 0.75, & 0 < t < \dfrac{1}{4\nu_0}, \\ -1.00, & \dfrac{1}{4\nu_0} < t < \dfrac{1}{2\nu_0}. \end{cases}$$

(a) Plot the amplitude profile for $-1/2\nu_0 < t < 1/2\nu_0$.

(b) Write symbolic code for the coefficients in the Fourier expansion of $f(t)$ on the interval of periodicity and for the expansion at an adjustable truncation level. By examining plots of the truncated expansions, select a truncation level that gives a good representation of the waveform.

(c) Compute the intensities of the first 14 harmonics, and scale them so that the fundamental has unit intensity. Identify the harmonic of greatest intensity; this will define the perceived pitch of the musical note.

Additional Readings

Arfken, G. B., Weber, H. J. & Harris, F. E. (2013). *Mathematical methods for physicists* (7th ed.). New York: Academic Press.

Carslaw, H. S. (1921). *Introduction to the theory of Fourier's series and integrals* (2nd ed.). London: Macmillan (3rd revised and enlarged ed., New York: Dover (1950). A detailed and classic work).

Churchill, R. V. (1993). *Fourier series and boundary value problems* (5th ed.). New York: McGraw-Hill.

Folland, G. B. (1992). *Fourier analysis and its applications.* Providence, RI: American Mathematical Society.

Lighthill, M. J. (1958). *Introduction to Fourier analysis and generalized functions.* Cambridge, UK: Cambridge University Press.

Oberhettinger, F. (1973). *Fourier expansions, a collection of formulas.* New York: Academic Press.

Chapter 13

INTEGRAL TRANSFORMS

13.1 INTRODUCTION

An **integral transform** $\mathcal{L}$ connects functions $f(t)$ with corresponding functions $g(s)$ by an integral of the general form

$$g(s) = [\mathcal{L}f(t)](s) = \int_a^b K(s,t)\, f(t)\, dt\,. \qquad (13.1)$$

Here a, b, and $K(s,t)$ (called the **kernel** of the transform) are fixed, i.e., independent of $f(t)$ and $g(s)$. Thus, Eq. (13.1) is a recipe for converting functions $f(t)$ into their transforms $g(s)$, and we can regard this equation as describing the action of an operator $\mathcal{L}$ on $f(t)$. Note that the operator $\mathcal{L}$ is linear, since

$$\mathcal{L}[f(t) + h(t)] = \mathcal{L}f(t) + \mathcal{L}h(t) \qquad \text{and} \qquad \mathcal{L}[cf(t)] = c\mathcal{L}f(t)\,. \qquad (13.2)$$

We can regard $\mathcal{L}$ as creating a relationship between functions $f(t)$ in "direct space" with counterpart functions $g(s)$ in an "image space." This relationship makes integral transforms useful in a variety of contexts in both pure and applied mathematics.

An important application of the direct-space/image-space duality is that if a mathematical relationship (in particular, a differential equation) is satisfied by $f(t)$, there will be an analogous relationship satisfied by $g(s)$. For many transforms the operation of differentiation in direct space corresponds to a simple algebraic operation in the image space, thereby causing direct-space ODEs to become image-space algebraic equations. These observations suggest the following strategy for solving ODEs (and other similar problems) using integral transforms:

1. Convert the quantities entering our problem into image space, writing the transforms of the known quantities in the problem, and representing any unknown quantities in the problem by symbols denoting their transforms;

2. Simplify and solve the problem in the image space, thereby obtaining the transform of the solution;

3. Obtain the inverse of the transformed solution so as to get the direct-space solution to our problem.

This strategy will be more fully illustrated with specific examples later in this chapter.

Mathematics for Physical Science and Engineering.
http://dx.doi.org/10.1016/B978-0-12-801000-6.00013-4

It is obvious that for the outlined strategy to work it will be necessary that our transform have an inverse and (of at least equal importance) that we have a practical method of finding the inverse. The entire content of this chapter, then, is to be understood as applicable (for a given transform) only to classes of functions for which transforms and their inverses exist.

In this book we consider the two integral transforms that have been of greatest interest in science and engineering: the **Fourier transform**,

$$g(\omega) = \frac{1}{\sqrt{2\pi}} \int_{-\infty}^{\infty} f(t) e^{i\omega t}\, dt, \tag{13.3}$$

and the **Laplace transform**, for which

$$g(s) = \int_{0}^{\infty} e^{-st} f(t)\, dt\,. \tag{13.4}$$

13.2 FOURIER TRANSFORM

The Fourier transform can be thought of as a generalization of the Fourier series to an interval of periodicity that is infinite in length, and therefore becomes appropriate for the representation of functions that are in fact not periodic. This approach leads naturally to formulas both for the Fourier transform and for its inverse, thereby providing also a method for "undoing" a transform.

FOURIER SERIES, INFINITE INTERVAL

Let's start with an exponential Fourier series for a function $g(\omega)$ on the interval $(-L, L)$, for which

$$g(\omega) = \sum_{n=-\infty}^{\infty} c_n\, e^{i\pi n\omega/L}, \qquad c_n = \frac{1}{2L} \int_{-L}^{L} g(\omega)\, e^{-i\pi n\omega/L}\, d\omega\,. \tag{13.5}$$

Since it is our intention to let L become large, successive n will have closely spaced values of $n\pi/L$, suggesting that we may define $n\pi/L = t$ and make the replacement

$$\sum_{n=-\infty}^{\infty} \longrightarrow \frac{L}{\pi} \int_{-\infty}^{\infty} dt\,.$$

If we also make the notational change $L\,c_n = f(t)\sqrt{\pi/2}$ and take the limit $L \to \infty$, Eq. (13.5) then takes the form

$$g(\omega) = \frac{1}{\sqrt{2\pi}} \int_{-\infty}^{\infty} f(t)\, e^{i\omega t}\, dt\,, \qquad f(t) = \frac{1}{\sqrt{2\pi}} \int_{-\infty}^{\infty} g(\omega)\, e^{-i\omega t}\, d\omega\,. \tag{13.6}$$

The relationships shown in Eq. (13.6) are known as the **Fourier integral theorem**, and a more rigorous proof of that theorem confirms those equations for any $f(t)$ that satisfies the Dirichlet conditions on every finite interval, providing that $\int_{-\infty}^{\infty} |f(t)|\, dt$ is also finite.

We regard the first equation of Eq. (13.6) as giving the **Fourier transform** of $f(t)$, and the second equation is therefore a formula for the **inverse** of the transform. Note that the inversion formula is nearly the same as the transform formula, differing therefrom only in the sign of i.

DIRAC DELTA FUNCTION

Another instructive approach to the Fourier integral theorem can be developed from the observation that the following integral (in the limit of large n) is a representation of the Dirac delta function:

$$\delta_n(x) = \frac{1}{2\pi}\int_{-n}^{n} e^{ixt}\,dt\,. \tag{13.7}$$

Evaluating the integral for δ_n, we get

$$\delta_n(x) = \frac{e^{ixn} - e^{-ixn}}{2\pi i x} = \frac{\sin nx}{\pi x}\,, \tag{13.8}$$

which at $x = 0$ is indeterminate, but with the limiting value n/π. Thus, at large n, δ_n is strongly peaked at $x = 0$. It is also evident that at large n the quantity δ_n becomes localized, with all its significant contributions in an interval of width proportional to $1/n$. The result is that for arbitrary "well-behaved" functions $f(x)$ we have

$$\int_{-\infty}^{\infty} f(x)\,\delta_n(x)\,dx \approx f(0)\int_{-\infty}^{\infty}\frac{\sin nx}{\pi x}\,dx = f(0)\,,$$

using the known definite integral $\displaystyle\int_{-\infty}^{\infty}\frac{\sin x}{x}\,dx = \pi$.

FOURIER INTEGRAL THEOREM AGAIN

We can now rederive the Fourier integral theorem by simply combining the integrals of Eq. (13.6) and identifying therein a delta function:

$$\begin{aligned} f(u) &= \frac{1}{\sqrt{2\pi}}\int_{-\infty}^{\infty} e^{-i\omega u} g(\omega)\,d\omega = \frac{1}{\sqrt{2\pi}}\int_{-\infty}^{\infty} e^{-i\omega u}\left[\frac{1}{\sqrt{2\pi}}\int_{-\infty}^{\infty} e^{i\omega t} f(t)\,dt\right] d\omega, \\ &= \frac{1}{2\pi}\int_{-\infty}^{\infty} f(t)\left[\int_{-\infty}^{\infty} e^{i(t-u)\omega}\,d\omega\right] dt = \frac{1}{2\pi}\int_{-\infty}^{\infty} f(t)\Big[2\pi\delta(t-u)\Big]\,dt, \\ &= f(u)\,. \end{aligned} \tag{13.9}$$

The interchange of the order in which the integrations in Eq. (13.9) are carried out and the identification of the ω integral as a generalized function can be justified when $f(t)$ meets the conditions for application of the Fourier integral theorem.

The process leading to Eq. (13.9) is a typical use of the delta-function representation, based on Eq. (13.7), which we typically write in the somewhat imprecise form

$$\delta(x) = \frac{1}{2\pi}\int_{-\infty}^{\infty} e^{ixt}\,dt\,. \tag{13.10}$$

This equation can be used to reduce many expressions involving Fourier transforms, but it is well to remember that it has meaning only when inserted into an integral.

BASIC FOURIER TRANSFORM FORMULAS

Rewriting the above, and extending to the Fourier sine and cosine transforms that can be obtained by writing $e^{i\omega t} = \cos\omega t + i\sin\omega t$, we have the following formulas:

$$g(\omega) = \frac{1}{\sqrt{2\pi}} \int_{-\infty}^{\infty} f(t)\, e^{i\omega t}\, dt\,, \tag{13.11}$$

$$f(t) = \frac{1}{\sqrt{2\pi}} \int_{-\infty}^{\infty} g(\omega)\, e^{-i\omega t}\, d\omega\,, \tag{13.12}$$

$$g_c(\omega) = \sqrt{\frac{2}{\pi}} \int_{0}^{\infty} f_c(t)\, \cos\omega t\, dt\,, \tag{13.13}$$

$$f_c(t) = \sqrt{\frac{2}{\pi}} \int_{0}^{\infty} g_c(\omega)\, \cos\omega t\, d\omega\,, \tag{13.14}$$

$$g_s(\omega) = \sqrt{\frac{2}{\pi}} \int_{0}^{\infty} f_s(t)\, \sin\omega t\, dt\,, \tag{13.15}$$

$$f_s(t) = \sqrt{\frac{2}{\pi}} \int_{0}^{\infty} g_s(\omega)\, \sin\omega t\, d\omega\,. \tag{13.16}$$

A word of caution about Eqs. (13.11) through (13.16) may be in order. We have defined these transforms in a way that causes the transforms and their inverses to have the same factor premultiplying the integral. However, this choice of scaling, which makes the formulas as symmetric as possible, is not universal. Many authors define the transform $g(\omega)$ without a premultiplying factor, and that choice makes it necessary to define the formula for $f(t)$ (the inverse transform) with a factor $1/2\pi$ preceding the integral. It is advisable to determine the definitions in use before using tables of Fourier transforms from diverse sources.

Example 13.2.1. Fourier Transform of Pulse

As noted previously, Fourier transforms can appropriately be applied to functions that are not periodic and cannot be treated using Fourier series. Let's find the Fourier transform of the function

$$f(t) = \begin{cases} 1, & |t| < 1, \\ 0, & |t| > 1. \end{cases}$$

A compact way of writing $f(t)$ involves the **unit step function (Heaviside function)** $u(t)$, defined in Eq. (6.61). Noting that $u(t-a) = 1$ for $t > a$ and $u(t-a) = 0$ for $t < a$, we have

$$f(t) = u(t+1) - u(t-1)\,.$$

Using Eq. (13.11) and noting that $f(t)$ is nonzero only for $|t| < 1$, we have

$$g(\omega) = \frac{1}{\sqrt{2\pi}} \int_{-1}^{1} e^{i\omega t}\, dt = \frac{1}{\sqrt{2\pi}} \frac{e^{i\omega} - e^{-i\omega}}{i\omega} = \frac{1}{\sqrt{2\pi}} \frac{2\sin\omega}{\omega} = \sqrt{\frac{2}{\pi}} \frac{\sin\omega}{\omega}\,.$$

We can check that $g(\omega)$ has been obtained correctly by applying Eq. (13.12) and verifying that we properly recover $f(t)$. Doing so, we get

$$f(t) = \frac{1}{\sqrt{2\pi}} \sqrt{\frac{2}{\pi}} \int_{-\infty}^{\infty} \frac{\sin\omega}{\omega} e^{-i\omega t}\, d\omega = \frac{1}{\pi} \int_{-\infty}^{\infty} \frac{\sin\omega(\cos\omega t - i\sin\omega t)}{\omega}\, d\omega .$$

The imaginary part of the above integrand is an odd function and therefore integrates to zero; the real part is even and it can be written as two times its integral from zero to infinity. Thus, we have

$$f(t) = \frac{2}{\pi} \int_0^{\infty} \frac{\sin\omega \cos\omega t}{\omega}\, d\omega . \tag{13.17}$$

In principle we could now perform the indicated integration and thereby confirm our result for $g(\omega)$, but it is of more interest to recognize that the process leading to Eq. (13.17) has generated a formula for $f(t)$ that for some purposes may be more convenient than the original definition. This formula, which is an **integral representation** of $f(t)$, involves neither a two-range specification nor absolute value signs, and therefore may be easier to work with than the original formula. We note that the generation of integral representations is an important application of the Fourier transform.

We close this example with a few further observations. If we set $t = 0$ in the integral representation for $f(t)$, we obtain

$$1 = \frac{2}{\pi} \int_0^{\infty} \frac{\sin\omega}{\omega}\, d\omega,$$

from which we obtain the well-known result

$$\int_0^{\infty} \frac{\sin\omega}{\omega}\, d\omega = \frac{\pi}{2} . \tag{13.18}$$

The formula in Eq. (13.18) is essentially that which was used to establish the scaling of Eq. (13.8).

Then, if we set $t = 1$ and write $\sin\omega\cos\omega = \frac{1}{2}\sin 2\omega$, we can establish that the integral representation for $f(1)$ is

$$\frac{2}{\pi} \int_0^{\infty} \frac{\sin\omega \cos\omega}{\omega}\, d\omega = \frac{1}{2},$$

showing that the value reported at the discontinuity ($t = 1$) is the midpoint of the jump in $f(t)$, analogous to the corresponding result for Fourier series.

■

We continue by examining a few additional Fourier transforms.

Example 13.2.2. Fourier Transform of Gaussian

The **Gaussian function** e^{-at^2}, with $a > 0$, has Fourier transform

$$g(\omega) = \frac{1}{\sqrt{2\pi}} \int_{-\infty}^{\infty} e^{-at^2} e^{i\omega t}\, dt = \frac{1}{\sqrt{2\pi}} \int_{-\infty}^{\infty} e^{-at^2} \cos\omega t\, dt . \tag{13.19}$$

We obtained Eq. (13.19) by setting $e^{-at^2} e^{i\omega t} = e^{-at^2}(\cos\omega t + i\sin\omega t)$ and dropping the sine term from the integrand because it is odd and its integral vanishes.

This integral is easily evaluated using complex integration techniques, or, with more effort, by expanding the cosine in power series and manipulating the resultant expansion. The result is

$$g(\omega) = \frac{1}{\sqrt{2a}}\, e^{-\omega^2/4a}\,. \tag{13.20}$$

The demonstration of Eq. (13.20) is the topic of Exercise 13.2.6.

It is of interest to note that the Fourier transform of a Gaussian is also a Gaussian, but of a different width (i.e., with a different constant in the exponential). We may also observe that because the width parameter appears in the transform in a position reciprocal to its position in the original function, a narrower (more localized, larger a) Gaussian will have a wider (less localized, smaller $1/4a$) transform. This qualitative behavior is typical for Fourier transforms.

■

Example 13.2.3. Fourier Transform of Delta Function

To find the Fourier transform of the Dirac delta function is nearly trivial:

$$g(\omega) = \frac{1}{\sqrt{2\pi}} \int_{-\infty}^{\infty} \delta(t)\, e^{i\omega t}\, dt = \frac{1}{\sqrt{2\pi}}\,. \tag{13.21}$$

This example shows that a completely localized function, $\delta(t)$, has a completely delocalized (constant) Fourier transform.

The delta function does not satisfy the Dirichlet conditions, but a formal application of the Fourier inversion formula yields Eq. (13.10) when applied to this $g(\omega)$. However, again we point out that expressions involving delta functions have precise meanings only when the delta function is to be used within a convergent integral.

■

Example 13.2.4. Shifted Variable

Using the superscript T to denote the Fourier transform, and continuing with the convention that g is the Fourier transform of f, we have the following useful formula:

$$\Big[f(t-a)\Big]^T(\omega) = e^{ia\omega} g(\omega)\,. \tag{13.22}$$

Equation (13.22) makes it easy to write the transform of a function which is expressed relative to a displaced origin.

The proof of Eq. (13.22) is straightforward. We write the transform formula for $f(t-a)$ and change the variable of integration from t to $u = t - a$:

$$\frac{1}{\sqrt{2\pi}} \int_{-\infty}^{\infty} f(t-a)\, e^{i\omega t}\, dt = \frac{1}{\sqrt{2\pi}} \int_{-\infty}^{\infty} f(u)\, e^{i\omega(u+a)}\, du,$$

$$= e^{i\omega a} \left[\frac{1}{\sqrt{2\pi}} \int_{-\infty}^{\infty} f(u)\, e^{i\omega u}\, du \right] = e^{i\omega a} g(\omega)\,.$$

Illustrating Eq. (13.22) by applying it to a Gaussian, we have

$$\Big[\,|\, e^{-k(t-a)^2}\Big]^T = e^{i\omega a}\Big[e^{-kt^2}\Big]^T = \frac{1}{\sqrt{2k}}\, e^{i\omega a}\, e^{-\omega^2/4k}\,.$$

■

Exercises

13.2.1. Show in detail how the formulas of Eq. (13.6) follow from Eq. (13.5), including specifically the factors $1/\sqrt{2\pi}$ that premultiply the integrals.

13.2.2. Show that the value of $\delta_n(0)$, obtained as the limit $x \to 0$ of Eq. (13.8), is n/π.

13.2.3. Explain in detail how to obtain the Fourier sine and cosine transforms from the corresponding formulas for the (exponential) Fourier transform.

13.2.4. Using Eq. (13.18), show that $\dfrac{2}{\pi}\displaystyle\int_0^\infty \frac{\sin\omega\cos\omega}{\omega}\,d\omega = \frac{1}{2}$.

13.2.5. Evaluate the Fourier transform of a pulse of unit height (like that of Example 13.2.1 but extending from $t = 0$ to $t = 2$)

(a) By explicit computation of the transform integral, and

(b) By applying the shift formula, Eq. (13.22), to the transform given in the Example.

Verify that both methods give the same result.

13.2.6. Establish Eq. (13.20), the formula for the Fourier transform of a Gaussian, by carrying out the following process:

1. Rewrite Eq. (13.19), changing the integration variable to $u = at^2$ and the range to $(0, \infty)$.
2. Expand $\cos(\omega\sqrt{u/a})$ in power series and integrate term-by-term, identifying the integrals as gamma functions of half-integer arguments,
3. Apply the Legendre duplication formula, Eq. (9.12), to simplify the combinations of gamma functions,
4. Identify the resulting summation as $g(\omega)$.

13.3 FOURIER TRANSFORM: SYMBOLIC COMPUTATION

Both MAPLE and MATHEMATICA support the computation of Fourier transforms. MAPLE defines the transform (and therefore also its inverse) differently than this text:

$$g_{\text{maple}}(\omega) = \int_{-\infty}^{\infty} f(t)\,e^{-i\omega t}\,dt\,, \qquad f(t) = \frac{1}{2\pi}\int_{-\infty}^{\infty} g_{\text{maple}}(\omega)\,e^{i\omega t}\,d\omega\,. \qquad (13.23)$$

Note that the signs of the imaginary exponents are opposite from our definitions and that MAPLE and the text differ in the placement of the quantity 2π. Curiously, MAPLE's definitions of the Fourier sine and cosine transforms do agree with the definitions in this text.

To access Fourier transforms in MAPLE, one must first activate MAPLE's integral transform package by invoking the command `with(inttrans)`. Then the Fourier transform of an expression *expr*, where *expr* is a function of t and the transform is a function of w, is given by

`fourier(`*expr*`,t,w)`.

To obtain Fourier sine or cosine transforms, replace `fourier` by `fouriersin` or `fouriercos`.

In MAPLE,

`invfourier(`*expr*`,w,t)`

yields the inverse Fourier transform. Note that the positions of the arguments `w` and `t` are in reverse order as compared to their positions in `fourier`. The MAPLE sine and cosine transforms are defined in a way that makes them self-inverse, so no additional commands for their inversion are needed.

MATHEMATICA uses (as defaults) the same definitions as this text for Fourier, Fourier sine, and Fourier cosine transforms (and therefore also their inverses). Advanced users can change these definitions in various ways; we do not discuss the procedure for doing that.

The MATHEMATICA command for the transform of *expr*, where *expr* is a function of t and the transform a function of w, is

`FourierTransform[`*expr*`,t,w]`.

For sine or cosine transforms, use `FourierSinTransform` or `FourierCosTransform`. The inverse transforms are reached using

`InverseFourierTransform[`*expr*`,w,t]`,

`InverseFourierSinTransform`, or `InverseFourierCosTransform`. These latter two inverses are not really needed (the default sine and cosine transforms are self-inverse), but MATHEMATICA provides them because nondefault transform definitions may cause loss of the self-inverse property.

There is a frequent need for transform equations involving unit step functions or the Dirac delta function. Both MAPLE and MATHEMATICA can handle these generalized functions in Fourier-transform contexts.

MAPLE: `Heaviside(x)` and `Dirac(x)`.

MATHEMATICA: `HeavisideTheta[x]` and `DiracDelta[x]`.

Here `Heaviside` and `HeavisideTheta` refer to the unit step function $u(x)$, first encountered in this text at Eq. (6.61). This function is zero if $x < 0$ and unity if $x > 0$. The functions `Dirac` and `DiracDelta` are $\delta(x)$.

Example 13.3.1. Fourier Transforms Using Maple

Let's try to use MAPLE to obtain the transforms from the Examples in Section 13.2.

```
> with(inttrans):
> F := Heaviside(t+1) - Heaviside(t-1):
```

The pulse of Example 13.2.1.

```
> G := fourier(F,t,w);
```

$$G := \frac{2 \sin w}{w}$$

We must multiply this result by $1/\sqrt{2\pi}$ before comparing with the example; it checks. We can also check by inverting `G`:

```
> invfourier(G,w,t);
```

$$\text{Heaviside}(t+1) - \text{Heaviside}(t-1)$$

Note that the unit step function enables the pulse to be represented without having to specify different functional forms for different ranges of t. The commands `fouriersin` and `fouriercos` are unable to process `F`.

Moving on to the transform of the delta function, we find

```
> fourier(Dirac(t),t,w);
```
$$1$$

This checks after multiplication by $1/\sqrt{2\pi}$.

A straightforward attempt to evaluate the Fourier transform of e^{-at^2} fails. The problem is that the transform only exists if the real part of a is positive. We can resolve this impasse by specifying that a is to be assumed positive; the MAPLE command to do so is `assume(a>0)`. Thus, continuing our MAPLE session,

```
> fourier(exp(-a*t^2),t,w);
```
$$fourier\left(e^{-at^2}, t, w\right)$$

```
> assume(a>0);
```
(a is now written $a\sim$)

```
> fourier(exp(-a*t^2),t,w);
```
$$e^{-w^2/4a\sim}\sqrt{\frac{\pi}{a\sim}}$$

Here is one more example:

```
> fourier(1/(1+t^2),t,w);
```
$$\pi\left(e^{w}\,\text{Heaviside}(-w) + e^{-w}\,\text{Heaviside}(w)\right)$$

This result can be written in the more transparent form $\pi\, e^{-|w|}$. We should get an equivalent result if we take the Fourier cosine transform of this even function:

```
> GC := fouriercos(1/(1+t^2),t,w);
```
$$GC := \frac{1}{2}\sqrt{2}\sqrt{\pi}\, e^{-w}$$

After allowing for the difference in scaling of MAPLE's exponential and cosine transforms, this result checks. Because the cosine transform is only defined for $w \geq 0$, there is no need for MAPLE to generate a result that is valid for both signs of w, and it simplified the answer by assuming $w \geq 0$. We can also check by inverting `GC`:

```
> fouriercos(GC,w,t);
```
$$\frac{1}{1+t^2}$$

■

Example 13.3.2. Fourier Transforms Using Mathematica

Applying MATHEMATICA to the Examples in Section 13.2,

```
F = HeavisideTheta[t+1] - HeavisideTheta[t-1]
```
The pulse of Example 13.2.1.

```
G = FourierTransform[F,t,w]
```
$$\frac{\sqrt{\frac{2}{\pi}}\,\text{Sin}[w]}{w}$$

This should agree with the earlier example. It checks.

We look next at

```
GC = FourierCosTransform[F,t,w]
```
$$\frac{\sqrt{\frac{2}{\pi}}\,\text{Sin}[w]}{w}$$

```
GS = FourierSinTransform[F,t,w]
```
$$\frac{2\sqrt{\frac{2}{\pi}}\,\text{Sin}\left[\frac{w}{2}\right]^2}{w}$$

We note that GC is equal to G (as it should be, because F is an even function), while GS is, as expected, different. We can further check these results by inverting the transforms:

```
InverseFourierTransform[G,w,t]
```
$$\frac{1}{2}\,(\mathrm{Sign}[1-t]+\mathrm{Sign}[1+t])$$

```
InverseFourierCosTransform[GC,w,t]
```
$$\frac{1}{2}\,(1+\mathrm{Sign}[1-t])$$

```
InverseFourierSinTransform[GS,w,t]
```
$$\frac{1}{2}\,(1+\mathrm{Sign}[1-t])$$

The first of these results is equivalent to F; the second and third correspond to F only for positive values of t. Note also that instead of **HeavisideTheta**, MATHEMATICA used **Sign[x]**, which has the value +1 if $x > 0$ and -1 if $x < 0$.

Further examples:

```
FourierTransform[DiracDelta[t],t,w]
```
$$\frac{1}{\sqrt{2\pi}}$$

```
FourierTransform[E^(-a*t^2),t,w]
```
$$\frac{e^{-w^2/4a}}{\sqrt{2}\,\sqrt{a}}$$

We see that MATHEMATICA takes the Fourier transform of a Gaussian making the unstated assumption that $a > 0$.

Finally, look at

```
FourierTransform[1/(1+t^2),t,w]
```
$$e^{\mathrm{Abs}[w]}\sqrt{\frac{\pi}{2}}$$

■

Exercises

13.3.1. Using symbolic computation if helpful, find the Fourier transforms of the following:

(a) $e^{-|t|}$, (c) $e^{-|t|}\sin t$, (e) $e^{-t^2}\cos t$,

(b) te^{-t^2}, (d) $u(t)\,e^{-t}$, (f) $e^{-t^2}\sin t$.

Here $u(t)$ is the Heaviside (unit step) function.

13.3.2. Repeat Exercise 13.3.1, taking Fourier cosine transforms.

13.3.3. Repeat Exercise 13.3.1, taking Fourier sine transforms.

13.3.4. Find the inverse Fourier transforms of

(a) $\dfrac{1}{1-i\omega}$, (c) $\omega^2\,e^{-\omega^2}$,

(b) $\dfrac{1}{1+\omega^2}$, (d) $u(\omega)\,e^{-\omega}\cos\omega$.

Here $u(t)$ is the Heaviside (unit step) function.

13.3.5. Find the Fourier sine and Fourier cosine transforms of $\mathrm{erfc}(t)$ and verify that application of the respective inverse transforms recovers $\mathrm{erfc}(t)$.

13.4 FOURIER TRANSFORM: SOLVING ODEs

Fourier transforms apply to derivatives in a way that is useful for solving differential equations. In particular, we note the following, where the superscript T denotes the Fourier transform, and g is the Fourier transform of f:

$$\left[\frac{df(t)}{dt}\right]^T(\omega) = -i\omega g(\omega)\,, \tag{13.24}$$

$$\left[\frac{d^n f(t)}{dt^n}\right]^T(\omega) = (-i\omega)^n g(\omega)\,. \tag{13.25}$$

The proof of Eq. (13.24) is fairly direct. Starting from the formula defining the transform of the derivative, we integrate by parts, dropping the integrated terms because they must vanish if the transform is defined:

$$\begin{aligned}\frac{1}{\sqrt{2\pi}}\int_{-\infty}^{\infty}\frac{df(t)}{dt}\,e^{i\omega t}\,dt &= -\frac{1}{\sqrt{2\pi}}\int_{-\infty}^{\infty} f(t)\frac{d}{dt}\left[e^{i\omega t}\right]dt,\\ &= -\frac{1}{\sqrt{2\pi}}\int_{-\infty}^{\infty} f(t)(i\omega)\,e^{i\omega t}\,dt = -i\omega\, g(\omega)\,.\end{aligned}$$

Equation (13.25) now follows if Eq. (13.24) is applied n times.

Let's see how Eqs. (13.24) and (13.25) can help us to solve ODEs.

Example 13.4.1. Driven Classical Oscillator

Newton's Second Law for a particle of mass m subject to a force $-kx$ and free to move only in the x direction leads to the equation of motion

$$F = m\,a = m\,\frac{d^2x}{dt^2} = -k\,x\,.$$

This equation has the general solution $x(t) = A\sin\alpha t + B\cos\alpha t$, with $\alpha = \sqrt{k/m}$. If in addition there is an external force $f(t)$ applied to the particle, the system is referred to as **driven**, and the equation of motion becomes

$$m\,\frac{d^2x}{dt^2} = -k\,x + f(t)\,, \qquad \text{or} \qquad \frac{d^2x}{dt^2} + \alpha^2 x = \frac{f(t)}{m}\,. \tag{13.26}$$

Let's use Fourier transforms to solve this driven-oscillator ODE. In this and later examples involving several transforms, we adopt the practice of using uppercase letters to denote the transforms of quantities that are assigned corresponding lowercase letters. Thus, let $X(\omega)$ denote the (as yet unknown) transform of $x(t)$, define $F(\omega)$ to be the transform of $f(t)$, and rewrite Eq. (13.26) by transforming all its terms. Using Eq. (13.25) to handle d^2x/dt^2, we have

$$-\,\omega^2 X(\omega) + \alpha^2 X(\omega) = \frac{1}{m}\,F(\omega)\,. \tag{13.27}$$

Solving for $X(\omega)$, we find

$$X(\omega) = \frac{1}{m}\,\frac{F(\omega)}{\alpha^2 - \omega^2}\,. \tag{13.28}$$

Since $f(t)$ is known, its Fourier transform is in principle available, so our formula for $X(\omega)$ is completely specified, and we may take the inverse transform of X to obtain $x(t)$. A significant feature of the present analysis is that the original ODE has, in the image (transform) space, become an algebraic equation. The price for this simplification is that to complete the analysis we must take an inverse transform.

We illustrate the solution method of this example by taking $f(t) = m\cos\beta t$. We need $F(\omega)$, which is most easily obtained by writing $\cos\beta t$ in terms of complex exponentials. We have

$$F(\omega) = \frac{m}{\sqrt{2\pi}} \int_{-\infty}^{\infty} \frac{e^{i\beta t} + e^{-i\beta t}}{2}\, e^{i\omega t}\, dt = m\sqrt{\frac{\pi}{2}}\, \frac{1}{2\pi} \int_{-\infty}^{\infty} \left(e^{i(\omega+\beta)t} + e^{i(\omega-\beta)t} \right) dt,$$

$$= m\sqrt{\frac{\pi}{2}} \left(\delta(\omega+\beta) + \delta(\omega-\beta) \right). \tag{13.29}$$

Inserting this result into Eq. (13.28),

$$X(\omega) = \sqrt{\frac{\pi}{2}} \left[\frac{\delta(\omega+\beta) + \delta(\omega-\beta)}{\alpha^2 - \omega^2} \right]. \tag{13.30}$$

Taking now the inverse transform,

$$x(t) = \frac{1}{2} \int_{-\infty}^{\infty} \frac{\delta(\omega+\beta) + \delta(\omega-\beta)}{\alpha^2 - \omega^2}\, e^{-i\omega t}\, d\omega = \frac{1}{2}\, \frac{e^{i\beta t} + e^{-i\beta t}}{\alpha^2 - \beta^2}.$$

$$= \frac{\cos\beta t}{\alpha^2 - \beta^2}. \tag{13.31}$$

Note that although we have solved a second-order ODE we have obtained a unique answer despite the fact that the ODE has a general solution that has two independent constants. The uniqueness arises from the boundary conditions implied by the transform/inverse-transform process, and what we actually have here is a procedure for obtaining a single particular integral to our inhomogeneous ODE. Use of the Fourier transform can be viewed as an alternative to the methods introduced in Chapter 10 for solving inhomogeneous ODEs.

Commenting further, note that if $\beta = \alpha$ we do not have a solution to our driven ODE. In that case the driving force is at the natural frequency of the oscillator and the behavior, termed **resonance**, is that the oscillation amplitude increases over time (without limit) and is not suitable for a description using Fourier transforms.

■

SYMBOLIC COMPUTATION

Both MAPLE and MATHEMATICA know how to use Eqs. (13.24) and (13.25). In MAPLE,

```
> with(inttrans):
> fourier(diff(f(x),x),x,w);                    I w fourier(f(x),x,w)
> fourier(diff(1/(1-x^2),x),x,w);          Iπw sin(w)(−1 + 2Heaviside(w))
```

Because $-1 + 2\,$`Heaviside(w)` $=$ `sign(w)`, the output of the last line above is equivalent to $i\pi|w|\sin w$. Note also that the transform definition of MAPLE causes its derivative formulas to differ from Eqs. (13.22) and (13.24) by a sign change ($-i$ becomes $+i$). If one plans to use MAPLE and (by hand) inserts a factor $-i\omega$ to form the transform of a derivative, processing thereafter by MAPLE will be inconsistent, and may result in errors. See Example 13.4.2 for an illustration of this issue.

In MATHEMATICA:

```
FourierTransform[D[f[x],x],x,w]
```
$-Iw\,\texttt{FourierTransform}[f[x],x,w]$

```
FourierTransform[D[1/(1-x^2),x],x,w]
```
$-\frac{1}{2}e^{-iw}\left(-1+e^{2iw}\right)\sqrt{\frac{\pi}{2}}\,w\,\texttt{Sign}[w]$

```
ExpToTrig[Expand[%]]
```
$-i\sqrt{\frac{\pi}{2}}\,w\,\texttt{Sign}[w]\,\texttt{Sin}[w]$

The MATHEMATICA and MAPLE results for the last computation become consistent when we multiply the MAPLE transform by $1/\sqrt{2\pi}$ and change the sign of i. Note however that we had to use the simplification command **ExpToTrig** to force MATHEMATICA to rewrite the complex exponentials as a trigonometric function.

Example 13.4.2. Differential Equation

Here we solve an inhomogeneous ODE for which symbolic methods may be viewed as simpler than hand computation. Consider

$$-y''(x) + y'(x) + 2y(x) = h(x)\,, \qquad h(x) = \begin{cases} 6\,e^{x}, & x < 0, \\ 0, & x > 0. \end{cases}$$

Working first in MAPLE, letting $Y(\omega)$ be the Fourier transform of $y(x)$, and $H(\omega)$ the transform of $h(x)$, the ODE transforms to

$$\omega^2 Y + i\omega Y + 2Y = (\omega^2 + i\omega + 2)Y = H(\omega)\,. \qquad \text{(MAPLE only)} \qquad (13.32)$$

Note that the transform of y' must be written for the MAPLE transform definition, as $+i\omega Y$. Our symbolic computing session is therefore

```
> with(inttrans):
> h := Heaviside(-x)*6*exp(x):
> H := fourier(h,x,w);
```
$$\frac{6}{1 - Iw}$$

```
> Y := H/(w^2+I*w+2);
```
$$Y := \frac{6}{(1 - Iw)(w^2 + iw + 2)}$$

```
> y := invfourier(Y,w,x);
```
$$y := e^{-x}\text{Heaviside}(x) + \left(3e^{x} - 2e^{2x}\right)\text{Heaviside}(-x)$$

If we wanted to avoid having to consider how MAPLE treats the transforms of derivatives, we could have preceded the statement assigning Y by

```
> fourier(-diff(y(x),x,x) + diff(y(x),x) + 2*y(x), x, w);
```
$$(w^2 + Iw + 2)\,\mathit{fourier}(y(x), x, w).$$

This output provides the information given in Eq. (13.32).

In MATHEMATICA (and also in the notation of the present text), the transformed ODE is

$$\omega^2 Y - i\omega Y + 2Y = (\omega^2 - i\omega + 2)Y = H(\omega)\,,$$

and our symbolic computing session is

```
h = HeavisideTheta[-x]*6*E^x;
```

```
H = FourierTransform[h,x,w]
```

$$-\frac{3i\sqrt{\frac{2}{\pi}}}{-i+w}$$

```
Y = H/(w^2-I*w+2)
```

$$-\frac{3i\sqrt{\frac{2}{\pi}}}{(-i+w)(2-iw+w^2)}$$

```
y = InverseFourierTransform[Y,w,x]
```

$$-e^{-x}\left(e^{2x}(-3+2e^{x})\text{HeavisideTheta}[-x]-\text{HeavisideTheta}[x]\right)$$

```
Collect[Expand[%],HeavisideTheta[-x]]
```

$$(3e^{x}-2e^{2x})\text{HeavisideTheta}[-x]+e^{-x}\text{Heaviside}[x]$$

Note that despite the difference in transform definition, both computing systems produce the same final result.

■

The utility of the present procedure for solving inhomogeneous ODEs is limited to situations in which the coefficients of the unknown function and its derivatives are constants and the inhomogeneous term [$h(x)$ in the above example] has a Fourier transform. The procedure is clearly applicable when $h(x)$ is finite and of limited extent in x (and usually exhibiting discontinuities in $h(x)$ or in its slope). Such functions do possess Fourier transforms, and it is then that the procedure of this section is most valuable. However, often an inhomogeneous term without a Fourier transform will possess a Laplace transform, and a method for solving ODEs using Laplace transforms (described later in this chapter) is applicable to a wider class of $h(x)$ than the Fourier-transform method described here.

Exercises

13.4.1. Consider the following ODE in the independent variable t

$$y'' + 2y' + y = f(t)\,, \qquad \text{with} \quad f(t) = u(t+1) - u(t-1)\,.$$

Note that $f(t)$ is the pulse whose Fourier transform was obtained in Example 13.2.1.

(a) Obtain a particular solution to this inhomogeneous ODE by the Fourier-transform method described in the present section of this text.

(b) Characterize and check your solution

(i) By plotting it on the range $-2 < t < 10$ and

(ii) By writing symbolic code to compute the left-hand side of the ODE and plotting that result for $-1 < t < 2$.

13.4.2. Use Fourier-transform methods to obtain a particular solution to the inhomogeneous ODE

$$y''(x) - 2y'(x) + y(x) = e^{-|x|}\,.$$

Plot and check your solution as in Exercise 13.4.1.

13.4.3. (a) Use the unit step (Heaviside) function to describe a triangular pulse

$$f(x) = \begin{cases} 1-x, & 0 < x < 1, \\ 1+x, & -1 < x < 0, \\ 0, & |x| > 1, \end{cases}$$

and use your symbolic system to find its Fourier transform.

(b) Use Fourier-transform methods to obtain a particular solution to

$$y''(x) - 2y'(x) + y(x) = f(x)\,.$$

Plot and check your solution as in Exercise 13.4.1.

13.5 FOURIER CONVOLUTION THEOREM

It is useful to define a quantity known as the **convolution** of two functions $f_1(t)$ and $f_2(t)$, with notation $f_1 * f_2$, as the integral

$$(f_1 * f_2)(t) = \frac{1}{\sqrt{2\pi}} \int_{-\infty}^{\infty} f_1(u) f_2(t-u)\,du\,. \tag{13.33}$$

Note that the asterisk, which is not a superscript but is written in-line, is the symbol denoting the convolution operation. It must not be confused with the complex-conjugation symbol, nor with the use of the asterisk in computer programs to indicate multiplication. Sometimes a convolution is referred to as a **Faltung**, a German word that means "folding." By making the change of variable $v = t - u$, it is easily verified that

$$f_1 * f_2 = f_2 * f_1\,. \tag{13.34}$$

CONVOLUTION THEOREM

The importance of the convolution arises from the fact that it satisfies the **convolution theorem**. If $F_1(\omega)$ and $F_2(\omega)$ are respectively the Fourier transforms of $f_1(t)$ and $f_2(t)$, the convolution theorem states that

$$(f_1 * f_2)^T(\omega) = F_1(\omega) F_2(\omega)\,, \tag{13.35}$$

where $(f_1 * f_2)^T$ is the Fourier transform of $f_1 * f_2$. The proof of Eq. (13.35) is a nice exercise in the use of the delta-function formula, Eq. (13.10), and the formula for the inverse transform, Eq. (13.12). We start by writing

$$f_1(u) = \frac{1}{\sqrt{2\pi}} \int_{-\infty}^{\infty} F_1(\omega_1) e^{-i\omega_1 u}\,d\omega_1 \text{ and } f_2(t-u) = \frac{1}{\sqrt{2\pi}} \int_{-\infty}^{\infty} F_2(\omega_2) e^{-i\omega_2(t-u)}\,d\omega_2\,.$$

We next insert these expressions into the integral (in both u and t) that represents the left-hand side of Eq. (13.35):

$$\begin{aligned}(f_1 * f_2)^T(\omega) &= \frac{1}{\sqrt{2\pi}} \int_{-\infty}^{\infty} (f_1 * f_2)(t)\, e^{i\omega t}\,dt = \frac{1}{2\pi} \int_{-\infty}^{\infty} dt \int_{-\infty}^{\infty} du f_1(u) f_2(t-u)\, e^{i\omega t} \\ &= \frac{1}{(2\pi)^2} \int_{-\infty}^{\infty} dt \int_{-\infty}^{\infty} du \int_{-\infty}^{\infty} d\omega_1 \int_{-\infty}^{\infty} d\omega_2 F_1(\omega_1) F_2(\omega_2)\, e^{-i\omega_1 u - i\omega_2(t-u) + i\omega t}.\end{aligned} \tag{13.36}$$

Note now that the complex exponential has the form $e^{i(\omega_2-\omega_1)u}e^{i(\omega-\omega_2)t}$, so the u and t integrations, including the factor $1/(2\pi)^2$, evaluate to the delta-function product $\delta(\omega_2-\omega_1)\,\delta(\omega-\omega_2)$. Equation (13.36) therefore reduces to

$$(f_1 * f_2)^T(\omega) = \int_{-\infty}^{\infty} d\omega_1 \int_{-\infty}^{\infty} d\omega_2 F_1(\omega_1)F_2(\omega_2)\delta(\omega_2-\omega_1)\,\delta(\omega-\omega_2) = F_1(\omega)F_2(\omega)\,,$$

as claimed.

SIGNAL PROCESSING

Fourier transform methods are often used for problems in which the variable t represents time, and the inverse transform formula, Eq. (13.12), can be identified as an integral in which contributions $g(\omega)$ at all angular frequencies ω are summed to describe a function $f(t)$. Because our analysis is not restricted to periodic functions we need to describe $f(t)$ by an integral over ω rather than a sum. Incidentally, the negative ω provide a mechanism for dealing with the fact that the signals constituting $f(t)$ can occur at arbitrary relative phases.

One can think of the Fourier transform as providing a connection between the description of a signal as a function of time ("in the time domain") and its description in terms of the frequencies into which it can be decomposed (the "frequency domain"). As long as we stick to problems in which the behavior is linear, it is useful to characterize a physical system in terms of either its behavior at different times or its behavior at different frequencies. Often the frequency decomposition is more useful.

In many electrical and mechanical systems involving oscillations, an input signal $f(t)$ produces an output $y(t)$ that is related to the input by a differential equation with constant coefficients, of the generic form

$$Ay'' + By' + Cy = f(t)\,. \tag{13.37}$$

Following the procedure illustrated in Example 13.4.1 and again in Example 13.4.2, we can use Fourier transform methods to relate the transform $Y(\omega)$ of $y(t)$ to the ODE and the transform $F(\omega)$ of the **forcing function** $f(t)$. In the general case presently under discussion, we have

$$Y(\omega) = \frac{1}{-\omega^2 A - i\omega B + C}\,F(\omega) = \Phi(\omega)F(\omega)\,, \quad \text{with} \quad \Phi = \frac{1}{-\omega^2 A - i\omega B + C}\,. \tag{13.38}$$

The function $\Phi(\omega)$ is called a **transfer function**.

Equation (13.38) shows that (in the frequency domain) each frequency component of an incoming signal, i.e., each value of $F(\omega)$, is processed independently by being combined with the transfer function, which depends upon the ODE but not upon $f(t)$. Then, by taking the inverse transform, we translate the behavior into the time domain.

The convolution theorem is relevant to our present analysis. That statement may become more apparent if we observe that Φ is the Fourier transform of some quantity $\varphi(t)$. Then the convolution theorem lets us write

$$Y = (\varphi * f)^T\,, \quad \text{equivalent to} \quad y = \varphi * f \quad \text{or} \quad y(t) = \frac{1}{\sqrt{2\pi}}\int_{-\infty}^{\infty} \varphi(u)f(t-u)\,du\,. \tag{13.39}$$

Equation (13.39) is useful because φ is easy to obtain (i.e., Φ is easily inverted), and we never need to take the transform of f. These observations leave us with two ways of dealing with transfer-function problems: (1) invert ΦF directly (when it is easy to do so), or (2) use Eq. (13.39), in which the integral for $y(t)$ can be evaluated numerically when an analytical evaluation is difficult or impossible.

PARSEVAL RELATION

As a final topic of this section, consider the result of inverting Eq. (13.35) for the special case $t = 0$ with $f_1(u) = f(u)$ and $f_2(-u) = f^*(u)$. Then

$$(f_1 * f_2)(t=0) = \frac{1}{\sqrt{2\pi}} \int_{-\infty}^{\infty} f(u) f^*(u)\, du = \frac{1}{\sqrt{2\pi}} \int_{-\infty}^{\infty} F_1(\omega) F_2(\omega)\, d\omega \,. \tag{13.40}$$

Letting F be the Fourier transform of f, we have $F_1 = F$, but need to develop a formula for F_2 to proceed further. We write

$$F_2(\omega) = \frac{1}{\sqrt{2\pi}} \int_{-\infty}^{\infty} f_2(u)\, e^{i\omega u}\, du = \frac{1}{\sqrt{2\pi}} \int_{-\infty}^{\infty} f^*(-u)\, e^{i\omega u}\, du,$$

$$= \left[\frac{1}{\sqrt{2\pi}} \int_{-\infty}^{\infty} f(-u)\, e^{-i\omega u}\, du \right]^* = \left[\frac{1}{\sqrt{2\pi}} \int_{-\infty}^{\infty} f(t)\, e^{i\omega t}\, dt \right]^*,$$

where the last expression was obtained by changing the integration variable to $t = -u$. We now recognize $F_2(\omega) = F^*(\omega)$, so we can return to Eq. (13.40), where we cancel the factors $1/\sqrt{2\pi}$ and obtain

$$\int_{-\infty}^{\infty} f(u) f^*(u)\, du = \int_{-\infty}^{\infty} F(\omega) F^*(\omega)\, d\omega \,. \tag{13.41}$$

Equation (13.41) is called the **Parseval relation**. Its interpretation is that a function and its Fourier transform have the same norm, a result that is not surprising when one notes the similarity between the direct and inverse transform formulas.

Example 13.5.1. Integral Evaluation

The Parseval relation provides an opportunity for integral evaluation. As an example, let's evaluate

$$I = \int_{-\infty}^{\infty} \frac{\sin^2 x}{x^2}\, dx \,.$$

Noting that $\sin x/x$ is proportional to the Fourier transform of the unit pulse discussed in Example 13.2.1, we make the following identifications:

$$f(t) = u(t+1) - u(t-1), \qquad F(\omega) = \sqrt{\frac{2}{\pi}}\, \frac{\sin\omega}{\omega},$$

$$I = \frac{\pi}{2} \int_{-\infty}^{\infty} F^*(\omega) F(\omega)\, d\omega = \frac{\pi}{2} \int_{-\infty}^{\infty} f^*(t) f(t)\, dt \,.$$

The integral involving $f^*(t)f(t)$ is easy to evaluate because this product is equal to unity between $t = -1$ and $t = 1$, and zero elsewhere. We therefore have

$$\int_{-\infty}^{\infty} f^*(t) f(t)\, dt = 2, \quad \text{and} \quad I = \frac{\pi}{2}(2) = \pi \,.$$

■

Exercises

13.5.1. Given $f(t) = (t^2+1)^{-1}$ and $g(t) = \dfrac{t}{t^2+1}$, evaluate the following convolutions:

$$\text{(a)}\ f * f\,, \qquad \text{(b)}\ g * g\,, \qquad \text{(c)}\ f * g\,.$$

13.5.2. You are given the following Fourier transforms:

$$f(t) = \frac{1}{t^2+1}, \qquad f^T(s) = F(s) = \sqrt{\frac{\pi}{2}}\; e^{-|s|},$$

$$g(t) = \frac{t}{t^2+1}, \qquad g^T(s) = G(s) = i\sqrt{\frac{\pi}{2}}\; \text{sign}(s)\, e^{-|s|}\,,$$

where $\text{sign}(u) = 1$ if $u > 0$ and -1 if $u < 0$.

Form the convolution $f * g$ and show that its transform is the product $F(s)G(s)$, as required by the convolution theorem.

Note. The answers of Exercise 13.5.1 may be helpful.

13.5.3. Using symbolic computation, evaluate the integral

$$I = \int_{-\infty}^{\infty} \left[\frac{x\cos x - \sin x}{x^2}\right]^2 dx$$

in the following two ways:

(a) By a direct call to a symbolic integration procedure, and

(b) By using the Parseval relation.

13.6 LAPLACE TRANSFORM

We turn now to the **Laplace transform**, for which we find it convenient to use the notation

$$\mathcal{L}[f(t)] = F(s) = \int_0^{\infty} e^{-st} f(t)\, dt\,. \tag{13.42}$$

We are adopting the convention that a direct-space function is represented by a lowercase letter, with its Laplace transform denoted by the same letter, in uppercase. This notation is parallel to that we have been using for the Fourier transform. Like the Fourier transform, the Laplace transform operator $\mathcal{L}$ is linear, with

$$\mathcal{L}[f(t)+g(t)] = \mathcal{L}[f(t)] + \mathcal{L}[g(t)] \qquad \text{and} \qquad \mathcal{L}[kf(t)] = k\mathcal{L}[f(t)]\,. \tag{13.43}$$

The Laplace transform exists when the integral in Eq. (13.42) converges. If there exist constants t_0, s_0, and M such that for all $t > t_0$, then convergence at large t is assured if $s > s_0$, and

$$\left|e^{-s_0 t} f(t)\right| < M\,.$$

Note that a transform may exist even if $f(t)$ does not vanish at large t; in fact, a function such as $f(t) = Ce^{s_0 t}$ has a Laplace transform.

For a transform to exist it is also necessary that Eq. (13.42) converge at small t. In particular, if the small-t behavior of $f(t)$ is as t^n, we can have a transform only if

$n > -1$. Finally, note that only values of $f(t)$ with $t \geq 0$ enter the transform formula. We may therefore without error set $f(t) = 0$ for all $t < 0$.

Summarizing, we observe that these limitations are less severe than those applicable to the Fourier transform, and it is found that they are not a barrier to the effective use of Laplace transforms in a variety of problems.

It turns out that the inversion of Laplace transforms is not as easy as that of Fourier transforms, and a general transform inversion formula requires a fairly sophisticated complex integration procedure (to be deferred to Chapter 17). For that reason it has in the past proved useful to have tables giving a large number of transform pairs, and to carry out transform inversions by table lookup. Such tables can be supplemented by techniques that enable the construction of entries that extend the tables, and prior to the widespread deployment of digital computers many physicists and engineers became highly skilled in formal procedures for Laplace transform evaluation. Symbolic computing has now made it much easier to find inverse Laplace transforms, and users need not rely as much as before on transform tables.

A short list of Laplace transform pairs is given here in Table 13.1. More extensive lists are in AMS-55 and the two-volume work of Erdélyi et al. (see Additional Readings). Symbolic computing of Laplace transforms and their inverses is discussed later in this section of the text.

LAPLACE TRANSFORM EVALUATION

Here we review some elementary methods for identifying transforms and their inverses.

Example 13.6.1. Direct Evaluation

Sometimes the integral defining a Laplace transform is easily integrated. For example, the first two entries in Table 13.1 are

$$f(t) = 1; \qquad F(s) = \int_0^\infty 1 \cdot e^{-st}\, dt = \frac{1}{s}\,, \tag{13.44}$$

$$\begin{aligned} f(t) = t^n; \qquad F(s) &= \int_0^\infty t^n\, e^{-st}\, dt = \frac{1}{s^{n+1}} \int_0^\infty (st)^n\, e^{-st}\, d(st), \\ &= \frac{1}{s^{n+1}} \int_0^\infty u^n\, e^{-u}\, du = \frac{1}{s^{n+1}}\, \Gamma(n+1)\,. \end{aligned} \tag{13.45}$$

It is important when deriving a transform formula to identify the conditions or limitations that apply to it. The integrals in both the above formulas only converge if $s > 0$, and the second integral has the additional convergence condition $n > -1$. These limitations are attached to the table entries.

■

Example 13.6.2. Partial Fraction Decomposition

If we have a transform that is not in our table, we can sometimes decompose it into partial fractions that may be recognizable as table entries. Consider the following

Table 13.1: Laplace Transforms.[a]

	$F(s)$	$f(t)$	Limitation
1.	$\frac{1}{s}$	1	$s > 0$
2.	$\frac{\Gamma(n+1)}{s^{n+1}}$	t^n	$s > 0,\ n > -1$
3.	$\frac{1}{s-k}$	e^{kt}	$s > k$
4.	$\frac{1}{(s-k)^2}$	te^{kt}	$s > k$
5.	$\frac{s}{s^2+k^2}$	$\cos kt$	$s > 0$
6.	$\frac{k}{s^2+k^2}$	$\sin kt$	$s > 0$
7.	$\frac{s}{s^2-k^2}$	$\cosh kt$	$s > k$
8.	$\frac{k}{s^2-k^2}$	$\sinh kt$	$s > k$
9.	$\frac{s^2-k^2}{(s^2+k^2)^2}$	$t\cos kt$	$s > 0$
10.	$\frac{2ks}{(s^2+k^2)^2}$	$t\sin kt$	$s > 0$
11.	$\frac{e^{-as}}{s}$	$u(t-a)$	$s > 0,\ a \geq 0$
12.	e^{-as}	$\delta(t-a)$	$s > 0,\ a \geq 0$
13.	$\frac{\tanh(s/2)}{s}$	See Fig. 13.1	$s > 0$
14.	$Y(s-k)$	$e^{kt}y(t)$	$s > k$
15.	$e^{-as}Y(s)$	$y(t-a)$	$y(x<0) = 0$
16.	$\frac{1}{a}Y(s/a)$	$y(at)$	$a > 0$
17.	$Y^{(n)}(s)$	$(-t)^n\, y(t)$	
18.	$\int_s^\infty Y(p)\, dp$	$\frac{1}{t}\, y(t)$	Y is integrable
19.	$\frac{1}{s}Y(s)$	$\int_0^t y(x)\, dx$	y is integrable
20.	$X(s)\, Y(s)$	$(x * y)(t)$	
21.	$s\, Y(s) - y(0)$	$y'(t)$	
22.	$s^2Y(s) - sy(0) - y'(0)$	$y''(t)$	
23.	$s^nY(s) - \sum_{i=1}^{n} s^{n-i}\, y^{(i-1)}(0)$	$y^{(n)}(t)$	

[a] $u(t-a)$ is the unit step function; $x(t)$ and $y(t)$ are arbitrary, with transforms $X(s)$ and $Y(s)$. $(x * y)$ is a convolution.

inverse transform and its partial-fraction decomposition

$$\frac{1}{s^2(s^2-k^2)} = \frac{1}{k^2}\left[\frac{1}{s^2-k^2} - \frac{1}{s^2}\right].$$

Methods for working with partial fractions are reviewed in Appendix E.

The partial fractions correspond to entries in Table 13.1, so we have the result (not in the table)

$$\mathcal{L}^{-1}\left[\frac{1}{s^2(s^2-k^2)}\right] = \frac{\sinh kt - t}{k^2}. \tag{13.46}$$

Equation (13.46) is valid subject to the more restrictive of the limitations identified in the table for its individual terms, namely $s > k$.

■

Example 13.6.3. Translation of Transform

To replace a transform variable s by the translated form $s - k$, we append a factor e^{kt} to $f(t)$. To verify that this is so, note that the Laplace transform of $e^{kt} f(t)$ can be written

$$\mathcal{L}\left[e^{kt} f(t)\right](s) = \int_0^\infty f(t)\, e^{kt}\, e^{-ts}\, dt = \int_0^\infty f(t)\, e^{-t(s-k)}\, dt = \mathcal{L}[f(t)](s-k). \tag{13.47}$$

The result is $F(s-k)$. This general result corresponds to Table entry 14.

An example of the use of Eq. (13.47) is with $f(t) = 1$, the transform of which is $F(s) = 1/s$. The transform of $e^{kt} f(t)$ is therefore $F(s-k) = 1/(s-k)$. This is Table entry 3, and is valid when $s - k > 0$, i.e., when $s > k$.

■

Example 13.6.4. Differentiate Parameter

Differentiation of a transform pair with respect to a parameter may lead to additional useful formulas. This operation commutes with the transform integral formula and the result is therefore also a transform pair. Consider the pair (Table entry 3)

$$f(t) = e^{kt}, \qquad F(s) = \frac{1}{s-k}.$$

Differentiating both these formulas with respect to k, we get

$$\frac{df(t)}{dk} = t\, e^{kt}, \qquad \frac{dF(s)}{dk} = \frac{1}{(s-k)^2}. \tag{13.48}$$

This transform pair, valid for $s > k$, is Table entry 4.

■

Example 13.6.5. Transforms of Trigonometric Functions

Judicious use of the Euler identity $e^{iax} = \cos ax + i \sin ax$ can convert known transform pairs into expressions involving sines or cosines. Start from Table entry 3, but replace

k by ik and use the Euler identity. The result is

$$\mathcal{L}\left[e^{ikt}\right](s) = \mathcal{L}[\cos kt + i \sin kt] = \frac{1}{s - ik}\,.$$

Expanding into real and imaginary parts, this equation becomes

$$\mathcal{L}[\cos kt] + i\mathcal{L}[\sin kt] = \frac{s}{s^2 + k^2} + i\,\frac{k}{s^2 + k^2}\,.$$

The Laplace transforms of the sine and cosine are real, so we may identify

$$\mathcal{L}[\cos kt] = \frac{s}{s^2 + k^2} \quad \text{and} \quad \mathcal{L}[\sin kt] = \frac{k}{s^2 + k^2}\,. \tag{13.49}$$

These are Table entries 5 and 6, and are valid for $s > 0$.

■

Example 13.6.6. Discontinuous Functions

Expressions involving delta functions or the Heaviside (unit step) function $u(x)$ may have convenient Laplace transforms. We look first at the transform of $f(t) = u(t-a)$. We only consider $a \geq 0$ because all negative values of a produce the same transform as that with $a = 0$. Thus, for positive a we have

$$\mathcal{L}[u(t-a)] = \int_0^\infty u(t-a)\, e^{-st}\, dt = \int_a^\infty e^{-st}\, dt = \frac{e^{-as}}{s}\,. \tag{13.50}$$

This is Table entry 11.

Next let's look at the Laplace transform of the delta function $\delta(t-a)$, again with $a > 0$. By direct evaluation we get

$$\mathcal{L}[\delta(x-a)] = \int_0^\infty \delta(t-a)\, e^{-st}\, dt = e^{-as}\,. \tag{13.51}$$

This is Table entry 12.

A final, and more robust example, is provided by a function $f(t)$ that is a square wave of the form shown in Fig. 13.1. We can write this waveform as a series of unit step functions:

$$f(t) = u(t) - 2u(t-1) + 2u(t-2) - 2u(t-3) + \cdots = -1 + 2\sum_{n=0}^{\infty}(-1)^n u(t-n)\,.$$

Taking the Laplace transform term-by-term, we have, using Eq. (13.50),

$$\mathcal{L}[f(t)] = -\frac{1}{s} + 2\sum_{n=0}^{\infty}(-1)^n\,\frac{e^{-ns}}{s}\,. \tag{13.52}$$

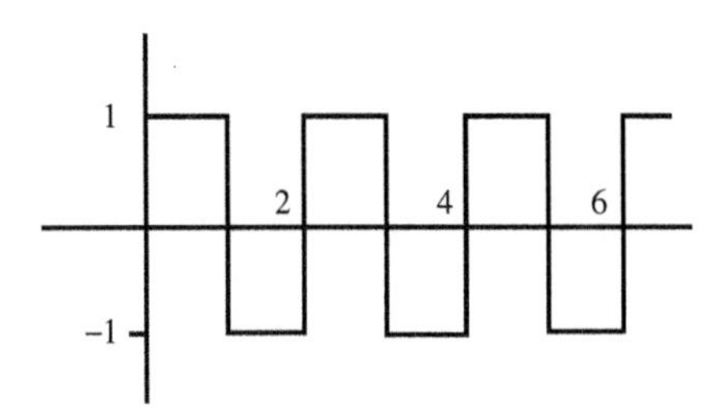

Figure 13.1: Square wave for Example 13.6.6.

The summation in Eq. (13.52) is a geometric series, so we can sum it to reach

$$\mathcal{L}[f(t)] = -\frac{1}{s} + \frac{2}{s}\frac{1}{1+e^{-s}} = \frac{1}{s}\left(\frac{1-e^{-s}}{1+e^{-s}}\right),$$

$$= \frac{1}{s}\left(\frac{e^{s/2}-e^{-s/2}}{e^{s/2}+e^{-s/2}}\right) = \frac{1}{s}\left(\frac{\sinh(s/2)}{\cosh(s/2)}\right) = \frac{\tanh(s/2)}{s}. \qquad (13.53)$$

This is Table entry 13.

We note that the discontinuous functions being studied here have well-behaved, continuous Laplace transforms, suggesting that use of Laplace transforms may alleviate problems that may arise when it is attempted to treat these discontinuities by direct-space methods.

■

PROPERTIES OF LAPLACE TRANSFORMS

We have already identified one general relationship satisfied by Laplace transforms, namely the translation property discussed in Example 13.6.3. We examine now some other general properties. These properties are included in Table 13.1 as entries 14–23.

Example 13.6.7. Translation of Function

Assume $f(t)$ to be such that $f(t) = 0$ when $t < 0$. Then the Laplace transform of $f(t-a)$ will have the form

$$\mathcal{L}[f(t-a)] = \int_0^\infty f(t-a)\, e^{-st}\, dt = \int_a^\infty f(t-a)\, e^{-st}\, dt = \int_0^\infty f(u)\, e^{-s(u+a)}\, du\,,$$

where we made a change of the integration variable from t to $u = t - a$. This equation can be rewritten as

$$\mathcal{L}[f(t-a)] = e^{-as}F(s)\,, \qquad (13.54)$$

and is subject to the aforementioned condition that $f(t) = 0$ when $t < 0$. This is Table entry 15.

■

Example 13.6.8. Change of Scale

The Laplace transform of $y(at)$ is given by the formula

$$\mathcal{L}[y(at)] = \int_0^\infty y(at)\, e^{-st}\, dt = \int_0^\infty y(u)\, e^{-su/a}\,\frac{du}{a} = \frac{1}{a}Y\left(\frac{s}{a}\right), \qquad (13.55)$$

where we made a change of variable to $u = at$. The resulting formula is Table entry 16, and is subject to the limitation $a > 0$ because it is assumed that $y(u) = 0$ if $u < 0$.

■

Example 13.6.9. Derivatives of Transform

To obtain a formula for the derivative of a Laplace transform we can differentiate the transform formula with respect to s. When applied to the transform integral, the differentiation affects only the factor e^{-st}. Thus,

$$\frac{d}{ds}\,\mathcal{L}[f(t)] = \frac{d}{ds}\int_0^\infty f(t)\,e^{-st}\,dt = \int_0^\infty (-t)\,f(t)\,e^{-st}\,dt\,,$$

which shows that

$$\frac{dF(s)}{ds} = \mathcal{L}\left[(-t)f(t)\right]. \tag{13.56}$$

Repeated differentiation inserts additional powers of $(-t)$ in the function being transformed. This is Table entry 17.

■

Example 13.6.10. Integral of Transform

If we integrate the transform, the integration (with respect to s) affects only the factor e^{-st}, and we have

$$\int_s^\infty \mathcal{L}[f(t)] = \int_0^\infty dt\,f(t)\int_s^\infty e^{-st}\,ds = \int_0^\infty dt\,f(t)\,\frac{e^{-st}}{t}\,,$$

corresponding to

$$\int_s^\infty F(s)\,ds = \mathcal{L}\left[\frac{f(t)}{t}\right]. \tag{13.57}$$

This is Table entry 18.

■

Example 13.6.11. Integral of Function

The Laplace transform under study here is

$$\mathcal{L}\left[\int_0^t f(x)\,dx\right] = \int_0^\infty dt\,e^{-st}\int_0^t f(x)\,dx\,.$$

Interchanging the order of the t and x integrations, we get

$$\mathcal{L}\left[\int_0^t f(x)\,dx\right] = \int_0^\infty dx\int_x^\infty dt\,e^{-st}\,f(x) = \int_0^\infty dx\,f(x)\left[\frac{e^{-sx}}{s}\right] = \frac{1}{s}\,F(s)\,. \tag{13.58}$$

This result corresponds to Table entry 19.

■

Transforms of the derivatives of a function are also of importance, and are discussed later in connection with the use of Laplace transforms to solve differential equations.

LAPLACE TRANSFORM CONVOLUTION

Like the Fourier transform, we can define for the Laplace transform a convolution operation and establish a convolution theorem. The Laplace convolution of two functions $f(t)$ and $g(t)$ is defined as follows:

$$(f*g)(t) = \int_0^t f(u)g(t-u)\,du\,. \tag{13.59}$$

To obtain a convolution theorem, we now write the Laplace transform of $(f * g)$ and interchange the order of the two integrations. We also write $e^{-st} = e^{-su}\,e^{-s(t-u)}$. Thus,

$$\begin{aligned}\mathcal{L}[(f*g)](s) &= \int_0^\infty dt\, e^{-st} \int_0^t f(u)\, g(t-u)\, du \\ &= \int_0^\infty du\, f(u)\, e^{-su} \int_u^\infty dt\, g(t-u)\, e^{-s(t-u)}\,.\end{aligned}$$

We now make the change of integration variable $t - u = v$ and note that the range of v is $(0, \infty)$. These observations bring us to

$$\begin{aligned}\mathcal{L}[(f*g)](s) &= \left[\int_0^\infty f(u)\, e^{-su}\, du\right] \left[\int_0^\infty g(v)\, e^{-sv}\, dv\right] \\ &= F(s)\, G(s)\,. \qquad (13.60)\end{aligned}$$

We note that the Laplace and Fourier convolution theorems are of essentially the same form. The relationship exhibited in Eq. (13.60) is Table entry 20.

One use of the Laplace convolution theorem is to provide a pathway toward the evaluation of the inverse transform of a product $F(s)G(s)$ in the case that $F(s)$ and $G(s)$ are individually recognizable as the transforms of known functions. The convolution theorem will also prove useful in connection with the use of the Laplace transform for the solution of differential equations (a topic we pursue in the next section of this text).

Example 13.6.12. A Laplace Transform Inversion

Let's use the convolution theorem to evaluate

$$\mathcal{L}^{-1}\left[\frac{1}{s^2(s^2+k^2)}\right].$$

We recognize the transform to be inverted as a product of two known transforms $F(s)G(s)$, where $F(s) = 1/s^2$ and $G(s) = 1/(s^2 + k^2)$. We note that $f(t) = t$ (Table entry 2) and that $g(t) = (\sin kt)/k$ (Table entry 6).

Applying now the convolution theorem, we identify $F(s)G(s)$ as the transform of $(f * g)$. Thus, we have

$$\mathcal{L}^{-1}\left[\frac{1}{s^2(s^2+k^2)}\right] = (f*g) = \int_0^t f(t-u)\, g(u)\, du = \frac{1}{k}\int_0^t (t-u)\sin ku\, du\,. \qquad (13.61)$$

Because the convolution theorem result is symmetric in F and G, we can associate either f or g with u (making the other factor in the convolution integrand a function of $t - u$). Our choice in this problem was to associate $t - u$ with the simpler of f and g, thereby obtaining a less complicated convolution integral. The integration in Eq. (13.61) can now be evaluated, leading to

$$\mathcal{L}^{-1}\left[\frac{1}{s^2(s^2+k^2)}\right] = \frac{kt - \sin kt}{k^3}\,. \qquad (13.62)$$

The present example is simple enough that we can check it easily. A partial-fraction decomposition of the transform is

$$\frac{1}{s^2(s^2+k^2)} = \frac{1}{k^2}\left[\frac{1}{s^2} - \frac{1}{s^2+k^2}\right],$$

and the inverse transform of this expansion confirms Eq. (13.62).

The significant point of this example is that its method can be used when other approaches are difficult, and even if the convolution integral cannot be evaluated analytically it provides a route to the numerical determination of the desired inverse transform.

■

Example 13.6.13. Use of Delta Function

Let's evaluate the inverse Laplace transform of $F(s) = e^{-2s}/s^2$. We write

$$F(s) = G(s)H(s), \quad \text{with } G(s) = e^{-2s}, \quad H(s) = \frac{1}{s^2}.$$

We recognize G and H as known transforms:

$$g(t) = \delta(t-2), \qquad h(t) = t.$$

We can now use the convolution theorem to find $f(t) = (g * h)$. Because g is a delta function, the computation is simple:

$$f(t) = \int_0^t h(u)g(t-u)\,du = \int_0^t u\,\delta(t-u-2)\,du$$

$$= \begin{cases} t-2, & t \geq 2, \\ 0, & t < 2. \end{cases}$$

Note that we automatically obtained a limitation of f to nonnegative values of its argument.

■

SYMBOLIC COMPUTATION

In MAPLE, Laplace transforms and their inverses are accessed using the `inttrans` package, whose commands only become available after executing first the statement `with(inttrans)`. The Laplace transform of $f(t)$, with transform variable s, and its inverse, are given as

`laplace(`$f(t)$`,t,s)` and `invlaplace(`$F(s)$`,s,t)`.

If $f(t)$ or $F(s)$ are such that the existence of the transform pair is not assured, MAPLE returns these commands unevaluated.

In MATHEMATICA, the corresponding commands (not requiring a prior package activation) are

`LaplaceTransform[`$f(t)$`,t,s]` and `InverseLaplaceTransform[`$F(s)$`,s,t]`.

MATHEMATICA normally processes these commands without regard to their possible limitations.

In both MAPLE and MATHEMATICA the transforms and their inverses have the same definitions as this text, and both languages have strong algorithms for finding inverse transforms. However, the results are sometimes written in ways that may seem opaque to many users. One reason for this is that neither symbolic system presents expressions in a multiple-range format, but uses instead expressions such as the Heaviside function to enable single-line specification of different formulas for different ranges of the independent variable.

Example 13.6.14. Symbolic Computation, Laplace Transforms

Here is a MAPLE session:

```
> with(inttrans);
> laplace(t^3,t,s);
```

$$\frac{6}{s^4}$$

```
> invlaplace(1/s^4,s,x);
```

$$\frac{1}{6}x^3$$

```
> laplace(y^n,y,u);
```

$$\mathit{laplace}(y^n, y, u)$$

```
> invlaplace(1/s^(n+1),s,t);
```

$$\mathit{invlaplace}(s^{-n-1}, s, t)$$

No processing of these last two commands occurs because the transform or inverse is subject to a limitation on the value of n. Thus, continue with

```
> assume(n > -1);
```

(n now written $n\sim$)

```
> laplace(y^n,y,u);
```

$$\Gamma(n\sim+1)u^{-n\sim-1}$$

```
> invlaplace(1/s^n,s,t);
```

$$\frac{t^{n\sim}}{\Gamma(n\sim+1)}$$

The corresponding session in MATHEMATICA is:

```
LaplaceTransform[t^3,t,s]
```

$$\frac{6}{s^4}$$

```
InverseLaplaceTransform[1/s^4,s,t]
```

$$\frac{t^3}{6}$$

```
LaplaceTransform[t^n,t,u]
```

$$u^{-1-n}\text{Gamma}[1+n]$$

```
InverseLaplaceTransform[1/s^(n+1),s,t]
```

$$\frac{t^n}{\text{Gamma}[1+n]}$$

The availability of transforms via symbolic computation greatly reduces the urgency of having extensive transform tables. Here are some examples showing that MAPLE and MATHEMATICA can handle transforms involving a variety of special functions. In MAPLE,

```
> with(inttrans):
> laplace(Ei(1,x),x,t);
```

$$\frac{\ln(s+1)}{s}$$

```
> invlaplace(ln(s+1)/s,s,t);
```
$-\mathrm{Ei}(-t)$

```
> laplace(erf(t),t,s);
```
$$\frac{e^{s^2/4}\,\mathrm{erfc}(\frac{1}{2}s)}{s}$$

```
> invlaplace(1/sqrt(s^2-b^2),s,t);
```
$BesselI(0, bt)$

In the above, `Ei(1,x)` is the exponential integral $E_1(x)$, defined in Section 2.10, `erf` and `erfc` are error functions defined in Eqs. (9.42) and (9.44), and `BesselI(0,bt)` is the modified Bessel function $I_0(bt)$, treated in Chapter 14. We need to observe that $-\mathrm{Ei}(-x) = E_1(x)$ to confirm the second of the above results.

In MATHEMATICA,

```
LaplaceTransform[ExpIntegralE[1,x],x,s]
```
$$\frac{\texttt{Log}[1+\texttt{s}]}{\texttt{s}}$$

```
InverseLaplaceTransform[Log[s+1]/s,s,t]
```
$-\texttt{ExpIntegralEi}[-\texttt{t}]$

```
LaplaceTransform[Erf[t],t,s]
```
$$\frac{e^{s^2/4}\,\texttt{Erfc}[\frac{s}{2}]}{s}$$

```
InverseLaplaceTransform[1/Sqrt[s^2-b^2],s,t]
```
$\texttt{BesselJ}[0, ibt]$

Here `ExpIntegralE[1,x]` is $E_1(x)$ and `ExpIntegralEi[-t]` is $\mathrm{Ei}(-t)$. The expression `BesselJ`$[0, ibt]$, which is $J_0(ibt)$, is equivalent to $I_0(bt)$.

■

Exercises

13.6.1. Take Laplace transforms by hand computation, and check your work by symbolic computation.

$$\text{(a)}\ \sin 4(t-3), \qquad \text{(b)}\ (2t-1)^2, \qquad \text{(c)}\ t^{3/2}.$$

13.6.2. Take inverse Laplace transforms, using symbolic computation if helpful.

$$\text{(a)}\ \frac{s^2}{(s^2+4)^2}, \qquad \text{(b)}\ \frac{e^{-6s}}{s-1}, \qquad \text{(c)}\ \frac{1}{s(s+1)(s+2)}.$$

13.6.3. By hand computation, take the Laplace transforms of the six pulses shown in Fig. 13.2. Check your work using symbolic computation.

Hint. For symbolic computation describe the pulses using the unit step function (`Heaviside(x)` or `HeavisideTheta[x]`).

13.6.4. Check the Laplace convolution theorem by comparing $\mathcal{L}[(f * g)]$ and FG, with $f = t^2 - 4$ and $g = \sin 2t$.

13.6.5. Find the inverse Laplace transform of $H(s) = e^{-as}/s$ by writing it as FG, where $F = e^{-as}$ and $G = 1/s$, and then applying the convolution theorem.

13.6.6. Apply the Laplace convolution theorem to $e^{-as}Y(s)$ and thereby obtain a confirmation of entry 15 in Table 13.1.

Note. The vanishing of the result for $t < a$ can be made explicit by writing $u(t-a)y(t-a)$, where u is the unit step function.

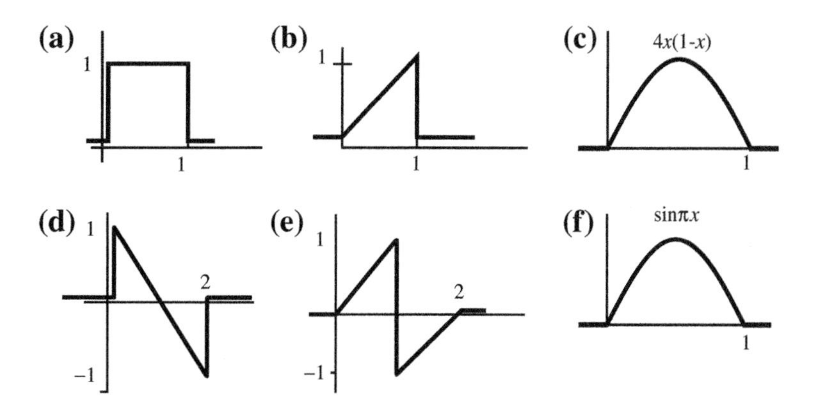

Figure 13.2: Pulse wave forms for Exercise 13.6.3.

13.7 LAPLACE TRANSFORM: SOLVING ODEs

Laplace transforms are useful for solving ODEs because the transforms of derivatives of a function have convenient forms. The formula for the transform of $df(t)/dt$ can be integrated by parts as follows:

$$\mathcal{L}[f'(t)] = \int_0^\infty e^{-st}\frac{df}{dt}\,dt = f(t)e^{-st}\Big|_{t=0}^{\infty} + \int_0^\infty se^{-st}f(t)\,dt = -f(0) + sF(s)\,. \tag{13.63}$$

This is entry 21 in Table 13.1.

If we apply Eq. (13.63) to $f'(t)$, we get a corresponding formula for the Laplace transform of f'':

$$\begin{aligned}\mathcal{L}[f''(t)] &= -f'(0) + s\mathcal{L}[f'(t)] = -f'(0) + s[-f(0) + sF(s)]\\ &= -f'(0) - sf(0) + s^2F(s)\,.\end{aligned} \tag{13.64}$$

This is entry 22 in the Table. Extension to higher derivatives follows analogously, with the result given as entry 23 in the Table.

We are now ready to solve inhomogeneous ODEs such as the following:

Example 13.7.1. Solving an ODE

Let's solve $y'' - 4y' + 4y = 4\,,$ with $y(0) = 1,\ y'(0) = 4\,.$

Taking Laplace transforms of all terms in the ODE, we get

$$s^2Y - sy(0) - y'(0) - 4\Big(sY - y(0)\Big) + 4Y = \frac{4}{s}\,,$$

equivalent to $(s^2 - 4s + 4)Y - s = \frac{4}{s}\,,$ with solution $Y = \frac{s^2+4}{s(s-2)^2}\,.$

We now take the inverse transform (can be done with a partial-fraction expansion, but is easier via symbolic computation). The result is

$$y(t) = 1 + 4t\,e^{2t}\,.$$

The reader should check that $y(t)$ is a solution to the ODE and that it satisfies the conditions $y(0) = 1$, $y'(0) = 4$.

Notice that we get a particular solution to the ODE that is determined by the values specified for $y(0)$ and $y'(0)$. In this respect the situation differs from that when the Fourier transform is used.

■

Example 13.7.2. Another ODE

Consider now $y'' - 4y = 4e^{2t}$, with $y(0) = 0,\ \ y'(0) = 1$.

This ODE cannot be solved using Fourier transforms because its inhomogeneous term does not have a Fourier transform. But there is no problem here because that term does have a Laplace transform. Transforming the ODE,

$$s^2Y - sy(0) - y'(0) - 4Y = \frac{4}{s-2} \quad \Longrightarrow \quad (s^2-4)Y = 1 + \frac{4}{s-2} = \frac{s+2}{s-2},$$

which can be solved to yield $Y = \dfrac{1}{(s-2)^2}$.

Inverting this transform, we get $y(t) = te^{2t}$. Again the reader should check that this solves the ODE and satisfies the conditions at $t = 0$.

■

Example 13.7.3. Simultaneous ODEs

Because the Laplace transform converts a linear ODE into an algebraic equation, a set of simultaneous inhomogeneous linear ODEs is reduced by transformation into a set of algebraic equations (which will be linear if the ODEs have constant coefficients). Look at this equation system, where y and z are both unknown functions of x, and

$$y'(x) - y(x) + 2z(x) = 0\,,$$

$$z'(x) - 2y(x) + z(x) = 0\,,$$

with $y(0) = 0$ and $z(0) = 1$. Letting Y and Z be respectively the Laplace transforms of y and z, these equations transform into

$$sY - y(0) - Y + 2Z = 0\,,$$

$$sZ - z(0) - 2Y + Z = 0\,,$$

which we solve for Y and Z, obtaining first

$$Y(s) = -\frac{2}{s^2+3}\,, \qquad\qquad Z(s) = \frac{s-1}{s^2+3}\,,$$

and then

$$y = -\frac{2}{\sqrt{3}}\sin(\sqrt{3}\,x)\,, \qquad\qquad z = \cos(\sqrt{3}\,x) - \frac{1}{\sqrt{3}}\sin(\sqrt{3}\,x)\,.$$

These results can be checked by substitution into the ODE system and verifying that $y(0) = 0$ and $z(0) = 1$.

■

CONVOLUTIONS

In our discussion of signal processing earlier in this chapter we noted that the Fourier transform of the solution of an inhomogeneous linear ODE could be written as a product with one factor coming from the left-hand side of the ODE (the homogeneous part), while the other factor was the Fourier transform of the right-hand side (the inhomogeneous term). A similar observation applies when ODEs are solved using Laplace transforms, even though signal processing is ordinarily not involved. In particular, an ODE of the form

$$ay''(t) + by'(t) + cy(t) = f(t)\,, \tag{13.65}$$

with initial conditions $y_0 = y(0)$, $y'_0 = y'(0)$, will have a solution $y(t)$ whose Laplace transform is

$$Y(s) = \Phi(s)\Big(F(s) + (as+b)y_0 + ay'_0\Big)\,, \quad \text{with} \quad \Phi(s) = \frac{1}{as^2+bs+c}\,. \tag{13.66}$$

We therefore identify $Y(s)$ as the Laplace transform of a convolution, in which one factor is the inverse transform of $\Phi(s)$, denoted $\varphi(t)$, and the other factor is the inverse transform of $F(s) + (as+b)y_0 + ay'_0$, and write

$$\begin{aligned} y(t) &= (\varphi * f) + ay_0\mathcal{L}^{-1}[s\Phi(s)] + (ay'_0 + by_0)\mathcal{L}^{-1}[\Phi(s)] \\ &= (\varphi * f) + ay_0\varphi'(t) + (ay'_0 + by_0)\varphi(t)\,. \end{aligned} \tag{13.67}$$

The inverse transform of $s\Phi$ has the indicated form because $\varphi(0)$ vanishes.

To make Eq. (13.67) useful we need to evaluate $\varphi(t)$ and $\varphi'(t)$. Letting α and β denote the roots of $as^2 + bs + c$, so $as^2 + bs + c = a(s-\alpha)(s-\beta)$, the transform inversion is straightforward, leading to

$$\varphi(t) = \begin{cases} \dfrac{1}{a}\left(\dfrac{e^{\alpha t} - e^{\beta t}}{\alpha - \beta}\right), & \alpha \neq \beta, \\[2ex] \dfrac{t}{a}\,e^{\alpha t}, & \alpha = \beta. \end{cases} \qquad \varphi'(t) = \begin{cases} \dfrac{1}{a}\left(\dfrac{\alpha e^{\alpha t} - \beta e^{\beta t}}{\alpha - \beta}\right), & \alpha \neq \beta, \\[2ex] \left(\dfrac{1+\alpha t}{a}\right)e^{\alpha t}, & \alpha = \beta. \end{cases} \tag{13.68}$$

One advantage of using the convolution theorem is that we do not need to obtain the Laplace transform of $f(t)$. A disadvantage is that we must evaluate the integral corresponding to $\varphi * f$, with φ the form given in Eq. (13.68).

Example 13.7.4. Solving ODE using Convolution

Consider the following ODE with the initial conditions $y(0) = 1$, $y'(0) = 0$:

$$2y'' + y' - 6y = e^t\,.$$

Let's solve this ODE using Eq. (13.67). The polynomial $2s^2 + s - 6$ has roots $\alpha = 3/2$, $\beta = -2$, and the coefficient of s^2 and s in the polynomial are $a = 2$ and $b = 1$. Then $\alpha - \beta = 7/2$, and

$$\varphi = \frac{1}{2}\left[\frac{e^{3t/2} - e^{-2t}}{7/2}\right],$$

$$\varphi' = \frac{1}{2}\left[\frac{\frac{3}{2}\,e^{3t/2} + 2e^{-2t}}{7/2}\right],$$

and, noting that $y_0 = 1$, $y_0' = 0$, Eq. (13.67) yields

$$y(t) = (\varphi * e^t) + 2\varphi'(t) + \varphi(t)$$
$$= (\varphi * e^t) + \frac{4}{7}\,e^{3t/2} + \frac{3}{7}\,e^{-2t}. \tag{13.69}$$

We next evaluate the convolution:

$$(\varphi * e^t) = \int_0^t \frac{1}{2}\left[\frac{e^{3u/2} - e^{-2u}}{7/2}\right] e^{t-u}\,du = \frac{e^t}{7}\int_0^t \left[e^{u/2} - e^{-3u}\right] du$$
$$= \frac{2}{7}\,e^{3t/2} - \frac{1}{3}\,e^t + \frac{1}{21}\,e^{-2t}.$$

Substituting this result into Eq. (13.69), we reach

$$y(t) = \frac{6}{7}\,e^{3t/2} - \frac{1}{3}\,e^t + \frac{10}{21}\,e^{-2t}. \tag{13.70}$$

We can verify the correctness of this solution by substituting into the original ODE and by confirming that $y(0)$ and $y'(0)$ have the prescribed values.

If we had chosen not to use the convolution theorem, we could write the Laplace transform of our ODE:

$$2s^2Y - 2sy_0 - 2y_0' + sY - y_0 - 6Y = \frac{1}{s-1},$$

where $1/(s-1)$ is the transform of e^t. Inserting $y_0 = 1$, $y_0' = 0$ and solving for Y,

$$Y = \left(\frac{1}{2s^2 + s - 6}\right)\left(\frac{1}{s-1} + (2s+1)\right) = \frac{s(2s-1)}{(2s-3)(s+2)(s-1)}, \tag{13.71}$$

which has an inverse transform that agrees with Eq. (13.70).

■

RESPONSE FUNCTIONS

Suppose that a dynamical system is described by an ODE of the general form

$$ay''(t) + by'(t) + cy(t) = f(t),$$

where $f(t)$, sometimes called a **forcing function**, is zero when the system is not subject to external forces or source terms. It is of interest to be able to describe the response of a system, initially in the time-independent state $y = 0$, when a sudden impulse is applied at $t = t_0$. This situation corresponds (for a unit impulse at t_0) to the equation

$$ay'' + by' + cy = \delta(t - t_0), \tag{13.72}$$

with $y(t) = y'(t) = 0$ for $t \le t_0$.

The function $y(t)$ that solves this problem is sometimes called the **response function** of the dynamical system. It can also be identified as an initial-value Green's function and used in a similar fashion. Even though the impulse causes a discontinuity in y' at $t = t_0$, it can be shown that Eq. (13.72) can be solved using Laplace transforms with $y(t_0)$ and $y'(t_0)$ set to zero in the transform formula.

Example 13.7.5. Response Function

Taking $t - t_0$ as the independent variable in the ODE

$$y'' + k^2 y = \delta(t - t_0)\,,$$

the Laplace transform of the equation (with $y(0) = y'(0) = 0$) has the form

$$s^2 Y + k^2 Y = 1, \qquad \text{leading to } \; Y = \frac{1}{s^2 + k^2}\,.$$

Inverting the transform, we get

$$y = \begin{cases} \dfrac{1}{k}\,\sin k(t - t_0)\,, & t > t_0, \\ 0\,, & t < t_0. \end{cases}$$

As expected, the impulse has caused a discontinuous change in y'.

■

Exercises

13.7.1. Verify the steps in the solution of the ODE in Example 13.7.1, including the use of symbolic computation to confirm that the inverse transform of $Y(s)$ is the form given for $y(t)$. Then check that $y(t)$ satisfies both the ODE and the initial conditions.

13.7.2. Repeat the procedure of Exercise 13.7.1 to check the solution of the ODE in Example 13.7.2.

13.7.3. Verify that the simultaneous ODEs of Example 13.7.3 have the solution given in that Example. Include a check of the initial conditions on $y(0)$ and $z(0)$.

Use Laplace transforms to solve the following ODEs (in the independent variable t) subject to the indicated initial conditions. Use symbolic computing methods where helpful. Verify that your solutions satisfy both the ODE and the initial conditions.

13.7.4. $y'' - 4y = 3e^{-t}$, with $y(0) = 1$, $y'(0) = -1$.

13.7.5. $y'' + 2y' + 5y = 10\cos t$, with $y(0) = 1$, $y'(0) = 2$.

13.7.6. $y'' + 2y' + 5y = 10\cos t$, with $y(0) = 2$, $y'(0) = 1$.

13.7.7. $y'' + y = \sin t$, with $y(0) = 0$, $y'(0) = 1$.

13.7.8. $y'' + y = \sin t$, with $y(0) = 1$, $y'(0) = 0$.

13.7.9. $y'' - y = \sin t$, with $y(0) = 0$, $y'(0) = 0$.

13.7.10. $y'' - y = \sin t$, with $y(0) = 1$, $y'(0) = -1$.

13.7.11. Use symbolic computing to check that $y(x)$ and $z(x)$ as found in Example 13.7.3 are the inverse Laplace transforms of Y and Z and that y and z constitute a solution to the ODE system of that Exercise (including its initial conditions).

Use Laplace transforms to solve the following sets of simultaneous ODEs subject to the indicated initial conditions. Use symbolic computing methods where helpful. Verify that your solutions satisfy both the ODEs and the initial conditions.

13.7.12. $y'(t) + z(t) = 2\cos t\,,$ $\quad z'(t) - y(t) = 1\,,$ with $y(0) = 0$, $z(0) = 0$.

13.7.13. $y''(t) + z''(t) - z'(t) = 0\,,$ $\quad y'(t) + z'(t) - 2z(t) = 1 - e^t\,,$ with $y(0) = 1$, $y'(0) = 0$, $z(0) = 0$, $z'(0) = 1$.

13.7.14. Use symbolic computing to verify that the inverse transform given in Eq. (13.71) yields the expression for $y(t)$ given in Eq. (13.70).

13.7.15. Use the convolution theorem to solve

$$2y'' + y' - 6y = \sin x\,,$$

with $y(0) = 2$ and $y'(0) = -1$.

13.7.16. By solving the following ODEs subject to $y(t - t_0) = y'(t - t_0) = 0$, find the response functions for

(a) $y'' - 4y' - 5y = \delta(t - t_0)\,,$

(b) $y'' + 7y' + 10y = \delta(t - t_0)\,,$

(c) $y'' - 4y = \delta(t - t_0)\,.$

Additional Readings

Abramowitz, M., & Stegun, I. A. (Eds.). (1972). *Handbook of mathematical functions with formulas, graphs, and mathematical tables (AMS-55).* Washington, DC: National Bureau of Standards (Reprinted, Dover (1974). Contains a table of Laplace Transforms).

Erdélyi, A., Magnus, W., Oberhettinger, F., & Tricomi, F. G. (1954). *Tables of integral transforms* (Vol. 2). New York: McGraw-Hill.

McCollum, P. A., & Brown, B. F. (1965). *Laplace transform tables and theorems.* New York: Holt, Rinehart and Winston.

Prudnikov, A. P., Brychkov, Yu. A., & Marichev, O. I. (1992). Integrals and series. *Direct Laplace transforms* (Vol. 4), and *Inverse Laplace transforms* (Vol. 5). New York: Gordon and Breach. (Table erratum: Math. Comp. 66:1766 (1997)).

Roberts, G. E., & Kaufman, H. (1966). *Table of Laplace transforms.* Philadelphia: Saunders.

Sneddon, I. N. (1951). *Fourier transforms.* New York: McGraw-Hill (Reprinted, Dover (1995). A detailed treatment with many applications).

Sneddon, I. N. (1974). *The use of integral transforms.* New York: McGraw-Hill (More elementary than Sneddon's 1951 treatise).

Chapter 14

SERIES SOLUTIONS: IMPORTANT ODEs

In this chapter we introduce methods for solving second-order ODEs through the introduction of power series, and apply the techniques to several differential equations that are of great importance in the physical sciences.

The method of series solution, also known as the **method of Frobenius**, enables us to obtain, as series expansions, the solutions to a wide class of linear homogeneous second-order ODEs. Most such ODEs do not have series solutions whose sums can be identified in terms of the well-known elementary functions, but if the ODE is of frequent use its series solution may have been defined as a **special function**. One advantage of appropriate special-function definitions is that the properties of the special functions can be studied and used whenever those functions are encountered.

This situation is not really that different from some that you are already familiar with. The trigonometric functions $\cos x$ and $\sin x$, which arise from the relationships between the sides of a right triangle with acute angle x, are defined algebraically as power series. The function e^x is expressible as a power series. Frequent encounters create the feeling that these are known quantities; their use is made convenient because of relationships you have learned. For example, you know that $\sin^2 x + \cos^2 x = 1$, that $\sin 2x = 2 \sin x \cos x$, that $e^{x+y} = e^x\, e^y$, and (perhaps more recently) that $e^{ix} = \cos x + i \sin x$. Your familiarity with these functions is enhanced by plotting them as a function of x; you thereby see the periodicity (and the difference in phase) of the sine and cosine, and the "exponential growth" or "exponential decay" of the exponential.

This chapter therefore has the dual objective of (1) showing how to obtain the solutions to some ODEs of practical importance, and (2) surveying the properties of the special functions that are the solutions to the ODEs.

14.1 SERIES SOLUTIONS OF ODEs

When the clever methods of Chapter 10 do not yield a closed-form solution to a linear homogeneous second-order ODE, our back-up plan is ordinarily to seek a solution given as an infinite series. The method of Frobenius attempts a series solution of the form

$$y(x) = x^s \sum_{j=0}^{\infty} a_j x^j = \sum_{j=0}^{\infty} a_j x^{s+j} \,. \tag{14.1}$$

Mathematics for Physical Science and Engineering.
http://dx.doi.org/10.1016/B978-0-12-801000-6.00014-6

The factor x^s (where s may be of either sign and is not restricted to integer values) provides a flexibility that is needed to make the method maximally useful. Once we have adopted the form of Eq. (14.1), we can without loss of generality require the coefficient a_0 to be nonzero. This means that (for small x) the dominant term of the expansion is a_0x^s. For simplicity we restrict discussion here to expansions about $x = 0$. One can, of course, expand about any point $x = a$, in which case every x on the right-hand sides of Eq. (14.1) is replaced by $x - a$.

The conditions under which a series solution can work is the subject of a theorem by Fuchs. Fuch's theorem states that when the ODE is written in the standard form

$$y''(x) + p(x)y'(x) + q(x)y(x) = 0\,,$$

at least one series solution about $x = a$ will exist if $x = a$ is either an **ordinary point** or a **regular singular point** of the ODE, but no series solution is guaranteed to exist if $x = a$ is an **irregular singular point** of the ODE. An ordinary point of the ODE is one at which the coefficients $p(x)$ and $q(x)$ are nonsingular. We define $x = a$ to be a **regular** singular point of the ODE if $p(x)$ or $q(x)$ is singular, but the two limits

$$\lim_{x=a}(x-a)p(x) \qquad \text{and} \qquad \lim_{x=a}(x-a)^2q(x)$$

exist. In other words, if $x = a$ is a regular singular point, $p(x)$ diverges no more strongly than $1/(x-a)$ and $q(x)$ is no more divergent that $1/(x-a)^2$. Singular points that are not regular are termed **irregular**.

We will not discuss it at length in this text, but one can also seek the solution to an ODE in x by expansion about $x = \infty$, meaning that we expand in powers of $1/x$. To determine regularity or irregularity at $x = \infty$, one rewrites the ODE in terms of $t = 1/x$ and tests the resulting equation for regularity at $t = 0$.

Example 14.1.1. Regular and Irregular Singular Points

To find the singular points of the ODE

$$xy'' - \frac{y'}{x} + \frac{y}{1-x^2} = 0$$

we first bring it to the form $y'' + p(x)y' + q(x)y = 0$. Here

$$p(x) = -\frac{1}{x^2} \qquad \text{and} \qquad q(x) = \frac{1}{x(1-x)(1+x)}\,.$$

At $x = 0$, $p(x)$ is singular, and so is $xp(x)$. Thus $x = 0$ is an irregular singular point of the ODE. At $x = \pm 1$ $p(x)$ is regular but $q(x)$ is singular. However, $(x-1)q(x)$ is nonsingular at $x = 1$, and $(x+1)q(x)$ is nonsingular at $x = -1$, so $x = +1$ and $x = -1$ are both regular singular points of the ODE.

■

The following example illustrates the application of the method of Frobenius, and in particular discusses the role of the parameter s in the exponents of the expansion.

Example 14.1.2. Frobenius Method

Consider the ODE

$$y''(x) + k^2y(x) = 0\,,$$

which we have already solved several times; its general solution is $C_1 \sin kx + C_2 \cos kx$. However, assuming that we do not remember how to solve the ODE, we proceed to a series solution by expansion about $x = 0$. We expect to obtain at least one solution because $x = 0$ is a regular point of the ODE.

We begin the solution process by writing the expansions for y and y''. In expanded form (not using summation signs), we have

$$\begin{aligned} y(x) &= a_0 x^s + a_1 x^{s+1} + a_2 x^{s+2} + a_3 x^{s+3} + \cdots , \\ y''(x) &= s(s-1) a_0 x^{s-2} + (s+1)s\, a_1 x^{s-1} + (s+2)(s+1) a_2 x^s \\ & \quad + (s+3)(s+2) a_3 x^{s+1} + \cdots . \end{aligned}$$

Using these expansions, we now form $y'' + k^2 y$:

$$y'' + k^2 y = s(s-1) a_0 x^{s-2} + (s+1)s\, a_1 x^{s-1} + \Big[(s+2)(s+1) a_2 + k^2 a_0 \Big] x^s + \Big[(s+3)(s+2) a_3 + k^2 a_1 \Big] x^{s+1} + \cdots . \quad (14.2)$$

Since $y'' + k^2 y = 0$ for all values of x, and since power-series expansions are unique, we can set the coefficient of each power of x separately to zero, thereby getting a set of equations involving the parameter s and the coefficients a_j. From the above equation, we get

$$\begin{aligned} s(s-1) a_0 &= 0 , \\ (s+1)s\, a_1 &= 0 , \\ (s+2)(s+1) a_2 + k^2 a_0 &= 0 , \\ (s+3)(s+2) a_3 + k^2 a_1 &= 0 , \\ & \cdots . \end{aligned}$$

It would be more efficient and would also give us the general members of the equation set if we carried out the above analysis using summation signs and index summations. Working directly from Eq. (14.1), we have

$$k^2 y(x) = \sum_{j=0}^{\infty} k^2 a_j x^{s+j} ,$$

$$y''(x) = \sum_{j=0}^{\infty} (s+j)(s+j-1) a_j x^{s+j-2} .$$

It would be easier to add these two summations together if the indicated powers of x in the summations were the same, so we rewrite the summation for y'' by replacing j by $j+2$, so as to change the power of x to $s+j$. We can do this because j is a dummy index; its function is only to indicate the values to be used when performing the summations. We must only be careful to adjust the summation limits so that the summation includes the correct set of index values. The replacement $j \to j+2$ causes the lower summation limit to be $j = -2$, not zero. To avoid confusion, we write

$$y''(x) = \sum_{j=-2}^{\infty} (s+j+2)(s+j+1)a_{j+2}x^{s+j}$$

$$= s(s-1)a_0 x^{s-2} + (s+1)s\,a_1 x^{s-1} + \sum_{j=0}^{\infty}(s+j+2)(s+j+1)a_{j+2}x^{s+j}\,.$$

Now it is easy to form $y'' + k^2 y$:

$$y'' + k^2 y = s(s-1)a_0 x^{s-2} + (s+1)s\,a_1 x^{s-1}$$

$$+ \sum_{j=0}^{\infty}\Big[(s+j+2)(s+j+1)a_{j+2} + k^2 a_j\Big]\,x^{s+j}\,. \quad (14.3)$$

Equation (14.3) contains, in a general form, the same information as Eq. (14.2), so we continue from Eq. (14.3). The coefficient of each power of x must vanish, so in particular we have (from the coefficient of x^{s-2})

$$s(s-1)a_0 = 0\,. \quad (14.4)$$

Since we have already specified that $a_0 \neq 0$ (so that the series actually starts with x^s), Eq. (14.4) can only be satisfied if $s(s-1) = 0$, in which case a_0 is arbitrary. This first condition, which depends upon s but not on the a_j, is called the **indicial equation** of the method, and it determines the possible values for s. In the present case, those values are $s = 0$ and $s = 1$. Whenever Fuchs' theorem is satisfied, at least one of these s values will correspond to a solution to the ODE. Sometimes, but not always, the other s value will correspond to another ODE solution; in later examples we encounter both possibilities.

We proceed further by choosing one of the s values that satisfy Eq. (14.4); we take first $s = 0$. We now continue to the coefficient of x^{s-1} in Eq. (14.3). It gives

$$(s+1)s\,a_1 = 0\,. \quad (14.5)$$

Because we have already chosen $s = 0$, this equation is satisfied for any value of a_1. The simplest way to handle this situation is to set $a_1 = 0$. If necessary we can revisit that choice later.

We are now ready to look at the general term, the coefficient of x^{s+j} in the summation. That coefficient has the form (for $s = 0$)

$$(j+2)(j+1)a_{j+2} + k^2 a_j = 0, \qquad \text{or} \quad a_{j+2} = -\frac{k^2}{(j+2)(j+1)}\,a_j\,. \quad (14.6)$$

Equation (14.6) shows that the a_j satisfy a **recurrence relation**, so that once we know a_j we can compute a_{j+2}. The overall pattern of the solution process now becomes clear; our choice $a_1 = 0$ will, by virtue of Eq. (14.6), cause a_3, a_5, and all a_{2n+1} to also vanish, and our solution for $s = 0$ will contain only even nonnegative powers of x. In particular, from Eq. (14.6),

$$a_2 = -\frac{k^2}{1 \cdot 2}\, a_0\,,$$

$$a_4 = -\frac{k^2}{3 \cdot 4}\, a_2 = +\frac{k^4}{1 \cdot 2 \cdot 3 \cdot 4}\, a_0\,,$$

$$a_{2n} = (-1)^n \frac{k^{2n}}{(2n)!}\, a_0\,.$$

Inserting these coefficients into the power series for $y(x)$, we get (remembering that this is for $s = 0$)

$$y(x) = a_0 \sum_{n=0}^{\infty} \frac{(-1)^n k^{2n} x^{2n}}{(2n)!} = a_0 \cos kx\,. \tag{14.7}$$

We can get a second solution by repeating the above analysis with the other solution to the indicial equation, $s = 1$. With that choice of s, Eq. (14.5) tells us that $a_1 = 0$, and the general term of Eq. (14.3) yields

$$(j+3)(j+2)a_{j+2} + k^2 a_j = 0\,, \qquad \text{or} \quad a_{j+2} = -\frac{k^2}{(j+3)(j+2)}\, a_j\,. \tag{14.8}$$

Thus, once again $a_{2n+1} = 0$ for all n, and

$$a_{2n} = (-1)^n \frac{k^{2n}}{(2n+1)!}\, a_0\,. \tag{14.9}$$

With these coefficients (and remembering that $s = 1$), we get

$$y(x) = a_0 \sum_{n=0}^{\infty} \frac{(-1)^n k^{2n} x^{2n+1}}{(2n+1)!} = \frac{a_0}{k} \sin kx\,. \tag{14.10}$$

The two values of s satisfying the indicial equation have produced the two linearly independent solutions to our ODE (at arbitrary scale).

The features of this problem that transfer to other ODEs include the role of the indicial equation in determining the values that can be assigned to the parameter s, the setting to zero of a_1 if its value is arbitrary, and the use of recurrence to determine the coefficients a_j. What is special about this example is the fact that we obtained two series that could be identified as elementary functions. More often a series solution will be that of a known special function or of a function that has not been named and studied.

■

Exercises

Obtain series solutions for the following ODEs.

14.1.1. $y' = 3x^2 y\,.$

14.1.2. $2xy'' - y' + 2y = 0\,.$

14.1.3. $2xy'' + y' + 2y = 0\,.$

14.2 LEGENDRE EQUATION

Our first practical example of series solution is the Legendre ODE,

$$(1-x^2)y''(x) - 2xy'(x) + \lambda y(x) = 0\,. \tag{14.11}$$

This ODE arises when the Laplace or Helmholtz partial differential equation is written in spherical polar coordinates. In those applications x is the cosine of the polar angle, and we require solutions that are nonsingular for $-1 \le x \le +1$. This requirement is equivalent to requiring that our series solution, an expansion about $x = 0$, be convergent over that entire range.

Before proceeding, let's check that Fuchs' theorem is satisfied for the Legendre ODE. Writing the ODE in the standard form

$$y'' - \frac{2x}{(1+x)(1-x)}\,y' + \frac{\lambda}{(1+x)(1-x)}\,y = 0\,,$$

we see that $x = 0$ is a regular point of the ODE, and that the ODE has regular singular points at $x = \pm 1$. These singularities will not prevent us from finding a solution by expansion about $x = 0$, but they suggest (at least to experienced mathematicians) that the expansions may present difficulties at $x = \pm 1$.

Applying the method of Frobenius, expanding about $x = 0$, we have (before adjusting the summation indices)

$$\sum_{j=0}^{\infty}(s+j)(s+j-1)a_j x^{s+j-2} - \sum_{j=0}^{\infty}(s+j)(s+j-1)a_j x^{s+j}$$
$$- 2\sum_{j=0}^{\infty}(s+j)a_j x^{s+j} + \lambda\sum_{j=0}^{\infty} a_j x^{s+j} = 0\,.$$

Performing an index shift on the first summation and combining the last three, we get

$$\sum_{j=-2}^{\infty}(s+j+2)(s+j+1)a_{j+2}x^{s+j} + \sum_{j=0}^{\infty}\Big[-(s+j)(s+j-1) - 2(s+j) + \lambda\Big]a_j x^{s+j} = 0\,.$$

From this equation we set the coefficient of each power of x to zero:

$$s(s-1)a_0 = 0\,, \tag{14.12}$$

$$s(s+1)a_1 = 0\,, \tag{14.13}$$

$$(s+j+2)(s+j+1)a_{j+2} + \Big[-(s+j)(s+j-1) - 2(s+j) + \lambda\Big]a_j = 0\,. \tag{14.14}$$

Equation (14.12) tells us that the possible values of s are $s = 0$ and $s = 1$, and we can satisfy Eq. (14.13) by setting $a_1 = 0$. Then, for $s = 0$ our solution will consist of a series containing only even powers of x, while the $s = 1$ solution will consist entirely of odd powers of x.

For $s = 0$, the recurrence relation, used only for even j, is

$$a_{j+2} = \frac{j(j+1) - \lambda}{(j+1)(j+2)}\,a_j\,. \tag{14.15}$$

For $s = 1$, also for even j, we have

$$a_{j+2} = \frac{(j+1)(j+2) - \lambda}{(j+2)(j+3)}\, a_j \,. \tag{14.16}$$

Using Eqs. (14.15) and (14.16), the general solution to the Legendre ODE can be written as the following expansion about $x = 0$:

$$\begin{aligned} y(x) &= \\ & C_1\left[1 - \frac{\lambda\, x^2}{1\cdot 2} - \frac{\lambda(2\cdot 3-\lambda)x^4}{1\cdot 2\cdot 3\cdot 4} - \frac{\lambda(6-\lambda)(4\cdot 5-\lambda)x^6}{6!} - \frac{\lambda(6-\lambda)(20-\lambda)(6\cdot 7-\lambda)x^8}{8!} - \cdots\right] \\ & + C_2\left[x + \frac{(1\cdot 2-\lambda)\, x^3}{2\cdot 3} + \frac{(2-\lambda)(3\cdot 4-\lambda)x^5}{2\cdot 3\cdot 4\cdot 5} + \frac{(2-\lambda)(12-\lambda)(5\cdot 6-\lambda)x^7}{7!} + \cdots\right] . \end{aligned} \tag{14.17}$$

For unrestricted values of λ, it can be shown that both the series in Eq. (14.17) diverge at $x = 1$. But we can nevertheless obtain valid solutions at $x = 1$ (and at $x = -1$) by setting either C_1 or C_2 to zero, and then choosing a value of λ that causes the remaining series to terminate, leaving a polynomial (which presents no convergence issues).

Carrying out the prescription of the preceding paragraph, we first set $C_2 = 0$, and then look for values of λ that cause the C_1 sum to terminate. The first possibility is $\lambda = 0$, which leaves the zero-degree polynomial $y(x) = C_1$. The next possibility is $\lambda = 2\cdot 3 = 6$, which leaves

$$y(x) = C_1\left[1 - \frac{6\,x^2}{2}\right] = -2C_1\left[-\frac{1}{2} + \frac{3}{2}x^2\right] .$$

Note that $\lambda = 2\cdot 3$ produced an even polynomial of degree 2. Another choice for λ is $4\cdot 5$, which leaves an even polynomial of degree 4. In general, the C_1 sum can be terminated, leaving a polynomial of some **even** degree n, by choosing $\lambda = n(n+1)$. The constant C_1 is arbitrary because the Legendre equation is homogeneous. It is customary to scale the polynomial solutions to make $y(1) = 1$.

Turning now to the case in which we set $C_1 = 0$ and choose λ to terminate the C_2 sum, we find that the choice $\lambda = n(n+1)$, but now with n odd, leaves us with an **odd** polynomial of degree n.

The two cases (C_1 or C_2 zero) lead to the unified result that

The Legendre ODE, when written in the form

$$(1-x^2)y''(x) - 2xy'(x) + n(n+1)y(x) = 0\,, \tag{14.18}$$

with n a nonnegative integer, has a polynomial solution of degree n, with the same parity as n. This polynomial solution, designated $P_n(x)$ and scaled such that $P_n(1) = 1$, is called the **Legendre polynomial** *of degree n.*

We will shortly identify approaches for evaluating Legendre polynomials that are easier than the direct use of Eq. (14.17). However, that equation (with the scaling

Table 14.1: Legendre Polynomials.

n	$P_n(x)$
0	1
1	x
2	$\frac{3}{2}x^2 - \frac{1}{2}$
3	$\frac{5}{2}x^3 - \frac{3}{2}x$
4	$\frac{35}{8}x^4 - \frac{15}{4}x^2 + \frac{3}{8}$
5	$\frac{63}{8}x^5 - \frac{35}{4}x^3 + \frac{15}{8}x$
6	$\frac{231}{16}x^6 - \frac{315}{16}x^4 + \frac{105}{16}x^2 - \frac{5}{16}$
7	$\frac{429}{16}x^7 - \frac{693}{16}x^5 + \frac{315}{16}x^3 - \frac{345}{15}x$
8	$\frac{6435}{128}x^8 - \frac{3003}{232}x^6 + \frac{3465}{64}x^4 - \frac{315}{32}x^2 + \frac{35}{128}$

$P_n(1) = 1$ and λ set to $n(n+1)$) does define the P_n. Using that information we construct the explicit formulas given for P_0 through P_8 in Table 14.1.

It may be worth mentioning that although we call ODEs with different values of λ (or n) a "Legendre equation," each different λ value really defines a different ODE, with its own solution set, so the "solutions of the Legendre equation" are really solutions of a parameterized set of related ODEs. Since the Legendre ODE is second-order, it will (for each n value) have a second solution that is not a polynomial. These second solutions can be found by the method presented in Section 10.6, and, when scaled conventionally, are designated $Q_n(x)$. The solutions Q_n all have singularities at $x = \pm 1$ and are not appropriate for use in place of the P_n.

SYMBOLIC COMPUTATION

The Legendre polynomials can be accessed using symbolic computing. The syntax

`LegendreP(n,x)` (MAPLE) or `LegendreP[n,x]` (MATHEMATICA)

produces $P_n(x)$. If the arguments n and x are numeric, both symbolic systems return $P_n(x)$ as a numeric quantity (in decimal form if either argument is a decimal). If both n and x are given only as symbols, the unevaluated function call is returned. However, if n is integral (and x is a symbol), MATHEMATICA returns the explicit form of the Legendre polynomial, but MAPLE does not. The sample coding shows how to see the form of $P_n(x)$ in MAPLE.

MAPLE:

```
> LegendreP(2,1/2);
```
$$-\frac{1}{8}$$

```
> LegendreP(2,0.5);
```
-0.1250000000

```
> LegendreP(n,x);
```
LegendreP(n, x)

	`> LegendreP(2,x);`	LegendreP$(2, x)$
	`> simplify(%);`	$-\frac{1}{2}+\frac{3}{2}x^2$
MATHEMATICA:	`LegendreP[2,1/2]`	$-\frac{1}{8}$
	`LegendreP[2,0.5]`	-0.125
	`LegendreP[n,x]`	`LegendreP[n,x]`
	`LegendreP[2,x]`	$\frac{1}{2}(-1+3x^2)$

We do not make much use of them in this book, but for reference we note that the second solution to the Legendre equation, denoted $Q_n(x)$, is accessed in the symbolic systems as

`LegendreQ(n,x)` (MAPLE) or `LegendreQ[n,x]` (MATHEMATICA).

Exercises

14.2.1. Plot $P_n(x)$ on the range $(-1, +1)$ for integer n from 0 through 8 (recommend that you do this on separate plots).

(a) For each P_n, count the number of zeros on the plotted range and compare this number to n.

(b) Identify the point (or points) where each P_n has its maximum magnitude, and note that maximum value.

(c) Determine from your plots whether each P_n has a definite parity, and if so, identify that parity (even or odd).

14.2.2. Find the three roots of $P_3(x)$.

Hint. One way to do this is by plotting P_3 and then restricting the range of the plot to the neighborhood of a zero until its location can be read out to sufficient precision.

14.2.3. Express $f(x) = 3x^3 - 2x^2 - 3x + 5$ as a linear combination of Legendre functions.

Hint. Start by finding the coefficient of P_3, denoted c_3. Then observe that $f(x) - c_3P_3(x)$ is a polynomial of degree 2. Continue with the coefficients of P_2, P_1, and finally P_0.

GENERATING FUNCTION

The notion of a **generating function** was introduced in Section 2.9, where it was used to define the Bernoulli numbers. Here we identify a generating function for the Legendre polynomials P_n; it is useful for deriving recurrence formulas that are efficient ways of obtaining the P_n.

Our generating function is

$$g(x,t) = \frac{1}{(1-2xt+t^2)^{1/2}} = \sum_{n=0}^{\infty} P_n(x)t^n \,. \tag{14.19}$$

As we shortly see in more detail, Eq. (14.19) determines not only the functional form, but also the scale of $P_n(x)$.

We need to verify that Eq. (14.19) gives results consistent with our earlier definitions of P_n. To start, note that the coefficient of t^n must be a polynomial of degree n in x (because t occurs both in xt and in t^2, the power of x multiplying t^n can be at most n). To check the scale for P_n given from the generating function, set $x=1$ and note that

$$g(1,t) = \frac{1}{(1-2t+t^2)^{1/2}} = \frac{1}{1-t} = 1+t+t^2+t^3+\cdots \,. \tag{14.20}$$

Since $g(1,t) = P_0(1) + P_1(1)t + P_2(1)t^2 + \cdots$, we see that the generating function formula yields $P_n(1) = 1$ for all n.

Now that we have established that the generating function formula produces $P_n(x)$ that are polynomials of the expected degree and scale, it remains to show that the generated P_n are solutions of the Legendre ODE. As a first step in that direction, note that

$$t\,\frac{\partial^2}{\partial t^2}t^{n+1} = n(n+1)t^n \,, \qquad \text{so} \qquad t\,\frac{\partial^2}{\partial t^2}\big[tg(x,t)\big] = \sum_{n=0}^{\infty} n(n+1)P_n(x)t^n \,. \tag{14.21}$$

We next observe that if the functions P_n are solutions of the Legendre equation, then

$$\sum_{n=0}^{\infty}\left[(1-x^2)\frac{\partial^2}{\partial x^2} - 2x\frac{\partial}{\partial x} + n(n+1)\right] P_n(x)t^n = 0\,, \tag{14.22}$$

and we can check that the generating function yields this result. Rewriting Eq. (14.22), using Eq. (14.21) for the $n(n+1)$ term, we should have

$$(1-x^2)\frac{\partial^2 g(x,t)}{\partial x^2} - 2x\frac{\partial g(x,t)}{\partial x} + t\,\frac{\partial^2}{\partial t^2}\big[tg(x,t)\big] = 0\,. \tag{14.23}$$

We leave it as an exercise to show that Eq. (14.23) is an identity.

Example 14.2.1. Potential of Charge not at Origin

The generation-function formula, Eq. (14.19), has an interesting physical interpretation. If we place a charge q at $z=a$ on the polar axis of a spherical coordinate system, the potential it produces at (r,θ) (irrespective of the value of φ) is

$$V(r,\theta) = \frac{q}{4\pi\varepsilon_0}\,\frac{1}{r_1}\,,$$

where r_1 is the distance shown in Fig. 14.1. Applying the law of cosines, we find

$$V(r,\theta) = \frac{q}{4\pi\varepsilon_0}\left(r^2+a^2-2ar\cos\theta\right)^{-1/2}. \tag{14.24}$$

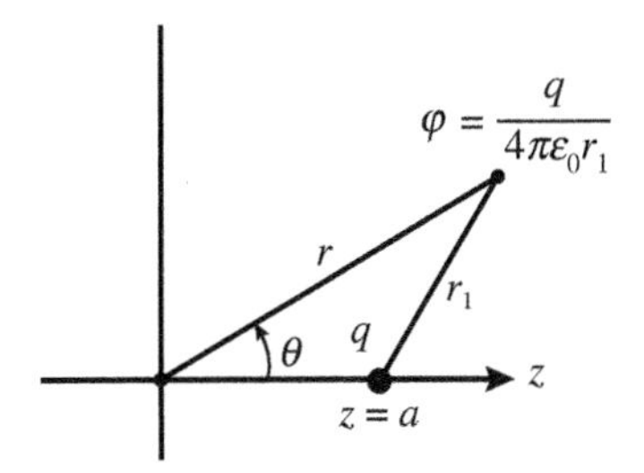

Figure 14.1: Potential at (r,θ) of a charge q at $z=a$.

What we do next depends on whether or nor $r>a$. If $r>a$, we rewrite Eq. (14.24) to the form

$$V(r,\theta)=\frac{q}{4\pi\varepsilon_0 r}\left(1-\frac{2a}{r}\cos\theta+\frac{a^2}{r^2}\right)^{-1/2}=\frac{q}{4\pi\varepsilon_0 r}\,g(\cos\theta,a/r)\,,$$

which we recognize as containing the Legendre generating function $g(\cos\theta,a/r)$. We can therefore write

$$V(r,\theta)=\frac{q}{4\pi\varepsilon_0 r}\sum_{n=0}^{\infty}P_n(\cos\theta)\left(\frac{a}{r}\right)^n=\frac{q}{4\pi\varepsilon_0 a}\sum_{n=0}^{\infty}P_n(\cos\theta)\left(\frac{a}{r}\right)^{n+1},\quad (r>a). \tag{14.25}$$

Since it can be shown that for $|x|\le 1$ all $P_n(x)$ satisfy $|P_n(x)|\le 1$, the series in Eq. (14.25) will be convergent because we have specified that $r>a$.

On the other hand, if $r<a$, the series in Eq. (14.25) does not converge, and we instead manipulate Eq. (14.24) to

$$V(r,\theta)=\frac{q}{4\pi\varepsilon_0 a}\left(\frac{r^2}{a^2}+1-\frac{2r}{a}\cos\theta\right)^{-1/2}=\frac{q}{4\pi\varepsilon_0 a}\sum_{n=0}^{\infty}P_n(\cos\theta)\left(\frac{r}{a}\right)^n,\quad (r<a), \tag{14.26}$$

which converges because of our restriction to $r<a$.

This Example illustrates the central role of the Legendre polynomials in the electrostatics of charge distributions. Suppose now that we have two charges q_1 and q_2 at points $z=a_1$ and $z=a_2$ on the polar axis, with r greater than both a_1 and a_2. Then, using Eq. (14.24) for each charge, we get the convergent expansion

$$V(r,\theta)=\frac{1}{4\pi\varepsilon_0}\left[(q_1+q_2)\frac{P_0(\cos\theta)}{r}+(a_1q_1+a_2q_2)\frac{P_1(\cos\theta)}{r^2}\right.$$
$$\left.+(a_1^2q_1+a_2^2q_2)\frac{P_2(\cos\theta)}{r^3}+\cdots\right]. \tag{14.27}$$

This is an expansion in **multipole moments** of the charge distribution; the nth moment is identified as the 2^n-pole moment, using Greek prefixes: monopole, dipole, quadrupole, etc. The first term contains the 2^0, or **monopole moment**, i.e., $Q=q_1+q_2$, the sum of the charges, the next term contains the $2^1 =$ **dipole moment** of the charge distribution (relative to the origin), $\mu=a_1q_1+a_2q_2$, and the third term contains the **quadrupole moment** $\omega=a_1^2q_1+a_2^2q_2$. Inserting these identifications and evaluating the Legendre polynomials, Eq. (14.27) becomes

$$V(r,\theta)=\frac{1}{4\pi\varepsilon_0}\left[\frac{Q}{r}+\frac{\mu\cos\theta}{r}+\frac{\omega}{r^2}\left(\frac{3}{2}\cos^2\theta-\frac{1}{2}\right)+\cdots\right]. \tag{14.28}$$

This expansion is valid in its present form only for charges that lie on the polar axis (i.e., for linear multipoles), with all charges closer to the origin than r. Expansions for charge distributions of more general geometries are treated in Section 15.4.

■

RECURRENCE FORMULAS

Our motivation for introducing a generating function is that it can be used to obtain recurrence formulas. The general approach is to differentiate $g(x,t)$ with respect to t or x, thereby obtaining expressions that contain contiguous index values. We illustrate with the most important of these formulas. Start with

$$\frac{\partial g}{\partial t} = \frac{x-t}{(1-2xt+t^2)^{3/2}} = \sum_{n=0}^{\infty} nP_n(x)t^{n-1}.$$

We rearrange this equation to

$$(x-t)g(x,t) = (1-2xt+t^2)\sum_{n=0}^{\infty} nP_n(x)t^{n-1},$$

which expands to

$$\sum_{n=0}^{\infty}\left[xP_n(x)t^n - P_n(x)t^{n+1} - nP_n(x)t^{n-1} + 2nxP_n(x)t^n - nP_n(x)t^{n+1}\right] = 0.$$

Collecting similar terms, this equation simplifies to

$$\sum_{n=0}^{\infty}\left[(2n+1)xP_n(x)t^n - (n+1)P_n(x)t^{n+1} - nP_n(x)t^{n-1}\right] = 0. \tag{14.29}$$

Combining the coefficients of t^n from the individual terms of Eq. (14.29), we obtain for each n the recurrence formula

$$(2n+1)xP_n(x) - nP_{n-1}(x) - (n+1)P_{n+1}(x) = 0. \tag{14.30}$$

Recurrence formulas involving $P'_n = dP_n/dx$ result if we differentiate the generating function with respect to x. Of the many formulas that can be obtained in this way, we list the following:

$$(2n+1)xP_n(x) = nP_{n-1}(x) + (n+1)P_{n+1}(x), \tag{14.31}$$

$$(2n+1)P_n(x) = P'_{n+1}(x) - P'_{n-1}(x), \tag{14.32}$$

$$(n+1)P_n(x) = P'_{n+1}(x) - xP'_n(x), \tag{14.33}$$

$$(1-x^2)P'_n(x) = nP_{n-1}(x) - nxP_n(x), \tag{14.34}$$

$$(1-x^2)P'_n(x) = (n+1)xP_n(x) - (n+1)P_{n+1}(x). \tag{14.35}$$

Equation (14.31) provides an efficient route to the P_n, starting from the initial values $P_0(x) = 1$ and $P_1(x) = x$. The other recurrence formulas are also useful; for example, we employ Eq. (14.32) later in this Section.

Example 14.2.2. Recurrence Formula

Let's use symbolic computing to generate some P_n recursively. Methods for using recurrence formulas are discussed in Appendix D; if this Example seems opaque, the material in that Appendix may be helpful.

Our plan is to use Eq. (14.31) to obtain the functional forms of $P_n(x)$, writing a procedure that uses $P_0(x) = 1$ and $P_1(x) = x$ as starting values. Note that if our actual need is for a table of P_j rather than just P_n, such a procedure would not be optimal, as calling it repeatedly for a succession of j values would cause the same P_j to be generated (and discarded) repeatedly. Note also that generating the P_n by a newly written procedure has only pedagogical value; the built-in procedure `LegendreP` already does this quite efficiently.

That said, we start this Example by writing a formula for P_n in terms of P_{n-1} and P_{n-2}:

$$P_n(x) = \frac{(2n-1)xP_{n-1}(x) - (n-1)P_{n-2}(x)}{n}.$$

In MAPLE, we first define an array `P` of sufficient size, with indexing starting from zero:

```
> P := Array(0 .. 10):
```

We initialize with P_0 and P_1:

```
> P[0] := 1:   P[1] := x:
```

We next write a loop that contains a straightforward formula for P_n,

```
> for j from 2 to n do
>    P[j]:=((2*j-1)*x*P[j-1]-(j-1)*P[j-2])/j; end do;
```

Examining the output from this loop, we see that there is a need to simplify the results; to do so, including a sort to make the powers of x occur in order, we **expand**, and then `sort` the expression for `P[n]`. Incorporating our modified loop into a procedure, we have

```
> LegP := proc(n,x) local P,j;
>    P := Array(0,n);
>    P[0] := 1;    P[1] := x;
>    for j from 2 to n do
>      P[j] :=sort(expand(((2*j-1)*x*P[j-1]-(j-1)*P[j-2])/j));
>    end do;
> end proc;
```

Since the last statement executed in this procedure was the computation of P_n, that quantity will be the output of the procedure.

`LegP(8,z);` $\qquad \frac{6435}{128}z^8 - \frac{3003}{32}z^6 + \frac{3465}{64}z^4 - \frac{315}{32}z^2 + \frac{35}{128}$

In MATHEMATICA, we collect our polynomials under a symbol that does not conflict with any MATHEMATICA definitions. We initialize with P_0 and P_1:

```
p[0] = 1; p[1] = x;
```

Note that the syntax is `[j]` and not `[[j]]`. This syntax implies that `p[j]` is a **function** that has $P_j(x)$ as its value for integer `j`; it is **not** an indexed array. More discussion of this point can be found in Appendix D.

We next write a loop that contains a straightforward formula for P_n,

```
Do[p[j] = ((2*j-1)*x*p[j-1] - (j-1)*p[j-2])/j, {j, 2, n}]
```

Examining the output from this loop, we see that there is a need to simplify the results. To do so, we `Expand` the expression for `p[j]`. Incorporating our modified loop into a procedure, we have

```
legP[n_,x_] := Module[{j,p},
  p[0]=1; p[1]=x;
  Do[p[j] = Expand[((2*j-1)*x*p[j-1]-(j-1)*p[j-2])/j], {j, 2, n}];
p[n] ]
```

The `Do` statement produces no output; to get the value of P_n returned we need to make a call to it the last statement that is executed. Executing the procedure,

`legP[8,z]` $\qquad \dfrac{35}{128} - \dfrac{315}{32}z^2 + \dfrac{3465}{64}z^4 - \dfrac{3003}{32}z^6 + \dfrac{6435}{128}z^8$

■

Exercises

14.2.4. Differentiate the Legendre generating function, Eq. (14.19), with respect to x, and thereby establish Eqs. (14.32) and (14.33).

Hint. It may be necessary to use Eq. (14.31), which was derived in the text.

14.2.5. Show that given either of Eqs. (14.34) and (14.35), one can derive the other.

14.2.6. By explicitly generating the first few terms in the expansion of the Legendre generating function, obtain an expression for $P_4(x)$.

Hint. This may require first an expansion of $(1-2xt+t^2)^{-1/2}$ and then a further expansion of some of the resulting terms.

14.2.7. Verify that Eq. (14.23) is an identity.

14.2.8. Show that $P_n'(1) = \dfrac{n(n+1)}{2}$.

14.2.9. Referring to the discussion between Eqs. (14.27) and (14.28), show that for a two-charge system, if the total charge Q is zero, the dipole moment μ remains the same if the two charges are moved the same distance along the z-axis (meaning that $a_1 \to a_1 + d$ and $a_1 \to a_2 + d$, with d arbitrary). Then show that the quadrupole moment ω remains unchanged under this position shift if both Q and μ are zero.

14.2.10. Enter into your symbolic system the recurrence-formula coding of Example 14.2.2 and verify that it works properly. Then remove the command `expand` or `Expand` (but of course keeping the expression that was to be expanded), and note how rapidly the complexity of the final result escalates as n is increased.

LEGENDRE EXPANSIONS

As the reader may by now suspect, the polynomial solutions of the Legendre equation are (when given the conventional scaling) identical with the polynomials we obtained in Example 11.2.1 by orthogonalizing, with unit weight, the set $\{x^n\}$ on the interval $(-1, 1)$. But now that we have found the P_n to be solutions of a second-order ODE, we are poised to obtain further insight. If we write the Legendre ODE in the form

$$\mathcal{L}y = n(n+1)y\,, \qquad \text{with} \quad \mathcal{L} = -(1-x^2)\,\frac{d^2}{dx^2} + 2x\,\frac{d}{dx}\,,$$

we can identify its solutions P_n as eigenfunctions of the operator $\mathcal{L}$, with respective eigenvalues $n(n+1)$. This operator $\mathcal{L}$ was identified as defining a self-adjoint ODE in Example 11.6.1, and it was also pointed out in that Example that its eigenfunctions would be orthogonal with a scalar product of definition

$$\langle\varphi|\chi\rangle = \int_{-1}^{1} \varphi(x)^*\chi(x)\,dx\,. \tag{14.36}$$

Example 11.6.1 did not show how to find the orthogonal eigenfunctions, but we have now found that they are the Legendre polynomials.

The orthogonality property, and the fact that the P_n are polynomials of successive degrees, enable us to conclude that the P_n, found as ODE solutions, must be equivalent to the similarly named functions obtained by the Gram-Schmidt process in Section 11.2.

It is instructive to plot some Legendre polynomials on the range $(-1, 1)$. All the zeros of each P_n lie in this range; that feature is necessary to yield the orthogonality. Figure 14.2 contains plots of P_3 and P_{10}. Note the odd or even symmetry, both of which cause the zeros to be symmetrically distributed. The symmetry and scaling also causes the P_n to have the special values

$$P_n(1) = 1\,, \qquad P_n(-1) = (-1)^n\,, \qquad P_{2n+1}(0) = 0\,. \tag{14.37}$$

Since the P_n form an orthogonal set of functions, they can be used as a basis for the expansion of general functions on the range of orthogonality, $(-1, 1)$. To make such expansions we need the normalization integral $\langle P_n|P_n\rangle$. To evaluate this integral

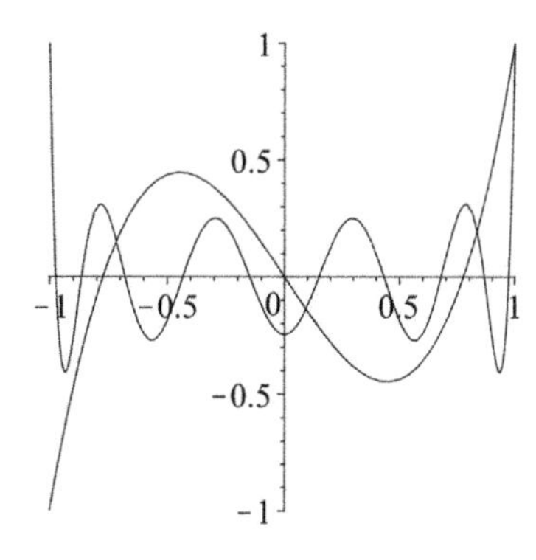

Figure 14.2: Legendre polynomials $P_3(x)$ and $P_{10}(x)$. Note that P_n has n zeros in the plotted range of x, and that over the entire range, $|P_n| \leq 1$.

we replace one of the factors P_n by its equivalent as given by Eq. (14.32):

$$\langle P_n|P_n\rangle = \frac{1}{2n+1}\left[\int_{-1}^{1} P_n(x)P'_{n+1}(x)\,dx - \int_{-1}^{1} P_n(x)P'_{n-1}(x)\,dx\right].$$

We now note that $P'_{n-1}(x)$ is a polynomial of degree $n-2$ and therefore can be written as a linear combination of P_j in which the largest value of j is $n-2$; the integral containing P'_{n-1} must therefore vanish due to orthogonality. We next integrate the remaining integral by parts; the result is

$$\langle P_n|P_n\rangle = \frac{1}{2n+1}\int_{-1}^{1} P_n(x)P'_{n+1}(x)\,dx,$$

$$= \frac{1}{2n+1}\left[P_n(x)P_{n+1}(x)\Big|_{-1}^{1} - \int_{-1}^{1} P'_n(x)P_{n+1}(x)\,dx\right].$$

Since P'_n is a polynomial of degree $n-1$ the integral in this equation vanishes; noting also that $P_n(1)P_{n+1}(1) = 1$ and $P_n(-1)P_{n+1}(-1) = -1$, our final result reduces to

$$\langle P_n|P_n\rangle = \frac{1-(-1)}{2n+1} = \frac{2}{2n+1}. \tag{14.38}$$

Example 14.2.3. A Legendre Series

We have already seen (Example 14.2.1) that the electrostatic potential of a charge not at the coordinate origin can be expanded in Legendre polynomials P_n, but the expansion was not specifically derived as an orthogonal expansion. Here we examine a situation in which we use the orthogonality properties of the P_n to develop an expansion.

Orthogonal expansions can handle functions that have discontinuities either in the function itself or in its derivatives; our present example, exhibiting a slope discontinuity, is

$$f(x) = e^{-2|x|}, \quad f(x) = \sum_{n=0}^{\infty} a_n P_{2n}(x), \quad a_n = \frac{\langle P_{2n}|f\rangle}{\langle P_{2n}|P_{2n}\rangle} = \frac{2\displaystyle\int_0^1 e^{-2x}P_{2n}(x)\,dx}{\dfrac{2}{2(2n)+1}}.$$

The above was written in a way that recognizes that $f(x)$ is an even function and its expansion involves only the even P_n.

Let's do this example using symbolic computing. A first step might be to evaluate the integral in the numerator of a_n; code to do this for a given n might take the form

MAPLE:
```
> s[n]:=evalf(int(exp(-2*x)*LegendreP(2*n,x), x=0 .. 1)):
```

MATHEMATICA:
```
s[n]=N[Integrate[Exp[-2*x]*LegendreP[2*n, x], {x, 0, 1}] ];
```

If one uses the above coding for small n (e.g., $n \le 6$), satisfactory results are obtained. If larger n values are used, both symbolic systems exhibit catastrophic numerical error unless the precision of the computations is increased. To retain enough precision to obtain a good plot with an expansion through P_{60}, in MAPLE one needs to set `Digits:=100`; in MATHEMATICA one needs to insert 20 as a second argument of the `N` command.

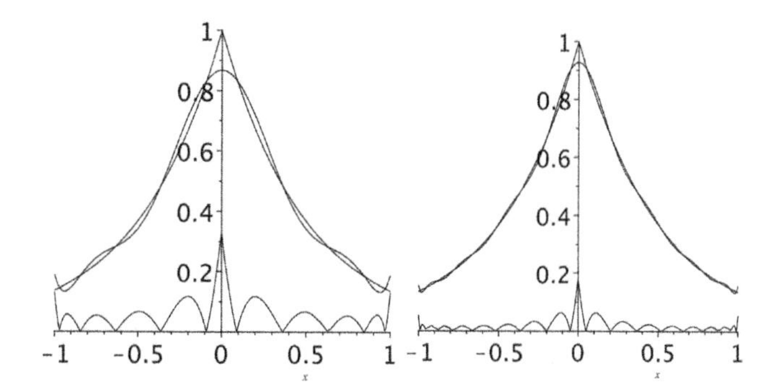

Figure 14.3: Legendre series expansions of $e^{-2|x|}$. Left: Truncated after P_8; right: Truncated after P_{16}. Also shown (at 2.5 times magnification) is the magnitude of the error in the expansion.

Assuming that `s[n]` through $P_{\rm nmax}$ have been generated by one of the following sets of commands,

MAPLE:
```
> Digits := 100:
>  for n from 0 to nmax by 2 do
>     s[n/2]:=evalf(int(exp(-2*x)*LegendreP(n,x),x=0.. 1))
   end do;
```

MATHEMATICA:
```
Do[ s[k/2] = N[Integrate[Exp[-2*x]*LegendreP[k, x], {x, 0, 1}],20],
          {k, 0, nmax, 2}]
```

we can then form and plot a series truncated at any `nn` not larger than `nmax` from the coding

MAPLE:
```
> T := add((4*n+1)*s[n]*LegendreP(2*n,x), n=0 .. nn/2):
> plot([exp(-abs(2*x)), T, abs(2.5*(T-exp(-abs(2*x))))],
          x=-1 .. 1, 0 .. 1);
```

MATHEMATICA:
```
T = Sum[(4*n+1)*s[n]*LegendreP[2*n,x], {n,0,nn/2}];
Plot[{Exp[-2*Abs[x]], T, Abs[2.5*(T-Exp[-2*Abs[x]])]}, {x, -1, 1},
      PlotRange -> {0, 1}]
```

The plots include the absolute value of the error (multiplied by 2.5) along with the exact and truncated forms of $f(x)$. We illustrate in Fig. 14.3 for `nn` = 8 and 16.

■

RODRIGUES FORMULA

The Legendre polynomials can also be represented by a compact formula due to Rodrigues:

$$P_n(x) = \frac{1}{2^n\, n!}\, \frac{d^n}{dx^n}\, (x^2 - 1)^n\,. \tag{14.39}$$

Adopting the temporary notations

$$D = \frac{d}{dx} \qquad \text{and} \qquad y = D^n (x^2 - 1)^n \,,$$

one way to show that P_n as defined in Eq. (14.39) satisfies the Legendre ODE is to compute $D^{n+2}(x^2-1)^{n+1}$ in two different ways and set to zero the difference of the two results. Our first evaluation, which makes use of Leibniz' formula for differentiating a product, Eq. (2.62), is

$$\begin{aligned} D^{n+2}(x^2-1)^{n+1} &= (x^2-1)D^{n+2}(x^2-1)^n + (n+2)\left[D(x^2-1)\right] D^{n+1}(x^2-1)^n \\ &\quad + \frac{(n+2)(n+1)}{2}\left[D^2(x^2-1)\right] D^n(x^2-1)^n \\ &= (x^2-1)D^2y + (n+2)(2x)Dy + (n+2)(n+1)y \,. \end{aligned}$$

The second evaluation, which also uses the Leibniz formula, is

$$\begin{aligned} D^{n+1}\left[D(x^2-1)^{n+1}\right] &= D^{n+1}\left[2(n+1)x(x^2-1)^n\right] \\ &= 2(n+1)\left[xD^{n+1}(x^2-1)^n + (n+1)D^n(x^2-1)^n\right] \\ &= 2(n+1)xDy + 2(n+1)^2 y \,. \end{aligned}$$

Subtracting the first of these equations from the second, we reach

$$(1-x^2)y'' - 2xy' + n(n+1)y = 0 \,,$$

the Legendre equation.

We now know that P_n, as given in Eq. (14.39), satisfies the Legendre equation. It is also obvious that the Rodrigues formula for P_n defines a polynomial of degree n. It remains only to show that the P_n are generated at the proper scale. At $x = 1$ the only nonzero contribution to P_n is that for which each factor $x^2 - 1$ is differentiated; there are $n!$ terms of that sort ($n!$ is the number of ways the operators D can be applied, one to each of the n factors). Each application of D produces $2x$, or at $x = 1$ an overall factor 2^n. The $n!$ and 2^n are exactly what is needed to leave us with $P_n(1) = 1$.

One can use the Rodrigues formula to derive the orthogonality and normalization of the P_n; we do not pursue that approach because we have already established those properties.

Exercises

14.2.11. Show that $P_n(-1) = (-1)^n$.

14.2.12. Show that $\displaystyle\int_{-1}^{1} x^m P_n(x)\,dx = 0$ if $m < n$.

14.2.13. Show that $\displaystyle\int_{-1}^{1} P_n(x)P'_{n-1}(x)\,dx = 0$.

14.2.14. Evaluate $\displaystyle\int_{-1}^{1} P_n(x)\,dx$.

14.2.15. Using symbolic computing, write code that produces the Legendre expansion of the step function

$$f(x) = \begin{cases} 0, & -1 \le x < 0, \\ 1, & 0 < x \le 1, \end{cases}$$

and make calculations for enough expansion lengths to give a good indication of the rate at which the expansion converges.

14.2.16. Using symbolic computing, obtain the Legendre expansion of the function $\sin^{-1} x$ (this is not $1/\sin x$).

14.2.17. The Legendre expansion of a function $f(x)$, truncated after P_N, which can be written

$$f(x) = \sum_{n=0}^{N} a_n P_n(x) \,,$$

is a **least-squares-error** approximation to $f(x)$. This means that

$$I = \int_{-1}^{1} \left[f(x) - \sum_{n=0}^{N} b_n P_n(x) \right]^2 dx$$

is a minimum when the b_n are the coefficients in the Legendre expansion of $f(x)$, i.e., $b_n = a_n$, $n = 0, \ldots, N$.

Expanding I, we have

$$I = \langle f|f \rangle - 2 \sum_n b_n \langle P_n|f \rangle + \sum_{mn} b_m b_n \langle P_m|P_n \rangle \,.$$

Evaluating some of the expressions in I and then adding to I the zero quantity

$$\sum_{n=0}^{N} a_n^2 \langle P_n|P_n \rangle - \sum_{n=0}^{N} a_n^2 \langle P_n|P_n \rangle \,,$$

show that I is indeed a minimum when $b_n = a_n$ for $n = 0, \ldots, N$.

14.3 ASSOCIATED LEGENDRE FUNCTIONS

The **associated Legendre functions** are solutions to an ODE that is a generalization of the Legendre equation; this generalization is needed when solving the Laplace or Helmholtz partial differential equation in polar coordinates for situations not exhibiting symmetry about the polar axis. The associated Legendre ODE has the form

$$(1 - x^2)y'' - 2xy' + \left[n(n+1) - \frac{m^2}{1 - x^2} \right] y = 0 \,. \tag{14.40}$$

In almost all problems of physical interest, m is an integer. The value $m = 0$ corresponds to the Legendre equation.

The most useful approach to the associated Legendre ODE is to exploit its similarity to the Legendre equation. We can remove the potentially difficult occurrence of $1 - x^2$ in the denominator of the m^2 term in Eq. (14.40) by making the substitution

$$y = (1 - x^2)^{m/2} u \,. \tag{14.41}$$

In applications this substitution does not greatly complicate matters; if x represents $\cos\theta$ (where θ is the polar angle), $(1-x^2)^{m/2}$ becomes $\sin^m\theta$.

Using Eq. (14.41), we reduce Eq. (14.40) to the more benign form

$$(1-x^2)u'' - 2(m+1)xu' + \big[n(n+1) - m(m+1)\big]\, u = 0\,. \tag{14.42}$$

We now make the key observation of the current analysis: we notice that if we differentiate Eq. (14.42) we recover a similar ODE, but with $m \to m+1$ and u' in place of u:

$$-2x(u')' + (1-x^2)(u')'' - 2(m+1)u' - 2(m+1)x(u')' + \big[n(n+1) - m(m+1)\big]\, u'$$

$$= (1-x^2)(u')'' - 2(m+2)x(u')' + \big[-2(m+1) + n(n+1) - m(m+1)\big]\, u'$$

$$= (1-x^2)(u')'' - 2(m+2)x(u')' + \big[n(n+1) - (m+2)(m+1)\big]\, u' = 0\,.$$

This equation tells us that if u is a solution of Eq. (14.42) for a given value of m, then u' is a solution for $m \to m+1$. Since we already know that $P_n(x)$ is a solution for $m = 0$, we may conclude that $d^m P_n/dx^m$ is a solution for positive integer m. Combining with Eq. (14.41), we can state that the associated Legendre equation, Eq. (14.40), has the solution

$$y = (1-x^2)^{m/2} \frac{d^m}{dx^m} P_n(x)\,. \tag{14.43}$$

It is customary to identify these solutions as $P_n^m(x)$ and to call them **associated Legendre functions**, with the precise definition

$$P_n^m(x) = (-1)^m (1-x^2)^{m/2} \frac{d^m}{dx^m} P_n(x)\,. \tag{14.44}$$

When $m = 0$, it is usually omitted, as $P_n^0 = P_n$. Also, note the factor $(-1)^m$. Some authors prefer not to include this factor, but we retain it to maintain agreement with the most widely used compilations of mathematical data (in Additional Readings, see Abramowitz or Olver et al.).

If we use the Rodrigues formula for P_n, we can write

$$P_n^m(x) = \frac{(-1)^m}{2^n\, n!} (1-x^2)^{m/2} \frac{d^{n+m}}{dx^{n+m}} (x^2-1)^n\,. \tag{14.45}$$

Equation (14.45) is valid for both positive and negative m in the range $|m| \le n$, and it can be shown that

$$P_n^{-m}(x) = (-1)^m \frac{(n-m)!}{(n+m)!}\, P_n^m(x)\,. \tag{14.46}$$

That P_n^m and P_n^{-m} are proportional is no surprise, because the associated Legendre ODE depends only on m^2. Some authors define P_n^{-m} to be equal to P_n^m; a problem with that definition is that recurrence formulas then have forms that depend upon the signs of m in their individual terms.

The first few associated Legendre **functions** (they are not polynomials) are shown in Table 14.2. Because these functions are frequently used with $x = \cos\theta$, the table gives values in terms of both x and θ.

The P_n^m are useful in a variety of problems because they are nonsingular in the entire range $|x| \le 1$ (the range appropriate for $\cos\theta$, where θ is the polar angle in spherical coordinates). Corresponding to each P_n^m there is a second solution Q_n^m, but these see less use in physics and engineering because they are all singular at $x = \pm 1$.

Table 14.2: Associated Legendre Functions.

n	m	$P_n^m(x) = P_n^m(\cos\theta)$
0	0	$1 = 1$
1	1	$-(1-x^2)^{1/2} = -\sin\theta$
1	0	$x = \cos\theta$
1	−1	$\frac{1}{2}(1-x^2)^{1/2} = \frac{1}{2}\sin\theta$
2	2	$3(1-x^2) = 3\sin^2\theta$
2	1	$-3x(1-x^2)^{1/2} = -3\sin\theta\cos\theta$
2	0	$\frac{3}{2}x^2 - \frac{1}{2} = \frac{1}{2}\left(3\cos^2\theta - 1\right)$
2	−1	$\frac{1}{2}x(1-x^2)^{1/2} = \frac{1}{2}\sin\theta\cos\theta$
2	−2	$\frac{1}{8}(1-x^2) = \frac{1}{8}\sin^2\theta$
3	3	$-15(1-x^2)^{3/2} = -15\sin^3\theta$
3	2	$15x(1-x^2) = 15\sin^2\theta\cos\theta$
3	1	$-\frac{3}{2}\left(5x^2-1\right)(1-x^2)^{1/2} = -\frac{3}{2}\sin\theta\left(5\cos^2\theta - 1\right)$
3	0	$\frac{5}{2}x^3 - \frac{3}{2}x = \frac{1}{2}\left(5\cos^3\theta - 3\cos\theta\right)$
3	−1	$\frac{1}{8}(5x^2-1)(1-x^2)^{1/2} = \frac{1}{8}\sin\theta\left(5\cos^2\theta - 1\right)$
3	−2	$\frac{1}{8}x(1-x^2) = \frac{1}{8}\sin^2\theta\cos\theta$
3	−3	$\frac{1}{48}(1-x^2)^{3/2} = \frac{1}{48}\sin^3\theta$

SYMBOLIC COMPUTING

The associated Legendre functions $P_n^m(x)$ are available in our symbolic computing systems using the syntax

`LegendreP(n,m,x)` (MAPLE) or `LegendreP[n,m,x]` (MATHEMATICA).

These functions have the same names as those representing $P_n(x)$ but the symbolic systems can detect that they have three (and not two) arguments. If the syntax shown here is used with `m` explicitly shown but set to zero, the polynomials $P_n(x)$ are recovered.

In MATHEMATICA, `LegendreP[n,m,x]` for both positive and negative `m` produces output consistent with Eqs. (14.45) and (14.46). Unfortunately, MAPLE has adopted conventions that for nonzero `m` cannot be made consistent with those equations (and in some cases are in error).[1] We therefore recommend that MAPLE users use instead the function `LegenP(n,m,x)` that is provided in Appendix K. That appendix also contains examples of the use of both MAPLE and MATHEMATICA to evaluate associated Legendre functions.

We do not discuss the second solutions of the associated Legendre equation, but they are available for symbolic computing as `LegendreQ(n,m,x)` or `LegendreQ[n,m,x]`.

PROPERTIES OF THE P_n^m

Procedures similar to those used for the Legendre polynomials can be employed to develop recurrence relations for the associated Legendre functions. Because these functions have two indices, many formulas are possible. We present only three here. For others, consult the Additional Readings.

$$(2n+1)xP_n^m(x) = (n+m)P_{n-1}^m(x) + (n-m+1)P_{n+1}^m(x)\,, \tag{14.47}$$

$$(2n+1)(1-x^2)^{1/2}P_n^m(x) = P_{n-1}^{m+1}(x) - P_{n+1}^{m+1}(x)\,, \tag{14.48}$$

$$P_n^{m+1}(x) + \frac{2mx}{(1-x^2)^{1/2}}P_n^m(x) + (n+m)(n-m+1)P_n^{m-1}(x) = 0\,. \tag{14.49}$$

For nonzero m, the P_n^m have the special values

$$P_n^m(1) = P_n^m(-1) = 0\,,\quad P_n^m(0) = \begin{cases} (-1)^{(n+m)/2}\,\dfrac{(n+m-1)!!}{(n-m)!!}\,, & n+m \text{ even},\\ 0\,, & n+m \text{ odd}. \end{cases} \tag{14.50}$$

The parity of the P_n^m is given by

$$P_n^m(-x) = (-1)^{n+m}\,P_n^m(x)\,. \tag{14.51}$$

The functions P_m^m have a simple form due to the fact that in the Rodrigues formula the quantity $(x^2-1)^m$ is differentiated $2m$ times; if $(x^2-1)^m$ is expanded, the only term that does not become zero after the $2m$ differentiations is x^{2m}, and we have

$$P_m^m(x) = \frac{(-1)^m}{2^m\,m!}\,(1-x^2)^{m/2}\frac{d^{2m}x^{2m}}{dx^{2m}} = (-1)^m(2m-1)!!\,(1-x^2)^{m/2}\,. \tag{14.52}$$

Combining this result with Eq. (14.46), we obtain

$$P_m^{-m}(x) = \frac{1}{(2m)!!}\,(1-x^2)^{m/2}\,. \tag{14.53}$$

The P_n^m of equal m but unequal n are orthogonal on the interval $(-1,1)$, with the orthogonality and normalization corresponding to the integral

$$\int_{-1}^{1} P_n^m(x)P_{n'}^m(x)\,dx = \frac{2\delta_{nn'}}{2n+1}\,\frac{(n+m)!}{(n-m)!}\,. \tag{14.54}$$

Equation (14.54) provides the information necessary to make expansions (for any fixed m) in P_n^m. Such expansions will generally converge rapidly only for functions that approach zero at $x=\pm1$ at least as fast as $(1-x^2)^{m/2}$.

[1]The error referred to occurs in `Maple13` and in earlier MAPLE releases when an environment variable is given a setting appropriate to the use of $P_n^m(x)$ for the range $|x| \le 1$. We do not know whether this error will be corrected in future releases.

Example 14.3.1. Checking a Recurrence Formula

Let's verify that the definitions we have adopted for the associated Legendre functions are consistent with the recurrence formula, Eq. (14.48). Our approach will be to write a procedure that makes the check for input values of n and m, and then prints either OK, ERROR, or BAD INPUT, whichever is indicated.

A suitable MAPLE procedure might be the following

```
> RecTest := proc(n,m) local S;
>   if (n<0 or abs(m) > n) then
>      RETURN(print("BAD INPUT")) end if;
>   if (abs(m+1) > n-1) then S := 0
>     else S := LegenP(n-1,m+1,x) end if;
>   S := S - LegenP(n+1,m+1,x)-(2*n+1)*(1-x^2)^(1/2)*LegenP(n,m,x);
>   S := simplify(S);
>   if S = 0 then print("(n,m), (",n,m,")  OK")
>               else print("(n,m), (",n,m,")  ERROR") end if;
> end proc;
```

A MATHEMATICA procedure for testing this recurrence relation might be

```
recTest[n_, m_] := Module[{S},
  If[n < 0 || Abs[m] > n, Return[Print["BAD INPUT"]] ];
  If[Abs[m+1] > n-1, S = 0, S = LegendreP[n-1, m+1, x] ];
  S = S - LegendreP[n+1, m+1, x]
               - (2*n+1)*(1-x^2)^(1/2)*LegendreP[n, m, x];
  S = Simplify[S];
  If[S == 0, Return[Print["n,m = ", n, ",", m, "  OK"]]];
  Print["n,m = ", n, ",", m, "  ERROR"] ]
```

The ERROR condition cannot be made the *else-statement* of the If command because MATHEMATICA regards a relation that can be true for some values of an undefined variable as indeterminate, in which case neither the *then-statement* nor the *else-statement* is executed.

■

Exercises

14.3.1. Using LegenP (MAPLE; Appendix K) or LegendreP (MATHEMATICA), plot $P_n^m(x)$ for several n and m values on the range $-1 \le x \le 1$, and from the plots

(a) Count the number of zeros on the plotted range and relate this number to n and m.

(b) Determine whether each P_n^m has a definite parity, and if so, identify that parity (even or odd).

14.3.2. Verify, using symbolic computing, that the recurrence formula connecting P_n^m of successive m values, Eq. (14.49), is valid for arbitrary n. Your check must include m values of both signs and cases where $m+1$ and $m-1$ are of opposite sign.

14.3.3. Show that the substitution $y(x) = (1-x^2)^{m/2}u(x)$ converts the associated Legendre equation, Eq. (14.40), to the form

$$(1-x^2)u'' - 2(m+1)xu' + \big[n(n+1) - m(m+1)\big]\, u = 0\,.$$

14.3.4. Show that the substitution $x = \cos\theta$ converts the following ODE into the associated Legendre equation.

$$\frac{1}{\sin\theta}\frac{d}{d\theta}\left(\sin\theta\,\frac{dy}{d\theta}\right) + \left[l(l+1) - \frac{m^2}{\sin^2\theta}\right] y = 0\,.$$

14.3.5. Prove Eqs. (14.52) and (14.53).

14.3.6. Consider the expansion of $P_n^{-m}(x)$ in the orthogonal functions $P_j^m(x)$:

$$P_n^{-m}(x) = \sum_j c_j P_j^m(x)\,, \qquad c_j = \frac{\langle P_j^m | P_n^{-m}\rangle}{\langle P_j^m | P_j^m\rangle}\,.$$

Evaluate the numerator of c_j by inserting the Rodrigues formulas for P_j^m and P_n^{-m} and carrying out repeated integrations by parts; use Eq. (14.54) to evaluate the denominator. Thereby establish the validity of Eq. (14.46) for P_n^{-m}.

14.4 BESSEL EQUATION

Our next adventure in series solution of ODEs is provided by the Bessel equation, with solutions that are termed **Bessel functions**. It is difficult to overstate the importance of Bessel functions in science and engineering. They are sometimes called **cylinder functions**, because they are encountered when the Laplace or Helmholtz equation is written in cylindrical coordinates. But the importance of Bessel functions is broader than that; they occur also in many spherical-coordinate problems in expressions that are called **spherical Bessel functions**, they are used to describe the propagation of electromagnetic waves (both in free space and in waveguides), and they are ubiquitous in problems involving damped sinusoidal oscillations. In addition, they are found as solutions to many problems that have no obvious geometric connection to a sphere or cylinder.

Bessel functions can be thought of as generalizations of the familiar trigonometric functions sine and cosine, and in common with those functions, they are connected by an extensive array of formulas and relationships. The importance and rich features of Bessel functions have inspired mathematicians to investigate them in great detail, and there exist a number of books entirely dedicated to these functions and their properties. The material in the present text is but a short introduction to these versatile functions; far more detail is available in the Additional Readings.

SERIES SOLUTION

Our starting point for the study of Bessel functions is that they are solutions to the Bessel ODE,

$$x^2y''(x) + xy'(x) + (x^2 - n^2)y(x) = 0\,. \tag{14.55}$$

Like the Legendre equation, the Bessel equation is a set of ODEs whose members are distinguished by the parameter n, which is called the **order** of the Bessel equation and its solutions. It is not necessary that n be an integer.

Dividing Eq. (14.55) through by x^2, we note that $x = 0$ is a regular singular point of the ODE, so we expect the method of Frobenius to yield at least one solution. We therefore try

$$y(x) = \sum_{j=0}^{\infty} a_j x^{s+j}\,, \tag{14.56}$$

obtaining initially

$$x^2\sum_{j=0}^{\infty}(s{+}j)(s{+}j{-}1)a_jx^{s+j-2}+x\sum_{j=0}^{\infty}(s{+}j)a_jx^{s+j-1}+x^2\sum_{j=0}^{\infty}a_jx^{s+j}-n^2\sum_{j=0}^{\infty}a_jx^{s+j}=0\,.$$

Combining the first, second, and fourth sums, but keeping the third by itself, we reach

$$\sum_{j=0}^{\infty}\Big[(s+j)^2-n^2\Big]a_jx^{s+j}+\sum_{j=0}^{\infty}a_jx^{s+j+2}=0\,. \tag{14.57}$$

We next take the terms with $j = 0$ and $j = 1$ out of the first sum, and then replace the dummy index j by $j + 2$. These steps convert the first term of Eq. (14.57) to

$$\begin{aligned}(s^2-n^2)a_0x^s &+ \big[(s+1)^2-n^2\big]a_1x^{s+1}+\sum_{j=2}^{\infty}\big[(s+j)^2-n^2\big]a_jx^{s+j}\\ &= (s^2-n^2)a_0x^s+\big[(s+1)^2-n^2\big]a_1x^{s+1}+\sum_{j=0}^{\infty}\big[(s+j+2)^2-n^2\big]a_{j+2}x^{s+j+2}.\end{aligned} \tag{14.58}$$

Inserting the expression in Eq. (14.58) in place of the first term of Eq. (14.57) and combining terms with the same power of x, we get

$$(s^2-n^2)a_0x^s+\big[(s{+}1)^2-n^2\big]a_1x^{s+1}+\sum_{j=0}^{\infty}\Big(\big[(s{+}j{+}2)^2-n^2\big]a_{j+2}+a_j\Big)x^{s+j+2}=0\,. \tag{14.59}$$

Since Eq. (14.59) must be satisfied for all x, the coefficient of each power of x must vanish. We therefore have

$$(s^2-n^2)\,a_0=0\,, \tag{14.60}$$

$$\big[(s+1)^2-n^2\big]\,a_1=0\,, \tag{14.61}$$

$$\big[(s+j+2)^2-n^2\big]\,a_{j+2}+a_j=0\,. \tag{14.62}$$

From the indicial equation, Eq. (14.60), we note that because a_0 must be nonzero, we find $s^2 = n^2$, i.e., the possible values of s are n and $-n$. With either of these values of s, we now note that in Eq. (14.61) the coefficient of a_1 is nonzero, so we must set $a_1 = 0$. Then, from Eq. (14.62), we see that a_3, a_5, and all a_{2j+1} must vanish, and that a_2, a_4, and all a_{2j} can be computed recursively, starting from an arbitrary value of a_0.

Since our power series has contributions only for even values of the index j, it is convenient to substitute $2k$ for j, thereby causing the summation to run over all integer values of k. Then the series solution has the form

$$y(x) = \sum_{k=0}^{\infty} b_k x^{s+2k}\,, \tag{14.63}$$

where $b_k = a_{2k}$, and Eq. (14.62) becomes

$$\left[(s+2k+2)^2 - n^2\right] b_{k+1} + b_k = 0\,. \tag{14.64}$$

Continuing, with s set to $+n$, Eq. (14.64) can be rearranged to the form

$$b_{k+1} = -\frac{b_k}{4(k+1)(n+k+1)}\,. \tag{14.65}$$

Let's now use Eq. (14.65) to compute the first few b_k. Starting from an arbitrary value of b_0, we get (assuming for now that n is a nonnegative integer)

$$b_1 = -\frac{b_0}{4(1)(n+1)}\,,$$

$$b_2 = -\frac{b_1}{4(2)(n+2)} = +\frac{b_0}{4^2(1)(2)(n+1)(n+2)}\,,$$

$$b_3 = -\frac{b_2}{4(3)(n+3)} = -\frac{b_0}{4^3(1)(2)(3)(n+1)(n+2)(n+3)}\,,$$

from which we see that the general formula for the b_k is

$$b_k = \frac{(-1)^k n!\, b_0}{2^{2k} k!\,(n+k)!}\,. \tag{14.66}$$

Substituting this formula for b_k into Eq. (14.63), giving $y(x)$ the name $J_n(x)$, and setting $b_0\, n!\, 2^n = 1$ to cause J_n to have its conventional scaling, we have

$$J_n(x) = \sum_{k=0}^{\infty} \frac{(-1)^k}{k!\,(n+k)!} \left(\frac{x}{2}\right)^{n+2k}\,, \qquad \text{(nonnegative integer } n\text{)}. \tag{14.67}$$

J_n is called a **Bessel function** of order n, or sometimes, to distinguish it from the other solution to the Bessel equation of the same n^2, a **Bessel function of the first kind**.

Since we really want a formula that is valid even if n is not an integer, we rewrite $(n+k)!$ in Eq. (14.66) as a gamma function, changing n to ν to reinforce the notion that the order of the Bessel function is not necessarily integral. The result is

$$J_\nu(x) = \sum_{k=0}^{\infty} \frac{(-1)^k}{k!\,\Gamma(\nu+k+1)} \left(\frac{x}{2}\right)^{\nu+2k}\,. \tag{14.68}$$

In the development leading to Eq. (14.68) we have assumed n (and now ν) to be nonnegative. However, the indicial equation $s^2 - \nu^2 = 0$ has solution $-\nu$ as well as $+\nu$. If we repeat the analysis with ν assumed negative, we still reach Eq. (14.68), but now without a restriction on the sign of ν.

So long as ν is not an integer, Eq. (14.68) produces an expansion for $J_{-\nu}$ that starts with a different power of x and is certainly linearly independent of J_ν. $J_{-\nu}$ is a second solution to the same ODE as J_ν (the same ODE because the Bessel ODE of order ν depends only on ν^2). However, for $\nu = -n$, a negative integer, we have the problem that the first n terms of the expansion have infinite denominators (due to the singularities of the gamma function) and all are therefore zero. The first nonzero term of the expansion for J_{-n} contains $(x/2)^n$. Examining the resulting series in more detail, we find that

$$J_{-n}(x) = (-1)^n J_n(x)\,, \qquad \text{(integer } n). \tag{14.69}$$

In other words, J_n and J_{-n} are linearly dependent, so, for integer n, we have only one solution to the Bessel ODE.

For reasons we do not explain here, the standard form adopted for a second solution to the Bessel ODE is a linear combination of J_ν and $J_{-\nu}$:

$$Y_\nu(x) = \frac{\cos(\pi\nu)J_\nu(x) - J_{-\nu}(x)}{\sin \pi\nu}\,. \tag{14.70}$$

This functional form is clearly a solution of the Bessel ODE and can be evaluated in a straightforward fashion if ν is not an integer. However, if ν is integral, the formula for Y_ν becomes an indeterminate form of the type 0/0 with a limit that is identified as Y_ν. The function Y_ν is sometimes called a **Bessel function of the second kind.**[2]

The key point of the current discussion is that a well-defined mathematical procedure has been used to define a second solution to the Bessel ODE, so that for a given value of the parameter n^2 in that ODE, the ODE will have the linearly independent solutions J_n and Y_n, with $n \geq 0$. The fact that it may be tedious to evaluate these functions by hand is of limited relevance; their numerical values are readily available via symbolic computing.

SYMBOLIC COMPUTING

Bessel functions are known to our symbolic computing systems. To access $J_n(x)$ or $Y_n(x)$, use

`BesselJ(n,x)` or `BesselY(n,x)` (MAPLE),

`BesselJ[n,x]` or `BesselY[n,x]` (MATHEMATICA).

The symbol `n` need not be an integer and may be of either sign.

A call to `BesselJ` or `BesselY` does not produce numerical output unless both arguments are numerical and at least one is a decimal quantity. However, numerical evaluation can be forced using `evalf` or `N`. Thus,

MAPLE:

```
> BesselJ(1,2);              BesselJ(1,2)
> BesselJ(1,2.);             0.5767248078
> evalf(BesselJ(1,2));       0.5767248078
```

[2]The notation Y_ν for these Bessel functions is that most widely used now. Sometimes one encounters the older notation N_ν and the name **Neumann function**.

MATHEMATICA:	`BesselJ[5/4,3]`	$\text{BesselJ}\left[\frac{5}{4},\ 3\right]$
	`BesselJ[1.25,3]`	0.426601
	`N[BesselJ[5/4,3],8]`	0.42660130

Exercises

14.4.1. Write symbolic code that evaluates a Bessel function from its power series, truncated after 10 terms. How successful is this truncated expansion at representing $J_0(1)$, $J_0(10)$, $J_{16}(1)$, and $J_{16}(10)$?

14.4.2. Exercise 14.4.1 shows that the power series will not always be a practical way to evaluate Bessel functions. Fortunately, the extensive study received by Bessel functions has let to many alternate methods for their evaluation. A useful integral representation (for integer n) is

$$J_n(x) = \frac{1}{\pi}\int_0^{\pi} \cos(x\sin\theta - n\theta)\,d\theta\,.$$

Carrying out numerical integrations, check this formula against the values given by your symbolic program for $J_0(10)$ and $J_{16}(10)$.

PLOTS OF J_n and Y_n

The series for $J_n(x)$ in Eq. (14.67) converges more rapidly than that for the exponential and is therefore convergent for all n and for all finite values of x. Thus, our series for J_n provides a complete description of that function, and Eq. (14.70) then fully defines Y_n. Though we cannot reduce these functions to anything more elementary than the sums that define them, we need to regard them as known functions. After all, while we cannot reduce the expansions of $\sin x$, $\cos x$, or e^x to simpler objects, nor can we easily evaluate them to numerical values without a computer or tables, we nevertheless regard them as known, familiar quantities. It is the objective of the present subsection to develop some familiarity and ease with respect to our new friends, the Bessel functions.

A starting point for gaining an understanding of $J_n(x)$ and $Y_n(x)$ is to plot them for various values of n. The four panels of Fig. 14.4 show that for all n, both J_n and Y_n are oscillatory, with (for $x \gg n$) equal oscillation amplitudes that decay gradually as x increases. The oscillation is (for both functions) of approximate period 2π, with the phases of the two oscillations differing by approximately 90°. The period and phase difference approach those respective values in the limit of large x. In that limit, these functions are essentially slightly damped sines and cosines. The singularity at $x = 0$ in the Bessel ODE shows up in the behavior of its solutions for small x: $J_0(0) = 1$, while for $n > 0$, $J_n(0) = 0$; but $Y_n(x)$ is singular at $x = 0$ for all n.

The behavior shown in the plots can be rationalized if we write the Bessel ODE in the form

$$y'' + \frac{1}{x}\,y' + \left(1 - \frac{n^2}{x^2}\right) y = 0\,. \tag{14.71}$$

Comparing with the ODE for a damped oscillator (see Example 10.4.2), we note that Eq. (14.71) looks like the equation of such an oscillator as soon as n^2/x^2

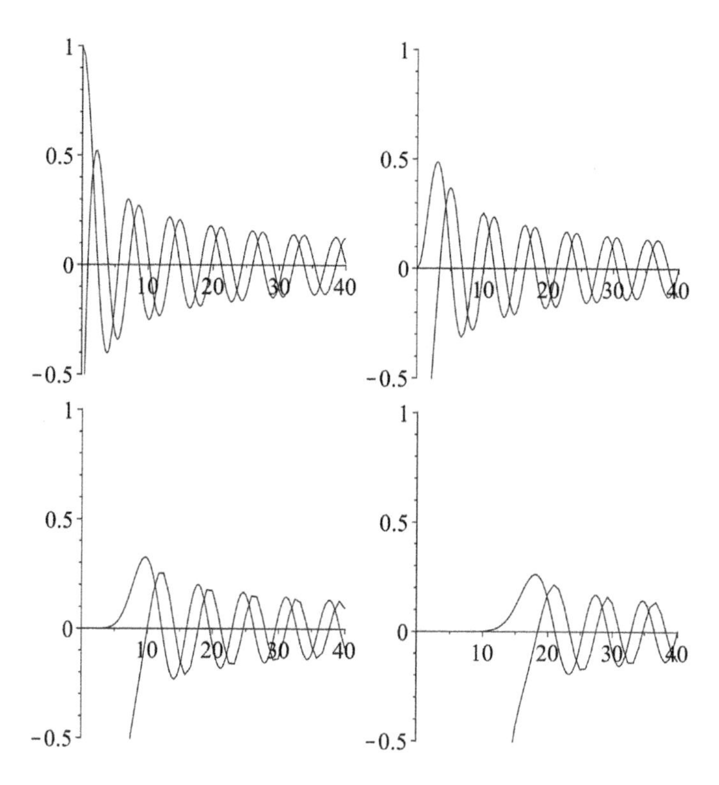

Figure 14.4: Bessel functions $J_n(x)$ and $Y_n(x)$, for $n = 0$, 2, 8, and 16.

becomes small, with a damping (the y' term) that diminishes as x increases. The main difference relative to the damped oscillator is at small x, the neighborhood of the singularity in the ODE.

RECURRENCE FORMULAS

Bessel functions of contiguous orders satisfy recurrence relations; those of most importance are the following:

$$\frac{2\nu}{x}\,J_\nu(x) = J_{\nu-1}(x) + J_{\nu+1}(x)\,, \tag{14.72}$$

$$2J'_\nu(x) = J_{\nu-1}(x) - J_{\nu+1}(x)\,, \tag{14.73}$$

$$\frac{d}{dx}\left[x^\nu J_\nu(x)\right] = x^\nu J_{\nu-1}(x)\,, \tag{14.74}$$

$$\frac{d}{dx}\left[x^{-\nu} J_\nu(x)\right] = -x^{-\nu} J_{\nu+1}(x)\,. \tag{14.75}$$

These equations are satisfied also by the $Y_\nu(x)$. They are written for order ν to emphasize that they apply whether or not the Bessel functions are of integer order, and for either sign of ν. The recurrence formulas may be proved for J_ν using the series expansion, Eq. (14.68), and for Y_ν using also Eq. (14.70).

Exercises

14.4.3. Use a recurrence formula and the relation between J_n and J_{-n} to show that $J_0'(x) = -J_1(x)$.

14.4.4. Use the result from Exercise 14.4.3 to show that

$$\int_0^\infty J_1(x)\,dx = 1\,.$$

14.4.5. It can be shown that there exists a generating function for Bessel functions of integral order corresponding to the formula

$$g(x,t) = e^{x(t-t^{-1})/2} = \sum_{n=-\infty}^{\infty} J_n(x)t^n\,.$$

Note that the expansion of $g(x,t)$ contains an infinite number of terms with each power of t, so the quantities defined by the generating function will not be polynomials.

By differentiating the formula for $g(x,t)$, derive the recurrence formulas given in Eqs. (14.72) and (14.73).

14.4.6. The Bessel generating function given in Exercise 14.4.5 has the property that $g(-x,1/t) = g(-x,-t) = g(x,t)$. Use those relationships to show that $J_{-n}(x) = (-1)^n J_n(x)$.

ODEs WITH BESSEL FUNCTION SOLUTIONS

There are many ODEs that can be transformed in a way that makes their solutions identifiable in terms of Bessel functions. We mention here two important cases of this type. First, consider

$$x^2y''(x) + xy'(x) + \left(k^2x^2 - n^2\right)y(x) = 0\,. \qquad (14.76)$$

This equation has the solutions $J_n(kx)$ and $Y_n(kx)$. To verify these solutions, write $u = kx$ and $y(x) = z(u)$, which requires us to make the following changes in Eq. (14.76):

$$k^2x^2 = u^2, \qquad x\,\frac{dy}{dx} = \frac{u}{k}\frac{dz}{du}\frac{du}{dx} = u\,z', \qquad x^2\,\frac{d^2y}{dx^2} = \frac{u^2}{k^2}\frac{d^2z}{du^2}\left(\frac{du}{dx}\right)^2 = u^2z''\,.$$

These changes bring us to

$$u^2z''(u) + uz'(u) + (u^2 - n^2)z(u) = 0\,,$$

which has the solutions $z(u) = J_n(u)$ and $Y_n(u)$, i.e., equivalent to the stated solutions $J_n(kx)$ and $Y_n(kx)$.

Now look at

$$y'' + \frac{2}{x}\,y' + \left[k^2 - \frac{l(l+1)}{x^2}\right]y = 0\,, \qquad (14.77)$$

which we will show has solutions of the form

$$x^{-1/2}J_{l+1/2}(kx) \quad \text{and} \quad x^{-1/2}Y_{l+1/2}(kx). \qquad (14.78)$$

Our proof starts by rewriting Eq. (14.77) in the form

$$x^2y'' + 2xy' + \left[k^2x^2 - l(l+1)\right] y = 0 . \tag{14.79}$$

This equation would look more like a Bessel ODE if it did not have the factor 2 in the y' term; that factor can be removed by the substitution $y = x^{-1/2}z$. We then have

$$y' = x^{-1/2}z' - \tfrac{1}{2}\, x^{-3/2}z\,, \qquad y'' = x^{-1/2}z'' - x^{-3/2}z' + \tfrac{1}{2}\tfrac{3}{2}\, x^{-1/2}z\,,$$

and Eq. (14.79) can be reduced to

$$x^2z'' + xz' + \left[k^2x^2 - \left(l + \tfrac{1}{2}\right)^2\right] z = 0 . \tag{14.80}$$

Comparing with Eq. (14.76), we see that Eq. (14.80) has the solutions $z(x) = J_{l+1/2}(kx)$ and $z(x) = Y_{l+1/2}(kx)$, consistent with the solutions to Eq. (14.77) given in Eq. (14.78).

Exercises

14.4.7. Verify that the substitution $y = x^{-1/2}z$ converts Eq. (14.79) into Eq. (14.80).

14.4.8. By making the substitution $y = x^c Z_\nu(ax^b)$ into the ODE

$$y'' + \frac{1-2c}{x}y' + \left[\left(abx^{b-1}\right)^2 + \frac{c^2 - \nu^2b^2}{x^2}\right] y = 0$$

verify that y is a solution when Z_ν is a Bessel function of order ν.

ORTHOGONALITY

Bessel functions of different orders do not satisfy a useful orthogonality relation; instead there is an important orthogonality among Bessel functions of the same order, but with arguments of different scales. This behavior is analogous to that of the trigonometric functions, for which (with m and n integers)

$$\int_0^1 \sin n\pi x \sin m\pi x \, dx = 0\,, \qquad m \neq n.$$

The relationship we shall establish is of the form (valid only for certain α and β)

$$\int_0^1 J_\nu(\alpha x) J_\nu(\beta x) w(x)\, dx = 0\,, \qquad \alpha \neq \beta. \tag{14.81}$$

Because our situation is more general that involving the sine functions, we need a weight factor in the scalar product and we must recognize that the scale factors α and β may not be proportional to integers.

The way to proceed is provided by the analysis in Section 11.6 (Sturm-Liouville theory), which deals with the orthogonality properties of the eigenfunctions of second-order differential operators. If we write the ODE of Eq. (14.76) in the form

$$\mathcal{L}y = \left[\frac{d^2}{dx^2} + \frac{1}{x}\frac{d}{dx} - \frac{\nu^2}{x^2}\right] y = -k^2 y\,, \tag{14.82}$$

we see that for any fixed ν, Eq. (14.82) is an eigenvalue equation with $-k^2$ as eigenvalues. Since the ODE in question has solutions $J_\nu(kx)$, we have identified the Bessel functions of any given ν, but various k, as eigenfunctions of the same $\mathcal{L}$.

Next we look to Section 11.6 to tell us the conditions under which eigenfunctions of different eigenvalues (here $J_\nu(kx)$ of different k) are orthogonal. The first requirement is that the orthogonality must be determined using a scalar product

$$\langle\varphi|\chi\rangle = \int_a^b \varphi^*(x)\chi(x)\,w(x)\,dx$$

containing the weight factor needed to make $\mathcal{L}$ a self-adjoint ODE operator. That factor is x; to check this claim we note that

$$x\left(y'' + \frac{1}{x}y'\right) = (xy')'.$$

Second, the endpoints a and b of the scalar-product integral must be such that for any two eigenfunctions $u = J_\nu(\alpha x)$ and $v = J_\nu(\beta x)$, where α and β are the k values corresponding to two eigenfunctions,

$$w(x)(uv' - u'v)\Big|_a^b = x(uv' - u'v)\Big|_a^b = 0. \tag{14.83}$$

We plan to take $a = 0$ and $b = 1$, and the presence of the factor x in Eq. (14.83) causes that equation to reduce to the condition

$$u(1)v'(1) - u'(1)v(1) = J_\nu(\alpha)J'_\nu(\beta) - J'_\nu(\alpha)J_\nu(\beta) = 0. \tag{14.84}$$

One way to satisfy Eq. (14.84) is to choose α and β such that $J_\nu(\alpha) = J_\nu(\beta) = 0$. We refer to these values of α or β as **zeros of** J_ν. We note in passing that there are other possibilities for satisfying Eq. (14.84) that we do not consider in this text.

Summarizing our findings to this point, we have made the observation that the functions $J_\nu(\alpha x)$ and $J_\nu(\beta x)$ are orthogonal if α and β are zeros of J_ν, when the scalar product is defined as

$$\langle\varphi|\chi\rangle = \int_0^1 \varphi^*(x)\chi(x)\,x\,dx\,,$$

i.e., that

$$\int_0^1 J_\nu(\alpha x)J_\nu(\beta x)\,x\,dx = 0 \tag{14.85}$$

if α and β are different zeros of J_ν.

Since we will want to use the $J_\nu(\alpha x)$ in orthogonal expansions, we will also need the normalization integral $\langle J_\nu(\alpha x)|J_\nu(\alpha x)\rangle$; when α is a zero of J_ν (**and only then**), it can be shown that

$$\langle J_\nu(\alpha x)|J_\nu(\alpha x)\rangle = \int_0^1 \Big[J_\nu(\alpha x)\Big]^2 x\,dx = \frac{1}{2}\Big[J_{\nu+1}(\alpha)\Big]^2. \tag{14.86}$$

ZEROS OF BESSEL FUNCTIONS

To use the orthogonality relations we have just derived we will need to have available the zeros of the Bessel function J_ν of the order ν we plan to use. While the zeros of Bessel functions occur at intervals of approximate length π, there are no closed formulas for them, and before the advent of computers the zeros were available only in tables that were produced by laborious hand computations. We can now get the zeros of Bessel functions directly from symbolic computing systems; to do so, use

`BesselJZeros(n,j)` or `BesselYZeros(n,j)` (MAPLE),

`BesselJZero[n,j]` or `BesselYZero[n,j]` (MATHEMATICA).

Note that in these names, **Zeros** is plural in MAPLE but **Zero** is singular in MATHEMATICA. In both systems, these calls produce the x value of the jth positive zero of $J_n(x)$ or $Y_n(x)$ (if $x = 0$ is a zero it is not regarded as a positive zero). In MAPLE, j must be an integer (a floating-point integer is not accepted); in MATHEMATICA, a floating-point integer is accepted for j and nonintegral j produces (without notice) an erroneous result. In both symbolic systems, only a symbolic result is returned when neither argument is a decimal quantity.

MAPLE:	`> BesselJZeros(1,2);`	`BesselJZeros(1,2)`
	`> BesselJZeros(1,2.);`	(error)
	`> BesselJZeros(1.,2);`	7.015586670
	`> evalf(BesselJZeros(1,2));`	7.015586670
MATHEMATICA:	`BesselJZero[5/4,3]`	`BesselJZero` $\left[\frac{5}{4}, 3\right]$
	`BesselJZero[1.25,3]`	10.5408
	`BesselJZero[1.25,3.5]`	12.1197 (this is nonsense)
	`N[BesselJZero[5/4,3],8]`	10.540833

BESSEL SERIES

If we let $\alpha_{\nu j}$ stand for the jth positive zero of J_ν, then the functions $J_\nu(\alpha_{\nu j}x)$, $j = 1, 2, \ldots$ form an orthogonal basis that can be used to expand arbitrary functions. The successive values of j correspond to basis members with increasing numbers of nodes within the range $(0, 1)$. See Fig. 14.5. We might expect that the convergence rate of a Bessel series might be comparable to that of a Fourier sine series.

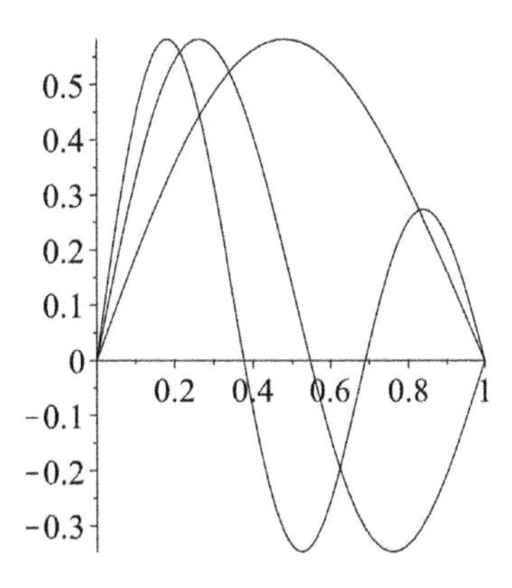

Figure 14.5: First three orthogonal Bessel functions of order 1.

Using the usual rules for orthogonal expansions, we write the expansion of $f(x)$ in terms of Bessel functions of order ν as

$$f(x) = \sum_{j=1}^{\infty} c_j J_\nu(\alpha_{\nu j} x), \qquad \text{with} \quad c_j = \frac{\langle J_\nu(\alpha_{\nu j} x) | f(x) \rangle}{\langle J_\nu(\alpha_{\nu j} x) | J_\nu(\alpha_{\nu j} x) \rangle}. \tag{14.87}$$

Inserting the formula for the normalization integral, Eq. (14.86), and writing the numerator of Eq. (14.87) as an integral, we have

$$c_j = \frac{2}{[J_{\nu+1}(\alpha_{\nu j})]^2} \int_0^1 J_\nu(\alpha_{\nu j} x)\, f(x)\, x\, dx. \tag{14.88}$$

We may want to use a Bessel series on an interval other than $(0, 1)$, and it is easy to change the interval to $(0, a)$. Replacing x by x/a in the basis functions and recalculating the normalization integral, we reach the more general formulas

$$f(x) = \sum_{j=1}^{\infty} c_j J_\nu\left(\frac{\alpha_{\nu j} x}{a}\right), \qquad c_j = \frac{2}{a^2 [J_{\nu+1}(\alpha_{\nu j})]^2} \int_0^a J_\nu\left(\frac{\alpha_{\nu j} x}{a}\right) f(x)\, x\, dx. \tag{14.89}$$

Example 14.4.1. A Bessel Series

Let's expand $F(x) = x(1 - x)$ in a Bessel series on the interval $(0, 1)$, using Bessel functions of order 1. Obtaining this expansion would require considerable effort in the absence of symbolic computing availability, but is straightforward in both MAPLE and MATHEMATICA.

Taking MAPLE first, we write a procedure to generate the coefficients by evaluation of Eq. (14.88). Letting n be the maximum number of coefficients we want to compute, we write

```
> CC := proc(C,n) local j,Z,x,F;
>   F := x*(1-x);
>   for j from 1 to n do
>     Z := BesselJZeros(1,j);
>     C[j] := evalf(int(x * F * BesselJ(1,Z*x), x=0 .. 1)
>                * 2/BesselJ(2,Z)^2)   end do
> end proc;
```

We next execute our procedure for some value of n; here we choose $n = 5$, and then build and plot FX, the expansion of F:

```
> CC(C,5):
> FX := add(C[j] * BesselJ(1,BesselJZeros(1,j)*x), j = 1 .. 5);
> plot([FX, x*(1-x), 2.5*(FX-x*(1-x))], x = 0 .. 1);
```

Already at $n = 2$ the expansion is fairly accurate, as can be seen in Fig. 14.6, which also contains (at 2.5 times magnification) the error in the expansion.

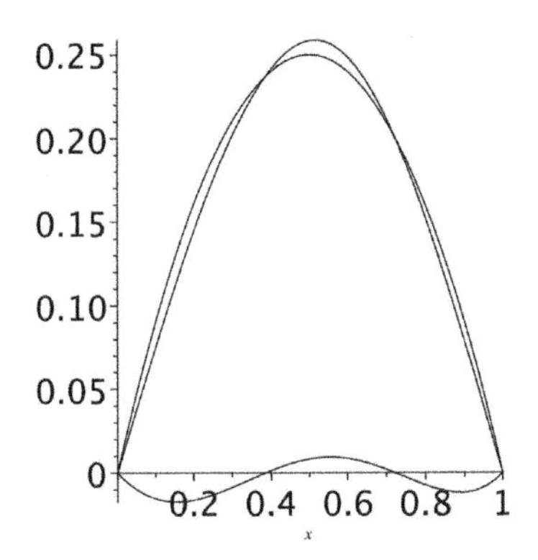

Figure 14.6: The function $x(1-x)$ and its two-term expansion in Bessel functions of order 1. Also shown (at 2.5 times magnification) is the error in the expansion.

In MATHEMATICA, we write a procedure to generate the coefficients by evaluation of Eq. (14.88). Letting n be the maximum number of coefficients we want to compute, we write

```
cc[c_,n_] := Module[{f,x,z,j},
  f = x*(1-x);
  Do[z = BesselJZero[1,j];
     c[j] = N[Integrate[x * f * BesselJ[1,z*x], {x,0,1}]
           * 2/BesselJ[2,z]^2],
   {j,1,n}] ]
```

We next execute our procedure for some value of n; here we choose $n = 5$, and then build and plot fx, the expansion of f:

```
cc(c,5);
fx = Sum[c[j] * BesselJ[1, BesselJZero[1,j]*x], {j,1,5}]
Plot[{fx, x*(1-x), 2.5*(fx-x*(1-x))}, {x,0,1}]
```

Already at $n = 2$ the expansion is fairly accurate, as can be seen in Fig. 14.6, which also contains (at 2.5 times magnification) the error in the expansion.

At $n = 5$ the curves are virtually indistinguishable.

■

Exercises

14.4.9. Find the first two positive zeros of $J_1(x)$ by plotting it and then restricting the range of the plot to the neighborhood of a zero until to its location can be read out to sufficient precision. Check your results by comparison with the output of BesselJZero or BesselJZeros.

14.4.10. Show that

(a) $J_{\nu-1}(x) = J_{\nu+1}(x)$ at every maximum or minimum of $J_\nu(x)$.

(b) $J_{\nu-1}(x) = -J_{\nu+1}(x)$ at every zero of $J_\nu(x)$.

14.4.11. Expand $\sin x$ on the interval $0 \le x \le \pi$ in the Bessel series

$$\sin x = \sum_{j=1}^{\infty} a_j J_1(\alpha_{1j} x/\pi)\,,$$

where α_{1j} is the jth zero of J_1. Make a plot showing the accuracy of your expansion; do not truncate before $\sin x$ and its expansion are nearly indistinguishable on your plot.

Hint. Remember to use the normalization integral appropriate for the interval $(0, \pi)$.

14.4.12. Expand the step function

$$f(x) = \begin{cases} 1, & 0 \le x \le \frac{1}{2}, \\ 0, & \frac{1}{2} \le x \le 1, \end{cases}$$

as a series in Bessel functions of order zero.

LIMITING VALUES

Limiting values of $J_\nu(x)$ and $Y_\nu(x)$ for small x can be obtained from the power-series expansion of $J_{\pm\nu}$ and the definition of Y_ν. Asymptotic values in the large-x limit can also be established. These limiting values are (for $\nu \ge 0$)

$$\begin{aligned}
x \to 0^+: &\quad J_\nu(x) = \frac{1}{\Gamma(\nu+1)}\left(\frac{x}{2}\right)^\nu, \\
x \to \infty: &\quad J_\nu(x) \sim \sqrt{\frac{2}{\pi x}}\cos\left(x - \frac{2\nu+1}{4}\pi\right), \\
x \to 0^+: &\quad Y_\nu(x) = \begin{cases} \dfrac{2}{\pi}\ln x, & \nu = 0, \\ -\dfrac{\Gamma(\nu)}{\pi}\left(\dfrac{2}{x}\right)^\nu, & \nu > 0, \end{cases} \\
x \to \infty: &\quad Y_\nu(x) \sim \sqrt{\frac{2}{\pi x}}\sin\left(x - \frac{2\nu+1}{4}\pi\right).
\end{aligned} \tag{14.90}$$

Exercises

14.4.13. Show that $\displaystyle\lim_{x\to 0}\frac{J_1(x)}{x} = \frac{1}{2}$.

14.4.14. Find $\displaystyle\lim_{x\to 0} J_\nu(x)/Y_\nu(x)$.

14.5 OTHER BESSEL FUNCTIONS

Bessel's equation, being of second order, has only two independent solutions, and by studying $J_\nu(x)$ and $Y_\nu(x)$ we have already discussed both. However, one can form expressions involving these solutions that are convenient for various applications, and it is useful to be able to recognize these alternate functional forms.

HANKEL FUNCTIONS

Continuing the analogy with the trigonometric functions, we have many times pointed out that $\sin x$ and $\cos x$ are solutions of the ODE $y'' + y = 0$, and we also know that $e^{\pm ix} = \cos x \pm i \sin x$ form an alternate, but equivalent solution set. In the Bessel-function world, the corresponding alternate solution set is the **Hankel functions**, sometimes called **Bessel functions of the third kind**, defined as

$$\begin{aligned} H_\nu^{(1)}(x) &= J_\nu(x) + iY_\nu(x)\,, \\ H_\nu^{(2)}(x) &= J_\nu(x) - iY_\nu(x)\,. \end{aligned} \tag{14.91}$$

These functions see use in problems involving incoming or outgoing waves, because the oscillation of J_ν and Y_ν is converted into a large-x behavior of e^{ix} for $H^{(1)}(x)$ and e^{-ix} for $H^{(2)}(x)$. For real x, $H_\nu^{(1)}(x)$ and $H_\nu^{(2)}(x)$ are complex conjugates.

The Hankel functions are known to our symbolic computing systems as

`HankelH1(n,x)` and `HankelH2(n,x)` (MAPLE)

`HankelH1[n,x]` and `HankelH2[n,x]` (MATHEMATICA)

Sample computations:

MAPLE:	`> HankelH1(2,1.23);`	0.1663693838 − 1.220292439 I
	`> BesselJ(2,1.23);`	0.1663693838
	`> BesselY(2,1.23);`	−1.220292439
MATHEMATICA:	`HankelH2[2,1.23]`	0.166369 + 1.22029 i
	`BesselJ[2,1.23]`	0.166369
	`BesselY[2,1.23]`	−1.22029

Exercises

14.5.1. (a) Using the results in Eq. (14.90), find the limiting forms of $H_\nu^{(1)}(x)$ and $H_\nu^{(2)}(x)$ at $x \to \infty$.

(b) If these expressions are multiplied by $e^{-i\omega t}$, explain why the resulting function of x and t is identified as an outgoing or incoming wave. Determine which combination is "outgoing" and which is "incoming."

MODIFIED BESSEL FUNCTIONS

Modified (also called hyperbolic) Bessel functions are solutions of the ODE

$$x^2y''(x) + xy'(x) - (x^2 + n^2)y(x) = 0\,, \tag{14.92}$$

which corresponds to Eq. (14.76) with $k = i$. Since the solutions to Eq. (14.76) are $J_n(kx)$ and $Y_n(kx)$, the solutions to Eq. (14.92) are $J_n(ix)$ and $Y_n(ix)$. These quantities are analogous to the hyperbolic functions $\sinh x$ and $\cosh x$, which are related to the circular functions by $\sin(ix) = i \sinh x$ and $\cos(ix) = \cosh x$. It is customary to identify the solutions to Eq. (14.92) as $I_n(x)$ and $K_n(x)$, with definitions

$$I_n(x) = i^{-n} J_n(ix)\,,$$
$$K_n(x) = \frac{\pi}{2}\, i^{n+1} H_n^{(1)}(ix)\,. \tag{14.93}$$

The powers of i in Eq. (14.93) have been inserted to make $I_n(x)$ and $K_n(x)$ real when x is real.

Looking at the series expansion of I_n, i.e., of $J_n(ix)$, we see that its dependence on ix causes all the terms to have the same sign, so I_n is not oscillatory, but increases without limit as x is increased. The large-x behavior of K_n can be deduced from that of $H_n^{(1)}$. We already noted that $H_n^{(1)}(x)$ behaves asymptotically as e^{ix}, which means that $H_n^{(1)}(ix)$ has asymptotic dependence $e^{i(ix)} = e^{-x}$. In other words, $K_n(x)$ goes exponentially to zero at large x.

Summarizing, the usual definitions of the modified Bessel functions make I_n an analog of $\sinh x$ and K_n analogous to e^{-x}. The modified Bessel functions occur in many physical problems.

The modified Bessel functions are known to our symbolic computing systems as

`BesselI(n,x)` and `BesselK(n,x)` (MAPLE),

`BesselI[n,x]` and `BesselK[n,x]` (MATHEMATICA).

Both symbolic systems use the same definitions of I_n and K_n as this text.

Sample computations:

MAPLE:	`> BesselI(2,1.23);`	0.2141109561
	`> I^(-2)*BesselJ(2,1.23*I);`	0.2141109561
	`> BesselK(2,1.23);`	0.9801528316
	`> (Pi/2)*I^3*HankelH1(2,1.23*I);`	$(0.31199123363 - 0.\text{I})\pi$
	`> evalf(Pi/2)*I^3*HankelH1(2,1.23*I);`	$0.9801528318 - 0.\text{I}$
MATHEMATICA:	`BesselI[2,1.23]`	0.214111
	`I^(-2)*BesselJ[2,1.23*I]`	$0.24111 + 0.\text{i}$
	`BesselK[2,1.23]`	0.980153
	`(Pi/2)*I^3*HankelH1[2,1.23*I]`	$0.980153 + 0.\text{i}$

Exercises

14.5.2. Show that as $x \to \infty$, $I_\nu(x)$ and $K_\nu(x)$ have the limiting forms

$$I_\nu(x) \sim \frac{1}{\sqrt{2\pi x}}\, e^x\,, \qquad K_\nu(x) \sim \sqrt{\frac{\pi}{2x}}\, e^{-x}\,.$$

14.5.3. Develop recurrence relations for I_n and K_n that are parallel to those given for J_n in Eqs. (14.72) and (14.73). Check your work by substitution of values of I_n and K_n obtained by symbolic computing.

14.5.4. Referring to the answer to Exercise 14.4.8, show that the Airy differential equation,

$$y'' - xy = 0\,,$$

has solutions that can be written in the standard forms

$$\mathrm{Ai}(x) = \frac{1}{\pi}\sqrt{\frac{x}{3}}\,K_{1/3}\left(\tfrac{2}{3}\,x^{3/2}\right),$$

$$\mathrm{Bi}(x) = \sqrt{\frac{x}{3}}\left[I_{-1/3}\left(\tfrac{2}{3}\,x^{3/2}\right) + I_{1/3}\left(\tfrac{2}{3}\,x^{3/2}\right)\right].$$

SPHERICAL BESSEL FUNCTIONS

The ODE

$$x^2y'' + 2xy' + \left[x^2 - n(n+1)\right]y = 0 \tag{14.94}$$

arises when the Helmholtz equation is solved in spherical polar coordinates; its solutions are known as **spherical Bessel functions**. This equation is the case $k = 1$ of the ODE in Eq. (14.79), and its solutions are conventionally written as

$$j_n(x) = \sqrt{\frac{\pi}{2x}}\,J_{n+1/2}(x)\,, \tag{14.95}$$

$$y_n(x) = \sqrt{\frac{\pi}{2x}}\,Y_{n+1/2}(x)\,. \tag{14.96}$$

It turns out that for integral n both the j_n and y_n become closed expressions involving $\sin x$, $\cos x$, and negative powers of x. Our first step in confirming this behavior is to write $j_0(x)$ using the series expansion, Eq. (14.68), for $J_{1/2}$. We get

$$j_0(x) = \sqrt{\frac{\pi}{2x}}\sum_{k=0}^{\infty}\frac{(-1)^k}{k!\,\Gamma(k+3/2)}\left(\frac{x}{2}\right)^{2k+1/2}. \tag{14.97}$$

Inserting

$$\Gamma(k+3/2) = \frac{\sqrt{\pi}\,(2k+1)!!}{2^{k+1}}\,,$$

and simplifying, using the relation

$$2^k\,k!\,(2k+1)!! = (2k+1)!\,,$$

we find

$$j_0(x) = \sum_{k=0}^{\infty}\frac{(-1)^k\,x^{2k}}{(2k+1)!} = \frac{1}{x}\sum_{k=0}^{\infty}\frac{(-1)^k\,x^{2k+1}}{(2k+1)!} = \frac{\sin x}{x}\,. \tag{14.98}$$

We now see why the factor $\sqrt{\pi/2x}$ was included in the definition of j_n; that factor causes $j_0(0)$ to be unity (the limit as $x \to 0$ of $\sin x/x$).

A process similar to that which led to Eq. (14.98) can also be applied to $j_{-1}(x)$. The result is

$$j_{-1}(x) = \sqrt{\frac{\pi}{2x}}\sum_{k=0}^{\infty}\frac{(-1)^k}{k!\,\Gamma(k+1/2)}\left(\frac{x}{2}\right)^{2k-1/2},$$

which simplifies to

$$j_{-1}(x) = \frac{(-1)^k\, x^{2k-1}}{(2k)!} = \frac{\cos x}{x}\,. \tag{14.99}$$

From j_0 and j_{-1} we can make all other j_n recursively. To do so, we use Eq. (14.72) with $\nu = n + 1/2$, corresponding to the formula

$$\frac{2n+1}{x}\, j_n(x) = j_{n-1}(x) + j_{n+1}(x)\,. \tag{14.100}$$

It is obvious that repeated use of Eq. (14.100) will produce a linear combination of sines and cosines, each multiplied by a negative power of x. Despite the fact that the closed forms for $j_n(x)$ (for $n \geq 0$) seem to be singular at $x = 0$ (see Table 14.3), they actually behave for small x as x^n.

To analyze the spherical Bessel functions y_n, we note that when Eq. (14.70) is used to find $Y_{1/2}$, we have $\cos(\pi/2) = 0$ and $\sin(\pi/2) = 1$, so we get $Y_{1/2} = -J_{-1/2}$, corresponding to

$$y_0(x) = -j_{-1}(x) = -\frac{\cos x}{x}\,. \tag{14.101}$$

We can also establish that $y_{-1}(x) = +j_0(x)$ and use Eq. (14.100), which is also valid for y_n, to obtain $y_n(x)$ for general positive integers n. Some y_n are given in Table 14.3.

There are also spherical versions of the other Bessel functions:

$$h_n^{(1)}(x) = \sqrt{\frac{\pi}{2x}}\, H_{n+1/2}^{(1)}(x) = j_n(x) + i y_n(x)\,, \tag{14.102}$$

$$h_n^{(2)}(x) = \sqrt{\frac{\pi}{2x}}\, H_{n+1/2}^{(2)}(x) = j_n(x) - i y_n(x)\,, \tag{14.103}$$

$$i_n(x) = \sqrt{\frac{\pi}{2x}}\, I_{n+1/2}(x)\,, \tag{14.104}$$

$$k_n(x) = \sqrt{\frac{2}{\pi x}}\, K_{n+1/2}(x)\,. \tag{14.105}$$

Table 14.3: Spherical Bessel functions.

$j_0(x) = \dfrac{\sin x}{x}$	$y_0(x) = -\dfrac{\cos x}{x}$
$j_1(x) = \dfrac{\sin x}{x^2} - \dfrac{\cos x}{x}$	$y_1(x) = -\dfrac{\cos x}{x^2} - \dfrac{\sin x}{x}$
$h_0^{(1)}(x) = -\dfrac{i}{x}\, e^{ix}$	$h_0^{(2)}(x) = \dfrac{i}{x}\, e^{-ix}$
$h_1^{(1)}(x) = \left(-\dfrac{1}{x} - \dfrac{i}{x^2}\right) e^{ix}$	$h_1^{(2)}(x) = \left(-\dfrac{1}{x} + \dfrac{i}{x^2}\right) e^{-ix}$
$i_0(x) = \dfrac{\sinh x}{x}$	$k_0(x) = \dfrac{e^{-x}}{x}$
$i_1(x) = \dfrac{\cosh x}{x} - \dfrac{\sinh x}{x^2}$	$k_1(x) = \left(\dfrac{1}{x} + \dfrac{1}{x^2}\right) e^{-x}$

Note that the factor premultiplying the Bessel function is different for k_n than for the other spherical Bessel functions. Some of each of these functions are also given in Table 14.3.

The behavior of the spherical Bessel functions for small arguments can be deduced from their series expansions, while the behavior for large arguments follow from the asymptotic formulas for J_ν and Y_ν. Useful specific results include the following:

$$\begin{aligned}
x \to 0^+ : &\qquad j_n(x) = \frac{x^n}{(2n+1)!!}, \\
x \to \infty : &\qquad j_n(x) \sim \frac{1}{x} \sin\left(x - \frac{n\pi}{2}\right), \\
x \to 0^+ : &\qquad y_n(x) = -\frac{(2n-1)!!}{x^{n+1}}, \\
x \to \infty : &\qquad y_n(x) \sim -\frac{1}{x} \cos\left(x - \frac{n\pi}{2}\right).
\end{aligned} \tag{14.106}$$

All these spherical Bessel functions can be accessed using symbolic programming, but not all have built-in definitions in either or both of our symbolic languages. In MAPLE, $j_n(x)$, $y_n(x)$, $h_n^{(1)}(x)$, $h_n^{(2)}(x)$ can be obtained from the procedures

```
> SphBesselJ:= proc(n,x);
>       simplify(sqrt(Pi/(2*x))*BesselJ(n+1/2,x),symbolic) end proc:
> SphBesselY:= proc(n,x);
>       simplify(sqrt(Pi/(2*x))*BesselY(n+1/2,x),symbolic) end proc:
> SphHankelH1:= proc(n,x);
>       simplify(sqrt(Pi/(2*x))*HankelH1(n+1/2,x),symbolic) end proc:
> SphHankelH2:= proc(n,x);
>       simplify(sqrt(Pi/(2*x))*HankelH2(n+1/2,x),symbolic) end proc:
```

The extra argument `symbolic` of `simplify` causes the two instances of fractional powers of x to be combined.

In MATHEMATICA, these functions have built-in definitions:

```
SphericalBesselJ[n,x]     SphericalBesselY[n,x]
SphericalHankelH1[n,x]    SphericalHankelH2[n,x]
```

In MATHEMATICA, the forms of the above built-in functions may not be explicitly displayed. To display them, use the command `FunctionExpand`. For example,

```
SphericalHankelH1[0,x]                          SphericalHankelH1[0,x]
FunctionExpand[SphericalHankelH1[0,x]]
```

$$-\frac{ie^{ix}}{x}$$

Neither symbolic system has built-in definitions for $i_n(x)$ or $k_n(x)$. MAPLE procedures for them are

```
> SphBesselI:= proc(n,x);
>       simplify(sqrt(Pi/(2*x))*BesselI(n+1/2,x),symbolic) end proc:
```

```
> SphBesselK:= proc(n,x);
>       simplify(sqrt(2/(Pi*x))*BesselK(n+1/2,x),symbolic) end proc:
```

In MATHEMATICA, the corresponding procedures are

```
SphericalBesselI[n_,x_]:=
          Simplify[Sqrt[Pi/(2*x)]*BesselI[n+1/2,x], x>0]
SphericalBesselK[n_,x_]:=
          Simplify[Sqrt[2/(Pi*x)]*BesselK[n+1/2,x], x>0]
```

The additional argument `x>0` of `Simplify` causes the fractional powers of x to be combined by MATHEMATICA.

Exercises

14.5.5. Plot $j_n(x)$ and $y_n(x)$ on the same graph for $n = 0$ and (on a different graph) for $n = 8$. Extend the range of x enough that the asymptotic behavior for large x is apparent. Enlarge a portion of each graph so that the asymptotic phases of j_n and y_n can be read out accurately and compared with the formulas in Eq. (14.106).

14.5.6. Verify the recurrence formula, Eq. (14.100), for both j_n and y_n, by using symbolic computing to check various cases.

14.5.7. Derive recurrence formulas for i_n and k_n, and use symbolic computing to check your formulas.

14.5.8. Prove that $y_{-1}(x) = j_0(x)$.

14.5.9. (a) Develop a recurrence formula for $j_n(x)$ parallel to Eq. (14.75).

(b) Use the formula you found in part (a) to write j_{n+1} in terms of a derivative of j_n, then in terms of two differentiations involving j_{n-1}, then in terms of three differentiations involving j_{n-2}, etc. Generalize to prove that

$$j_n(x) = x^n \left(-\frac{1}{x}\frac{d}{dx}\right)^n j_0(x) = x^n \left(-\frac{1}{x}\frac{d}{dx}\right)^n \left(\frac{\sin x}{x}\right).$$

14.5.10. Obtain a repeated-differentiation formula parallel to Exercise 14.5.9 for $y_n(x)$.

14.6 HERMITE EQUATION

The Hermite ODE is

$$y'' - 2xy' + 2ny = 0\,. \tag{14.107}$$

Here n is a parameter that, if restricted to integral values, causes this ODE to have polynomial solutions.

We attempt a series solution of the usual form,

$$y_n(x) = \sum_{j=0}^{\infty} a_j x^{s+j}\,. \tag{14.108}$$

Substituting this assumed form into the Hermite ODE and collecting the coefficients of successive powers of x, we find

$$s(s-1)a_0 = 0\,,$$

$$(s+1)s\,a_1 = 0\,, \tag{14.109}$$

$$(s+j+2)(s+j+1)a_{j+2} + (2n-2s-2j)a_j = 0\,, \qquad j = 0,\ 1,\ \ldots$$

From Eq. (14.109), we conclude that the possible values of s are $s = 0$ and $s = 1$, that a_0 is arbitrary, that $a_1 = 0$, and

$$a_{j+2} = \frac{2(s+j-n)\,a_j}{(s+j+1)(s+j+2)}\,. \tag{14.110}$$

We therefore conclude that $a_3 = a_5 = a_{2j+1} = 0$. For $s = 0$, Eq. (14.110) leads to the explicit formulas

$$a_2 = -\frac{2n}{2!}\,a_0\,, \qquad a_4 = \frac{2^2 n(n-2)}{4!}\,a_0\,, \qquad a_6 = -\frac{2^3 n(n-2)(n-4)}{6!}\,a_0\,, \quad \ldots$$

A similar analysis for $s = 1$ yields

$$a_2 = -\frac{2(n-1)}{3!}\,a_0\,, \qquad a_4 = \frac{2^2(n-1)(n-3)}{5!}, \quad \ldots,$$

and the general series solution for y_n then takes the form

$$\begin{aligned} y_n(x) = C_1 &\left[1 - \frac{2n}{2!}\,x^2 + \frac{2^2 n(n-2)}{4!}\,x^4 - \frac{2^3 n(n-2)(n-4)}{6!}\,x^6 + \cdots\right] \\ &+ C_2\left[x - \frac{2(n-1)}{3!}\,x^3 + \frac{2^2(n-1)(n-3)}{5!}\,x^5 - \cdots\right]. \end{aligned} \tag{14.111}$$

Just as for the Legendre ODE, we can obtain a polynomial solution for y_n only by choosing n to be an even integer and setting $C_2 = 0$ or by choosing n to be an odd integer and setting $C_1 = 0$.

When $y_n(x)$ is assigned its conventional scaling (that which makes the coefficient of x^n equal to 2^n), the resulting expressions are called the **Hermite polynomials** and traditionally denoted $H_n(x)$. The first few Hermite polynomials are shown in Table 14.4.

SYMBOLIC COMPUTING

The Hermite polynomial $H_n(x)$ can be accessed symbolically as

`HermiteH(n,x)` (MAPLE) or `HermiteH[n,x]` (MATHEMATICA).

In MATHEMATICA, if n is a nonnegative integer and x is a symbol this function call returns the explicit form of $H_n(x)$; in MAPLE, the explicit form is obtained only by application of the command `simplify`. As for other functions, decimal evaluation occurs automatically only if at least one of the arguments is a decimal quantity. Thus,

Table 14.4: Hermite polynomials.

n	$H_n(x)$
0	1
1	$2x$
2	$4x^2 - 2$
3	$8x^3 - 12x$
4	$16x^4 - 48x^2 + 12$
5	$32x^5 - 160x^3 + 120x$
6	$64x^6 - 480x^4 + 720x^2 - 120$
7	$128x^7 - 1344x^5 + 3360x^3 - 1680x$
8	$256x^8 - 3584x^6 + 13440x^4 - 13440x^2 + 1680$

MAPLE:

```
> HermiteH(2,x);                HermiteH(2,x)
> simplify(HermiteH(2,x);       -2 + 4x^2
> HermiteH(3,1.4);              5.152000000
```

MATHEMATICA:

```
HermiteH[2,x]                   -2 + 4x^2
HermiteH[2, 4/3]                46/9
```

PROPERTIES

Earlier we saw that various properties of the Legendre polynomials could be developed starting from a generating function or a Rodrigues formula. This approach is available for general sets of polynomials derived from an ODE, and (without proof) we identify the corresponding formulas for the Hermite polynomials. Their generating function formula is

$$g(x,t) = e^{2xt-t^2} = \sum_{n=0}^{\infty} H_n(x)\,\frac{t^n}{n!}\,, \tag{14.112}$$

and from this starting point one easily derives the recurrence formulas

$$H_{n+1}(x) = 2xH_n(x) - 2nH_{n-1}(x)\,, \tag{14.113}$$

$$H_n'(x) = 2nH_{n-1}(x)\,. \tag{14.114}$$

At $x = 0$ we have the following special values:

$$H_{2n}(0) = (-1)^n\,\frac{(2n)!}{n!}\,, \qquad H_{2n+1}(0) = 0\,. \tag{14.115}$$

The Hermite polynomials can be thought of as eigenfunctions of the operator

$$\mathcal{L} = \frac{d^2}{dx^2} - 2x\,\frac{d}{dx}\,,$$

with the eigenfunction H_n having eigenvalue $-2n$. This observation indicates that there should be conditions under which the H_n are orthogonal. We therefore note that the Hermite ODE is not self-adjoint, but becomes so if multiplied by e^{-x^2}. Then, following the general discussion in Section 11.6, the H_n will be orthogonal if the scalar product is defined as

$$\langle \varphi | \chi \rangle = \int_a^b \varphi^*(x)\chi(x)\, e^{-x^2}\, dx\,,$$

and the endpoints a and b are chosen such that

$$\Big[H_n(x)H'_m(x) - H'_n(x)H_m(x) \Big] e^{-x^2} \Big|_a^b = 0\,.$$

The choice $a = -\infty$, $b = \infty$ satisfies this condition because of the presence of the factor e^{-x^2}, and we have the final result

$$\int_{-\infty}^{\infty} H_n(x)H_m(x)\, e^{-x^2}\, dx = \begin{cases} 0\,, & n \neq m\,, \\ 2^n n! \sqrt{\pi}\,, & n = m\,. \end{cases} \tag{14.116}$$

Evaluation of the normalization integral shown as a case of Eq. (14.116) is the topic of Exercise 14.6.4.

Exercises

14.6.1. Plot $H_n(x)$ (suggested range $-n/2 \le x \le n/2$, vertical range $-2^{2n} \cdots 2^{2n}$) for $n = 2,\ 6,\ 9$, and 10.

14.6.2. Determine the values of $H_n(0)$ for general n by carrying out suitable operations on the generating-function formula.

14.6.3. Establish the recurrence formulas, Eqs. (14.113) and (14.114), by processes that begin by differentiating the generating-function formula.

14.6.4. Derive the normalization integral for the H_n given in Eq. (14.116) by the following steps:

1. Form a sum of orthogonality/normalization integrals:
$$\int_{-\infty}^{\infty} g(x,t)g(x,u)e^{-x^2}\, dx = \sum_{nm} \frac{t^n u^m}{n!m!} \langle H_n | H_m \rangle\,.$$
2. Eliminate terms on the right-hand side of this equation that vanish due to orthogonality.
3. Organize the left-hand side of this equation so that one of its factors within the integral is $e^{-(x-t-u)^2}$.
4. Evaluate the left-hand-side integral.
5. Do whatever is necessary to equate the coefficients of $(tu)^n$ on both sides of the equation, thereby finding $\langle H_n | H_n \rangle$.

14.6.5. For $g(x,t)$ as given in Eq. (14.112), note that

$$\left(\frac{d}{dx}\right)^m g(x,t) = (2t)^m g(x,t)\,.$$

Show that if the coefficient of x^m in $H_m(x)$ is c_m, then $\left(\frac{d}{dx}\right)^m H_m(x) = m!\,c_m$ and $c_m = 2^m$.

14.6.6. Expand $f(x) = 1/\cosh 2x$ in Hermite polynomials, using symbolic computing to evaluate the coefficients. Compare plots of $f(x)$ and its expansions truncated after H_4 and H_{10}. Note that even keeping terms through H_{10} the expansion is very poor beyond about $|x| = 2$.

A RELATED ODE

The ODE

$$\mathcal{L}\,\varphi(x) = \frac{1}{2}\left(-\frac{d^2}{dx^2} + x^2\right)\varphi(x) = \left(n + \frac{1}{2}\right)\varphi(x) \tag{14.117}$$

arises when the simple harmonic oscillator is studied in quantum mechanics. The substitution

$$\varphi(x) = e^{-x^2/2}y(x) \tag{14.118}$$

converts the original ODE into an equation for y that we can identify as the Hermite equation, with the **same** value of n as that appearing in Eq. (14.117). We therefore see that Eq. (14.117) has a solution

$$\varphi_n(x) = H_n(x)\,e^{-x^2/2}\,. \tag{14.119}$$

Noting that Eq. (14.117) is a self-adjoint ODE, and that its solutions $\varphi_n(x)$ vanish at $x = \pm\infty$, we conclude that the $\varphi_n(x)$ are orthogonal with the (unweighted) scalar product

$$\langle u|v\rangle = \int_{-\infty}^{\infty} u^*(x)v(x)\,dx\,. \tag{14.120}$$

We therefore have

$$\langle\varphi_n|\varphi_m\rangle = \int_{-\infty}^{\infty} H_n(x)H_m(x)e^{-x^2}\,dx = 0\,, \qquad n \neq m, \tag{14.121}$$

which is really the same as Eq. (14.114).

AN OPERATOR APPROACH

An interesting alternative approach leading to the φ_n starts by writing Eq. (14.117) in the form

$$\mathcal{L}\varphi(x) = \tfrac{1}{2}\left(-D^2 + x^2\right)\varphi(x) = \left(n + \tfrac{1}{2}\right)\varphi(x)\,, \tag{14.122}$$

where we have introduced D as a shorthand for d/dx. We then note the following identities related to quantities appearing in the operator $\mathcal{L}$:

$$\begin{aligned}(x + D)(x - D)f(x) &= -D^2 f(x) + x^2 f(x) + f(x) = (2\mathcal{L} + 1)f(x)\,,\\ (x - D)(x + D)f(x) &= -D^2 f(x) + x^2 f(x) - f(x) = (2\mathcal{L} - 1)f(x)\,.\end{aligned} \tag{14.123}$$

To derive these identities, note that, for arbitrary $f(x)$,

$$D[xf(x)] = f(x) + xDf(x) = (1 + xD)f(x)\,.$$

Applying these identities to the solution φ_n of Eq. (14.117), and making the further definitions $\hat{\mathbf{a}} = (x + D)/\sqrt{2}$ and $\hat{\mathbf{a}}^\dagger = (x - D)/\sqrt{2}$, we reach

$$\hat{\mathbf{a}}\,\hat{\mathbf{a}}^\dagger\,\varphi_n = \tfrac{1}{2}(-D^2 + x^2 + 1)\varphi_n = (\mathcal{L} + \tfrac{1}{2})\varphi_n = (n+1)\varphi_n\,, \tag{14.124}$$

$$\hat{\mathbf{a}}^\dagger\,\hat{\mathbf{a}}\,\varphi_n = \tfrac{1}{2}(-D^2 + x^2 - 1)\varphi_n = (\mathcal{L} - \tfrac{1}{2})\varphi_n = n\,\varphi_n\,. \tag{14.125}$$

In the last members of these equations, we have replaced $\mathcal{L}\varphi_n$ by $(n + \frac{1}{2})\varphi_n$. We also find it useful to rearrange these equations to obtain

$$\mathcal{L}\varphi_n = \left[\hat{\mathbf{a}}\hat{\mathbf{a}}^\dagger - \tfrac{1}{2}\right]\varphi_n = \left[\hat{\mathbf{a}}^\dagger\hat{\mathbf{a}} + \tfrac{1}{2}\right]\varphi_n\,. \tag{14.126}$$

We now compute $\mathcal{L}\left[\hat{\mathbf{a}}^\dagger\varphi_n\right]$. The result is

$$\begin{aligned}\mathcal{L}\left[\hat{\mathbf{a}}^\dagger\varphi_n\right] &= \left(\hat{\mathbf{a}}^\dagger\hat{\mathbf{a}} + \tfrac{1}{2}\right)\hat{\mathbf{a}}^\dagger\varphi_n = \hat{\mathbf{a}}^\dagger\left(\hat{\mathbf{a}}\hat{\mathbf{a}}^\dagger + \tfrac{1}{2}\right)\varphi_n = \hat{\mathbf{a}}^\dagger\left(\mathcal{L} + 1\right)\varphi_n,\\ &= \left(n + \tfrac{3}{2}\right)\left[\hat{\mathbf{a}}^\dagger\varphi_n\right].\end{aligned} \tag{14.127}$$

Equation (14.127) shows that $\hat{\mathbf{a}}^\dagger\varphi_n$ is also an eigenfunction of $\mathcal{L}$, but with an eigenvalue corresponding to $n \to n + 1$. Since the effect of $\hat{\mathbf{a}}^\dagger$ is to convert φ_n into φ_{n+1}, it is called a **raising** operator (or in field-theory contexts, a **creation** operator). Of importance here is the fact that we can use $\hat{\mathbf{a}}^\dagger$ to generate φ_{n+1} if φ_n is already known.

Turning now to $\mathcal{L}\left[\hat{\mathbf{a}}\varphi_n\right]$, an analysis similar to that just performed leads to the equation

$$\mathcal{L}\left[\hat{\mathbf{a}}\varphi_n\right] = \left(n - \tfrac{1}{2}\right)\left[\hat{\mathbf{a}}\varphi_n\right]\,, \tag{14.128}$$

with the interpretation that $\hat{\mathbf{a}}$ is a **lowering**, or **annihilation** operator that converts φ_n into an eigenfunction with $n \to n-1$. Raising and lowering operators are referred to collectively as **ladder operators**, as they enable passage up or down (in eigenvalue) through a succession of eigenfunctions.

Up to this point, the discussion has not told us how we might find the functions φ_n. But if φ_0 is the eigenfunction of smallest n (an assertion that is true but which we do not prove), we can then find φ_0 from the equation

$$\sqrt{2}\,\hat{\mathbf{a}}\varphi_0 = (x + D)\varphi_0 = \left(\frac{d}{dx} + x\right)\varphi_0 = 0\,. \tag{14.129}$$

This is a first-order separable equation with solution $\varphi_0 = Ce^{-x^2/2}$, the same as our earlier solution, $H_0(x)\,e^{-x^2/2}$. Additional φ_n can now be constructed in a straightforward fashion by evaluating

$$\varphi_n = (x - D)^n\varphi_0\,. \tag{14.130}$$

From a rearrangement of Eq. (14.130) (and multiplication by $e^{x^2/2}$ to recover the Hermite polynomials) we can obtain the Rodrigues formula for them:

$$H_n(x) = (-1)^n e^{x^2}\,\frac{d^n}{dx^n}\,e^{-x^2}\,. \tag{14.131}$$

Details of this rearrangement are the topic of Exercise 14.6.7.

Exercises

14.6.7. To obtain the Rodrigues formula for H_n, Eq. (14.131),

(a) Start by showing that the following two operators $\hat{A}$ and $\hat{B}$ are identical:

$$\hat{A} = e^{x^2/2}(x - D)e^{-x^2/2} \qquad \text{and} \qquad \hat{B} = e^{x^2}(-D)e^{-x^2}\,.$$

Hint. Show that they lead to the same result when applied to the same arbitrary function $f(x)$.

(b) Now form $\hat{A}^n$ and $\hat{B}^n$ and apply both to $f(x) = 1$. Relate $\hat{A}^n H_0$ to Eq. (14.130) and $\hat{B}^n H_0$ to Eq. (14.131).

(c) Complete the demonstration of Eq. (14.131) by showing that it produces $H_n(x)$ at the conventional scale.

14.6.8. The operators $\hat{\mathbf{a}} = (x + D)/\sqrt{2}$ and $\hat{\mathbf{a}}^\dagger = (x - D)/\sqrt{2}$ were introduced in the text just before Eq.(14.124).

(a) Show that the notations $\hat{\mathbf{a}}$ and $\hat{\mathbf{a}}^\dagger$ are appropriate because they are adjoints of each other in a Hilbert space for which the scalar product is defined as in Eq. (14.120).

(b) Writing

$$\langle \hat{\mathbf{a}}\varphi_n | \hat{\mathbf{a}}\varphi_n \rangle = \langle \varphi_n | \hat{\mathbf{a}}^\dagger \hat{\mathbf{a}} | \varphi_n \rangle\,,$$

where φ_n are the functions defined in Eq. (14.119), show that this scalar product is positive if $n > 0$, zero when $n = 0$, and negative if $n < 0$.

(c) Since the scalar product of part (b) can never be negative (explain why!), show how the result of part (b) leads to the conclusion that eigenfunctions φ_n exist only if n is a nonnegative integer.

14.7 LAGUERRE EQUATION

The Laguerre ODE is

$$xy''(x) + (1 - x)y'(x) + ny = 0\,, \qquad (14.132)$$

with polynomial solutions (the **Laguerre polynomials**) $L_n(x)$. The conventional scaling of the L_n corresponds to the Rodrigues formula

$$L_n(x) = \frac{1}{n!}\, e^x \,\frac{d^n}{dx^n}\left(x^n e^{-x}\right)\,. \qquad (14.133)$$

From Eq. (14.133) one can carry out the indicated differentiations, thereby obtaining the explicit form

$$L_n(x) = \sum_{j=0}^{n} (-1)^j \binom{n}{j} \frac{x^j}{j!}\,. \qquad (14.134)$$

By substitution into the ODE one can (with some effort) confirm the validity of Eqs. (14.133) and (14.134). The first few Laguerre polynomials are listed in Table 14.5. A somewhat simpler form has been obtained by making the tabulated quantity $n!L_n(x)$ rather than just L_n. A few authors omit the factor $1/n!$ from Eq. (14.133); one should check the definition of $L_n(x)$ before using material from other sources.

Table 14.5: Laguerre polynomials.

n	$n!\,L_n(x)$
0	1
1	$1-x$
2	$2-4x+x^2$
3	$6-18x+9x^2-x^3$
4	$24-96x+72x^2-16x^3+x^4$
5	$120-600x+600x^2-200x^3+25x^4-x^5$
6	$720-4320x+5400x^2-2400x^3+450x^4-36x^5+x^6$

PROPERTIES

The Laguerre polynomials have a generating function:

$$g(x,t) = \frac{e^{-xt/(1-t)}}{1-t} = \sum_{n=0}^{\infty} L_n(x)\,t^n\,, \tag{14.135}$$

and with its help one can establish the recurrence formulas

$$(n+1)L_{n+1}(x) = (2n+1-x)L_n(x) - nL_{n-1}(x)\,, \tag{14.136}$$

$$xL_n'(x) = nL_n(x) - nL_{n-1}(x)\,. \tag{14.137}$$

The Laguerre ODE is not self-adjoint but can be brought to self-adjoint form by multiplication of Eq. (14.132) by e^{-x}. The L_n are therefore orthogonal with scalar product

$$\int_0^{\infty} L_n(x)L_m(x)\,e^{-x}\,dx = \delta_{mn} \tag{14.138}$$

because xe^{-x} is the coefficient of y'' after the multiplication by e^{-x} and

$$\left[L_n(x)L_m'(x) - L_n'(x)L_m(x)\right] xe^{-x}\Big|_0^{\infty} = 0\,.$$

This is an application of the general discussion in Section 11.6. The fact that the L_n are normalized to unity is the topic of Exercise 14.7.6.

ASSOCIATED LAGUERRE POLYNOMIALS

Related sets of polynomials can be obtained by differentiating the Laguerre ODE. After a k-fold differentiation of the ODE with n replaced by $n+k$, i.e.,

$$\frac{d^k}{dx^k}\Big[xy'' + (1-x)y' + (n+k)y\Big] = 0\,,$$

the result can be written in the form

$$xz'' + (k+1-x)z' + nz = 0, \qquad \text{with } z = \frac{d^k}{dx^k}L_{n+k}(x)\,. \tag{14.139}$$

Table 14.6: Associated Laguerre polynomials.

n	$n!\,L_n^k(x)$
0	1
1	$(k+1)-x$
2	$(k+1)_2-2(k+2)x+x^2$
3	$(k+1)_3-3(k+2)_2x+3(k+3)x^2-x^3$
4	$(k+1)_4-4(k+2)_3x+6(k+3)_2x^2-4(k+4)x^3+x^4$
5	$(k+1)_5-5(k+2)_4x+10(k+3)_3x^2-10(k+4)_2x^3+5(k+5)x^4-x^5$
6	$(k+1)_6-6(k+2)_5x+15(k+3)_4x^2-20(k+4)_3x^3+15(k+5)_2x^4$ $-6(k+6)x^5+x^6$

The notations $(k+n)_m$ are Pochhammer symbols, defined in Eq. (2.53).

The ODE of Eq. (14.139) is called the **associated Laguerre** equation, and the standard notation for its solutions, the **associated Laguerre polynomials**, is

$$L_n^k(x) = (-1)^k \frac{d^k}{dz^k} L_{n+k}(x)\,. \tag{14.140}$$

This notational convention, which makes L_n^k a polynomial of degree n, has not seen universal adoption. It is therefore important to check the definitions when combining material from various sources. The first few L_n^k are given in Table 14.6. Again note that the quantity listed is $n!\,L_n^k$. It is possible to present results for general values of the index k; the formulas are kept compact by making use of the Pochhammer symbol, defined in Eq. (2.53).

The associated Laguerre polynomials have the Rodrigues formula

$$L_n^k(x) = \frac{x^{-k}e^x}{n!}\frac{d^n}{dx^n}\left(x^{n+k}e^{-x}\right), \tag{14.141}$$

and they satisfy many recurrence formulas, of which we cite only two:

$$(n+1)L_{n+1}^k(x) - (2n+k+1-x)L_n^k(x) + (n+k)L_{n-1}^k(x) = 0\,, \tag{14.142}$$

$$(n+1)L_{n+1}^{k-1}(x) = (k+n)L_n^{k-1}(x) - xL_n^k(x)\,, \tag{14.143}$$

and the orthogonality relation

$$\int_0^\infty x^k e^{-x} L_n^k(x) L_m^k(x)\,dx = \frac{(n+k)!}{n!}\,\delta_{nm}\,. \tag{14.144}$$

SYMBOLIC COMPUTING

MAPLE and MATHEMATICA use the same definitions as this text for both L_n and L_n^k. In MAPLE, these functions are

`LaguerreL(n,x)` and `LaguerreL(n,k,x)`,

In MATHEMATICA, they are

LaguerreL[n,x] and LaguerreL[n,k,x].

If n does not reduce to a numeric quantity, both systems return only the unevaluated function; if n is numeric, an explicit evaluation is automatically provided by MATHEMATICA, while evaluation in MAPLE only takes place with application of simplify.

MAPLE:

```
> LaguerreL(3,x);
```

LaguerreL(3,x)

```
> simplify(LaguerreL(3,x));
```

$$1-3x+\frac{3}{2}x^2-\frac{1}{6}x^3$$

```
> simplify(LaguerreL(3,k,x));
```

$$\frac{1}{6}k^3+k^2+\frac{11}{6}k+1-\frac{1}{2}x\,k^2-\frac{5}{2}x\,k-3x+\frac{1}{2}x^2k+\frac{3}{2}x^2-\frac{1}{6}x^3$$

MATHEMATICA:

```
LaguerreL[3,x]
```

$$\frac{1}{6}\left(6-18x+9x^2-x^3\right)$$

```
LaguerreL[3,k,x]
```

$$\frac{1}{6}\left(6+11k+6k^2+k^3-18x-15k\,x-3k^2x+9x^2+3kx^2-x^3\right)$$

These formulas for $L_3^k(x)$, though correct, are not optimal in that they fail to reveal the structure for this function that is shown in Table 14.6. We can improve the presentation by collecting the expressions according to powers of x and then causing each term to be factored. However, we need to trick the symbolic system into collecting the terms that are independent of x into a single term that will also get factored. In MAPLE, we can proceed as follows (with u undefined):

```
> u*x*simplify(LaguerreL(3,k,x):
> S := collect(%,x);
```

$$S:=-\frac{1}{6}\,ux^4+u\left(\frac{1}{2}\,k+\frac{3}{2}\right)x^3+u\left(-\frac{1}{2}\,k^2-\frac{5}{2}\,k-3\right)x^2+u\left(\frac{1}{6}\,k^3+k^2+\frac{11}{6}\,k+1\right)x$$

The extra factor x has caused collection of the terms that were originally independent of x.

Next we factor the individual terms of S using the command map, which causes its first argument to be applied to each term of the expression represented by its second argument:

```
> S := map(factor,S):
```

$$S:=-\frac{1}{6}\,ux^4+\frac{1}{2}u(k+3)x^3-\frac{1}{2}\,u(k+2)(k+3)x^2+\frac{1}{6}\,u(k+1)(k+2)(k+3)x$$

Having now completed the termwise factoring, we set $u=1/x$ to restore the algebraic value to that representing L_3^k:

```
> u := 1/x;
> S;
```

$$\frac{1}{6}(k+1)(k+2)(k+3)-\frac{1}{2}(k+2)(k+3)x+\frac{1}{2}(k+3)x^2-(1/6)x^3$$

In MATHEMATICA, the simplification of L_3^k proceeds as follows (with u an undefined quantity). The multiplication by u*x causes the terms that were originally independent of x to be collected into a single term.

```
S = Collect[u*x*LaguerreL[3,k,x],x]
```

$$\frac{1}{6}(6+11k+6k^2+k^3)ux+\frac{1}{6}(-18-15k-3k^2)ux^2+\frac{1}{6}(9+3k)ux^3-ux^4/6$$

We now apply the command Map, which causes its first argument to be applied to each term of the expression represented by its second argument:

```
Map[Factor,S]
```

$$\frac{1}{6}(1+k)(2+k)(3+k)ux-\frac{1}{2}(2+k)(3+k)ux^2+\frac{1}{2}(3+k)ux^3-\frac{ux^4}{6}$$

Finally, we remove both u and the extra power of x by setting $u=1/x$:

```
u = 1/x;
S
```

$$\frac{1}{6}(1+k)(2+k)(3+k)-\frac{1}{2}(2+k)(3+k)x+\frac{1}{2}(3+k)x^2-\frac{x^3}{6}$$

Exercises

14.7.1. Plot $L_n(x)$ for $n=2$ over a range of x sufficient to include all its zeros. In particular, determine whether all the zeros occur for positive values of x. Then repeat this exercise for $n=8$.

14.7.2. Starting from the Rodrigues formula for $L_n(x)$, show that $L_n(0)=1$.

14.7.3. Using Leibniz' rule for the nth derivative of a product, show how Eq. (14.134) follows from Eq. (14.133).

14.7.4. Verify that $L_n(x)$ as given in Eq. (14.134) satisfies the Laguerre ODE.

14.7.5. Use the Laguerre generating function, Eq. (14.135), to establish the recurrence formulas, Eqs. (14.136) and (14.137).

14.7.6. Establish the normalization of the Laguerre polynomials as follows:

1. Using the Laguerre generating function, Eq. (14.135), write

$$\int_0^\infty e^{-x}g(x,u)g(x,v)\,dx=\sum_{nm}\langle L_n|L_m\rangle u^n v^m\,.$$

2. Change the integration variable from x to $(1-uv)x/(1-u)(1-v)$ and carry out the left-hand-side integration.

3. Expand the result of that integration in powers of uv, and identify the normalization integrals.

14.7.7. In **Hartree atomic units** the differential equation for the radial wave functions of the hydrogen atom is

$$-\frac{1}{2}y''-\frac{1}{r}y'-\frac{y}{r}+\frac{L(L+1)}{2r^2}y=Ey\,,$$

where L is an integer (an angular quantum number), and E is an eigenvalue which for bound states must be negative. The solutions y must be finite at $r=0$ and approach zero asymptotically as $r\to\infty$.

(a) Show that the term with r^{-2} dependence is removed from the ODE by the substitution $y = r^L u$, yielding the equation

$$-\frac{1}{2}u'' - \frac{L+1}{r}u' - \frac{u}{r} = Eu\,.$$

(b) Show that the further substitutions $u = e^{-\alpha r}v$ and $E = -\alpha^2/2$ remove the terms of the ODE that are dominant at large r, leaving

$$rv'' + (2L + 2 - 2\alpha r)v' + [2 - 2\alpha(L+1)]v = 0\,.$$

(c) Show that the change of variable $x = 2\alpha r$, with $v(r)$ renamed $z(x)$, yields the equation

$$xz'' + (2L + 2 - x)z' + \left(\frac{1}{\alpha} - L - 1\right)z = 0\,.$$

(d) The ODE of part (c) is an associated Laguerre equation, and it will have solutions consistent with the requirements of the present problem only if the coefficient of z, denoted n in Eq. (14.139), is a nonnegative integer. We must therefore set $1/\alpha$ to an integer value, N, and require that $N - L - 1$ be nonnegative. Based on the above, write (in terms of associated Laguerre polynomials) the bound-state hydrogenic wave functions for general N and L, and state the ranges of values these parameters may assume.

(e) Write explicit formulas for the six hydrogenic radial wave functions of smallest E, identifying each by its values of N, L, and E. Write the Laguerre functions in polynomial form.

14.8 CHEBYSHEV POLYNOMIALS

There are two sets of Chebyshev polynomials, identified as Type I and Type II. We discuss here only those of Type I, which are solutions of the ODE

$$(1 - x^2)y''(x) - xy'(x) + n^2 y(x) = 0. \tag{14.145}$$

Equation(14.145) has a solution that is a polynomial of degree n, denoted $T_n(x)$; it has the Rodrigues representation

$$T_n(x) = \frac{(-1)^n \pi^{1/2}(1-x^2)^{1/2}}{2^n \Gamma(n+1/2)} \frac{d^n}{dx^n}\left[(1-x^2)^{n-1/2}\right] \tag{14.146}$$

and the generating function

$$g(x,t) = \frac{1-t^2}{1-2xt+t^2} = T_0(x) + 2\sum_{n=1}^{\infty} T_n(x)t^n\,. \tag{14.147}$$

Table 14.7: Chebyshev Type I polynomials.

n	$T_n(x)$
0	1
1	x
2	$2x^2 - 1$
3	$4x^3 - 3x$
4	$8x^4 - 8x^2 + 1$
5	$16x^5 - 20x^3 + 5x$
6	$32x^6 - 48x^4 + 18x^2 - 1$

From the generating function one can establish the recurrence formula

$$T_{n+1}(x) - 2xT_n(x) + T_{n-1}(x) = 0 \tag{14.148}$$

and the initial values

$$T_0(x) = 1\,, \qquad T_1(x) = x\,. \tag{14.149}$$

We also have the special values

$$T_n(1) = 1\,, \quad T_n(-1) = (-1)^n\,, \quad T_{2n}(0) = (-1)^n\,, \quad T_{2n+1}(0) = 0\,. \tag{14.150}$$

Explicit expressions for the first few T_n are presented in Table 14.7.

SYMBOLIC COMPUTING

The Chebyshev polynomials are available in our symbolic computing systems:

`ChebyshevT(n,x)` (MAPLE), `ChebyshevT[n,x]` (MATHEMATICA).

Explicit display of the functional form of $T_n(x)$ occurs automatically in MATHEMATICA but in MAPLE only upon application of `simplify`.

MAPLE: `> ChebyshevT(3,x);` → `ChebyshevT(3,x)`

`> simplify(ChevyshevT(3,x);` → $4x^3 - 3x$

MATHEMATICA: `ChebyshevT[3,x]` → $-3x + 4x^3$

FURTHER PROPERTIES

If we make the change of variable $x = \cos\theta$, we get the interesting result

$$T_n = \cos n\theta = \cos(n\cos^{-1} x)\,. \tag{14.151}$$

An important consequence of Eq. (14.151) is that all the extrema of T_n are seen to have the same magnitude, namely unity. This feature is illustrated in Fig. 14.7.

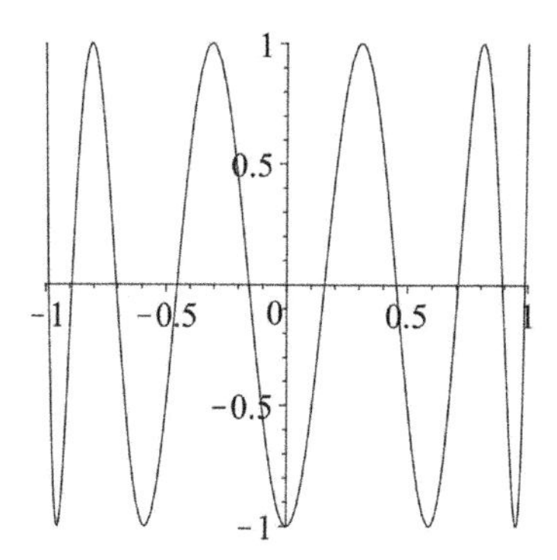

Figure 14.7: The Chebyshev polynomial T_{10}.

Following the discussion of Section 11.6, one can establish that the T_n satisfy the orthogonality relation

$$\int_{-1}^{1} T_n(x)T_m(x)(1-x^2)^{-1/2}\,dx = \begin{cases} 0, & m \neq n, \\ \dfrac{\pi}{2}, & m = n \neq 0, \\ \pi, & m = n = 0\,. \end{cases} \tag{14.152}$$

The main use of the Chebyshev polynomials is in numerical work, as a basis for the expansion of functions on a finite range that can be mapped onto $(-1, 1)$. Because of Eq. (14.152) we can use the T_n in an orthogonal expansion, and for many such expansions the convergence rate is optimized because of the property established by Eq. (14.151). When near enough convergence that the contributions from successive T_n are decreasing rapidly, the dominant part of the error from truncating the expansion will be from the first term not included, and in that term the maximum error has been minimized by making all the extrema of the same magnitude.

Example 14.8.1. A Chebyshev Expansion

Let's illustrate the numerical virtue of a Chebyshev expansion by comparing power-series and Chebyshev expansions of $e^{x/2}$ on the range $(-1, 1)$.

Using MAPLE first, we write a procedure to get the expansion coefficients through T_n:

```
> CC := proc(C,n) local x,j,f,q;
>    f := exp(x/2);
>    for j from 0 to n do
>      if (j = 0) then q := Pi else q := Pi/2 end if;
>      C[j]:=evalf(int(f*ChebyshevT(j,x)/(1-x^2)^(1/2), x=-1 .. 1)/q)
>    end do
> end proc:
```

Then we form the Chebyshev series (through T_4) and compare it with a power-series expansion truncated at x^4. The plot displays the errors in the two expansions.

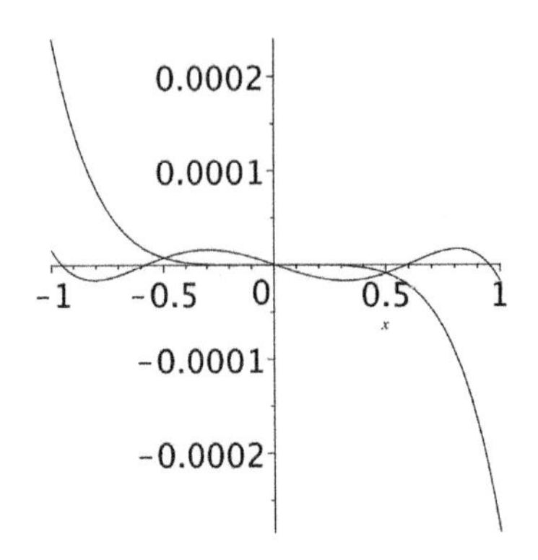

Figure 14.8: Errors in Chebyshev and power-series expansions, Example 14.8.1. The Chebyshev error has five zeros in the plotted interval.

```
> CC(C,4):
> fx := add(C[j] * ChebyshevT(j,x), j = 0 .. 4):
> h := add((x/2)^j/j!, j = 0 .. 4):
> plot([fx-exp(x/2), h-exp(x/2)], x = -1 .. 1);
```

The plot is shown in Fig. 14.8.

In MATHEMATICA, we start by writing a procedure to get the expansion coefficients through T_n:

```
cc[c,n] := Module[{x,j,f,q},
            f = E^(x/2);
            Do[If[j == 0, q = Pi, q = Pi/2];
               c[j] = N[Integrate[f*ChebyshevT[j,x]/(1-x^2)^(1/2),
                                  {x,-1,1}]/q ],
              {j,0,n} ] ]
```

Then we form the Chebyshev series (through T_4) and compare it with a power-series expansion truncated at x^4. The plot displays the errors in the two expansions.

```
cc[c,4]
fx = Sum[c[j]*ChebyshevT[j,x], j = 0 .. 4]
h = Sum[(x/2)^j/j!, {j, 0, 4}]
Plot[{fx-E^(x/2), h-E^(x/2)}, {x, -1, 1}]
```

The plot is shown in Fig. 14.8.

Both symbolic systems show that the Chebyshev expansion (through T_4) has maximum errors of about ± 0.00002, and that the error pattern exhibits **minimax** behavior, meaning that the maximum error has been approximately minimized by causing all the error extrema to be of about equal magnitude.

■

Exercises

14.8.1. (a) Expand x^8 in Chebyshev polynomials T_n. Do so in two ways:

1. Find the coefficient of T_8, denoted c_8, that causes $x^8 - c_8 T_8(x)$ to have a vanishing x^8 term. Then find the coefficient of T_6 that causes

$x^8 - c_8T_8(x) - c_6T_6(x)$ to have a vanishing x^6 term. Continue in the same way to find c_4, c_2, and c_0.

2. Using symbolic computing if helpful, use the orthogonality of the T_n to develop the expansion.

(b) Check the accuracy of the expansion of x^8 if truncated after T_6 by plotting x^8 and its expansion on the same graph. Also plot the error in the truncated expansion.

Additional Readings

Abramowitz, M., & Stegun, I. A. (Eds.). (1964). *Handbook of mathematical functions*, Applied Mathematics Series-55 (AMS-55). Washington, DC: National Bureau of Standards (Paperback edition, New York: Dover (1974). Chapter 22 is a detailed summary of the properties and representations of orthogonal polynomials. Other chapters summarize properties of Bessel, Legendre, hypergeometric, and confluent hypergeometric functions and much more. See also Olver et al. (2010)).

Byron, F. W., Jr., & Fuller, R. W. (1969). *Mathematics of classical and quantum physics*. Reading, MA: Addison-Wesley (Reprinted, Dover (1992)).

Dennery P., & Krzywicki, A. (1996). *Mathematics for physicists*. New York: Dover (Reprinted).

Erdelyi, A., Magnus, W., Oberhettinger, F., & Tricomi, F. G. (1953). *Higher transcendental functions* (Vol. 3). New York: McGraw-Hill (Reprinted Krieger (1981). A detailed, almost exhaustive listing of the properties of the special functions of mathematical physics).

Fox, L., & Parker, I. B. (1968). *Chebyshev polynomials in numerical analysis*. Oxford: Oxford University Press (A detailed, thorough, but very readable account of Chebyshev polynomials and their applications in numerical analysis).

Gradshteyn, I. S., & Ryzhik, I. M. (2007). *Table of integrals, series and products* (7th ed., A. Jeffrey & D. Zwillinger, Eds.). New York: Academic Press.

Kreyszig, E. (1988). *Advanced engineering mathematics* (6th ed.). New York: Wiley.

Margenau, H., & Murphy, G. M. (1956). *The mathematics of physics and chemistry* (2nd ed.). Princeton, NJ: Van Nostrand.

Olver, F. W. J., Lozier, D. W., Boisvert, R. F., & Clark, C. W. (Eds.). (2010). *NIST handbook of mathematical functions*. Cambridge, UK: Cambridge University Press (Update of AMS-55 (Abramowitz and Stegun, 1964), but links to computer programs are provided instead of tables of data).

Sansone, G. (1959). *Orthogonal functions* (A. H. Diamond, Trans.). New York: Interscience (Reprinted Dover (1991)).

Sneddon, I. N. (1980). *Special functions of mathematical physics and chemistry* (3rd ed.). New York: Longman.

Watson, G. N. (1952). *A treatise on the theory of bessel functions* (2nd ed.). Cambridge: Cambridge University Press (This is the definitive text on Bessel functions and their properties. Although difficult reading, it is invaluable as the ultimate reference).

Whittaker, E. T., & Watson, G. N. (1927). *A course of modern analysis*. Cambridge, UK: Cambridge University Press (Reprinted (2002). The classic text on special functions and real and complex analysis. Relatively advanced).

Chapter 15

PARTIAL DIFFERENTIAL EQUATIONS

15.1 INTRODUCTION

Because physical space is three-dimensional and many important problems in physics depend upon the spatial derivatives of a physical quantity, partial differential equations (PDEs) arise frequently in physics problems. Although a very wide variety of PDEs can be written down, it is both fortunate and interesting that a great deal of physics is described by just a few PDEs, and even much of that physics can be discussed in terms of homogeneous linear PDEs that are to be solved for specific regions of space, often with values of the dependent variable specified on the boundaries of the region defined by the problem under study. Let's illustrate by identifying a few PDEs and some of the problems in which they arise.

LAPLACE AND POISSON EQUATIONS

The **Laplace equation**,

$$\nabla^2\psi(\mathbf{r}) = 0\,, \tag{15.1}$$

is second-order, linear, and homogeneous. One place it occurs is in the description of the electrostatic potential ψ in a region containing no charges, currents, or time-dependent electromagnetic phenomena. Under these conditions the electric field $\mathbf{E}$ is irrotational and can be written as the negative gradient of a scalar potential: $\mathbf{E} = -\nabla\psi$, and the Maxwell equation (for points at which there is no charge density), $\nabla\cdot\mathbf{E} = 0$, becomes equivalent to $\nabla\cdot\nabla\psi = 0$, i.e., the Laplace equation. The Laplace equation also describes steady-state distributions of temperature in a material of constant thermal conductivity, so its importance is not limited to electricity and magnetism. In problems that are two-dimensional, either inherently so or because there is no variation of ψ in the third dimension, the Laplace equation reduces to

$$\frac{\partial^2\psi}{\partial x^2} + \frac{\partial^2\psi}{\partial y^2} = 0 \tag{15.2}$$

or its equivalent in other coordinate systems. In a later chapter we will also encounter this equation (in two dimensions) as satisfied by the real and the imaginary parts of so-called **analytic functions** of a complex variable.

Mathematics for Physical Science and Engineering.
http://dx.doi.org/10.1016/B978-0-12-801000-6.00015-8

The inhomogeneous equation related to the Laplace equation, called the **Poisson equation**, has the form

$$\nabla^2\psi(\mathbf{r}) = f(\mathbf{r})\,. \tag{15.3}$$

In electrostatics, ψ is the scalar potential, and $f(\mathbf{r}) = -\rho(\mathbf{r})/\varepsilon_0$, with ρ the charge density. In general, $f(\mathbf{r})$ is referred to as a **source density**, or simply a **source**.

WAVE EQUATION

In a homogeneous medium within which the speed of wave propagation is v (or in free space, where electromagnetic waves propagate at the velocity of light, with $v = c$), the wave amplitude satisfies the **wave equation**

$$\nabla^2\psi(\mathbf{r},t) = \frac{1}{v^2}\frac{\partial^2\psi}{\partial t^2}\,. \tag{15.4}$$

This equation is also linear, homogenous, and second-order and is still a PDE if the waves travel in one dimension, i.e.,

$$\frac{\partial^2\psi(x,t)}{\partial x^2} = \frac{1}{v^2}\frac{\partial^2\psi}{\partial t^2}\,. \tag{15.5}$$

The wave equation differs from the Laplace equation in the important way that, if written in the form

$$\nabla^2\psi - \frac{\partial^2\psi}{\partial\tau^2} = 0$$

(with $\tau = vt$), not all the partial derivatives occur with the same sign. This difference alters greatly the number, nature, and even the existence of PDE solutions.

DIFFUSION EQUATION

The **diffusion equation** differs from the wave equation in that it contains a first derivative of the time variable:

$$\nabla^2\psi(\mathbf{r},t) = \frac{1}{k^2}\frac{\partial\psi}{\partial t}\,. \tag{15.6}$$

The diffusion can be that of thermal energy (in which case $\psi = T$, the temperature), or of matter (with ψ then representing the concentration of some species). If a steady-state condition exists, then the time derivative vanishes and we have, as mentioned earlier, a Laplace equation. Part of the physical content of the diffusion equation is that the temperature or concentration changes with time when its local value deviates from straight-line spatial dependence (i.e., it has curvature).

An important generalization of the diffusion equation is the **time-dependent Schrödinger equation**, which for a single particle in dimensionless coordinates can take the form

$$-\frac{1}{2}\nabla^2\psi(\mathbf{r},t) + V(\mathbf{r})\psi = i\,\frac{\partial\psi}{\partial t}\,. \tag{15.7}$$

HELMHOLTZ EQUATION AND ITS GENERALIZATIONS

The **Helmholtz equation**,

$$\nabla^2\psi(\mathbf{r}) + k^2\psi = 0\,, \tag{15.8}$$

does not often arise in the initial formulation of physics problems, but it is frequently encountered when wave-propagation or diffusion problems are solved by various methods. Sometimes this equation occurs with a minus sign before the k^2; it may be helpful to note that this is a Helmholtz equation with an imaginary value of k.

Generalizations of the Helmholtz equation in which k^2 is replaced by a more general function of $\mathbf{r}$ do occur frequently in physical science; an example is the time-independent **Schrödinger equation**, which for a single particle subject to a potential $V(\mathbf{r})$ can be written (in dimensionless units)

$$-\frac{1}{2}\nabla^2\psi(\mathbf{r}) + \big[V(\mathbf{r}) - E\big]\psi = 0. \tag{15.9}$$

SOLUTIONS AND BOUNDARY CONDITIONS

We already know that a second-order ODE in y has a general solution containing two independent constants, which can often permit us to obtain a unique particular solution with specified values of y at two points (perhaps the ends of an interval on which we desire a solution to the ODE). Alternatively, sometimes we can obtain a unique solution to a second-order ODE with the values of y' specified at the ends of an interval. These specifications are called **boundary conditions**.

In two dimensions a second-order PDE in $\psi = \psi(x,y)$ defined for a rectangular area might be solvable subject to a boundary condition on each of its x boundaries (a and b), i.e., specifications of $\psi(a,y) \equiv \psi_a(y)$ and $\psi(b,y) \equiv \psi_b(y)$, where, as shown, ψ_a and ψ_b are functions of y. We might also have boundary conditions on the y boundaries (c and d), e.g., specifications of $\psi(x,c) \equiv \psi_c(x)$ and $\psi(x,d) \equiv \psi_d(x)$, with ψ_c and ψ_d functions of x. Because ψ_a through ψ_d can be arbitrary functions, we see that the set of solutions to a two-dimensional PDE could span an extremely wide variety of functional forms.

The discussion of the preceding paragraph can be generalized to a two-dimensional region defined by an arbitrary boundary curve, with the result that a unique solution to our 2-D PDE may perhaps be obtained subject to the requirement that $\psi(x,y)$ have specified values everywhere on the boundary. Further generalizing to three dimensions, a PDE in $\psi(\mathbf{r})$ for a given volume may perhaps only have a unique solution when $\psi(\mathbf{r})$ is specified everywhere on the surface bounding the volume.

Although the boundary conditions discussed in the two preceding paragraphs were all in terms of values of ψ, it is also possible to specify boundary conditions in terms of the first derivative of ψ in the direction normal to the boundary (the analogous situation for an ODE is a derivative directed toward the interior of the interval on which the ODE is defined).

When one of the independent variables in a PDE is the time t, we often encounter problems in which the dependent variable ψ has its spatial dependence (everywhere) specified at $t = 0$. This **initial condition** is a boundary condition for one end of an infinite time interval, and often appears with $\partial\psi/\partial t$ also specified everywhere at $t = 0$.

It is important to know whether a set of boundary conditions is sufficient to determine a unique solution to a PDE or whether too many conditions are specified (with the result that no solution satisfying the conditions can exist). A complete analysis of this topic is beyond the scope of the present text, but we can summarize the results of such an analysis for the PDEs to be considered here.

The solution to the Laplace equation (e.g., an electrostatic potential) is completely determined when the potential is specified everywhere on the boundary (∂V) of the spatial region (V) for which the PDE is to be solved. This specification is termed

Dirichlet boundary conditions. The sufficiency of Dirichlet boundary conditions for the Laplace equation is to be expected, since it is in essence a statement that if a distribution of charge external to V produces a specific distribution of potential on ∂V, it must also produce a unique potential at each point within V. Note also that a suitable external distribution of charge can produce a specific distribution of the normal derivative of potential, $\partial V/\partial \mathbf{n}$, at each point of ∂V, a specification called **Neumann boundary conditions**. Neumann boundary conditions are also sufficient to determine completely (except for an additive constant) a solution to the Laplace equation. However, if we attempt to impose both Dirichlet and Neumann boundary conditions (the combination is called **Cauchy boundary conditions**), the attempt will in general produce inconsistent results and therefore not lead to a solution to the Laplace PDE.

For wave equations, either Dirichlet or Neumann boundary conditions suffice to define PDE solutions, but the solutions are in general, not unique. Cauchy boundary conditions are usually too restrictive. However, we are sometimes interested in wave-propagation problems in which we impose Cauchy boundary conditions, but not for all positions and times. This situation arises when waves are propagated from a fully specified source but their behavior elsewhere is not specified.

Diffusion equations, having only a first time derivative, can accommodate fewer boundary conditions than the PDEs already discussed. In general they have no solutions if either Dirichlet or Neumann conditions are imposed at all spatial points at multiple times, but unique results are possible if either Dirichlet or Neumann conditions (but not both) are imposed at all spatial points at an initial time $t = 0$.

One observation regarding boundary conditions may be helpful: physical insight is a practical guide for understanding what to expect; one usually knows intuitively whether sufficient (but not excessive) conditions defining a problem have been imposed.

Exercises

15.1.1. Prove that if the electrostatic potential ψ is given at all points of a surface ∂V enclosing a charge-free region V, the value of ψ, given as the solution of the Laplace equation for the region V subject to the given boundary values, is unique. Start by assuming that there are two distinct solutions, ψ_1 and ψ_2; their difference, $u = \psi_1 - \psi_2$, then must satisfy the Laplace equation $\boldsymbol{\nabla}^2 u = 0$ for the region V, with $u = 0$ on ∂V. Next apply the divergence theorem to $\int \boldsymbol{\nabla} \cdot (u \boldsymbol{\nabla} u)\, d\tau$ to show that

$$\int_V u \boldsymbol{\nabla}^2 u \, d\tau + \int_V \boldsymbol{\nabla} u \cdot \boldsymbol{\nabla} u \, d\tau = 0 \,,$$

and (explain why) we can then conclude that $u = 0$ throughout V.

15.2 SEPARATION OF VARIABLES

A powerful method for solving homogeneous linear PDEs subject to boundary conditions is the method of **separation of variables**. The essence of this method is that we search for PDE solutions of the separated-variable form, e.g., for $\psi(x, y, z)$ the form $X(x)Y(y)Z(z)$; or for $\psi(x, t)$, the form $X(x)U(t)$ (we choose U to reserve T as the symbol for temperature). The method in no way requires that our solution (subject to the boundary conditions) have this factored form. Because our PDE is linear and homogeneous, we may obtain overall solutions that satisfy the boundary conditions by making expansions using the factored forms as a basis.

The essential feature of a PDE that will allow its solution by the method of separation of variables is that we will be able to write it in a form that lets us solve separately for some or all of $X(x)$, $Y(y)$, $Z(z)$, $U(t)$, with the boundary conditions also separating to at least some extent. Let's illustrate with some examples that are, for simplicity, in only two dimensions.

Example 15.2.1. Vibrating String

Our first example is a vibrating string that is tautly stretched between two fixed endpoints located at positions $x = 0$ and $x = L$. We assume that the string is free to execute small transverse vibrations in a single plane, with an instantaneous amplitude $\psi(x,t)$ (relative to the equilibrium position) at positions x and times t, and that the string is plucked from its midpoint at $t = 0$. The PDE governing the oscillatory motion of the string is the wave equation, which for the present problem reduces to

$$\frac{\partial^2 \psi(x,t)}{\partial x^2} = \frac{1}{v^2}\frac{\partial^2 \psi}{\partial t^2}. \tag{15.10}$$

Separating the variables—Irrespective of our thoughts as to what the solution to this problem should look like, we search for solutions to our PDE of the form $\psi = X(x)U(t)$. Substituting into Eq.(15.10) we get

$$\frac{\partial^2}{\partial x^2}\Big[X(x)U(t)\Big] = \frac{1}{v^2}\frac{\partial^2}{\partial t^2}\Big[X(x)U(t)\Big] \quad \longrightarrow \quad U(t)X''(x) = \frac{1}{v^2}X(x)U''(t). \tag{15.11}$$

Because all the differentiations now act on functions of a single variable we have written them as ordinary derivatives.

We now carry out a key step of the method of separation of variables: we divide Eq. (15.11) by $X(x)U(t)$ to reach

$$\frac{X''}{X} = \frac{1}{v^2}\frac{U''}{U}. \tag{15.12}$$

Equation (15.12) must be satisfied for all x and t, but it has the feature that its left-hand side depends only on x, while its right-hand side depends only on t. If we change x keeping t constant (which we may certainly do because these two variables are independent of each other), Eq. (15.12) can only remain satisfied if its left-hand side, X''/X, actually remains constant. Likewise, Eq. (15.12) can only remain satisfied when t is varied at constant x if U''/U remains constant. Furthermore, the constant values of X''/X and U''/U must be related in a way that satisfies Eq. (15.12). These requirements correspond to the single-variable equations

$$\frac{X''}{X} = C, \qquad \frac{U''}{U} = Cv^2,$$

where we as yet have no information as to possible values for C. Writing these equations in more conventional forms, we see that we have a pair of ODEs, one in x and one in t, that are coupled by virtue of having a parameter in common:

$$X'' = CX, \qquad U'' = Cv^2U. \tag{15.13}$$

Equations (15.13) must now be solved subject to at least some of the boundary conditions, and it becomes practical to do so because the boundary conditions can also be separated. The fixed endpoints are equivalent to the requirement that $X(0)$ and $X(L)$ vanish, so that $\psi(0,t)$ and $\psi(L,t)$ will be zero for all t. We thus not only have two ODEs in place of our original PDE, but we also have a full set of boundary (endpoint) conditions for the x equation.

Our problem also provides two boundary conditions for the t equation (we could alternatively call them initial conditions since they apply at $t = 0$). The plucking of the string at its center corresponds to a stationary triangular shape that, starting at $t = 0$, then moves in a way determined by the PDE. We therefore have the boundary conditions

$$\psi(x,0) = \psi_0(x)\,, \qquad \left.\frac{\partial\psi(x,t)}{\partial t}\right|_{t=0} = 0\,, \tag{15.14}$$

where ψ_0 is the original shape. The second of these conditions states that ψ is stationary at $t = 0$.

The X ODE—The x equation and its boundary conditions correspond to a problem we have already solved, in Example 11.4.1. Its solutions consist of sinusoidal oscillations of wavelengths such that nodes can be placed at $x = 0$ and $x = L$, and with C values that depend upon the wavelengths. Since nodes, like children, occur in integer numbers, the solutions and the associated C values can be classified according to their numbers of nodes. Algebraically, our current problem in x has solutions

$$X_n = \sin\frac{n\pi x}{L}\,, \quad \text{with} \quad C_n = -\frac{n^2\pi^2}{L^2} \quad \text{and } n = 1,\ 2, \ldots \tag{15.15}$$

One can easily verify that Eq. (15.15) satisfies both the x equation and its boundary conditions. The x ODE also has solutions for positive values of C, but solutions with $C > 0$ (exponentials) cannot be zero at two values of x and therefore cannot satisfy the boundary conditions.

Normal modes and their time dependence—A solution to the vibrating-string PDE that consists of a spatial factor multiplied by a time factor is of a type known as a **standing wave** because that functional form describes a nodal pattern that does not change with time. The standing waves are also known as **normal modes** of oscillation of the string; when we solve the equation for $U(t)$ for a given normal mode, i.e., for a given value of C_n from Eq. (15.15), we are finding the time dependence of that normal mode. With C_n specified, the t equation becomes

$$U'' = -\frac{n^2\pi^2 v^2}{L^2}\,U\,, \tag{15.16}$$

with general solution

$$U_n(t) = a_n \cos\frac{n\pi vt}{L} + b_n \sin\frac{n\pi vt}{L}\,, \tag{15.17}$$

thereby defining the overall solution for the nth normal mode:

$$\psi_n(x,t) = \sin\frac{n\pi x}{L}\left[a_n \cos\frac{n\pi vt}{L} + b_n \sin\frac{n\pi vt}{L}\right]. \tag{15.18}$$

As discussed in Chapter 12, the standing wave of Eq. (15.18) has wavelength λ, period τ, and frequency ν found from the equations

$$\frac{n\pi\lambda}{L} = 2\pi \longrightarrow \lambda = \frac{2L}{n}\,, \qquad \frac{n\pi v\tau}{L} = 2\pi \longrightarrow \tau = \frac{2L}{nv}\,, \qquad \nu = \frac{1}{\tau} = \frac{nv}{2L}\,. \tag{15.19}$$

Combining the normal modes—The wave equation is linear and homogeneous, so any linear combination of its solutions will also be a solution. Moreover, all the solutions ψ_n that we have found satisfy the boundary conditions $X(0) = X(L) = 0$, and (because these boundary values are zero) any linear combination of the ψ_n will also satisfy these boundary conditions. At this point we can therefore state that a general solution satisfying the boundary conditions at $x = 0$ and L is

$$\psi(x,t) = \sum_{n=1}^{\infty} \sin\frac{n\pi x}{L}\left[a_n \cos\frac{n\pi vt}{L} + b_n \sin\frac{n\pi vt}{L}\right]. \tag{15.20}$$

In the current problem, Eq. (15.20) describes a combination of the normal modes, each at arbitrary amplitude and phase. Thus, Eq. (15.20) indicates that any superposition of the normal modes is possible.

There remains the task of choosing the coefficients a_n and b_n (i.e., the amplitudes and phases of the normal modes) in a way that satisfies the initial conditions. To see how to accomplish this, let's start by writing down general expressions for ψ and $\partial\psi/\partial t$ at $t = 0$. From Eq. (15.20),

$$\psi(x,0) = \sum_{n=1}^{\infty} a_n \sin\frac{n\pi x}{L}, \tag{15.21}$$

$$\left.\frac{\partial\psi(x,t)}{\partial t}\right|_{t=0} = \sum_{n=1}^{\infty} b_n\left(\frac{n\pi v}{L}\right)\sin\frac{n\pi x}{L}. \tag{15.22}$$

The condition that $\partial\psi/\partial t$ vanish at $t = 0$ is easily handled by setting all the b_n to zero. We also note that $\psi(x,0)$ is given by Eq. (15.21) as a Fourier sine series with arbitrary coefficients, so we can choose those coefficients in a way that makes this series properly represent the required form for $\psi(x,0)$.

The Fourier sine series for a plucked string was the topic of Example 12.5.1. Adapting the result found there to a displacement of arbitrary magnitude, we have

$$a_n = \begin{cases} A\dfrac{(-1)^{(n-1)/2}}{n^2}, & n \text{ odd}, \\ 0, & n \text{ even}, \end{cases}$$

and the complete solution to our vibrating string problem is

$$\psi(x,t) = A\sum_{n\ \mathrm{odd}} \frac{(-1)^{(n-1)/2}}{n^2} \sin\frac{n\pi x}{L} \cos\frac{n\pi vt}{L}. \tag{15.23}$$

■

The ideas to be carried forward from the foregoing example include the following, relevant when the method of separation of variables is applicable:

- *After introducing a separated-variable product, the PDE is written in a way that makes its individual terms each depend upon only one independent variable.*
- *Each term is then set equal to a constant, with the constants of the individual terms related.*
- *The boundary conditions are, to the extent possible, written in separated form.*
- *The separated terms of the PDE are written as ODEs and solved, where possible subject to the separated boundary conditions.*

- *Products of these ODE solutions are then identified as solutions of the original PDE. (In the example, each of these product solutions is a normal mode of oscillation of the string.)*
- *Because the original PDE is linear and homogeneous, any linear combination of its solutions is also a solution. These linear expansions may continue to satisfy some of the boundary conditions, and any remaining boundary conditions are now satisfied by making a proper choice of the coefficients in the expansion.*

Exercises

15.2.1. Solve the vibrating string problem of Example 15.2.1 subject to the initial conditions that the string be undisplaced at time $t = 0$ (i.e., $\psi(x, 0) = 0$ for all x), but that its initial velocity distribution be of the form

$$\left.\frac{\partial\psi(x,t)}{\partial t}\right|_{t=0} = \begin{cases} Ax, & 0 \le x \le L/2, \\ A(L-x), & L/2 \le x \le L. \end{cases}$$

15.2.2. Repeat Exercise 15.2.1 but with the initial velocity a delta function at the point x_0 (the result of striking the string with a hammer at $x = x_0$).

15.2.3. Compare the results of Exercises 15.2.1 and 15.2.2 (the latter for the case $x_0 = L/2$) by computing the intensities b_n^2 of the first 6 harmonics (the values of n from 1 to 6), and compare these intensities with the a_n^2 that were found as the solution of Example 15.2.1.

Example 15.2.2. Laplace Equation: Rectangle

We next use the method of separation of variables to determine the steady-state temperature distribution $T(x, y)$ of a rectangular plate with boundaries at $x = 0$ and $x = L_x$ and at $y = 0$ and $y = L_y$, when the boundary at $x = L_x$ is kept at a nonzero temperature T_0 and the other three boundaries are kept at $T = 0$. The temperature distribution is governed by a two-dimensional Laplace equation with the above-described boundary conditions. Our strategy will be to obtain a set of solutions each of which satisfies some of the boundary conditions, and then from that set to construct a unique solution that satisfies the remaining boundary conditions.

Separating the variables—Our starting point is to search for solutions of the form $T(x, y) = X(x)Y(y)$, with the objective of separating our PDE into ODEs in the individual dimensions. Applying the Laplace operator, we get initially

$$\left[\frac{\partial^2}{\partial x^2} + \frac{\partial^2}{\partial y^2}\right] X(x)Y(y) = Y(y)\,\frac{d^2X(x)}{dx^2} + X(x)\,\frac{d^2Y(y)}{dy^2} = 0\,. \tag{15.24}$$

Again we note that because the partial derivatives are now operating on functions of a single variable we can write them as ordinary derivatives. Our next step is to divide Eq. (15.24) through by $X(x)Y(y)$, leading to

$$\frac{1}{XY}\left[Y(y)\,\frac{d^2X(x)}{dx^2} + X(x)\,\frac{d^2Y(y)}{dy^2}\right] = \frac{X''}{X} + \frac{Y''}{Y} = 0\,. \tag{15.25}$$

As in Example 15.2.1, the requirement that Eq. (15.25) must be satisfied for all x and y leads us to the conclusion that X''/X and Y''/Y must each be constant, and the constant values of X''/X and Y''/Y must sum to zero. We therefore have the great simplification that for some value of C

$$\frac{X''}{X} = C\,, \qquad \frac{Y''}{Y} = -C\,,$$

equivalent to the pair of ODEs

$$X'' = CX\,, \qquad Y'' = -CY\,. \tag{15.26}$$

Equations (15.26) must now be solved subject to the boundary conditions, and it becomes practical to do so because three of the four boundary conditions separate into simple requirements:

$$\begin{aligned}
T(x,0) = 0 &\to X(x)Y(0) = 0 \ \ \text{(for all } x) \ \to Y(0) = 0\,,\\
T(x,L_y) = 0 &\to X(x)Y(L_y) = 0 \ \ \text{(for all } x)\to Y(L_y) = 0\,,\\
T(0,y) = 0 &\to X(0)Y(y) = 0 \ \ \text{(for all } y) \ \to X(0) = 0\,.
\end{aligned}$$

Solving the Y ODE—Our previous example suggests (at least to the author) that the homogeneous boundary conditions on Y might permit solutions to the y equation to be combined without disturbing the boundary conditions, so our next step is to look for solutions to the y equation subject to $Y(0) = Y(L_y) = 0$. This is an ODE with boundary conditions that we have already investigated several times; it has the following solutions satisfying the boundary conditions:

$$Y_n(y) = \sin\frac{n\pi y}{L_y}\,, \quad \text{with } C_n = \frac{n^2\pi^2}{L_y^2} \quad \text{and } n = 1,\ 2,\ 3,\ldots \tag{15.27}$$

Solving the X ODE—For each value of C_n we can now solve the x equation subject to $X(0) = 0$, leaving for later the question of how to satisfy the boundary condition at $x = L_x$. From the sign of C_n (and the sign with which it appears in the x equation), we see that the relevant solutions to that equation are exponentials, or alternatively and equivalently the hyperbolic functions sinh and cosh. That is,

$$X'' = C_n X = \frac{n^2\pi^2}{L_y^2}X \ \text{ has solution } \ X_n = a_n \sinh\frac{n\pi x}{L_y} + b_n \cosh\frac{n\pi x}{L_y}\,. \tag{15.28}$$

The solution corresponding to $X_n(0) = 0$ is the sinh function, and the corresponding $T_n = X_nY_n$ is

$$T_n(x,y) = \sinh\frac{n\pi x}{L_y}\,\sin\frac{n\pi y}{L_y}\,. \tag{15.29}$$

Each T_n from Eq. (15.29) is a solution to the Laplace equation that satisfies three of the four boundary conditions.

Satisfying the remaining boundary condition—Because the Laplace equation is homogeneous, any linear combination of the T_n is also a solution, and the most general solution satisfying the first three boundary conditions has the form

$$T(x,y) = \sum_{n=1}^{\infty} a_n T_n(x,y) = \sum_{n=1}^{\infty} a_n \sinh\frac{n\pi x}{L_y}\,\sin\frac{n\pi y}{L_y}\,. \tag{15.30}$$

Our final step in solving the current problem is to choose the set of coefficients a_n that yields (for all y)

$$T(L_x, y) = T_0 = \sum_{n=1}^{\infty} \left[a_n \sinh \frac{n\pi L_x}{L_y} \right] \sin \frac{n\pi y}{L_y} . \tag{15.31}$$

While Eq. (15.31) may seem unusual, it is simply the Fourier sine series for the constant function $f(y) = T_0$ on the interval $0 \le y \le L_y$, and the quantity in square brackets is the nth coefficient in the expansion. Referring to Eq. (12.27), we see that

$$T_0 = \sum_{n \text{ odd}} \frac{4T_0}{n\pi} \sin \frac{n\pi y}{L_y} , \tag{15.32}$$

and we can make the identification

$$a_n \sinh \frac{n\pi L_x}{L_y} = \begin{cases} \dfrac{4T_0}{n\pi}, & n \text{ odd}, \\ 0, & n \text{ even}. \end{cases} \tag{15.33}$$

Solving Eq. (15.33) for a_n and inserting the result into Eq. (15.30), we reach our final result

$$T(x, y) = \frac{4T_0}{\pi} \sum_{n \text{ odd}} \left[n \sinh \frac{n\pi L_x}{L_y} \right]^{-1} \sinh \frac{n\pi x}{L_y} \sin \frac{n\pi y}{L_y} . \tag{15.34}$$

To give an idea of the physical content of Eq. (15.34), a contour plot of the temperature distribution is shown in Fig. 15.1. Such plots are the topic of Exercise 15.2.5.

■

The relatively simple problem of Example 15.2.2 can easily be generalized to deal with an arbitrary temperature distribution at $x = L_x$. We need only to replace the Fourier sine series of the constant T_0 with the series representing the boundary values $T(L_x, y)$.

The fact that linear combinations of solutions to the Laplace equation are also solutions enables us to generalize the present problem in an even more interesting way. Suppose that instead of a given distribution at the boundary $x = L_x$, we had another arbitrary distribution at the boundary $x = 0$. If we solve the problem for that second distribution (with $T = 0$ on all the other boundaries), we can add it to our

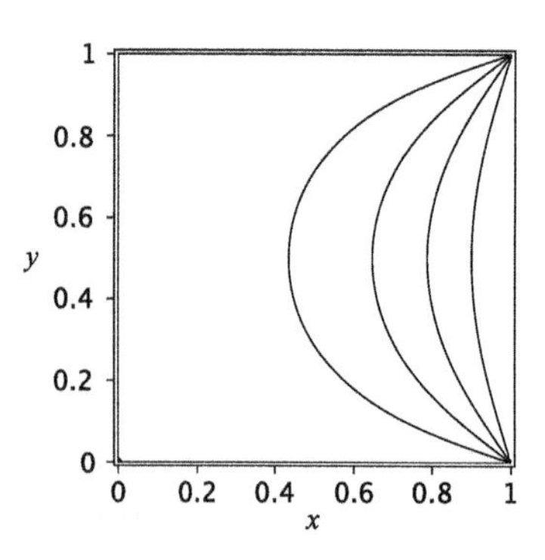

Figure 15.1: Temperature distribution of a square plate of unit dimensions with the boundary conditions given in Example 15.2.2. Contours at 0.2, 0.4, 0.6, and 0.8 T_0.

original solution (for which the boundary temperature is nonzero only at $x = L_x$), and thereby get a temperature distribution whose boundary conditions are the sum of those of both problems. Continuing this process by solving two more problems, each with nonzero temperatures on one of the other boundaries ($y = 0$ or $y = L_y$), we can solve the Laplace equation for a rectangle with arbitrary boundary values, i.e., for arbitrary Dirichlet boundary conditions. We have thus parlayed our simple problem into a general treatment of the Laplace equation for rectangular geometry.

Example 15.2.2 has identified one more feature of the solution process for linear homogeneous PDEs. It is

- *If we are able to solve a linear homogeneous PDE in T when T is given on one portion of the boundary and zero on the remainder of the boundary, and can do the same when each other part of the boundary has a specified value of T (with the remainder of the boundary then zero), we can add the solutions of these relatively simple problems to obtain the solution of a more complicated problem in which ψ has arbitrary values for all portions of the boundary.*

Exercises

15.2.4. Show that a change of variable from k to ik causes the space spanned by $\sin kx$ and $\cos kx$ to become a space spanned by $\sinh kx$ and $\cosh kx$. Thus, changing the sign of k^2 in the ODE $\psi'' + k^2\psi = 0$ converts its solutions from trigonometric to hyperbolic functions.

15.2.5. MAPLE and MATHEMATICA both support contour plots. Create a procedure `tt` that computes $T(x, y)$ as given in Eq. (15.34) and plot the temperature distribution for a unit square plate, with contours 0, 0.2, 0.4, 0.6, and 0.8 (in units of T_0). Basic coding to produce the plots is:

MAPLE:
```
> with(plots):
> contourplot(tt(x,y), x = 0 .. 1, y = 0 .. 1,
>                 contours = [0.,.2,.4,.6,.8] );
```

MATHEMATICA:
```
ContourPlot[tt[x,y], {x,0,1}, {y,0,1},
                 Contours -> {.2,.4,.6,.8} ]
```

(The desired contours are the default values for MATHEMATICA and hence need not have been specified.)

More detail regarding contour plots, including a discussion as to how to control a plot's aspect ratio, can be found in Appendix A.

15.2.6. Modify the solution for the two-dimensional Laplace equation given in Eq. (15.34) so that it applies for $T = T_0$ on the boundary at $x = 0$, with $T = 0$ on the other boundaries of the rectangle $x = (0, L_x)$, $y = (0, L_y)$. Then (referring to Exercise 15.2.5 if helpful), for $L_x = L_y = 1$,

(a) Make a contour plot of $T(x, y)$ for this new problem.

(b) Form the solution $T(x, y)$ for a problem in which $T = T_0$ on both the boundaries $x = 0$ and $x = L_x$ (with $T = 0$ on $y = 0$ and $y = L_y$).

(c) Make a contour plot of $T(x, y)$ for the problem of part (b).

15.2.7. What steady-state temperature distribution within the rectangle of Example 15.2.2 do you expect if the edges at $x = 0$, $y = 0$, and $y = L_y$ are all kept at $T = T_0$ while that at $x = L_x$ is set to zero?

(a) Obtain the temperature distribution $T(x, y)$ for this problem, using equations motivated by Eq. (15.34).

(b) For $L_x = L_y = 1$, make a contour plot for the distribution of part (a). (See Exercise 15.2.5 if helpful.)

(c) Compare your plot with that obtained in Exercise 15.2.5.

Example 15.2.3. Diffusion Equation

A long thin cylindrical rod of material is insulated at its curved boundaries and extends longitudinally from $x = 0$ to $x = L$. Its temperature at position x and time t is denoted $T(x, t)$. Prior to $t = 0$ the entire rod is at temperature $T = 0$. Starting at $t = 0$ the end of the rod at $x = 0$ is maintained at $T = 0$ while the end at $x = L$ is kept at temperature T_0. Our task is to find the distribution of temperature on the rod as a function of t. The insulation and the constancy of the boundary conditions over the cross section of the rod cause the heat flow to be only in the x direction, so our diffusion equation takes the relatively simple form

$$\nabla^2 T = \frac{1}{k^2}\frac{\partial T}{\partial t} \qquad \longrightarrow \qquad \frac{\partial^2 T(x,t)}{\partial x^2} = \frac{1}{k^2}\frac{\partial T}{\partial t}\,, \tag{15.35}$$

where k depends upon the thermal conductivity of the material.

Making our problem solvable—We would like to apply the method of separation of variables to solve this problem, but a direct approach does not work because the boundary conditions at $x = 0$ and $x = L$ do not lead us to a set of functions of x that we can use, as in previous examples, to write the overall solution as an expansion. But all is not lost; we can be more inventive and describe our problem as a difference of two simpler problems, each of which we **can** solve. More specifically, we reframe the problem in terms of its time evolution toward the final steady-state temperature distribution, which has the simple form

$$T(x) = \frac{xT_0}{L}\,. \tag{15.36}$$

Though Eq. (15.36) may be reasonably obvious, it is the topic of Exercise 15.2.8.

Because the diffusion equation is linear and homogeneous, the difference of two of its solutions, along with the difference of their boundary conditions, will also be a solution. Our first solvable problem, with solution $T_1(x, t)$, will be the time-independent temperature distribution obtained if we start from the final steady-state temperature and boundary conditions:

$$T_1(x,0) = T_1(x,t) = \frac{xT_0}{L}\,, \qquad T_1(0,t) = 0\,, \qquad T_1(L,t) = T_0\,. \tag{15.37}$$

Our second solvable problem, with solution $T_2(x, t)$, will also start from the final steady-state temperature, but with different boundary conditions:

$$T_2(x,0) = \frac{xT_0}{L}\,, \qquad T_2(0,t) = 0\,, \qquad T_2(L,t) = 0\,. \tag{15.38}$$

This problem is solvable because the boundary condition on x is that $T = 0$ at both $x = 0$ and $x = L$. Now, from Eqs. (15.37) and (15.38) we see that $T_1 - T_2$ is the solution to a problem in which

$$[T_1 - T_2](x,0) = 0\,, \qquad [T_1 - T_2](0,t) = 0\,, \qquad [T_1 - T_2](L,t) = T_0\,, \qquad (15.39)$$

i.e., it is the problem we wish to solve.

Solving for T_2—To complete our work we must now actually find the solution T_2. Qualitatively, you probably already know what will happen: keeping both ends of the rod at $T = 0$, it will over time equilibrate to $T_2 = 0$ for all x. But we need a quantitative description of the equilibration process, which we obtain using the method of separation of variables.

Assuming the separated-variable form $T = X(x)U(t)$ and proceeding as in earlier examples, we reach

$$X'' = CX\,, \qquad U' = Ck^2U\,. \qquad (15.40)$$

Our next step is to solve the x equation subject to the boundary conditions derived from Eq. (15.38), namely $X(0) = 0$ and $X(L) = 0$. That equation is similar (including boundary conditions) to the x equation in Example 15.2.2, so we know that its relevant solutions are

$$X_n(x) = \sin\frac{n\pi x}{L}\,, \quad \text{with} \quad C_n = -\frac{n^2\pi^2}{L^2} \quad \text{and} \quad n = 1,\ 2,\ 3,\ldots \qquad (15.41)$$

Using these values of C_n, the U equation then assumes the form

$$U_n' = -\frac{n^2\pi^2k^2}{L^2}\,U_n\,, \quad \text{with general solution} \quad U_n = e^{-(n^2\pi^2k^2/L^2)t} + C'\,. \qquad (15.42)$$

We must set the constant C' to zero to make U_n go to zero at infinite t, so our work up to this point has identified the following solutions for $T_2(x,t)$:

$$T_{2,n}(x,t) = \sin\frac{n\pi x}{L}\,e^{-(n^2\pi^2k^2/L^2)t}\,. \qquad (15.43)$$

Equation (15.43) is a reasonable result because thermal equilibration of a sinusoidal temperature distribution can be expected to be faster when it has more nodes: the regions of positive and negative temperature are then closer together.

To complete the determination of T_2, we must now take the linear combination of the $T_{2,n}(x,t)$ that satisfies the boundary condition $T_2(x,0) = xT_0/L$, so we write

$$T_2(x,t) = \sum_n a_n \sin\frac{n\pi x}{L}\,e^{-(n^2\pi^2k^2/L^2)t}\,, \qquad (15.44)$$

and therefrom get

$$\frac{xT_0}{L} = T_2(x,0) = \sum_n a_n \sin\frac{n\pi x}{L}\,. \qquad (15.45)$$

This is a Fourier sine series for an interval of length L, so we know from Eq. 12.23 that the a_n are given by the formula

$$a_n = \frac{2}{L}\int_0^L \frac{T_0x}{L}\,\sin\frac{n\pi x}{L}\,dx = \frac{2(-1)^{n+1}T_0}{n\pi}\,, \qquad (15.46)$$

and therefore

$$T_2(x,t) = \frac{2T_0}{\pi}\sum_{n=1}^{\infty}\frac{(-1)^{n+1}}{n}\sin\frac{n\pi x}{L}\,e^{-(n^2\pi^2k^2/L^2)t}\,. \tag{15.47}$$

Putting it all together—To complete the solution of this Example, we now form $T = T_1 - T_2$, giving the final result

$$T(x,t) = \frac{xT_0}{L} - \frac{2T_0}{\pi}\sum_{n=1}^{\infty}\frac{(-1)^{n+1}}{n}\sin\frac{n\pi x}{L}\,e^{-(n^2\pi^2k^2/L^2)t}\,. \tag{15.48}$$

■

Example 15.2.3 illustrates one additional point worthy of mention:

- *Sometimes the solution to a PDE that is needed only becomes accessible by the method of separation of variables if it is identified as a linear combination of solvable problems which are then solved separately and combined. This strategy is similar to that illustrated earlier for the Laplace equation but may be less obvious in the context of initial-value problems such as that of the most recent example.*

Exercises

15.2.8. Verify that Eq. (15.36) is a solution to $X'' = 0$ such that $X(0) = 0$ and $X(L) = T_0$.

15.2.9. Either by hand or using symbolic computing, verify the integration in Eq. (15.46).

15.2.10. Two large slabs, each of thickness d, are caused to come to a steady-state temperature distribution with one side at $T = 0$ and the other side at $T = T_0$. At time $t = 0$ they are stacked with the $T = T_0$ sides touching each other and the outer ($T = 0$) sides are maintained (at all times) at temperature $T = 0$. Applying the diffusion equation, Eq. (15.35),

(a) Find the temperature distribution within the slabs as a function of position and time.

(b) Make plots of the temperature profile at the times at which the interface between the two slabs is at temperatures (i) $0.8\ T_0$ and (ii) $0.6\ T_0$.

15.2.11. Repeat the calculation and plotting of Exercise 15.2.10 if the outer sides of the two slabs (initially at $T = 0$) are insulated (so no heat flows between them and the surroundings). The boundary condition for an insulated end at $x = x_0$ is that $\partial T/\partial x$ vanish at $x = x_0$.

Hint. The final temperature of both slabs will be constant, at $T = T_0/2$.

Example 15.2.4. Schrödinger Equation

This example deals with the time-independent Schrödinger equation for a particle confined to a box bounded by $x = 0$ and $x = a$, by $y = 0$ and $y = b$, and by $z = 0$

and $z = c$, but subject to no additional forces. In dimensionless units (unit particle mass, Planck's constant h set to 2π), the PDE governing this problem is

$$-\frac{1}{2}\nabla^2\psi(x,y,z) = -\frac{1}{2}\left[\frac{\partial^2\psi}{\partial x^2} + \frac{\partial^2\psi}{\partial y^2} + \frac{\partial^2\psi}{\partial z^2}\right] = E\psi(x,y,z)\,, \tag{15.49}$$

subject to the conditions that $\psi(x,y,z) = 0$ on all the boundaries of the box. The quantity E is an eigenvalue, and solutions satisfying the boundary conditions exist only for certain values of E. Our task is to determine the possible values of E and the corresponding eigenfunctions $\psi(x,y,z)$.

Separating the variables—We start by assuming that $\psi = X(x)Y(y)Z(z)$, an appropriate choice because the boundary conditions also separate:

$$X(0) = X(a) = 0\,, \qquad Y(0) = Y(b) = 0\,, \qquad Z(0) = Z(c) = 0\,.$$

Inserting the separated-variable form, we have

$$\nabla^2\psi = YZX'' + XZY'' + XYZ'' = -2E\,XYZ$$

$$\longrightarrow \quad \frac{X''}{X} + \frac{Y''}{Y} + \frac{Z''}{Z} = -2E\,. \tag{15.50}$$

Equation (15.50) is a bit more complicated than the separations we have previously encountered, but has most of the same features. Each of the terms X''/X, Y''/Y, Z''/Z must individually be a constant (we name these constants $-2E_x$, $-2E_y$, $-2E_z$), but the new wrinkle is that instead of being equal, they must sum to $-2E$. Thus, we have (in a form quantum physicists might like)

$$-\frac{1}{2}X'' = E_x X\,, \quad -\frac{1}{2}Y'' = E_y Y\,, \quad -\frac{1}{2}Z'' = E_z Z\,, \tag{15.51}$$

with the additional condition

$$E_x + E_y + E_z = E\,. \tag{15.52}$$

The boundary conditions make each of Eqs. (15.51) a fully-defined one-dimensional eigenvalue problem.

Solving the one-dimensional problems—Taking the X equation first, we solve it for $0 \le x \le a$ subject to $X(0) = X(a) = 0$. The solutions, used earlier in this section, are

$$X_n(x) = \sin\frac{n\pi x}{a}, \qquad \text{with } E_x = \frac{n^2\pi^2}{2a^2} \qquad \text{and } n = 1,\ 2,\ 3,\ldots \tag{15.53}$$

The Y and Z equations are similar; they yield (with m and p positive integers)

$$Y_m(y) = \sin\frac{m\pi y}{b}, \quad Z_p(z) = \sin\frac{p\pi z}{c}, \qquad E_y = \frac{m^2\pi^2}{2b^2}, \quad E_z = \frac{p^2\pi^2}{2c^2}, \tag{15.54}$$

and we therefore have

$$\psi_{nmp}(x,y,z) = \sin\frac{n\pi x}{a}\sin\frac{m\pi y}{b}\sin\frac{p\pi z}{c}, \qquad \text{with}$$

$$E_{nmp} = E_x + E_y + E_z = \frac{\pi^2}{2}\left[\frac{n^2}{a^2} + \frac{m^2}{b^2} + \frac{p^2}{c^2}\right]. \tag{15.55}$$

Each set of (n,m,p) defines a different solution to our problem, and (unless they happen to have the same value of E_{nmp}) we cannot add solutions together and still have a solution to our PDE. To see this, look at $\Psi = \psi_{nmp} + \psi_{n'm'p'}$. We get (if $E_{nmp} \neq E_{n'm'p'}$)

$$-\frac{1}{2}\nabla^2\Psi = E_{nmp}\psi_{nmp} + E_{n'm'p'}\psi_{n'm'p'} \neq (\text{constant}) \times \Psi .$$

Thus, in contrast with our earlier examples, we cannot combine individual solutions to produce an overall solution meeting some condition, but instead have individual solutions, each with its own value of E_{mnp} (the eigenvalue). Note also that the value of E is not arbitrary, but can only have specific values that are determined from the boundary conditions.

The only exception to the above observations arises if two different E_{mnp} happen to be equal. If $E_{nmp} = E_{n'm'p'}$, then any linear combination of ψ_{nmp} and $\psi_{n'm'p'}$ is also an eigenfunction, and we have (in that case) a two-dimensional subspace all of whose members are eigenfunctions of the same E_{nmp}. The eigenvalue is then referred to as **degenerate** because it is associated with more than a single eigenfunction.

In the present problem there are no limitations on the values of n, m, and p other than that they be positive integers, so we are finding that this PDE has an infinite number of solutions consistent with the boundary conditions. This behavior is an indication that the quantum particle in a box can be in any of an infinite number of states; simply being in the box is not sufficient to determine its state completely. ■

A take-home lesson from the foregoing Example is

- *Because each separated-variable solution to this eigenvalue problem is associated with its respective eigenvalue, the solutions cannot in general be added together and remain solutions. We therefore obtain a* **set** *of solutions to our eigenvalue problem. Because we cannot form linear combinations of solutions to satisfy boundary conditions, the method of separation of variables only works when the separated equations can be solved in a way that satisfies all the boundary conditions of the problem.*

Exercises

15.2.12. Solve the quantum problem of a particle in a cube of side a. Use dimensionless units, as in Example 15.2.4.

(a) Show that the energy eigenvalues are of the form

$$E_{nmp} = \frac{\pi^2}{2a^2}\left[n^2 + m^2 + p^2\right],$$

and tabulate all the solutions for which $n^2 + m^2 + p^2 \leq 14$.

(b) Show that the degenerate eigenfunctions can be interconverted by carrying out symmetry operations on the cube, and that an eigenvalue is unique only when the eigenfunction has (except possibly for sign) the full cubic symmetry.

15.2.13. Consider a quantum particle in a two-dimensional rectangular box with sides a and $2a$. Using dimensionless units, as in Example 15.2.4, find all the eigenfunctions with $E \leq 3\pi^2/a^2$. Identify any degeneracies.

15.2.14. Consider a three-dimensional quantum harmonic oscillator, with wave functions described by the time-independent Schrödinger equation (in dimensionless units), with $V = (x^2 + y^2 + z^2)/2$:

$$-\frac{1}{2}\nabla^2\psi(x,y,z) + \frac{1}{2}(x^2+y^2+z^2)\psi = E\psi\,. \tag{15.56}$$

This equation must be solved subject to the condition that ψ approach zero for points far from the coordinate origin (i.e., when any of $\pm x$, $\pm y$, or $\pm z$ become large).

(a) We plan to solve this equation by separating the variables. Letting $\psi = X(x)Y(y)Z(z)$, show that the separation leads to the ODEs

$$-\frac{X''}{2}+\frac{x^2}{2}X = E_xX\,, \quad -\frac{Y''}{2}+\frac{y^2}{2}Y = E_yY\,, \quad -\frac{Z''}{2}+\frac{z^2}{2}Z = E_zZ\,, \tag{15.57}$$

with $E = E_x + E_y + E_z$.

(b) The equation for X has solutions

$$X_n(x) = e^{-x^2/2}H_n(x), \qquad n = 0,\ 1,\ 2,\ \ldots, \tag{15.58}$$

where $H_n(x)$ is the Hermite polynomial of degree n. Note that this solution vanishes in the limit of large $|x|$. Insert the formula for $X_n(x)$ into the X equation and, comparing with the Hermite ODE, Eq. (14.107), verify the solution and show that the value of E_x corresponding to X_n is $E_x = n + \frac{1}{2}$.

Hint. It may be helpful to look at Eqs. (14.117) and (14.119).

(c) Extending the work of part (b) to the Y and Z equations, show that the eigenfunctions of ψ can be denoted ψ_{nmp} with $E_{nmp} = n + m + p + \frac{3}{2}$, and give a formula for ψ_{nmp}.

(d) List all the states with $E \le 9/2$, for each giving its values of the indices n, m, p, its eigenfunction ψ_{nmp} (expanded as an explicit polynomial times an exponential), and its value of E.

Note. Save your answer to part (d); it will be needed in connection with a later Exercise.

15.3 SEPARATION OF VARIABLES IN CYLINDRICAL COORDINATES

The method of separation of variables can be used when PDEs are written in coordinate systems other than Cartesian, but the procedure must take into account that operators such as the Laplacian are not in an entirely separated form. As an initial two-dimensional example, let's find the characteristic vibrational frequencies of a circular membrane (such as a drumhead) and the oscillation of the membrane given specific initial conditions.

Example 15.3.1. Circular Membrane

A uniformly stretched membrane, fastened to a fixed circular boundary of radius h, has a displacement ψ that is a function of position on the membrane and the time t.

Let's find the normal modes of vibration of the membrane and the corresponding characteristic vibration frequencies.

The displacement is governed by the wave equation, which before specializing to the present problem has the general form given in Eq. (15.5). For two spatial dimensions, and using polar coordinates (a choice motivated by the boundary conditions), the wave equation reduces to

$$\frac{\partial^2\psi(\rho,\varphi,t)}{\partial\rho^2} + \frac{1}{\rho}\frac{\partial\psi}{\partial\rho} + \frac{1}{\rho^2}\frac{\partial^2\psi}{\partial\varphi^2} = \frac{1}{v^2}\frac{\partial^2\psi}{\partial t^2}. \tag{15.59}$$

Equation (15.59) is obtained from Eq. (15.5) using the formula for the Laplacian in cylindrical coordinates, Eq. (7.91), removing the dependence on z, and simplifying the ρ derivative as shown in Exercise 7.6.11.

Separating the variables—We attempt to find solutions of the separated-variable form, $\psi = P(\rho)\Phi(\varphi)U(t)$. (Cultural note: P is an uppercase ρ.) Inserting this form for ψ and dividing through by $P\Phi U$, we have

$$\frac{P''}{P} + \frac{P'}{\rho P} + \frac{1}{\rho^2}\frac{\Phi''}{\Phi} = \frac{1}{v^2}\frac{U''}{U}. \tag{15.60}$$

As in our earlier examples, the right-hand side of this equation depends only on t, while its left-hand side is independent of t. Therefore each side is equal to a constant, which we call $-C^2$. We take this constant as a negative number because we anticipate (correctly) that the time behavior will be oscillatory. Thus, at this point we have

$$U''(t) = -C^2v^2U(t), \qquad \frac{P''}{P} + \frac{P'}{\rho P} + \frac{1}{\rho^2}\frac{\Phi''}{\Phi} = -C^2. \tag{15.61}$$

We now rearrange the second of Eqs. (15.61) to

$$\rho^2\frac{P''}{P} + \rho\frac{P'}{P} + C^2\rho^2 = -\frac{\Phi''}{\Phi}, \tag{15.62}$$

and it now becomes obvious that we can separate the ρ and φ dependence. Setting each side to the constant value m^2 (this implied choice of sign will shortly be seen to be appropriate), Eq. (15.62) can be separated into the two equations

$$\rho^2\frac{P''}{P} + \rho\frac{P'}{P} + C^2\rho^2 = m^2, \qquad -\frac{\Phi''}{\Phi} = m^2,$$

which in more conventional forms can be written

$$\rho^2P''(\rho) + \rho P'(\rho) + (C^2\rho^2 - m^2)P(\rho) = 0, \tag{15.63}$$

$$\Phi''(\varphi) = -m^2\Phi(\varphi). \tag{15.64}$$

Implied boundary conditions—Our statement of the present problem has thus far only identified one explicit boundary condition, namely that ψ must vanish on the circular boundary at $\rho = h$. In the separated variables, this boundary condition is $P(h) = 0$. However, there are additional implied boundary conditions that arise from the requirement that ψ be continuous and differentiable at all points interior to the circular boundary. These implied boundary conditions become significant because of the properties of the curvilinear coordinate system. In particular, there must be conditions on Φ that take account of the fact that $\varphi = 0$ and $\varphi = 2\pi$ are actually the same coordinate point, so we must require Φ to be periodic, with period 2π.

Solving the Φ ODE—We are now ready to solve the φ equation subject to the aforementioned boundary (connection) conditions:

$$\Phi'' = -m^2\Phi \quad \text{has general solution} \quad \Phi_m = a_m \cos m\varphi + b_m \sin m\varphi\,, \tag{15.65}$$

and this solution will have period 2π only if $m = 0,\ \pm 1,\ \pm 2,\ \ldots$

Note that we designated the separation constant $-m^2$, suggesting that we expect it to be negative. If we had written $+m^2$, our formula for Φ_m would have contained hyperbolic functions, which could only be periodic if m were chosen to be imaginary.

Solving the P ODE—We turn now to the P ODE, which we identify as having solutions that are Bessel functions. Note that if we had made a different choice for the sign of C^2, the solutions to the P ODE would have been **modified** Bessel functions, which do not have the oscillatory character that we will need to satisfy the boundary conditions of the present problem.

Comparing with Eq. (14.76), we see that Eq. (15.63) has the solutions $J_m(C\rho)$ and $Y_m(C\rho)$; we reject Y_m because it puts an unphysical singularity at $\rho = 0$, the center of the membrane.

We are now ready to identify possible values for C; they are determined by requiring that $J_m(C\rho)$ vanish at $\rho = h$. Letting α_{mj} stand for the jth positive zero of J_m, this boundary condition leads to a set of C (depending on both j and m) such that $C_{nj}h = \alpha_{mj}$, so we have

$$P_{mj}(\rho) = J_m\left(\frac{\alpha_{mj}\,\rho}{h}\right), \qquad \text{with} \qquad C_{mj} = \frac{\alpha_{mj}}{h}\,. \tag{15.66}$$

Normal modes—With both $\Phi_m(\varphi)$ and $P_{mj}(\rho)$ in hand, and with C_{mj} determined, we can solve the equation for $U(t)$ and thereby obtain a set of separated-variable solutions that describe the normal modes of oscillation of our membrane. In particular,

$$U'' = -C_{mj}^2 v^2 U \quad \longrightarrow \quad U_{mj} = a_{mj}\cos C_{mj}vt + b_{mj}\sin C_{mj}vt\,. \tag{15.67}$$

Since it is our intention to identify a linearly independent set of normal modes, we can combine the spatial and time factors of the separated-variable solutions by writing

$$\psi_{mj}(\rho,\varphi,t) = J_m\left(\frac{\alpha_{mj}\rho}{h}\right)\left\{\begin{array}{c}\cos m\varphi\\ \\ \sin m\varphi\end{array}\right\}\left\{\begin{array}{c}\cos\dfrac{\alpha_{mj}vt}{h}\\ \sin\dfrac{\alpha_{mj}vt}{h}\end{array}\right\}. \tag{15.68}$$

Equation (15.68) shows that the characteristic frequencies of the circular membrane are proportional to the zeros α_{mj} of the Bessel function of order m. When we studied Bessel functions in Chapter 14, we found that those zeros are not integer multiples of each other, and that they only approach equal spacing in the limit of large j. The result is that the harmonics of a drum do not have musically related frequencies, and therefore produce sound that is less "harmonious" than that of, say, a violin.

It is sometimes useful to enumerate the normal modes ψ_{mj} by adopting the convention that if $m \geq 0$ we will choose the $\cos m\varphi$ function of Eq. (15.68), but if $m < 0$ we will use Eq. (15.68) for $|m|$ but choose the $\sin m\varphi$ function.

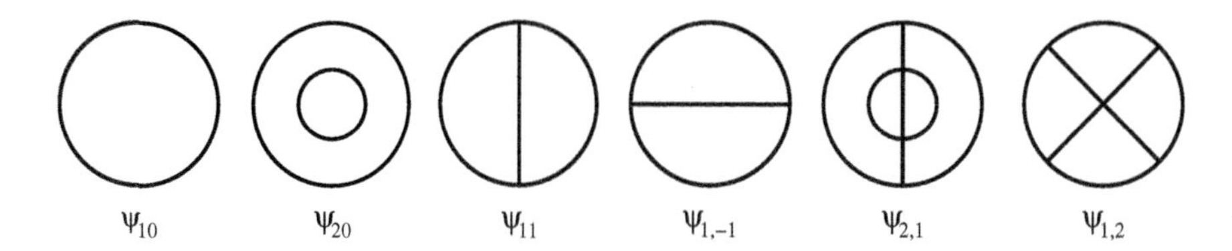

Figure 15.2: Nodal patterns of some normal modes ψ_{mj} of a circular membrane.

A feature of the set of normal modes that becomes apparent in the present problem is that individual modes may exhibit less symmetry than the problem in which they occur. For example, our membrane has circular symmetry, but a mode with $m = 1$ will have a node (which in a two-dimensional problem is a line, rather than a point) in a particular angular direction. (With our enumeration convention $m = 1$ corresponds to a factor $\cos\varphi$, which is zero at the angles $\varphi = \pi/2$ and $3\pi/2$ and therefore defines a straight line.) However, some nodes (e.g., at ρ values that are interior zeros of J_m) are circular. Several nodal possibilities for the circular membrane are shown in Fig. 15.2.

The symmetry of the membrane is consistent with the existence of angular nodes because such nodes can occur at any angular orientation. Again for $m = 1$, the normal modes given in Eq. (15.68) include (for each j) one with the angular orientation $\cos\varphi$ and another with orientation $\sin\varphi$. These are the two linearly independent modes of vibration for $m = 1$, but we can take linear combinations of these modes (which have the same time dependence) to describe similar behavior at any other orientation; note that

$$\cos(\varphi + \delta) = \cos\delta\cos\varphi - \sin\delta\sin\varphi\,,$$

and that the two functions $\cos(\varphi + \delta)$ and $\sin(\varphi + \delta)$ span the same space as $\cos\varphi$ and $\sin\varphi$. In other words, Eq. (15.68) remains equally valid if we change the direction relative to which we measure the angle φ.

Vibrations with initial conditions—Though we will not do so here, one can find a solution for the membrane vibration subject to initial conditions by combining the normal modes with arbitrary coefficients, and then matching the coefficients to expansions of the initial distributions of displacement and velocity of the membrane. The coefficients in these expansions can be found using the formulas for Bessel series that were developed in Section 14.4. In principle, it is possible to apply initial conditions in a way that excites only a specific normal mode; in practice that situation may not be easy to achieve.

In orchestras, kettledrums (timpani) are constructed and played (hit) in a way that causes much (but by no means all) of the sound energy to be at the fundamental frequency. The nature of the normal mode distribution then produces a characteristic sound quality that is quite different from that obtained from other orchestral instruments.

■

Our first example of the separation of variables in curvilinear coordinates indicates the following:

- *It may still be possible to separate the variables in a curvilinear coordinate system, but the process may be nested or sequential. For example, in the circular membrane problem the value of the constant m, determined from the Φ equation,* *enters as a parameter in the P equation. Only when the P equation is solved with a given value of m are we able to determine the separation constant that is needed to solve the equation for $U(t)$.*

- *Individual normal modes for a multidimensional vibration problem may have less symmetry than the underlying problem, but in that case the set of all normal modes can reproduce the overall problem symmetry.*

Exercises

15.3.1. Cylindrical problems do not always involve Bessel functions. Find the steady-state temperature distribution if the surface of a long cylinder of material (radius a) is kept at the temperature distribution $\psi_0(\varphi)$ (independent of z). The PDE governing this situation, expressed in cylindrical coordinates, is

$$\frac{\partial^2\psi}{\partial\rho^2} + \frac{1}{\rho}\frac{\partial\psi}{\partial\rho} + \frac{1}{\rho^2}\frac{\partial^2\psi}{\partial\varphi^2} = 0\,.$$

(a) Obtain the solution $\psi(\rho,\varphi)$ as an expansion valid for general ψ_0.

(b) Obtain a closed-form solution for the case $\psi_0 = T_0\cos\varphi$.

(c) Obtain an expansion giving the solution for the case

$$\psi_0(\varphi) = \begin{cases} T_0, & -\frac{\pi}{2} < \varphi < +\frac{\pi}{2}, \\ 0, & \text{otherwise.} \end{cases}$$

(d) Write a symbolic procedure `psi(rho,phi)` or `psi[rho_,phi_]` that evaluates the expansion of part (c) (with $T_0 = 1$), keeping at least the first 20 nonzero terms. Check your expansion by plotting $\psi(a,\varphi)$ for $-\pi \le \varphi \le \pi$ and compare with ψ_0.

(e) Make a contour plot of your solution to part (c). See Exercise 15.2.5 for coding that produces contour plots.

Example 15.3.2. Electrostatic Potential Inside a Cylinder

Consider a charge-free circular cylinder of radius a and length L, with the curved surface and one end cap grounded (at potential $\psi = 0$) and with the other end cap held at the constant potential V_0. Our task is to find the potential ψ at all points interior to the cylinder. The geometry of the problem suggests the use of cylindrical coordinates (ρ,φ,z); we place the cylinder with its centerline along the z axis, with the grounded end cap at $z = 0$ and the cap with potential V_0 at $z = L$. In these coordinates, the Laplace equation takes the form

$$\nabla^2\psi(\rho,\varphi,z) = \frac{\partial^2\psi}{\partial\rho^2} + \frac{1}{\rho}\frac{\partial\psi}{\partial\rho} + \frac{1}{\rho^2}\frac{\partial^2\psi}{\partial\varphi^2} + \frac{\partial^2\psi}{\partial z^2} = 0\,. \qquad (15.69)$$

Because of the angular symmetry of both the cylinder and the boundary potentials, the solution we seek will be independent of the coordinate φ. We therefore apply the method of separation of variables by taking solutions of the form $\psi = P(\rho)Z(z)$, and Eq. (15.69) reduces to the two ODEs

$$\rho^2 P''(\rho) + \rho P'(\rho) + C^2\rho^2 P(\rho) = 0\,, \qquad Z''(z) - C^2 Z(z) = 0\,. \qquad (15.70)$$

Because we have a zero boundary condition for the entire boundary in ρ and a nonzero boundary condition for part of the boundary in z, namely $Z(L) = V_0$, our strategy will be to first obtain a set of solutions to the ρ ODE and then to solve the z ODE as a series expansion.

Solving the P ODE—The P ODE is similar to that of Example 15.3.1 except that here it does not contain the parameter m^2. We recognize it as having solutions that are Bessel functions of order zero. The relevant solution is regular at $\rho = 0$, namely $J_0(C\rho)$. As in the earlier example, we require solutions that place a zero of the Bessel function at $\rho = a$; letting α_{0j} be the jth positive zero of J_0, we choose

$$C_j = \frac{\alpha_{0j}}{a} \quad \text{and} \quad P_j(\rho) = J_0\left(\frac{\alpha_{0j}\rho}{a}\right). \tag{15.71}$$

Note that the sign of C^2 was chosen so that the ρ equation would be a (regular, not modified) Bessel equation. The modified Bessel functions are not oscillatory and cannot satisfy the boundary condition at $\rho = a$.

Solving the Z ODE—With the positive sign assigned to C_j^2, the Z ODE will have exponentials, or equivalently, hyperbolic functions as solutions. We choose the sinh solution to satisfy the boundary condition $Z(0) = 0$, and we therefore have

$$Z_j(z) = \sinh C_j z = \sinh \frac{\alpha_{0j} z}{a}$$

and an overall solution of the form

$$\psi(\rho, z) = \sum_{j=1}^{\infty} b_j J_0\left(\frac{\alpha_{0j}\rho}{a}\right) \sinh \frac{\alpha_{0j} z}{a}. \tag{15.72}$$

Satisfying the boundary condition at $z = L$—The expansion of Eq. (15.72) satisfies all the boundary conditions except that at $z = L$, where we must have $\psi(\rho, L) = V_0$. Writing

$$\psi(\rho, L) = V_0 = \sum_{j=1}^{\infty} \left[b_j \sinh \frac{\alpha_{0j} L}{a} \right] J_0\left(\frac{\alpha_{0j}\rho}{a}\right), \tag{15.73}$$

we see that we now need the coefficients in the Bessel series of order zero for V_0 on the range $(0, a)$. That series is known; it is

$$V_0 = \sum_{j=1}^{\infty} \frac{2V_0}{\alpha_{0j} J_1(\alpha_{0j})} J_0\left(\frac{\alpha_{0j}\rho}{a}\right). \tag{15.74}$$

However, Eq. (15.74) may be unknown to you, and in many problems of practical interest the necessary expansion may not have closed-form coefficients. We therefore proceed based on the assumption that we lack an analytical expansion with which to fit our Bessel series.

Accordingly, we use symbolic computing to obtain a numerical evaluation of the formula for the coefficients, Eq.(14.89), which for the current problem becomes

$$V_0 = \sum_j c_j J_0\left(\frac{\alpha_{0j}\rho}{a}\right), \qquad c_j = \frac{2V_0}{a^2 [J_1(\alpha_{0j})]^2} \int_0^a J_0\left(\frac{\alpha_{0j} x}{a}\right) x\, dx. \tag{15.75}$$

MAPLE code for the first N coefficients c_j is

```
> V0:=1: a:=1: L:=2: N:=100:                    (set to desired values)
> for j from 1 to N do
>   alpha := BesselJZeros(0,j);
>   c[j] := evalf(2 * V0/a^2/BesselJ(1,alpha)^2)
>          * evalf(int(BesselJ(0,alpha*x/a)*x, x=0 .. a))
> end do;
```

In MATHEMATICA, the first n coefficients c_j can be obtained from

```
v0=1; a=1; l=2; n=100;                          (set to desired values)
Do[alpha = BesselJZero[0,j];
   c[j] = 2. * v0/a^2/BesselJ[1,alpha]^2
        * Integrate[BesselJ[0,alpha*x/a]*x, {x,0,a}],
   {j,1,n} ]
```

We can now obtain the coefficients b_j that are needed for Eq. (15.72) from the formula

$$b_j = c_j \left[\sinh \frac{\alpha_{0j} L}{a} \right]^{-1}$$

and build $\psi(\rho, z)$ by the following procedure: In MAPLE,

```
> psi := proc(rho,z) local S,j,alpha;
>   S := 0;
>   for j from 1 to N do
>     alpha := BesselJZeros(0,j);
>     S := S + c[j]/sinh(alpha * L/a) * BesselJ(0, alpha * rho/a)
>                * sinh(alpha * z/a)
>   end do; S
> end proc;
```

In MATHEMATICA, the corresponding procedure is

```
psi[rho_,z_] := Module[{s,j,alpha},
    s = 0;
    Do[alpha = BesselJZero[0,j];
       s = s + c[j]/sinh[alpha * l/a] * BesselJ[0, alpha * rho/a]
                  * sinh[alpha * z/a],
       {j,1,n}];
    s]
```

The above procedure enables computation of $\psi(\rho, z)$ for arbitrary input, but we may get a better feeling for the result by looking at a contour plot. Code for such a plot can be constructed using symbolic coding as illustrated in Exercise 15.2.5. Your

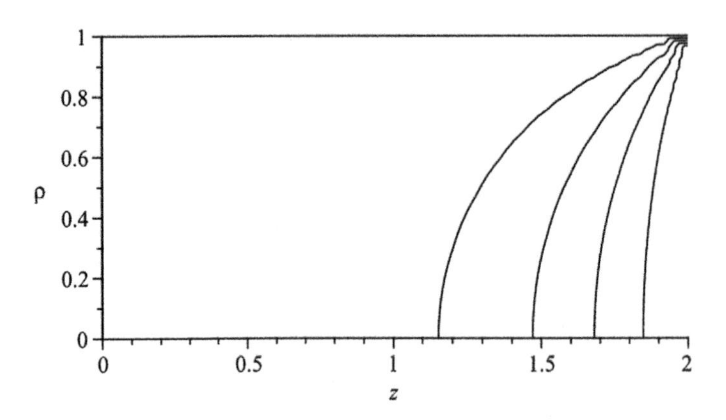

Figure 15.3: Contour plot for Example 15.3.2.

plot may look something like that in Fig. 15.3; to save space in the book it is printed with the z-axis horizontal.

■

Exercises

15.3.2. Verify that the separation of variables in Eq. (15.69) leads to Eq. (15.70).

15.3.3. Find the normal modes of oscillation of a circular cylindrical cavity of radius a and length h if the amplitude $\psi(\rho, \varphi, z, t)$ of the oscillation is governed by the PDE

$$\nabla^2\psi = \frac{1}{v^2}\frac{\partial^2\psi}{\partial t^2},$$

subject to the boundary condition that ψ vanish on all surfaces of the cavity.

Find the functional form of each linearly independent normal mode in terms of ρ, φ, and z, and identify its frequency of oscillation.

15.3.4. A semi-infinite circular cylinder of radius a is placed on the positive z-axis, with its finite end at $z = 0$. Find the steady-state temperature distribution within the cylinder if its curved surface is maintained at temperature $T = 0$ and its end at $z = 0$ is maintained at the temperature distribution $\psi_0 = T_0 y = T_0 \rho \sin\varphi$. The PDE governing the temperature distribution is $\nabla^2\psi = 0$.

Hint. You may use symbolic methods to evaluate any necessary integrals.

15.4 SPHERICAL-COORDINATE PROBLEMS

A special, but highly important set of problems has the feature that both the relevant PDE and its boundary conditions have convenient forms when written in spherical polar coordinates. A nonexclusive list of examples includes

- **Central-force problems** in quantum mechanics, including the hydrogen atom and the most widely used approximate methods for the treatment of larger atoms.

- Electrostatics problems with boundary conditions specified on a spherical surface.
- Dissipation of thermal energy from a localized or point source.
- Wave propagation into an infinite medium (or electromagnetic waves in free space), with arbitrary angular distribution.

The above and other similar problems have in common that they can be addressed using the method of separation of variables in spherical polar coordinates. The most important problems of this type involve PDEs of the form

$$\nabla^2\psi(r,\theta,\varphi) + G(r)\psi = 0\,. \tag{15.76}$$

Equation (15.76) includes not only the Laplace equation (for which $G = 0$), but also the Helmholtz equation, with $G = k^2$, or the Schrödinger equation for a central force, with $G = 2[E - V(r)]$. The separation-of-variables method will be applicable when the boundary conditions are also to some extent separable.

SEPARATION OF VARIABLES

As usual, we set $\psi(r,\theta,\varphi) = R(r)\Theta(\theta)\Phi(\varphi)$, use the form for the Laplacian in spherical coordinates as given in Eq. (7.97) and Exercise 7.6.12, and thereby get

$$\Theta\Phi\left(R'' + \frac{2}{r}R'\right) + \frac{R\Phi}{r^2\sin\theta}\frac{d}{d\theta}\Big(\sin\theta\,\Theta'\Big) + \frac{R\Theta}{r^2\sin^2\theta}\Phi'' + G(r)R\Theta\Phi = 0\,. \tag{15.77}$$

Dividing through by $R\Theta\Phi$, multiplying by r^2, and grouping terms as shown,

$$\frac{r^2}{R}\left(R'' + \frac{2}{r}R' + G(r)R\right) + \frac{1}{\sin^2\theta}\left[\frac{\sin\theta}{\Theta}\frac{d}{d\theta}\Big(\sin\theta\,\Theta'\Big) + \frac{\Phi''}{\Phi}\right] = 0\,. \tag{15.78}$$

We could rearrange Eq. (15.78) to put Φ''/Φ on one side of the equation, leaving quantities only dependent on r and θ on the other side. Equivalently, we can simply observe that all the φ dependence in Eq. (15.78) is localized in Φ''/Φ, and Eq. (15.78) can be satisfied for all r, θ, and φ only if Φ''/Φ is actually a constant. Therefore, we set

$$\frac{\Phi''}{\Phi} = -m^2, \qquad \text{or} \qquad \Phi'' = -m^2\,\Phi\,. \tag{15.79}$$

We now return to Eq. (15.78), writing it with Φ''/Φ replaced by $-m^2$. That equation is then in separated form:

$$\frac{r^2}{R}\left(R'' + \frac{2}{r}R' + G(r)R\right) + \frac{1}{\sin^2\theta}\left[\frac{\sin\theta}{\Theta}\frac{d}{d\theta}\Big(\sin\theta\,\Theta'\Big) - m^2\right] = 0\,, \tag{15.80}$$

and we can therefore divide it into the two equations

$$\frac{r^2}{R}\left(R'' + \frac{2}{r}R' + G(r)R\right) = \lambda\,, \tag{15.81}$$

$$\frac{1}{\Theta\sin\theta}\frac{d}{d\theta}\Big(\sin\theta\,\Theta'\Big) - \frac{m^2}{\sin^2\theta} = -\lambda\,, \tag{15.82}$$

with the possible values of λ determined by the boundary conditions.

SOLUTION OF THE ANGULAR EQUATIONS

At this point it is useful to notice that the Θ and Φ equations are independent of $G(r)$ and that for all central-force problems their solutions will be subject to the same boundary conditions, namely that

(1) Φ must be periodic, with period 2π (remember that $\varphi = 0$ and $\varphi = 2\pi$ are the same angle), and
(2) Θ, the range of which is $(0 \le \theta \le \pi)$, must be finite on its entire range. This is not a trivial requirement because $\theta = 0$ and $\theta = \pi$ are singular points of the ODE defined by Eq. (15.82).

Of great practical importance is that once we have identified the solutions of the Θ and Φ equations for any central-force problem, they will be applicable, without change, to all such problems.

The periodicity of Φ leads, as in problems with cylindrical symmetry, to the requirement that m be an integer (or zero), and we have

$$\Phi_m = \left\{ \begin{array}{c} \cos m\varphi \\ \sin m\varphi \end{array} \right\}, \qquad m = 0,\ 1,\ 2,\ \ldots \tag{15.83}$$

If we are willing to deal with complex solutions, we can simplify Eq. (15.83) by writing instead

$$\Phi_m = e^{im\varphi}, \qquad m = 0,\ \pm 1,\ \pm 2,\ \ldots \tag{15.84}$$

The form given in Eq. (15.84) is ideal for applications in quantum mechanics; in other contexts we can make use of the fact that the real and imaginary parts of Φ_m are individually ODE solutions.

Turning now to the Θ equation, we find that although it is superficially somewhat complicated, it assumes a simpler form if we make the substitution $t = \cos\theta$. The range of t is $(-1, 1)$.

To write the derivatives in terms of t, note that $\sin^2\theta = 1 - t^2$ and

$$\frac{d}{d\theta} = \left(\frac{dt}{d\theta}\right)\frac{d}{dt} = -\sin\theta\,\frac{d}{dt}.$$

We then have

$$\frac{1}{\sin\theta}\frac{d}{d\theta}\left(\sin\theta\frac{d}{d\theta}\right) = \frac{1}{\sin\theta}\left(-\sin\theta\frac{d}{dt}\right)\left[\sin\theta\left(-\sin\theta\frac{d}{dt}\right)\right],$$

$$= \frac{d}{dt}\left[(1-t^2)\frac{d}{dt}\right],$$

and making the notational change $\Theta(\theta) \to P(t)$, the Θ equation becomes

$$\frac{d}{dt}\left[(1-t^2)\frac{dP(t)}{dt}\right] - \frac{m^2}{1-t^2}P(t) + \lambda P(t) = 0. \tag{15.85}$$

Expanding the derivative that is the first term of Eq. (15.85), we recognize the Θ equation, written in terms of $t = \cos\theta$, as the associated Legendre equation:

$$(1-t^2)P'' - 2tP' - \frac{m^2}{1-t^2}P + \lambda P = 0. \tag{15.86}$$

From our earlier analysis of that equation, which is Eq. (14.40) of Section 14.3, we have already learned that it has solutions, designated $P_l^m(t)$, where l is a positive integer or zero, and m is an integer such that $|m| \le l$. The $P_l^m(t)$ are finite for the entire interval $-1 \le t \le +1$, and the λ value corresponding to P_l^m is $l(l+1)$. We have changed the subscript n, used in Eq. (14.40), to l to agree with the usual notational choice for central-force problems in quantum mechanics; in that domain l is identified as an angular-momentum quantum number.

Summarizing our findings to this point: the general central-force PDE has as the angular part of its solutions the functions

$$P_l^{\pm m}(\cos\theta)\left\{\begin{array}{c}\cos m\varphi\\ \\ \sin m\varphi\end{array}\right\} \qquad \text{or} \qquad P_l^{\pm m}(\cos\theta)\, e^{im\varphi}\,. \tag{15.87}$$

Remembering that P_l^m and P_l^{-m} are proportional to each other, we see that for any given l, Eq. (15.87) defines $2l+1$ linearly independent angular solutions. A convenient way to identify these angular solutions is to let m have the range $-l \le m \le l$, and define

$$S_l^m(\theta,\varphi) = \left\{\begin{array}{ll} P_l^m(\cos\theta)\cos m\varphi, & 0 \le m \le l \\ \\ P_l^m(\cos\theta)\sin m\varphi, & -l \le m \le -1 \end{array}\right\}, \quad \text{or} \tag{15.88}$$

$$= P_l^m(\cos\theta)\, e^{im\varphi}\,, \qquad -l \le m \le l\,. \tag{15.89}$$

Radial equation—In contrast to the situation for the angular equations, the radial ODEs for spherical-coordinate problems will differ in ways that depend upon the definition of $G(r)$. We limit the current general discussion to a presentation of the ODE that results from assigning a value to the separation constant λ. That ODE, which will be different for each l, has the form

$$R_l'' + \frac{2}{r}\,R_l' + \left[G(r) - \frac{l(l+1)}{r^2}\right] R_l = 0\,. \tag{15.90}$$

In addition to differences arising from choices of $G(r)$ and the value of l, spherical-coordinate problems can also differ in their radial boundary conditions. More specific situations are the topics of later examples.

SPHERICAL HARMONICS

The individual solutions to the angular equations can describe the normal modes of vibration of a spherical surface (e.g., a balloon), or, of more importance, the angular parts of the standing-wave distributions in quantum mechanics problems. For the moment let's discuss the angular solutions written as real functions, as defined in Eq. (15.88). The solution S_0^0, which has the same value at all θ and φ, describes a spherically symmetric angular distribution. If S_0^0 were multiplied by a sinusoidal function of the time, it could describe a standing-wave vibrational motion in which a spherically symmetric object (e.g., a balloon) uniformly expands and contracts, maintaining its spherical symmetry. This situation differs from that of a vibrating string in that there are no fixed points (such as the ends, for a string); the entire sphere oscillates about its equilibrium configuration, alternately moving outward (in the case of a balloon, working against the elastic force that tends to shrink its area) and inward (increasing the air pressure, which opposes shrinkage).

Less symmetric modes of vibration are also possible. The solution S_1^0, which reduces to $\cos\theta = z/r$, corresponds to displacement toward positive z in the hemisphere $z > 0$ and displacement toward negative z in the hemisphere $z < 0$, with a nodal line at the equator $\theta = \pi/2$ ($z = 0$). This mode of vibration corresponds to motion in which the balloon alternately lengthens and shortens in the z direction. However, there is nothing unique about the z direction. The solution S_1^1, with explicit form

$$S_1^1 = P_1^1(\cos\theta)\cos\varphi = -\sin\theta\cos\varphi = -x/r\,,$$

can describe a standing wave with amplitude in the x direction, while

$$S_1^{-1} = P_1^{-1}(\cos\theta)\sin\varphi = \frac{1}{2}\sin\theta\sin\varphi = y/2r$$

has amplitude in the y direction. Note that the three angular solutions with $l = 1$ all have one nodal line, each corresponding to a different angular direction. There are three solutions because the value of m can take each integer value in the range $(-1, +1)$, i.e., -1, 0, and $+1$. At this point the reader may ask: "What about vibrations in oblique directions?" They correspond to linear combinations of the standing waves we have already identified; those standing waves can be mixed because the symmetry of the problem assures that they will vibrate at the same frequency.

Going on to S_l^m with $l = 2$, we find that there are five independent normal modes ($m = -2, -1, 0, 1, 2$); these modes all have two nodal lines. We do not prove it, but it is possible to show that linear combinations of these five modes can produce the two-node patterns in any orientation. The generalization that is useful to note here is that the $2l + 1$ functions S_l^m of any given l (but different m) form a complete set of l-node angular solutions to a PDE that is separable in spherical polar coordinates.

We could have carried out an analysis using the complex form of the S_l^m as given in Eq. (15.89), but in that case, instead of finding nodes corresponding to specific values of φ, we would have observed that as φ moves from 0 to 2π, the function $e^{im\varphi}$ cycles around the unit circle (in the complex plane) $|m|$ times instead of passing through $|m|$ pairs of points where $\cos m\varphi$ or $\sin m\varphi$ is zero.

It is also important to note that, because of the properties of the trigonometric or complex exponential functions, solutions S_l^m with different m are orthogonal when integrated in φ over the range $(0, 2\pi)$. Moreover, because of the properties of the associated Legendre functions, solutions S_l^m of the same m but different l are orthogonal when integrated in $t = \cos\theta$ over the range $-1 \le t \le 1$, equivalent to an integration in θ from 0 to π with weight factor $|dt/d\theta| = \sin\theta$.

Because the S_l^m form a complete orthogonal set of angular functions, we can use them to make expansions of arbitrary functions of the angles θ and φ. For this purpose it is convenient to scale the S_l^m in a way that makes them normalized. It can be shown that a set of normalized angular functions, known as **spherical harmonics** and denoted Y_l^m, with definition

$$Y_l^m(\theta,\varphi) = \sqrt{\left(\frac{2l+1}{4\pi}\right)\frac{(l-m)!}{(l+m)!}}\,P_l^m(\cos\theta)\,e^{im\varphi}\,, \quad \left\{ \begin{array}{c} l = 0,\ 1,\ 2,\ldots \\ -l \le m \le l \end{array} \right\}, \tag{15.91}$$

with m restricted to integer values, are orthonormal under the scalar product definition

$$\langle F|G\rangle \equiv \int_0^{2\pi} d\varphi \int_0^{\pi} \sin\theta\, d\theta\, F^*(\theta,\varphi)\,G(\theta,\varphi) = \int d\Omega\, F^*(\theta,\varphi)\,G(\theta,\varphi)\,, \tag{15.92}$$

where $d\Omega$ is the element of solid angle: $d\Omega = \sin\theta\, d\theta\, d\varphi$. The orthonormality condition can be stated

$$\left\langle Y_l^m \middle| Y_{l'}^{m'} \right\rangle = \delta_{ll'}\delta_{mm'} \,. \tag{15.93}$$

The importance of the orthonormality condition is that an arbitrary angular distribution $f(\theta, \varphi)$ can be expanded in spherical harmonics according to the formulas

$$f(\theta,\varphi) = \sum_{l=0}^{\infty} \sum_{m=-l}^{l} c_{lm} Y_l^m(\theta,\varphi) \,, \tag{15.94}$$

$$c_{lm} = \langle Y_l^m | f \rangle = \int_0^{2\pi} d\varphi \int_0^{\pi} \sin\theta\, d\theta\, [Y_l^m(\theta,\varphi)]^* \, f(\theta,\varphi) \,. \tag{15.95}$$

Even though the individual Y_l^m are complex, they can still be used to expand real functions. Y_l^{-m} is proportional to $(Y_l^m)^*$, and these quantities can appear with coefficients that make the overall expansion real.

For reference we tabulate in Table 15.1 the first few spherical harmonics, giving also their forms when the angular functions are expressed in terms of x/r, y/r, and z/r.

Exercises

15.4.1. (a) Show that the space spanned by the three spherical harmonics Y_1^m $(m = 1,\ 0,\ -1)$ is also spanned by the real basis x/r, y/r, z/r.

(b) Regarding each member of the real basis identified in part (a) as a function in three-dimensional space, characterize its nodal surface(s), i.e., the set of points where the function is zero.

(c) Let (x', y', z') be the point to which (x, y, z) is transformed when a Cartesian coordinate system is subjected to a rotation. Show that x'/r', y'/r', z'/r' span the same space as x/r, y/r, z/r. This means that each

Table 15.1: Spherical harmonics.

$$\begin{aligned}
Y_0^0(\theta,\varphi) &= \frac{1}{\sqrt{4\pi}} \\
Y_1^1(\theta,\varphi) &= -\sqrt{\frac{3}{8\pi}}\sin\theta\, e^{i\varphi} = -\sqrt{\frac{3}{8\pi}}\,(x+iy)/r \\
Y_1^0(\theta,\varphi) &= \sqrt{\frac{3}{4\pi}}\cos\theta = \sqrt{\frac{3}{4\pi}}\, z/r \\
Y_1^{-1}(\theta,\varphi) &= +\sqrt{\frac{3}{8\pi}}\sin\theta\, e^{-i\varphi} = \sqrt{\frac{3}{8\pi}}\,(x-iy)/r \\
Y_2^2(\theta,\varphi) &= \sqrt{\frac{15}{32\pi}}\,\sin^2\theta\, e^{2i\varphi} = \sqrt{\frac{15}{32\pi}}\,(x^2-y^2+2ixy)/r^2 \\
Y_2^1(\theta,\varphi) &= -\sqrt{\frac{15}{8\pi}}\,\sin\theta\cos\theta\, e^{i\varphi} = -\sqrt{\frac{15}{8\pi}}\, z(x+iy)/r^2 \\
Y_2^0(\theta,\varphi) &= \sqrt{\frac{5}{16\pi}}\,(3\cos^2\theta - 1) = \sqrt{\frac{5}{16\pi}}\,(3z^2-r^2)/r^2 \\
Y_2^{-1}(\theta,\varphi) &= \sqrt{\frac{15}{8\pi}}\,\sin\theta\cos\theta\, e^{-i\varphi} = +\sqrt{\frac{15}{8\pi}}\, z(x-iy)/r^2 \\
Y_2^{-2}(\theta,\varphi) &= \sqrt{\frac{15}{32\pi}}\,\sin^2\theta\, e^{-2i\varphi} = \sqrt{\frac{15}{32\pi}}\,(x^2-y^2-2ixy)/r^2
\end{aligned}$$

member of the primed basis can be written as a linear combination of the unprimed basis functions.

15.4.2. (a) Form a real basis for the space spanned by the spherical harmonics with $l = 2$ by forming linear combinations of Y_2^m and Y_2^{-m}.

(b) Multiplying each member of your real basis by r^2 to produce a set of functions in three-dimensional space, describe the nodal structure of each basis member. Characterize each nodal surface by its shape (planar, spherical, conical, etc.) and position.

(c) If we rotate a Cartesian coordinate system so that $x' = z$, $y' = y$, $z' = -x$, state how the nodal structure of each basis member (in its original position) is described in the primed coordinate system.

(d) Show that it is possible to write each original (unprimed) basis member as a linear combination of basis members constructed for the primed coordinate system.

15.4.3. Find the seven spherical harmonics with $l = 3$ that correspond to the definition in Eq. (15.91). Then repeat Exercise 15.4.2 for $l = 3$.

15.4.4. (a) There is one monomial in x, y, z, of degree 0, namely 1. There are three such monomials of degree 1, namely x, y, and z. There are six such monomials of degree 2: x^2, y^2, z^2, xy, xz, and yz. Find the set of such monomials of degree 3.

(b) The monomial of degree zero (divided by r^0) is a basis for the Y_l^m with $l = 0$. The monomials of degree 1 (divided by r^1) are a basis for the Y_l^m with $l = 1$. From the six monomials of degree 2 (divided by r^2), form a five-membered basis for the Y_l^m with $l = 2$ and identify the function built from this six-monomial set that is linearly independent of that five-membered basis. Is this sixth function a basis for any set of spherical harmonics?

(c) Using the ideas developed in part (b), analyze the set of monomials of degree three and use them to make bases for sets of spherical harmonics.

(d) Obtain the monomials of degree four and find the sets of spherical harmonics for which they can form bases. Can you extrapolate your findings to monomial sets of arbitrary degree?

15.4.5. The angular parts of the wave functions that are solutions to quantum central-force problems are traditionally labeled by alphabetic symbols that indicate the l values of their spherical harmonics. The code is: "s" for $l = 0$; "p" for $l = 1$; "d" for $l = 2$; "f" for $l = 3$. Higher l values are assigned symbols in alphabetic order: "g," "h,"...

(a) Explain why "s" refers to a unique spatial function, while "p" refers to one or more of three functions with different spatial orientations.

(b) How many functions are in a set designated "d"? In a set designated "f"?

(c) Explain why the different members of a function set with a given l value will all lead to the same radial ODE in a central-force problem.

SYMBOLIC COMPUTING

Spherical harmonics—The spherical harmonics are accessible in both our symbolic computing systems. In MATHEMATICA, they are defined with the same conventions as are used in this book, and called using the command `SphericalHarmonicY`. However, in MAPLE, the conventions that are used for the associated Legendre functions make MAPLE's Y_l^m different than our definition, and the difference cannot be compensated by changing settings within MAPLE. We have therefore included in Appendix K a procedure `SphY`, which we recommend using instead of the built-in MAPLE procedure `SphericalY`.

The spherical harmonic $Y_L^M(\theta, \varphi)$ is called by invoking

```
SphY(L, M, theta, phi)                  (MAPLE),
SphericalHarmonicY[L, M, theta, phi]    (MATHEMATICA).
```

Here are some examples of the spherical-harmonic procedures:

MAPLE:

```
> SphY(3, -1, theta, phi);
```

$$\frac{1}{8}\frac{\sqrt{21}(5\cos(\theta)^2 - 1)\ \sin(\theta)e^{-I\phi}}{\sqrt{\pi}}$$

```
> SphY(2, 1, Pi/3, 1.23);
```

$$\frac{(-0.02088985794 - 0.05890555012\ \mathrm{I})\ \sqrt{30}\ \sqrt{3}}{\sqrt{\pi}}$$

This output can be further simplified:

```
> evalf(%):
```

$$-0.1118102980 - 0.3152844377\ \mathrm{I}$$

MATHEMATICA:

```
SphericalHarmonicY[3, -1, theta, phi]
```

$$\frac{1}{8}\mathrm{e}^{-\mathrm{i\,phi}}\sqrt{\frac{21}{\pi}}\ \left(-1 + 5\,\mathrm{Cos[theta]}^2\right)\ \mathrm{Sin[theta]}$$

```
SphericalHarmonicY[2, 1, Pi/3, 1.23]
```

$$-0.11181 - 0.315284\ \mathrm{i}$$

In MATHEMATICA, this function call simplifies completely.

Angular Scalar Products—For making spherical harmonic expansions, it is desirable to have symbolic code for evaluating scalar products of the form given in Eq. (15.92).

In MAPLE, the procedure

```
SProd(U,V,w1,w2)
```

carries out the evaluation, with `U` and `V` expressions in the two arbitrarily named variables `w1` and `w2`, with `w1` the variable corresponding to θ and `w2` that corresponding to φ:

```
> SProd := proc(U,V,w1,w2) local UU, VV, _theta , _phi;
>   assume( _theta, real); assume( _phi, real);
>   UU := subs(w1 = _theta, w2 = _phi, U);
>   VV := subs(w1 = _theta, w2 = _phi, V);
>   evalf(int(int(conjugate(UU)*VV*sin(_theta), _phi = 0 .. 2*Pi),
>             _theta = 0 .. Pi))
> end proc:
```

Some of the apparently unnecessary statements in the above coding are needed to keep MAPLE from getting confused when carrying out the double integral. The two `assume` statements assure MAPLE that `_theta` and `_phi` are real when they appear in the `conjugate` command; the `subs` command replaces (in copies of `U` and `V`) `w1` by `_theta` and `w2` by `_phi`. (The underscores preceding `theta` and `phi` make these quantities different symbols than `theta` and `phi`; they were chosen to make it easy to understand the content of the procedure while avoiding a collision with variable names that the user is likely to choose.)

In MATHEMATICA, one can write a somewhat simpler procedure

`SProd[U, V]` ,

but the price paid for the simplification is that the angular variables in `U` and `V` are required to be `theta` and `phi`:

```
SProd[u_, v_] := N[ Integrate[
          Integrate[ Conjugate[u] * v * Sin[theta], {phi, 0, 2*Pi} ],
          {theta, 0, Pi} ] ]
```

Here are some illustrations of the use of the scalar product.

MAPLE:

```
> SProd(cos(theta)^2,sin(theta)^2,theta,phi);     1.675516082  (= 8π/15)
> SProd(1,1,u,v);                                 12.56637062  (= 4π)
> SProd(SphY(2,1,theta,phi),SphY(2,1,theta,phi),theta,phi);     1.
> SProd(SphY(2,1,x,y),SphY(2,-1,x,y),x,y);                      0.
```

MATHEMATICA:

```
SProd[Cos[theta]^2,Sin[theta]^2]                  1.67552  (= 8π/15)
SProd[1,1]                                        12.5664  (= 4π)
SProd[SphericalHarmonicY[2,1,theta,phi],
               SphericalHarmonicY[2,1,theta,phi]]         1.
SProd[SphericalHarmonicY[2,1,theta,phi],
               SphericalHarmonicY[2,-1,theta,phi]]        0.
```

Example 15.4.1. Spherical Harmonic Expansions

As a first example, take $F(\theta, \phi) = \cos^2\theta$. From the form of F we know that the only spherical harmonics in its expansion will be Y_L^M with $M = 0$ and $L \le 2$. Finding the corresponding expansion coefficients, in MAPLE we have

```
> F := cos(theta)^2:
> C0 := SProd(SphY(0,0,theta,phi),F,theta,phi);        C0 := 1.181635901
> C1 := SProd(SphY(1,0,theta,phi),F,theta,phi);        C1 := 0.
> C2 := SProd(SphY(2,0,theta,phi),F,theta,phi);        C2 := 1.056887279
```

We can check the expansion by examining it:

```
> C0 * SphY(0,0,theta,phi) + C2 * SphY(2,0,theta,phi);
```

$$\frac{0.5908179505}{\sqrt{\pi}} + \frac{0.5284436395\,\sqrt{5}\left(-\frac{1}{2}+\frac{3}{2}\cos(\theta)^2\right)}{\sqrt{\pi}}$$

```
> evalf(%);
```

$$2.\,10^{-10} + 0.9999999999\,\cos(\theta)^2$$

In MATHEMATICA,

```
F = Cos[theta]^2;
C0 = SProd[SphericalHarmonicY[0,0,theta,phi], F]          1.18164
C1 = SProd[SphericalHarmonicY[1,0,theta,phi], F]          0.
C2 = SProd[SphericalHarmonicY[2,0,theta,phi], F]          1.05689
```

Checking the expansion,

```
C0 * SphericalHarmonic[0,0,theta,phi]
                  + C2 * SphericalHarmonicY[2,0,theta,phi]
```

$$0.333333 + 0.333333\,(-1 + 3\,\texttt{Cos}[\texttt{theta}]^2)$$

```
Simplify[%]
```

$$-1.11022 \times 10^{-16} + 1.\,\texttt{Cos}[\texttt{theta}]^2$$

A slightly more demanding expansion is that of $G(\theta, \phi) = \cos\theta + \sin\theta \sin\varphi$. Proceeding without identifying the nonzero coefficients, we first compute a coefficient array and then form the expansion. We assume it is acceptable to truncate the expansion after $L = N$.

In MAPLE,

```
> G := sin(theta)*sin(phi) + cos(theta):
> for L from 0 to N do for M from -L to L do
>    C[L,M] := SProd(SphY(L,M,theta,phi),G,theta,phi)
> end do; end do;
```

Assuming the above code was run with $N \geq 2$, one could look at individual coefficients by entering commands such as

```
> C[1,1];          1.447202509 I
> C[2,0];          0.
```

Let's check the expansion, keeping six significant figures, setting `N` to at least 2:

```
> Digits := 6:
> S := 0: for L from 0 to N do for M from -L to L do
>    S := S + C[L,M] * SphY[L,M,theta,phi]
> end do end do;
```

$$\frac{0.361800\,\mathrm{I}\sqrt{6}\,\sin(\theta)\,e^{-\mathrm{I}\,\phi}}{\sqrt{\pi}}+\frac{1.02332\sqrt{3}\,\cos(\theta)}{\sqrt{\pi}}-\frac{0.361800\,\mathrm{I}\sqrt{6}\,\sin(\theta)\,e^{\mathrm{I}\,\phi}}{\sqrt{\pi}}$$

To bring this expression to a more convenient form, we apply `evalf` and convert the exponentials to trigonometric form, using `convert(..,trig)` and `simplify`:

```
> evalf(%};
```

$$0.500000\,\mathrm{I}\,\sin(\theta)e^{-1.\,\mathrm{I}\phi}+0.999995\cos(\theta)-0.500000\,\mathrm{I}\,\sin(\theta)\,e^{1.\mathrm{I}\phi}$$

```
> simplify(convert(%,trig));
```

$$\sin(\theta)\sin(\phi)+0.999995\cos(\theta)$$

There is some round-off error in the above check, but the expansion is clearly correct.

In MATHEMATICA,

```
g = Sin[theta] * Sin[phi] + Cos[theta]
Do[ Do[ c[L, M] =
        SProd[SphericalHarmonicY[L, M, theta, phi], g], {M, -L, L} ],
    {L, 0, 2} ]
```

Note that we write `c[L,M]` and not `c[[L,M]]`, so `c` is not an array but a function defined for arguments `L,M`. We cannot make `c` an array because MATHEMATICA does not permit arrays to have zero or negative indices. We can now examine individual coefficient values:

```
c[1,1]          0. + 1.4472 i
c[2,0]          0.
```

Let's check the expansion, setting `N` to at least 2:

```
SS = 0;
Do[ Do[ SS = SS +
        c[L, M]*SphericalHarmonicY[L, M, theta, phi], {M, -L, L} ],
    {L, 0, 2}]; SS
```

$$0.+1.\,\mathrm{Cos[theta]}+(0.+0.5\,\mathrm{i})\,\mathrm{e}^{-\mathrm{i}\,\mathrm{phi}}\,\mathrm{Sin[theta]}-(0.+0.5\,\mathrm{i})\,\mathrm{e}^{\mathrm{i}\,\mathrm{phi}}\,\mathrm{Sin[theta]}$$

The complex exponentials can be converted to trigonometric functions using the command `ExpToTrig`. We get

```
ExpToTrig[SS]
```

$$(0.+0.\,\mathrm{i})+1.\,\mathrm{Cos[theta]}+(1.+0.\,\mathrm{i})\,\mathrm{Sin[phi]}\,\mathrm{Sin[theta]}$$

This is clearly the correct result.

■

LAPLACE EQUATION

With the angular functions now in hand, let's return to the radial equation, considering here the Laplace equation, for which the function $G(r)$ is zero. Equation (15.90) then reads

$$R_l'' + \frac{2}{r}\,R_l' - \frac{l(l+1)}{r^2}\,R_l = 0\,. \tag{15.96}$$

The form of Eq. (15.96) indicates that its solutions will be of the form $R_l = r^p$, and we can determine the possible values of p by substituting that form into the ODE. (If not apparent initially, this observation would become obvious when trying to set up the indicial equation of the Frobenius method.) Using the suggested form for R_l, we obtain

$$p(p-1)\,r^{p-2} + 2p\,r^{p-2} + l(l+1)\,r^{p-2} = 0\,,$$

which is an algebraic equation with solutions $p = l$ and $p = -(l+1)$. The radial equation therefore has solutions $R(r) = r^l$ and $R(r) = r^{-l-1}$.

Combining our newly found radial solution with the angular solution, and remembering that there are angular solutions for all integer $l \geq 0$ and $|m| \leq l$, the general solution for the Laplace equation, when written in spherical polar coordinates, takes the form

$$\psi(r,\theta,\varphi) = \sum_{l=0}^{\infty}\sum_{m=-l}^{l}\left[a_{lm}r^l + \frac{b_{lm}}{r^{l+1}}\right]Y_l^m(\theta,\varphi)\,, \tag{15.97}$$

$$= \sum_{l=0}^{\infty}\sum_{m=-l}^{l}\left[a'_{lm}r^l + \frac{b'_{lm}}{r^{l+1}}\right]P_l^m(\cos\theta)\,e^{im\varphi}\,. \tag{15.98}$$

The coefficients a_{lm} and b_{lm}, or a'_{lm} and b'_{lm} if Eq. (15.98) is used, are determined from the boundary conditions.

Example 15.4.2. Two Electrostatics Problems

Our first problem is to find the potential inside an empty sphere of radius r_0, with a specified distribution of potential on the sphere. When we place the center of the sphere at the origin of a spherical polar coordinate system, our boundary condition is that the potential is described on the sphere by the function $f(\theta,\varphi) = A\cos^2\theta$. In other words, we require a solution $\psi(r,\theta,\varphi)$ such that

$$\psi(r_0,\theta,\varphi) = A\cos^2\theta\,. \tag{15.99}$$

Our knowledge of the physical situation tells us that the potential cannot become infinite within an empty sphere on which the potential is finite, so our solution cannot include any terms with negative powers of r. Taking our solution in the form given by Eq. (15.98), we can immediately set all the b'_{lm} to zero. Equation (15.99) then becomes equivalent to

$$\psi(r_0,\theta,\varphi) = \sum_{l=0}^{\infty}\sum_{m=-l}^{l} a'_{lm}r_0^l P_l^m(\cos\theta)\,e^{im\varphi} = A\cos^2\theta\,. \tag{15.100}$$

It is clear that we do not need terms of nonzero m to satisfy Eq. (15.100); it is also obvious that it would be helpful to write $\cos^2\theta$ in terms of Legendre polynomials.

Either by carrying out the expansion of $\cos^2\theta$ in terms of P_l or (perhaps easier here) to note by inspection that

$$\cos^2\theta = \frac{2}{3}\,P_2(\cos\theta) + \frac{1}{3}\,P_0(\cos\theta)\,.$$

Equation (15.100) can then be rewritten

$$\psi(r_0,\theta,\varphi) = \sum_{l=0}^{\infty} a'_{l0} r_0^l P_l(\cos\theta) = A\left[\frac{2}{3}\,P_2(\cos\theta) + \frac{1}{3}\,P_0(\cos\theta)\right]. \qquad (15.101)$$

We now see that all the a'_{l0} vanish except a'_{00} and a'_{20}, and that

$$a'_{00} = \frac{A}{3} \qquad \text{and} \qquad a'_{20} r_0^2 = \frac{2A}{3}\,.$$

Solving for the a'_{l0} and substituting into Eq. (15.98), we have the final result

$$\psi(r,\theta,\varphi) = \frac{A}{3}\,P_0(\cos\theta) + \frac{2A}{3}\left(\frac{r}{r_0}\right)^2 P_2(\cos\theta)\,,$$

$$= A\left[\frac{1}{3} + \frac{r^2}{r_0^2}\left(\cos^2\theta - \frac{1}{3}\right)\right].$$

Our second problem is to find the potential outside a sphere of radius r_0 at the origin of a spherical polar coordinate system given that the potential on the sphere is $f(\theta,\varphi)$, where the explicit form of f is not specified. Based on our experience from the previous problem of this Example, we start by obtaining the spherical harmonic expansion of f:

$$f(\theta,\varphi) = \sum_{lm} c_{lm} Y_l^m(\theta,\varphi)\,, \qquad c_{lm} = \langle Y_l^m | f\rangle\,.$$

The potential we seek can be given by Eq. (15.97), but with all the coefficients a_{lm} set to zero to avoid an unphysical divergence of the potential at large r. In other words, we have a solution of the form

$$\psi(r,\theta,\phi) = \sum_{l=0}^{\infty}\sum_{m=-l}^{l} \frac{b_{lm}}{r^{l+1}}\,Y_l^m(\theta,\varphi)\,. \qquad (15.102)$$

We then have, on the boundary,

$$\psi(r_0,\theta,\varphi) = \sum_{lm} \frac{b_{lm}}{r_0^{l+1}}\,Y_l^m(\theta,\varphi) = \sum_{lm} c_{lm}\,Y_l^m(\theta,\varphi)\,,$$

from which we have, for each l and m,

$$\frac{b_{lm}}{r_0^{l+1}} = c_{lm}\,.$$

Inserting into Eq. (15.102) the coefficients obtained for the present problem, we reach

$$\psi(r,\theta,\varphi) = \sum_{lm} c_{lm}\left(\frac{r_0}{r}\right)^{l+1} Y_l^m(\theta,\varphi)\,. \qquad (15.103)$$

Except in simple cases it may be advisable to use symbolic computing to evaluate the scalar products $\langle Y_l^m | f\rangle$.

■

Exercises

15.4.6. A complete three-dimensional space contains no charge outside a sphere of radius r_0; within that sphere are charges that cause the electrostatic potential on the sphere to be given by $\psi(r_0,\theta,\varphi) = \sin^2(2\theta)\sin^2\varphi$. Find the potential at all points external to the sphere.

15.4.7. A sphere of radius r_0 contains no charge, but charges external to the sphere cause the electrostatic potential on it to be given by $\psi(r_0,\theta,\varphi) = \sin^2(2\theta)$ $\sin^2\varphi$. Find the potential at all points within the sphere.

15.4.8. A sphere of radius r_0 contains no charge. On the sphere the derivative of the potential in the outward normal direction has the position-dependent value $\psi_r(\theta,\varphi) = \cos^2\theta$. Find the potential at all points within the sphere. To what extent is your answer not unique?

15.4.9. A complete three-dimensional space contains no charge outside a sphere of radius r_0; within that sphere are charges that cause the electrostatic potential on the sphere to be given by $\psi(r_0,\theta,\varphi) = A\cos^2\theta\sin^2\varphi$. Using symbolic computation if helpful, find the terms through r^6 in the spherical harmonic expansion of the potential within the sphere.

HELMHOLTZ EQUATION

The Helmholtz equation often appears when the Laplacian occurs in an eigenvalue problem. In that case, the Helmholtz equation may only have a solution consistent with the boundary conditions for certain values of the parameter appearing in the equation.

Example 15.4.3. Quantum Particle in a Spherical Box

While stated as a quantum-mechanics problem, from a purely mathematical viewpoint this problem is simply that of finding the standing-wave modes of a spherical box when the wave amplitude is required to vanish on the boundary. For this box problem, the Schrödinger equation (in dimensionless units) is

$$-\nabla^2\psi = k^2\,\psi\,, \tag{15.104}$$

where $k^2 = 2E$ is an eigenvalue (with E interpreted as the energy of the quantum state represented by the standing wave). Our mission is to find all the solutions to Eq. (15.104) within a sphere of radius a centered at the coordinate origin, subject to the condition that $\psi = 0$ at $r = a$.

Writing Eq. (15.104) in spherical polar coordinates, our PDE becomes

$$\left[\frac{\partial^2}{\partial r^2} + \frac{2}{r}\frac{\partial}{\partial r} + \frac{1}{r^2\sin\theta}\frac{\partial}{\partial\theta}\left(\sin\theta\frac{\partial}{\partial\theta}\right) + \frac{1}{r^2\sin^2\theta}\frac{\partial^2}{\partial\varphi^2} + k^2\right]\psi = 0\,. \tag{15.105}$$

Separating the variables by introducing $\psi = R(r)\Theta(\theta)\Phi(\varphi)$, the angular equations reduce in the usual way to yield solutions

$$\Theta(\theta)\Phi(\varphi) = Y_l^m(\theta,\varphi) \quad \text{with} \quad l = 0,\ 1,\ 2,\ldots, \quad m = -l,\ \ -l+1,\ \ \ldots,+l, \tag{15.106}$$

and the equation for $R(r)$, which depends upon l, is

$$R_l'' + \frac{2}{r}R_l' + \left[k^2 - \frac{l(l+1)}{r^2}\right]R_l = 0\,. \tag{15.107}$$

This ODE has been discussed at Eq. (14.79) and at Eq. (14.94), where it was shown to have solutions

$$j_l(kr) = \sqrt{\frac{\pi}{2kr}}\, J_{l+1/2}(kr) \quad \text{and} \quad y_l(kr) = \sqrt{\frac{\pi}{2kr}}\, Y_{l+1/2}(kr)\,. \tag{15.108}$$

For the present problem, we require solutions that are nonsingular at $r = 0$, so we consider further only the solutions $j_l(kr)$.

To satisfy the boundary condition at $r = a$, we need to have $j_l(ka) = 0$, showing that for any given l we must choose k in such a way that ka is a zero of j_l. Comparing with Eq. (15.108), we see that the zeros we need are those of $J_{l+1/2}$, so the pth such zero, which we denote β_{lp}, can be obtained using symbolic computing as `BesselJZeros(l+1/2, p)` or `BesselJZero[l+1/2, p]`.

Summarizing the above analysis, we find that the standing-wave modes for the region interior to a sphere of radius a have the amplitude distribution

$$\psi_{plm}(r,\theta,\varphi) = j_l\left(\frac{\beta_{lp}r}{a}\right) Y_l^m(\theta,\varphi)\,, \qquad \begin{cases} p = 1,\ 2,\ \ldots, \\ l = 0,\ 1,\ \ldots, \\ m = -l,\ \ldots,\ l\,, \end{cases} \tag{15.109}$$

and the value of k associated with ψ_{plm} is β_{lp}/a. In the quantum-mechanics problem, $E = k^2/2 = \beta_{lp}^2/2a^2$.

■

Example 15.4.4. Enumeration of Standing-Wave Modes

We continue here the analysis started in Example 15.4.3, making a numerical tabulation of the standing waves. The input data that we need for the tabulation are the values of the zeros of the spherical Bessel functions. Using one of the commands `BesselJZeros(l+1/2,p)` or `BesselJZero[l+1/2,p]` we start by building the following table:

β_{lp}	$l = 0$	$l = 1$	$l = 2$	$l = 3$
$p = 1$	3.1416	4.4934	5.7635	6.9879
$p = 2$	6.2832	7.7253	9.0950	10.4171
$p = 3$	9.4248	10.9041	12.3229	13.6980

The argument of the spherical Bessel function j_l is $\beta_{lp}r/a$, which causes the boundary at $r = a$ to coincide with the pth positive zero of j_l. Thus, if $p = 1$ there is no zero between the origin and the boundary; if $p = 2$ there is one interior value of r where j_l is zero; if $p = 3$ there are two interior values of r where j_l vanishes, etc. These interior zeros define spheres on which j_l is zero; they are **nodal surfaces** of ψ_{nlm}. We have previously discussed the nodal pattern of Y_l^m; its real or imaginary part has l nodes in directions that are determined by the m value. Since the nodes of Y_l^m do not depend upon r, they will apply for all r and therefore also define nodal surfaces.

Let's now use the data from the table of β_{lp} to obtain more explicit descriptions of the ψ_{plm}, for simplicity setting the sphere radius a to unity. For each ψ_{plm} we also

Table 15.2: Eigenfunctions ψ_{plm} and eigenvalues E_{plm} for particle in spherical box, Example 15.4.4.

	E		E
$\psi_{1,0,0} = j_0(3.14r)\,Y_0^0(\theta,\varphi)$	4.90	$\psi_{1,3,3} = j_3(6.99r)\,Y_3^3(\theta,\varphi)$	24.42
$\psi_{1,1,1} = j_1(4.49r)\,Y_1^1(\theta,\varphi)$	10.10	$\psi_{1,3,2} = j_3(6.99r)\,Y_3^2(\theta,\varphi)$	24.42
$\psi_{1,1,0} = j_1(4.49r)\,Y_1^0(\theta,\varphi)$	10.10	$\psi_{1,3,1} = j_3(6.99r)\,Y_3^1(\theta,\varphi)$	24.42
$\psi_{1,1,-1} = j_1(4.49r)\,Y_1^{-1}(\theta,\varphi)$	10.10	$\psi_{1,3,0} = j_3(6.99r)\,Y_3^0(\theta,\varphi)$	24.42
$\psi_{1,2,2} = j_2(5.76r)\,Y_2^2(\theta,\varphi)$	16.61	$\psi_{1,3,-1} = j_3(6.99r)\,Y_3^{-1}(\theta,\varphi)$	24.42
$\psi_{1,2,1} = j_2(5.76r)\,Y_2^1(\theta,\varphi)$	16.61	$\psi_{1,3,-2} = j_3(6.99r)\,Y_3^{-2}(\theta,\varphi)$	24.42
$\psi_{1,2,0} = j_2(5.76r)\,Y_2^0(\theta,\varphi)$	16.61	$\psi_{1,3,-3} = j_3(6.99r)\,Y_3^{-3}(\theta,\varphi)$	24.42
$\psi_{1,2,-1} = j_2(5.76r)\,Y_2^{-1}(\theta,\varphi)$	16.61	$\psi_{2,1,1} = j_1(7.73r)\,Y_1^1(\theta,\varphi)$	29.84
$\psi_{1,2,-2} = j_2(5.76r)\,Y_2^{-2}(\theta,\varphi)$	16.61	$\psi_{2,1,0} = j_1(7.73r)\,Y_1^0(\theta,\varphi)$	29.84
$\psi_{2,0,0} = j_0(6.28r)\,Y_0^0(\theta,\varphi)$	19.74	$\psi_{2,1,-1} = j_1(7.73r)\,Y_1^{-1}(\theta,\varphi)$	29.84

calculate the energy $E_{plm} = \beta_{lp}^2/2$. The results (for all solutions with $E \le 32$) are given (to two decimal places) in Table 15.2, in the order of their energies.

Note that each value of the index set (p,l,m) defines a different solution to our k^2-dependent PDE; a linear combination of these solutions will not also be a solution unless all the terms have the same value of k^2. Solutions of the same k^2 will only occur when the solutions have the same values of p and l, meaning that they differ only in the value of m. The physical significance of the equal values of k^2 (called **degeneracy**) is that different m values correspond to different spatial orientations within a set of angular distributions.

■

Exercises

15.4.10. The three-dimensional quantum harmonic oscillator was the topic of Exercise 15.2.14, where it was solved by the method of separation of variables in Cartesian coordinates. This problem makes use of the answer to part (d) of that earlier Exercise. If you have not done that Exercise and saved your answer, you should do it now in preparation for the present problem.

We return to the 3-D quantum oscillator, but now attack it using spherical polar coordinates. This problem uses the time-independent Schrödinger equation for the potential $V = (x^2+y^2+z^2)/2$, which we now write as $V = r^2/2$. Our Schrödinger equation therefore takes the form

$$-\frac{1}{2}\nabla^2\psi + \frac{r^2}{2}\psi = E\psi\,, \tag{15.110}$$

with $\psi = \psi(r,\theta,\varphi)$. This is a central-force problem, and we set $\psi = R(r)\,Y(\theta,\varphi)$. We know that the angular part of the solution will be a spherical harmonic $Y_l^m(\theta,\varphi)$.

(a) Show that the radial equation for this problem when the angular solution is the spherical harmonic Y_l^m is

$$-\frac{1}{2}R'' - \frac{1}{r}R' + \frac{r^2}{2}R + \frac{l(l+1)}{2r^2}R = E\,R\,. \qquad (15.111)$$

(b) Show that the substitution $R(r) = r^l e^{-r^2/2}U(r^2)$ converts Eq. (15.111) into

$$r^2U''(r^2) + \left(l + \tfrac{3}{2} - r^2\right)U'(r^2) = -\frac{1}{2}\left(E - l - \tfrac{3}{2}\right)U(r^2)\,. \quad (15.112)$$

Hint. Keep in mind that the notation U' indicates the derivative of U with respect to its argument. Thus, for example, $dU(r^2)/dr = 2rU'(r^2)$.

(c) We now take note that Eq. (14.139), the associated Laguerre ODE, has the following form when its dependent variable (denoted z in that equation but here called L_n^k) is a function of x^2:

$$x^2(L_n^k)''(x^2) + (k + 1 - x^2)(L_n^k)'(x^2) = -nL_n^k(x^2)\,, \qquad (15.113)$$

where n is any nonnegative integer and L_n^k is a polynomial of degree n in its argument (and therefore here an even polynomial of degree $2n$ in x).

A comparison of the left-hand sides of Eqs. (15.112) and (15.113) shows that Eq. (15.112) becomes an identity (and is thereby solved) if $U(r^2)$ is identified as $L_n^{l+1/2}(r^2)$. Show that this identification leads to the equation

$$-n = -\frac{1}{2}\left(E - l - \tfrac{3}{2}\right)\,,$$

which rearranges to

$$E = 2n + l + \frac{3}{2}\,. \qquad (15.114)$$

(d) Based on parts (a) through (c), write expressions for the PDE solution $\psi_{nlm}(r,\theta,\varphi)$ and for its associated eigenvalue E_{nlm}.

(e) Make a list of all the eigenfunctions and their eigenvalues for $E \le 9/2$. Compare with your answer to part (d) of Exercise 15.2.14, and verify that the two methods of solving this problem lead to solution sets with the same degree of degeneracy (i.e., that the two solution sets have equal numbers of eigenfunctions for each E value).

(f) Complete the task of verifying that the Cartesian and spherical-coordinate solution sets are equivalent by obtaining the explicit forms of all the individual solutions and showing that they span the same solution spaces. For the spherical-coordinate solutions, the necessary associated Laguerre polynomials are most easily obtained by symbolic computation. The fractional upper index is accepted by the symbolic procedures

`LaguerreL(n,l+1/2,r^2)` or `LaguerreL[n,l+1/2,r^2]`.

15.4.11. The hydrogen atom is a quantum central-force problem that (in dimensionless units) is described by the Schrödinger equation

$$-\frac{1}{2}\nabla^2\psi - \frac{1}{r}\psi = E\psi\,. \qquad (15.115)$$

Here r is the distance from the origin (the position of the hydrogen nucleus).

(a) Using spherical polar coordinates, we know that this central-force problem has angular solutions $Y_l^m(\theta,\varphi)$. Show, writing $\psi = R(r)\,Y_l^m(\theta,\varphi)$, that the radial equation for any nonnegative integer l takes in spherical polar coordinates the form

$$-\frac{1}{2}R'' - \frac{1}{r}R' - \frac{1}{r}R + \frac{l(l+1)}{2r^2}R = E\,R\,. \qquad (15.116)$$

(b) Show that the substitution $R(r) = r^l e^{-r/n}\,U(2r/n)$ (where n is to be determined later) converts Eq. (15.116) into

$$-\frac{2r}{n}U''(2r/n) - \left(2l+2-\frac{2r}{n}\right)U'(2r/n) - (n-l-1)\,U(2r/n)$$
$$-\frac{r}{2n}U(2r/n) = nrE\,U(2r/n)\,. \qquad (15.117)$$

(c) The associated Laguerre equation, Eq. (14.139), takes the following form when its independent variable is $2r/n$, the n in that equation is renamed p, and the dependent variable is identified as the associated Laguerre polynomial $L_p^k(2r/n)$:

$$\frac{2r}{n}(L_p^k)''(2r/n) + \left(k+1-\frac{2r}{n}\right)(L_p^k)'(2r/n) + p\,L_p^k = 0\,. \qquad (15.118)$$

Show by comparison of the left-hand side of Eq. (15.118) with the first three terms of Eq. (15.117) that those three terms will combine to yield zero if $U(2r/n)$ is taken to be $L_{n-l-1}^{2l+1}(2r/n)$.

We will not prove it, but the bound-state solutions of the hydrogen atom correspond to L_{n-l-1}^{2l+1} with n a positive integer that is at least as large as $l+1$. These associated Laguerre functions are polynomials of degree $n-l-1$, and were discussed in Section 14.7.

(d) Show that, after canceling the terms that combine to yield zero, the remaining terms of Eq. (15.117) become an identity if $E = -1/2n^2$.

(e) Summarize the work of parts (a) through (d) by showing that

(i) The smallest value of n consistent with the present development is $n = 1$, and in that case, l must be zero;

(ii) All larger integer values of n are acceptable, but the possible values of l range from zero to $n-1$, and

(iii) The value of E depends only on n.

(f) List all the hydrogen-atom solutions through $n = 3$, including all relevant l and m values, and (using symbolic computing if helpful) write each wave function ψ_{nlm} as an explicit radial function times a spherical harmonic (which can be left as a symbol Y_l^m). Identify the value of E for each wave function.

Note. The Laguerre functions can be accessed symbolically as

`LaguerreL(p,q,x)` or `LaguerreL[p,q,x]`.

(g) Using the notations s, p, d, etc. to represent l values (see Exercise 15.4.5), label the wave functions of part (f) as n followed by the l code, as illustrated by $1s$ or $3d$. Explain why the following labels will not occur, even in a more complete list of hydrogenic wave functions: $1p$, $2d$, $4g$.

15.5 INHOMOGENEOUS PDEs

Many of the ideas developed for inhomogeneous ODEs are also applicable to inhomogeneous PDEs. A linear inhomogeneous PDE will have as its general solution the sum of a particular solution and the general solution to the corresponding homogeneous PDE. We can also generalize to two- and three-dimensional problems the notions corresponding to the introduction and use of Green's functions that was developed for ODEs in Section 10.7. A general exposition is beyond the scope of this text, but we can illustrate the essential ideas by looking at the Poisson equation (the inhomogeneous equation corresponding to the Laplace equation).

POISSON EQUATION

As pointed out earlier, the Poisson equation is satisfied by the potential $\psi(\mathbf{r})$ in an electrostatics problem in which $\rho(\mathbf{r})$ is the charge density:

$$\nabla^2\psi(\mathbf{r}) = -\frac{\rho(\mathbf{r})}{\varepsilon_0}. \tag{15.119}$$

If the only charge density is that of a point charge q at a point $\mathbf{r}'$, you probably already know that $\psi(\mathbf{r})$ is proportional to $q/|\mathbf{r}-\mathbf{r}'|$; in the unit system used in this book, the relationship is

$$\psi(\mathbf{r}) = \frac{q}{4\pi\varepsilon_0}\,\frac{1}{|\mathbf{r}-\mathbf{r}'|}. \tag{15.120}$$

Equations (15.119) and (15.120) are consistent because $\rho = q\delta(\mathbf{r}-\mathbf{r}')$ and, referring to Eq. (7.68),

$$\nabla^2\left(\frac{1}{|\mathbf{r}-\mathbf{r}'|}\right) = -4\pi\delta(|\mathbf{r}-\mathbf{r}'|). \tag{15.121}$$

Our present interest in these equations arises from the fact that they enable us to define a **Green's function** $G(\mathbf{r},\mathbf{r}')$ for the Laplace/Poisson equation (which we will call a **Coulomb Green's function**) as

$$G(\mathbf{r},\mathbf{r}') = -\frac{1}{4\pi|\mathbf{r}-\mathbf{r}'|} \tag{15.122}$$

with the property that

$$\nabla^2 G(\mathbf{r}, \mathbf{r}') = \delta(\mathbf{r} - \mathbf{r}') \tag{15.123}$$

(where ∇^2 operates on $\mathbf{r}$ but not on $\mathbf{r}'$), and that the Poisson equation, Eq. (15.119), has solution

$$\psi(\mathbf{r}) = -\frac{1}{\varepsilon_0} \int G(\mathbf{r}, \mathbf{r}')\rho(\mathbf{r}')\, d^3r' \,. \tag{15.124}$$

At this point we do not have anything surprising or new: Eq. (15.124) simply states that the potential at $\mathbf{r}$ is the sum of the Coulombic potentials of all elements of the charge distribution. Note however that the present development assumes that there are no boundary conditions that imply the existence of additional charges. For example, such additional charges occur when the Poisson equation is solved for the region within a grounded enclosure (i.e., a boundary on which $\psi = 0$); the maintenance of this zero potential is brought about by the presence of induced charges on the boundary.

Thus, what we have now is a "fundamental" Coulomb Green's function applicable for an infinite unbounded region. However, starting from that $G(\mathbf{r}, \mathbf{r}')$ we can (at least in principle) add a solution to the Laplace equation that will produce the desired boundary conditions. Some instances of that approach may already be familiar to you. Here are two examples.

Example 15.5.1. Method of Images

As an initial example, suppose our region of interest is that on one side of a grounded plane; for definiteness let $\psi = 0$ on the plane $z = 0$ and consider the region $z > 0$. To determine the Coulomb Green's function, consider a unit point source in this region, at $\mathbf{r}' = (x', y', z')$ (with $z' > 0$). Our fundamental Green's function G_0 will be

$$G_0(\mathbf{r}, \mathbf{r}') = -\frac{1}{4\pi|\mathbf{r} - \mathbf{r}'|} \,.$$

If to this G_0 we add the potential from a source of unit magnitude and opposite sign at the mirror-image point $\bar{\mathbf{r}}' = (x', y', -z')$, as shown in the left panel of Fig. 15.4, the sum of the potentials of the two sources will add to zero on the plane $z = 0$, and the potential from the image source will satisfy the Laplace equation everywhere except

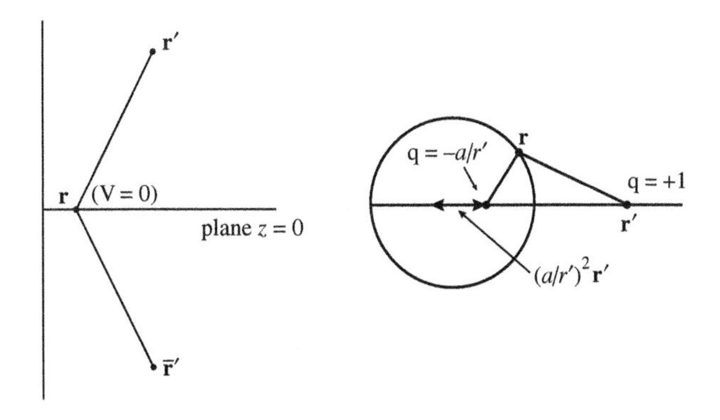

Figure 15.4: Image charges. Left, zero potential on plane. Right, zero potential on sphere.

at $\bar{\mathbf{r}}'$, which is not within the region under study. Thus, a Green's function suitable for our semi-infinite region will be

$$G(\mathbf{r},\mathbf{r}') = -\frac{1}{4\pi|\mathbf{r}-\mathbf{r}'|} + \frac{1}{4\pi|\mathbf{r}-\bar{\mathbf{r}}'|}\,.$$

As a second example, consider the space external to a sphere of radius a centered at the coordinate origin, with a unit source at the point $\mathbf{r}'$, with $r' > a$. It can be shown (see Exercise 15.5.1) that a suitable image of this source will have strength $-a/r'$ and be located at $(a/r')^2\mathbf{r}'$ (which is inside the sphere). The source and its image are shown in the right panel of Fig. 15.4. Our Green's function for the region external to the sphere can then be seen to have the form

$$G(\mathbf{r},\mathbf{r}') = -\frac{1}{4\pi|\mathbf{r}-\mathbf{r}'|} + \frac{a}{4\pi r'}\,\frac{1}{|\mathbf{r}-(a/r')^2\mathbf{r}'|}\,.$$

It may not always be easy to find a solution to the Laplace equation that can be added to our fundamental Green's function and thereby satisfy specific boundary conditions, but in principle a suitable solution always exists.

■

Exercises

15.5.1. A charge outside a sphere of radius a and its image within the sphere lie on a line that passes through the center of the sphere. Using that line as the z-axis of a Cartesian coordinate system, place a unit charge at $(x,y,z) = (0,0,r')$ and place its image, of strength $-a/r'$, at $(x,y,z) = (0,0,a^2/r')$. Show that the sum of the potentials produced by these charges is zero at an arbitrary point on the sphere, at $(x,y,z) = (a\sin\theta\cos\varphi, a\sin\theta\sin\varphi, a\cos\theta)$, where θ and φ are the usual spherical polar angular coordinates.

15.5.2. A unit charge is located at the point $\mathbf{r}'$ **within** a sphere of radius a. Find the position and magnitude of an image charge that will cause the combined potential of the charge and its image to vanish on the sphere.

15.5.3. Find the Coulomb Green's function $G(\mathbf{r},\mathbf{r}')$ for the portion of three-dimensional space with $x > 0$ and $y > 0$, subject to the boundary condition that G vanish when $\mathbf{r}$ is on the boundary of that region (i.e., on the xz- and yz-planes).

15.6 INTEGRAL TRANSFORM METHODS

In Chapter 13 we saw how integral transforms can be used to convert a linear ODE into an algebraic equation. If these techniques are applied to one of the variables in a PDE, the dependence on that variable becomes algebraic. If the original PDE had two independent variables it would be reduced to an ODE. We illustrate this procedure with two examples: one using the Laplace transform, and one using the Fourier transform.

Example 15.6.1. Thermal Diffusion

An insulated rod with one end at $x = 0$ extends to infinity in the positive x direction and is initially at temperature $T = 0$. Starting at time $t = 0$ the $x = 0$ end of the rod is maintained at temperature T_0. The temperature distribution in the rod is governed by the PDE

$$\frac{\partial^2\psi(x,t)}{\partial x^2} = \frac{1}{k^2}\frac{\partial\psi(x,t)}{\partial t}. \tag{15.125}$$

This problem is the large-L limit of that studied in Example 15.2.3.

We proceed by taking the Laplace transform of Eq. (15.125), regarding t as the variable to be transformed. We denote the transform $\Psi(x,s)$. Noting that when we take the transform x is just a parameter, Eq. (15.125) becomes

$$\frac{d^2\Psi(x,s)}{dx^2} = \frac{1}{k^2}\mathcal{L}\left[\frac{\partial\Psi}{\partial t}\right] = \frac{1}{k^2}\left[s\Psi(x,s) - \psi(x,0)\right],$$

where we have used the formula for the Laplace transform of a derivative, Eq. (13.63). Since $\psi = 0$ for all x at $t = 0$, this equation simplifies to

$$\frac{d^2\Psi(x,s)}{dx^2} = \frac{s}{k^2}\Psi(x,s). \tag{15.126}$$

We now have an ODE instead of the original PDE, and we can solve it subject to the boundary conditions $\psi(0,t) = T_0$ and $\psi(\infty,t) = 0$. Translating to the corresponding conditions on Ψ,

$$\psi(0,t) = T_0 \longrightarrow \Psi(0,s) = \frac{T_0}{s} \qquad \text{and} \qquad \psi(\infty,t) = 0 \longrightarrow \Psi(\infty,s) = 0.$$

Here we have used the fact that $\mathcal{L}[1] = 1/s$.

We now solve Eq. (15.126) subject to the boundary conditions on Ψ, obtaining

$$\Psi(x,s) = \frac{T_0}{s}e^{-(\sqrt{s}/k)x}. \tag{15.127}$$

To get the temperature distribution $\psi(x,t)$, our only remaining step is to find the inverse transform of Ψ. The difficulty of this final step often limits the usefulness of the Laplace transform method of PDE solution, but in the present case the inverse is easy to find. Resorting to symbolic computation,

MAPLE:
```
> with(inttrans):
> assume(x/k > 0);
> invlaplace(T0/s*exp(-(x/k)*sqrt(s)),s,t);
```

$$T0\ \mathrm{erfc}\left(\frac{1}{2}\frac{x\sim}{k\sim\sqrt{t}}\right)$$

MATHEMATICA:
```
InverseLaplaceTransform[T0/s*E^(-x/k*Sqrt[s]), s, t]
```

$$\frac{\mathrm{T0}\sqrt{\frac{\mathrm{k^2t}}{\mathrm{x^2}}}\ \mathrm{x\,Erfc}\left[\frac{1}{2\sqrt{\frac{\mathrm{k^2t}}{\mathrm{x^2}}}}\right]}{\mathrm{k}\sqrt{\mathrm{t}}}$$

```
Simplify[%, k/x > 0]
```

$$\mathrm{T0\ Erfc}\left[\frac{\mathrm{x}}{2\mathrm{k}\sqrt{\mathrm{t}}}\right]$$

In MAPLE the inverse transform evaluation will not proceed unless x/k is first identified as positive; in MATHEMATICA the initial result will not simplify unless we assume $k/x > 0$.

Since $\mathrm{erfc}(z) = 1$ at $z = 0$ and decreases toward zero at large z, we see that our solution

$$\psi(x,t) = T_0 \,\mathrm{erfc}\left(\frac{x}{2k\sqrt{t}}\right) \tag{15.128}$$

will decrease from T_0 at $x = 0$ toward zero at large x, with the rate of decrease smaller at larger times t.

Since the inversion of Fourier transforms is easier than that of Laplace transforms, one might ask whether this problem could be treated using Fourier transforms. That option is not possible here, because the condition $\psi(0,t) = T_0$ does not have a Fourier transform.

■

Example 15.6.2. Two-Dimensional Laplace Equation

We require the solution of the Laplace equation

$$\frac{\partial^2\psi}{\partial x^2} + \frac{\partial^2\psi}{\partial y^2} = 0 \tag{15.129}$$

for the entire first quadrant of a Cartesian coordinate system, with boundary conditions $\psi = 0$ on the positive y-axis and $\psi = x/(x^2+1)$ on the positive x-axis. Our method of solution will be to apply a Fourier transform to the x-dependence of ψ. Because $\psi = 0$ at $x = 0$ and our region of interest is $x \geq 0$, we choose to use the Fourier sine transform.

Letting $\Psi(k,y)$ be the transform of $\psi(x,y)$, and using Eq. (13.25) in the form

$$\left[\frac{\partial^2\psi(x,y)}{\partial x^2}\right]^T(k) = -k^2\Psi(k,y)\,,$$

the Laplace equation becomes the ODE

$$-\,k^2\Psi + \frac{\partial^2\Psi}{\partial y^2} = 0\,, \tag{15.130}$$

while the boundary condition $\psi(x,0) = x/(x^2+1)$ has Fourier sine transform

$$\Psi(k,0) = \sqrt{\frac{\pi}{2}}\,e^{-k}\,. \tag{15.131}$$

Equation (15.131) is easily checked using symbolic computation.

The general solution to Eq. (15.130) is

$$\Psi(k,y) = Ae^{ky} + Be^{-ky}\,,$$

where A and B must not depend upon y but can be functions of anything else, e.g., k. A solution to Eq. (15.130) that is consistent with Eq. (15.131) and which approaches zero as $y \to \infty$ is

$$\Psi(k,y) = \sqrt{\frac{\pi}{2}}\,e^{-k(y+1)}\,. \tag{15.132}$$

To complete the solution of our original problem we need only to find the inverse Fourier sine transform of $\Psi(k, y)$. Using symbolic computation or other means, we obtain

$$\psi(x, y) = \frac{x}{x^2 + (y+1)^2} . \tag{15.133}$$

It is clear that this formula for ψ properly reproduces the boundary condition at $y = 0$. One can also check that it satisfies the Laplace equation.

■

Exercises

15.6.1. Verify that $\Psi(x, s)$ as given by Eq. (15.127) is a solution to Eq. (15.126) that satisfies the boundary conditions on $\Psi(0, s)$ and $\Psi(\infty, s)$.

15.6.2. (a) Using symbolic computation or otherwise, verify that $\Psi(k, 0)$ as given in Eq. (15.131) is the Fourier sine transform of $\psi(x, 0) = x/(x^2 + 1)$.

(b) Verify that $\psi(x, y)$ as given in Eq. (15.133) is the inverse Fourier sine transform of $\Psi(k, y)$, Eq. (15.132).

(c) Verify that $\psi(x, y)$, as given in Eq. (15.133), is a solution of the two-dimensional Laplace equation, Eq. (15.129).

15.6.3. Use the Laplace transform method to solve (for all $t > 0$) the PDE

$$\frac{\partial y(x, t)}{\partial x} + x \frac{\partial y(x, t)}{\partial t} = x$$

subject to the conditions $y(0, t) = 1$ and $y(x, 0) = 1$.

Hint. Both because we need a solution only for $t > 0$ and because the dependence of the PDE on t is simpler than its dependence on x, take the transform with respect to the variable t. Feel free to use symbolic methods to obtain the inverse transform.

15.6.4. Use the Laplace transform method to solve (for all positive x and t) the PDE

$$\frac{\partial y}{\partial x} + \frac{\partial y}{\partial t} = 0$$

subject to the conditions $y(0, t) = 0$ and $y(x, 0) = \sin x$.

This PDE describes a source that had been transmitting a sinusoidal oscillation toward positive x for an indefinitely long interval before it was turned off at $t = 0$. Explain how your answer describes what then happens.

Additional Readings

Arfken, G. B., Weber, H. J., & Harris, F. E. (2013). *Mathematical methods for physicists* (7th ed.). New York: Academic Press.

Cohen, H. (1992). *Mathematics for scientists and engineers.* Englewood Cliffs, NJ: Prentice-Hall.

Folland, G. B. (1995). *Introduction to partial differential equations* (2nd ed.). Princeton, NJ: Princeton University Press.

Margenau, H., & Murphy, G. M. (1956). *The mathematics of physics and chemistry* (2nd ed.). Princeton, NJ: Van Nostrand (Chapter 5 covers curvilinear coordinates and 13 specific coordinate systems).

Chapter 16

CALCULUS OF VARIATIONS

16.1 INTRODUCTION

In an elementary calculus course you have doubtless encountered the problem of finding the point where a function has its minimum (or maximum) value. If the function is smooth (meaning that it has a continuous derivative) you also know that its **stationary points** (minima, maxima, and points of inflection) can all be found by setting the derivative of the function to zero. In the present chapter we address a generalization of that problem: How can we find a function connecting two given endpoints that minimizes or maximizes some property of that function? In other words, instead of finding a **point** at which a function is stationary, we seek to vary a **function** to make one of its properties stationary. Perhaps for that reason the branch of mathematics under study in this chapter is known as the **calculus of variations**.

The central idea of this chapter becomes clearer if we give a few concrete examples of problems within its scope:

1. What is the shortest curve connecting two given points: (a) on a plane, (b) on a sphere, (c) on an ellipsoid? (The shortest curve is called a **geodesic** of the space in question, and is relevant in general relativity as well as in ordinary geometry.)

   You already know the answer for a plane: the straight line connecting the endpoints. You probably also know the answer for a sphere: the shorter of the two great-circle arcs connecting the endpoints, and perhaps you can also develop the equation for that curve. But unless you have a formal method for obtaining the result, you will probably have great difficulty in solving this problem for an ellipsoid or any other more general surface.

2. What shape will a flexible chain of given length take (under the influence of gravity) if it is suspended between two fixed points?

   The curve that is the solution to this problem, called a **catenary** (it is not a parabola or other simple algebraic curve), is relevant to the design of suspension bridges.

Mathematics for Physical Science and Engineering.
http://dx.doi.org/10.1016/B978-0-12-801000-6.00016-X

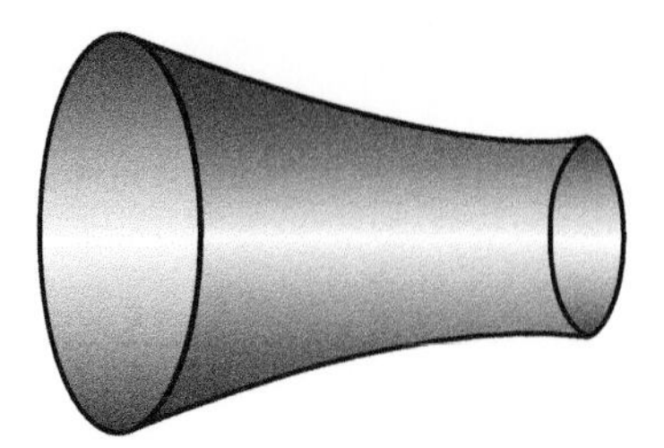

Figure 16.1: Soap film.

3. If a bead is caused to slide (without friction) along a wire under the influence of gravity, what shape must the wire have to minimize the time of travel between two fixed points?

   This is the famous **brachistochrone** problem; attempts to solve it contributed to the development of the methods studied in this chapter.

4. What shape is assumed by a soap film that is suspended between two coaxial wire circles (see Fig. 16.1)?

   Neglecting the influence of gravity, the surface tension of the film will cause it to be in the shape of a surface of revolution of minimum area.

5. **Fermat's principle** of optics states that the actual path of a light ray between two given points is that for which the time of travel is stationary.

   This problem becomes most meaningful when the ray travels through media with different transmission speeds, and solution of the problem only requires the methods of this chapter when the speed of light varies continuously with position.

All the above problems have in common that they can be framed in terms of a line or surface integral that is to be made stationary (in some cases subject to constraints such as the fixed path length in Example 2 above). Let's illustrate how this works for a simple case of each of the above examples. Note that we are only providing more complete definitions of these problems; we are not yet discussing methods for their solution.

1. **Geodesic.**– Let a function $y(x)$ define a path in the xy-plane connecting the endpoints (x_1, y_1) and (x_2, y_2), meaning that $y(x_1) = y_1$ and $y(x_2) = y_2$. The differential ds of path length corresponding to a change in x from x to $x + dx$ and to the accompanying change of y along the path, namely $dy = (dy/dx)\,dx$, is given by

$$ds = \sqrt{dx^2 + dy^2} = \left(\sqrt{1 + \frac{dy^2}{dx^2}}\right) dx = \sqrt{1 + y'^2}\, dx\,. \tag{16.1}$$

   The shortest path connecting the endpoints is that which minimizes

$$W = \int_{x_1}^{x_2} ds = \int_{x_1}^{x_2} \sqrt{1 + y'^2}\, dx\,,$$

   with respect to all curves $y(x)$ that pass through (x_1, y_1) and (x_2, y_2).

2. **Catenary.**– Again we have a path $y(x)$ that connects points (x_1, y_1) and (x_2, y_2), and we are to minimize the gravitational potential energy associated

with the path, but with the additional requirement that the path be of length L. This fixed-length requirement is called a **constraint**. Using Eq. (16.1), our constraint corresponds to the equation

$$L = \int_{x_1}^{x_2} \sqrt{1 + y'^2}\, dx \,, \tag{16.2}$$

with a fixed value of L.

Taking y as the vertical coordinate and assuming our chain to be uniform (so its mass per unit length is constant), the gravitational potential energy for an element of length ds at x is proportional to $y\, ds = y\sqrt{1 + y'^2}\, dx$. We therefore need to minimize

$$W = \int_{x_1}^{x_2} y\sqrt{1 + y'^2}\, dx \tag{16.3}$$

with respect to paths through the endpoints but also subject to the constraint given in Eq. (16.2).

3. **Brachistochrone.**– Because we want to minimize the travel time of the sliding bead, we are interested in the differential of time, dt, when the bead moves an amount ds along its path at the speed v. For simplicity we consider a case where the initial x and y values are zero, the direction of positive y is downward, and the bead starts from rest, so that its velocity can be calculated, using energy conservation, as

$$\frac{mv^2}{2} = mgy \,, \qquad \text{or} \qquad v = \sqrt{2gy}\,.$$

We therefore have

$$dt = \frac{ds}{v} = \frac{\sqrt{1 + y'^2}}{\sqrt{2gy}}\, dx \,,$$

and we need to minimize

$$W = \int_{x_1}^{x_2} dt = \int_{x_1}^{x_2} \sqrt{\frac{1 + y'^2}{y}}\, dx \tag{16.4}$$

with respect to paths through the endpoints.

4. **Soap film.**– In the soap-film problem let r_1 and r_2 be the radii of the two wire circles, centered at the respective points x_1 and x_2 on a common axis, and let the soap film at x have radius $r(x)$. Then the contribution to the soap-film area for interval dx will be $dA = 2\pi r(x)\, ds = 2\pi r(x)\sqrt{1 + r'(x)^2}\, dx$, and the integral to be minimized is

$$W = 2\pi \int_{x_1}^{x_2} r\sqrt{1 + r'^2}\, dx \,. \tag{16.5}$$

5. **Fermat's principle.**– In many cases light-ray problems do not require the special methods to be developed in this chapter. We illustrate by using Fermat's principle to confirm the law of specular reflection.

   Consider a mirror in the xz-plane and a light ray that passes from $(x_1, y_1, 0)$ to $(x_2, y_2, 0)$ along a path that includes reflection by the mirror. If we assume that the velocity of light is constant at all points of the light path, we require the point on the mirror that connects the endpoints by straight-line segments

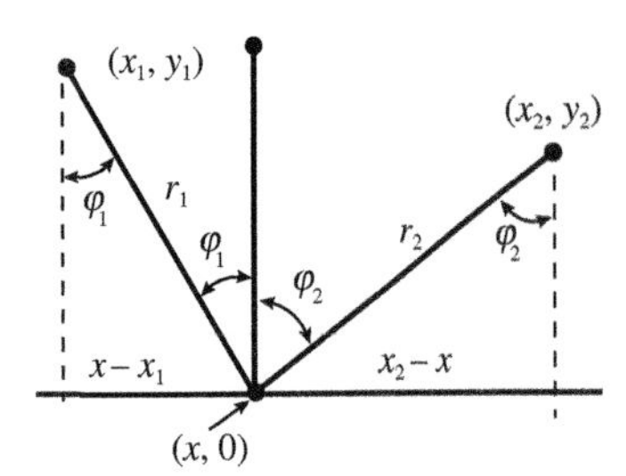

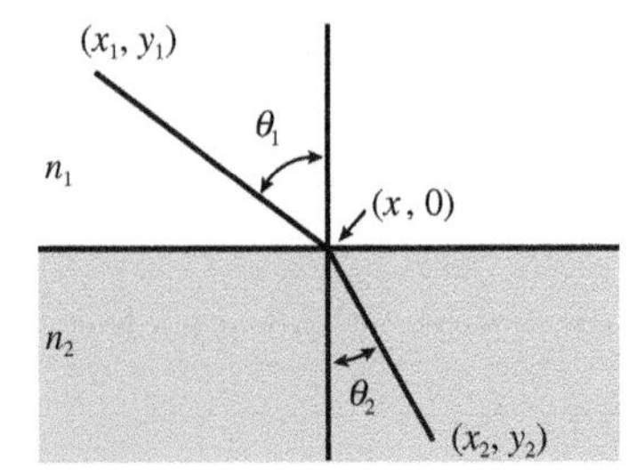

Figure 16.2: Light rays: Left, reflection; right, refraction.

of shortest total length. It is obvious that the shortest path will reflect from the mirror at some point $(x, 0, 0)$, with the value of x that minimizes

$$w(x) = \sqrt{(x - x_1)^2 + y_1^2} + \sqrt{(x - x_2)^2 + y_2^2} .$$

This is a problem in ordinary differential calculus. Setting $w'(x) = 0$, we find

$$\frac{x - x_1}{\sqrt{(x - x_1)^2 + y_1^2}} + \frac{x - x_2}{\sqrt{(x - x_2)^2 + y_2^2}} = 0 .$$

Writing $r_1 = \sqrt{(x - x_1)^2 + y_1^2}$, $r_2 = \sqrt{(x - x_2)^2 + y_2^2}$, and referring to the left panel of Fig. 16.2 for the definitions of φ_1 and φ_2, we have

$$\frac{x - x_1}{r_1} = \frac{x_2 - x}{r_2}, \quad \text{or} \quad \sin\varphi_1 = \sin\varphi_2 , \quad \text{or} \quad \varphi_1 = \varphi_2 .$$

The equality of φ_1 and φ_2 is the law of specular reflection.

Exercises

16.1.1. The velocity of light in vacuum is c; in a medium of refractive index n, it is c/n. When a ray of light reaches a boundary between two media of different refractive indices n_1 and n_2, the direction of the ray is bent at the boundary in a way that depends upon n_1, n_2, and the angle of incidence of the light ray (relative to the direction normal to the media boundary). See the right panel of Fig. 16.2. Apply Fermat's principle (that the path actually taken requires a travel time that is stationary with respect to infinitesimally neighboring paths) to find the formula connecting n_1, n_2, θ_1, and θ_2. This relationship is known as **Snell's Law of refraction**.

Hint. The path within each medium is a straight line.

16.2 EULER EQUATION

Now that we have identified a number of problems that involve minimizing an integral that contains an initially unknown function, let's move toward developing general methods for dealing with this class of problems. This effort will lead to the derivation of the **Euler equation** of the calculus of variations.

We need to find a way to cast our present problems in a form that permits us to use our knowledge of differential calculus to find extrema. Our route in this direction starts by assuming that there exists (for a prototype problem in one dimension) a differentiable function $y(x)$ passing through the endpoints (x_1, y_1) and (x_2, y_2) that makes the integral

$$W = \int_{x_1}^{x_2} F\left(y(x), \frac{dy(x)}{dx}, x\right) dx \tag{16.6}$$

stationary with respect to all infinitesimal variations in $y(x)$. What is new here is that W is not just a function of some variable(s); it depends upon the entire range (i.e., all values) of the function $y(x)$. For this reason W is sometimes called a **functional** of y, and occasionally one sees the dependence of W upon y indicated by the notation $W[y]$. We are assuming that the integrand of our integral may depend explicitly upon a function y with specified endpoint values, and/or upon the derivative of y, and/or it may have an explicit dependence on x. F need not depend upon all these quantities. The requirement that W be stationary is sometimes written $\delta W = 0$; this notation is used instead of $dW = 0$ as an additional indicator that the variation is with respect to a function, and not just that of a variable. It is also customary to write y_x to denote the derivative of y with respect to x. This notation will be particularly useful when we later encounter expressions with more than one independent variable. The notational observations of this paragraph enable our variational problem to be written in the more compact form

$$\delta W = \int_{x_1}^{x_2} F(y,\, y_x,\, x)\, dx = 0\,. \tag{16.7}$$

A description of functional variations can start by defining a function $\eta(x)$ that is entirely arbitrary except that it is differentiable and with the property that $\eta(x_1) = \eta(x_2) = 0$. This definition means that the function $y(x) + \alpha\eta(x)$, with α an infinitesimal quantity, describes a function with the same endpoints as $y(x)$, but otherwise deformed infinitesimally in an arbitrary fashion. Note also that

$$\frac{\partial}{\partial x}\left[y(x) + \alpha\eta(x)\right] = y_x + \alpha\eta_x,$$

where η_x is shorthand for $\partial\eta/\partial x$.

The condition that W be stationary when y is the function $y(x)$ can now be stated in a more conventional fashion as the requirement

$$\frac{dW}{d\alpha} = \int_{x_1}^{x_2} \frac{d}{d\alpha} F(y + \alpha\eta,\, y_x + \alpha\eta_x,\, x)\, dx = 0\,, \quad \text{for all } \eta \text{ and for } \alpha = 0. \tag{16.8}$$

Our next step is to manipulate Eq. (16.8) by evaluating the derivative within the integral. We have

$$\frac{dF}{d\alpha} = \frac{\partial F}{\partial y}\eta + \frac{\partial F}{\partial y_x}\eta_x\,,$$

where we are treating F as dependent on the three formal variables y, y_x, and x. Since we are interested in an evaluation at $\alpha = 0$ these partial derivatives are to be evaluated at that α value. However, note that η and its derivative η_x are still in our formula, so we have

$$\left.\frac{dW}{d\alpha}\right|_{\alpha=0} = \int_{x_1}^{x_2} \left(\frac{\partial F}{\partial y}\eta + \frac{\partial F}{\partial y_x}\eta_x\right) dx = 0\,, \qquad \text{all } \eta. \tag{16.9}$$

We now subject the second term within the integral to an integration by parts, integrating η_x and differentiating $\partial F/\partial y_x$. Because η vanishes at both x_1 and x_2, there are no integrated endpoint contributions, and Eq. (16.9) becomes

$$\left.\frac{dW}{d\alpha}\right|_{\alpha=0} = \int_{x_1}^{x_2} \eta(x) \left(\frac{\partial F}{\partial y} - \frac{d}{dx}\frac{\partial F}{\partial y_x} \right) dx = 0\,, \qquad \text{all } \eta. \tag{16.10}$$

It is important to note that while the partial derivatives indicate only the differentiation of F with respect to the formal variables y or y_x, the ordinary derivative d/dx in Eq. (16.10) applies to **all** the dependence on x in the quantity on which it acts (not just on any explicit dependence on x that might be present in F). This interpretation is correct because the ordinary derivative appeared when we carried out an integration by parts.

We are now ready for the key step in the current analysis. In order for Eq. (16.10) to be satisfied for all functions η, it is necessary that the remainder of the integrand be identically zero. If the parenthesized factor in that equation were nonzero in the neighborhood of some value of x, the equation would be violated when η is chosen to be nonzero in that neighborhood and zero everywhere else. Note that in general the vanishing of an integral does not imply the vanishing of its integrand. But here we have the additional feature that one of the factors in the integrand is arbitrary; that fact enables us to conclude that the other factor must vanish. Thus, we reach the important and valuable result known as the **Euler equation**,

$$\frac{\partial F}{\partial y} - \frac{d}{dx}\frac{\partial F}{\partial y_x} = 0\,. \tag{16.11}$$

Example 16.2.1. Geodesic on a Plane

Our first example of the Euler equation will be to verify something we already know, namely that the shortest distance between two points in a plane is the straight line connecting them. The path length from (x_1, y_1) to (x_2, y_2) defined by the function $y(x)$ is given by L in Eq. (16.2), corresponding to choosing F in Eq. (16.6) as $F = \sqrt{1+y_x^2}$. We then have

$$\frac{\partial F}{\partial y} = 0\,, \qquad \text{and} \qquad \frac{\partial F}{\partial y_x} = \frac{y_x}{\sqrt{1+y_x^2}}\,.$$

The Euler equation for this example is therefore

$$\frac{d}{dx}\frac{y_x}{\sqrt{1+y_x^2}} = 0\,,$$

which is equivalent to $y_x/\sqrt{1+y_x^2} =$ constant, or the even simpler result $y_x =$ constant. This simple ODE can be integrated to yield $y = ax + b$, and the constants a and b are to be chosen so that $y(x_1) = y_1$ and $y(x_2) = y_2$.

The above analysis actually only shows that the straight line through a pair of points is stationary in length relative to infinitesimal variations. It is clear, however, that the stationary $y(x)$ is of minimum length.

■

Exercises

16.2.1. Find and solve the Euler equation that makes each of the following integrals stationary.

Hint. Note that y is missing in all these problems. Use that fact to identify the ODEs you generate as first-order equations in the dependent variable y_x.

(a) $\displaystyle\int_{x_1}^{x_2} (y_x^2 + y_x)\,dx$ (b) $\displaystyle\int_{x_1}^{x_2} \frac{\sqrt{1+y_x^2}}{x}\,dx$

(c) $\displaystyle\int_{x_1}^{x_2} x\sqrt{1-y_x^2}\,dx$ (d) $\displaystyle\int_{x_1}^{x_2} \sqrt{x(1+y_x^2)}\,dx$

ALTERNATE FORMS OF EULER EQUATION

There are several possibilities for bringing the Euler equation to forms that may be convenient for various problems. First, Eq. (16.11) was written under the assumption that x is the independent variable and that y is a function of x. However, either x or y could have been chosen as the independent variable (or, equivalently, there are alternative choices as to which variable in a physical problem is to be deemed independent). These choices may vary in the ease with which the Euler equation is constructed and solved. Moreover, the derivation of the Euler equation did not make use of any geometric properties of the variables; that point is illustrated and commented upon in Example 16.2.2.

It can also be shown that from the Euler equation we can obtain

$$\frac{\partial F}{\partial x} - \frac{d}{dx}\left(F - y_x\frac{\partial F}{\partial y_x}\right) = 0\,, \tag{16.12}$$

which when F has no explicit dependence on x reduces to

$$F - y_x\frac{\partial F}{\partial y_x} = \text{constant}. \tag{16.13}$$

In general Eq. (16.11) or (16.12) will be a second-order differential equation, involving $y_{xx} = d^2y/dx^2$. Equation (16.13) is only first order, and is available in the frequently occurring case that F has no explicit dependence on x.

Example 16.2.2. Geodesic on a Cone

Our task here is to find the shortest path between two points on the cone defined (in Cartesian coordinates) by the equation $z^2 = 3(x^2 + y^2)$. It is convenient to use cylindrical coordinates for this problem: setting $\rho^2 = x^2 + y^2$, the cone has equation $z = \rho\sqrt{3}$. In the cylindrical coordinates, the differential ds of path length on the cone satisfies the equation

$$ds^2 = d\rho^2 + dz^2 + \rho^2 d\theta^2 = 4d\rho^2 + \rho^2 d\theta^2\,, \quad \text{equivalent to} \quad ds = \sqrt{4d\rho^2 + \rho^2 d\theta^2}\,,$$

where we have used the fact that $z^2 = 3\rho^2$, and that $dz = \sqrt{3}\,d\rho$.

We can write the path length using either ρ or θ as the independent variable (i.e., the variable with respect to which we integrate). A look at the Euler equation indicates that it will be simpler if the dependent variable, which is denoted y in Eq. (16.11), is absent from the function F in the integrand. That observation motivates the choice to choose θ as the dependent variable and to integrate (and differentiate) with respect to ρ. With this choice, we define the path length between points (ρ_1, θ_1) and (ρ_2, θ_2) as

$$L = \int ds = \int_{\rho_1}^{\rho_2} \sqrt{4 + \rho^2\theta_\rho^2}\,d\rho\,. \tag{16.14}$$

We are now ready to write the Euler equation corresponding to $\delta L = 0$. With $F = \sqrt{4 + \rho^2\theta_\rho^2}$, we have

$$\frac{\partial F}{\partial \theta} - \frac{d}{d\rho}\frac{\partial F}{\partial \theta_\rho} = 0\,, \tag{16.15}$$

the first term of which vanishes, and the second term integrates immediately to yield

$$\frac{\partial F}{\partial \theta_\rho} = C_1 \quad \text{(a constant)}. \tag{16.16}$$

At this point we note that the form of Eq. (16.15) contains no factors associated with the metric properties of the angular coordinate; the Euler equation has the same form irrespective of the geometric meanings of its variables.

Inserting into Eq. (16.16) the functional form of F, carrying out the indicated differentiation, and solving for θ_ρ, we get

$$\frac{\rho^2\theta_\rho}{\sqrt{4 + \rho^2\theta_\rho^2}} = C_1\,, \qquad \theta_\rho = \frac{d\theta}{d\rho} = \frac{2C_1}{\rho\sqrt{\rho^2 - C_1^2}}\,. \tag{16.17}$$

Our next step is to integrate the expression for θ_ρ, i.e., to evaluate

$$\theta(\rho) = \int \frac{2C_1}{\rho\sqrt{\rho^2 - C_1^2}}\, d\rho = 2\cos^{-1}\left(\frac{C_1}{\rho}\right) + C_2\,. \tag{16.18}$$

The integral evaluation shown above could have been carried out using symbolic computation. A somewhat simpler expression can be obtained by transforming Eq. (16.18) to

$$\rho\,\cos\left(\frac{\theta - C_2}{2}\right) = C_1\,. \tag{16.19}$$

To complete the solution of our problem we now need to find the values of C_1 and C_2 that cause the curve defined by Eq. (16.19) to pass through the specified points. That task is the topic of Exercise 16.2.3.

■

Here is another example that illustrates different ways of using the Euler equation.

Example 16.2.3. Soap Film

As indicated by the discussion leading to Eq. (16.5), the problem of finding a soap film of minimum area connecting two coaxial circular loops can lead to the variational equation

$$\delta W = 2\pi \int r\sqrt{1 + r_x^2}\, dx = 0\,. \tag{16.20}$$

However, we could with equal validity have written an integral in which r was chosen as the independent variable. With that choice, the element of area corresponding to changes in r and x becomes

$$dA = 2\pi r\sqrt{dx^2 + dr^2} = 2\pi r\sqrt{\left(\frac{dx}{dr}\right)^2 + 1}\; dr = 2\pi r\sqrt{x_r^2 + 1}\, dr\,,$$

and our variational equation would then be

$$\delta W = 2\pi \int r\sqrt{x_r^2 + 1}\, dr = 0\,. \tag{16.21}$$

We can use either Eq. (16.20) or Eq. (16.21). Taking first Eq. (16.21), the Euler equation, Eq. (16.11), becomes (with x and y in that equation respectively replaced by r and x)

$$\frac{\partial F}{\partial x} - \frac{d}{dr}\frac{\partial F}{\partial x_r} = 0\,, \qquad \text{with} \quad F = r\sqrt{x_r^2 + 1}\,. \tag{16.22}$$

One advantage of the choice of Eq. (16.21) is that $\partial F/\partial x = 0$, so the Euler equation reduces to

$$\frac{\partial F}{\partial x_r} = C_1 \text{ (a constant)}, \qquad \text{i.e.,} \quad \frac{r x_r}{\sqrt{1 + x_r^2}} = C_1\,.$$

Solving for x_r, we get

$$x_r = \frac{C_1}{\sqrt{r^2 - C_1^2}} \quad \text{and} \quad x = \int \frac{C_1\, dr}{\sqrt{r^2 - C_1^2}} = C_1 \cosh^{-1}\left(\frac{r}{C_1}\right) + C_2\,, \tag{16.23}$$

where C_2 is another constant. Equation (16.23) can be rearranged to

$$r = C_1 \cosh\left(\frac{x - C_2}{C_1}\right). \tag{16.24}$$

If we were to complete the discussion of this problem, we would need to find values of C_1 and C_2 that are consistent with the specified endpoints (the positions x_1 and x_2 of the loops and their respective radii r_1 and r_2).

We turn now to the alternate formulation of the soap-film problem in which x is the independent variable, namely Eq. (16.20). Because $F = r\sqrt{1 + r_x^2}$ has no explicit dependence on x, we can use the alternative form of the Euler equation given in Eq. (16.12) (with r replacing y), and continue to Eq. (16.13), reaching

$$F - r_x \frac{\partial F}{\partial r_x} = r\sqrt{1 + r_x^2} - r_x r \frac{r_x}{\sqrt{1 + r_x^2}} = C_1\,. \tag{16.25}$$

Equation (16.25) can be solved for r_x, with the relatively simple result

$$r_x = \frac{dr}{dx} = \frac{\sqrt{r^2 - C_1^2}}{C_1}, \quad \text{yielding} \quad x = C_1 \int \frac{dr}{\sqrt{r^2 - C_1^2}} = C_1 \cosh^{-1}\left(\frac{r}{C_1}\right) + C_2\,. \tag{16.26}$$

This is the same as the result that we obtained starting from Eq. (16.21).

It would take us too far afield to complete the solution of this problem; a more detailed discussion can be found in the texts by Arfken et al., by Bliss, and by Courant and Robbins (see the Additional Readings). The soap-film problem also exhibits several pitfalls that await the unwary in using the calculus of variations. For some parameter values there are multiple solutions, while for others there are none (indicating conditions for which the film cannot be stable). The problem also admits a special solution that cannot be found by solving the general Euler equation (separate circular disks of film on the individual loops). Euler's method fails to find this special solution because it cannot be reached by continuous deformation of a general functional form and as a consequence violates the assumption that the deformation function η is differentiable.

■

Example 16.2.4. Brachistochrone

The brachistochrone is of enough historical interest that it should be included even in an elementary exposition of the calculus of variations. As indicated in Eq. (16.4), the function F for this problem is $F = \sqrt{1+y_x^2}/\sqrt{y}$. Since F has no explicit dependence on x, we can write the Euler equation in the form given in Eq. (16.13), in the present case reaching

$$\sqrt{\frac{1+y_x^2}{y}} - y_x\left(\frac{y_x}{\sqrt{y(1+y_x^2)}}\right) = C_1\,,$$

which we can solve for y_x, obtaining

$$y_x = \frac{dy}{dx} = \sqrt{\frac{1-C_1^2 y}{C_1^2 y}}\,.$$

This is a separable ODE, and can be written

$$dx = \sqrt{\frac{C_1^2 y}{1-C_1^2 y}}\;dy\,. \tag{16.27}$$

It turns out that a useful form for the solution of Eq. (16.27) is obtained if we make the substitution

$$C_1^2 y = \sin^2(\theta/2) = \frac{1}{2}\,(1-\cos\theta)\,. \tag{16.28}$$

Then

$$C_1^2\,dy = \sin(\theta/2)\cos(\theta/2)\,d\theta,$$

$$\sqrt{\frac{C_1^2 y}{1-C_1^2 y}} = \frac{\sin(\theta/2)}{\cos(\theta/2)}\,,$$

$$dx = \frac{1}{C_1^2}\,\sin^2(\theta/2)\,d\theta\,,$$

and

$$\begin{aligned} x &= \frac{1}{C_1^2}\int \sin^2(\theta/2)\,d\theta = \frac{1}{2C_1^2}\int (1-\cos\theta)\,d\theta \\ &= \frac{1}{2C_1^2}\,(\theta-\sin\theta) + C_2\,. \end{aligned} \tag{16.29}$$

We can now view Eqs. (16.28) and (16.29) as giving x and y in terms of the parameter θ, and a plot of the x–y trajectory produces a **cycloid**, namely the curve traced out by a point on the circumference of a circle when the circle is rolled along the x axis. This curve has the general appearance shown in Fig. 16.3; its scale, but not its shape, is determined by the radius $1/(2C_1^2)$ of the rolling circle. The constant

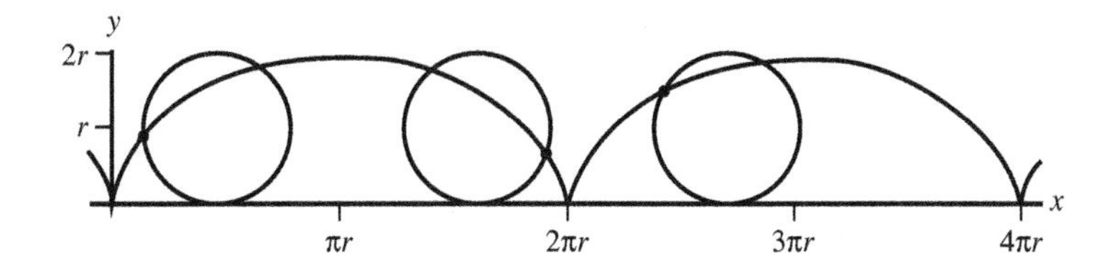

Figure 16.3: Cycloid: Path traced by dot on circumference of circle.

C_2 determines the starting position in which the marked point is on the x-axis; the radius of the rolling circle is equal to $1/(2C_1^2)$. In Exercise 16.2.4 you are asked to verify this interpretation of the solution given by Eqs. (16.28) and (16.29).

In the current problem, we set $C_2 = 0$ (so the cycloid starts at $x = 0$) and roll our circle on the under side of the x-axis. The proper value of C_1 is that which causes the endpoint (x, y) to lie on the first lobe of the cycloid. This value of C_1 is most easily obtained numerically.

One way to obtain a numeric solution is to note that y/x on the cycloid is a function of θ only, given by the equation

$$\frac{1-\cos\theta}{\theta-\sin\theta} - \frac{y}{x} = 0\,. \tag{16.30}$$

As we move through the first lobe of the cycloid, its value of y/x decreases monotonically from ∞ to zero; we know this because the cycloid is vertical at $x = 0$ and is everywhere convex (as a function of y). For a given endpoint (x, y) there will therefore be a unique value of θ for which Eq. (16.30) is satisfied; this value is easily found graphically. The procedure is illustrated in Fig. 16.4 for $y = 1$, $x = 4$, i.e., for $y/x = 1/4$.

Thus a solution procedure can be

1. Given the endpoint coordinates x and y, compute y/x.
2. Solve Eq. (16.30) numerically for θ at the endpoint.
3. Obtain C_1^2 by inserting the known values of y and θ into Eq. (16.28).

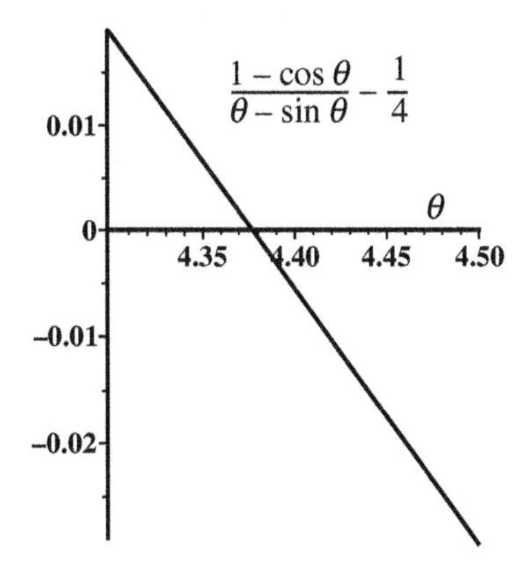

Figure 16.4: Left-hand side of Eq. (16.30) as a function of θ, for $y/x = 1/4$.

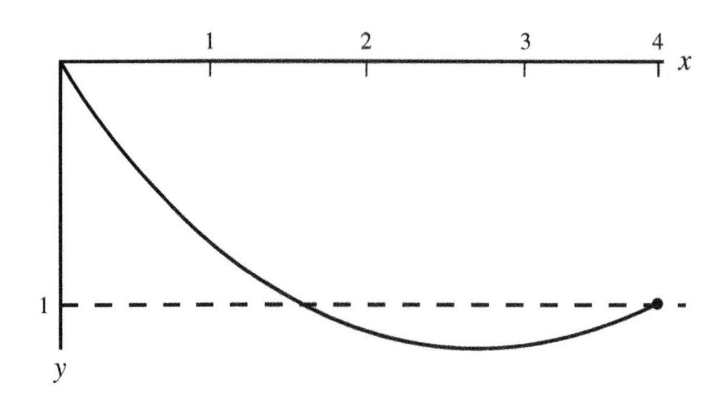

Figure 16.5: Brachistochrone, travel from $(x, y) = (0, 0)$ to $(4, 1)$.

Note that if y/x is small, it will intersect the cycloid during the latter half of its initial lobe, meaning that the shortest-time path passes through points lower than the endpoint. This behavior is seen for our test endpoint $(x, y) = (4, 1)$. See Fig. 16.5. Further discussion of the brachistochrone is in Exercise 16.2.4.

■

Exercises

16.2.2. Prove Eq. (16.12).

Hint. Be careful to distinguish between $\partial/\partial x$ and d/dx and watch for an opportunity to use the original form of the Euler equation to make a simplification.

16.2.3. Complete the solution of the geodesic for the cone of Example 16.2.2 by specializing to the situation that the path is from (ρ_1, θ_1) to (ρ_2, θ_2) with $\theta_1 = 0$, $0 < \theta_2 < \pi$, and $\rho_1 > \rho_2$. With those parameter choices, the value of C_2 is in the range $(0, \pi)$ and the arccos function of Eq. (16.18) must be assigned a value in the range $(-\pi, 0)$. Proceed as follows:

(a) Write Eq. (16.19) for the two endpoints of the path and solve for C_2 as a function of ρ_1, ρ_2, and θ_2.

(b) Determine C_1 and thereby have a useful expression for $\theta(\rho)$.

(c) Plot θ as a function of ρ for $\rho_1 = 2$, $\rho_2 = 1$, and several values of θ_2.

16.2.4. (a) Plot a cycloid that passes through the coordinate origin, with x the horizontal axis and y the vertical axis. Use the parametric description given as Eqs. (16.28) and (16.29).

(b) Explain how the parametric equations of the cycloid correspond to the trajectory of a point on the circumference of a rolling circle.

(c) To make the cycloid of this problem describe a brachistochrone whose endpoints are at the origin and $(x, y) = (5, 1)$ (with the y axis in the downward direction), solve Eq. (16.30) numerically for the θ value of its nonzero endpoint, and thereby determine the parameter C_1. Make an x–y plot of the cycloid, with the graph restricted to the x values between the endpoints.

(d) The graph of part (c) shows that the fastest trajectory passes through y values lower than the endpoint. Find another endpoint that will cause the trajectory to only pass through y values above the endpoint.

(e) Find the condition on y/x that determines whether the trajectory will pass through a minimum y value lower than the endpoint.

16.2.5. Find algebraic equations whose solution makes stationary the integral

$$\int_{\theta_1}^{\theta_2} \sqrt{r^2 r_\theta^2 + r^4}\, d\theta,$$

16.2.6. Apply Fermat's principle to obtain an algebraic equation for the path traveled by a light ray when the index of refraction is proportional to

(a) e^x, (b) $\dfrac{1}{2y+1}$, (c) $(y+2)^{1/2}$.

16.3 MORE GENERAL PROBLEMS

The variational problems of the preceding sections can be generalized in several ways:

1. There may be more than one dependent variable, but only one independent variable, x. The simplest such case, with the dependent variables denoted $y(x)$ and $z(x)$, corresponds to

$$\delta W = \int F(y, y_x, z, z_x, x)\, dx = 0\,, \qquad \text{all } y(x) \text{ and } z(x). \tag{16.31}$$

2. There may be more than one independent variable (here x and y), but only a single dependent variable, $u(x, y)$. Our prototype variational problem is then

$$\delta W = \int F(u, u_x, u_y, x, y)\, dx\, dy = 0\,, \qquad \text{all } u(x, y). \tag{16.32}$$

3. Our variational problem may involve higher derivatives, as in

$$\delta W = \int F(y, y_x, y_{xx}, x)\, dx\,, \qquad \text{all } y(x), \tag{16.33}$$

where y_{xx} denotes the second derivative of $y(x)$. This generalization is the topic of Exercise 16.3.1.

Further generalizations (which we do not develop here) correspond to situations with two or all three of the features enumerated above.

SEVERAL DEPENDENT VARIABLES

We can develop Euler equations for this case by introducing a different arbitrary deformation function for each dependent variable, then using integrations by parts to reach a form in which the integrand consists of terms each of which contains a different arbitrary factor. The result is that we get an Euler equation for each dependent variable. For the problem defined by Eq. (16.31), we have

$$\frac{\partial F}{\partial y} - \frac{d}{dx}\frac{\partial F}{\partial y_x} = 0\,,$$
$$\frac{\partial F}{\partial z} - \frac{d}{dx}\frac{\partial F}{\partial z_x} = 0\,. \tag{16.34}$$

Example 16.3.1. Lagrangian Mechanics

An important case of Eqs. (16.34) is provided by the Lagrangian formulation of mechanics, which starts from what is known as **Hamilton's principle**, namely that the time integral of the **Lagrangian** $L \equiv T - V$ (where T and V are respectively the kinetic and potential energies) is a stationary function of the path in the coordinate space. It can be shown that Hamilton's principle is equivalent to Newton's law of mechanics; a demonstration to that effect can be found in many advanced mechanics texts. For a single particle of mass m in ordinary three-dimensional space, the "path" is the trajectory of the particle, given as a function of time t. Thus, in Cartesian coordinates, $x(t)$, $y(t)$, $z(t)$ are dependent variables that are functions of the independent variable t, the potential energy V is some function $V(x, y, z)$, and the kinetic energy T is

$$T = \frac{1}{2}mv^2 = \frac{m}{2}\left(\dot{x}^2 + \dot{y}^2 + \dot{z}^2\right). \tag{16.35}$$

Here we are using the convention (common in mechanics) that dots over a variable indicate its time derivatives (so $\dot{x}$ is the same as x_t). The connection to our earlier formulas is that

$$L = \frac{m}{2}\left(\dot{x}^2 + \dot{y}^2 + \dot{z}^2\right) - V(x, y, z) \tag{16.36}$$

plays the role of the quantity originally designated F. (In the present discussion, L has no explicit dependence on t, but such a dependence is a formal possibility.)

Continuing, Eqs. (16.34) for the dependent variable x takes in our current notation the form

$$\frac{\partial L}{\partial x} - \frac{d}{dt}\frac{\partial L}{\partial \dot{x}} = -\frac{\partial V}{\partial x} - \frac{d}{dt}\frac{\partial T}{\partial \dot{x}} = -\frac{\partial V}{\partial x} - \frac{d}{dt}m\dot{x} = -\frac{\partial V}{\partial x} - m\ddot{x} = 0\,,$$

and the complete set of three Euler equations is

$$-\frac{\partial V}{\partial x} = m\ddot{x}\,, \qquad -\frac{\partial V}{\partial y} = m\ddot{y}\,, \qquad -\frac{\partial V}{\partial z} = m\ddot{z}\,. \tag{16.37}$$

We have recovered Newton's law, perhaps a result that could be characterized as overkill.

However, the Lagrangian formulation of mechanics has much to recommend it. First, it enables classical mechanics to be described as depending on a single scalar function, the Lagrangian. Second, the form of the Euler equations does not depend on the coordinate system; for example, in spherical polar coordinates these equations are

$$\frac{\partial L}{\partial r} - \frac{d}{dt}\frac{\partial L}{\partial \dot{r}} = 0\,,$$
$$\frac{\partial L}{\partial \theta} - \frac{d}{dt}\frac{\partial L}{\partial \dot{\theta}} = 0\,, \qquad (16.38)$$
$$\frac{\partial L}{\partial \varphi} - \frac{d}{dt}\frac{\partial L}{\partial \dot{\varphi}} = 0\,.$$

Of course, the form of L does depend upon the coordinate system. But note that once T and V are written in the chosen coordinates (which are not even required to be orthogonal), the Euler equations provide a straightforward recipe leading to the equations of motion. The reader may find that approach to be easier than an attempt to make a direct transformation of Newton's law to a general coordinate system.

■

SEVERAL INDEPENDENT VARIABLES

A prototype problem involving two independent variables was introduced at Eq. (16.32). In that problem, the dependent variable u is a function of both x and y, as must also be its deformation function $\eta(x, y)$. When we generalize the procedure leading to the Euler equation, we get

$$\frac{\partial F}{\partial u} - \frac{d}{dx}\frac{\partial F}{\partial u_x} - \frac{d}{dy}\frac{\partial F}{\partial u_y} = 0\,. \qquad (16.39)$$

We have written d/dx and d/dy as ordinary derivatives despite the fact that the quantities to which they are applied are functions of the two variables x and y. The intent is to avoid any potential confusion as to whether or not these derivatives apply only to the explicit appearance of x or y in F. The derivation of Eq. (16.39) makes it clear that these differentiations are with respect to **all** the dependence on x (or y), both explicit and implicit.

Example 16.3.2. Electrostatics

An electrostatic field $\mathbf{E}$ can be shown to have an energy density given in terms of the electrostatic potential φ by

$$\frac{\varepsilon}{2}\mathbf{E}^2 = \frac{\varepsilon}{2}(\boldsymbol{\nabla}\varphi)^2 = \frac{\varepsilon}{2}(\varphi_x^2 + \varphi_y^2 + \varphi_z^2)\,.$$

A variational principle applicable to the electrostatic field is that, for given values of φ on the boundary of a charge-free region and for φ continuous and differential within the region, the electrostatic energy must be a minimum. This principle defines a problem to which an Euler equation applies, with

$$\delta\,(\text{Energy}) = \int F(\varphi_x, \varphi_y, \varphi_z)\,dx\,dy\,dz = 0\,, \quad \text{with} \quad F = \varphi_x^2 + \varphi_y^2 + \varphi_z^2\,. \qquad (16.40)$$

The Euler equation (noting that F does not depend explicitly upon φ) is

$$\frac{d}{dx}\frac{\partial F}{\partial \varphi_x} + \frac{d}{dy}\frac{\partial F}{\partial \varphi_y} + \frac{d}{dz}\frac{\partial F}{\partial \varphi_z} = 0\,.$$

Evaluating the derivatives, this equation reduces to

$$\frac{d}{dx}\varphi_x + \frac{d}{dy}\varphi_y + \frac{d}{dz}\varphi_z = \varphi_{xx} + \varphi_{yy} + \varphi_{zz} = \nabla^2\varphi = 0\,,$$

which is just the Laplace equation.

■

Exercises

16.3.1. Following the method used for the original derivation of the Euler equation, find its generalization for the situation that the integrand of Eq. (16.7) also contains an explicit dependence on the second derivative of y, y_{xx}. It will be necessary to carry out two integrations by parts on the equation analogous to Eq. (16.9).

Solve the following mechanics problems using the stationary property of the time integral of the Lagrangian, introduced in Example 16.3.1. Use the fact that in Cartesian coordinates the kinetic energy has the form $m[\dot{x}^2 + \dot{y}^2 + \dot{z}^2]/2$, using the dot to indicate differentiation with respect to time.

16.3.2. (a) Find the equation of motion for a particle of mass m moving without friction on the surface of the paraboloid $z = x^2 + y^2$ and under the influence of gravity, which exerts a force mg in the $-z$ direction. This problem is easiest if solved in cylindrical coordinates.

(b) Find a solution to the equation of motion corresponding to circular motion at constant z, and determine the angular velocity of the particle.

16.3.3. Two particles of mass m are connected by a string of fixed length that passes through a small hole in a horizontal circular table. One of the particles can move on the surface of the table; the other is subject to the gravitational force mg and can move vertically up or down. Assume all motion is without friction. See Fig. 16.6.

(a) Find the equations of motion for this system, expressed in terms of the polar coordinates of the particle on the table.

(b) Find a solution to the equations of motion corresponding to uniform circular motion of the particle on the table.

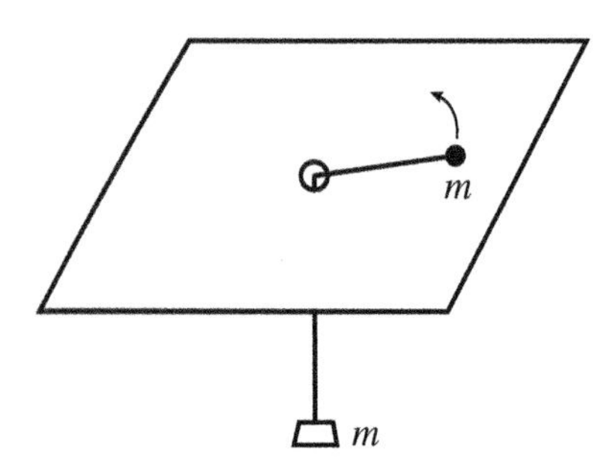

Figure 16.6: Mechanical system of Exercise 16.3.3.

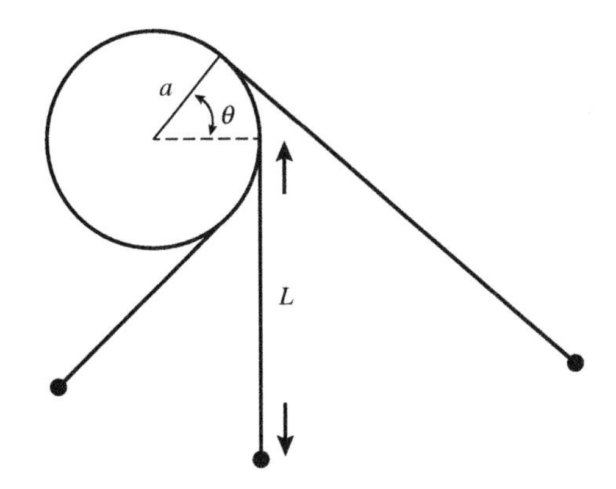

Figure 16.7: Mechanical system of Exercise 16.3.4.

16.3.4. One end of a string is wound around a fixed horizontal cylinder of radius a. A mass m, attached to the free end of the string, executes oscillations in a vertical plane, with the string winding or unwinding as the mass swings back and forth. Assume that the unwound part of the string forms a straight line tangent to the surface of the cylinder, and that when the unwound string is vertical, its unwound length is L. See Fig. 16.7.

Write the Lagrangian in terms of the angular coordinate θ (in the figure) and obtain the equation of motion for this pendulum.

16.4 VARIATION WITH CONSTRAINTS

We have already identified a need to be able to minimize a functional W subject to the constraint that another functional have a fixed value; our prototype for this situation was the determination of the minimum-energy configuration of a flexible chain of fixed length. Now that we have developed a method for finding an unconstrained extremum, we can use Lagrange's method of undetermined multipliers to handle the constraint. If you have not already had occasion to study the method of undetermined multipliers (it was used earlier in this text), it is recommended that you examine its presentation in Appendix G before continuing with the current discussion.

The essence of Lagrange's method is that a function W can be minimized subject to a constraint $L =$ constant by

1. Finding an unconstrained minimum of $W + \lambda L$, where λ is an **undetermined parameter** (i.e., one whose value has not yet been specified), and
2. Assigning λ a value that causes the constraint to be satisfied.

What is relevant here is that the procedure for finding the unconstrained minimum is not dictated by Lagrange's method, so we can use that method in the context of the calculus of variations as legitimately as in the context of algebraic function minimization. The key message of this paragraph is:

> *If W and L are integrals involving an unknown function $y(x)$, we can minimize W subject to a constraint on L simply by applying the Euler equation to $W+\lambda L$ and then choosing λ in a way that causes the constraint to be satisfied.*

Example 16.4.1. Catenary

The shape assumed by a uniform flexible chain suspended from two fixed endpoints will be that which minimizes its gravitational potential. In Eq. (16.3) we saw that this potential is proportional to the integral

$$W = \int y\sqrt{1+y_x^2}\,dx\,,$$

and we must choose $y(x)$ such that $\delta W = 0$ subject to the constraint that the length of the chain is fixed. In terms of $y(x)$, this length is

$$L = \int \sqrt{1+y_x^2}\,dx\,. \tag{16.41}$$

Using the method of Lagrange multipliers, we seek the unconstrained minimum of $U = W - \lambda L$, where λ is an undetermined multiplier. We therefore have

$$\delta U = \int F(y, y_x)\,dx = 0\,, \qquad \text{with} \quad F = (y+\lambda)\sqrt{1+y_x^2}\,. \tag{16.42}$$

We note that F does not contain x, and we therefore use the Euler equation in the alternate form given in Eq. (16.13):

$$F - y_x\frac{\partial F}{\partial y_z} = C_1 \quad \longrightarrow \quad (y+\lambda)\sqrt{1+y_x^2} - y_x\left[\frac{(y+\lambda)y_x}{\sqrt{1+y_x^2}}\right] = C_1\,, \tag{16.43}$$

where C_1 is a constant to be determined later. Equation (16.43) can be solved for y_x:

$$y_x = C_1^{-1}\sqrt{(y+\lambda)^2 - C_1^2}\,. \tag{16.44}$$

This is a separable ODE with general solution

$$y + \lambda = C_1 \cosh\left(\frac{x - C_2}{C_1}\right)\,. \tag{16.45}$$

To find solutions of specific problems, we need to choose C_1, C_2, and λ in a way such that (x_1, y_1) and (x_2, y_2) are the coordinates of the ends of the chain, that Eq. (16.45) is satisfied, and that Eq. (16.41) gives the proper value for the length L of the chain.

It can be cumbersome to find the parameter values for general conditions, but let's consider the relatively simple case that the two ends are at vertical position $y = 0$ and at the horizontal locations $\pm x_0$. The solution must then be an even function of x, so we can set $C_2 = 0$ and $\lambda = C_1 \cosh(x_0/C_1)$, reaching

$$y(x) = C_1\left[\cosh\left(\frac{x}{C_1}\right) - \cosh\left(\frac{x_0}{C_1}\right)\right]\,. \tag{16.46}$$

It remains to choose C_1 in a way that satisfies Eq. (16.41) for L.

From Eq. (16.46) we find

$$y_x = \sinh(x/C_1)\,, \qquad 1 + y_x^2 = 1 + \sinh^2(x/C_1) = \cosh^2(x/C_1)\,,$$

which when substituted into Eq. (16.41) yields

$$L = \int_{-x_0}^{x_0} \cosh(x/C_1)\,dx$$
$$= 2C_1 \sinh(x_0/C_1)\,. \tag{16.47}$$

Equation (16.47) has no solution unless $L \geq 2x_0$; proof of that statement is left to Exercise 16.4.1, and corresponds to the physical requirement that the chain be at least as long as the distance between the points at which its ends are connected. If the condition $L > 2x_0$ is met, Eq. (16.47) can be solved graphically for any desired input, by plotting $L - 2C_1 \sinh(x_0/C_1)$ against C_1. For example, if we take $x_0 = 0.5$ and $L = 2$, a plot like that in Fig. 16.8 indicates that $C_1 = 0.2297$, and using that value we can plot $y(x)$ as given in Eq. (16.46). The result, shown in Fig. 16.9, is a curve called a **catenary**.

■

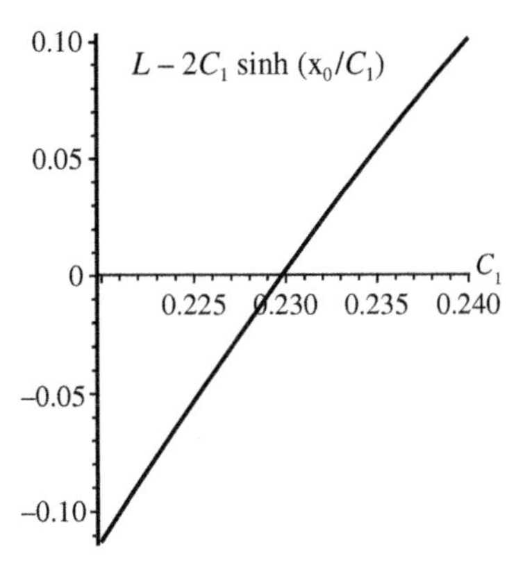

Figure 16.8: Plot of $L - 2C_1 \sinh(x_0/C_1)$ against C_1, Example 16.4.1.

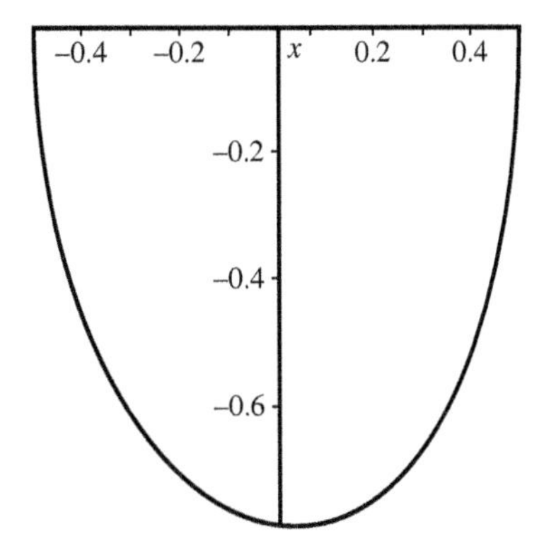

Figure 16.9: Catenary of Example 16.4.1.

Exercises

16.4.1. Prove that there is no real value of C_1 permitting Eq. (16.47) to be satisfied if $L < 2x_0$. We are therefore prevented from obtaining a "solution" when the problem is physically unrealizable.

16.4.2. Plot the catenary with the parameters used for Fig. 16.9 and add to the plot a parabola that passes through the same endpoints and value of $y(0)$.

16.4.3. (a) Find an algebraic equation (not an ODE) for the curve of arc length L that connects the points x_1 and x_2 on the x-axis and encloses the maximum area above the x-axis.

(b) Show that this problem has no solution if $L < |x_2 - x_1|$.

(c) Find and plot a specific solution for $x_1 = 1$, $x_2 = 2$, and $L = 3$.

16.4.4. A fixed volume of water is in a vertical cylindrical tube of radius r_0; the tube (and the water inside) are rotating about the rotation axis of the tube at a constant angular frequency ω. The rotation generates a centrifugal potential energy per unit mass $-\omega^2 r^2/2$, where r is the distance from the rotation axis. Find the curve that describes the location of the water surface as a function of r, considering the combined effect of the centrifugal and gravitational energy.

16.4.5. Show that for a perimeter of fixed length the plane figure of maximum area is a circle.

Additional Readings

Arfken, G. B., Weber, H. J., & Harris, F. E. (2013). *Mathematical methods for physicists* (7th ed.). New York: Academic Press.

Bliss, G. A. (1925). *Calculus of variations*. LaSalle, IL: The Mathematical Association of America, Open Court Publishing Co. (As one of the older texts, this is still a valuable reference for details of problems such as minimum-area problems).

Courant, R., & Robbins, H. (1996). *What is mathematics?* (2nd ed.). New York: Oxford University Press (Chapter VII contains a fine discussion of the calculus of variations, including soap film solutions to minimum-area problems).

Chapter 17

COMPLEX VARIABLE THEORY

In Chapter 3 we encountered complex numbers, complex variables, complex functions, and a variety of formulas. However, the material of that chapter does not give a full indication of the interesting and useful properties associated with the applications of calculus in the complex domain. We start the present chapter by considering the derivative and what is implied by its existence, and then continue to a remarkable series of theorems that underlie the great power of complex analysis. These and related topics are often referred to as **complex variable theory**.

A comprehensive exposition of complex variable theory is far outside the scope of this text. More detailed presentations focused on applications to physical science can be found in the works by Kurala and by Levinson and Redheffer in the Additional Readings. See also the text by Arfken et al. Also included is a text by Churchill et al. that takes a more purely mathematical viewpoint, and an excellent summary at a practical level by Spiegel.

17.1 ANALYTIC FUNCTIONS

When we consider a function $f(x)$ of a real variable x, we define its derivative as the limit

$$\frac{df(x)}{dx} = \lim_{\Delta x \to 0} \frac{f(x + \Delta x) - f(x)}{\Delta x},$$

providing (at least at points in the interior of the region of x under consideration) that this limit exists and has the same value as Δx approaches zero from either direction (i.e., from either positive or negative Δx values). If this requirement is not met at some point x_0, a plot of $f(x)$ will either have a slope discontinuity at x_0 or even be discontinuous there, and we conclude that the derivative does not exist at that point.

In the complex domain, the derivative of a function $f(z)$ is defined as

$$\frac{df(z)}{dz} = \lim_{\Delta z \to 0} \frac{f(z + \Delta z) - f(z)}{\Delta z}, \tag{17.1}$$

and a function possessing a derivative is termed **analytic**; it can also be referred to as **regular** or **nonsingular**.

For the derivative to exist (i.e., for the function to be analytic), the limit in Eq. (17.1) must have the same value for all directions in which Δz can approach zero.

Mathematics for Physical Science and Engineering.
http://dx.doi.org/10.1016/B978-0-12-801000-6.00017-1

Thus, analyticity at a point z is associated with behavior in a small two-dimensional region surrounding z (called a **neighborhood** of z); behavior on a curve through z does not suffice.

Analyticity is a rather restrictive requirement, but it is met by a wide variety of complex functions. If $f(z)$ is specified explicitly as a function of z alone, i.e., not as $u(x, y) + iv(x, y)$ (with $z = x + iy$ and u and v real), functions with derivatives in the real domain will also have them as complex functions. As a near-trivial example, consider z^2. We have

$$\frac{dz^2}{dz} = \lim_{\Delta z \to 0} \frac{(z + \Delta z)^2 - z^2}{\Delta z} = \lim_{\Delta z \to 0} \frac{2z\Delta z + (\Delta z)^2}{\Delta z} = 2z\,.$$

This result is clearly independent of the direction in which Δz approaches zero, and is the same as what we would get if z were restricted to real values.

If one considers a more general complex function $w(z) = u(x, y) + iv(x, y)$, by comparing the results of the choices $\Delta z = \Delta x$ and $\Delta z = i\Delta y$, it can be shown that the derivatives in the x and y directions are equal only if

$$\frac{\partial u}{\partial x} = \frac{\partial v}{\partial y} \quad \text{and} \quad \frac{\partial u}{\partial y} = -\frac{\partial v}{\partial x}\,. \tag{17.2}$$

It can also be proved that Eqs. (17.2) suffice to make the derivative entirely independent of the direction of Δz. These relations are called the **Cauchy-Riemann** equations.

By differentiating the Cauchy-Riemann equations we can obtain

$$\frac{\partial^2 u}{\partial x^2} = \frac{\partial^2 v}{\partial x \partial y}\,,$$

$$\frac{\partial^2 u}{\partial y^2} = -\frac{\partial^2 v}{\partial y \partial x}\,.$$

Adding these equations, we reach the useful result

$$\frac{\partial^2 u}{\partial x^2} + \frac{\partial^2 u}{\partial y^2} = 0\,, \tag{17.3}$$

showing that **any** $u(x, y)$ that is the real part of an analytic function satisfies the two-dimensional Laplace equation. Similar steps enable a proof that $v(x, y)$ also satisfies the Laplace equation, and it is further possible to show that $u(x, y)$ and $v(x, y)$ are complementary in that the lines of constant u (in a Cartesian plot) intersect the lines of constant v at right angles. These observations have made complex variable theory useful for studying electrostatics problems that can be reduced to two dimensions; then u can represent equipotentials and v stream lines (lines of force), or with equal validity, the reverse.

A typical function of a complex variable will have points where it is not analytic (i.e., points where the limit in Eq. (17.1) does not have the same finite value for approach to that point from all directions). Such points are identified as **singular**. For example, the function $1/z$ is singular at $z = 0$, as are the more complicated functions $\exp(1/z)$ and $\ln z$.

Finally, there are functions that are not analytic anywhere in the complex plane. The simplest example of such a function is $z^* = x - iy$, for which the Cauchy-Riemann equations are never satisfied.

Example 17.1.1. Completing an Analytic Function

If we are given the real (or imaginary) part of an analytic function we can complete its definition by using the Cauchy-Riemann equations. Let's find $w(x,y) = u(x,y) + iv(x,y)$, given that $v = 2xy$. From the two Cauchy-Riemann equations,

$$\frac{\partial u}{\partial x} = \frac{\partial v}{\partial y} = \frac{\partial}{\partial y} 2xy = 2x \quad \longrightarrow \quad u = x^2 + g(y)\,,$$

$$\frac{\partial u}{\partial y} = -\frac{\partial v}{\partial x} = -\frac{\partial}{\partial x} 2xy = -2y \quad \longrightarrow \quad u = -y^2 + h(x)\,,$$

where $g(y)$ and $h(x)$ are arbitrary. The most general function satisfying these conditions is

$$u(x,y) = x^2 - y^2 + C\,, \quad \text{and} \quad f(z) = x^2 - y^2 + C + 2ixy = z^2 + C\,.$$

■

Exercises

17.1.1. Use the Cauchy-Riemann equations to determine whether each of the following functions is analytic.

(a) $y + ix$ (b) $x^2 - y^2 + 2ixy$ (c) $x^2 - y^2 - 2ixy$

(d) e^{x+iy} (e) $\dfrac{y - ix}{x^2 + y^2}$ (f) $\dfrac{x - iy}{x^2 + y^2}$

(g) $\ln|z|$ (h) $1/z$ (i) $z^2 - (z^*)^2$

17.1.2. Use the two-dimensional Laplace equation to determine whether each of the following can be the real part of an analytic function.

(a) xy (b) x^3 (c) $x^2 - y^2$

(d) $\sinh y \sin x$ (e) y (f) $\dfrac{y}{(1-x)^2 + y^2}$

(g) $x^2 + y^2$ (h) $\sin x \sin y$ (i) $x^2 y - y^2 x$

17.1.3. For each of the quantities in Exercise 17.1.2 that can be the real part of an analytic function, find the corresponding imaginary part and write the analytic function in terms of $z = x + iy$ alone.

17.1.4. By introducing Δz and letting it approach zero, use techniques similar to those the text used to show that z^2 is analytic, prove that if $f(z)$ and $g(z)$ are analytic,

(a) $\dfrac{d}{dz}\Big[f(z)g(z)\Big] = f'(z)g(z) + f(z)g'(z)$, (b) $\dfrac{d}{dz}\left[\dfrac{f(z)}{g(z)}\right] = \dfrac{f'g - fg'}{g^2}$.

17.1.5. Find the Cauchy-Riemann equations in polar coordinates.

Hint. Set $z = re^{i\theta}$ and work with $f(z) = u(r,\theta) + iv(r,\theta)$. The chain rule may be helpful.

17.1.6. Using the Cauchy-Riemann equations in polar coordinates, determine whether the following functions are analytic.

(a) $|z|$ (b) $\ln z$ (c) $\sqrt{z}$

(d) $e^{|z|}$ (e) $|z|^2$ (f) $|z|^{1/2}e^{i\theta/2}$

17.1.7. If $u(x,y)$ and $v(x,y)$ are the real and imaginary parts of the same analytic function of $z = x + iy$, show that in a plot using Cartesian coordinates, the lines of constant u intersect the lines of constant v at right angles.

17.2 SINGULARITIES

The previous section illustrated some ways in which a function of a complex variable can become singular. It is useful to classify and name the various possible types of singularities.

POLES

A singularity in a function is called a **pole** if the function diverges at a point, at a rate corresponding to a finite negative integer power of z. Poles are classified by the strengths of their divergences (their **orders**) and are also identified by the value of z at which they diverge. Thus, $1/z$ is said to have a pole of **first order**, located at $z = 0$. First-order poles are sometimes called **simple poles**. The function $1/(z-3)^2$ has a **second-order** pole at $z = 3$, and $(z-a)^{-n}$, with n an integer, has an nth order pole at $z = a$.

The examples we have already given are easily recognized, but we need a general definition to enable the resolution of more complicated cases. That definition is the following:

A function $f(z)$ has a pole of order n at $z = z_0$ (where n is a positive integer) if and only if, for all integers m less than n,

$$\lim_{z\to z_0} (z-z_0)^m f(z) \qquad \textit{does not exist (is divergent),}$$

but this limit is finite and nonzero for $m = n$.

Example 17.2.1. Some Poles

Let's find the poles and their orders, of the following functions:

1. $f(z) = z/\sin z\,.$

   The only points at which there can be a pole are those at which $\sin z = 0$; the candidates are $z = n\pi$, where n is an integer. However, we can rule out $z = 0$, because

$$\lim_{z\to 0} f(z) = \lim_{z\to 0} \frac{z}{\sin z} = 1 \quad \text{(if in doubt, check using l'Hôpital's rule).}$$

   For nonzero integer n of either sign, using l'Hôpital's rule,

$$\lim_{z\to n\pi} (z-n\pi)f(z) = \lim_{z\to n\pi} \frac{z(z-n\pi)}{\sin z} = \lim_{z\to n\pi} \frac{2z-n\pi}{\cos z} = \frac{n\pi}{\cos n\pi} = (-1)^n n\pi\,,$$

which is finite and nonzero. We have first-order poles at $z = n\pi$ for all nonzero integers n.

2. $f(z) = \dfrac{1}{z} + \dfrac{1}{z^2}$.

   Note that $f(z)$ and $zf(z)$ both diverge at $z = 0$, but $z^2 f(z)$ is finite and nonzero. Therefore $f(z)$ has a pole of order 2 at $z = 0$. It does **not** have poles of both order 1 and 2.

3. $f(z) = z^{-1/2}$.

   Here $f(z)$ diverges at $z = 0$, but $zf(z) = 0$ at $z = 0$. Thus, this function is singular at $z = 0$, but the singularity is not a pole.

4. $f(z) = \dfrac{1}{2z^{1/2} + 3}$.

   The function $f(z)$ has poles where $z^{1/2} = -3/2$; these are at $z_0 = i\sqrt{3/2}$ and at $-z_0$. For z_0, we examine (using l'Hôpital's rule)

$$\lim_{z \to z_0} \frac{z - z_0}{2z^{1/2} + 3} = \lim_{z \to z_0} \frac{1}{z^{-1/2}} = z_0^{1/2},$$

   which is nonzero. Therefore $f(z)$ has a simple pole at z_0. A similar analysis reveals another simple pole at $-z_0$. There is also a singularity of $f(z)$ at $z = 0$ (see below), but it is not a pole.

■

BRANCH POINTS

Suppose that a function $f(z)$ is multiple-valued and that we assign to it any one of its values at a nonsingular point z_0. For example, if $f(z) = z^{1/2}$ and $z_0 = 1$ we might take $f(z_0) = +1$. This function $f(z)$ is analytic for all nonzero z, and the values we assign to it at points near z_0 must be those that correspond to continuity in $f(z)$; in the current example $f(1+\varepsilon) \approx 1+\varepsilon/2$ or $f(1+i\varepsilon) \approx 1+i\varepsilon/2$, and **not** approximately $-1 + O(\varepsilon)$.

In this way, once we have chosen a specific value for $f(z_0)$, we can identify unique values for $f(z)$ for z near z_0, and by repeating this process we can identify a region on which we have made $f(z)$ single-valued. The single-valued function identified in this way is called a **branch** of $f(z)$. Another branch of $f(z)$ can be obtained by starting from a different value for $f(z_0)$; in the present example an alternate choice is $f(z_0) = -1$.

The process of mapping out a single-valued region for $f(z)$ has limitations. Suppose, with $f(z) = z^{1/2}$, we move in z on the unit circle, where $z = e^{i\theta}$, from $z_0 = 1$, for which $\theta = 0$ and $f(z_0) = 1$, by increasing θ to 2π, which is z_0 again because $e^{2\pi i} = 1$. In Table 17.1 we list z and θ in steps of 90°. We also list values of $f(z)$, first as $e^{i\theta/2}$, and then in the form $x + iy$. An important thing to notice is that when z traverses the entire circle, $f(z) = e^{i\theta/2}$ will have changed from 1 only to $e^{i\pi}$, which is -1, the value of $f(z_0)$ for the other branch of our function. Thus, $f(z)$ is no longer single-valued. Some thought should enable the reader to see that single-valuedness in $f(z)$ cannot be maintained for any region in which we can take a path that encircles $z = 0$, and for that reason $z = 0$ is called a **branch point** of $f(z) = z^{1/2}$. Because we can reach either branch of $f(z)$ by infinitesimal changes in z from $z = 0$, we cannot have a well-defined derivative at that point, and branch points are singularities.

Table 17.1: Values of $z^{1/2}$ on the unit circle.

z	1	i	-1	$-i$	1
θ	0	$\pi/2$	π	$3\pi/2$	2π
$f(z) = e^{i\theta/2}$	e^0	$e^{i\pi/4}$	$e^{i\pi/2}$	$e^{3i\pi/4}$	$e^{i\pi}$
$f(z)$ (simplified)	1	$\frac{1+i}{\sqrt{2}}$	i	$\frac{-1+i}{\sqrt{2}}$	-1

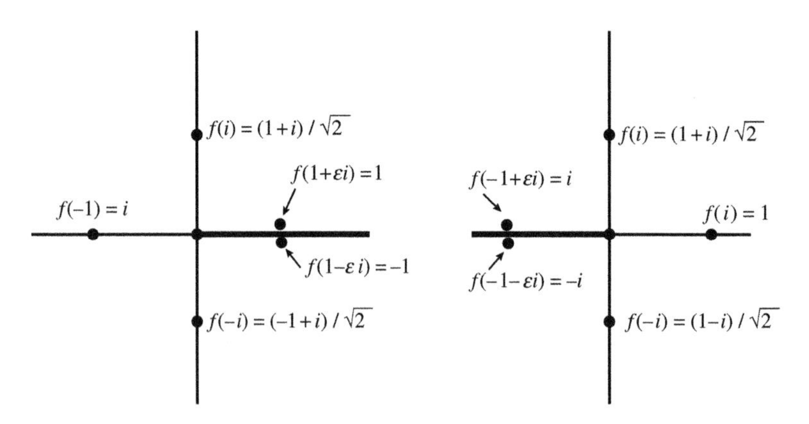

Figure 17.1: Two possible branch cuts for $f(z) = z^{1/2}$ and some corresponding values of $f(z)$.

For the function $f(z) = z^{1/2}$, a single circuit of the branch point at $z = 0$ causes a change in the sign of $f(z)$; a second circuit restores the original sign. Since the original value of $f(z)$ is restored after two circuits of the branch point, the branch point is said to be of order 2. The function $\ln z$ has a branch point at $z = 0$ with the property that a circuit of $z = 0$ adds $2\pi i$ to the value of $\ln z$. Further circuits each add an additional $2\pi i$, so the initial value is never restored. That branch point is said to be of infinite order.

One way to define a region in which a function $f(z)$ with branch point(s) is single-valued is to draw line(s) that it is understood cannot be crossed, thereby preventing the circuit of any branch point. Such lines are called **cut lines** or **branch cuts**. To accomplish their intended purpose, cut lines must start at branch points. For our function $f(z) = z^{1/2}$, two possible choices for a branch cut are shown in Fig. 17.1. The branch cut would have been equally effective if drawn on any path that started at $z = 0$ and continued to infinity; the two choices in the figure are equally valid. However, for some points z we see that $f(z)$ will depend upon the location of the branch cut; note the values of $f(z)$ identified in Fig. 17.1.

Example 17.2.2. Branch Cuts

The function $f(z) = (z^2 - 1)^{1/2}$ has two branch points, at $z_1 = 1$ and at $z_2 = -1$. For further discussion it may be helpful to make the definitions $z - z_1 = r_1\, e^{i\theta_1}$ and $z - z_2 = r_2\, e^{i\theta_2}$, so

$$f(z) = (z - z_1)^{1/2}(z - z_2)^{1/2} = (r_1 r_2)^{1/2}\, e^{i(\theta_1+\theta_2)/2}\,. \tag{17.4}$$

The relationships between the quantities in Eq. (17.4) are shown in Fig. 17.2.

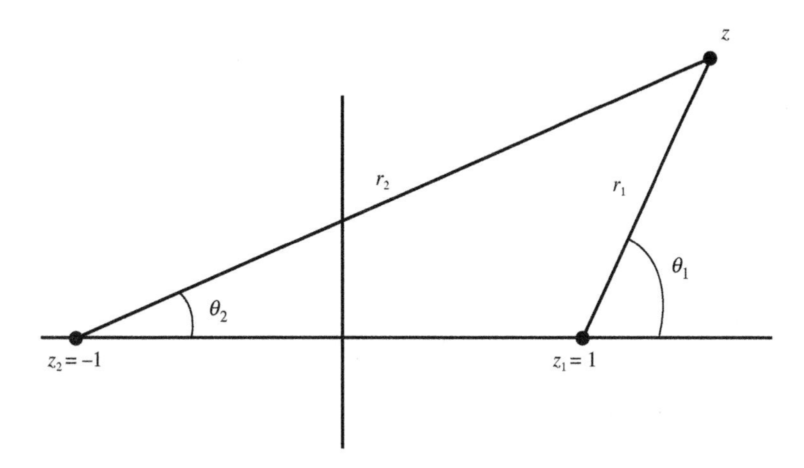

Figure 17.2: Geometry for Example 17.2.2.

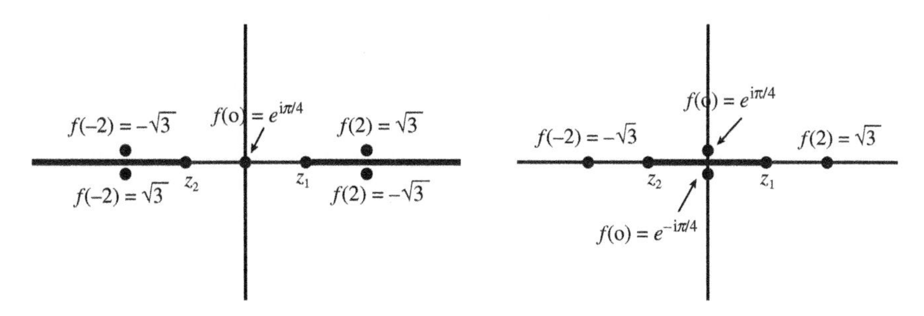

Figure 17.3: Possible branch cuts for $f(z)$ of Example 17.2.2 and some corresponding values of $f(z)$.

Let's choose a branch for $f(z)$ by setting $f(2) = +\sqrt{3}$. Starting from $z = 2$, we can maintain single-valuedness if we vary z along any closed path that encircles neither z_1 nor z_2, because both θ_1 and θ_2 will then return to their original values at the end of the circuit. Somewhat less obvious is that we retain single-valuedness if we encircle **both** z_1 and z_2; for such a circuit $(\theta_1 + \theta_2)/2$ increases by 2π. However, if we encircle z_1 but not z_2, or z_2 but not z_1, a complete circuit will change the sign of $f(z)$.

With the above data in mind, we see that to keep $f(z)$ single-valued, we need to draw branch cuts that prevent the encirclement of either branch point alone, while it does not matter whether we can encircle both branch points. Two different, equally valid ways of accomplishing this objective are shown in Fig. 17.3. They differ in the values assigned to $f(z)$ in parts of the complex plane.

The reader should be cautioned that just because a function has two branch points one cannot conclude that a branch cut connecting them will suffice. Consider, for example, $f(z) = (z^2 - 1)^{1/3}$. This function is not single-valued if one makes a circuit of either, or of both branch points.

■

ESSENTIAL SINGULARITIES

It may happen that a function $f(z)$ is singular at $z = z_0$ (where z_0 is not a branch point), and $(z - z_0)^n f(z)$ is also singular for all finite integers n. Thus, we have a "pole of infinite order," which is officially called an **essential singularity**. An example is

$e^{1/z}$, which has the power-series expansion

$$e^{1/z} = \sum_{n=0}^{\infty} \frac{1}{n! z^n} \,. \tag{17.5}$$

The essential singularity is at $z = 0$.

POINT AT INFINITY

In the complex plane, points infinitely removed from the coordinate origin can be reached by displacement in any direction. The most useful way to consider the behavior of complex functions at infinity is to make a change of variable from z to $Z = 1/z$ and then to study the situation at $Z = 0$. With this transformation, the passage of z toward $+\infty$ corresponds to an approach of Z toward zero from positive values. Passage of z toward $-\infty$ corresponds to Z approaching zero from negative values, and passage of z toward infinity through complex values corresponds to complex-valued approaches to zero by Z. In terms of Z, infinity, however approached, is at $Z = 0$, and it is customary in complex variable theory to refer to $Z = 0$ as the **point at infinity**.

A function may be regular (nonsingular) at the point at infinity; $f(z) = 1/z$, corresponding to $F(Z) = Z$, is regular at $Z = 0$. A function may have a pole at infinity, as for example $f(z) = z^2$; here we have $F(Z) = Z^{-2}$, which has a pole of second order at $Z = 0$.

Multiple-valued functions may have branch points at infinity; for example, $f(z) = z^{1/2}$ corresponds to $F(Z) = Z^{-1/2}$, which has a second-order branch point at $Z = 0$. Thus, either choice of the branch cut for this $f(z)$ (Fig. 17.1) can now be said to connect the two branch points: $z = 0$ and $z = \infty$.

Let's return to Example 17.2.2, where we found two branch points, at $z_1 = 1$ and $z_2 = -1$. In the limit of large $|z|$, $f(z)$, as given by Eq. (17.4), becomes $f(z) \approx z$, so $f(z)$ has a first-order pole at infinity. We now understand why we can make a branch cut that did not go to infinity: there is no branch point there. The alternate pattern with two cut lines approaching infinity from different directions can now be interpreted as having a single cut line that connects the two branch points of our problem via infinity rather than by the shorter direct path.

Some functions $f(z)$ have no singularities at finite values of z; they are sometimes called **entire functions**. However, it can be shown that unless merely a constant, every entire function will be singular at infinity. We already noted that $f(z) = z^2$, which is entire, has a pole at infinity. Other entire functions, such as $f(z) = e^z$, have an essential singularity at infinity: $F(Z) = e^{1/Z}$.

Exercises

17.2.1. Identify all the singularities of the following functions, including any that may be at infinity. For each pole or branch point, also specify its order.

(a) $z^2 + \dfrac{1}{2}$ (b) $\cosh z$ (c) $(z-1)^{-1/2}$

(d) $ze^{1/z}$ (e) $\dfrac{\tan z}{z}$ (f) $\ln(1+z)$

(g) $\left(\dfrac{3}{z+z^{-1}}\right)^2$ (h) $\dfrac{1}{z^3 - 3z^2 + 2z}$ (i) $\dfrac{1}{z^4 + 2z^2 + 1}$

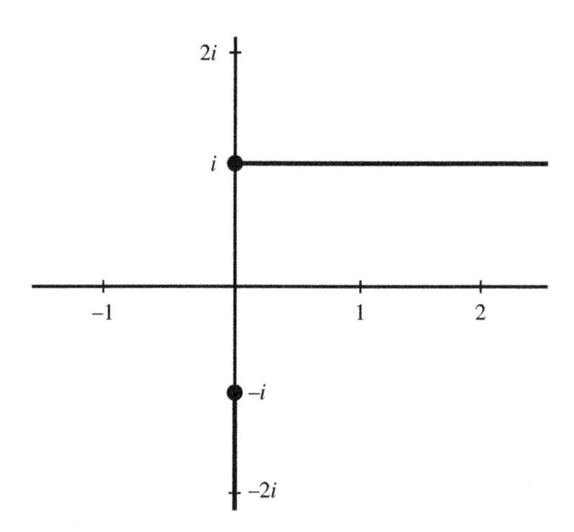

Figure 17.4: Branch cuts for Exercise 17.2.2.

17.2.2. If the function $f(z) = \dfrac{(z^2+1)^{1/2}}{z}$

is made single-valued by making the branch cuts shown in Fig. 17.4, and we take a branch on which $f(1) = \sqrt{2}$, find the values of $f(-1)$, $f(2i)$, and (on each side of the branch cut) $f(-2i)$.

17.2.3. (a) Show that the function $f(z) = \dfrac{1}{[z^2(1-z)]^{1/3}}$

is regular at $z = \infty$, and prove that $f(z)$ can be made single-valued by a branch cut that connects its singularities.

(b) Using the branch cut you found in part (a) and choosing a branch such that $f(-1)$ is real, find the values of $f(i)$, $f(-i)$, and $f(2)$.

17.3 POWER-SERIES EXPANSIONS

Functions of a complex variable can be expanded in Taylor series and in a generalization thereof known as a **Laurent series.**

TAYLOR SERIES

In preparation for making Taylor series expansions of a function $f(z)$ about a point z_0 at which $f(z)$ is analytic, we need a theorem (which we do not prove) stating that an analytic function possesses derivatives of all orders. This result is in contrast with the situation for functions of real variables, where the existence of a first derivative does not guarantee the existence of higher derivatives. The behavior of the complete set of derivatives can also be shown to guarantee the existence of a unique Taylor expansion for $f(z)$ about any point at which $f(z)$ is analytic, and the Taylor expansion can be viewed as a definition of $f(z)$ wherever it converges.

We already know the conditions under which a Taylor series converges. If the series has the form

$$f(z) = \sum_{n=0}^{\infty} a_n (z - z_0)^n ,$$

convergence is assured if

$$S|z - z_0| < 1, \qquad \text{where} \qquad S = \lim_{n\to\infty} \left| \frac{a_{n+1}}{a_n} \right| ,$$

and may be possible for some z for which $S|z - z_0| = 1$. The series will surely diverge if $S|z - z_0| > 1$. We can therefore state that the Taylor series will converge within a circular disk in the complex plane of radius $1/S$, and may converge on, but not beyond the boundary of that disk. It is obvious that a Taylor series for $f(z)$ cannot converge on a disk that contains a singularity of $f(z)$, and it can be shown that the disk of convergence of the Taylor expansion of $f(z)$ about z_0 extends to the singularity of $f(z)$ that is closest to z_0.

The observations of the foregoing paragraph can elucidate a situation that may seem perplexing when viewed through the lens of real-variable theory. The function

$$f(x) = \frac{1}{x^2 + 1}$$

has a power-series expansion about $x = 0$ that only converges for $|x| < 1$, and it may not be obvious why that is so. But now we know (from a complex-variable perspective) that the region of convergence is limited by the poles of $f(z)$ at $\pm i$, corresponding to a disk of convergence of unit radius.

LAURENT SERIES

A Laurent series about a point z_0 includes negative as well as perhaps positive powers of $z - z_0$ and is useful for expanding a function $f(z)$ about a point at which it is singular. Laurent's theorem states that if $f(z)$ is analytic between two concentric circles centered at z_0, it can be expanded in a series of the general form

$$f(z) = \cdots + a_{-3}\,(z - z_0)^{-3} + a_{-2}\,(z - z_0)^{-2} + a_{-1}\,(z - z_0)^{-1}$$
$$+ a_0 + a_1\,(z - z_0) + a_2\,(z - z_0)^2 + a_3\,(z - z_0)^3 + \cdots , \qquad (17.6)$$

with the series convergent in the interior of the annular region between the two circles. The portion of the series with negative powers of $z - z_0$ is called the **principal part** of the expansion.

It is important to realize that if a function $f(z)$ has several singularities at different distances from the expansion point z_0, there will be several annular regions, each with its own Laurent expansion about z_0. Figure 17.5 illustrates a situation in which there are singularities at z_0, z_1, and z_2, so there is a different Laurent expansion valid for each of the three regions R_1, R_2, and R_3.

If the largest negative power of $z - z_0$ in a Laurent expansion for the region immediately around $z = z_0$ is for some finite value of $-n$, then $f(z)$ can be identified as having a pole of order n at z_0. If the values if n extend to $-\infty$, then $f(z)$ has an essential singularity at z_0. In the special case that all the coefficients a_{-n} vanish, we have a Taylor series for $f(z)$ and $f(z)$ is analytic in the neighborhood of z_0.

It is not necessary that the expansion point of a Laurent series be at a singularity, and if there are multiple singularities there may be several different annular regions of analyticity, each with its own Laurent expansion. For a given function, expansion point, and annular region, the Laurent expansion is unique, and may be found by

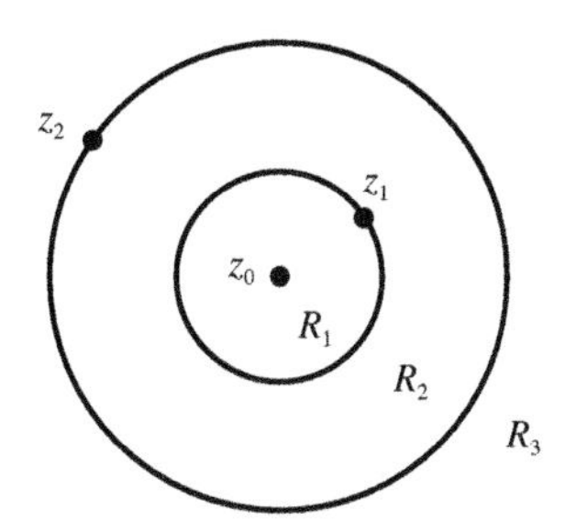

Figure 17.5: A function with these singularities has different Laurent expansions about z_0 for the regions R_1, R_2, and R_3.

any legitimate mathematical method. The coefficients in the expansion are given by general formulas that involve integrals, but it is often easier to proceed in other ways.

Example 17.3.1. Laurent Expansions

1. Frequently occurring instances of Laurent expansions are for functions that are analytic everywhere except for a single singularity at a point $z = z_0$. The region of convergence is then bounded by an infinitesimal circle about z_0 and a circle of infinite radius, as in

$$\frac{e^z}{z^2} = \frac{1}{z^2} + \frac{1}{z} + \frac{1}{2!} + \frac{z}{3!} + \cdots .$$

The expansion, obtained by dividing that for e^z termwise by z^2, shows $z = 0$ to be a pole of order 2. This expansion converges for $0 < |z| < \infty$, the "annular region" of the present problem.

2. Consider expansions about $z = 0$ of

$$f(z) = \frac{1}{z(z-2)} .$$

This $f(z)$ has singularities at $z = 0$ and at $z = 2$, so there will be two Laurent expansions about $z = 0$, one for the region $0 < |z| < 2$, and another for $2 < |z| < \infty$. Our first step for finding both expansions is to write a partial-fraction decomposition of $f(z)$:

$$f(z) = -\frac{1}{2z} + \frac{1}{2(z-2)} .$$

For the region $|z| < 2$, we expand the second partial fraction in powers of z,

$$\frac{1}{2(z-2)} = -\frac{1}{4\left(1-\dfrac{z}{2}\right)} = -\frac{1}{4}\left[1 + \frac{z}{2} + \left(\frac{z}{2}\right)^2 + \cdots\right],$$

which when combined with the first partial fraction yields

$$f(z) = -\frac{1}{2z} - \frac{1}{4}\left[1 + \frac{z}{2} + \left(\frac{z}{2}\right)^2 + \cdots\right].$$

The negative powers of this expansion consist only of a single term and there is therefore no issue as to its convergence; however, that term is singular at $z = 0$

and shows $z = 0$ to be a simple pole. The positive powers form an infinite series that converges for $|z| < 2$, so our Laurent series is convergent for $0 < |z| < 2$.

For the region $|z| > 2$, we expand the second partial fraction in powers of $1/z$, obtaining

$$\frac{1}{2(z-2)} = \frac{1}{2z}\left[1 + \frac{2}{z} + \left(\frac{2}{z}\right)^2 + \cdots\right],$$

leading to

$$f(z) = \frac{1}{4}\left[\left(\frac{2}{z}\right)^2 + \left(\frac{2}{z}\right)^3 + \cdots\right].$$

This series converges for $|z| > 2$, as required. This result tells us nothing about the behavior of $f(z)$ near $z = 0$, as the expansion we are now considering is only applicable for $|z| > 2$.

■

Exercises

17.3.1. For each of the following functions,

- Identify its singularity nearest to $z = 0$,
- Obtain (by methods such as were developed in Chapter 2) its power-series expansion about $z = 0$, and
- Determine the radius of its disk of convergence, verifying that this radius is equal to the distance from the origin to the nearest singularity.

(a) $\dfrac{1}{z - 3i}$ (b) $\dfrac{1}{1+z}$ (c) $\dfrac{z}{z^2+16}$

(d) $\sin z$ (e) $\ln(1-z)$ (f) e^{-iz}

(g) $(1+z^2)^{1/3}$ (h) $\cosh(z-1)$ (i) $\tan^{-1} z$

17.3.2. Find Laurent expansions about $z = 0$ for each of the following functions (valid for a region near $z = 0$), and determine the range of z for which each expansion converges.

(a) $\dfrac{z-1}{z^2}$ (b) $\dfrac{e^{z+1}}{z^3}$ (c) $\dfrac{1}{z(z-3)^2}$

(d) $(z+2)\sin\left(\dfrac{1}{z}\right)$ (e) $\dfrac{\cosh z}{z^5}$ (f) $\dfrac{1}{z(z+2)}$

17.3.3. For each of the expansions in Exercise 17.3.2 that does not converge for $z \to \infty$, find another Laurent expansion about $z = 0$ that converges for larger $|z|$ than the expansion found in that Exercise.

17.4 CONTOUR INTEGRALS

Much of the power of complex analysis stems from a set of integral theorems, largely due to Cauchy. These theorems deal with line integrals of analytic functions $f(z)$ over closed curves denoted by symbols such as C, generically written in notations like

$$\oint_C f(z)\,dz\,.$$

Here the curve C is often called a **contour**, and the line integral is referred to as a **contour integral**. As in earlier sections of this book, the circle on the integral sign serves as a reminder that the curve C is a closed loop. Unless it is specifically indicated otherwise, it is understood that **the contour is to be traversed in the counterclockwise (mathematically positive) direction**. A reversal of the direction of traverse changes the sign of the integral.

CAUCHY'S THEOREM

Cauchy's theorem states that if C is any simple closed curve[1] in the complex plane and $f(z)$ is analytic on C and everywhere in the area enclosed by C, then

$$\oint_C f(z)\,dz = 0\,. \tag{17.7}$$

To prove Cauchy's theorem we use Green's theorem in the plane, as given in Eq. (6.46). Restating that theorem,

$$\int_{\partial A} \left[P(x,y)\,dy + Q(x,y)\,dx\right] = \int_A \left[\frac{\partial P}{\partial x} - \frac{\partial Q}{\partial y}\right] dA\,. \tag{17.8}$$

As before, the notation ∂A denotes the closed curve that is the boundary of the area A; the line integral is to be evaluated along that curve.

Writing $f(z) = u(x,y) + iv(x,y)$ and $dz = dx + idy$, the line integral of Eq. (17.7) takes the form

$$\oint_C f(z)\,dz = \oint_C (u + i\,v)(dx + i\,dy) = \oint_C \left[u\,dx - v\,dy\right] + i\oint_C \left[u\,dy + v\,dx\right]\,. \tag{17.9}$$

We have separated the real and the imaginary contributions to Eq. (17.9).

Our next step is to use Eq. (17.8) to rewrite the line integrals in the right-hand member of Eq. (17.9) as integrals over the area A enclosed by C:

$$\oint_C f(z)\,dz = \int_A \left[-\frac{\partial u}{\partial y} - \frac{\partial v}{\partial x}\right] dA + i\int_A \left[\frac{\partial u}{\partial x} - \frac{\partial v}{\partial y}\right] dA\,. \tag{17.10}$$

A rigorous proof of Cauchy's theorem includes a demonstration that the functions u and v must be differentiable if $f(z)$ is analytic; assuming that to be the case, we accept Eq. (17.10) and continue by noting that each integrand in Eq. (17.10) is an expression that vanishes by virtue of the Cauchy-Riemann equations, Eqs. (17.2). Note that these integrands vanish at all points within A because it has been assumed that $f(z)$ is analytic throughout A. If that analyticity is absent anywhere within A, even if only at a single point, the theorem does not apply.

[1] A "simple closed curve" is one that does not cross itself; a more precise statement of the theorem also excludes curves C with an infinite number of derivative discontinuities (corners), but that is a refinement largely irrelevant for us.

IMPLICATIONS OF CAUCHY'S THEOREM

Line integral independent of path–Consider two different paths (denoted C_1 and C_2), both starting at a point A and ending at a point B, where A, B, both paths, and the region between them are all in the same region of analyticity of a function $f(z)$. See Fig. 17.6. Then

$$\int_{C_1} f(z)\,dz - \int_{C_2} f(z)\,dz = \oint f(z)\,dz = 0\,. \tag{17.11}$$

We have used the fact that taking minus the integral over C_2 is equivalent to traversing C_2 starting at B and ending at A. We can rearrange Eq. (17.11) to

$$\int_{C_1} f(z)\,dz = \int_{C_2} f(z)\,dz\,. \tag{17.12}$$

Equation (17.12) states that the line integral of $f(z)\,dz$ between A and B is invariant with respect to deformation of the path C, so long as the deformation is entirely within a region in which $f(z)$ is analytic.

Deformed path enclosing singularities–Consider now the situation illustrated in the left panel of Fig. 17.7, in which we are integrating $f(z)\,dz$ on either of the closed paths C_1 or C_2 in a region of analyticity that encloses one or more singularities. These singularities can be of any type (poles, branch points, or essential singularities), but if there are branch cuts, they must lie entirely within the "island" containing the singularities. Cauchy's theorem does not let us conclude that the value of this integral is zero, but we are nevertheless able to show that its value is unchanged by deformations that can be reached without leaving the region of analyticity of $f(z)$, and that therefore the integral over C_2 has the same value as that over C_1. Note that even if the singularity is only at a single point, the deformations under discussion here do not include those that move the contour through the singularity.

To prove that the integrals on paths C_1 and C_2 of Fig. 17.7 have the same value (if traversed in the same direction), we modify those paths as shown in the right panel of the figure, by cutting them open at the points marked A and reconnecting them by the lines marked B and B'. We then integrate over C_1, B, C_2, and B' in the directions shown by the arrows, thereby forming a closed loop about a region in which $f(z)$ is entirely analytic, and to which Cauchy's theorem applies. Because the segments B

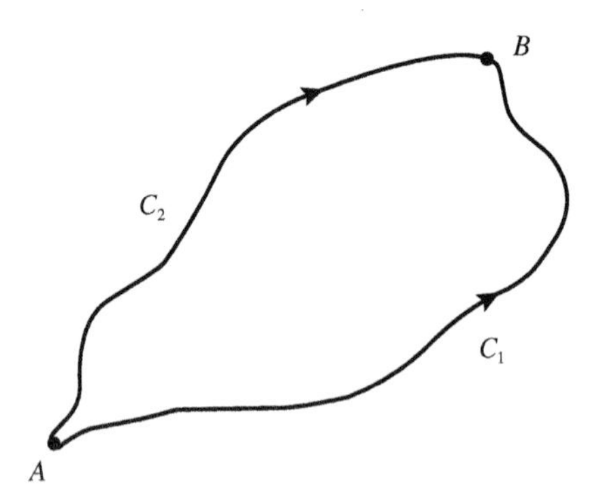

Figure 17.6: Two paths from A to B.

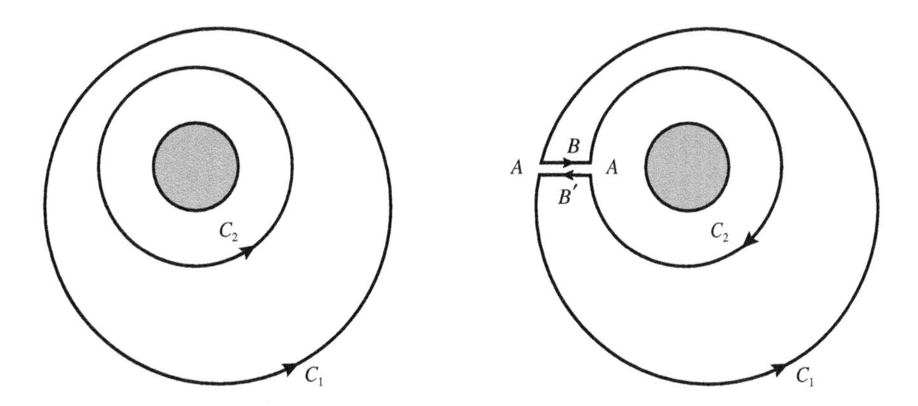

Figure 17.7: Left: Paths enclosing singularities. Right: Modified path for use of Cauchy's theorem.

and B' lie in a region where $f(z)$ is analytic and are close together and traversed in opposite directions, they make no net contribution to our contour integral, while the contributions of C_1 and C_2 (which are traversed in opposite directions) add to zero, as required by Cauchy's theorem. If both C_1 and C_2 are traversed in the same direction, their contributions are therefore shown to be equal.

Summarizing, we have shown that a contour integral enclosing singularities can be subjected to an arbitrary continuous deformation within its region of analyticity without changing its value. For example, a contour surrounding a pole or an essential singularity can be contracted to a circle of arbitrarily small radius without changing the value of the integral.

CAUCHY'S INTEGRAL FORMULA

We now examine a specific contour integral enclosing a singularity. **Cauchy's integral formula** states that

$$f(z_0) = \frac{1}{2\pi i} \oint_C \frac{f(z)}{z - z_0}\, dz\,, \tag{17.13}$$

where $f(z)$ is a function that is analytic on and within C, and z_0 is a complex number that is in the region enclosed by C. Let's prove this formula.

The integrand in Eq. (17.13) is analytic everywhere on and within C except for a simple pole at $z = z_0$, so our first step in analyzing the integral in that equation will be to contract C to a circle of some small radius r about z_0 and to make a change of variables to $z - z_0 = re^{i\theta}$, where the integration is over $0 \le \theta \le 2\pi$. We have

$$\frac{1}{z - z_0} = \frac{e^{-i\theta}}{r} \qquad \text{and} \qquad dz = \frac{\partial\left(z_0 + re^{i\theta}\right)}{\partial\theta}\, d\theta = ire^{i\theta}\, d\theta\,, \tag{17.14}$$

and therefore

$$\oint_C \frac{f(z)}{z - z_0}\, dz = \int_{\theta=0}^{2\pi} f(z_0 + re^{i\theta})\, \frac{e^{-i\theta}}{r}\, ire^{i\theta}\, d\theta = i \int_0^{2\pi} f(z_0 + re^{i\theta})\, d\theta$$

$$= if(z_0) \int_0^{2\pi} d\theta = 2\pi i\, f(z_0)\,.$$

The second line of this equation is obtained by letting r approach zero; $f(z_0+re^{i\theta})$ can then be replaced by $f(z_0)$ because $f(z)$ is analytic at z_0. The final result is equivalent to Eq. (17.13).

A trivial generalization of Eq. (17.13) is that its contour integral evaluates to zero if z_0 is not enclosed by C. In that case we simply have an instance of Cauchy's theorem.

An interesting aspect to Cauchy's integral formula is that it shows that the value of $f(z)$ at any point z_0 within the curve C is entirely determined by the values of $f(z)$ on C. There is no corresponding result in real-variable theory; real functions are not determined over an entire interval by their values at the ends of the interval.

ANOTHER INTEGRAL FORMULA

The technique used to derive Cauchy's integral formula can be used to evaluate integrals of the generic type

$$I_n = \frac{1}{2\pi i} \oint_C (z - z_0)^n \, dz \,, \tag{17.15}$$

where C encloses z_0 and n is an integer (of either sign) or zero. If $n \geq 0$, the integrand is analytic and, by Cauchy's theorem, $I_n = 0$. If n is a negative integer, we make the substitution $z = z_0 + re^{i\theta}$, use the expression for dz from Eq. (17.14), and integrate over a circle of radius r. Equation (17.15) becomes

$$I_n = \frac{1}{2\pi i} \int_0^{2\pi} \left(r^n e^{in\theta}\right) ire^{i\theta} \, d\theta = \frac{r^{n+1}}{2\pi} \int_0^{2\pi} e^{i(n+1)\theta} \, d\theta \,. \tag{17.16}$$

If n is any integer other than -1, the integral over θ evaluates to zero (irrespective of the value of r); we have

$$\int_0^{2\pi} e^{i(n+1)\theta} \, d\theta = \left. \frac{e^{i(n+1)\theta}}{i(n+1)} \right|_0^{2\pi} = 0 \,, \qquad (n \neq -1),$$

while for $n = -1$ we have a result corresponding to Cauchy's integral formula for $f(z) = 1$, namely $I_{-1} = 1$. Summarizing this important result,

$$I_n = \frac{1}{2\pi i} \oint_C (z - z_0)^n \, dz = \delta_{n,-1} \,. \tag{17.17}$$

We can now use Eq. (17.17) to derive a formula for the coefficients in a Laurent (or Taylor) series. If the contour C is chosen to lie entirely within the region where a Laurent (or Taylor) series converges, and we write the series in the form

$$f(z) = \sum_n a_n (z - z_0)^n \,, \tag{17.18}$$

where the range of n is whatever is needed to represent $f(z)$, we can then obtain the coefficient a_m (for any integer m, positive, negative, or zero) by evaluating the following contour integral:

$$a_m = \frac{1}{2\pi i} \oint_C (z - z_0)^{-m-1} f(z) \, dz \,. \tag{17.19}$$

Remember that in using Eq. (17.19), C must be in the region of analyticity for the series whose coefficients are to be calculated.

The proof of Eq. (17.19) is direct; substitution of the series for $f(z)$, Eq. (17.18), produces a series of integrals all of which vanish according to Eq. (17.17) except that with $n = m$, for which the integral (including its prefactor) evaluates to a_m.

As observed previously, direct evaluation of the integral in Eq. (17.19) will often not be the easiest way to establish a Laurent or Taylor series; it is of course legitimate to obtain the series in other ways.

Exercises

17.4.1. Compute by explicit evaluation of the line integrals, for a square path connecting (in the order given) the points $(x = 0, y = 0)$, $(1, 0)$, $(1, 1)$, $(0, 1)$, $(0, 0)$:

$$\text{(a)} \quad \oint (2iy - 1)\, dx\,, \qquad \text{(b)} \quad \oint (x - iy)(dx + idy)\,, \qquad \text{(c)} \quad \oint (z^2 - 1)\, dz\,.$$

17.4.2. Comment on the relation between your answers to Exercise 17.4.1 and the analyticity of the integrands involved.

17.4.3. Compute by explicit evaluation the following contour integrals, for a counterclockwise traverse of the unit circle. Work in polar coordinates, for which $z = re^{i\theta}$:

$$\text{(a)} \;\; x^3 - 3xy^2\,, \qquad \text{(b)} \;\; 3x^2y - y^3\,, \qquad \text{(c)} \;\; z^3\,.$$

(d) Find the real and imaginary parts of z^3 and comment on the relevance of those quantities to your answers for parts (a) through (c).

17.4.4. The integral $\displaystyle\oint_C \frac{dz}{z - 2}$

is zero when the contour C is the square defined in Exercise 17.4.1. Confirm that this is the case by explicit evaluation of the line integral, and explain why this result is to be expected.

Note. Assume all the logarithms you may encounter are to be evaluated on the same branch.

17.4.5. For C the unit circle, evaluate (in any legitimate way) the following contour integrals:

$$\text{(a)} \quad \oint_C \frac{3z^2\, dz}{z - 3i} \qquad \text{(b)} \quad \oint_C \frac{4z^2 - 1}{(2z - 1)^2}\, dz \qquad \text{(c)} \quad \oint_C (z^2 - 3z + 2)\, dz$$

$$\text{(d)} \quad \oint_C \frac{(z + 1)\sin z}{z^2}\, dz \quad \text{(e)} \quad \oint_C \frac{z^2\, dz}{(6z - 1)(z + 6i)} \quad \text{(f)} \quad \oint_C \frac{\sinh z}{z(2z + i)}\, dz$$

17.5 THE RESIDUE THEOREM

Consider now a contour integral in which the integrand is analytic everywhere on and within the contour except for one or more point singularities (i.e., the region enclosed contains no branch cuts). We may deform the contour so that it consists of small circles about the individual singularities, connected by segments whose contributions to the integral are equal and opposite and therefore cancel. We have thereby reduced

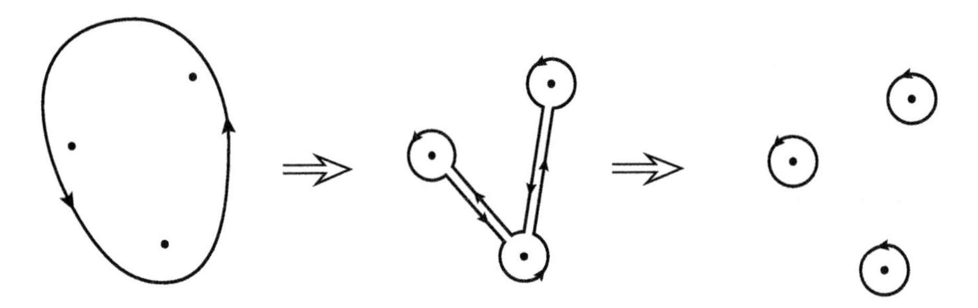

Figure 17.8: Deforming a contour to isolate the enclosed point singularities.

the contour integral into separate integrals that enclose the individual singularities of the integrand. See Fig. 17.8.

Now, for each of the isolated singularities z_i of $f(z)$, introduce a Laurent series of the form given in Eq. (17.18), valid for the immediately surrounding region. Then, integrating $f(z)$ term by term about z_i, we have

$$\oint_C \sum_n a_n(z-z_i)^n = a_{-1}\oint_C \frac{dz}{z-z_i} = 2\pi i\, a_{-1}\,, \tag{17.20}$$

where the integrals of the individual terms of the sum are evaluated using Eq. (17.17). We have used the fact that all these integrals vanish except that involving $1/(z-z_i)$, leaving the interesting and useful result that our overall integral is simply $2\pi i$ times the coefficient of $1/(z-z_i)$ in the Laurent expansion of $f(z)$. This coefficient, a_{-1}, therefore has specific importance for the integral of $f(z)$ around its isolated singularity at z_i, and is called the **residue** of $f(z)$ at z_i.

Combining the integrals about all the singularities z_i, we have the relatively simple final result:

If a curve C encloses a region in which $f(z)$ is analytic except for isolated singularities (poles or essential singularities), the integral of $f(z)$ over C has the value

$$\oint_C f(z)\,dz = 2\pi i \sum (\text{residues of } f(z) \text{ at its singularities within } C). \tag{17.21}$$

The residue of $f(z)$ about $z=z_0$ is defined as the coefficient of $(z-z_0)^{-1}$ in its Laurent series immediately about z_0. Equation (17.21) is the **residue theorem**. *This theorem assumes that the contour is traversed in the positive (counterclockwise) direction. If the traverse is clockwise, the signs of the residues must be reversed.*

FINDING RESIDUES

The residue theorem has great value because, even for relatively complicated functions, it is often easy to identify singular points and the corresponding residues. We also have the simplifying feature that the residue is the only coefficient in the Laurent series that is needed to evaluate contour integrals.

A systematic procedure for obtaining the residue is available when the singularity in question is a pole. If $f(z)$ has a pole of order n at $z = z_0$, then

$$F(z) = \lim_{z\to z_0} (z-z_0)^n f(z)$$

will be analytic and nonzero at z_0, and the residue we seek will be the coefficient of $(z - z_0)^{n-1}$ in the Taylor expansion of $F(z)$ about z_0. From Taylor's theorem, that coefficient will be

$$\frac{1}{(n-1)!} \frac{d^{n-1}}{dz^{n-1}} F(z) ,$$

so an overall formula for the residue of a pole of order n at $z = z_0$ is

$$\text{Residue, pole of order } n \text{ at } z_0 = \frac{1}{(n-1)!} \lim_{z \to z_0} \frac{d^{n-1}}{dz^{n-1}} \Big[(z - z_0)^n f(z) \Big] . \tag{17.22}$$

In fact, a close look at the derivation of Eq. (17.22) would enable us to generalize it to read

$$\text{Residue, pole of any order} \le n \text{ at } z_0 = \frac{1}{(n-1)!} \lim_{z \to z_0} \frac{d^{n-1}}{dz^{n-1}} \Big[(z - z_0)^n f(z) \Big] . \tag{17.23}$$

One feature of Eq. (17.23) is that it works even if one does not know the order of the pole, providing that n is chosen to be large enough.

While Eq. (17.22) or (17.23) is always available to find the residue at a pole, it is sometimes easier just to make a Laurent expansion and read out its residue term. However, we cannot use Eq. (17.22) for an essential singularity because there is no finite n for which that equation applies. Then the Laurent expansion may be the easiest route to the residue. If all else fails, one can (at least in principle) use the integral formula for a_{-1}, Eq. (17.19).

Example 17.5.1. Some Residues

Let's compute residues for a number of typical situations:

1. $f(z) = \dfrac{z^2 - 5}{z - 3}$.

   We start by observing that $z^2 - 5$ is nonsingular and nonzero at $z = 3$. We therefore conclude that the only singularity of $f(z)$ for finite z is a simple pole at $z = 3$. From Eq. (17.22), we have

$$(\text{Residue at } z = 3) = \lim_{z \to 3} \left[(z - 3) \frac{z^2 - 5}{z - 3} \right] = (z^2 - 5)\Big|_{z=3} = 4 . \tag{17.24}$$

2. $f(z) = \dfrac{g(z)}{z - 3}$, with $g(z)$ any function that is nonsingular at $z = 3$.

   Note that in general the formula for the residue simply returns $g(3)$. In fact, this result is to be expected because the integral giving the residue corresponds to Cauchy's integral formula.

3. $f(z) = \dfrac{g(z)}{2z - 5}$, with $g(z)$ nonsingular.

   Here the pole is at $z = 5/2$; note that a useful way to write $f(z)$ is

$$f(z) = \frac{g(z)}{2\left(z - \frac{5}{2}\right)} ,$$

   showing that the residue at $z = 5/2$ is $\dfrac{1}{2} g(5/2)$.

4. $f(z) = \dfrac{z-5}{\tan z}$.

   Here the singularities of $f(z)$ are at $n\pi$, $n = 0,\ \pm 1,\ \pm 2, \dots$ Using Eq. (17.22) and applying l'Hôpital's rule,

$$(\text{Residue at } z = n\pi) = \lim_{z \to n\pi} \frac{(z - n\pi)(z-5)}{\tan z} = \frac{n\pi - 5}{\sec^2 n\pi} = n\pi - 5 \,. \qquad (17.25)$$

5. $f(z) = \dfrac{z^2 + 3z + 2}{(z-1)^2}$.

   Here $f(z)$ has a pole of order 2 at $z = 1$. Applying Eq. (17.22),

$$(\text{Residue at } z = 1) = \frac{d}{dz}\left[(z-1)^2 \frac{z^2+3z+2}{(z-1)^2}\right]_{z=1} = 2z + 3\Big|_{z=1} = 5 \,. \quad (17.26)$$

6. $f(z) = \dfrac{\sin z}{z^4}$.

   In this case it is much easier to insert an expansion for $\sin z$ than to use the general formula for the residue. We write

$$f(z) = \frac{1}{z^4}\left[z - \frac{z^3}{3!} + \frac{z^5}{5!} - \cdots\right], \qquad (17.27)$$

   from which we read out the coefficient of z^{-1}, namely $-1/3!$.

7. $f(z) = \dfrac{z+1}{(z^{1/2}-1)^2}$.

   Here $f(z)$ has a branch point at $z = 0$ and a pole of order 2 at $z = 1$; the branch cut can be chosen in a way that does not go through $z = 1$. One point illustrated by this example is that the presence of a multiple-valued quantity does not necessarily prevent a function from having pole(s).

   We now proceed to find the residue. Note that in order for there to be a pole, we must be on the branch of $z^{1/2}$ and of $f(z)$ for which $z^{1/2} = +1$ at the pole. We must keep this information in mind when we evaluate the residue.

   Applying Eq. (17.22) and remembering that all powers of z evaluate to unity at $z = 1$, we find that it is necessary to apply l'Hôpital's rule twice:

$$(\text{Residue at } z = 1) = \lim_{z \to 1}\left[\frac{(z-1)^2(z+1)}{(z^{1/2}-1)^2}\right]$$

$$= \lim_{z \to 1}\left[\frac{3z^2 - 2z - 1}{1 - z^{-1/2}}\right] = \frac{6z-2}{z^{-3/2}/2}\Bigg|_{z=1} = 8 \,. \qquad (17.28)$$

■

SYMBOLIC COMPUTATION

In practice, one would not often use symbolic computing to obtain residues as a step toward the evaluation of a contour integral, because it would be more natural to proceed directly to the symbolic computation of the desired integral. However, it is possible that your symbolic program cannot evaluate the integral, or you may want to check a residue computation.

Both our symbolic systems know how to find residues. The residue of *expr*, a function of z, at the point $z0$, is given by

MAPLE:	`residue(` *expr*`, z = z0 )`
MATHEMATICA:	`Residue[` *expr*`, {z, z0} ]`

Let's check some of the results from Example 17.5.1:

MAPLE:	`residue( (z-5)/tan(z), z=Pi);`	$\pi - 5$
	`residue( (z+1)/(z^(1/2)-1)^2, z=1);`	8
MATHEMATICA:	`Residue[ Sin[z]/z^4, {z,0} ]`	$-\dfrac{1}{6}$
	`Residue[ (z^2+3*z+2)/(z-1)^2, {z,1} ]`	5

CONTOUR INTEGRAL EVALUATIONS

We proceed now to some preliminary examples of the use of the residue theorem for the evaluation of contour integrals.

Example 17.5.2. Multiple Poles

We are to evaluate

$$I = \oint_C \frac{e^{i\pi z}}{z(2z-1)(z-2)}\, dz\,,$$

with C the unit circle.

The integrand has poles at $z = 0$, $z = 1/2$, and $z = 2$. Those at 0 and 1/2 lie within the contour; that at $z = 2$ is outside. Our integral I is therefore given as

$$I = 2\pi i \Big[(\text{residue at } z = 0) + (\text{residue at } z = 1/2) \Big] .$$

The residue at $z = 2$ does not contribute to I.

$$(\text{residue at } z = 0) = \left. \frac{e^{i\pi z}}{(2z-1)(z-2)} \right|_{z=0} = \frac{1}{2}\,,$$

$$(\text{residue at } z = 1/2) = \left. \frac{e^{i\pi z}}{z(2)(z-2)} \right|_{z=1/2} = \frac{e^{i\pi/2}}{-3/2} = -\frac{2i}{3}\,.$$

Using these residues, we find $I = \pi\left(\dfrac{4}{3} + i\right)$.

■

Example 17.5.3. Contour Avoiding Branch Cut

We next evaluate

$$I = \oint_C \frac{\ln z}{z^2+1}\, dz\,,$$

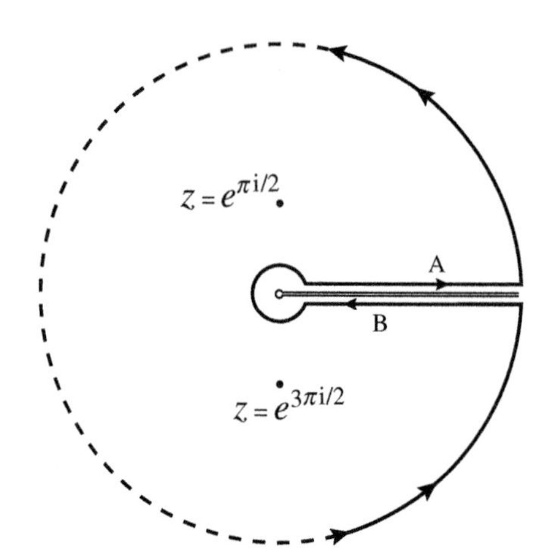

Figure 17.9: Contour that avoids a branch cut.

where the contour has the complicated shape given in Fig. 17.9. It is assumed that the large circular arc is "at infinity," i.e., at a very large distance from the origin, so it will encircle all the poles of the integrand at finite z.

The branch cut causes $\ln z$ to be single-valued, and for definiteness we choose the branch of the integrand for which $\ln z$ is real just above the cut (i.e., it does not contain $2n\pi i$ with nonzero n). Note that the contour has been drawn in a way that causes the branch cut to be outside the area that is enclosed, so we may use the residue theorem to evaluate the integral. (The reason such a contour integral may be of interest will become apparent in the next section of the text. For now, just view this Example as an abstract problem.)

If we identify z^2+1 as $(z-i)(z+i)$, we can see that the integrand of I has simple poles at $z=\pm i$; both poles are enclosed by the contour. Our contour integral therefore has the value

$$I = 2\pi i\Big[(\text{residue at } z=i) + (\text{residue at } z=-i)\Big].$$

When we calculate these residues, we must be careful to use the correct value of $\ln z$. Writing $z = re^{i\theta}$, we have already stated that on the real axis above the cut we are using the branch of $\ln z$ that is real. This choice corresponds for complex z (above the cut) to $\ln z = \ln r + i\theta$. We must therefore use this same form for all evaluations of $\ln z$, inserting the θ values that can be reached continuously from $\theta = 0$ on the positive real axis. In particular, at $z = i = e^{i\pi/2}$, the appropriate value of $\ln z$ is $i\pi/2$. At $z=-i=e^{3i\pi/2}$, we must take $\ln z = 3i\pi/2$, and **not** $-i\pi/2$ (the latter can only be reached by a path that goes outside the region of integration and crosses the branch cut).

Now that we have established the way in which $\ln z$ is to be calculated, we can compute the residues needed to evaluate I:

$$(\text{residue at } z=i) = \left.\frac{\ln z}{z+i}\right|_{z=i} = \frac{\pi i/2}{2i} = \frac{\pi}{4},$$

$$(\text{residue at } z=-i) = \left.\frac{\ln z}{z-i}\right|_{z=-i} = \frac{3\pi i/2}{-2i} = -\frac{3\pi}{4},$$

leading to $I = 2\pi i\left(\frac{\pi}{4} - \frac{3\pi}{4}\right) = -\pi^2 i$.

■

Exercises

17.5.1. Using any convenient method of hand computation, find the residues of the following functions at all their finite singularities. Then check your work using symbolic computing.

(a) $\dfrac{1}{z(2z-3)}$ (b) $\dfrac{1}{(z-1)(z-5)}$ (c) $\dfrac{2z+1}{z^2-z-2}$

(d) $\dfrac{z-3}{z^2+1}$ (e) $\left(\dfrac{1+z}{1-z}\right)^2$ (f) $\dfrac{(z-1)^2}{(z+2)^3}$

(g) $\dfrac{z^2}{(z^2+1)^2}$ (h) $\dfrac{z^2}{(1-3z)(z-3)}$ (i) $\dfrac{\sin z}{z^3}$

(j) $\dfrac{1}{e^{2z}+1}$ (k) $\dfrac{\sin z}{\pi - 2z}$ (l) $\dfrac{z^6+1}{z^3(z-2)}$

(m) $\dfrac{\cosh z - 1}{z^7}$ (n) $\dfrac{e^{-z}}{3\cosh z - 5}$ (o) $\dfrac{e^{2z}-2z-1}{z^3}$

(p) $\dfrac{e^z}{z^3+1}$ (q) $\dfrac{\cos z + 1}{(\pi - z)^3}$ (r) $\dfrac{e^{iz}}{4z^2+1}$

17.5.2. Use the residue theorem to evaluate the contour integral of each of the functions in Exercise 17.5.1, with the contour a circle of radius 5/2 with center at $z = 0$.

17.5.3. A multiple-valued function may have singularities that only occur on specific branches. Functions containing $z^{1/2}$ have two branches, which we can distinguish by their values of $z^{1/2}$ at $z = 1$.

Identify the value of z and the branch or branches at which the following functions have poles, and for each pole compute its residue.

(a) $\dfrac{z^2}{2z^{1/2}+1}$, (b) $\dfrac{z^{3/2}}{z - z^{1/2} - 2}$, (c) $\dfrac{z^{1/2}}{3z-1}$.

17.5.4. The only finite-z singularities of the function $\Gamma(z)$ are at zero and the negative integers. Find a general formula for the residues of $\Gamma(z)$ at $z=-n$, $n=0$, $1, 2, \ldots$

Hint. Relate $\Gamma(-n+z)$ to $\Gamma(1+z)$.

17.6 EVALUATION OF DEFINITE INTEGRALS

An important application of complex variable theory is to the evaluation of definite integrals with real integrands and real integration limits. In fact, many such integrals of importance in physical science are extremely difficult to evaluate by any other method. In most cases, evaluations become practical because the integral under study can be related to a contour integral whose value can be established using the residue theorem.

A convenient way to develop this topic is by presenting a series of examples that illustrate the techniques involved.

INTEGRALS ON RANGE $(0,\,2\pi)$

The integrals under consideration here are on the range $\theta = (0, 2\pi)$, with integrands that contain trigonometric functions of θ. The basic idea of the evaluation method is to make the substitution $z = e^{i\theta}$, so values of θ correspond to points z on the unit circle in the complex plane. An integral in θ from 0 to 2π then corresponds in z to a complete counterclockwise traverse of the unit circle. Our real integral over θ has been converted into a closed contour integral over z in the complex plane.

The basic relationships needed to carry out the conversion from θ to z are the following:

$$z = e^{i\theta}, \qquad dz = ie^{i\theta}\,d\theta \quad \text{or} \quad d\theta = \frac{dz}{iz}, \tag{17.29}$$

and the trigonometric identities (from Chapter 3)

$$\begin{aligned} \cos\theta &= \frac{e^{i\theta}+e^{-i\theta}}{2} = \frac{z+z^{-1}}{2}, \\ \sin\theta &= \frac{e^{i\theta}-e^{-i\theta}}{2i} = \frac{z-z^{-1}}{2i}. \end{aligned} \tag{17.30}$$

Trigonometric functions of integer multiples of θ also have convenient forms in terms of z:

$$\cos n\theta = \frac{z^n+z^{-n}}{2}, \qquad \sin n\theta = \frac{z^n-z^{-n}}{2i}. \tag{17.31}$$

Example 17.6.1. Trigonometric Integral

Let $I = \displaystyle\int_0^{2\pi} \frac{\cos 2\theta\, d\theta}{5-4\cos\theta}$.

Using the relationships in Eqs. (17.29) and (17.30), we rewrite I as

$$I = \oint \frac{\frac{1}{2}\left(z^2+z^{-2}\right)}{5-2\left(z+z^{-1}\right)}\,\frac{dz}{iz}, \tag{17.32}$$

where the contour is the unit circle. Equation (17.32) simplifies to

$$I = \frac{i}{4}\oint \frac{(z^4+1)\,dz}{z^2(z-2)(z-\frac{1}{2})}.$$

This expression for I can now be evaluated using the residue theorem.

The integrand has simple poles at $z = 2$ and $z = 1/2$, and a pole of order 2 at $z = 0$. The poles at $z = 0$ and $z = 1/2$ are inside the contour and we will need the residues of the integrand at those points, but the pole at $z = 2$ is outside the contour and need not be considered further. Thus, we compute

$$(\text{residue at } z=0) = \frac{d}{dz}\left[\frac{z^4+1}{(z-2)\left(z-\frac{1}{2}\right)}\right]_{z=0} = \frac{5}{2},$$

$$(\text{residue at } z=1/2) = \left[\frac{z^4+1}{z^2(z-2)}\right]_{z=1/2} = -\frac{17}{6}.$$

We then apply the residue theorem and evaluate the integral:

$$I = 2\pi i\left(\frac{i}{4}\right)\left[\frac{5}{2}-\frac{17}{6}\right] = \frac{\pi}{6}.$$

■

Exercises

17.6.1. Evaluate the following definite integrals:

(a) $\displaystyle\int_0^{2\pi}\frac{d\theta}{5+4\cos\theta}$ (b) $\displaystyle\int_0^{2\pi}\frac{d\theta}{13-5\sin\theta}$

(c) $\displaystyle\int_0^{2\pi}\frac{d\theta}{3+2\cos\theta+\sin\theta}$ (d) $\displaystyle\int_0^{2\pi}\frac{d\theta}{(5-3\sin\theta)^2}$

(e) $\displaystyle\int_0^{2\pi}\frac{\cos 3\theta\, d\theta}{5-4\cos\theta}$ (f) $\displaystyle\int_0^{2\pi}\frac{d\theta}{(2-\cos\theta)^2}$

17.6.2. Consider the integral

$$I = \int_0^{2\pi}\frac{d\theta}{1+b\cos\theta},$$

with b a real constant restricted to the range $0 < b < 1$.

(a) Show that when converted to a contour integral in $z = e^{i\theta}$, the integrand has two poles. Prove that one pole lies within the unit circle, while the other is outside.

(b) Show that the integral has the value $I = \dfrac{2\pi}{\sqrt{1-b^2}}$.

INTEGRALS ON RANGE $(-\infty, \infty)$

In contrast to the trigonometric integrals considered earlier, we do not make a change of variables to convert a real integral on $(-\infty, \infty)$ to a complex integral on a closed contour. Instead, we regard our real integral as a corresponding complex integral along the real axis, and seek to close the contour by an arc which contributes a known (preferably zero) value to the contour integral.

Example 17.6.2. A Simple Integral

We want to evaluate the integral

$$I = \int_0^{\infty}\frac{dx}{1+x^2} = \frac{1}{2}\int_{-\infty}^{\infty}\frac{dx}{1+x^2}$$

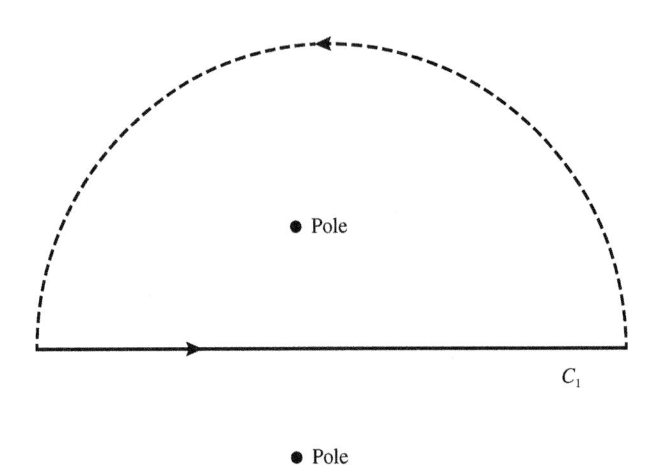

Figure 17.10: Contour closed by arc in upper half-plane, for function with poles at $\pm i$.

by identifying it as part of a contour integral that can be treated using the residue theorem. As is often the case, our original integral was for the range $(0, \infty)$ but we use the fact that its integrand is even to extend the range to $(-\infty, \infty)$, which is a better starting point for setting up a tractable contour integral.

We therefore consider the contour integral

$$J = \frac{1}{2} \oint_{C_1} \frac{dz}{1+z^2}$$

for the contour shown in Fig. 17.10. The arc in the upper half-plane has radius R, with $R \to \infty$. The portion of the contour along the real axis therefore corresponds to the integral I, and I and the contour integral J are related by

$$J = I + \frac{1}{2} \int_{\text{arc at } R} \frac{dz}{1+z^2} \,.$$

The integral over the arc at radius R is best written in polar coordinates: $z = Re^{i\theta}$, with $dz = iRe^{i\theta} d\theta$. We get (in the limit of large R)

$$\int_{\text{arc at } R} \frac{dz}{1+z^2} = \int_0^\pi \frac{iRe^{i\theta}\, d\theta}{1+R^2 e^{2i\theta}} \approx \frac{i}{R} \int_0^\pi e^{-i\theta}\, d\theta \quad \longrightarrow \quad 0\,. \tag{17.33}$$

We see that the integrand approaches zero at large $|z|$ rapidly enough that there is no contribution to the contour integral from the large arc, and we have the simple result that $I = J$.

Our remaining task is to evaluate the contour integral. Writing

$$\frac{1}{1+z^2} = \frac{1}{(z-i)(z+i)}\,,$$

we see that the integrand of J has two poles, one at $+i$ and the other at $-i$. The locations of these poles are marked in Fig. 17.10. We note that only the pole at $z = i$

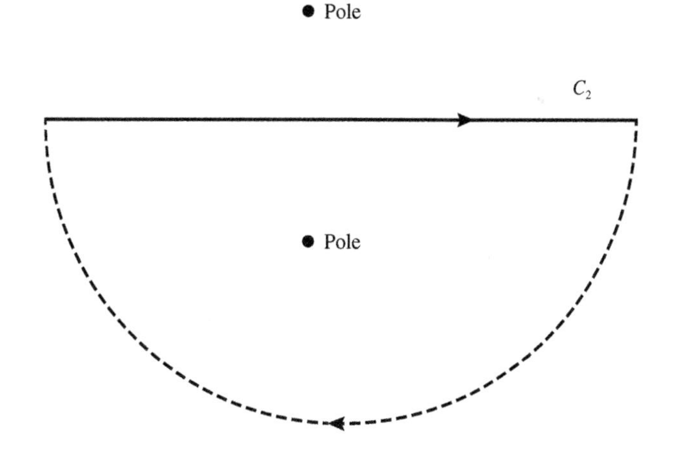

Figure 17.11: Contour closed by arc in lower half-plane, for function with poles at $\pm i$.

is within the contour; we have

$$(\text{residue at } z = i) = \frac{1}{2}\left[\frac{1}{z+i}\right]_{z=i} = \frac{1}{2}\frac{1}{2i} = \frac{1}{4i}.$$

Now applying the residue theorem,

$$I = J = 2\pi i\left(\frac{1}{4i}\right) = \frac{\pi}{2}. \tag{17.34}$$

We could have closed the contour for this example with an arc in the lower half-plane, as shown in Fig. 17.11. The contribution from that arc also vanishes. The contour C_2 encloses only the pole at $z = -i$, and that pole is encircled in the **clockwise** direction, so its residue should be taken with a minus sign. The reader can verify that when these differences are taken into consideration the lower-arc contour yields (as it must) the same result for the integral I, namely $\pi/2$.

The integral I is simple enough that it can easily be evaluated directly, thereby checking our foray into complex analysis. We have

$$\int_0^\infty \frac{dx}{1+x^2} = \tan^{-1} x\Big|_0^\infty = \frac{\pi}{2},$$

in agreement with Eq. (17.34).

The importance of this example, of course, is that the contour-integral evaluation can be applied when elementary integrations are impractical or impossible.

■

ZERO CONTRIBUTIONS FROM LARGE ARCS

In the foregoing example we found that the large arc closing the contour made no contribution to the integral. Similar situations arise for many contour integrals, and it is useful to identify the general conditions under which integrals over such arcs vanish. If we look again at Eq. (17.33), we can see that the property causing the arc integral to vanish is that the integrand approaches zero for large $|z|$ faster than $1/|z|$. That is to be expected because the arc length is proportional to $|z|$ and the arc length times the magnitude of the integrand then tends to zero for large $|z|$. This observation is worth capturing as a theorem:

If C is a full or partial circular arc of radius R with center at the origin and $\lim_{R\to\infty} Rf(z) = 0$ for all points z on C, then

$$\lim_{R\to\infty} \int_C f(z)\,dz = 0\,. \tag{17.35}$$

A companion result, known as **Jordan's lemma**, deals with the frequently occurring situation when a large arc closes the contour for an integral whose integrand contains a complex exponential. We state the lemma without proof.

If C is a full or partial semicircle of radius R in the upper half-plane with center at the origin, and $\lim_{R\to\infty} f(z) = 0$ for all points z on C, and $a > 0$, then

$$\lim_{R\to\infty} \int_C e^{iaz} f(z)\,dz = 0\,. \tag{17.36}$$

The theorem as stated applies only when C is in the upper half-plane because e^{iaz} becomes negligible only when the imaginary part of z is large and positive. Note that if $a < 0$, then Jordan's lemma applies when C is a full or partial semicircle in the **lower** *half-plane because that is where e^{iaz} becomes small.*

The statements of both these theorems make it clear that they continue to apply if the arc is less than a full circle or semicircle.

Example 17.6.3. Jordan's Lemma

Consider the integral

$$I = \int_0^\infty \frac{\cos x}{x^2+1}\,dx = \frac{1}{2}\int_{-\infty}^\infty \frac{\cos x}{x^2+1}\,dx\,.$$

Now write $\cos x$ in terms of complex exponentials, getting

$$I = \frac{1}{4}\left[\int_{-\infty}^\infty \frac{e^{ix}\,dx}{x^2+1} + \int_{-\infty}^\infty \frac{e^{-ix}\,dx}{x^2+1}\right].$$

Next, convert each of these integrals along the real axis into a contour integral. The contour can be closed for the first integral by an arc in the upper half-plane; for the second integral we must close the contour by an arc in the lower half-plane. Since these arcs do not contribute to their respective integrals, we have, writing $z^2+1 = (z-i)(z+i)$

$$I = \frac{1}{4}\oint_{C_1} \frac{e^{iz}\,dz}{(z-i)(z+i)} + \frac{1}{4}\oint_{C_2} \frac{e^{-iz}\,dz}{(z-i)(z+i)}\,,$$

where C_1 and C_2 are the contours shown in Figs. 17.10 and 17.11. We note that C_1 encircles the simple pole at $z = i$ (but not the pole at $z = -i$), while C_2 encircles $z = -i$ but not $z = i$. Remembering that C_2 is traversed in the negative direction, we apply the residue theorem, obtaining

$$I = \frac{1}{4}(2\pi i)\left\{\left[\frac{e^{iz}}{z+i}\right]_{z=i} - \left[\frac{e^{-iz}}{z-i}\right]_{z=-i}\right\} = \frac{\pi}{2e}\,. \tag{17.37}$$

Note that to use Jordan's lemma we had to close the contour for each term in the half-plane for which the exponential becomes small when z is far from the real axis. We also had to pay attention to the direction in which each pole is encircled.

■

Exercises

17.6.3. Verify the final formula, Eq. (17.37), given for the integral I in Example 17.6.3.

17.6.4. Evaluate, using contour integration methods:

(a) $\displaystyle\int_0^\infty \frac{dx}{x^4+1}$ (b) $\displaystyle\int_0^\infty \frac{x^2\,dx}{x^4+16}$

(c) $\displaystyle\int_0^\infty \frac{dx}{(x^2+1)^2}$ (d) $\displaystyle\int_0^\infty \frac{dx}{x^4+x^2+1}$

(e) $\displaystyle\int_{-\infty}^\infty \frac{dx}{(x^2+4x+5)^2}$ (f) $\displaystyle\int_0^\infty \frac{\cos 2x\,dx}{9x^2+4}$

(g) $\displaystyle\int_0^\infty \frac{x\sin x}{x^2+1}\,dx$ (h) $\displaystyle\int_0^\infty \frac{\cos x\,dx}{1+x^2+x^4}$

17.6.5. One can evaluate $I = \displaystyle\int_0^\infty \frac{\ln(x^2+1)}{x^2+1}\,dx$

by writing it as

$$I = \frac{1}{2}\int_{-\infty}^\infty \frac{\ln(x+i)}{x^2+1}\,dx + \frac{1}{2}\int_{-\infty}^\infty \frac{\ln(x-i)}{x^2+1}\,dx\,,$$

and then converting each of the above integrals into a contour integral in which the branch point of its integrand is outside the contour (and the branch cut can also be chosen to be entirely outside). In addition, for both integrals, one should choose the branch of the logarithm that causes it to have real values on the real line.

Following the above suggestions, show that

$$I = \pi \ln 2\,.$$

INTEGRATIONS ON CIRCULAR SECTORS

Integrals that cannot be extended to the range $(-\infty, \infty)$ require different evaluation approaches. One technique that is frequently useful is to identify a contour on which two lines at different angles make related contributions to an integral.

Example 17.6.4. Circular Sector Contour

Consider

$$I = \int_0^\infty \frac{dx}{x^3+1}\,.$$

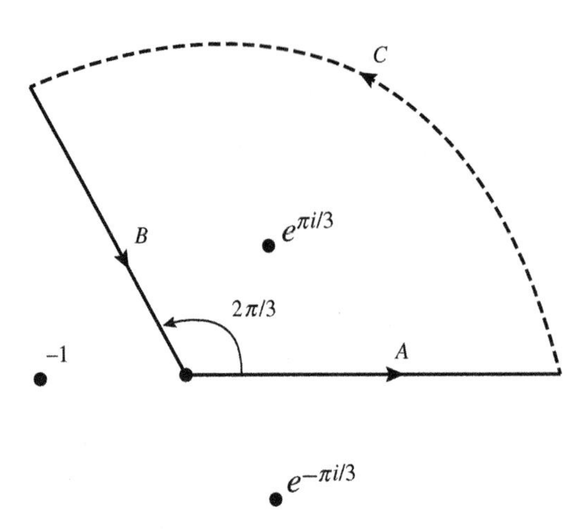

Figure 17.12: Contour for Example 17.6.4.

The integrand of I is not an even function, and in fact the entire integrand does not have definite parity. However, it does have the feature that z^3 assumes the same set of values on the ray at angle $\theta = 2\pi/3$ as it does on the positive real axis. We therefore look closely at the integral

$$J = \oint_C \frac{dz}{z^3+1}$$

for the contour C shown in Fig. 17.12. The contribution to J from the segment A of the figure is just the integral I whose value we seek. To find the contribution from segment B, we note that for this segment $z = re^{2\pi i/3}$, so $dz = e^{2\pi i/3}\,dr$, and the integration of this segment is from $r = \infty$ to $r = 0$. We also observe that

$$z^3 = (r\,e^{2\pi i/3})^3 = r^3 e^{2\pi i} = r^3\,,$$

and we get

$$\int_B \frac{dz}{z^3+1} = \int_\infty^0 \frac{e^{2\pi i/3}\,dr}{r^3+1} = -e^{2\pi i/3}\int_0^\infty \frac{dr}{r^3+1} = -e^{2\pi i/3}\,I\,.$$

From our first theorem on circular arcs, Eq. (17.35), the large arc of Fig. 17.12 makes no contribution to J, so we have

$$J = I - e^{2\pi i/3} I\,, \qquad \text{or} \quad I = \frac{J}{1 - e^{2\pi i/3}}\,.$$

All that remains to finish this example is to use the residue theorem to evaluate J.

The integrand of J has poles at the three values of z where $z^3 = -1$; they are at $z = e^{\pi i/3}$, $z = -1$, and $z = e^{-\pi i/3}$. The locations of these poles are marked in Fig. 17.12. Only the pole at $e^{\pi i/3}$ lies within the contour. We therefore need only

$$(\text{residue at } z_1 = e^{\pi i/3}) = \lim_{z\to z_1} \frac{z - z_1}{z^3+1} = \left.\frac{1}{3z^2}\right|_{z=z_1} = \frac{1}{3e^{2\pi i/3}}\,.$$

We then find

$$J = \frac{2\pi i}{3e^{2\pi i/3}} \qquad \text{and} \qquad I = \frac{2\pi i}{3e^{2\pi i/3}\left(1 - e^{2\pi i/3}\right)} = \frac{\pi}{3}\left(\frac{2i}{e^{2\pi i/3} - e^{4\pi i/3}}\right)\,.$$

We know that this expression must simplify to something real (I is a real integral), and that may motivate us to notice that $e^{4\pi i/3} = e^{-2\pi i/3}$, so our expression for I reduces to

$$I = \frac{\pi}{3}\,\frac{2i}{e^{2\pi i/3} - e^{-2\pi i/3}} = \frac{\pi}{3}\,\frac{1}{\sin 2\pi/3} = \frac{\pi}{3\sin 120^\circ} = \frac{2\pi}{3\sqrt{3}}\,.$$

■

Exercises

17.6.6. Use contour integration to evaluate the following:

$$\text{(a)}\quad \int_0^\infty \frac{x\,dx}{x^3+1}\,, \qquad\qquad \text{(b)}\quad \int_0^\infty \frac{dx}{x^5+1}\,.$$

17.6.7. The Fresnel integrals of infinite argument,

$$F_c(\infty) = \int_0^\infty \cos u^2\,du\,, \qquad F_s(\infty) = \int_0^\infty \sin u^2\,du\,,$$

arise in optics problems. One way to obtain $F_c(\infty)$ and $F_s(\infty)$ starts by recognizing that

$$I = F_c(\infty) + iF_s(\infty) = \int_0^\infty \frac{e^{ix}}{2x^{1/2}}\,dx\,.$$

This integral can be evaluated using the contour shown in Fig. 17.13. The integration along the positive real axis evaluates to I; the integration along the imaginary axis can be identified as proportional to $\Gamma(1/2)$.

Verify that $I = F_c(\infty) + iF_s(\infty)$, show that the small and large arcs of the contour do not contribute to I, and complete the evaluation of $F_c(\infty)$ and $F_s(\infty)$.

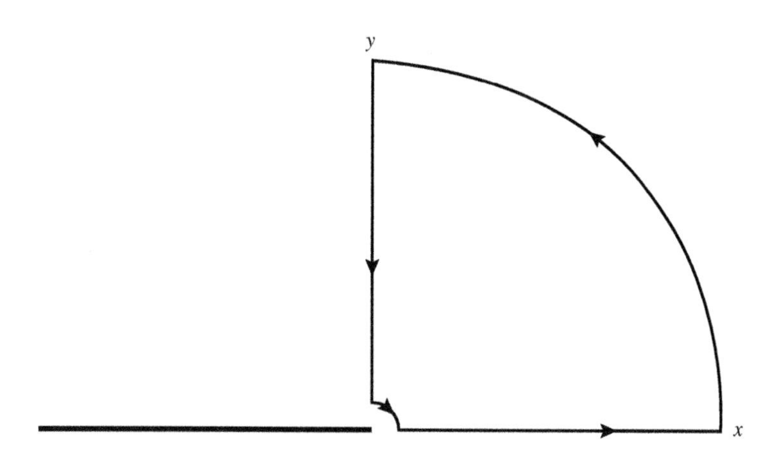

Figure 17.13: Contour for Exercise 17.6.7.

USE OF PERIODICITY

Sometimes we can close a contour by taking advantage of the periodicity of an integrand.

Example 17.6.5. Periodic Integrand

Consider

$$I = \int_{-\infty}^{\infty} \frac{dx}{\cosh x}\,.$$

We cannot usefully close the contour for this integral by a circular arc either in the upper or the lower half-plane because the integrand is periodic in the imaginary part of z, with an oscillation magnitude comparable to unity for small $\mathcal{R}e(z)$. However, we can exploit the periodicity by closing the contour in a different way. A useful contour for this problem is the rectangle shown in Fig. 17.14, which we refer to as C. The horizontal lines of the contour are the real axis and at y (the imaginary part of z) equal to π.

In the limit of large R (see the figure), the vertical sides of the contour C make negligible contributions, and the traversal (toward positive x) along the real axis yields the integral I. To analyze the contribution to the integral of the line at $y = \pi$, note that

$$\cosh(x + \pi i) = \cosh x \cosh \pi i + \sinh x \sinh \pi i = -\cosh x\,.$$

This result tells us that integration (toward negative x) along the line $y = \pi$ will also yield the integral I. The considerations of the present paragraph correspond to the formula

$$J = \oint_C \frac{dz}{\cosh z} = 2I\,.$$

The rest of the story is familiar: we need now to evaluate J using the residue theorem. Writing $\cosh z = \cos iz$, we see that $\cosh z$ has zeros at $iz = (n + 1/2)\pi$ for all integer n, which means that there will be one zero of $\cosh z$ (and therewith a pole of the integrand) within the contour, namely at $z = i\pi/2$. This pole is of first order, with a residue computed (using l'Hôpital's rule) as

$$(\text{residue at } z_0 = i\pi/2) = \lim_{z\to z_0} \frac{z - z_0}{\cosh z} = \left.\frac{1}{\sinh z}\right|_{z=i\pi/2} = \frac{1}{i}\,.$$

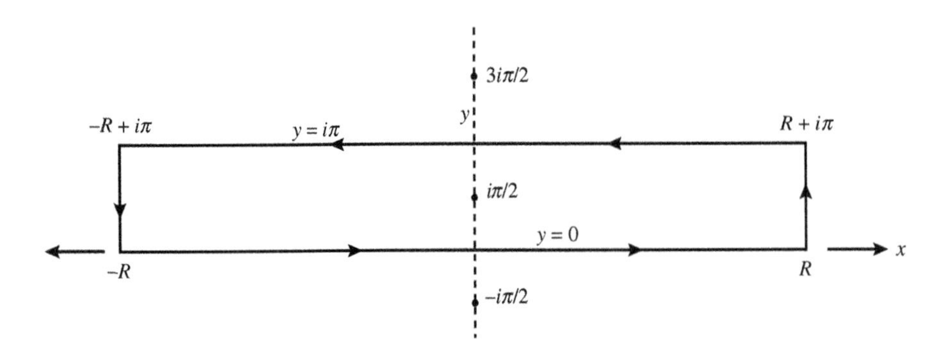

Figure 17.14: Contour for Example 17.6.5.

We conclude the calculation:

$$J = 2\pi i \left(\frac{1}{i}\right) = 2\pi\,; \qquad I = \frac{J}{2} = \pi\,.$$

■

Exercises

17.6.8. Evaluate by contour integration, using a contour similar to that of Fig. 17.14:

$$I = \int_{\infty}^{\infty} \frac{e^{sx}\,dx}{1+e^{x}}\,,$$

where $0 < s < 1$.

Hint. Place the upper horizontal line of the contour at a value of y that makes the integral on this line proportional to the integral along the real axis.

17.6.9. Evaluate by contour integration, using the contour of Fig. 17.14:

$$\int_0^{\infty} \frac{\cos px}{\cosh x}\,dx\,,$$

where $-1 < p < 1$.

DEALING WITH BRANCH POINTS

You may think that branch points are a nuisance because we must avoid integrating through the branch cuts that accompany them. While it is true that we cannot integrate through any branch cuts, their presence sometimes aids us in evaluating integrals.

Example 17.6.6. Gamma Function Reflection Formula

When we studied the beta function in Chapter 9, we developed the following formula, presented there as Eq. (9.39):

$$B(p,q) = \int_0^{\infty} \frac{r^{p-1}}{(1+r)^{p+q}}\,dr\,.$$

We would now like to evaluate the integral corresponding to $B(p,1-p)$:

$$B(p,\,1-p) = \int_0^{\infty} \frac{r^{p-1}}{1+r}\,dr\,, \tag{17.38}$$

which converges for $0 < p < 1$. We are interested in this integral because it will help us to establish the reflection formula for the gamma function, presented without proof as Eq. (9.11).

Our approach to the integral of Eq. (17.38) is to consider the contour integral

$$J = \oint_C \frac{z^{p-1}}{1+z}\,dz\,, \tag{17.39}$$

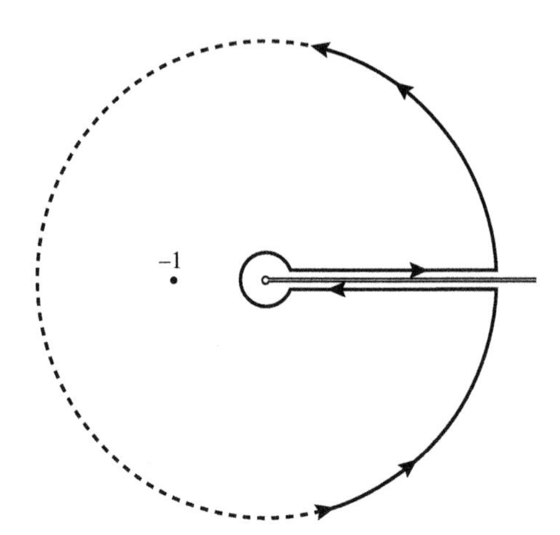

Figure 17.15: Contour for Example 17.6.6.

where C is the relatively complicated contour shown in Fig. 17.15. Because $p-1$ is not an integer the integrand of Eq. (17.39) has a branch point at $z=0$, and we have chosen to make the necessary branch cut along the positive real axis. The integrand also has a pole at $z=-1$, and the contour we have chosen encircles a region that is entirely analytic except for that pole. We shall now show that the arcs at infinity and near zero do not contribute to the contour integral, and that the lines on the two sides of the branch cut can be related to the real integral $B(p,1-p)$.

Dealing first with the two arcs, that at infinity does not contribute because, at large $|z|$, the integrand scales asymptotically as $|z|^{p-2}$, and $p-2<-1$. The small arc does not contribute because at small $|z|$ the integrand scales as $|z|^{p-1}$, while the circumference of the circle is proportional to $|z|$ and the product of these two quantities scales as $|z|^p$, a positive power of $|z|$.

We turn now to the contributions of the lines above and below the branch cut. Above the branch cut, we take z^{p-1} to be on the branch given in polar coordinates as $r^{p-1}e^{i(p-1)\theta}$, i.e., a choice that causes z^{p-1} to have the real value r^{p-1} on the upper edge of the branch cut, where $\theta=0$. Noting that z on the upper edge of the cut is just r, the integral for the segment above the branch cut can therefore be recognized as $B(p,1-p)$.

To determine what is going on below the branch cut, we need to vary z^{p-1} continuously within the region of analyticity. That means we must reach the lower side of the branch cut by moving counterclockwise in z around $z=0$, so when this travel is complete the value of z^{p-1} will have become $r^{p-1}e^{i(p-1)2\pi}$. Of course, z will then again just be r. The integral for the segment below the branch cut therefore takes the form (noting that its direction is from $r=\infty$ to $r=0$)

$$\text{(Integral, line below cut)} = e^{i(p-1)2\pi}\int_\infty^0 \frac{r^{p-1}}{1+r}\,dr = -e^{i(p-1)2\pi}B(p,1-p)\,.$$

Combining the two nonzero contributions to the contour integral, we get

$$J=\left(1-e^{2\pi i(p-1)}\right)B(p,1-p)\,.$$

The branch cut has played a valuable role here. In its absence the integrals for these two lines would have added to zero.

We proceed, as usual, by using the residue theorem to evaluate the contour integral. The residue at the pole within the contour is z^{p-1} at $z = -1$, computed using the branch of the integrand that is consistent with our earlier choice. We therefore take

$$(\text{residue at } z = e^{+\pi i}) = e^{\pi i(p-1)}\,,$$

leading to $J = 2\pi i\, e^{\pi i(p-1)}$ and

$$B(p, 1-p) = \frac{2\pi i\, e^{\pi i(p-1)}}{1 - e^{2\pi i(p-1)}} = \frac{\pi}{\sin(1-p)\pi}\,. \tag{17.40}$$

To make a connection to the usual form of the gamma function reflection formula, note that

$$B(p, 1-p) = \frac{\Gamma(p)\Gamma(1-p)}{\Gamma(1)} = \Gamma(p)\Gamma(1-p)\,,$$

and use the fact that $\sin(1-p)\pi = \sin p\pi$. We recover Eq. (9.11):

$$\Gamma(p)\Gamma(1-p) = \frac{\pi}{\sin p\pi}\,.$$

■

Exercises

17.6.10. Use contour integration to evaluate the following integrals. The parameter s can be assumed to be in the range $0 < s < 1$.

(a) $\displaystyle\int_0^\infty \frac{x^s\,dx}{x^2+1}$ (b) $\displaystyle\int_0^\infty \frac{x^s\,dx}{(x+1)^2}$

(c) $\displaystyle\int_0^\infty \frac{\ln x\,dx}{x^2+1}$ (d) $\displaystyle\int_0^\infty \frac{x^s \ln x}{x^2+1}\,dx$

17.7 EVALUATION OF INFINITE SERIES

The residue theorem can be used to evaluate infinite series, providing that the series meets several conditions, of which the most important are that the series be for the index range $(-\infty, \infty)$ and that the general term of the series (viewed as a function of a complex variable) have only point singularities. It then often turns out to be possible to make the series correspond to a sum of residues that can be evaluated by application of the residue theorem.

The key to this method for series evaluation is a theorem, due to Mittag-Leffler, that indicates how to make **pole expansions** of functions that possess only simple poles, with each term in the expansion containing an individual pole. The use of Mittag-Leffler's theorem leads to the following expansions, useful for dealing with summations:

$$\pi \cot \pi z = \sum_{n=-\infty}^{\infty} \frac{1}{z-n}\,, \tag{17.41}$$

$$\pi \csc \pi z = \sum_{n=-\infty}^{\infty} \frac{(-1)^n}{z-n}\,, \tag{17.42}$$

$$\pi \tan \pi z = -\sum_{n=-\infty}^{\infty} \frac{1}{z - (n + \frac{1}{2})}, \tag{17.43}$$

$$\pi \sec \pi z = -\sum_{n=-\infty}^{\infty} \frac{(-1)^n}{z - (n + \frac{1}{2})}. \tag{17.44}$$

Let's begin our analysis by seeing how we can use Eq. (17.41) to evaluate the sum

$$S = \sum_{n=-\infty}^{\infty} f(n), \tag{17.45}$$

where the complex version of $f(n)$, namely $f(z)$, is analytic for all finite z except possibly for poles, which cannot be at real integers if S is to converge. Our approach is to consider the function

$$F(z) = [\pi \cot \pi z] f(z).$$

If we integrate $F(z)$ over a circular contour C of very large radius R (where R is not an integer), the contour will encircle all the poles of $\pi \cot \pi z$ for $|n| < R$, and all the poles of $f(z)$. The poles of $\pi \cot \pi z$ can be seen from Eq. (17.41) to be at the integers n, each corresponding to a residue of $F(z)$ equal to $f(n)$. Using the information thus far developed, we have

$$\oint_C F(z)\,dz = 2\pi i \sum_{n=-\infty}^{\infty} f(n) + 2\pi i \sum \text{residues of } F(z) \text{ at the poles of } f(z)$$

$$= 2\pi i \left[S + \sum \text{residues of } F(z) \text{ at the poles of } f(z) \right]. \tag{17.46}$$

We next note that S will certainly be a convergent sum if the values of $f(n)$ go to zero for large $|n|$ faster than $|1/n|$, and the integral of $F(z)$ on a circular contour of noninteger R will then go to zero at large R. Under those conditions we have

$$2\pi i \left[S + \sum \text{residues of } F(z) \text{ at the poles of } f(z) \right] = 0,$$

equivalent to the very useful formula

$$\sum_{n=-\infty}^{\infty} f(n) = -\sum \text{residues of } [\pi \cot \pi z] f(z) \text{ at the poles of } f(z). \tag{17.47}$$

Example 17.7.1. A Simple Sum

Let's evaluate the sum

$$S = \sum_{n=1}^{\infty} \frac{1}{n^2 + a^2},$$

in which we assume that $a \neq 0$. We need to relate S to a sum for $n = (-\infty, \infty)$, which is easily accomplished because the summands of S are even functions of n. More specifically, we note that

$$\sum_{n=-\infty}^{-1} \frac{1}{n^2 + a^2} = S,$$

so we can write

$$S' = \sum_{n=-\infty}^{\infty} \frac{1}{n^2 + a^2} = \frac{1}{a^2} + 2S\,.$$

We now evaluate S' using Eq. (17.47), with $f(z) = (z^2 + a^2)^{-1}$. The function $f(z)$ has simple poles at $\pm ia$, so we need the following residues of $[\pi \cot \pi z](z^2 + a^2)^{-1}$:

$$(\text{residue at } z = ia) = \pi \lim_{z \to ia} \left[\frac{z - ia}{z^2 + a^2} \cot \pi z \right] = \frac{\pi}{2ia} \cot i\pi a = -\frac{\pi}{2a} \coth \pi a\,,$$

$$(\text{residue at } z = -ia) = \pi \lim_{z \to -ia} \left[\frac{z + ia}{z^2 + a^2} \cot \pi z \right] = \frac{\pi}{-2ia} \cot(-i\pi a) = -\frac{\pi}{2a} \coth \pi a\,.$$

Therefore,

$$S' = -\Big[(\text{residue at } z = ia) + (\text{residue at } z = -ia) \Big] = \frac{\pi}{a} \coth \pi a\,,$$

$$S = \frac{1}{2} \left(S' - \frac{1}{a^2} \right) = \frac{\pi}{2a} \coth \pi a - \frac{1}{2a^2}\,.$$

■

The other quoted series stemming from Mittag-Leffler's theorem lead to related summation formulas, so we repeat Eq. (17.47) and add to it these additional formulas to make a unified set of equations useful for series evaluations:

$$\sum_{n=-\infty}^{\infty} f(n) = -\sum \text{residues of } [\pi \cot \pi z] f(z) \text{ at the poles of } f(z), \quad (17.48)$$

$$\sum_{n=-\infty}^{\infty} (-1)^n f(n) = -\sum \text{residues of } [\pi \csc \pi z] f(z) \text{ at the poles of } f(z), \quad (17.49)$$

$$\sum_{n=-\infty}^{\infty} f(n + \tfrac{1}{2}) = \sum \text{residues of } [\pi \tan \pi z] f(z) \text{ at the poles of } f(z), \quad (17.50)$$

$$\sum_{n=-\infty}^{\infty} (-1)^n f(n + \tfrac{1}{2}) = \sum \text{residues of } [\pi \sec \pi z] f(z) \text{ at the poles of } f(z). \quad (17.51)$$

Example 17.7.2. An Alternating Sum

Consider the series

$$S = \frac{1}{1^3} - \frac{1}{3^3} + \frac{1}{5^3} - \cdots = \sum_{n=0}^{\infty} \frac{(-1)^n}{(2n+1)^3}\,.$$

We note that

$$S = \frac{1}{2} S'\,, \qquad \text{with} \qquad S' = \sum_{n=-\infty}^{\infty} \frac{(-1)^n}{(2n+1)^3}\,.$$

The summation S' is of the type represented by Eq. (17.51), with $f(z) = 1/(2z)^3$. We note that the only singularity of $f(z)$ is a pole at $z = 0$, and therefore

$$S' = \text{residue of } [\pi \sec \pi z]/8z^3 \text{ at } z = 0.$$

We compute this residue, noting that the singularity is a pole of order 3:

$$S' = \frac{\pi}{8}\frac{1}{2!}\left.\frac{d^2 \sec \pi z}{dz^2}\right|_{z=0} = \frac{\pi^3}{16}.$$

Thus,

$$S = \sum_{n=0}^{\infty} \frac{(-1)^n}{(2n+1)^3} = \frac{1}{2}S' = \frac{\pi^3}{32}.$$

■

SUMMATIONS WITH MISSING TERMS

Sometimes we may have a summation in which a singular term has been omitted. For illustrative purposes let's assume the missing term is for the index value $n = 0$. Returning to the discussion at Eqs. (17.46) and (17.47), we rewrite the statement that all the residues of $F(z)$ sum to zero as

$$\left(\sum_{n\neq 0} \text{residue of } F(z) \text{ at } z = n\right) + \Big(\text{all other residues of } F(z)\Big) = 0. \qquad (17.52)$$

This equation makes it clear that a summation that does not include the term with $n = 0$ will have as its value (minus) the sum of the other residues of F, i.e., those at the singularities of $f(z)$, plus (if not already included) the residue at $z = 0$.

Example 17.7.3. Summation with Missing Term

Consider $\zeta(2)$, a zeta function defined by the summation $\zeta(2) = \sum_{n=1}^{\infty} \frac{1}{n^2}$.

A summation related to $\zeta(2)$ is

$$S = \sum_{n\neq 0} \frac{1}{n^2}, \quad \text{so} \quad \zeta(2) = \frac{1}{2}S.$$

The sum S corresponds to the nonzero integer values of $f(z) = z^{-2}$, and we note that $f(z)$ has only one singularity, a pole at $z = 0$. Thus, based on the discussion at Eq. (17.52), we have

$$S = -\left(\text{residue of } \frac{\pi \cot \pi z}{z^2} \text{ at } z = 0\right).$$

The pole of $[\pi \cot \pi z]/z^2$ at $z = 0$ is of order 3. We could use the general formula for its residue, but in the present problem a simpler approach is to use the power-series expansion for the cotangent, namely

$$\cot u = \frac{1}{u} - \frac{1}{3}u - \frac{1}{45}u^3 - \cdots,$$

leading to

$$\frac{\pi \cot \pi z}{z^2} = \frac{\pi}{z^2}\left[\frac{1}{\pi z} - \frac{\pi z}{3} - \frac{(\pi z)^3}{45} - \cdots\right],$$

from which we read out the residue: $-\pi^2/3$.

Thus, $S = \pi^2/3$ and we get the well-known result

$$\zeta(2) = \sum_{n=1}^{\infty} \frac{1}{n^2} = \frac{S}{2} = \frac{\pi^2}{6}. \tag{17.53}$$

■

Exercises

17.7.1. Use contour integration methods to evaluate the following sums.

(a) $\displaystyle\sum_{n=0}^{\infty} \frac{1}{(2n+1)^2} = \frac{1}{1^2} + \frac{1}{3^2} + \frac{1}{5^2} + \frac{1}{7^2} + \cdots$

(b) $\displaystyle\sum_{n=1}^{\infty} \frac{1}{n^4} = \frac{1}{1^4} + \frac{1}{2^4} + \frac{1}{3^4} + \frac{1}{4^4} + \cdots$

(c) $\displaystyle\sum_{n=1}^{\infty} \frac{1}{n^6} = \frac{1}{1^6} + \frac{1}{2^6} + \frac{1}{3^6} + \frac{1}{4^6} + \cdots$

(d) $\displaystyle\sum_{n=0}^{\infty} \frac{1}{(2n+1)^4} = \frac{1}{1^4} + \frac{1}{3^4} + \frac{1}{5^4} + \frac{1}{7^4} + \cdots$

(e) $\displaystyle\sum_{n=1}^{\infty} \frac{1}{n(n+2)} = \frac{1}{1\cdot 3} + \frac{1}{2\cdot 4} + \frac{1}{3\cdot 5} + \frac{1}{4\cdot 6} + \cdots$

(f) $\displaystyle\sum_{n=1}^{\infty} \frac{(-1)^{n-1}}{n^2} = \frac{1}{1^2} - \frac{1}{2^2} + \frac{1}{3^2} - \frac{1}{4^2} + \cdots$

(g) $\displaystyle\sum_{n=1}^{\infty} \frac{1}{(n^2+1)^2} = \frac{1}{2^2} + \frac{1}{5^2} + \frac{1}{10^2} + \frac{1}{17^2} + \cdots$

17.7.2. Check your answers for Exercise 17.7.1 using your symbolic computing system. The command for the formal evaluation of the sum of terms u_n for n from `n1` to `n2` is

(MAPLE)	`sum(` u_n`, n=n1 .. n2 )`
(MATHEMATICA)	`Sum[` u_n`, {n,n1,n2} ]`

Possible values for summation limits include `infinity` (MAPLE) or `Infinity` (MATHEMATICA).

17.8 CAUCHY PRINCIPAL VALUE

We introduce the topic of this section by considering the integral

$$I = \int_0^\infty \frac{\sin x}{x}\,dx\,. \tag{17.54}$$

Our approach here is to write $\sin x$ as complex exponentials:

$$\sin x = \frac{e^{ix} - e^{-ix}}{2i}$$

obtaining initially

$$I = \int_0^\infty \frac{e^{ix} - e^{-ix}}{2ix}\,dx\,. \tag{17.55}$$

We would like to separate this into two individual integrals, but if we do so, each integral will diverge at $x = 0$ because the integrand behaves there like $1/x$. However, the integral as written in Eq. (17.55) does not diverge, and we can without changing its value write it as

$$I = \lim_{\varepsilon\to 0+} \int_\varepsilon^\infty \frac{e^{ix} - e^{-ix}}{2ix}\,dx = \lim_{\varepsilon\to 0+} \left[\int_\varepsilon^\infty \frac{e^{ix}}{2ix}\,dx - \int_\varepsilon^\infty \frac{e^{-ix}}{2ix}\,dx \right] .$$

The notation $0+$ indicates that ε is restricted to approach zero through positive values, so that both integrals will be individually convergent for nonzero ε. If we next make a change of variable from x to $-x$ in the second of the above integrals, we get

$$I = \lim_{\varepsilon\to 0+} \left[\int_{-\infty}^{-\varepsilon} \frac{e^{ix}}{2ix}\,dx + \int_\varepsilon^\infty \frac{e^{ix}}{2ix}\,dx \right] , \tag{17.56}$$

still a convergent result.

Expressions like that of Eq. (17.56) arise often enough that they have been given a name and notation. The right-hand side of that expression for I is called a **Cauchy principal value** integral, and in this book such integrals are written with a line through the integral sign, indicating that the integration range is cut open at the point where the integrand is singular. Thus, we write

$$I = \fint_{-\infty}^\infty \frac{e^{ix}}{2ix}\,dx\,. \tag{17.57}$$

Instead of drawing a horizontal line through the integral sign, some authors write P or PV before the integral sign, a choice this author finds potentially confusing. In any case, it is usually left to the reader to identify the singular point at which the integration range is cut.

It is important to realize that the notation of Eq. (17.57) does not cause the convergence of a divergent integral, in the present example

$$I' = \int_{-\infty}^\infty \frac{e^{ix}}{2ix}\,dx\,, \qquad \text{(divergent)}$$

but instead defines a useful related quantity that does converge to a definite value. Convergence is achieved because the two partial ranges end at the same distance from the singularity; a plot of $1/x$ near $x = 0$, Fig. 17.16, illustrates how the infinite contributions near the singularity cancel.

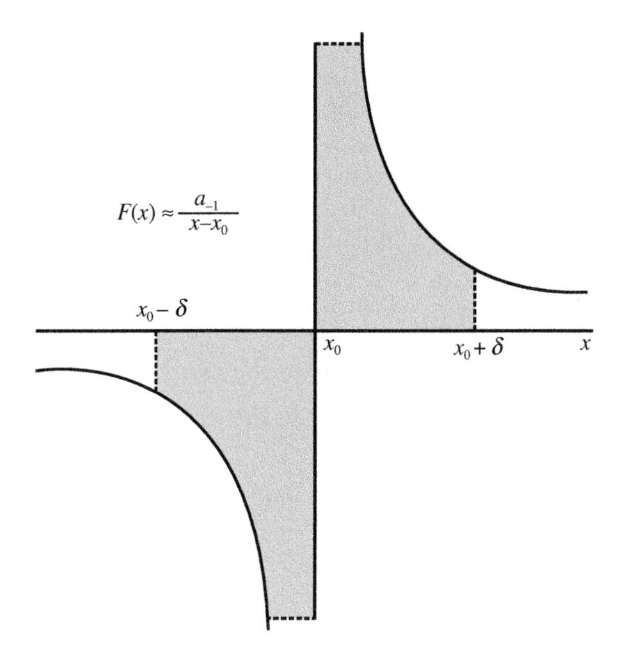

Figure 17.16: Principal-value cancellation in $1/x$.

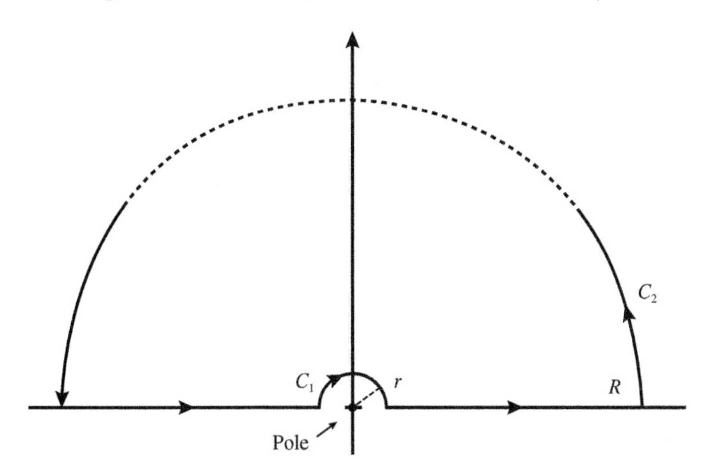

Figure 17.17: Contour for Cauchy principal value integral.

We can evaluate the integral I by considering the contour integral

$$J = \int_C \frac{e^{iz}}{2iz}\, dz \,, \tag{17.58}$$

where C is the contour shown in Fig. 17.17. The two pieces of the straight line along the real axis together evaluate to I; note that to have the integrand analytic everywhere on the contour we cannot connect these lines through the singular point $z = 0$, but must go around it. Invoking Jordan's lemma, the large arc C_2 closing the contour in the upper half-plane does not contribute to J. However, the small arc C_1 near $z = 0$ does make a contribution to J and must be analyzed further.

Noting that there is a simple pole at $z = 0$, we see that the semicircular arc about that point constitutes half the circuit about that pole, and therefore, since we are integrating [(residue) $i\,d\theta$], we get $\pi i\times$ (residue) instead of $2\pi i\times$ (residue). This is a general result when we go half-way around a first-order pole. We point out however that the same result is not obtained for the half-circuit of a higher-order pole.

In the present problem, the contour of Fig. 17.17 is a half-circuit in the negative direction (and it leaves the pole outside the contour). There are therefore no singularities encircled by our contour and $J = 0$. Taking these observations into consideration and noting that the residue of the integrand is $1/2i$, we find that

$$J = I - \pi i \left(\frac{1}{2i}\right) = 0\,, \qquad \text{so} \qquad I = \frac{\pi}{2}\,.$$

This is an important integral, so we write it more explicitly:

$$I = \int_0^\infty \frac{\sin x}{x}\,dx = \frac{\pi}{2}\,. \tag{17.59}$$

We could have drawn the contour for J to make the small half-circle go below $z = 0$, with the result that the pole is half-enclosed in the positive direction, and that it is now within the contour. With these changes, J is now nonzero but we get the same result for I.

DISPERSION RELATIONS

As we have already seen from our study of the Cauchy-Riemann equations, the real and the imaginary parts of an analytic function are related by expressions involving their derivatives. It is therefore not surprising that these quantities are also connected by relations involving integrals. Such integral relations were first applied in physical science to the study of the refractive and absorption properties of light, and for that reason have become known, irrespective of their area of application, as **dispersion relations**.

To develop this topic, consider a Cauchy principal value integral

$$I = \fint_{-\infty}^{\infty} \frac{f(x)}{x-a}\,dx\,, \tag{17.60}$$

and a related contour integral of the form

$$J = \int_C \frac{f(z)}{z-a}\,dz\,, \tag{17.61}$$

where C is a contour similar to Fig. 17.17, $f(z)$ (with $z = x + iy$) is analytic in the upper half-plane, and $f(z)$ approaches zero for large y sufficiently rapidly that the large arc closing the contour makes no contribution to the integral. We also note that the integrand is analytic in the entire region enclosed by the contour, so $J = 0$. Examining the various pieces of the contour, we see that the small clockwise arc makes a contribution corresponding to $(-1/2)$ times the residue at $z = a$ (compare with the earlier discussion leading to Eq. (17.59) if helpful), namely $-\pi i f(a)$. Moreover, the straight-line portions of the integral correspond to the Cauchy principal value integral I of Eq. (17.60). We therefore get

$$J = I - \pi i f(a) = 0\,,$$

which leads to the formula needed here:

$$\fint_{-\infty}^{\infty} \frac{f(x)}{x-a}\,dx = \pi i f(a)\,. \tag{17.62}$$

If we set $f(x) = u(x)+iv(x)$ and change the name of the quantity a to x_0, Eq. (17.62) can be decomposed into the two real equations

$$u(x_0) = \frac{1}{\pi} \fint_{-\infty}^{\infty} \frac{v(x)}{x - x_0}\, dx\,, \qquad v(x_0) = -\frac{1}{\pi} \fint_{-\infty}^{\infty} \frac{u(x)}{x - x_0}\, dx\,. \tag{17.63}$$

These are the formulas known as dispersion relations. As is obvious from their form, they enable determination of the real part of $f(x)$ from its imaginary part, and vice versa. Equations (17.63) are also known as those defining **Hilbert transforms**; if the first of these equations is interpreted as giving the transform of v, the second provides a formula for the inverse transform of u. When written in a form applicable to the range $x = (0, \infty)$, these equations are called the Kronig-Kramers relations (see Exercise 17.8.4).

Exercises

17.8.1. Evaluate the following principal-value integrals. Assume the quantities k and a to be real, and restrict s to the range $0 < s < 1$.

(a) $\displaystyle \fint_0^{\infty} \frac{\cos kx}{x^2 - a^2}\, dx$ (b) $\displaystyle \fint_0^{\infty} \frac{x \sin kx}{x^2 - a^2}\, dx$

(c) $\displaystyle \fint_0^{\infty} \frac{x^{-s}\, dx}{x - 1}$ (d) $\displaystyle \fint_{-\infty}^{\infty} \frac{\sin x\, dx}{(3x - a)(x^2 + a^2)}$

17.8.2. Write symbolic code that provides a numerical verification of the result of part (c) of Exercise 17.8.1. Your code should calculate directly

$$\int_0^{1-\varepsilon} \frac{x^{-s}\, dx}{x - 1} + \int_{1+\varepsilon}^{\infty} \frac{x^{-s}\, dx}{x - 1}\,,$$

and you should run it for a value of ε sufficiently small that the integral sum approaches a limit. Your result should then be compared numerically for several values of s with the principal values obtained when you carried out part (c).

17.8.3. Use contour integral methods to evaluate

$$I = \int_0^{\infty} \frac{x\, dx}{\sinh x}\,.$$

Use a contour similar to that illustrated in Fig. 17.14, but modify it to allow for the fact that there is a pole at $z = i\pi$. Explain why there is not a pole at $z = 0$.

17.8.4. The dispersion relations can be expressed as integrals on the range $(0, \infty)$ if $u(x)$ is an even function and $v(x)$ is odd. Assuming that u and v have these properties, show that Eqs. (17.63) can be expressed as integrals on the range $(0, \infty)$ and brought to the following forms:

$$u(x_0) = \frac{1}{\pi} \fint_0^{\infty} \frac{2xv(x)}{x^2 - x_0^2}\, dx\,, \qquad v(x_0) = -\frac{1}{\pi} \fint_0^{\infty} \frac{2x_0 u(x)}{x^2 - x_0^2}\, dx\,.$$

These equations are known as the Kronig-Kramers relations.

17.9 INVERSE LAPLACE TRANSFORM

Recall the formula for the Laplace transform $F(s)$ of a function $f(\omega)$:

$$F(s) = \int_0^\infty e^{-s\omega} f(\omega)\, d\omega\,. \tag{17.64}$$

We have chosen the symbol ω in place of that (t) which was used earlier for reasons that will soon become apparent.

The objective of the present section is to develop a formula for the inverse of the Laplace transform. A clue as to how to proceed is provided by the recognition that if the transform variable in Eq. (17.64) is permitted to become imaginary, we then have an integral somewhat similar to that of the Fourier transform, for which a simple inversion formula is available. Therefore, for reference in the present discussion we write here the formula for the Fourier transform $\Phi(\omega)$ of a function $\varphi(y)$ and for the inverse transformation:

$$\Phi(\omega) = \frac{1}{\sqrt{2\pi}} \int_{-\infty}^\infty \varphi(y)\, e^{i\omega y}\, dy \quad \text{and} \quad \varphi(y) = \frac{1}{\sqrt{2\pi}} \int_{-\infty}^\infty \Phi(\omega)\, e^{-iy\omega}\, d\omega\,. \tag{17.65}$$

A crucial difference between Laplace and Fourier transforms is that the latter are convergent only when a function and its transform vanish at infinity, while Laplace transforms also exist for functions $f(\omega)$ that vary for large ω as $e^{k\omega}$, where k is a positive constant. A function related to $f(\omega)$ that has a Fourier transform will therefore be $e^{-x\omega} f(\omega)$, where x is any real number larger than k.

Accordingly, to exploit the similarities of the Fourier and Laplace transforms, consider a function

$$\Phi(\omega) = \begin{cases} e^{-x\omega} f(\omega)\,, & \omega > 0\,, \\ 0\,, & \omega < 0\,, \end{cases} \tag{17.66}$$

where $f(\omega)$ has a Laplace transform and x has a value sufficiently large that $\Phi(\omega)$ can be a Fourier transform. We now take the inverse Fourier transform of $\Phi(\omega)$:

$$\begin{aligned} \varphi(y) &= \frac{1}{\sqrt{2\pi}} \int_{-\infty}^\infty e^{-iy\omega}\, \Phi(\omega)\, d\omega = \frac{1}{\sqrt{2\pi}} \int_0^\infty e^{-(x+iy)\omega} f(\omega)\, d\omega \\ &= \frac{1}{\sqrt{2\pi}} F(x+iy)\,, \end{aligned} \tag{17.67}$$

where F is the Laplace transform of f, occurring here with an argument $x+iy$ that is complex. The relevance of the choice of Φ and its identification as an inverse Fourier transform has become obvious: We have now found that the Fourier transform of $F(x+iy)$ is equal to $e^{-x\omega} f(\omega)$.

We write now the formula for the Fourier transform of $F(x+iy)$ for $\omega > 0$,

$$e^{-x\omega} f(\omega) = \frac{1}{\sqrt{2\pi}} \int_{-\infty}^\infty e^{i\omega y} \frac{1}{\sqrt{2\pi}} F(x+iy)\, dy\,,$$

and rearrange it to the form

$$f(\omega) = \frac{1}{2\pi} \int_{-\infty}^\infty e^{(x+iy)\omega}\, F(x+iy)\, dy\,. \tag{17.68}$$

We conclude our present derivation by making several notational changes. Starting from Eq. (17.68), we write $z = x+iy$, we note that $dy = dz/i$ and that the integral in

z is over a straight line in the complex plane from $x - i\infty$ to $x + i\infty$, and we replace the symbol ω by t. It can also be shown that the condition on x, introduced at the beginning of the derivation, is equivalent to a requirement that the vertical contour of z pass to the right of all the singularities of $F(z)$.

These changes and observations lead us to the following:

If $F(z)$ is the Laplace transform of $f(t)$ and all the singularities of $F(z)$ are at points whose real part is less than x, then we can invert the transform by the formula

$$f(t) = \frac{1}{2\pi i} \int_{x-i\infty}^{x+i\infty} e^{tz} F(z)\, dz\,. \tag{17.69}$$

This formula, showing how a function $f(t)$ can be obtained by inverting its Laplace transform $F(z)$, is known as the **Bromwich integral**. *The value of x, a real number, must be sufficiently large that $F(z)$ has no singularities at points whose real part is equal to or larger than x. Otherwise, the value of x is arbitrary.*

The reader may ask why the Bromwich integral is (for sufficiently large x) independent of x. The answer to this query is that because $F(z)$ is analytic to the right of the contour, the contour can be moved to larger x values without changing the value of the integral.

We close this section by remarking that, at least in principle, the Bromwich integral frees us from a limitation to transforms whose inverses are in tables, as the integral can be evaluated by any legitimate method, including often the closure of its contour by a large arc to the left and the use of the residue theorem.

Example 17.9.1. Laplace Transform Inversion

Let's use the Bromwich integral to find the function whose Laplace transform is

$$F(s) = \frac{1}{(s-a)^2 + k^2}\,,$$

with a and k real. The Bromwich integral for this problem is

$$f(t) = \frac{1}{2\pi i} \int_{x-i\infty}^{x+i\infty} \frac{e^{tz}}{(z-a)^2 + k^2}\, dz\,.$$

Since it is our intention to evaluate this integral using the residue theorem, we start by writing it in a form that exhibits the singularities of its integrand:

$$f(t) = \frac{1}{2\pi i} \int_{x-i\infty}^{x+i\infty} \frac{e^{tz}}{(z-a-ik)(z-a+ik)}\, dz\,.$$

This equation shows that in order for all singularities of the integrand to be to the left of the contour, we must choose $x > a$. Accordingly, we consider the integral

$$J = \frac{1}{2\pi i} \oint_C \frac{e^{tz}}{(z-a-ik)(z-a+ik)}\, dz\,,$$

where the contour is that shown in Fig. 17.18. The straight line at the right of the contour contributes $f(t)$ to the integral.

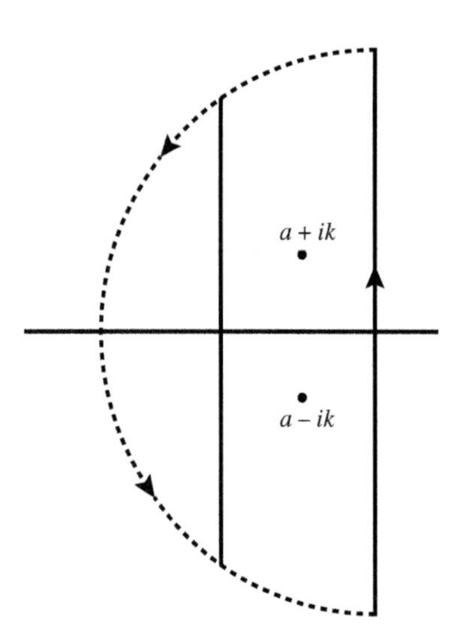

Figure 17.18: Contour for Example 17.9.1.

On the large arc closing the contour for J to the left, the integrand is everywhere smaller (in absolute value) than $e^{ta}/|(z-a)^2+k^2|$, which for large $|z|$ goes to zero as $|z|^{-2}$. The large arc therefore does not contribute to the integral, and we have $J = f(t)$.

We complete this problem by evaluating J. We note that the contour encloses two simple poles, at $a \pm ik$. From the residue theorem (which produces a factor $2\pi i$ that cancels the similar factor in the definition of J), we get

$$f(t) = J = (\text{residue at } z = a + ik) + (\text{residue at } z = a - ik)\,.$$

These residues have the respective values

$$(\text{residue at } z = a + ik) = \frac{e^{(a+ik)t}}{2ik}\,,$$

$$(\text{residue at } z = a - ik) = \frac{e^{(a-ik)t}}{-2ik}\,,$$

and therefore

$$F(s) = \frac{1}{(s-a)^2+k^2} \quad \Longrightarrow \quad f(t) = \frac{e^{at}}{k}\left[\frac{e^{ikt}}{2i} - \frac{e^{-ikt}}{2i}\right] = \frac{e^{at}}{k}\sin kt\,.$$

We can check this result against the entries in Table 13.1. If we take Entry 6 of the table and use Entry 14 to replace s by $s - a$, we confirm our result for $f(t)$.

■

We close this section with a further example that shows how we can use the Bromwich integral even when the Laplace transform $F(z)$ has a branch point.

Example 17.9.2. Inversion, Laplace Transform with Branch Cut

Let's use the Bromwich integral to find the function $f(t)$ whose Laplace transform is $F(s) = s^{-1/2}$. We therefore need to evaluate

$$f(t) = \frac{1}{2\pi i}\int_{x-i\infty}^{x+i\infty} \frac{e^{zt}}{z^{1/2}}\,dz\,.$$

A contour integral that includes $f(t)$ as part of its evaluation is

$$J = \frac{1}{2\pi i}\int_C \frac{e^{zt}}{z^{1/2}}\,dz\,,$$

where the contour is that shown in Fig. 17.19. For this contour, $f(t)$ is the contribution of the vertical line to the right, the small and large arcs do not contribute (proof of that claim is the topic of Exercise 17.9.1), and we have nonzero contributions from the horizontal lines above and below the branch cut.

Above the branch cut, write $z = re^{i\pi}$, so

$$dz = -dr\,, \quad z^{-1/2} = r^{-1/2}e^{-i\pi/2} = -ir^{-1/2}\,, \quad \text{and} \quad e^{tz} = e^{-tr}\,.$$

Keeping in mind that the integral above the branch point is from ∞ to zero in r, we have

$$\begin{aligned}
(\text{integral above branch cut}) &= \frac{1}{2\pi i}\int_\infty^0 \left(-ir^{-1/2}\right) e^{-tr}\,(-dr) = -\frac{1}{2\pi}\int_0^\infty r^{-1/2}e^{-tr}\,dr \\
&= -\frac{1}{2\pi t^{1/2}}\int_0^\infty u^{-1/2}e^{-u}\,du = -\frac{1}{2\pi t^{1/2}}\,\Gamma(1/2) \\
&= -\frac{1}{2(\pi t)^{1/2}}\,.
\end{aligned}$$

Below the branch cut, write $z = re^{-i\pi}$, and

$$dz = -dr\,, \quad z^{-1/2} = r^{-1/2}e^{i\pi/2} = ir^{-1/2}\,, \quad \text{and} \quad e^{tz} = e^{-tr}\,.$$

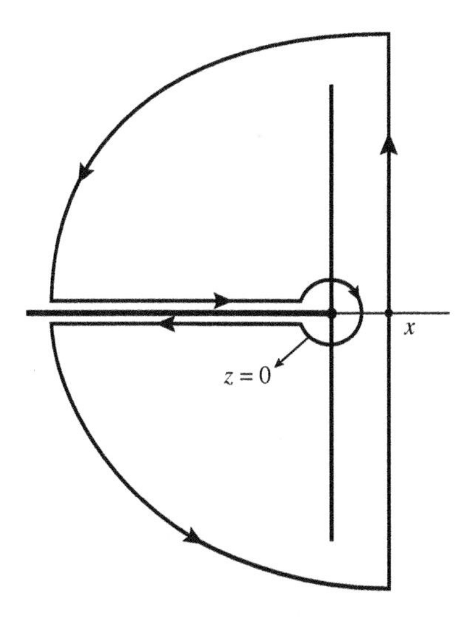

Figure 17.19: Contour for Example 17.9.2.

We therefore get

$$\text{(integral below branch cut)} = \frac{1}{2\pi i}\int_0^\infty \left(ir^{-1/2}\right) e^{-tr}\,(-dr)\,,$$

which reduces to the same result as the integral above the branch cut.

Adding the three contributions to J, we get

$$J = -\frac{1}{2(\pi t)^{1/2}} - \frac{1}{2(\pi t)^{1/2}} + f(t) = 0\,.$$

We have set $J = 0$ because the contour surrounds a region in which the integrand is analytic everywhere. Our final step is to solve for $f(t)$. The result,

$$F(s) = s^{-1/2} \quad \Longrightarrow \quad f(t) = \frac{1}{(\pi t)^{1/2}}\,,$$

is consistent with Entry 2 of Table 13.1 for $n = -1/2$.

■

Exercises

17.9.1. Show that Jordan's lemma, Eq. (17.36), can be identified as applicable to the large arc to the left that is needed to close the contour for the Bromwich integral.

17.9.2. Use the Bromwich integral to find the inverse Laplace transforms of the following functions:

(a) $\dfrac{1}{s(s+1)}$ (b) $\dfrac{1}{s^4-1}$ (c) $\dfrac{s}{s^4-1}$

(d) $\dfrac{1}{s^2-3s+2}$ (e) $\dfrac{s}{(s^2+4)^2}$ (f) $\dfrac{s^2}{(s^2-1)(s^2-4)}$

(g) $\dfrac{s^5}{s^6+2^6}$ (h) $\dfrac{s}{(s+1)(s^2+4)}$ (i) $\dfrac{(s-1)^2}{s(s+1)^2}$

17.9.3. Use the Bromwich integral to find the inverse Laplace transform of

$$F(s) = \frac{1}{(s^2+a^2)^{1/2}}\,.$$

Proceed as follows: Close the contour for the Bromwich integral by a large arc to the left, and note that the contour then encloses two branch points (at $z = \pm ia$) which can be connected by a branch cut. To evaluate the contour integral, shrink the contour so that it goes up one side of the branch cut and down the other, and reduce the contour integral to an ordinary integral from $-a$ to a. Use your symbolic computing system to evaluate this integral, and organize the result in a way that solves this exercise.

Additional Readings

Arfken, G. B., Weber, H. J., & Harris, F. E. (2013). *Mathematical methods for physicists* (7th ed.). New York: Academic Press.

Churchill, R. V., Brown, J. W., & Verkey, R. F. (1989). *Complex variables and applications* (5th ed.). New York: McGraw-Hill (This is an excellent text for both the beginning and advanced student. It is readable and quite complete).

Kurala, A. (1972). *Applied functions of a complex variable.* New York: Wiley (Interscience) (An intermediate-level text designed for scientists and engineers. Includes many physical applications).

Levinson, N., & Redheffer, R. M. (1970). *Complex variables.* San Francisco: Holden-Day (This text is written for scientists and engineers who are interested in applications).

Spiegel, M. R. (1964). Complex variables. In *Schaum's outline series.* New York: McGraw-Hill ((Reprinted 1995). An excellent summary of the theory of complex variables for scientists).

Chapter 18

PROBABILITY AND STATISTICS

Quantitative notions of probability and statistics are important in the physical sciences. Quantum mechanics describes physical systems in terms of probability distributions. Because systems containing large numbers of molecules cannot, even in principle, be described deterministically at the microscopic level (they are, after all, quantum-mechanical systems), their behavior can only be described statistically. Most scientific experiments involve data derived from measurements that are subject to random, and possibly also systematic errors, and we need methods for drawing conclusions based on our data sets and for assessing the degree of certainty of our findings. These kinds of practical problems are best addressed by a study of the theory of probability and of statistics.

A knowledge of the theory of probability is also of importance to those who undertake activities involving financial risk, and even to those who participate in games of chance or other activities in which chance is a factor.

Probability theory and statistics can involve sophisticated mathematical analysis, but our presentation here is limited to relatively elementary topics. Readers wanting more depth may wish to consult the Additional Readings.

Before embarking on the topic of this chapter, it may be useful to point out that mathematical analyses identify and quantify conclusions that follow from a set of initial assumptions. The applicability of results in probability and statistics depends upon the validity of the assumptions under which a system is described. For example, most analyses of probabilities involving dice assume that the six numbers on a die will appear with equal probability, and one usually assumes that a tossed coin will show heads and tails with equal probability. Analyses involving statistics have similar, but often more complicated, implied assumptions. Put simply, results in probability and statistics have a validity dependent on that of the relevant input assumptions.

18.1 DEFINITIONS AND SIMPLE PROPERTIES

The simplest situations involving probability deal with the description of "events" that either do, or do not occur (i.e., we rule out partial occurrences). A tossed coin (assuming it does not stay balanced on edge) either shows "heads" or "tails." A tossed

Mathematics for Physical Science and Engineering.
http://dx.doi.org/10.1016/B978-0-12-801000-6.00018-3

die shows exactly one of the integers from the set $\{1, 2, 3, 4, 5, 6\}$. Defining a **trial** as an observation that either does or does not detect a specific event (e.g., a coin toss showing "heads" or a die showing 6), the probability $p(x)$ of an event x is defined as the limit (for a large number of trials) of

$$p(x) = \frac{\text{number of times } x \text{ occurs}}{\text{number of trials}} . \tag{18.1}$$

If this limit does not exist, we cannot use the methods of probability theory to analyze the phenomenon. As a simple example, suppose that a trial consists of tossing a 1¢ and a 5¢ coin and recording the result (writing the 1¢ result first), and that our x is the occurrence of tt (for "tail, tail"). We then get a sequence of results such as

$$ht,\ tt,\ th,\ tt,\ hh,\ th,\ hh,\ ht,\ ht,\ hh,\ th,\ hh,\ tt,\ th,\ tt,\ \ldots,$$

from which we count the number of occurrences of tt and divide by the number of items in the sequence.

Sometimes the situation is such that we can form a set containing all the **mutually exclusive** possible outcomes. The notion of "mutually exclusive" is important here; we want each trial to produce exactly one of the possible outcomes. In the current example, an obvious set of mutually exclusive outcomes is

$$hh,\ ht,\ th,\ tt.$$

It is reasonable to assume that when a coin is tossed, heads and tails will occur with equal probability. This assumption may of course not be precisely correct if the coin is bent or not sufficiently symmetric, and is **input** to the mathematics (**not** derived therefrom). It is known how to study coin tossing problems with unsymmetric coins, but that is a topic we will come back to later. Right now, we make the additional reasonable assumption that the result from the tossing of one coin does not affect the result of the tossing of another, and these two assumptions lead us to conclude that the probability of each outcome of the two-coin toss is the same, namely 1/4. Note that the probabilities of all outcomes sum to unity.

The set of mutually exclusive outcomes, together with their probabilities of occurrence, is referred to as a **sample space**, and the sample space $\{hh, ht, th, tt\}$ is described as **uniform** because all its members occur with the same probability.

Once we have a sample space for a given problem, we can answer a variety of questions about the probabilities associated therewith. Because the members of a sample space refer to mutually exclusive outcomes, the probability of any one of several outcomes will be the sum of their individual probabilities. Thus, in our current example, the probability that both coins will have the same side up is the sum of $p(hh)$ and $p(tt)$, namely $1/4 + 1/4 = 1/2$. The probability that at least one coin will be heads is

$$p(1+\text{ heads}) = p(hh) + p(ht) + p(th) = 1/4 + 1/4 + 1/4 = 3/4, \tag{18.2}$$

or (because the probabilities of the members of the sample space sum to unity)

$$p(1+\text{ heads}) = 1 - p(tt) = 1 - 1/4 = 3/4. \tag{18.3}$$

A problem can have a variety of sample spaces that may be useful in different contexts, but all valid sample spaces must include a complete set of mutually exclusive events. If, in the current problem, we have agreed to pay a colleague 1 zot (a fictitious unit of currency) for each head, another relevant sample space might be

$$x_0(\text{no heads}),\ x_1\ (\text{one head}),\ x_2\ (\text{two heads}).$$

By identifying x_0 with tt, x_1 with th and ht, and x_2 with hh, and the elements of our new sample space would occur with probabilities $p(x_0) = 1/4$, $p(x_1) = 1/4 + 1/4 = 1/2$, $p(x_2) = 1/4$. These probabilities add to unity, but this sample space is not uniform. From our new sample space, we can read out that the probability of having to pay 1 zot is 1/2, the probability of paying 2 zot is 1/4, and the probability of getting off "zot free" is 1/4.

The foregoing example illustrates two concepts of general importance, which we state as theorems:

Theorem 1. *The probability that either of two mutually exclusive events occur is the sum of their individual probabilities, and*

Theorem 2. *The probability that two independent events both occur is the product of their individual probabilities.*

Example 18.1.1. Pair of Dice

Consider now the possible results when we throw a pair of dice. Most often the quantity of interest is the sum of the two numbers shown. We suspect (correctly) that different sums may not be equally likely, and that it might be easiest to start from a sample space that we have reason to believe is uniform. Such a space can be constructed by enumerating all the possibilities for the numbers shown. Supposing for the moment that one die is white and the other red (we correctly assume that the results cannot depend upon the colors of the dice), we list the possible outcomes in detail, with the number on the white die first:

$$\begin{array}{cccccc} 1,1 & 1,2 & 1,3 & 1,4 & 1,5 & 1,6 \\ 2,1 & 2,2 & 2,3 & 2,4 & 2,5 & 2,6 \\ 3,1 & 3,2 & 3,3 & 3,4 & 3,5 & 3,6 \\ 4,1 & 4,2 & 4,3 & 4,4 & 4,5 & 4,6 \\ 5,1 & 5,2 & 5,3 & 5,4 & 5,5 & 5,6 \\ 6,1 & 6,2 & 6,3 & 6,4 & 6,5 & 6,6 \end{array}$$

Figure 18.1: Sample space, two dice.

Since it is reasonable to assume that the probability of the number on the red die is completely independent of the number on the white die and that all numbers on either die are equally probable (with probability 1/6), each of the above tabular entries occurs with a probability $(1/6) \times (1/6) = 1/36$, a result corresponding to Theorem 2 above.

We are now ready to analyze the sample space we really need for the present problem, namely that whose entries correspond to the sum of the numbers on the two dice. Since the results in our outcome table are mutually exclusive, we can invoke Theorem 1 and state that $p(n)$, the probability that the two numbers add to n, is

$$p(n) = (1/36) \times (\text{the number of entries that add to } n). \tag{18.4}$$

Using Eq. (18.4), the members of our new sample space have the following probabilities:

n	2	3	4	5	6	7	8	9	10	11	12
$p(n)$	$\frac{1}{36}$	$\frac{2}{36}$	$\frac{3}{36}$	$\frac{4}{36}$	$\frac{5}{36}$	$\frac{6}{36}$	$\frac{5}{36}$	$\frac{4}{36}$	$\frac{3}{36}$	$\frac{2}{36}$	$\frac{1}{36}$

These probabilities add up to unity, reflecting the fact that every throw of a pair of dice must correspond to exactly one table entry.

The construction of this sample space provides an opportunity to make one further observation. We have reduced the size and complexity of the sample space in a way that is useful for our current purposes, but at the cost of losing control of some information. For example, the new sample space does not enable computation of the probability that a throw of the dice will produce a "double" (both dice showing the same number).

■

CONDITIONAL PROBABILITIES

If events are not entirely independent, the probability of an event B may depend upon whether or not some other event A is known to have occurred. We may be able to use this idea to simplify various probability computations.

Example 18.1.2. Balls in a Box—Brute Force

Suppose we put 4 white and 2 black balls into a box and then draw out at random two of the 6 balls, one after the other. We want to know the probabilities of the possible color sequences of the balls we draw out: ww (both white), wb (white, then black), bw (black, then white), and bb (both black).

One straightforward approach to this problem would be to count the possibilities in a sample space in which each member is an ordered pair of balls to be removed. Let's think of the balls as numbered (1–4 white, 5 and 6 black), and place the members of our sample space in a 6×6 array in which the row denotes the first ball removed and the column denotes the second ball removed. Each element in the array then denotes a possible removal, except for the diagonal elements, which would describe the impossible act of removing the same ball twice. After removing the diagonal elements, each remaining array position corresponds to an equally likely two-ball draw, with different portions of the array corresponding to different color patterns. We have

	1	2	3	4	5	6
1		ww	ww	ww	wb	wb
2	ww		ww	ww	wb	wb
3	ww	ww		ww	wb	wb
4	ww	ww	ww		wb	wb
5	bw	bw	bw	bw		bb
6	bw	bw	bw	bw	bb	

We now count the two-ball events of each color pattern, finding 12 ww, 8 wb, 8 bw, and 2 bb. There are 30 possibilities in all. Therefore,

$$p(ww) = \frac{12}{30} = \frac{2}{5}, \qquad p(wb) = p(bw) = \frac{8}{30} = \frac{4}{15}, \qquad p(bb) = \frac{2}{30} = \frac{1}{15}. \tag{18.5}$$

■

Computations of the type illustrated in Example 18.1.2 can become complicated and cumbersome. An alternate approach is to use the notion of **conditional probability**, defined as the probability of an event B, given that an event A is known to have occurred. The notation we use[1] to denote this probability is $p(B|A)$. With this notation, the probability of A, then B, is

$$p(AB) = p(A)p(B|A)\,. \tag{18.6}$$

Here $P(AB)$ indicates that A and B occur in the order A, then B. For example, this formula states that if A occurs with 60% probability, and if, when A is known to have occurred, B then occurs with 50% probability, the event sequence A, B has probability (60%)(50%) = (0.6)(0.5)=0.30, which is 30%.

Example 18.1.3. Balls in a Box—Conditional Probability

The advantage of using the concept of conditional probability is that it may be easier to compute $p(A)$ and $p(B|A)$ than to make a direct computation of $p(AB)$. Let's reexamine the problem of Example 18.1.2 from that perspective. Recall that we want the probabilities for the color patterns when two balls are drawn at random from a box originally containing 4 white and 2 black balls.

Considering initially the case that we first draw a white ball (which occurs with probability $p(w) = 4/6 = 2/3$), after which the box contains 3 white and 2 black balls. The conditional probabilities for the draw of a second ball are

$$p(w|w) = \frac{3}{5} \quad \text{and} \quad p(b|w) = \frac{2}{5}\,.$$

Now, using Eq. (18.6), we easily calculate

$$p(ww) = \frac{2}{3}\frac{3}{5} = \frac{2}{5} \quad \text{and} \quad p(wb) = \frac{2}{3}\frac{2}{5} = \frac{4}{15}\,. \tag{18.7}$$

Consider now what happens if we first draw a black ball (with probability $p(b) = 1/3$), leaving behind 4 white balls and 1 black ball, leading to the conditional probabilities

$$p(w|b) = \frac{4}{5} \quad \text{and} \quad p(b|b) = \frac{1}{5}\,.$$

Again using Eq. (18.6),

$$p(bw) = \frac{1}{3}\frac{4}{5} = \frac{4}{15} \quad \text{and} \quad p(bb) = \frac{1}{3}\frac{1}{5} = \frac{1}{15}\,. \tag{18.8}$$

Equations (18.7) and (18.8) give the same results as Eq. (18.5), but required less effort.

■

Exercises

18.1.1. For the two-coin tosses of the present section,

(a) Explain why the set $\{ht$, (at least one head), $tt\}$ is not a valid sample set.

[1] There are several notations in common use for conditional probability. What we are calling $p(B|A)$ is sometimes written by others as $p_A(B)$ or $p(B, A)$. We do not use this last form because it suggests to the unwary "B, then A."

(b) Find the probability of each member of the following sample set:

$$\{hh, \text{(at least one tail)}\}.$$

18.1.2. A standard deck of playing cards has four suits (spades, hearts, diamonds, clubs), each with 13 cards: (2–10, jack, queen, king, ace). If two cards are drawn "at random" (i.e., from a well-shuffled deck), what is the probability that they are of the same suit?

Hint. Don't forget that when the second card is drawn, the deck is then missing the first card.

18.1.3. For a shuffled standard deck of playing cards, compute the probability that

(a) A single card that is drawn is a spade.

(b) A single card that is drawn is an ace.

(c) A single card that is drawn is a spade or an ace.

(d) When three cards are drawn, exactly two are of the same suit.

(e) When three cards are drawn, none is an ace.

18.1.4. A card is drawn from a standard deck of playing cards, its identity is noted and it is returned to the deck, which is then reshuffled and a second drawing is carried out. Compute the probability that

(a) Both cards are spades.

(b) Both cards are of the same suit.

(c) The two cards are of different suits.

(d) The same card was drawn twice.

(e) Can you obtain the result for part (b) or for part (c) from that for part (a)? If so, explain why.

MORE PROBABILITY THEOREMS

The theorems to be discussed here involve sets of points in a sample space. An example of a set within the 36-member sample space for two dice (see Fig. 18.1) might be $A = \{(1,1),\ (2,2),\ (3,3),\ (4,4),\ (5,5),\ (6,6)\}$. Another set within this sample space might be $B = \{(1,5),\ (2,4),\ (3,3),\ (4,2),\ (5,1)\}$. Because A contains 6 members of the 36-member sample space, its probability of occurrence is $p(A) = 6/36 = 1/6$, while the probability of B is $p(B) = 5/36$.

One thing we might want to know is the probability that both A and B occur. In this context what we mean when we state "A occurs" is that some one of the mutually exclusive events in the event set A occurs. The statement "B occurs" means that one of the mutually exclusive events in the event set B occurs. Thus, the statement "A and B occur" means the occurrence of some mutually exclusive event that is a member of both the set A and the set B. In the present illustration, the only element A and B have in common is (3,3). Since we have previously determined that (3,3) occurs with probability $1/36$, the probability that "A and B occur" is $1/36$.

It is useful to express this result in a more compact notation. The set whose elements are common to A and B is called the **intersection** of A and B and is denoted $A \cap B$. We thus have $p(A \cap B) = 1/36$.

Another object of some interest is the probability that "A or B occur." As a statement in every-day speech, this phrase is somewhat ambiguous; does it mean "exactly one of A and B," or does it mean "A, B, or both?" In mathematics we cannot tolerate such an ambiguity, and it is generally accepted that in mathematical contexts, "A or B" means "**any one or both** of A and B." We write $A \cup B$ to denote this definition, with the precise meaning that $A \cup B$ is a set that contains all the members of A and all the members of B. This set is called the **union** of A and B.

The specification "exactly one of A or B" is identified as an **exclusive or**, sometimes written **xor**. There is nothing approaching universal agreement regarding a symbol for **xor**, and this text is written in a way that avoids the use of such a symbol.

In the current illustration, $A \cup B$ corresponds to the occurrence of any member of A or B, so its probability is the sum of the probabilities of the distinct members when the two sets are combined. By direct enumeration, and allowing for the fact that (3,3) appears in both sets, we find that $A \cup B$ has ten members, all mutually exclusive and each with probability 1/36. Therefore, $p(A \cup B) = 10/36 = 5/18$.

The relationship of A, B, $A \cap B$, and $A \cup B$ is easily represented pictorially; see Fig. 18.2. We start by drawing a circle labeled A, placing therein a set of points, each of which represents a member of the set A. The precise location of each point within the circle has no meaning; the points can be placed at any convenient locations. Now we draw another circle labeled B, containing points that represent members of B. If A and B have any members in common, we must position the circles so that they intersect and place the common members within the intersection region. Then the points encircled by curve A (or curve B) represent members of A (or members of B), those within the outer boundary of both circles are the members of the **union** $A \cup B$, and those within the intersection of the two circles are in the (aptly named) **intersection** $A \cap B$.

We can now use Fig. 18.2 to develop an equation connecting the probabilities of the sets involved in the figure. The probability of the union $A \cup B$ will be the sum of the probabilities for all the points in the figure. If our first step in this computation of this quantity is to add the probabilities for the points in A [this is $p(A)$] to the probabilities for the points in B [this is $p(B)$], we will double-count the probabilities for the points in the intersection of A and B. Thus,

Theorem 3. *The formula for the probability of $A \cup B$ (i.e., the probability of the occurrence of any of the events in $A \cup B$) is*

$$p(A \cup B) = p(A) + p(B) - p(A \cap B)\,. \tag{18.9}$$

Equation (18.9) is a generalization of our earlier Theorem 1, and reduces to that theorem if A and B are mutually exclusive, as they then have no members in common and have an empty intersection (obviously of zero probability). But our understanding

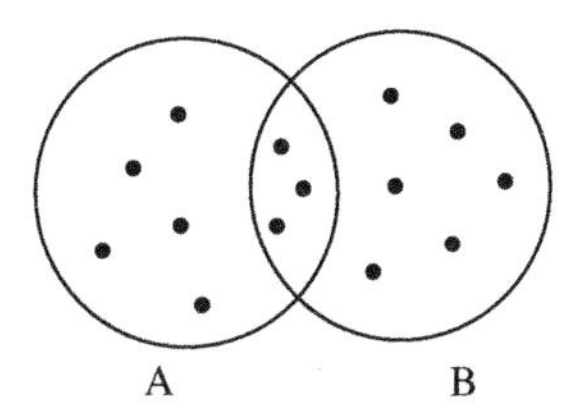

Figure 18.2: Union and intersection of two sets.

is now more nuanced; we have a formula that remains applicable if A and B have elements in common and can therefore both occur.

Example 18.1.4. Checking Theorem 3

The sets A and B introduced earlier in this section have respectively 6 and 5 members; we computed their individual probabilities as $p(A) = 6/36$ and $p(B) = 5/36$. We also determined that the intersection $A \cap B$ had one member, so $p(A \cap B) = 1/36$, and that the union $A \cup B$ had 10 distinct members, so $p(A \cup B) = 10/36$. Inserting these probabilities into Eq. (18.9), we get

$$p(A \cup B) = \frac{6}{35} + \frac{5}{36} - \frac{1}{36} = \frac{10}{36},$$

as required.

■

BAYES' THEOREM

Let's return to Eq. (18.7) and rewrite it as

$$p(A \cap B) = p(A)p(B|A),$$

and then write the similar equation involving the conditional probability that A also occurs if it is known that B has occurred, namely

$$p(B \cap A) = p(B)p(A|B).$$

The intersection of two sets has the same meaning regardless of which is listed first, so $A \cap B = B \cap A$, and the two preceding equations can be combined to yield

Theorem 4 (Bayes' theorem).[2]

$$p(A)p(B|A) = p(B)p(A|B) = p(A \cap B). \tag{18.10}$$

Example 18.1.5. Bayes' Theorem

A large employer is considering making a credit score of 650 an employment requirement. In reviewing its data from previous hires, it is determined that

(1) 80% of the previous hires were satisfactory employees, 20% were not.

(2) 85% of the satisfactory employees had credit scores $\geq$650.

(3) 10% of the unsatisfactory employees had credit scores $\geq$650.

Our task is to use the above data to determine the probability that a prospective employee with a credit score less than 650 will turn out to be satisfactory.

Letting L (for "low") stand for a credit score below 650 and letting S and U respectively stand for satisfactory and unsatisfactory employees, the probability needed here is $p(S|L)$. The input data correspond to the equations

$$p(S) = 0.80, \quad p(U) = 0.20, \quad p(L|S) = 0.15, \quad p(L|U) = 0.90.$$

[2]Be careful in trying to apply this theorem to the situation of Example 18.1.3. In that problem, $p(AB)$ denotes the probability that the **first ball drawn** is in A and that the **second ball drawn** is in B, so AB is not the same space as BA.

We do not have $p(S|L)$, but Bayes' theorem provides a way to obtain it. Using Eq. (18.10), we write

$$p(S|L) = \frac{p(S)p(L|S)}{p(L)}. \tag{18.11}$$

Equation (18.11) is likely to be useful because $p(L|S)$ and $p(S)$ are available in our data. However, we still need $p(L)$. But we can write (because $S \cap L$ and $U \cap L$ are mutually exclusive, with a union equal to L)

$$p(L) = p(S)p(L|S) + p(U)p(L|U) = (0.80)(0.15) + (0.20)(0.90) = 0.30,$$

and we then have everything needed to evaluate Eq. (18.11):

$$p(S|L) = \frac{(0.80)(0.15)}{0.30} = 0.40. \tag{18.12}$$

Our finding is that job candidates with lower credit scores have a 40% chance of success in this company.

■

Exercises

18.1.5. A box contains a white balls and b black balls; a second box contains c white balls and d black balls. A ball is transferred at random (without observing its color) from the first box to the second box. If a ball is then withdrawn from the second box, what is the probability that it will be white?

18.1.6. Two boxes are of identical appearance. The first contains 2 white balls and 3 red balls; the second contains 6 white balls and 4 red balls. One box is selected at random and a ball is withdrawn from it. What is the probability that this ball is white?

18.1.7. A for-profit school prepares high-school students for a college admission test. At the end of the preparatory process the students take a practice test and then go on to take the college admission test. The school reports that its students have 80% success rates for both the practice test and the college admission test, and that 60% of its students who fail the practice test also fail the college admissions test. What fraction of the students who pass the preparatory test fail the college admission test?

PERMUTATIONS AND COMBINATIONS

Many problems in probability include a counting of the number of ways in which data can be selected or sorted. The basic idea involved here is that if something (such as selecting an item, marking it, or whatever) can be done in n_1 ways, and that thereafter a second thing can be done in n_2 ways, the total number of ways of doing these two things (in the indicated order) is $n_1 n_2$. It is important to note that this analysis is applicable no matter how (or whether) the doing of the first thing changes the number of ways the second thing can be done; we must determine n_2 based on the state of our universe as it exists **after** the first thing is done. The same idea extends to a sequence of k successive acts; the total number of ways N that a succession of things can be done is

$$N = n_1 n_2 \cdots n_k\,, \quad \text{where } n_{j+1} \text{ is the number of ways the } (j+1)\text{th thing can be done, computed from the state of the system after the } j\text{th thing is completed.} \tag{18.13}$$

An important example of the application of this counting principle is the calculation of the number of distinct orderings of a set of objects. Taking three objects as an example, we can obtain an ordering of them by selecting a first object (this can be done in three ways), then a second object (two ways possible after the first object is removed), and then a third (one way; there is only one object left). Each set of these choices leads to a different ordering, so there are $3 \cdot 2 \cdot 1 = 3! = 6$ orderings of the three objects. Each ordering is called a **permutation**, so we have shown that there are 3! distinct permutations of three objects. For n objects, the number of permutations is $n!$ The notation we will use for the number of permutations of n objects is $P(n,n)$; we have just noted that

$$P(n,n) = n! \tag{18.14}$$

Sometimes we will start with n objects and make an ordered selection of r of them, where r is less than n. The number of ways of doing this, which we denote $P(n,r)$, is

$$P(n,r) = n(n-1)\cdots(n-r+1) = \frac{n!}{(n-r)!}\,. \tag{18.15}$$

We also encounter situations in which a set of items is selected, but the items are not required to be in any particular order. In that case, our interest is in the number of ways an unordered set of r items can be chosen from a set originally containing n members. The unordered set of chosen items is referred to as a **combination** selected from the original set of objects, and we use the notation $C(n,r)$ to stand for the number of ways in which r unordered objects can be chosen from an original population of size n; $C(n,r)$ is sometimes described as the "number of combinations of n objects, taken r at a time."

We have already developed a formula for $P(n,r)$. We note that the permutations counted by $P(n,r)$ can be sorted into sets, with all members of the same set containing a common collection of r items (in all possible orders). There are $r!$ members in each of these sets, and every member of such a set corresponds to the same item population. Thus, $P(n,r)$ represents a counting in which each **combination** is included $r!$ times, and we conclude that

$$C(n,r) = \frac{P(n,r)}{r!} = \frac{n!}{(n-r)!r!} = \binom{n}{r}\,. \tag{18.16}$$

The numbers of combinations are given by the binomial coefficients.

One way of viewing the counting of combinations is to identify $C(n,r)$ as the number of ways of dividing n items by putting r of them into a labeled box (Box A) and the remaining $n-r$ into a second box, Box B. It should be reasonably obvious that the number of ways of doing this would be the same if the r items had been put into Box B, leaving $n-r$ items for Box A. This observation corresponds to $C(n,r) = C(n,n-r)$, a relation satisfied by Eq. (18.16).

We can generalize the above combination problem by considering the placement of N items into m labeled boxes, where m can be larger than 2. Our objective is to find the number of distinct ways to put n_1 items in the first box, n_2 in the second, $\ldots$, and n_m in the mth. Of course, the n_i must sum to N. We denote this number of combinations $C_N(n_1, n_2, \ldots, n_m)$.

To find a formula for this general situation, we can proceed stepwise. We first choose n_1 items for the first box, which can be done in $C(N, n_1)$ ways. After that, we choose n_2 of the $N - n_1$ left-over items for the second box; this can be done in $C(N-n_1, n_2)$ ways. Continuing, the third box can be filled in $C(N-n_1-n_2, n_3)$ ways, etc. Let's look in more detail at this result. We first get

$$C_N = C(N, n_1)C(N-n_1, n_2)C(N-n_1-n_2, n_3)\cdots C(N-n_1-\cdots-n_{m-1}, n_m)$$

$$= \left[\frac{N!}{n_1!(N-n_1)!}\right]\left[\frac{(N-n_1)!}{n_2!(N-n_1-n_2)!}\right]\left[\frac{(N-n_1-n_2)!}{n_3!(N-n_1-n_2-n_3)!}\right]\cdots$$

$$\times\cdots\left[\frac{(N-n_1-\cdots-n_{m-2})!}{n_{m-1}!(N-n_1-\cdots-n_{m-1})!}\right]\left[\frac{(N-n_1-\cdots-n_{m-1})!}{n_m!(N-n_1-\cdots-n_m)!}\right].$$

With the exception of $N!$, all factors in the numerator cancel against identical factors in the denominator. We also note that $N-n_1-\cdots-n_m = 0$ and that $0! = 1$. We are left with

$$C_N(n_1, n_2, \ldots, n_m) = \frac{N!}{n_1!n_2!\cdots n_m!}. \tag{18.17}$$

The quantity C_n is called a **multinomial coefficient**.

STATISTICAL MECHANICS

Computations of the types presently under discussion are relevant in statistical mechanics. If we have N identical atoms or molecules, each of which is to be assigned one of m states (mathematically equivalent to being put into numbered boxes), the statistical-mechanical analysis depends on the number of distinct assignments that are possible. In classical theory, the particles, though identical, are regarded as distinguishable; in principle, we can follow their individual trajectories and thereby always know which is which. We can therefore get unique assignments by placing the first particle in some box (this can be done in m ways), then placing each following particle in any one of the m boxes (these assignments do not depend upon the placement of the earlier particles). The total number of possible assignments is therefore m^N. This classical counting procedure leads to what in statistical mechanics is called **Maxwell-Boltzmann statistics**.

In quantum theory, identical particles are regarded as inherently indistinguishable, so an interchange of the positions of two particles in different boxes does not produce a new assignment. This viewpoint is consistent with the Heisenberg uncertainty principle, which makes it impossible to trace particle trajectories with certainty and therefore maintain knowledge of individual particle identities. In particular, it has been found that counting only the distinguishable assignments (those where the numbers of particles in at least some of the boxes are different) leads to correct predictions of experimental results. It has also been found experimentally that the wave functions of quantum systems must be either symmetric or antisymmetric when the positions of two identical particles are interchanged. Antisymmetry is possible because physical phenomena depend quadratically on the wave function. Both symmetry and antisymmetry under particle exchange are observed in nature. Particles whose wave function is symmetric under particle interchange are said to obey **Bose-Einstein** statistics and are referred to as **bosons**; examples include the photon and the deuteron. Particles whose wave function is antisymmetric under particle exchange obey **Fermi-Dirac** statistics and are called **fermions**; examples include the proton, the neutron, and the electron. The indistinguishability and permutational symmetry of quantum particles must be considered when developing the formulas that count their numbers of distinguishable state assignments.

The permutational antisymmetry of fermions has as a consequence that a fermion wave function will vanish if more than one identical particle is assigned the same state (the interchange of two such particles will both leave the wave function invariant and change its sign, impossible unless the wave function is zero). Therefore, the number of distinct Fermi-Dirac states of N identical particles and m boxes is the number of ways N of the boxes can be chosen for occupancy by a single particle. This number is $C(m, N)$; it will be zero unless $m \geq N$. The limitation of fermions to singly-occupied states is the origin of shell structure and periodic properties in atoms; the world as we know it could not exist if the electrons were not fermions.

There is no restriction on the state occupancies of bosons; however, in counting the number of possible assignments we must keep in mind that the bosons are indistinguishable. We shall shortly show that the number of distinguishable assignments of N bosons among m states is $B_N(m)$, where

$$B_N(m) = \binom{N+m-1}{N} = \binom{N+m-1}{m-1}. \tag{18.18}$$

Example 18.1.6. Counting Boson Distributions

We want to know how many distinguishable ways there are for putting N indistinguishable objects into m numbered boxes. A cute way to approach this question is to consider the permutations of N balls and $m-1$ partitions, where both the balls and the partitions are regarded as objects to be permuted. After each permutation is identified, we identify the balls preceding the first partition as being in Box 1, those between the first and second partitions as in Box 2, etc., with those after the last (the $(m-1)$th) partition as in Box m. This number of permutations is $(N+m-1)!$.

The above-described permutations treat both the balls and the partitions as distinguishable. We do not want the balls to be distinguishable, and we do not need distinguishable partitions to associate the balls with numbered boxes. We therefore correct our result by dividing by both $N!$ and $(m-1)!$, leading to the result in Eq. (18.18).

■

Exercises

18.1.8. (a) What is the probability that two people have different birthdays (assume a 365-day year and exclude February 29)?

(b) What is the probability that three people all have different birthdays?

(c) What is the smallest number of people for which the probability that at least two have the same birthday exceeds 0.5?

Note. Consider writing symbolic code to make this computation.

18.1.9. You are dealt five cards from a standard 52-card deck of playing cards.

What is the probability that

(a) Four are aces?

(b) You have any four-of-a-kind?

(c) All five are the same suit (a flush)?

(d) All five are spades?

18.1.10. How many different ways can a committee consisting of a designated chair and four other members be selected from among 12 candidates?

18.1.11. Cards are drawn at random from a standard 52-card deck. What is the probability that exactly 8 cards are drawn before obtaining the first ace?

18.1.12. What is the probability that if a fair coin is tossed 100 times, it will land "heads" exactly 50 times?

Hint. Use symbolic computing or Stirling's formula.

18.1.13. A computer byte consists of n consecutive bits (a bit has two values, 0 and 1). Most modern computers have 8-bit bytes. There were once computers with 6-bit bytes.

(a) How many different 6-bit bytes are there?

(b) How many 6-bit bytes contain (anywhere) the sequence 1001?

18.1.14. By explicit enumeration, determine the number of ways two identical particles can be put into four states using Maxwell-Boltzmann, Fermi-Dirac, and Bose-Einstein statistics. Compare your results with the general formulas given earlier in this section of the text.

18.1.15. Each of n balls is placed at random in any one of n boxes (without regard for whether there is already another ball there). Show that in the limit of large n the probability that each box will contain one ball is $(2\pi n)^{1/2}e^{-n}$.

18.1.16. What is the probability of getting 20 points with 6 dice?

18.1.17. A red, a blue, a green, and a white ball are placed one each, at random, into boxes that are also colored red, blue, green, and white. What is the probability that no ball is in the box of the same color?

18.1.18. A box contains a white balls and b black balls. Balls are withdrawn until all the balls remaining in the box are the same color. What is the probability that this color is white?

18.1.19. Two dice were thrown 36 times. What is the probability that one or more of these throws produced a double-six?

18.2 RANDOM VARIABLES

In scientific and engineering contexts, we are often concerned with numerical data that may arise in connection with a problem involving probability. Perhaps the simplest example of this sort is that we may need the average of a set of numerical observations or measurements. For this and related purposes, we define the notion of a **random variable** X, meaning a quantity which can take on its various possible values x_i (which must be numerical), with specified probabilities $p(x_i)$, often abbreviated p_i.

These probabilities are usually determined by the physical problem in which X arises. The possible values of X can either be a discrete set (as in all the examples thus far discussed) or may fall on a continuous range (an alternative we discuss later). The requirement that X have numerical values does not prevent us from treating some of the problems already examined; we may often proceed usefully by introducing definitions such as "heads = 1" and "tails = 0" in place of symbols like h and t.

Among the reasons for requiring that a random variable have numerical values are that we may then introduce algebraic formulas for quantities such as the average of a set of x_i and that we can make diagrams in which quantities such as p_i are plotted against x_i.

EXPECTATION VALUES

Given the probabilities for the possible values of a random variable, we can compute the average of any function dependent on that random variable. That average, often referred to as an **expectation value** and identified by the use of angle brackets, corresponds for the function $g(x)$ to the following formula:

$$\langle g \rangle = \sum_i p_i g(x_i)\,. \tag{18.19}$$

Equation (18.19) weights the possible values of $g(x)$ according to their probabilities. We do not have to divide by a number characteristic of the size of the sample space because that notion is included in the p_i; they are defined so that they add to unity:

$$\sum_i p_i = 1\,. \tag{18.20}$$

As an example of the use of Eq. (18.19), we compute the expectation value of X^2 as

$$\langle X^2 \rangle = \sum_i p_i x_i^2\,.$$

Although this formula gives the "expectation value" of X^2, we cannot conclude that $\langle X^2 \rangle$ is the most probable value of x^2 or even that x^2 will ever have that value. Remember that "expectation value" is just a technical term meaning "average."

MEAN AND VARIANCE

Given a random variable X, one quantity of interest is its average value, technically called its **mean**, **mean value**, or **expectation value**. Several different symbols for the mean are in widespread use; some of them are $\overline{X}$, $\langle X \rangle$, and μ (when using μ the quantity being averaged must be obvious from the context). When X is described in a sample space described by x_i and p_i, its mean (the average of x) is given as a simple case of Eq. (18.19):

$$\overline{X} = \langle X \rangle = \mu = \sum_i p_i x_i\,. \tag{18.21}$$

If the points of the sample space all have equal probability, so that for a space of n members, $p_i = 1/n$ for all i, we can reduce Eq. (18.21) to

$$\mu = \frac{1}{n}\sum_i x_i\,. \tag{18.22}$$

There are three points about Eqs. (18.21) and (18.22) that may be worth mentioning.

(1) The mean is not a measure of the most likely result of an observation of X.
(2) The mean may not even be a possible value of X.
(3) There is no inherent requirement that half of the possible values of X fall on each side of the mean.

Example 18.2.1. Means of Two Probability Distributions

1. Find the mean of the following: 1.1, 2.2, 4.4, 5.5, 6.8, assuming all to be equally probable. Here we have five items, each with probability 1/5. The mean is

$$\mu = \frac{1}{5}(1.1 + 2.2 + 4.4 + 5.5 + 6.8) = \frac{20}{5} = 4.$$

Note that none of the data are at the mean.

2. The following integer data were obtained with the frequency indicated in the following table:

x_1	0	1	2	3	4	5	6
n_i	1	2	3	4	2	3	5

The total number of n_i is 20, so the probability p_i can be computed as $n_i/20$. We then compute

$$\mu = \frac{1}{20}(0 \cdot 1 + 1 \cdot 2 + 2 \cdot 2 + 3 \cdot 4 + 4 \cdot 2 + 5 \cdot 3 + 6 \cdot 5) = \frac{73}{20} = 3.65.$$

■

Also of frequent interest is a measure of the spread of the X values about the mean. Such a measure, called the **variance** of X and often denoted σ^2, is defined as

$$\sigma^2 = \left\langle \left(X - \langle X \rangle \right)^2 \right\rangle = \sum_i p_i \, (x_i - \mu)^2 \,. \tag{18.23}$$

It is obvious from the definition that σ^2 will be smaller if the x_i are clustered closely about their mean. It is also clear that, other things being equal, high-probability x_i will receive more weight in the computation of σ^2. Because the deviations from the mean occur as squares, the contributions from all deviations will be positive, and σ^2 will be zero only if $x_i = \mu$ for all i (i.e., that the probability distribution is entirely localized). If all points in our sample space occur with equal probability, Eq. (18.23) reduces to

$$\sigma^2 = \frac{1}{n}\sum_i (x_i - \mu)^2 \,. \tag{18.24}$$

If we expand the square in Eq. (18.23), we get

$$\sigma^2 = \sum_i p_i x_i^2 - 2\mu \sum_i p_i x_i + \mu^2 \sum_i p_i \,. \tag{18.25}$$

The first term on the right-hand side of Eq. (18.25) is the average of X^2, which we can write $\langle X^2 \rangle$; the summation following 2μ evaluates to μ, and the final summation of p_i evaluates to unity. Equation (18.25) therefore reduces to

$$\sigma^2 = \langle X^2 \rangle - \mu^2 = \langle X^2 \rangle - \langle X \rangle^2 . \tag{18.26}$$

Since σ^2 is an average of the squares of the deviations of X from the mean, a "root mean square" estimate of the average deviation is obtained by computing the **standard deviation** of X, defined as

$$\sigma = \sqrt{\sigma^2} = \sqrt{\sum_i p_i \,(x_i - \mu)^2} , \tag{18.27}$$

or, for a uniform-probability sample space, as

$$\sigma = \sqrt{\frac{1}{n} \sum_i (x_i - \mu)^2} . \tag{18.28}$$

When needed to avoid ambiguity, we write this quantity as $\sigma(X)$.

Suppose now that we rescale our random variable, replacing X by aX, where a is a constant. It is easy to show that

$$\langle aX \rangle = a\langle X \rangle , \qquad \sigma^2(aX) = a^2\sigma^2(X) , \qquad \sigma(aX) = a\,\sigma(X) . \tag{18.29}$$

Both the mean and the standard deviation scale linearly with a.

CHEBYSHEV INEQUALITY

A relation known as **Chebyshev's inequality** states that the probability that a random variable has a value more than $k\sigma$ from its mean is less than $1/k^2$. The proof is simple. If we start from Eq. (18.23) and limit the sum to points x_i such that $|x_i - \mu| > k\sigma$, the terms remaining in the sum will add to a value less than σ^2. We then further reduce the size of the sum by replacing $(x_i - \mu)^2$ by its minimum value, $k^2\sigma^2$. These steps correspond to the inequalities

$$\sigma^2 \;>\; \sum_{|x_i-\mu|>k\sigma} p_i(x_i - \mu)^2 \;>\; k^2\sigma^2 \sum_{|x_i-\mu|>k\sigma} p_i .$$

The summation in the last member of the above expression is the probability that x_i will be further from μ than $k\sigma$. Dividing through by $k^2\sigma^2$, we now obtain Chebyshev's inequality:

$$\sum_{|x_i-\mu|>k\sigma} p_i \;<\; \frac{1}{k^2} . \tag{18.30}$$

The bound provided by Chebyshev's inequality is often not very tight. As we will see later, the most used probability distribution (the Gauss normal distribution) indicates far smaller probabilities that a random variable will be further than $k\sigma$ from its mean. But Chebyshev's inequality does show that, irrespective of the applicable distribution, there is a low probability that a random variable will be more than a few standard deviations from its mean.

Example 18.2.2. Variance and Standard Deviation

Let's compute the variance and the standard deviation for the two data sets used in Example 18.2.1.

1. In the cited example, the mean for this data set was found to be $\mu = 4$. The values of $x_i - \mu$ are therefore -2.9, -1.8, 0.4, 1.5, and 2.8. We then compute
$$\sigma^2 = \frac{1}{5}\left[(-2.9)^2 + (-1.8)^2 + (0.4)^2 + (1.5)^2 + (2.8)^2\right] = 4.38,$$
corresponding to a standard deviation $\sigma = 2.09$.
2. Here the easiest approach may be to compute $\langle X^2 \rangle$ and use Eq. (18.26). We proceed as follows, using the data from Example 18.2.1:
$$\langle X^2 \rangle = \frac{1}{20}\left[0^2 \cdot 1 + 1^2 \cdot 2 + 2^2 \cdot 3 + 3^2 \cdot 4 + 4^2 \cdot 2 + 5^2 \cdot 3 + 6^2 \cdot 5\right] = 16.9.$$
Then, noting that the earlier example yielded $\mu = 3.65$,
$$\sigma^2 = 16.9 - (3.65)^2 = 3.58 \qquad \text{and} \qquad \sigma = 1.89.$$

■

Exercises

18.2.1. Find the mean, the variance, and the standard deviation for the sum obtained from throws of a pair of dice.

18.2.2. The following table gives n_i, the number of times each value x_i of a random variable X occurred in a collection of data:

x_i	0	1	2	3	4	5	6	7	8
n_i	1	0	1	10	20	5	1	1	1

(a) Find the mean, the variance, and the standard deviation of X.

(b) Find the number of data points further from the mean than 2σ and compare with the predictions of Chebyshev's inequality.

(c) Repeat part (b) for points further from the mean than 3σ.

COVARIANCE AND CORRELATION

Sometimes we will need to work with more than one random variable at a time, and we may need to know the extent to which our random variables are independent. Suppose that we have two random variables X and Y; a sample space for this two-variable system can have members each of which refers to some value x_i of X and some value y_j of Y, and occurs with a probability that depends upon both the choices of x_i and y_j. We denote this probability $p(x_i, y_j)$, which we henceforth write in the abbreviated notation p_{ij}. The probabilities of all members of the sample space add to unity:
$$\sum_{ij} p_{ij} = 1\,, \tag{18.31}$$
and the mean and variance of X (irrespective of the value of Y) can be computed as
$$\langle X \rangle = \sum_{ij} p_{ij} x_i\,, \tag{18.32}$$
$$\sigma^2(X) = \sum_{ij} p_{ij}(x_i - \langle X \rangle)^2\,. \tag{18.33}$$

The corresponding formulas for the mean and variance of Y are

$$\langle Y \rangle = \sum_{ij} p_{ij} y_j \,, \tag{18.34}$$

$$\sigma^2(Y) = \sum_{ij} p_{ij} (y_j - \langle Y \rangle)^2 \,. \tag{18.35}$$

The quantities defined above do not indicate whether X and Y are independent. However, a measure of the extent of their independence is provided by their **covariance**, which is defined as

$$\text{cov}(X, Y) = \Big\langle \, (X - \langle X \rangle) \, (Y - \langle Y \rangle) \, \Big\rangle \,. \tag{18.36}$$

This average value will be zero if the probabilities of the possible choices of X are independent of the choice of Y and vice versa; thus $\text{cov}(X, Y) = 0$ indicates that X and Y are independent.

To calculate $\text{cov}(X, Y)$, we rewrite Eq. (18.29) in terms of p_{ij}, leading to

$$\text{cov}(X, Y) = \sum_{ij} p_{ij} (x_i - \langle X \rangle)(y_j - \langle Y \rangle) \,. \tag{18.37}$$

If X and Y are independent, the probabilities p_{ij} will factor[3] into a product involving i and j separately, i.e., a product of the form $p(x_i)p(y_j)$. Our formula for cov then separates into

$$\text{cov}(X, Y) = \sum_{i} p(x_i)(x_i - \langle X \rangle) \sum_{j} p(y_j)(y_j - \langle Y \rangle) = 0. \tag{18.38}$$

Each of the sums in Eq. (18.38) can now be shown to vanish.

An interpretation of the covariance is most easily made if it is scaled so that its values lie between 1 and -1. It can be shown that this scaling is achieved by defining a quantity called **correlation**, defined as

$$\text{corr}(X, Y) = \frac{\text{cov}(X, Y)}{\sigma(X)\sigma(Y)} \,. \tag{18.39}$$

It is also useful to note that the covariance can be written in a form somewhat similar to Eq. (18.26) for σ^2. Expanding Eq. (18.36), we get

$$\text{cov}(X, Y) = \langle XY \rangle - \langle X \rangle \langle Y \rangle \,. \tag{18.40}$$

Let's summarize now some results for situations in which we have two random variables X and Y. To provide a context for our analysis, note that a possible scenario might be that the values of X and Y correspond to payoffs from games of chance, and our primary interest might be the expectation value of the overall payoff $Z = X + Y$, where Z is a random variable whose values are the sums of the individual values of X and Y. Given the probability data p_{ij}, we can compute

$$\langle Z \rangle = \sum_{ij} p_{ij} (x_i + y_j) = \sum_{ij} p_{ij} x_i + \sum_{ij} p_{ij} y_j = \langle X \rangle + \langle Y \rangle \,. \tag{18.41}$$

[3]One way to see this is to use Eq. (18.6), which we now write as $p_{ij} = p(x_i)p(y_j|x_i)$. If X and Y are independent $p(y_j|x_i)$ cannot depend on the choice of i, and can be denoted $p(y_j)$.

The mean of $X + Y$ is the sum of the means of X and Y whether or not X and Y are independent. A similar analysis leads to the more general result

$$\langle aX + bY \rangle = a\langle X \rangle + b\langle Y \rangle . \tag{18.42}$$

We might also want the variance of Z. We can proceed by writing

$$\begin{aligned}\sigma^2(Z) &= \langle Z^2 \rangle - \langle Z \rangle^2 = \langle (X+Y)^2 \rangle - \langle X+Y \rangle^2 \\ &= \langle X^2 \rangle + 2\langle XY \rangle + \langle Y^2 \rangle - \langle X \rangle^2 - 2\langle X \rangle \langle Y \rangle - \langle Y \rangle^2 .\end{aligned}$$

We now collect terms in a way permitting the use of Eqs. (18.26) and (18.40), reaching

$$\sigma^2(Z) = \sigma^2(X) + \sigma^2(Y) + 2\,\mathrm{cov}(X,Y) . \tag{18.43}$$

Equation (18.43) tells us that the variances of X and Y add only if these two variables are independent of each other.

Finally, remembering that the variance of aX is $a^2\sigma^2(X)$:

$$\sigma^2(aX) = \langle (aX)^2 \rangle - \langle aX \rangle^2 = a^2\sigma^2(X) ,$$

we compute the variance of $Z = aX + bY$, finding

$$\sigma^2(Z) = a^2\sigma^2(X) + b^2\sigma^2(y) + 2ab\,\mathrm{cov}(X,Y) . \tag{18.44}$$

Example 18.2.3. Covariance and Correlation

We consider two related problems, both of which deal with the drawing of two balls from a box that originally contained some black and some red balls. We define two random variables: X, which has value 1 if the first ball is black and zero if it is red; and Y, which has the value 1 if the second ball is black and zero if it is red.

Our first problem is the apparently trivial case that the box contains only two balls, one of each color. Our sample space contains only two members: $x_1 = 1$, $y_1 = 0$ and $x_2 = 0$, $y_2 = 1$, corresponding in colors to (black, red) and (red, black). These two members occur with equal probability, 1/2.

First we compute the means of X and Y:

$$\mu_x = \langle X \rangle = \frac{1}{2}x_1 + \frac{1}{2}x_2 = \frac{1}{2}, \qquad \mu_y = \langle Y \rangle = \frac{1}{2}y_1 + \frac{1}{2}y_2 = \frac{1}{2}.$$

Next we compute the variances of X and Y. Let's do so (for X) by computing

$$\langle X^2 \rangle = \frac{1}{2}x_1^2 + \frac{1}{2}x_2^2 = \frac{1}{2}, \qquad \sigma^2(X) = \langle X^2 \rangle - \mu_x^2 = \frac{1}{2} - \left(\frac{1}{2}\right)^2 = \frac{1}{4}.$$

It is obvious that we can also find $\sigma^2(Y) = 1/4$, and that $\sigma(X) = \sigma(Y) = 1/2$.

Then, to obtain the covariance, from $\mathrm{cov}(X,Y) = \langle XY \rangle - \mu_x\mu_y$, we note that

$$\langle XY \rangle = \frac{1}{2}x_1y_1 + \frac{1}{2}x_2y_2 = 0 ,$$

leading to

$$\mathrm{cov}(X,Y) = 0 - \left(\frac{1}{2}\right)\left(\frac{1}{2}\right) = -\frac{1}{4}.$$

Finally we compute the correlation,

$$\operatorname{corr}(X,Y) = \frac{-1/4}{(1/2)(1/2)} = -1\,.$$

This result is to be expected, since there is complete correlation between the choices of X and Y; once a red ball has been drawn, the other ball **must** be black, and vice versa. The correlation has value -1 because the values for X and Y must be opposite.

Let's now proceed to a second problem, in which the box originally contains two balls of each color. If we number the balls and regard each numbered pair as a member of our sample space, the space would have 12 members, each of equal probability. Instead, let's build a sample space whose members are distinguishable by color; this space has four members, but they are not equally probable. Using the same numerical values as in the earlier problem, our space is:

$$\left(x_1 = 1,\ y_1 = 1,\ p_1 = \frac{1}{6}\right), \quad \left(x_2 = 1,\ y_2 = 0,\ p_2 = \frac{1}{3}\right),$$

$$\left(x_3 = 0,\ y_3 = 1,\ p_3 = \frac{1}{3}\right), \quad \left(x_4 = 0,\ y_4 = 0,\ p_4 = \frac{1}{6}\right).$$

We now compute μ_x and $\langle X^2\rangle$:

$$\mu_x = \frac{1}{6}(x_1 + x_4) + \frac{1}{3}(x_2 + x_3) = \frac{1}{2}, \quad \langle X^2\rangle = \frac{1}{6}(x_1^2 + x_4^2) + \frac{1}{3}(x_2^2 + x_3^2) = \frac{1}{2}.$$

Continuing as in the earlier problem, we find

$$\mu_y = \frac{1}{2}, \quad \sigma^2(X) = \sigma^2(Y) = \frac{1}{4}, \quad \sigma(X) = \sigma(Y) = \frac{1}{2}.$$

Those results are the same as for the two-ball problem. But the covariance turns out differently. The only nonzero contribution to $\langle XY\rangle$ is from sample point 1, yielding

$$\langle XY\rangle = \frac{1}{6}x_1y_1 = \frac{1}{6}, \qquad \operatorname{cov}(X,Y) = \frac{1}{6} - \mu_x\mu_y = \frac{1}{6} - \frac{1}{4} = -\frac{1}{12}.$$

The correlation is $\dfrac{-1/12}{(1/2)(1/2)} = -\dfrac{1}{3}$. X and Y are less highly correlated than in our first problem. ■

Exercises

18.2.3. Repeat the analysis of Example 18.2.3 for a box that originally contained 4 black balls and 2 red balls.

18.2.4. Two cards are drawn from a standard 52-card deck of playing cards. The first card is not returned to the deck before drawing the second card. Let X be a random variable with value 1 if the first card is a spade and zero otherwise, let Y be a random variable with value 1 if the second card is a spade and zero otherwise, and let Z be a random variable with value 1 if the second card is red and zero otherwise.

(a) Calculate the mean values and variances of X, Y, and Z.

(b) Calculate the covariance and correlation of each pair of two variables (X,Y), (X,Z), and (Y,Z).

(c) Make comments that provide qualitative explanations for the results of part (b).

18.2.5. Repeat Exercise 18.2.4 for a situation in which the first card drawn is returned to the deck and the deck is reshuffled before drawing the second card. Your answer should include the comments demanded in part (c).

CONTINUOUS DISTRIBUTIONS

Up to this point, we have restricted attention to discrete random variables. But there are many problems in which a random variable can have any value in a continuous range. An example would be the time at which a radioactive particle decays; the decay can take place at any time and is caused by a nuclear reaction whose behavior is only predictable statistically.

Our method for dealing with continuous probability distributions must be by a limiting process similar to those used often used in calculus. Continuing with the radioactive decay problem, it makes no sense to associate a finite nonzero probability with each instant in time (there is an infinite number of such instants). It is more useful to recognize that for a short time interval Δt the decay probability will have some value proportional to Δt, and that the relevant quantity is the decay probability per unit time (computed in the limit of small Δt). Since this probability of decay per unit time may differ at different times, a general description for the decay within a time interval dt at time t must be of the form $f(t)\,dt$, where f can be called a **probability function**. Then the overall probability of decay during a time interval from t_1 to t_2 will be given by

$$\text{Probability of decay between times } t_1 \text{ and } t_2 = \int_{t_1}^{t_2} f(t)\,dt\,.$$

By making an analogy between decay per unit time and mass per unit volume (which is **density**), the probability function $f(t)$ is often referred to as a **probability density**.

We now restate and slightly expand the ideas of the preceding paragraph, framing the discussion in terms of a continuous random variable X whose probability density is $p(x)$. We assume the range of x to be $(-\infty, \infty)$. If part of that entire range is irrelevant, we simply set $p(x) = 0$ there. The probability distribution of our random variable has the following features:

(1) The probability that X has a value in the range $(x, x+dx)$ is $p(x)\,dx$.

(2) Because $p(x)\,dx$ must represent a probability, $p(x)$ must for all x be nonnegative.

(3) Because the total probability for all x must sum to unity, $p(x)$ must be such that

$$\int_{-\infty}^{\infty} p(x)\,dx = 1\,. \tag{18.45}$$

The reader may note that we did not impose a requirement that $p(x) \le 1$. In fact, $p(x)$ may even become infinite, providing it does so in a way that still permits the integral relation of Eq. (18.45) to be satisfied.

Now that we have a way of describing continuous probability distributions, let's see how to compute average values, including the mean and the variance. The sums of our earlier subsections now need to be replaced by integrals. The expectation value of a function $g(x)$ is given by

$$\langle g \rangle = \int_{-\infty}^{\infty} g(x) p(x)\, dx\,. \tag{18.46}$$

A first application of Eq. (18.46) is to the mean and variance of a continuous random variable X, which are described by the formulas

$$\langle X \rangle = \mu = \int_{-\infty}^{\infty} x p(x)\, dx\,, \tag{18.47}$$

$$\sigma^2 = \int_{-\infty}^{\infty} (x-\mu)^2\, p(x)\, dx\,. \tag{18.48}$$

Example 18.2.4. Classical Particle in Potential Well

A particle of unit mass moves (assuming classical mechanics) subject to the potential $V = x^2/2$, with a total energy $E = 1/2$ (dimensionless units). These conditions correspond to motion with kinetic energy $T = (1-x^2)/2$, so that when the particle is at x its velocity can be found from

$$\frac{v^2}{2} = \frac{1-x^2}{2} \qquad \text{leading to} \quad v(x) = \pm\sqrt{1-x^2}\,.$$

We see from this form for $v(x)$ that the particle will move back and forth between turning points at $x = \pm 1$, and that it will move fastest at $x = 0$ and momentarily become stationary at $x = \pm 1$.

The probability density for the particle's position will be proportional to the time spent in each element dx of its range, which in turn is (at x) proportional to $1/|v(x)|$. We therefore have

$$p(x) = \frac{C}{\sqrt{1-x^2}}\,, \tag{18.49}$$

where C must be assigned a value such that

$$\int_{-1}^{1} p(x)\, dx = \int_{-1}^{1} \frac{C\, dx}{\sqrt{1-x^2}} = 1\,.$$

Evaluating the integral, we find $C = 1/\pi$.

Let's comment briefly on the formula for $p(x)$ given in Eq. (18.49), and which we plot in Fig. 18.3.

The probability density becomes infinite at the turning points (the velocity is zero there) but does so in a way such that the overall probability for any region near $x = \pm 1$ remains finite.

Given $p(x)$, we can now compute the mean and variance of X, viewed as a random variable. We start by computing $\langle X \rangle$ and $\langle X^2 \rangle$. We get

$$\langle X \rangle = \int_{-1}^{1} x p(x)\, dx = \int_{-1}^{1} \frac{x\, dx}{\pi\sqrt{1-x^2}} = 0\,, \qquad \text{(a result due to symmetry)},$$

$$\langle X^2 \rangle = \int_{-1}^{1} x^2\, p(x)\, dx = \int_{-1}^{1} \frac{x^2\, dx}{\pi\sqrt{1-x^2}} = \frac{1}{2}\,, \qquad \text{(the reader can check this)}.$$

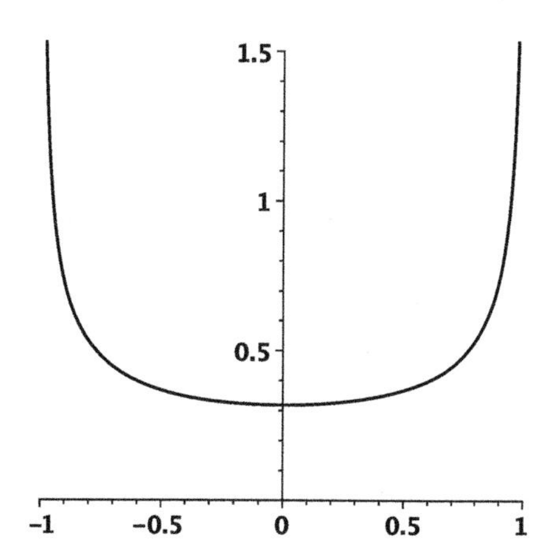

Figure 18.3: Probability density, classical particle in potential well. Turning points of the oscillation at ± 1.

From the first of these integrals we read out $\langle X \rangle = 0$; to compute the variance we can use Eq. (18.26), which is true for both discrete and continuous probability distributions:

$$\sigma^2 = \langle X^2 \rangle - \langle X \rangle^2 = \frac{1}{2} - 0 = \frac{1}{2}.$$

This result is equivalent to the standard deviation $\sigma = 1/\sqrt{2} = 0.71$. The standard deviation is larger than half way to the turning points because the particle spends a majority of the time in the outer half of its excursions.

■

CUMULATIVE DISTRIBUTION FUNCTIONS

Sometimes we are interested in the probability that the values of a continuous random variable X will fall between specified values. Calculations of that sort are most easily carried out if we first define a **cumulative distribution function** of definition

$$P(x) = \int_{-\infty}^{x} p(x)\, dx\,. \tag{18.50}$$

Then the probability that x is between x_1 and x_2 (with $x_2 > x_!$) is $P(x_2) - P(x_1)$. In favorable cases it may be possible to obtain an analytical expression for $P(x)$.

COVARIANCE FOR CONTINUOUS DISTRIBUTIONS

The covariance and correlation of continuous probability distributions X and Y are given by the formulas

$$\text{cov}(X,Y) = \int_{-\infty}^{\infty}\int_{-\infty}^{\infty} p(x,y)(x - \langle X \rangle)(y - \langle Y \rangle)\, dx\, dy\,, \tag{18.51}$$

$$\text{corr}(X,Y) = \frac{\text{cov}(X,Y)}{\sigma(X)\sigma(Y)}\,. \tag{18.52}$$

Exercises

18.2.6. For the continuous probability density

$$p(x) = \begin{cases} xe^{-x}, & x \geq 0, \\ 0, & x < 0, \end{cases}$$

(a) Make a plot of $p(x)$ for a range of x at least as large as $(-1, +5)$.

(b) Locate on your graph the most probable x value and verify that it corresponds to that found by differentiation of $p(x)$.

18.2.7. For the probability density of Exercise 18.2.6,

(a) Verify that $p(x)$ is normalized, i.e., that its integral over all x is unity.

(b) By evaluating appropriate integrals, find μ, σ^2, and the standard deviation σ.

(c) Obtain an integral representing the cumulative distribution function and use it to find the probabilities that

$$x > 1, \qquad |x - \mu| < \sigma, \qquad |x - \mu| < 2\sigma, \qquad |x - \mu| < 3\sigma.$$

(d) Verify that the results of part (c) are consistent with Chebyshev's inequality, Eq. (18.30).

18.2.8. For the probability density of Exercise 18.2.6, find the average of X for the portion of the probability distribution that lies outside the range $(\mu - \sigma, \mu + \sigma)$, i.e., for the x values more than one standard deviation from the mean.

Hint. Don't forget to allow for the fact that the total probability for X to be in only a part of its range is not unity.

18.2.9. The probability density for a pair of random variables X and Y has the following form:

$$p(x, y) = \begin{cases} 2\,e^{-(x+y)^2}, & x, y \geq 0, \\ 0, & \text{otherwise}. \end{cases}$$

(a) Verify that $p(x, y)$ is normalized, i.e., that its integral over all x and y is unity.

Hint. Use symbolic computing to evaluate the double integral. It is probably easiest to do so in two steps: integrate over x for an undefined value of y; then integrate the result over y.

(b) Compute the mean and variance of X and Y.

(c) Compute the covariance and correlation of X and Y.

(d) Comment on the qualitative significance of your result for part (c).

18.3 BINOMIAL DISTRIBUTION

This section is the first of three devoted to probability distributions of sufficiently frequent occurrence that they warrant detailed study even in a short presentation such as that of this chapter.

The **binomial distribution** is the discrete probability distribution resulting from repeated independent trials of individual events each of which can have two outcomes with constant (but not necessarily equal) probabilities. Such repeated trials are sometimes referred to as **Bernoulli trials**. Examples include repeated tosses of unbiased coins, or of the same biased coin, repeated attempts to obtain a specified number when a die is thrown, or the possibility of obtaining successful completion of any repeated experiment when individual successes can be treated as random with a fixed probability. Our interest in the binomial distribution is not limited to basic properties such as the mean and variance, but also extends to more detailed questions such as the probability of obtaining a specific result within a given number of attempts.

We start the analysis of the binomial distribution by developing a formula for $B(n, s, p)$, the probability of s successes when a process with an individual success probability p is carried out independently n times. We can determine $B(n, s, p)$ in two steps, of which the first is to count the number of ways s trials designated for success can be selected from the total of n trials, namely the number of combinations of s objects from a population of n, i.e., the quantity $C(n, s)$, which we showed in Eq. (18.16) to be the binomial coefficient $\binom{n}{s}$.

Each combination of trials marked for success (or failure) will occur with an overall probability that we can compute as the product of the individual probabilities for the mutually exclusive items contained therein: s successes, each at probability p, together contributing p^s, and $n - s$ failures, each at probability $q = 1 - p$, together contributing q^{n-s}. Combining these factors with the binomial coefficient, we get

$$B(n, s, p) = \binom{n}{s} p^s \, q^{n-s} = \binom{n}{s} p^s (1 - p)^{n-s} \, . \tag{18.53}$$

An overview of the properties of the probability distribution $B(n, s, p)$ can be obtained from plots of the set of s values for various values of n, the number of trials. Taking initially $p = 1/2$, Fig. 18.4 shows the probabilities of all s values for $n = 20$, $n = 200$, and $n = 2000$. In these plots the horizontal axis is s/n (the fraction of the trials that are successful); the vertical axis is (n times the probability of success), i.e., $nB(n, s, p)$. With the coordinates scaled in this way, the area under a curve defined

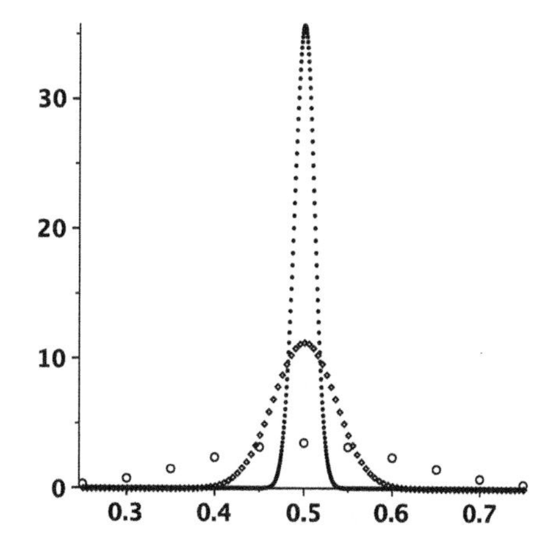

Figure 18.4: Binomial distributions for $p = 0.5$: $n = 20$, 200, 2000. Horizontal axis: Fraction of successes, s/n; vertical axis: n times the probability of s successes.

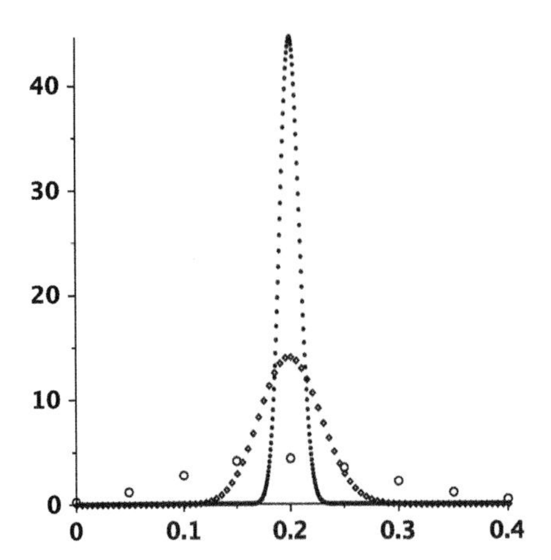

Figure 18.5: Binomial distributions for $p = 0.2$: $n = 20$, 200, 2000. Horizontal axis: Fraction of successes, s/n; vertical axis, n times the probability of s successes.

by each point set will be the same, corresponding to unit probability when one adds the probabilities of all s values.

All three of the graphs show distributions in which the s value of maximum probability corresponds to success in half the trials, i.e., at $s = n/2$. It is also clear that as n is increased, the fraction of successes clusters more tightly about the mean, so that for large n it becomes very unlikely that this fraction will deviate a large amount from 1/2. Keeping in mind the differing vertical scales of the graphs, we see that obtaining success in exactly half the trials becomes less likely as n increases; in fact, for large n it can be shown that the maximum value of $B(n, s, p)$ varies approximately as $n^{-1/2}$. However, this decrease is more than compensated by the fact that there are then more values of s/n that are close to 1/2.

Some similar features can be observed for individual trial probabilities p other than 1/2. Figure 18.5 shows the probability distributions for $p = 0.2$ and $n = 20$, 200, and 2000. The distributions are not entirely symmetric about the most probable s (a visual indication of this is that the rising and falling dot patterns are not at the same vertical positions). The most probable s is for all n is at 20% of the trials. The $p = 0.2$ distribution peaks more sharply with increasing n than for $p = 1/2$; in the next subsection we obtain a quantitative understanding of this observation.

MEAN AND VARIANCE

In principle we could compute the mean and variance for a set of n trials from the general formulas, Eqs. (18.21) and (18.26):

$$\langle S\rangle = \sum_s s\,B(n,s,p) = \sum_{s=0}^{n} s\binom{n}{s}p^s(1-p)^{n-s}\,, \tag{18.54}$$

$$\langle S^2\rangle = \sum_{s=0}^{n} s^2\binom{n}{s}p^s(1-p)^{n-s}\,, \tag{18.55}$$

$$\sigma^2(S) = \langle S^2\rangle - \langle S\rangle^2\,, \tag{18.56}$$

but that would require more effort than is necessary. It is easier to note, based on Eq. (18.41), that a sum of random variables will have a mean that is the sum of their individual means, and, based on Eq. (18.43), that a sum of random variables with zero covariance will have a variance that is the sum of their individual variances. Since a single trial has mean p and a variance that we can compute from

$$\langle X^2 \rangle = p(1^2) + (1-p)(0^2) = p \quad \text{and} \quad \langle X \rangle = p\,,$$

namely

$$\sigma^2(X) = \langle X^2 \rangle - \langle X \rangle^2 = p - p^2 = p(1-p) = pq\,,$$

we conclude (without computation) that

$$\langle S \rangle = np \quad \text{and} \quad \sigma^2(S) = np(1-p) = npq\,. \tag{18.57}$$

The first of Eqs. (18.57) confirms our expectation that the mean number of successes for n independent trials will be n times the individual success probability. The second of Eqs. (18.57) tells us that the standard deviation of $\langle S \rangle$ will vary as the square root of the number of trials, specifically as $\sqrt{npq}$.

For large numbers of trials we are more likely to be interested in the behavior of S/n than in that of S. We note that S/n is the fraction of trials that are successful and $\sigma^2(S/n)$ indicates its spread. From Eq. (18.57) and the general relations in Eqs. (18.42) and (18.29), we have

$$\langle S/n \rangle = p \qquad \text{and} \qquad \sigma^2(S/n) = \frac{pq}{n}\,. \tag{18.58}$$

The first of the above results is the expected result that the average probability of success in multiple trials is the same as the probability of success in a single trial. The second of the above results has a deeper significance; it reveals how the success fraction narrows about its average as the number of trials increases. More specifically, we see that the width of the S/n distribution varies as $\sigma \sim n^{-1/2}$, in agreement with the graphs previously presented as Figs. 18.4 and 18.5.

MORE DETAILED PROPERTIES

The probability distribution $B(n,s,p)$ provides enough information that we can answer more detailed questions about trials described by the binomial distribution. For example, $B(n,0,p)$ gives the probability that none of a set of n trials will succeed, and $B(n,n,p)$ gives the probability that all n trials succeed. The latter may be of importance in product distribution contexts; if an item has a nonzero probability of being defective, it may be useful to know how likely it is that a box of n similar items contains one or more that is defective.

SYMBOLIC COMPUTATION

Here is code to compute the binomial probability distribution given in Eq. (18.53). Because this code is needed for the Exercises, the reader may want to enter it into a workspace and preserve it for future use.

MAPLE:
```
> BinomialB := proc(n,s,p) binomial(n,s)*p^s*(1-p)^(n-s)
> end proc:
```

MATHEMATICA:
```
BinomialB[n_,s_,p_] := Binomial[n,s]*p^s*(1-p)^(n-s);
```

Here are some check values:

```
> BinomialB(3,1,1/4);
```

$$\frac{27}{64}$$

```
BinomialB[3,1,0.25]
```

0.421875

Exercises

18.3.1. Enter into a workspace the procedure BinomialB for your symbolic computing system and verify that it gives correct values for all s when $n = 3$ and $p = 1/4$.

18.3.2. Referring to the material in Appendix A on point plots and using the symbolic code for $B(n, s, p)$, make plots of the following binomial distributions (giving the probability of each s for given values of n and p). For $n = 20$, take $p = 0.1$, 0.4, and 0.7. For $n = 100$ and for $n = 1000$, take $p = 0.01$, 0.1, and 0.4.

18.3.3. Write a formula for $(x + y)^n$ using the binomial theorem, and from those results show how it can be concluded that the binomial distribution is normalized, i.e., that

$$\sum_{s=0}^{n} B(n, s, p) = 1\,,$$

irrespective of the value of p.

18.3.4. Form the difference $B(n, s + 1, p) - B(n, s, p)$ and from it show that the s value of maximum probability (for given n and p) is within one unit of np.

18.3.5. Write symbolic code that returns, as a decimal, the probability of the s value of a binomial distribution that is closest to np. Remember that s must be an integer. Don't worry about the special case in which np is exactly half way between two integers.

Hint. The integer closest to an arbitrary number is obtained by calling `round` (MAPLE) or `Round` (MATHEMATICA).

18.3.6. Write symbolic code that computes the cumulative probability distribution for n trials, single-trial probability p, and numbers of successes within the integer range $s_1 \le s \le s_2$. Test your code by seeing if it gives correct results for various cases that you can compute easily by hand.

18.3.7. (a) For $n = 20$ and $p = 0.4$, compute the mean and standard deviation of the binomial distribution.

(b) Identify the range of successes s_1 through s_2 that fall within one standard deviation of the mean (they must be integers), and compute the fraction of the distribution that is within this range of s. The code needed for this calculation was the topic of Exercise 18.3.6.

(c) Repeat the analysis of part (b) for regions within 2 and 3 standard deviations about the mean. Verify that these results are consistent with the Chebyshev inequality.

18.3.8. An advanced digital circuit board contains 100 individual microprocessors. The manufacturing process has been pushed toward its technical limits, with the result that each microprocessor has a 1% chance of being defective.

(a) What fraction of the circuit boards can be, on average, expected to be entirely defect-free?

(b) How low would the individual-microprocessor defect rate need to be to cause the fraction of defect-free circuit boards to reach 50%?

18.4 POISSON DISTRIBUTION

The Poisson distribution is applicable to repeated events that occur at a constant probability per unit time. A typical example is provided by the decay of a radioactive isotope, providing that the period of observation is short enough that a negligible proportion of the original sample has decayed. For species with decay lifetimes measured in thousands or millions of years, this is certainly a physically reasonable assumption.

We approach the Poisson distribution by considering a situation in which an average of ν events occur per unit time. Further, we define $F_n(t)$ (for $n = 0$, 1, 2, ...) as the probability that exactly n events occur during a time interval t. In addition, let dt be a time interval that is short enough that the probability of more than one event within dt is negligible, so the probability of an event within dt is $\nu\, dt$ and the probability of no event within dt is $1 - \nu\, dt$.

With the above notation and definitions, we now write down an expression for $F_n(t + dt)$, the probability that exactly n events occur in a time interval $t + dt$. Two mutually exclusive possibilities that add to give $F_n(t + dt)$ are

(n events in time t and none in the following dt),
with probability $F_n(t)(1 - \nu\, dt)$.

($n - 1$ events in time t and one in the following dt),
with probability $F_{n-1}(t)\ (\nu\, dt)$.

Combining these possibilities, we have

$$F_n(t + dt) = F_n(t)(1 - \nu\, dt) + F_{n-1}(t)\ (\nu\, dt)\,.$$

Rearranging the above, we get, in the limit of small dt,

$$\frac{F_n(t + dt) - F_n(t)}{dt} = \nu\, F_{n-1}(t) - \nu\, F_n(t) \quad \longrightarrow \quad \frac{dF_n(t)}{dt} = \nu\, F_{n-1}(t) - \nu\, F_n(t)\,. \tag{18.59}$$

Equation (18.59) is actually a set of recursive equations that for each n relates F_n and F_{n-1}. However, note that in the special case $n = 0$, there will be no F_{n-1} term, as the only possibility for $F_0(t + dt)$ is "no event in time t and none in time dt."

We can solve Eqs. (18.59) successively, starting with that for $n = 0$, which is

$$\frac{dF_0(t)}{dt} = -\nu\, F_0(t)\,, \quad \text{with solution} \quad F_0(t) = e^{-\nu t}\,. \tag{18.60}$$

In obtaining the solution shown here we have set the arbitrary constant in the general solution so as to make $F_0(0) = 1$, corresponding to the certainty that no events occur in a time interval of length zero.

Once we have $F_0(t)$, we can solve Eq. (18.59) for $n = 1$. We have

$$F_1' + \nu F_1 = \nu e^{-\nu t}, \quad \text{with general solution} \quad F_1(t) = (C + \nu t)\, e^{-\nu t}.$$

We must set $C = 0$ because $F_1(0)$ must vanish.

Continuing to larger n, it can be shown (Exercise 18.4.1) that

$$F_n(t) = \frac{(\nu t)^n}{n!}\, e^{-\nu t}. \tag{18.61}$$

We now define the **Poisson distribution** $P(n, \nu)$ as the probability that exactly n events occur in a unit time interval, given that ν, the average number of events per unit time, is a constant. Since $P(n, \nu)$ is the value of $F_n(t)$ for $t = 1$, we have from Eq. (18.61)

$$P(n, \nu) = \frac{\nu^n}{n!}\, e^{-\nu}. \tag{18.62}$$

Equation (18.62) describes the distribution of a random variable N whose possible values are $n = 0, 1, 2, \ldots$, with respective probabilities $P(n, \nu)$. In deriving the form for $P(n, \nu)$ we made implicit use of the fact that probabilities for a random variable add to unity, but we did not make that an explicit condition in the derivation. Let's check to verify that $P(n, \nu)$ is properly scaled. We get

$$\sum_{n=0}^{\infty} P(n, \nu) = \sum_{n=0}^{\infty} \frac{\nu^n}{n!}\, e^{-\nu} = e^{-\nu} \sum_{n=0}^{\infty} \frac{\nu^n}{n!} = 1. \tag{18.63}$$

The summation is just the power-series expansion of e^{ν}.

MEAN AND VARIANCE

It is also of interest to determine the mean and variance of the random variable N. These can be computed as follows:

$$\langle N \rangle = \sum_{n=0}^{\infty} n \left(\frac{\nu^n}{n!} \right) e^{-\nu} = \nu \sum_{n=0}^{\infty} \frac{\nu^{n-1}}{(n-1)!}\, e^{-\nu} = \nu e^{-\nu} \sum_{n=0}^{\infty} \frac{\nu^n}{n!} = \nu,$$

$$\langle N^2 \rangle = \sum_{n=0}^{\infty} n^2 \left(\frac{\nu^n}{n!} \right) e^{-\nu} = \sum_{n=0}^{\infty} \left[\frac{1}{(n-2)!} + \frac{1}{(n-1)!} \right] \nu^n e^{-\nu},$$

$$= \nu^2 e^{-\nu} \sum_{n=0}^{\infty} \frac{\nu^{n-2}}{(n-2)!} + \nu e^{-\nu} \sum_{n=0}^{\infty} \frac{\nu^{n-1}}{(n-1)!} = \nu^2 + \nu.$$

In evaluating the summations we have used the fact that negative integer factorials are infinite. From the above equations we can read out

$$\langle N \rangle = \nu, \tag{18.64}$$

$$\sigma^2 = \langle N^2 \rangle - \langle N \rangle^2 = (\nu^2 + \nu) - (\nu)^2 = \nu. \tag{18.65}$$

The mean and variance are both equal to ν, the event probability per unit time. It is also instructive to calculate the mean and variance of N/ν. Using Eqs. (18.42) and (18.29), we find

$$\langle N/\nu \rangle = 1 \quad \text{and} \quad \sigma^2(N/\nu) = \nu^{-1}. \tag{18.66}$$

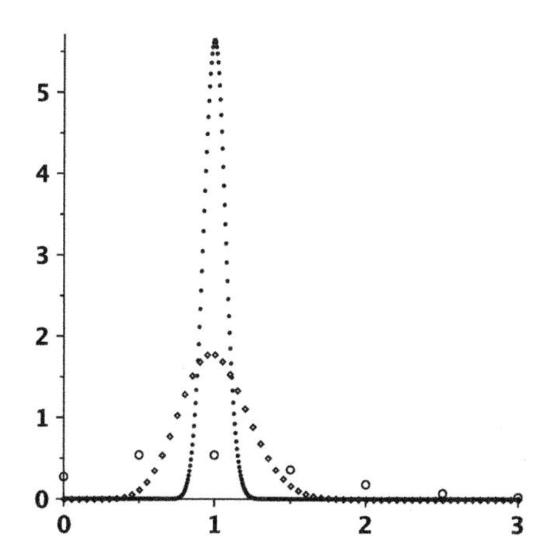

Figure 18.6: Poisson distributions: $\nu = 2$, 20, 200. Horizontal axis: Number of events n, plotted as n/ν; vertical axis: ν times the probability of n events.

To gain further perspective on the Poisson distribution, we show in Fig. 18.6 graphs of $P(n,\nu)$ for $\nu = 2$, 20, and 200. In these plots the horizontal axis is n/ν (the number of events per unit time expressed as a multiple of the average number of such events), and the vertical axis is $\nu P(n,\nu)$ (ν times the probability per unit time for exactly n events).

For small ν the plots are rather asymmetric, with significant values of $P(0,\nu)$, the probability that no events will occur in a unit time interval. For large ν the plots peak sharply about $n/\nu = 1$, become more nearly symmetric, and exhibit negligible probability of zero events per unit time. The peak location and width shown in the plots are those that are expected for the random variable N/ν, based on the results given in Eq. (18.66).

SYMBOLIC COMPUTATION

Here is code to compute the Poisson probability distribution given in Eq. (18.62). Because this code is needed for the Exercises, the reader may want to enter it into a workspace and preserve it for future use.

MAPLE:
```
> PoissonP := proc(n,nu); nu^n/n! * exp(-nu) end proc:
```

MATHEMATICA:
```
PoissonP[n_,nu_] := nu^n/n! * E^(-nu)
```

Here are some check values:

```
> PoissonP(4, 6);
```
$54\,e^{-6}$

```
> PoissonP(4, 6.);
```
0.1338526176

```
PoissonP[10, 8.]
```
0.0992615

Exercises

18.4.1. Establish the formula for $F_n(t)$ of general n, Eq. (18.61), by mathematical induction (see Appendix F). Values for $n = 0$ and $n = 1$ are available, so it is sufficient to show that F_n can be obtained from F_{n-1} for general integer $n \geq 2$.

18.4.2. Enter into a workspace the procedure `PoissonP` for your symbolic computing system and verify that it gives correct results for several different values of `n` and `nu`.

18.4.3. Referring to the material in Appendix A on point plots and using the symbolic code for $P(n, \nu)$, make plots of the following Poisson distributions (showing the probability of each n from zero to a value such that $P(n, \nu)$ is negligible). Take $\nu = 3$, 10, 100, and 1000.

18.4.4. If you receive spam e-mail at an average rate of five messages per day (an unrealistically small number), how many days per 30-day month would you expect to receive (a) exactly five spam messages; (b) more than five; (c) more than 10; (d) exactly one; (e) none at all?

18.4.5. Calculate the probability that exactly two of a group of 500 people have January 1 as their birthday, assuming a 365-day year and excluding February 29, in the following two ways:

1. Using the binomial distribution (which is exact) and
2. Using the Poisson distribution (which assumes that persons whose birthdays have already been identified are not removed from subsequent probability calculations).

18.5 GAUSS NORMAL DISTRIBUTION

In contrast to the binomial and Poisson distributions, the **Gauss normal** distribution is a continuous distribution on the range $(-\infty, \infty)$. It is of particular importance because it can be shown that in the limit of large numbers of trials or events, a wide class of other distributions approach a suitably defined normal distribution. It is beyond the scope of this text to prove that result, which is known as the **central limit theorem**. However, we do show, later in this section, that the binomial expansion for large n and the Poisson distribution for large ν approach normal distributions. The central limit theorem can be used as a justification for using the Gauss normal distribution when there is insufficient knowledge of the properties of a random variable to identify its actual distribution. For that reason, experimental data is often assumed to represent samples from a Gauss normal distribution.

The Gauss normal distribution is characterized by a bell-shaped curve corresponding to the probability density

$$g(x) = \frac{1}{\sigma\sqrt{2\pi}} \exp\left(-\frac{(x-\mu)^2}{2\sigma^2}\right). \tag{18.67}$$

It is apparent from its form that $g(x)$ is symmetric about μ, so that symbol has its usual meaning in probability theory, namely that it is the mean of the distribution. The constant preceding the exponential has the value needed to make the overall probability of the distribution add to unity; this claim can be checked by an examination of the integral

$$\int_{-\infty}^{\infty} g(x)\, dx\,.$$

If we make the substitution $(x - \mu)/\sigma\sqrt{2} = t$, then $dx/\sigma\sqrt{2} = dt$ and this integral can be recognized as one we have encountered before (see Section 9.5). We get

$$\int_{-\infty}^{\infty} g(x)\,dx = \frac{1}{\sqrt{\pi}} \int_{-\infty}^{\infty} e^{-t^2}\,dt = 1\,.$$

Similar operations confirm that the variance of $g(x)$ is indeed the quantity σ^2 appearing in Eq. (18.67).

Some Gauss normal probability distributions are shown (for $\mu = 0$) in Fig. 18.7. Note that as σ^2 increases, the curve broadens; since the total area under the curve remains unity, its maximum height must then decrease. If μ is assigned a nonzero value, the entire curve is uniformly shifted to the left or right to place its maximum at $x = \mu$.

Sometimes we are interested in the probability that x will fall between specified values. Calculations of that sort are most easily carried out using the notion of **cumulative distribution function**, defined in Eq. (18.50). For Gauss normal distributions, we define

$$G(x) = \int_{-\infty}^{x} g(x)\,dx\,, \tag{18.68}$$

so that the probability that x is between x_1 and x_2 (with $x_2 > x_1$) takes the form $G(x_2) - G(x_1)$.

If we scale a Gauss normal distribution by setting $\mu = 0$ and $\sigma^2 = 1$, we have what is called the **standard normal distribution**. In probability contexts, the probability density and the cumulative distribution function are then denoted $\varphi(z)$ and $\Phi(z)$, where

$$\begin{aligned} \varphi(z) &= \frac{1}{\sqrt{2\pi}}\, e^{-z^2/2}\,, \\ \Phi(z) &= \frac{1}{\sqrt{2\pi}} \int_{-\infty}^{z} e^{-t^2/2}\,dt\,. \end{aligned} \tag{18.69}$$

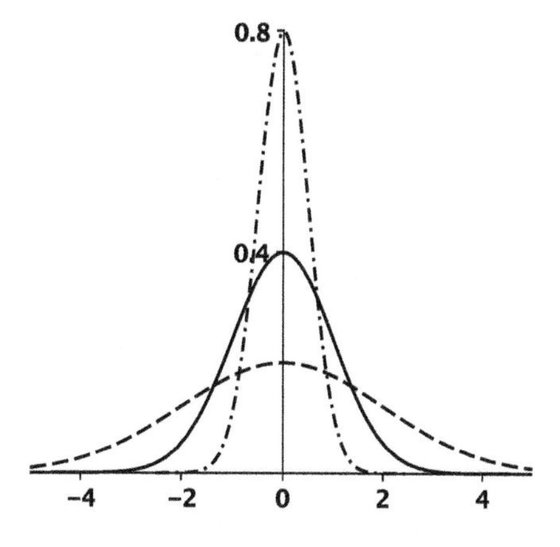

Figure 18.7: Gauss normal distributions for $\mu = 0$: $\sigma = 0.5$ (dash-dot), 1 (solid), 2 (dash).

The expression for Φ can be written in terms of error functions. As shown in Eq. (9.46), we have

$$\Phi(z) = \frac{1}{2}\left[1 + \text{erf}\left(\frac{z}{\sqrt{2}}\right)\right]. \tag{18.70}$$

Part of the value of the standard normal distribution is that we can write properties of a general Gauss distribution in terms of φ and Φ. For a Gauss distribution of mean μ and variance σ^2, we have

$$\begin{aligned} g(x) &= \frac{1}{\sigma}\,\varphi(z)\,, \\ G(x) &= \Phi(z)\,, \end{aligned} \qquad z = \frac{x-\mu}{\sigma}. \tag{18.71}$$

Example 18.5.1. Gauss Normal Distribution

Let's identify some properties of the Gauss normal distribution. First, note that the maximum of the distribution occurs at $x = \mu$, and that the height of the maximum is $1/\sigma\sqrt{2\pi}$. Next, it may be of interest to find the distance at which the probability density is half its maximum value. This point is at the values of x such that

$$\exp\left(-\frac{(x-\mu)^2}{2\sigma^2}\right) = \frac{1}{2},$$

the solution to which is

$$x - \mu = \pm\sigma\sqrt{2\ln 2}\,.$$

Of more importance than the above is the probability that x will be within 1, 2, or 3 standard deviations from its mean. These calculations involve the cumulative distribution function. Referring to Eq. (18.71) and the definition of G, the probability for $|x - \mu| < k\sigma$ is given (irrespective of the value of σ, as $\Phi(k) - \Phi(-k)$. From Eq. (18.70) and the relation $\text{erf}(-z) = -\text{erf}(z)$, we find

$$p(|x-\mu| < k\sigma) = \Phi(k) - \Phi(-k) = \text{erf}(k/\sqrt{2})\,. \tag{18.72}$$

Evaluating this expression by symbolic computing, we find

$$p(|x-\mu| < \sigma) = 0.683, \qquad p(|x-\mu| < 2\sigma) = 0.954, \qquad p(|x-\mu| < 3\sigma) = 0.997.$$

These computations show that the probability of being within a given number of standard deviations from the mean approaches certainty far more rapidly than is required by the Chebyshev inequality. For example, that inequality indicates that the probability of being outside 3σ is less than $1/9$ (so the probability of being inside exceeds $8/9 = 0.888$, much smaller than the corresponding probability for the Gauss normal distribution, which is 0.997.

Summarizing the above, we see that the probability of being within k standard deviations of the mean is independent of the value of σ, showing that σ is a useful and unique descriptor for Gauss normal distributions at all scales.

■

Example 18.5.2. Quality Control

Let's assume that the amounts of cereal dispensed by a packaging machine are described by a Gauss normal distribution and that the machine can be set to produce packages whose contents have a specified mean weight. A test reveals that (irrespective of the size of the packages) 10% of the cereal boxes contain at least 1 ounce less than the mean weight. The packager will be subject to a fine if too many of the cereal boxes are short of their nominal content. To what weight must the packager set the machinery so that 95% of the 16-ounce boxes actually contain at least 16 ounces?

Our first step toward solving this problem is to use the test data to determine the standard deviation σ of the assumed Gauss normal distribution. We observe that if 10% of the boxes are at least 1 ounce below the mean, there is another 10% that is at least 1 ounce above the mean, and we therefore want to determine the multiple of the standard deviation within which the cumulative probability is $1 - 0.20 = 0.80$. Applying Eq. (18.72),

$$0.80 = \text{erf}(k/\sqrt{2}), \qquad \text{with solution} \quad k = 1.282,$$

leading to the conclusion that 1 ounce $= 1.282\sigma$, or $\sigma = 0.780$ ounce.

We next ask how many standard deviations from the mean are needed to reach 90% probability ($90\% = 1-2\times 5\%$). This question is answered by solving the equation

$$0.90 = \text{erf}(k'/\sqrt{2}), \qquad \text{yielding} \quad k' = 1.645.$$

We then compute $k'\sigma = 1.645 \times 0.780 = 1.283$ ounce, and that is the amount by which we make the mean exceed the nominal weight. So the machinery needs to be set for a mean weight of 17.3 ounces.

■

SYMBOLIC COMPUTATION

Here is code to compute the Gauss normal probability distribution and the corresponding cumulative distribution, given in Eq. (18.71). Because this code is needed for the Exercises, the reader may want to enter it into a workspace and preserve it for future use.

MAPLE:

```
> GaussG := proc(mu,sigma,x);
>    evalf(1/sqrt(2*Pi)/sigma * exp(-(x-mu)^2/2/sigma^2))
> end proc:
> GaussPhi := proc(mu,sigma,x) local zz;
>              zz := (x-mu)/sigma/sqrt(2.); (1+erf(zz))/2
>            end proc:
```

MATHEMATICA:

```
GaussG[mu_,sigma_,x_] :=
        1/Sqrt[2*Pi]/sigma * E^(-(x-mu)^2/2/sigma^2)
GaussPhi[mu_,sigma_,x_] :=
   Module[ {zz}, zz = (x-mu)/sigma/Sqrt[2];
          (1 + Erf[zz])/2 ]
```

Here are some check values of the probability distribution:

```
> GaussG(0, 1., 1.)                          0.2419707244
  GaussG[0, 1., 1.]                          0.241971
```

The next two statements give the cumulative probability for x to be within $1.5\,\sigma$ of the mean at $x = 0$:

```
> GaussPhi(0, 1., 1.5) - GaussPhi(0, 1., -1.5)         0.9331927988
  GaussPhi[0, 1., 1.5] - GaussPhi[0, 1., -1.5]         0.933193
```

Exercises

18.5.1. Verify as follows that the Gauss normal probability distribution given in Eq. (18.67) has variance σ^2:

(a) Write integrals for $\langle X\rangle$ and $\langle X^2\rangle$.

(b) Rewrite the integrals of part (a) in terms of $t = (x-\mu)/\sigma\sqrt{2}$ and use symmetry to simplify the resulting expressions.

(c) Evaluate any integrals that remain after you have completed part (b). Integration by parts may be helpful here.

18.5.2. Enter into a workspace the procedure GaussG for your symbolic computing system and verify its correctness by comparing its output with a hand computation.

18.5.3. Using the procedure GaussG, make plots of the following Gauss normal distributions (showing a range of width at least 3σ on either side of the maximum at $x = \mu$):

$$g(1,\ 0.707,\ x),\qquad g(-1,\ 0.5,\ x),\qquad g(0,\ 0.2,\ x),\qquad g(4,\ 1,\ x),$$

where

$$g(\mu,\sigma,x) = \frac{1}{\sigma\sqrt{2\pi}}\,\exp\left(-\frac{(x-\mu)^2}{2\sigma^2}\right).$$

18.5.4. Enter into a workspace the procedure GaussPhi for your symbolic computing system and verify that it gives results consistent with Eq. (18.72) for one of the cases given after that equation.

18.5.5. Using the procedure GaussPhi, calculate the following Gauss normal cumulative probabilities:

(a) For $\mu = 2$, $\sigma = 2$, probability that x is in the interval $1 \le x \le 4$.

(b) For $\mu = 0$, $\sigma = 0.5$, probability that x is in the interval $0.25 \le x \le 0.75$.

(c) For $\mu = 1$, $\sigma = \sqrt{2}$, probability that x is in the interval $0 \le x \le 2$.

(d) Probability that x is in the interval $\mu \le x \le \mu + \sigma$.

18.5.6. Use your symbolic computing system to solve the transcendental equations for k and k' in Example 18.5.2.

Hint. It is easier to substitute trial values of k into $\text{erf}(k/\sqrt{2})$ or to solve graphically than it is to seek a formal inversion of the equation.

18.5.7. For a standard normal distribution ($\sigma = 1$, $\mu = 0$), find the value of x_0 (with $x_0 > \mu$) such that the probability that x is in the interval $\mu \le x \le \mu + x_0$ is (a) 0.1, (b) 0.25, (c) 0.5.

LIMITS OF BINOMIAL AND POISSON DISTRIBUTIONS

For binomial distributions of large n and Poisson distributions of large ν, Gauss normal distributions form good approximations in the regions where the probability is appreciable. We do not prove this statement, but illustrate it graphically. Figure 18.8 shows a binomial distribution with $n = 80$ and $p = 0.4$, which we compare with a Gauss normal distribution of the same mean, $\mu = np = 32$, and the same variance, $\sigma^2 = npq = 19.2$.

A comparison of the Poisson and Gauss normal distributions is shown in Fig. 18.9.

The Poisson distribution is for $\nu = 100$; the corresponding Gauss normal distribution has $\mu = \nu = 100$ and $\sigma^2 = \nu = 100$.

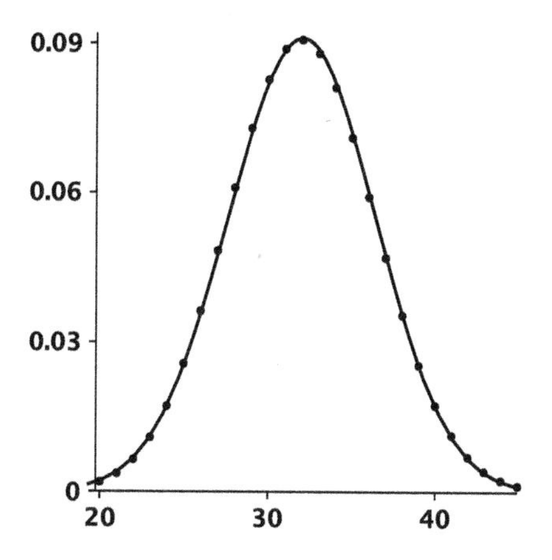

Figure 18.8: Comparison of binomial distribution for $p = 0.4$, $n = 80$ (dots) and Gauss normal distribution of the same mean and variance (solid curve).

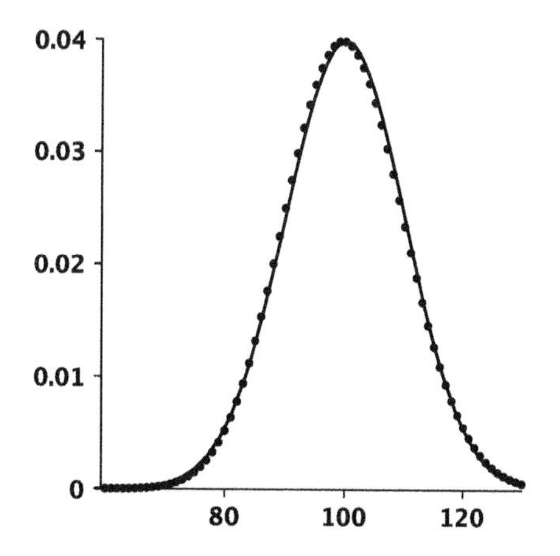

Figure 18.9: Comparison of Poisson distribution for $\nu = 100$ and Gauss normal distribution of the same mean and variance (solid curve).

Exercises

18.5.8. (a) For $n = 100$ and $p = 0.4$ create a binomial distribution plot that is assigned to the symbol `GraphB`. If needed, consult Exercises 18.3.1 and 18.3.2 for tips on making this plot.

(b) Determine the mean and variance of the binomial distribution that was plotted in part (a) and plot a Gauss normal distribution of that mean and variance, storing it by assignment to the symbol `GraphG`. If needed, consult Exercises 18.5.2 and 18.5.3 for tips on making this plot.

(c) Referring to the material in Appendix A on combination plots, combine `GraphB` and `GraphG` into a single graph so that they can be compared.

18.5.9. Proceed as in Exercise 18.5.8 to compare a Poisson distribution with $\nu = 80$ and a corresponding Gauss normal distribution.

18.6 STATISTICS AND APPLICATIONS TO EXPERIMENT

Two operations on experimental data that arise frequently are the following:

1. The determination of the most probable value and the degree of its certainty when we have repeated independent measurements of the same physical quantity.
2. The determination of the parameters describing a function of given form that yields an optimum fit to a set of data consisting of pairs (x_i, y_i).

We discuss these types of statistics problems in the present section of this chapter.

REPEATED MEASUREMENTS

We consider here a situation in which we do not have a functional characterization of the sample space corresponding to repeated measurements of the same physical quantity; instead, we have only a list of the results of a finite number of measurements. Our task is to deduce as much information as possible about the probability distribution that would result if we took an infinite number of measurements, using (of course) only the data that we have from the finite number of measurements that are actually available. Sometimes the hypothetical infinite set of data is referred to as the **parent population** of our problem; the data we actually have can then be called a **sample**. We will treat the experimental data as values of a random variable X, with the individual data points denoted x_i.

Our first task is to estimate the mean of the parent population, and the only relevant quantity available to us is the mean of our sample, which we assume contains n data points. We write $\langle X\rangle$ to denote the sample mean. Note that we do not identify this quantity with the symbol μ, because we reserve μ to stand for the unknown mean of the parent population. Following the practice illustrated throughout earlier sections of this chapter,

$$\langle X\rangle = \frac{1}{n}\sum_{i=1}^{n} x_i\,. \tag{18.73}$$

We next want to estimate the variance **of the mean**, i.e., the variance of $\langle X\rangle$. This is a meaningful quantity because a set of n measurements may not have the same average as the parent distribution, and different sets of n measurements may result in different averages. These observations are consistent with the notion that $\langle X \rangle$ is a value of a random variable (not just a fixed number).

To compute the variance of $\langle X \rangle$, we regard each measurement as an independent trial of X. Writing $\sigma^2(X)$ to denote the (as yet unknown) variance of the individual values of X, we can compute $\sigma^2(\langle X \rangle)$ using the relations in Eqs. (18.42) and (18.29):

$$\sigma^2(\langle X \rangle) = \frac{1}{n^2} \sum_{i=1}^{n} \sigma^2(X) = \frac{1}{n} \sigma^2(X). \tag{18.74}$$

Note carefully the difference in notation between $\sigma^2(X)$ and $\sigma^2(\langle X \rangle)$. Confusion between these quantities will lead to confusion about the present discussion.

Equation (18.74) confirms the intuitively expected result that the mean of a large set of measurements will be more narrowly distributed (subject to less error) than the individual measurements of which it is composed. This observation, in turn, suggests that the large-n limit of $\langle X \rangle$ will be a good measure of μ, the mean of the parent population.

The reader may note that we have up to now avoided a discussion of the computation of $\sigma^2(X)$; the problem with such a computation, which requires the evaluation of

$$\sigma^2(X) = \frac{1}{n} \sum_{i=1}^{n} (x_i - \mu)^2, \tag{18.75}$$

is that we do not know the value of μ, but only the approximation to it given by $\langle X \rangle$, which is itself a member of a probability distribution. A naive approach to this dilemma would be simply to approximate μ by $\langle X \rangle$, thereby creating a quantity we denote s^2, according to

$$s^2 = \frac{1}{n} \sum_{i=1}^{n} \Big(x_i - \langle X \rangle \Big)^2. \tag{18.76}$$

However, s^2 is an underestimate of $\sigma^2(X)$ because it fails to take account of the distribution of $\langle X \rangle$.

A more proper way to proceed starts by defining ε such that $\mu = \langle X \rangle + \varepsilon$ and writing Eq. (18.75) as

$$\begin{aligned} \sigma^2(X) &= \frac{1}{n} \sum_{i=1}^{n} \Big(x_i - \langle X \rangle - \varepsilon \Big)^2, \\ &= \frac{1}{n} \sum_{i=1}^{n} \Big(x_i - \langle X \rangle \Big)^2 - \frac{2\varepsilon}{n} \sum_{i=1}^{n} \Big(x_i - \langle X \rangle \Big) + \varepsilon^2. \end{aligned}$$

The first term on the right-hand side of this equation is s^2 and the second term vanishes, so we have

$$\sigma^2(X) = s^2 + \varepsilon^2. \tag{18.77}$$

To complete our analysis, we now note that ε is a random variable and that, invoking Eq. (18.74),

$$\langle \varepsilon^2 \rangle = \frac{1}{n} \sum_{i=1}^{n} \Big(\mu - \langle X \rangle \Big)^2 = \sigma^2(\langle X \rangle) = \frac{1}{n} \sigma^2(X). \tag{18.78}$$

Assigning to ε^2 its average value and solving Eq. (18.77), we get

$$\sigma^2(X) = \frac{n}{n-1}\, s^2 = \frac{1}{n-1}\sum_{i=1}^{n}\Big(x_i - \langle X\rangle\Big)^2 . \tag{18.79}$$

We now have the key results that are needed for the interpretation of repeated experimental measurements of a quantity X:

1. *The mean of n measurements of X (assumed to be independent and from the same probability distribution) is the best measure obtainable from the data for the value of X.*
2. *The individual data are distributed about the mean of the parent population (the exact value of X) with a variance $\sigma^2(X)$ given by Eq. (18.79) and therefore with a standard deviation*

$$\sigma(X) = \sqrt{\frac{n}{n-1}\, s^2} = \sqrt{\frac{1}{n-1}\sum_{i=1}^{n}\Big(x_i - \langle X\rangle\Big)^2} . \tag{18.80}$$

3. *The mean of the measurements, $\langle X\rangle$ is distributed about its value for the parent population with variance*

$$\sigma^2(\,\langle X\rangle\,) = \frac{1}{n(n-1)}\sum_{i=1}^{n}\Big(x_i - \langle X\rangle\Big)^2 . \tag{18.81}$$

The standard deviation of the mean is the square root of the above value given for $\sigma^2(\,\langle X\rangle\,)$.

Exercises

18.6.1. For each of the following data sets containing values of a random variable X, find an estimate of the mean value $\langle X\rangle$, the variance $\sigma^2(X)$ of the individual values of X relative to the mean of the parent distribution, and the variance of $\langle X\rangle$.

(a) The five data points 5.9, 6.0, 6.1, 6.2, 6.5.

(b) The data of part (a) plus additional data points 6.5, 6.2, 6.1, 6.0, 5.9.

(c) The data of part (a) plus five additional data points, all 5.9.

DISTRIBUTIONS OF DERIVED QUANTITIES

Often we are interested in a quantity which is not directly measured but is related functionally to one or more quantities that it is practical to measure. We may then want to know how the distribution of the quantity we calculate from the data is related to the distribution of the measured data. In particular, we may want to relate the mean and variance of a derived quantity to those of the measurements used in the computation. If our derived quantity depends upon the measurement of more than one random variable, we restrict the present discussion to the case that the measurement variables are independent (and therefore have zero covariance). Note that there are many practical situations in which this condition is not met.

We already have a partial answer to our present problem. In Section 18.2 we developed formulas for the mean and variance of linear combinations of random variables. For reference we repeat those formulas here for **independent** variables:

$$\langle aX + bY\rangle = a\langle X\rangle + b\langle Y\rangle\,,$$
$$\sigma^2(aX + bY) = a^2\sigma^2(X) + b^2\sigma^2(Y)\,. \tag{18.82}$$

We now want to generalize to the distribution of a quantity $f(X, Y)$ of more general form.

We assume that f can be expanded in Taylor series about the mean values $\langle X\rangle$ and $\langle Y\rangle$, and that the series converges rapidly enough that meaningful results are obtained when we retain only a few terms of the expansion. This assumption is most likely to be acceptable when the variances of X and Y are not too large. Letting $\Delta x = x - \langle X\rangle$ and $\Delta y = y - \langle Y\rangle$, the Taylor series for $f(x, y)$ takes the form

$$f(x,y) = f(\,\langle X\rangle, \langle Y\rangle\,) + \left(\frac{\partial f}{\partial x}\right)\Delta x + \left(\frac{\partial f}{\partial y}\right)\Delta y$$
$$+\frac{1}{2}\left(\frac{\partial^2 f}{\partial x^2}\right)(\Delta x)^2 + \frac{1}{2}\left(\frac{\partial^2 f}{\partial y^2}\right)(\Delta y)^2 + \left(\frac{\partial^2 f}{\partial x \partial y}\right)\Delta x \Delta y + \cdots\,. \tag{18.83}$$

The partial derivatives in Eq. (18.83) are to be evaluated at $x = \langle X\rangle$, $y = \langle Y\rangle$.

We now take the average value $\langle\, f(X,Y)\,\rangle$ to find the mean of f for the distribution of X and Y, assuming the availability of n_x data points for X and n_y data points for Y:

$$\langle\, f(X,Y)\,\rangle = \frac{1}{n_x n_y}\sum_{i,j}^{n_x,n_y} f(x_i, y_j) = f(\,\langle X\rangle, \langle Y\rangle\,) + \left(\frac{\partial f}{\partial x}\right)\langle\Delta x\rangle + \left(\frac{\partial f}{\partial y}\right)\langle\Delta y\rangle$$
$$+\frac{1}{2}\left(\frac{\partial^2 f}{\partial x^2}\right)\langle\,(\Delta x)^2\,\rangle + \frac{1}{2}\left(\frac{\partial^2 f}{\partial y^2}\right)\langle\,(\Delta y)^2\,\rangle + \left(\frac{\partial^2 f}{\partial x \partial y}\right)\langle\Delta x \Delta y\rangle + \cdots\,.$$

We can simplify this expression by recognizing that $\langle\Delta x\rangle = \langle\Delta y\rangle = 0$, that $\langle\,(\Delta x)^2\,\rangle = \sigma^2(X)$, $\langle\,(\Delta y)^2\,\rangle = \sigma^2(Y)$, and that $\langle\Delta x \Delta y\rangle$ is the covariance of X and Y, which vanishes because we have assumed these variables to be independent. We get

$$\langle\, f(X,Y)\,\rangle = f(\,\langle X\rangle, \langle Y\rangle\,) + \frac{1}{2}\left(\frac{\partial^2 f}{\partial x^2}\right)\sigma^2(X) + \frac{1}{2}\left(\frac{\partial^2 f}{\partial y^2}\right)\sigma^2(Y) + \cdots\,. \tag{18.84}$$

It is usually not useful to extend the expansion beyond this point.

We see from Eq. (18.84) that the best estimate for the value of $f(X, Y)$ is not the value of f when X and Y are assigned their mean values, but there are corrections that depend upon the spread of the measurements of X and Y. We leave it as an exercise to show that the variance of $f(X, Y)$ is related to those of X and Y by the formula

$$\sigma^2(f(X,Y)) = \left(\frac{\partial f}{\partial x}\right)^2\sigma^2(X) + \left(\frac{\partial f}{\partial y}\right)^2\sigma^2(Y) + \cdots\,. \tag{18.85}$$

Exercises

18.6.2. Prove Eq. (18.85). Remember that it is valid when X and Y are independent.

18.6.3. If X and Y are independent random variables, and $Z = XY$, then $\langle Z\rangle = \langle X\rangle\langle Y\rangle$. Show that

$$\frac{\sigma^2(Z)}{\langle Z\rangle^2} \approx \frac{\sigma^2(X)}{\langle X\rangle^2} + \frac{\sigma^2(Y)}{\langle Y\rangle^2}.$$

18.6.4. Prove that the result of Exercise 18.6.3 also applies if $Z = X/Y$.

CURVE FITTING

Here we consider the situation that we have a set of values y_i that have been obtained (possibly by measurement) for corresponding values of x, namely x_i. Our problem is to find the curve of a designated functional form that best represents our data points. In most cases, it is generally agreed that "best" means the minimization of the squares of the discrepancies between the data points and the curve that represents them, and these curve-fitting methods are called **least-squares** methods. The basic least-squares methods can be generalized to allow points to be assigned individual weights (i.e., importance) for influencing the fit; such generalizations include the **chi-square** methods that are of importance in statistics but are beyond the scope of the present text. More information regarding curve fitting, statistics, and data reduction can be found in the Additional Readings.

Even at an elementary level, the notion of "best" is in fact quite ambiguous; taking for discussion purposes the fitting of a point set by a straight line, here are several possibilities that are at least partially inconsistent with each other:

1. The discrepancies for the least-squares analysis are the vertical distances between the data point and the straight line.
2. The discrepancies for the least-squares analysis are the horizontal distances between the data point and the straight line.
3. The discrepancies for the least-squares analysis are the perpendicular distances between the data point and the straight line.
4. The straight-line fit to the data is restricted to lines that pass through a point (such as the coordinate origin) that the line should pass through, and this constraint is added to one of the other possibilities.

Clearly the above is not an all-inclusive list.

It is practical to fit data to a number of functional forms other than straight lines, but the practical complications of doing that take the discussion beyond what is reasonable here. However, the symbolic computing systems support curve fitting using a variety of functional forms, and we do provide in this text the information needed to access those capabilities.

The discussion here will not only be limited to linear fits without constraints, but will also deal only with discrepancies that are the vertical distances between the data points and the line that fits them. This choice will be most appropriate when one experimental property can be fairly precisely controlled (the "independent variable"), with more error-prone data taken for another quantity (the "dependent variable"). In the present discussion we name the independent variable x and the dependent variable y. The assignment of physical variables to x and y is the responsibility of the person analyzing the data.

LEAST SQUARES—LINEAR FIT

Our problem here is to find the linear function $y = ax + b$ that is the best fit to a data set

$$(X, Y) = \{(x_1, y_1), (x_2, y_2), \ldots, (x_n, y_n)\},$$

in the sense that a and b have been chosen to minimize

$$S^2 = \frac{1}{n}\sum_{i=1}^{n}(ax_i + b - y_i)^2 . \tag{18.86}$$

The quantity being summed is the square of the vertical distance between (x_i, y_i) and the line $y = ax + b$. We note that S^2 has the form of a variance, suggesting that it may be useful to write S^2 and other quantities we shall shortly encounter using the notations of the present chapter. Accordingly, we rewrite Eq. (18.86) as

$$S^2 = \langle\, (aX + b - Y)^2 \,\rangle . \tag{18.87}$$

To minimize S^2 we set to zero its partial derivatives with respect to a and b. We get

$$\frac{\partial S^2}{\partial a} = \langle 2(aX + b - Y)X \,\rangle = 2a\langle X^2\rangle + 2b\langle X\rangle - 2\langle XY\rangle = 0 , \tag{18.88}$$

$$\frac{\partial S^2}{\partial b} = \langle 2(aX + b - Y)\,\rangle = 2a\langle X\rangle + 2b - 2\langle Y\rangle = 0 . \tag{18.89}$$

Canceling the "2" from every term, we have a set of two simultaneous linear equations for a and b. The value of a that solves these equations can be written, using Cramer's rule, Eq. (4.75), as the following ratio of determinants:

$$a = \frac{\begin{vmatrix} \langle XY\rangle & \langle X\rangle \\ \langle Y\rangle & 1 \end{vmatrix}}{\begin{vmatrix} \langle X^2\rangle & \langle X\rangle \\ \langle X\rangle & 1 \end{vmatrix}} = \frac{\mathrm{cov}(X, Y)}{\sigma^2(X)} . \tag{18.90}$$

Then, from Eq. (18.89), we can obtain b as

$$b = \langle Y\rangle - a\langle X\rangle . \tag{18.91}$$

It can be shown that a good measure of the quality of the fit is given in terms of the **correlation** of X and Y, a quantity introduced at Eq. (18.39). A perfect fit corresponds to $\mathrm{corr}(X, Y) = \pm 1$, so a convenient quality measure is

$$\text{Fit quality} \sim \frac{\mathrm{cov}^2(X, Y)}{\sigma^2(X)\sigma^2(Y)} . \tag{18.92}$$

SYMBOLIC COMPUTING

Both MAPLE and MATHEMATICA support least-squares curve fitting, and the curves to which data can be fitted are not limited to straight lines. In both computing systems the data to be processed is in the form of a list of x, y pairs, where x is regarded as the "independent variable," meaning that the x values are regarded as known precisely, and the quantities whose sum of squares is to be minimized are the distances in y between the points and the curve. The data points must be in the following format:

MAPLE
```
> data := [ [x1,y1],[x2,y2],[x3,y3],...[xn,yn] ]:
```

MATHEMATICA
```
data = { {x1,y1},{x2,y2},{x3,y3},...{xn,yn} };
```

In MAPLE, a fit of data to the functional form $y = ax + b$ is obtained by the commands

```
> with(CurveFitting):
> LeastSquares(data,v):
```

Note that LeastSquares is in the CurveFitting package, and its use must be preceded by with(CurveFitting). In the LeastSquares command, v is the name the user wishes to name the independent variable; no matter what name is chosen, the first member of each data pair is the independent-variable coordinate. The output of LeastSquares will be an expression such as $6.25v + 9.05$; no name is given to the dependent variable.

If it is desired to fit to a form other than $y = ax+b$, one can add to LeastSquares an optional argument of the form curve=*form*, where *form* is a expression containing the independent variable and other variables which must occur linearly. Examples of possible optional arguments are (assuming that v has been identified as the independent variable)

```
curve=a*v^2+b*v+c   curve=a*sin(v)+b*cos(v)   curve=a*exp(-v^2/2)
```

MAPLE assumes that all symbols other than the independent variable are coefficients whose values are to be determined by the least-squares optimization.

In MATHEMATICA, a least-squares fit to a linear combination of basis functions can be obtained using coding illustrated by

```
Fit[ data, {f1, f2, f3}, v ]
```

Here f1, f2, and f3 are expressions in the independent variable v that are used to construct a fitting function of the form $c_1\,\mathtt{f1} + c_2\,\mathtt{f2} + c_3\,\mathtt{f3}$. No matter what name is chosen for the independent variable, the first member of each data pair is the independent-variable coordinate. The coefficients c_i are obtained by a least-squares procedure, and the output of the Fit command is a MATHEMATICA expression for the fitting function. For example, to fit data to a straight line with x as the independent variable, we write

```
Fit[ data, {1, x}, x ]
```

The output of this command will be a linear expression, such as $3.25x - 6.91$. As a second example, the fitting of data to a quadratic form is obtained from

```
Fit[data, {1, x, x^2}, x ]
```

Example 18.6.1. Least-Squares, Maple

Let's carry out least-squares fits to the following MAPLE data:

```
> T := [ [0,1.3], [1,2.1], [2,3.7], [3,4.2], [4,5.9], [5,6.4],
>        [6,7.4], [7,7.9], [8,8.7], [9,8.8], [10,9.6], [11,9.7],
```

```
>         [12,9.9], [13,10.2], [14,10.0], [15,10.2], [16,9.7],
>         [17,9.6], [18,8.9], [19,8.8], [20,7.8] ]:
```

To access the necessary packages and truncate to reasonable numbers of significant figures,

```
> with(CurveFitting):
> with(plots):
> Digits := 6:
```

We first try a linear fit (the default for MAPLE):

```
> T1 := LeastSquares(T, z);
```

$$T1 := 4.18831 + 0.346883\,z$$

To compare the point set and the fit given by $T1$, we make a graph of each; a further explanation of the plotting procedures and their use can be found in Appendix A. The points are graphed by entering

```
> G0 := pointplot(T);
```

$$G0 := PLOT(...)$$

The graph is stored under the name G0; when a graph is assigned it is not displayed but an abbreviated version of its coding is returned. We now graph the linear fit:

```
> G1 := plot(T1, z = 0 .. 20);
```

$$G1 := PLOT(...)$$

Finally, we show the result of combining the two graphs:

```
> display([G0,G1]);
```

The graphs are plotted on a common set of axes; the result (without color) is shown in the left panel of Fig. 18.10.

The fit to our data is not very satisfactory, mainly because a straight line is not an appropriate form for the fitting function. Fortunately, it is easy to cause MAPLE to use more general fitting functions. The curvature in the data suggests that we might do far better with a quadratic fitting function, so we now try

```
> T2 := LeastSquares(T, z, curve = a*z^2 + b*z  + c );
```

$$T2 := 1.05534 + 1.33624\,z - 0.0494680\,z^2$$

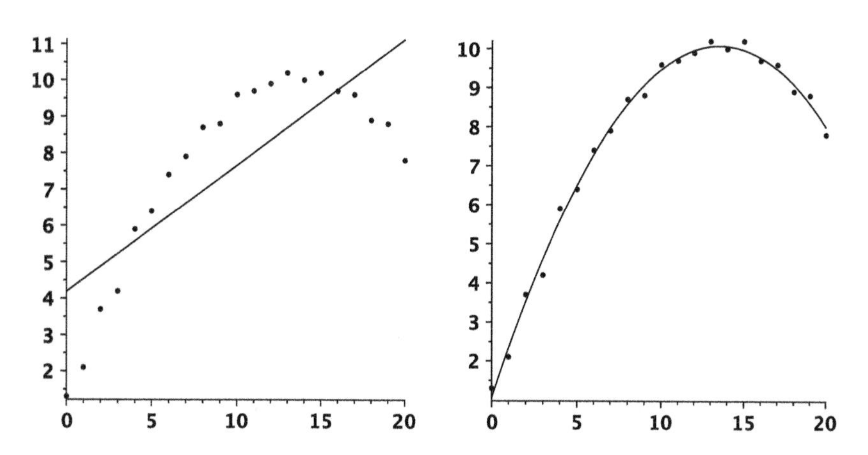

Figure 18.10: Linear (left) and quadratic (right) fits to the set of data points given in Example 18.6.1 (MAPLE) and in Example 18.6.2 (MATHEMATICA).

Because a, b, and c are undefined symbols other than the independent variable z, MAPLE identifies them as coefficients to be determined by the least-squares optimization. We graph the quadratic fit that was stored in T2:

```
> G2 := plot(T2, z = 0 .. 20);
```
$G2 := PLOT(...)$

We already have the plot of the point set stored as G0, so we combine the points and the quadratic fit by entering

```
> display([G0,G2]);
```

The result, which is much better, is shown (without color) as the right panel of Fig. 18.10.

■

Example 18.6.2. Least-Squares, Mathematica

Let's carry out least-squares fits to the following MATHEMATICA data:

```
t = { {0,1.3}, {1,2.1}, {2,3.7}, {3,4.2}, {4,5.9}, {5,6.4},
      {6,7.4}, {7,7.9}, {8,8.7}, {9,8.8}, {10,9.6}, {11,9.7},
      {12,9.9}, {13,10.2}, {14,10.0}, {15,10.2}, {16,9.7},
      {17,9.6}, {18,8.9}, {19,8.8}, {20,7.8} };
```

We first try a linear fit:

```
t1 = Fit[ t, {1, x}, x ]
```
$4.18831 + 0.346833\, x$

To compare the point set and the fit given by t1, we make a graph of each; a further explanation of the plotting procedures and their use can be found in Appendix A. The points are graphed by entering

```
g0 = ListPlot[ t ]
```
(plot not shown here)

We now graph the linear fit:

```
g1 = Plot[ t1, {x, 0, 20} ]
```
(plot not shown here)

The two graphs can now be combined:

```
Show[g0, g1]
```

This command produces a plot similar to that in the left panel of Fig. 18.10. It is important to make graph g0 the first argument of Show. As pointed out in Appendix A, the layout of the combined graph produced by Show is that of the graph that is its first argument. In the present case, the linear fit spans a much smaller vertical range than the point set, and many of the points will be outside the range of the plot if g1 is placed first.

The fit to our data displayed in Fig. 18.10 is not very satisfactory, mainly because a straight line is not an appropriate form for the fitting function. Fortunately, it is easy to cause MATHEMATICA to use more general fitting functions. The curvature in the data suggests that we might do far better with a quadratic fitting function, so we now try

```
t2 = Fit[ t, {1, x, x^2}, x ]
```
$1.05534 + 1.33624\, x - 0.049468\, x^2$

We graph the quadratic fit that was stored in t2:

```
g2 = Plot[ t2, {x, 0, 20} ]
```
(plot not shown here)

We already have the plot of the point set stored as g0, so we combine the points and the quadratic fit by entering

```
Show[g0,g2]
```

The result, which is much better, is shown (without color) as the right panel of Fig. 18.10.

■

Exercises

18.6.5. The following are values of $f(x)$ for integer values of x from 1 through 10:

$$3.00, 3.50, 4.33, 4.75, 5.80, 6.17, 6.14, 6.63, 6.78, 7.10$$

(a) Find the least-squares straight-line fit to these data.

(b) Find the quadratic function that is a least-squares fit to these data.

(c) Plot the original data and both fits on the same plot.

18.6.6. Repeat Exercise 18.6.5 for the following data set:

$$f(1) = 1,\ f(2) = 3,\ f(3) = 5,\ f(4) = 3,\ f(5) = 1.$$

Additional Readings

Bevington, P. R., & Robinson, D. K. (2003). *Data reduction and error analysis for the physical sciences* (3rd ed.). New York: McGraw-Hill.

Ramachandran, K. M., & Tsokos, C. P. (2009). *Mathematical statistics with applications.* New York: Academic Press (Relatively detailed but readable and self-contained).

Ross, S. M. (1997). *First course in probability* (5th ed.). (Vol. A). New York: Prentice Hall.

Ross, S. M. (1999). *Introduction to probability and statistics for engineers and scientists* (2nd ed.). New York: Academic Press.

Ross, S. M. (2000). *Introduction to probability models* (7th ed.). New York: Academic Press.

Suhir, E. (1997). *Applied probability for engineers and scientists*, New York: McGraw-Hill.

Uspensky, J. V. (1937). *Introduction to mathematical probability.* New York: McGraw-Hill (A clear and well-written exposition. The first few chapters are an exemplary introduction at a fully accessible level).

APPENDICES

A METHODS FOR MAKING PLOTS

There are a number of ways in which data can be plotted in our symbolic computing systems. The possibilities described here are the following;

1. **Function graphs.** A function of given form, $y(x)$, is plotted for a range of its arguments x. This type of plotting can be extended to permit the inclusion of two or more functions on the same graph.
2. **Parametric plots.** A pair of functions $y(t)$ and $x(t)$ are plotted against each other for a range of the parameter t. This type of plot describes the trajectory of the point (x, y) as t is varied.
3. **Contour plots.** A set of curves, each of which connects the points (x, y) with a different constant value of a function $f(x, y)$.
4. **Point plots.** A set of discrete points y_i corresponding to different values of x_i. This type of plot can be used to represent experimental data or discrete probabilities.
5. **Combination graphs**, containing simultaneous plots of any one or more of the above types. For example, this option can be used to compare a set of points with a corresponding functional form.

The details of these plotting methods differ greatly for different symbolic systems. We therefore describe them separately for MAPLE and MATHEMATICA.

MAPLE

Function graphs—The basic plotting command, `plot`, was introduced in the initial chapter of this text. We present here a more complete description, including various useful options. If `F(x)` is either a function of x or just an expression that depends on x, its plot for x in the range (x_1, x_2), with values of F shown for the vertical range (y_1, y_2), is obtained using a command such as

```
plot( F(x), x=x1 .. x2, y1 .. y2, color = red, linestyle = dash,
    thickness = 2, scaling = constrained )
```

The vertical range `y1 .. y2` and any or all the options following it can be omitted if not needed. Here are some comments about the options.

Mathematics for Physical Science and Engineering.
http://dx.doi.org/10.1016/B978-0-12-801000-6.00026-2

MAPLE recognizes a number of common color names (red, blue, green, yellow, orange, black, and several others); it also recognizes a very large number of colors by their names in the HTML language (the HTML names start with initial capitals). A list of all the available colors can be generated in a MAPLE session by entering the help command `?colornames`.

The function curves are by default rendered as solid lines (`solid`); additional possibilities, controlled by the option `linestyle`, are `dot`, `dash`, `dashdot`, `longdash`, and several others.

By default, plots are scaled differently in x and y when that will make them more useful or more attractive. This scaling freedom corresponds to the default value of `scaling`, namely `unconstrained`. To force the x and y scaling to be the same, set `scaling=constrained`.

The thickness of the function curves can be controlled by `thickness`, for which the default value is zero. Thicker curves are obtained by setting `thickness` to positive integer values.

The plot command can be used to plot more than one graph for the same horizontal and vertical ranges by changing `F` to a list containing the functions to be plotted. If it is desired to use the plot options to a significant extent, better results may often be obtained by defining the different curves as separate graphs, then combining them as a **combination graph**, as described later.

Here are two examples that include the use of options for function plots. The plot output are shown (without color) in the two panels of Fig. A.1.

```
> G := proc(x);  5/sqrt(2*Pi) * exp(-(x-3)^2/2  end proc:

> plot(G(x), x = 0 .. 6, 0 .. 2, linestyle = dash,
>                thickness = 2, color = red);

> plot(G(x), x = 0 .. 6, 0 .. 2, scaling = constrained);
```

Parametric plots—If `x(t)` and `y(t)` are functions or expressions for the coordinates in terms of a parameter `t`, then the curve defined by `x` and `y` for `t` on the range from `t1` to `t2` is generated by a command such as

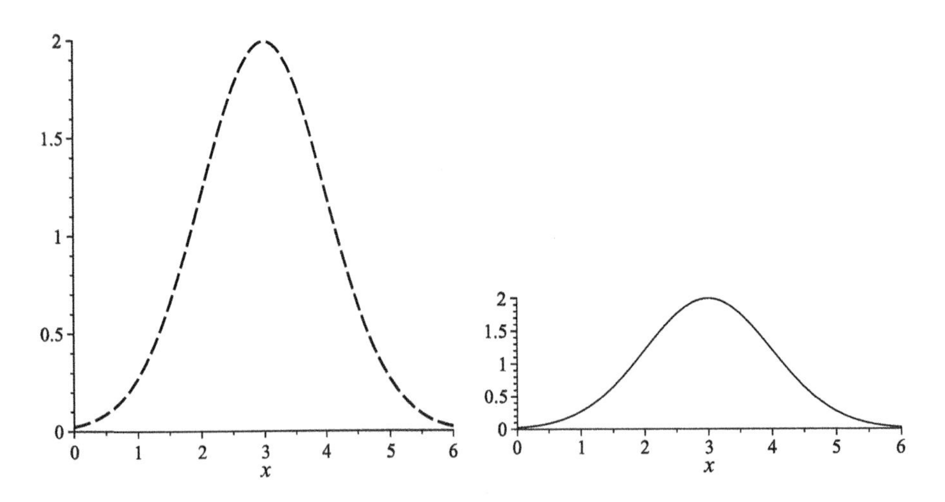

Figure A.1: Plots produced using `plot` with the options described in the text.

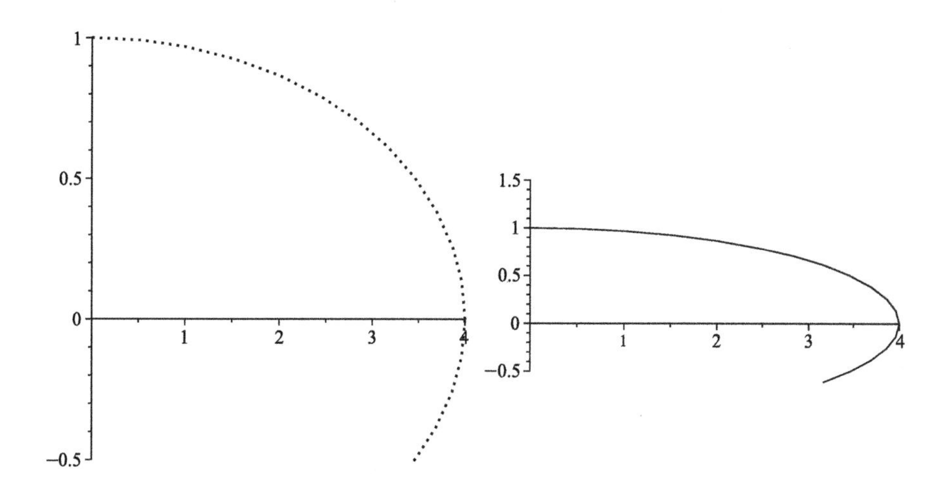

Figure A.2: Parametric plots, with the options described in the text.

```
> plot([x(t), y(t), t = t1 .. t2], view = [x1 .. x2, y1 .. y2],
    color=blue, linestyle=dot, thickness=2, scaling=constrained );
```

Note that `x(t)`, `y(t)`, and the range of `t` are the elements of a list. The list construction is required, and the range of `t` must be within the list. Another difference from our earlier function graphs is that we can no longer control the x and y ranges in different ways. To control the range that is displayed, we now need to use the option `view`. The other options work the same way as previously.

Here are two examples of the coding of a parametric plot. The plot output is shown (without color) in the two panels of Fig. A.2.

```
> plot([4*cos(t),sin(t), t = 0 .. 2*Pi], view = [0 .. 4, -0.5 .. 1],
        color = blue, linestyle = dot, thickness = 2 );
> plot([4*cos(t),sin(t), t = 0 .. 2*Pi], view = [0 .. 4, -0.5 .. 1],
        scaling = constrained );
```

Contour plots—The plot is of the contours of a function of two variables (or of an expression involving two variables), over a specified range of each variable. One can either specify the values of the function on the contours, or leave them to be determined by MAPLE. Note however, that if MAPLE chooses the contours, it does not indicate the function values it used. Also, be aware that MAPLE chooses the aspect ratio for contour plots just as it does for other plotting. Use the option `scaling=constrained` if you want the horizontal and vertical scales to be the same.

The command for generating a contour plot is `contourplot`. This command is part of MAPLE's `plots` package and its use must be preceded by the command `with(plots)`. Here is an example of coding to make a contour plot.

```
> with(plots):
> contourplot(x^2+2*y^2, x = -2 .. 2, y = -1 .. 1,
>    scaling = constrained, contours = [0.4, 0.8, 1.2, 1.6, 2.0] );
```

The plot output is shown (without color) in Fig. A.3.

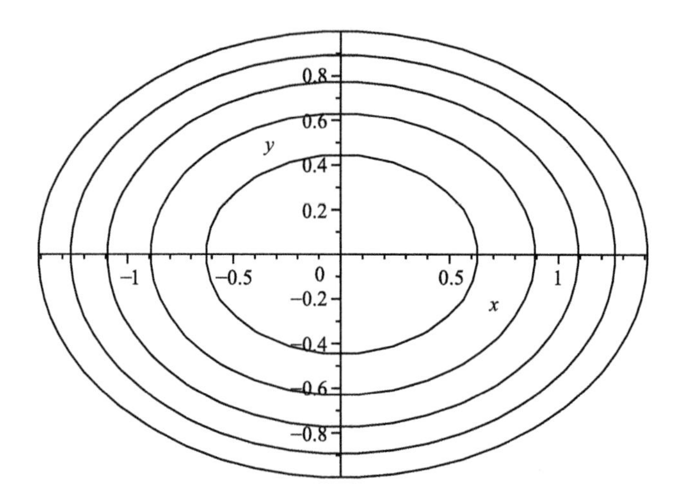

Figure A.3: Contour plot.

Point plots—The data to be plotted must be a numeric list of $[x_i, y_i]$ pairs. Any valid way of generating the data may be used; if the data originated from an experiment, it may have to be entered explicitly, e.g., as

```
> T := [ [0.10,0.384], [0.21,0.485],[0.44,0.777] ]:
```

Note that T is a **list**, each member of which is in turn a two-element list giving the horizontal and vertical coordinates of a point to be plotted.

If the data arise as values of a given function and are to be plotted for regularly-spaced values of its argument, we may use MAPLE's sequence-building procedure **seq**, which acts as illustrated here:

$$\texttt{seq(F(n), n = n1 .. n2)} \quad\longrightarrow\quad F(n_1),\ F(n_1+1), \cdots,\ F(n_2)\,.$$

F can either be a MAPLE function or simply an expression that may contain the variable **n**. For example, we can create a data structure whose elements are $[n, n^2]$ by the following coding:

```
> T := [ seq([n,n^2], n = 0 .. 4) ];        [ [0,0], [1,1], [2,4], [3,9], [4,16] ]
```

Note that in defining T the outer brackets are needed; T must be a list, not just a sequence.

Once we have an appropriate data set, we can make a plot using the command **pointplot**. However, that command is part of MAPLE's **plots** package and must be preceded by **with(plots)**. If T is a data set for point plotting, we may enter something such as

```
> with(plots):
> pointplot(T, color = red, symbol = solidcircle, symbolsize = 10):
```

The options shown here (color = red etc.) are optional. If the options are omitted, the points will be in the default color (black) using the default symbol and symbol size. There are also additional options that we leave the reader to discover (look at ?plot and the additional material to which it refers).

The options that we listed are those that are probably the most useful for displaying information from a point plot. The color options have already been described under **function graphs**. The choices available for symbols include the following: point, asterisk, box, circle, cross, diagonalcross, diamond, solidbox,

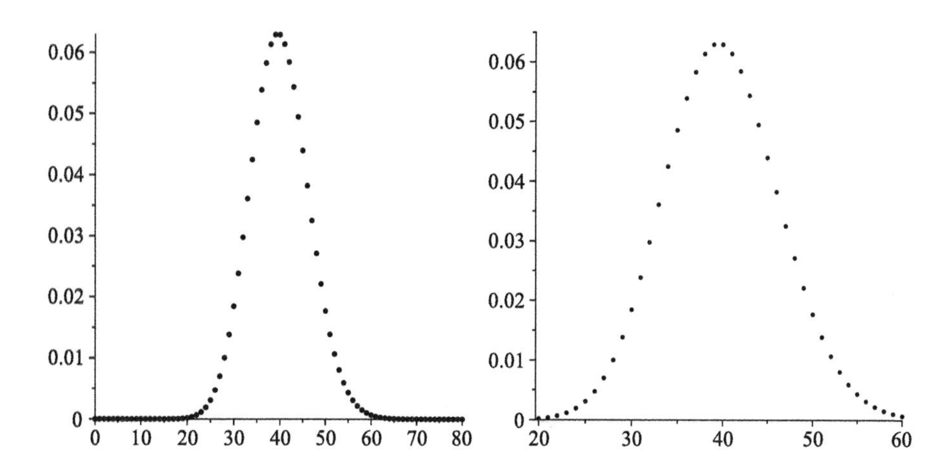

Figure A.4: Point plots, with the options described in the text.

`solidcircle`, `soliddiamond`. Except for `point`, which is always at the same size, the size at which the symbols are drawn can be controlled by entering a positive integer value for `symbolsize`. Larger values of `symbolsize` produce larger symbols (in units that are approximately printer points).

Here is the code to produce two point plots. The resulting graphs are shown (without color) in the two panels of Fig. A.4.

```
> with(plots):
> T := [ seq([n, 40^n*exp(-40)/n!], n = 0 .. 80) ]:
> pointplot(T, symbol = solidcircle, symbolsize = 12);
> pointplot(T, symbol = solidcircle, view = [20 .. 60, 0 .. 0.065] );
```

To display more than one point plot on the same graph, each with its own options, make them a **combination graph** as explained next.

Combination graphs—This option is useful for combining graphs that we have previously made separately. We proceed by designating a symbol to represent each individual graph, and then use the command `display` to present them together as a single graph; `display` is part of the `plots` package and its use must be preceded by `with(plots)`. Here is an example.

```
> with(plots):                    (needed for pointplot and display)
```

We first make the two individual graphs:

```
> G1 := pointplot([seq([n/20, 1-n^2/400], n = -20 .. 20)],
          symbol = solidcircle, symbolsize = 12);
```

$G1 := PLOT(\ldots)$

```
> G2 := plot(exp(-4*x^2), x = -1 .. 1, color = black,
          thickness = 2, linestyle = dash);
```

$G2 := PLOT(\ldots)$

Note that we specified a color only for `G2`. The assignment command causes the graph not to be drawn, with output only showing its coding in a highly abbreviated fashion.

We now combine the graphs using `display`. The graphs to be combined should be given as elements of a list; after that list can be added any global options we may

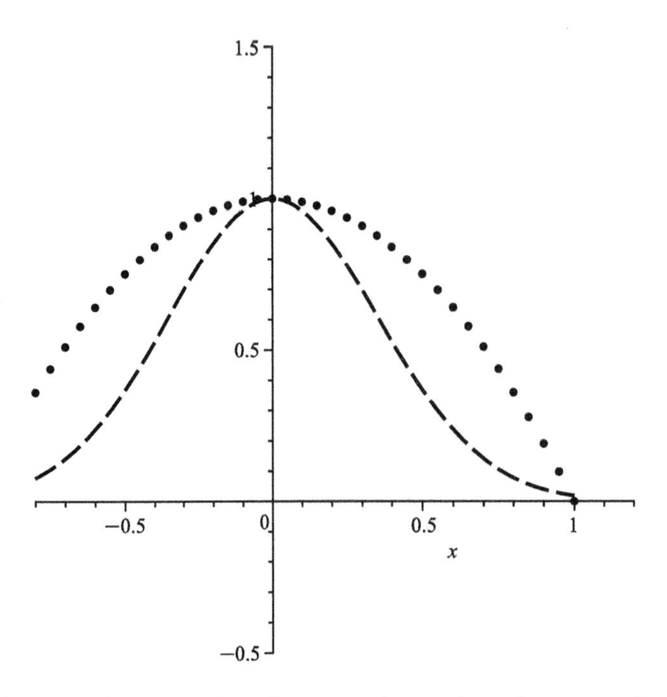

Figure A.5: Combination plot of two graphs, each with options (see text).

want to apply. The options we add in the present example are `color` (which applies only to graphs for which a color was not individually specified), and `view`, which controls the horizontal and vertical ranges of the combined graph. If `view` calls for a larger range than that of the input graphs, those graphs are of course only included to the extent of their original data. Continuing the present example,

```
> display( [G1,G2], color=blue, view = [-0.8 .. 1.2, -0.5 .. 1.5] );
```

This command produces the graphical output shown (without color) in Fig. A.5. Actual MAPLE output would show the point set $G1$ in blue and the continuous curve $G2$ in black.

MATHEMATICA

General—In the absence of user action to the contrary, MATHEMATICA adjusts the horizontal and vertical scaling of plots so that all (or almost all) the plot content is visible in a rectangular display window of pleasant proportions (with the horizontal dimension of the view window equal to the classical golden ratio times the vertical dimension of the window). This is usually a good choice for visualization of the salient properties of plots, but there are occasions on which one will want the same scaling in both coordinates, i.e., to have the plot at an aspect ratio of unity. However, MATHEMATICA defines the parameter `AspectRatio` to be the ratio of the Y to the X dimension of the displayed view window, not the scaling ratio of x and y. If the window displays values of x from x_1 to x_2 and values of y from y_1 to y_2, the scaling of x and y will be the same if the window dimensions have the ratio

$$\frac{Y_{\text{window}}}{X_{\text{window}}} = \frac{y_2 - y_1}{x_2 - x_1},$$

and therefore that is the value that `AspectRatio` must have to produce uniform scaling.

Function graphs—The basic plotting command, `Plot`, was introduced in the initial chapter of this text. If `F[x]` is either a function of x or just an expression that depends on x, its plot for x in the range (x_1, x_2), with values of F shown for the vertical range (y_1, y_2), is obtained using a command such as

```
Plot [F[x], {x,x1,x2}, PlotRange -> {y1,y2},
    PlotStyle -> { Red, Thickness[0.02], Dashing[Medium] },
    AspectRatio -> (y2-y1)/(x2-x1) ]
```

The option `PlotRange` is optional, but needs to be included if we are going to force a specific (i.e., **constrained**) scaling by setting `AspectRatio`. The value shown here for `AspectRatio` is that which produces equal scaling of the horizontal and vertical coordinates.

The options in `PlotStyle` are also optional. The color, if included, must be one of the many known to MATHEMATICA; these include `Red`, `Yellow`, `Green`, `Cyan`, `Blue`, `Magenta`, `Black`, and others. The color names all begin with capital letters. Consult the MATHEMATICA documentation if you want to use more subtle colors specified by their RGB (red/green/blue) values.

`Thickness` has an argument equal to the line width (in units of the horizontal plot dimension). `Dashing` accepts an argument permitting specification of the lengths of the dashes and the space between them when drawing function curves. Instead of that detailed specification it is permitted to use `Small`, `Medium`, or `Large` as the argument. Thickness can also alternatively be designated as `Small`, `Medium`, or `Large`.

It is permitted to provide a list of functions to be plotted together. In that event each option in `Plotstyle` must be a list, with elements that correspond, in order, with the functions being plotted. The options `PlotRange` and `AspectRatio`, if present, occur once and apply to the entire plot.

There are additional plot options; those already mentioned here are sufficient for many applications.

Here are two examples of coding for function plots. The resulting graphs are shown (without color) in the two panels of Fig. A.6.

```
f[x_] := 5/Sqrt[2*Pi] * E^(-(x-3)^2/2)                Defines f[x]
Plot[ f[x], {x, 0, 6},
     PlotStyle -> {Red, Thickness[0.02], Dashing[Large]} ]
Plot[ f[x], {x, 0, 6}, PlotRange -> {0,2}, AspectRatio -> 2/6 ]
```

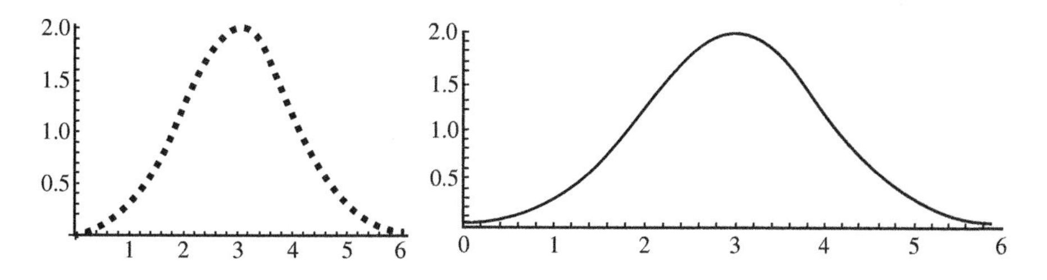

Figure A.6: Plots produced using `Plot` with the options described in the text.

Parametric Plots—If x[t] and y[t] are functions or expressions for the coordinates in terms of a parameter t, then the curve defined by x and y, for t on the range from t1 to t2 is generated by a command such as

```
ParametricPlot[ {x[t],y[t]}, {t, t1, t2},
    PlotStyle -> {Magenta, Dashing[Medium], Thickness[.05]},
    PlotRange -> {{x1,x2},{y1,y2}} ]
```

The AspectRatio, PlotStyle, and PlotRange options are optional; if PlotRange is present, it should indicate separate ranges for x and y. The options under PlotStyle work the same way as in the function plots considered earlier. MATHEMATICA's default aspect ratio for parametric plots is to make the horizontal and vertical scaling equal.

If there are no options, it is possible to replace a single x, y pair by a list of such pairs describing multiple graphs, but better results and more flexibility are usually obtained by defining different functions as separate plots and then combining them by the procedures described under **combination graphs**.

Here is the coding for two parametric plots.

```
ParametricPlot[ {Cos[t], 2*Sin[t]}, {t,0,2*Pi},
                PlotStyle -> Dashing[Medium] ]
ParametricPlot[ {Cos[t], 2*Sin[t]}, {t,0,2*Pi},
                PlotStyle -> {Red, Thickness[Large]},
                PlotRange -> {{0,1}, {0,2}} ]
```

The plots produced by this code are shown (without color) in the left and right panels of Fig. A.7.

Contour Plots—The command ContourPlot generates a set of contours representing constant values of the input function $f(x, y)$. Including its most important options, this command takes the form

```
ContourPlot[ f[x,y], {x,x1,x2}, {y,y1,y2},
    Contours -> {c1,c2,c3,c4,c5}, AspectRatio -> (y2-y1)/(x2-x1) ]
```

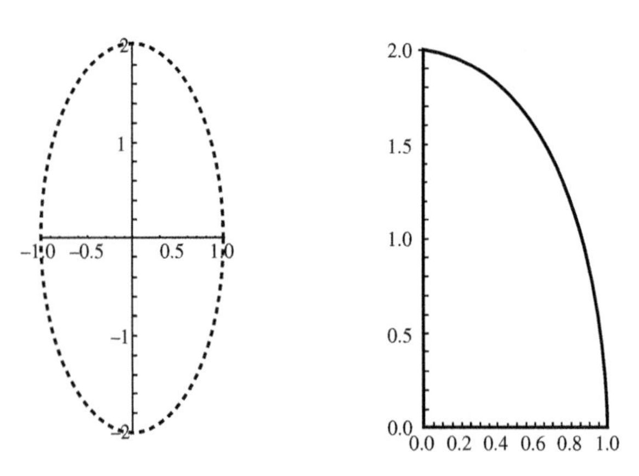

Figure A.7: Parametric plots, with the options described in the text.

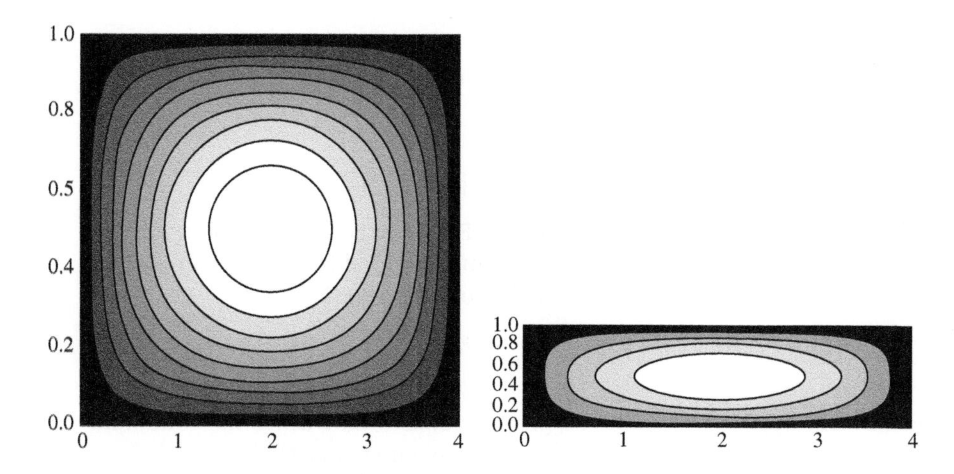

Figure A.8: Contour plots, with the options described in the text.

Mathematica supplies reasonable default values for the contours (of which there can be any number), and provides a nice color scheme for the spaces between the contours. However, the default aspect-ratio setting often distorts the physical system under study to an extent such that the option included here (constraining the scaling) may be useful. Here's the coding for two contour plots; the plots (without color, but with shading) are shown in Fig. A.8.

```
f[x_,y_ := x*(4-x)*y*(1-y)
ContourPlot[ f[x,y], {x,0,4}, {y,0,1} ]
ContourPlot[ f[x,y], {x,0,4}, {y,0,1},
             Contours -> {0.2, 0.4, 0.6, 0.8},
             AspectRatio  -> 1/4 ]
```

If it proves inconvenient to fix the aspect ratio as indicated above, a visual fix is possible by stretching or shrinking one of the coordinates in the displayed contour plot. You can drag one of the edges of the display; however, this will only change the aspect ratio if the `Shift` key is held down during the dragging process.

Point plots—The data to be plotted may either be a numeric list of $\{x_i, y_i\}$ pairs, or in the special case that the x_i are 1, 2, 3, ..., a numeric list $\{y_1, y_2, \ldots\}$. Any valid way of entering the data may be used; if the data originated from an experiment, it may have to be entered explicitly, e.g., as

```
T =  { {0.10,0.384}, {0.21,0.485}, {0.44,0.777} };
```

Note that `T` is a **list**, each member of which is in turn a two-element list giving the horizontal and vertical coordinates of a point to be plotted. If our data can usefully be identified with the positive integers, we can instead enter something like

```
T = {0.123, 0.234, 0.375};
```

If the data arise as values of a given function and are to be plotted for regularly-spaced values of its argument, we may use MATHEMATICA's `Table` command, which acts as illustrated here:

```
T = Table [F[n], {n,4} ]
```
{ F[1], F[2], F[3], F[4] }

or, if the table does not start from $n = 1$,

```
T = Table[ {n,F[n]}, {n,-1,1} ]
```
$\{\{-1, F[-1]\}, \{0, F[0]\}, \{1, F[1]\}\}$

The `Table` command produces a list as output. `F` can either be a MATHEMATICA function or simply an expression that may contain the variable `n`. For example, we can create a data structure for $F(n) = n^2$ by the following coding

```
T = Table[ {n,n^2}, {n,0,4}]
```
$\{\{0,0\},\{1,1\},\{2,4\},\{3,9\},\{4,16\}\}$

Once we have an appropriate data set, we can make a plot using the command `ListPlot`. If `T` is a data set for point plotting, we may enter something like

```
ListPlot[ T, PlotStyle -> {Red, PointSize[0.015]} ];
```

The style options shown here (`PlotStyle ->` $\cdots$) are optional. The colors that are known to MATHEMATICA are discussed under **function graphs**. If a color is not specified, the points will be plotted in the default color. The argument of `PointSize` is the diameter of the solid dots used as plotted points, expressed as a fraction of the width of the entire plot; if `PointSize` is not specified, it defaults to a small value. MATHEMATICA also understands `Small`, `Medium`, and `Large` as arguments for `PointSize` and these qualitative sizes often meet user needs.

Here is the coding for two point graphs:

```
f = Table[ {n, 40*n*E^(-40)/n!, {n, 0, 80} ]
ListPlot[f]
ListPlot[f, PlotStyle -> {Red, PointSize[Large]} ]
```

The graphs are shown (without color) in Fig. A.9.

The easiest way to display several point sets in the same graph, each with its own style options, is make them a **combination graph** as explained below.

Combination graphs—More than one graph can be displayed simultaneously by using the command `Show`. Each graph will be displayed subject to its style options

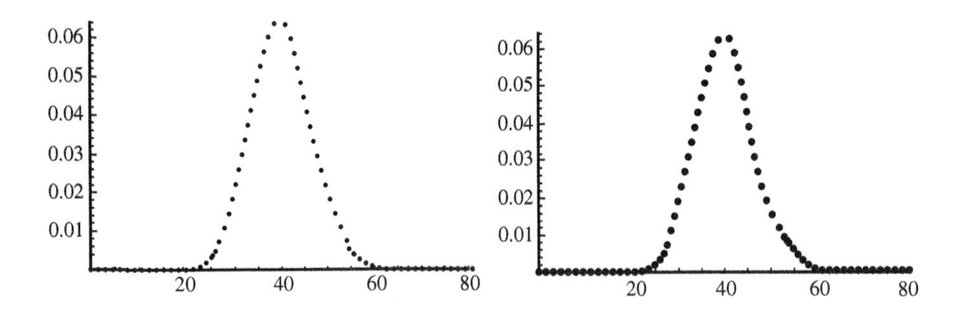

Figure A.9: Point plots, with the options described in the text.

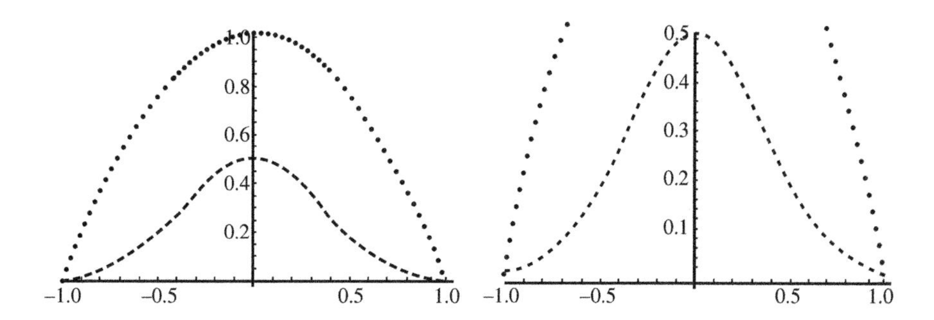

Figure A.10: Output of `Show` (see text). Left, for `[G1,G2]`; right, for `[G2,G1]`.

(or the defaults thereon), and the overall option parameters of the graph (its horizontal and vertical ranges, axis ticks, tick labels, etc.) will be those of the first graph in the input to `Show`.

Graphs to be displayed using `Show` should first be given names, as in the following examples:

```
T = Table[ {n/20, 1 - n^2/400}, {n,-20,20} ];
G1 = ListPlot[ T, PlotStyle -> {Red, PointSize[Medium]} ];
G2 = Plot[0.5 * E^(-4*x^2), {x,-1,1},
        PlotStyle -> {Blue, Dashing[Large], Thickness[Large]} ];
```

We can then combine these graphs by entering either

```
Show[ G1, G2 ]   or   Show[ G2, G1 ].
```

These two choices give different results, as can be seen from Fig. A.10. This figure is not in color; in actual MATHEMATICA output, the dots would be red and the dashed line blue.

The reason the two graphs of Fig. A.10 are different is that when the arguments of `Show` are `G1, G2` the range and scale of the combined graph is that of the individual graph `G1`, while the argument list `G2, G1` causes the combined-graph layout to be that of `G2`.

B PRINTING TABLES OF FUNCTION VALUES

This appendix contains a discussion of issues relevant to the generation of formatted tables of function values and presents code for generating such tables in both our symbolic languages. Because the printing procedures are rather different in the two languages, we discuss them separately. For each language we include code for a basic procedure in which the function arguments are presented as decimal numbers with a fixed number of decimal places, with the function values in **scientific notation** (numbers in the form $\pm 0.nn \cdots nn \times 10^p$ or possibly $\pm n.nn \cdots nn \times 10^p$).

Readers who wish to be able to change the details of tabular output will need to understand all the material in the part of this Appendix relevant to their symbolic language.

MAPLE

In MAPLE, formatted output is obtained by use of the command `printf`, whose first argument is a string that controls the formatting of the material, and whose remaining arguments contain the data to be printed. Unlike the command `print`, which displays its output centered and then moves the cursor to a new line, `printf` starts printing from the present position of the cursor (at the left margin if no prior `printf` command has left it positioned elsewhere) and only starts new lines if directed to do so in the format control string.

The MAPLE format control string has properties taken from the C computer language and its function is therefore already understood by C programmers. For the rest of us, the minimum one needs to know is fairly simple: the format control string contains three types of entries: (1) characters, sent to be printed "as is"; (2) specifications for formatting a number (more detail below); (3) control characters, of which the only one of current interest is `\n`, which causes the current print line to end and positions the cursor at the start of the next line.

We are interested here in two types of specifications for printing numbers (there are many others that we will ignore):

(1) A fixed decimal format whose total width is w columns and which presents right-adjusted numbers with d decimal places (therefore with decimal points aligned) is indicated by `%`w.d`f`. (The final period is not part of the specification.) This means that the percent sign is recognized as a special character and will not be part of what is printed. For example, if the specification `%8.3f` is applied to the number 4/3, the result will be *bbb*1.333, where *b* stands for a space (blank character). Note that one column is occupied by the decimal point. The same specification applied to $-4/3$ yields *bb*-1.333. The largest positive number representable in this format is 9999.999, while the largest representable negative number is -999.999.

(2) A scientific-notation format whose total width is w columns, with numbers given to d decimal places, indicated by `%`w.d`E`. For example, if `%12.4E` is used to format -123456, the result will be $b-1.2346E+05$. It is obviously necessary that w be large enough to hold the exponent and a possible minus sign.

Because `%` plays a special role, it can only be printed by doubling it: `%%` adds a percent sign to printed output.

Example B.1. Maple Print Lines

Suppose we want a line containing only the title of a table. MAPLE code to do that is illustrated here:

```
> printf("        Table of data, units are %%.\n");
```

This produces the following output, with the cursor moved after the printing to a position at the start of the next line:

bbbbbbbb`Table of data, units are %.`

The *b* characters are written (instead of actual blanks) to make it clear that the initial blanks in the format control string will cause blanks in the printed output.

Suppose next that we want the three numbers 1/7, $-2/7$, 3/7 to be printed to six decimal places with some space before and between them. To print these numbers we insert three numeric formats into our format control string; these will be used (in order) to format the three arguments of our `printf` command that follow the control string. We can get space to precede our printed numbers by making the field width bigger than is needed to hold the print, or by inserting blanks in the format control string. For example,

```
> printf("   %10.6f  %9.6f  %12.6f\n",1/7,-2/7,3/7);
```

*bbbbb*0.142857*bb* −0.285714*bbbbbb*0.428571

Before 0.142857 there are three blanks from the control string and two because the number requires eight of the ten spaces reserved for it; before −0.285714 the only blanks are from the control string; and before 0.428571 there are two blanks from the control string and four from the unused portion of the reserved field width.

Finally, we may have both fixed and scientific-notation data, as in this conversation about a circle:

```
> printf("Diameter: %4.2f, Circumference: %12.6E.\n",8.5,8.5*Pi);

  Diameter: 8.5, Circumference: 2.670354E+01.
```

Notice that the comma, colons, and period in the format control string are part of what is reproduced in the output.

■

With print control in hand, we now present code for the printing of a table of function values. The following coding is for a procedure that makes a table of values of a function whose MAPLE name is `func` for values of its argument x covering the range `x1 <= x <= x2` in steps of size `delta`. Because the function name and the range are procedure arguments, the entire procedure can be used for an arbitrary function; the main limitation is the number of digits of precision with which the data are printed.

Our coding is as follows:

```
> MakeTable := proc(func,x1,x2,delta)
>                 local fxstr,x1str,x2str,delstr,x;
>   fxstr := convert(func,string);
>   x1str := convert(x1,string);
>   x2str := convert(x2,string);
>   delstr := convert(delta,string);
>   printf("Values of "||fxstr||"(x) for x from "||x1str
>      ||" to "||x2str||" in steps of "||delstr||"\n\n");
>   printf("      x     "||fxstr||"\n");
>   x := x1;
>   while (x <= x2) do
>     printf("  %8.3f     %13.6E\n",x,func(x));
>     x := x + delta;
>   end do;
> end proc:
```

The MAPLE command `convert` is used to make the function name and range specification into strings, so that they can be inserted into the format control string. Notice that the control string is built from explicit string definitions (enclosed in double quotes) and variables (e.g., fxstr) that have strings as their values. These elements are connected into a single string using MAPLE's string concatenation operator ||. Note that material enclosed in double quotes cannot be broken between input lines; our overall string expression was long, so we broke it (legitimately) at a point not within any double quotes.

The use of this MAPLE procedure is illustrated in Example 1.5.2.

MATHEMATICA

In MATHEMATICA, the regular printing command `Print` is used for formatted printing. Each call of `Print`, however, starts a new line of output, with the print starting at the left margin of the line. It is therefore necessary to place all the content of each formatted line as arguments within a single `Print` command.

For our present purposes we consider three types of arguments for the `Print` command:

(1) Characters (or sequences of characters), called **strings**, which are written without modification into the output stream. MATHEMATICA does not put spaces or punctuation before or after strings, so two consecutive strings will print as adjacent character sequences with no intervening space. Strings can be specified by writing them explicitly between double quotes, or by causing a string to be the value of a variable whose name is then used as a `Print` argument.

(2) Numbers containing decimal points, which are to be placed right-adjusted with d decimal places into a field containing enough columns to represent a negative number with a total of w decimal digits and a decimal point. The quantity to be printed is converted (for output purposes only) to the specified format, padding with zeros as necessary to the right of the decimal point and padding with blanks as necessary to the left. The expression *expr*, formatted in this way, is obtained by the construction

    `PaddedForm[`*expr*`,{`w, d`}]`

(3) Numbers containing decimal points, to be presented in scientific notation with d digits to the right of the decimal point and with a total relative precision of p decimal digits. The value of p does not control the amount of space reserved for the output; it does, however, control the number of digits that are accurately represented in the output and therefore under ordinary circumstances should be $d+2$. The expression *expr*, formatted in this way, is obtained from

    `ScientificForm[`*expr*`, {`p, d`}, NumberPadding -> {"","0"}]`

Example B.2. Mathematica Print Lines

Suppose we want to build an output line (e.g., a table title) that contains no numeric data. We illustrate as follows, assuming that the variable `str` has as its value the string `functionname`,

```
Print["    ",str,"[x] values:"]
```

produces the printed output

bbbb`functionname[x] values:`

The b characters are written (instead of actual blanks) to make it clear that the blanks from the first argument of Print are copied into the print output.

Suppose next that we want the three numbers $1/7$, $-2/7$, $3/7$ to be printed to six decimal places with some space before and between them. We gain the desired spaces by designating widths larger than are necessary to represent our three numbers. We also need to prevent MATHEMATICA from treating these quantities as exact fractions, so we apply the N operator (with nominal precision of 10 digits):

```
Print[ PaddedForm[N[ 1/7, 10],  {9,6}],
       PaddedForm[N[-2/7, 10],  {9,6}],
       PaddedForm[M[ 3/7, 10], {10,6}]
     ]
```

produces $bbb0.142857bb-0.285714bbbb0.428571$

A padded format of width 9 occupies 11 print columns: 9 for digits, plus one each for the decimal point and a possible minus sign.

Finally, look at

```
Print["    ",ScientificForm[-0.1,{12,10},NumberPadding -> {"","0"}]]
```

This yields

$bbbb-1.0000000000 \times 10^{-1}$

MATHEMATICA's standard layout for scientific notation contains a single digit to the left of the mantissa's decimal point, plus room for a possible minus sign.

■

With print control in hand, we now present code for the printing of a table of function values. The following coding is for a procedure that makes a table of values of a function whose MATHEMATICA name is func for values of its argument x covering the range x1 <= x <= x2 in steps of size delta. Because the function name and the range are procedure arguments, the entire procedure can be used for an arbitrary function; the main limitation is the number of digits of precision with which the data are printed.

Our coding is as follows:

```
MakeTable[func_,x1_,x2_,delta_] := Module[{fxstr,x1str,x2str,delstr,x},
  fxstr = ToString[func];
  x1str = ToString[x1];
  x2str = ToString[x2];
  delstr = ToString[delta];
  Print["Values of ",fxstr,"[x] for x from ",x1str," to ",x2str,
                   " in steps of ",delstr];
  Print[ ];
  Print["     x          ",fxstr,"[x]"];
  x = x1;
  While[(x <= x2),
    Print[PaddedForm[x,{6,3}],"      ",
          ScientificForm[N[func[x],12],{8,6},NumberPadding
            -> {" ","0"}]];
    x = x + delta;] ]
```

The command `ToString` makes the function name and range specification into strings, so that they are more easily inserted into the table title line. Notice that the successive strings follow each other seamlessly when inserted as comma-separated arguments to the `Print` command, either identified by name or as explicit material enclosed in double quotes.

The use of this MATHEMATICA procedure is illustrated in Example 1.5.2.

Exercises

B.1. Modify the procedure `MakeTable` so that both the values of x and the function values are given as ordinary decimals, with the function values given to eight decimal places and in a field of width sufficient to accommodate all numbers with $|x| < 10^{12}$.

B.2. Modify the procedure `MakeTable` so that it can accept functions of two variables (x and y) with each variable running independently on a finite equally-spaced grid. It is not required to expend effort on the graph title or the column labeling, but if you do so it may be easier to use the code after you have forgotten how you constructed it.

C DATA STRUCTURES FOR SYMBOLIC COMPUTING

There are many situations in which we may need to work with mathematical quantities that are more complicated than individual numbers (or symbols that represent such numbers). Early in this book we encountered complex numbers, which from a fundamental viewpoint consist of ordered pairs of numbers (with associated rules for their manipulation). Then we encountered vectors (which can be represented by ordered n-tuples of numbers or symbols) and matrices (which correspond to two-dimensional numeric or symbolic arrays). Other objects, such as the data that control the extent and nature of a summation or series expansion, consist of several items not all of which are even of the same kind. It is also useful to define data arrays without specifying in advance what operational properties they possess. The designers of symbolic computing systems have found it useful to define data structures appropriate to all these purposes; some understanding of the definitions may make it easier to work with the objects that the data structures represent.

The data-structure philosophies of MAPLE and MATHEMATICA are markedly different. MAPLE defines different data structures for different types of mathematical objects, while MATHEMATICA defines one type of data structure but causes the way in which data are used to vary, depending upon the procedure that uses the data. While one might think these philosophies are only of importance to the developers of the symbolic languages, that view is not completely correct. The purpose of this appendix is to review those features of the symbolic data structures that are of importance to the system users and to identify issues that may cause problems if not understood.

MAPLE

The basic building block for data structures in MAPLE is a collection of elements, or entries, which can be grouped together either as an **ordered set**, which MAPLE calls a **list**, or as a set not possessing order, called by MAPLE a **set** (which is also its usual mathematical name). **Lists** are represented as their comma-separated elements enclosed in square brackets, and if the elements are to be referred to, they are considered indexed in the order of their appearance in the list, with the indexing starting from 1. **Sets** are represented by enclosing their comma-separated elements in curly braces. Thus, consider

```
> A := [3,4,2,3];        [3, 4, 2, 3]
> B := {3,4,2,3};        {2, 3, 4}
```

Here `A` has been defined as a list containing (in order) the four elements 3, 4, 2, 3; the duplication of "3" simply means that the first and fourth elements happen to have the same value. Because `B` is defined as a set, the duplication means that the element "3" has been listed redundantly; MAPLE removes the redundance and orders the set elements in whatever way the MAPLE developers found most convenient for later processing. We have little use for MAPLE sets in this book and will not discuss them further. By mentioning them here we call to attention that when a **list** is required, a brace-enclosed collection of elements would be accepted by MAPLE as a valid object (a **set**) but it would not properly designate an ordered list.

The individual elements of a list can be identified by placing after the list name the element's index number enclosed in square brackets (not parentheses, which are used to denote function arguments). Thus, with `A` as defined above, `A[2]` is 4, `A[1]` and `A[4]` are both 3, `A[3]` is 2, and a request for the value of `A[5]` generates an error. One can use the same notation to change the value of a list element:

```
> A[3] := 6;
```

will cause the "2" which was the third element of `A` to be replaced by "6."

MAPLE does not place restrictions as to the types of objects that can be members of lists. Thus, a list can have lists as elements, and there is no requirement that different elements of a list be of the same kind or dimension. For example,

```
C := [[a,b,c,d],[1,2,3,4],[w,x,y,z]];      F := [[a,b,c],17,x>4];
```

both define valid lists. Here `C` is a three-element list whose first element is the list `[a,b,c,d]`, and `F` is a list each of whose elements is a different type of quantity.

To obtain, say, the second element of the third element of `C`, one could write `C[3][2]` (its value is x), but MAPLE also accepts the simpler notation `C[3,2]`.

Another data structure in MAPLE, called a **sequence**, is a comma-separated set of items that is not surrounded by delimiters (parentheses, square brackets, or braces). Sequences are often produced in MAPLE output; one motivation for this output style is that a sequence can be inserted, perhaps along with other quantities, to build a list or define the range of a variable. We mention sequences here because they are not easily manipulated; they are not indexed quantities and their individual parts are not readily accessible. However, if a sequence is enclosed in square brackets it becomes a list, and its items can then be accessed using an index that indicates their position. For example, MAPLE may present the solutions to an equation as the sequence $4, 7, 11$. If we name this sequence and try to access its parts, we fail, but if we make it a list we can do so. Thus,

```
> S := 4,7,11;        4, 7, 11
> S[1];               (error)
> SS := [S];          [4, 7, 11]
> SS[1];              4
```

In MAPLE, the components of a vector are stored in a list, but the MAPLE data type produced by the command `Vector` contains additional information (e.g., whether the vector is a column or a row). That additional information is also used by MAPLE to speed up vector operations. Likewise, the elements of a matrix are stored in a list of lists, but `Matrix` generates a data type that also includes additional information. We therefore emphasize that the list construction (as used to make **A** and **C** above) is **not** an alternative way to construct a MAPLE vector or matrix; those should always be built using `Vector` or `Matrix`. Moreover, MAPLE uses the additional information associated with the `Vector` and `Matrix` data types to cause those quantities to be displayed as rows, columns, or two-dimensional arrays rather than in list notation. However, the underlying presence of the list construction enables the elements of vectors and matrices to be identified and changed using the list element notations `V[`i`]` and `M[`i,j`]`.

The observations of the preceding paragraph mean also that there is a difference between a list of vectors and a matrix. We illustrate that point by defining the following quantities:

```
> V1:=Vector([1,2,3]);
```

$$V1 := \begin{bmatrix} 1 \\ 2 \\ 3 \end{bmatrix}$$

```
> V2:=Vector([4,5,6]);
```

$$V2 := \begin{bmatrix} 4 \\ 5 \\ 6 \end{bmatrix}$$

We see that `V1` and `V2` are column vectors. Compare now a list `L` consisting of those two vectors and a matrix `M` made by placing them side-by-side:

```
> L := [V1,V2];
```

$$L := \left[\begin{bmatrix} 1 \\ 2 \\ 3 \end{bmatrix}, \begin{bmatrix} 4 \\ 5 \\ 6 \end{bmatrix} \right]$$

```
> M := Matrix([V1,V2]);
```

$$M := \begin{bmatrix} 1 & 4 \\ 2 & 5 \\ 3 & 6 \end{bmatrix}$$

Notice that the `Matrix` command can accept input that is a row of quantities that have been previously defined to be column vectors. That provides a way to enter a matrix by columns instead of by rows.

The above code shows how column vectors can be assembled into a matrix; one way to build a matrix from row vectors is to transform the vectors into columns, use the process illustrated above, and then transpose the resulting matrix.

We sometimes need a process opposite to that illustrated above. Given a matrix, we might want its columns as a list of vectors. MAPLE has a `LinearAlgebra` command, `Column`, that is useful for that purpose. We can proceed as follows, starting from the matrix `M` that we just built from `V1` and `V2`. `Column` has two arguments; the first is the matrix from which column(s) are to be extracted (as vectors), the second gives the columns to be extracted (either as the number of a single column or as a sequence of column numbers, e.g., notations like 1, 3, 5 (to mean columns 1, 3, and 5) or `n..m` (which means columns n through m). The result of the `Column` command is a **sequence** containing the requested column vectors. Thus,

```
> with(LinearAlgebra);
```

```
> Column(M,1..2);
```

$$\begin{bmatrix} 1 \\ 2 \\ 3 \end{bmatrix}, \begin{bmatrix} 4 \\ 5 \\ 6 \end{bmatrix}$$

```
> [%];
```

$$\left[\begin{bmatrix} 1 \\ 2 \\ 3 \end{bmatrix}, \begin{bmatrix} 4 \\ 5 \\ 6 \end{bmatrix}\right]$$

The last of the above coding lines converts the column sequence into a list for ease in further processing.

The manipulations illustrated above may be important when working with the solutions of eigenvalue problems. The MAPLE command `Eigenvectors` produces a set of eigenvectors as columns of a matrix. Conversion into a **list** of eigenvectors may be a first step toward further analysis. On the other hand, we may have a set of vectors we want to assemble into a transformation matrix so that we can carry out a similarity transformation (as in Section 5.3). The methods illustrated here may help to accomplish that.

MATHEMATICA

In MATHEMATICA there is only one compound data type of importance; it is an ordered **list** which is denoted by placing its comma-separated elements between curly braces. For example, a list `L` containing (in order) the elements a, b, c is formed by writing

```
L = {a,b,c}
```

$$\{a, b, c\}$$

That is exactly the construction we used to produce a vector; in MATHEMATICA, there is no difference between a vector and a list.

The elements of a list are indexed consecutively, starting from 1, and the ith element of a list `L` is identified as `L[[i]]`; that notation can be used both to obtain the value of a list element and to change it. Thus, for `L` as given above,

```
L[[2]]
```

$$b$$

```
L[[2]]=17; L
```

$$\{a, 17, c\}$$

Thus, the components of a vector `V` are accessed as `V[[i]]`.

There is no inherent limitation on the types of quantities that can be list elements, and it is possible for the elements of a list to be themselves lists. Thus, examples of lists are

```
C = {{a,b,c,d},{1,2,3,4},{w,x,y,z}}          F = {{a,b,c},17,x>4}
```

Thus, we see that the notation for a matrix is just a list of lists, with each sublist containing the elements of a row of our matrix. Given the list-of-lists formation of a matrix `M`, its ith row can be extracted as `M[[i]]`, and the jth element of the ith row (which is the matrix element M_{ij}) can be accessed as `M[[i]][[j]]`, or more conveniently as `M[[i,j]]`.

In MATHEMATICA, the commands `Eigenvectors` and `Eigensystem` produce the eigenvectors as a list of lists, with each sublist being an eigenvector. This list of lists corresponds to a matrix in which the eigenvectors are **rows**, whereas in the underlying mathematics as presented here (and almost everywhere else) the eigenvectors are actually columns. MATHEMATICA does this for a reason; if `M` is the matrix of row eigenvectors, then `M[[i]]` produces the ith eigenvector, and `M` is both a matrix and an eigenvector list. If a matrix of column eigenvectors is required, we can obtain it by transposing the matrix `M`: `Transpose[M]`.

Because the list notation does not itself indicate whether lists are intended to represent vectors or matrices, MATHEMATICA does not attempt to output any lists in the matrix display form. That is the reason for the existence of the command `MatrixForm`.

D SYMBOLIC COMPUTING OF RECURRENCE FORMULAS

The definitions used for vectors by both MAPLE and MATHEMATICA carry with them the built-in assumption that the numbering of the elements (the vector components) starts from 1. Moreover, the lower-level **list** construct has elements (operands) with numbering hard-wired to start with 1, and for both vectors and lists, "operand 0" is reserved for internal use by the symbolic system.

This rigidity in numbering most often becomes significant when list-type objects are to be used to store mathematical quantities whose natural numbering starts from zero; cases in point include the coefficients in a power series, the successive terms of many other summations, and series of functions whose initial member is traditionally numbered zero. The computational problems affected by the numbering system often involve situations where the nth member of a set of quantities can be generated recursively from prior members of the set. We illustrate by considering the **Hermite polynomials** $H_n(x), n = 0, 1, 2, \ldots$, with $H_n(x)$ a polynomial in x of degree n. As discussed in detail in Chapter 14, those polynomials are related by the recurrence formula

$$H_{n+1}(x) = 2xH_n(x) - 2nH_{n-1}(x)\,, \tag{D.1}$$

with the starting values $H_0(x) = 1$, $H_1(x) = 2x$.

If we wish to generate and store these polynomials, the symbolic code to do so will be more easily understood if a storage location associated with the index n can be associated with H_n. If that is not the case, then the coefficient n of "$2n$" and the storage index for H_n will be different, and the formula will perhaps be less transparent.

Due to the differences in the symbolic languages, this issue needs to be treated on a language-specific basis.

MAPLE

MAPLE has a data type `Array` which is useful in the present context. To define `A` as a one-dimensional array with indexing ranging from zero to `n`, write

```
> A := Array(0 .. n):
```

This construction can in fact be used for any integer range; possibilities that may be useful elsewhere are illustrated by `-3 .. 3` and `5 .. 12`.

The elements of `Array` (which the above command initializes to zero) are accessed using MAPLE's usual notation, and may be read or changed by code such as

```
> A[1] := 17:
> A[1];          17
> A[0];          0
```

For the Hermite polynomials, define an array H, re-initialize with values of H_0 and H_1, and check the result:

```
> H := Array( 0 .. 4);
> H[0] := 1:  H[1] := 2*x:
> ArrayElems(H);
```

$$\{\ (0) = 1, (1) = 2x\}$$

The output of the array elements is not pretty; note also that all elements that are zero are omitted from the output.

Now execute a loop to obtain the remaining H_n through H_4:

```
> for n from 1 to 3 do
>    H[n+1] := expand(2*x*H[n]-2*n*H[n-1]);
> end do;
```

$$H_2 := 4x^2 - 2$$
$$H_3 := 8x^3 - 12x$$
$$H_4 := 16x^4 - 48x^2 + 12$$

The command `expand` (or `simplify`) was needed to force similar terms to combine.

MATHEMATICA

While the construct `Array` also exists in MATHEMATICA, zero or negative indices are not processed in ways consistent with the evaluations of recurrence formulas, and a different approach is needed. One way to proceed, illustrated for the Hermite polynomials, is as follows:

- Do **not** define an array to hold the polynomials H_n.
- With `h` undefined, execute the following MATHEMATICA statements. (We use a lower-case `h` to emphasize the desirability of avoiding a clash with a quantity that MATHEMATICA might have defined):

```
h[0] = 1;    h[1] = 2*x;
For[ n=1, n <= 3, n = n+1, Expand[h[n+1] = 2*x*h[n] - 2*n*h[n-1]] ]
```

- The `For` command produces no output. Therefore, to check what we have done:

`h[0]`	1
`h[1]`	$2x$
`h[2]`	$4x^2 - 2$
`h[3]`	$8x^3 - 12x$
`h[4]`	$16x^4 - 48x^2 + 12$

Note that the `h[n]` are **not** elements of an array, and that they are identified using `[ ]` and not `[[ ]]`. From MATHEMATICA's viewpoint, **h** is a command which has the indicated values for arguments `n`. That means that the values of `h[n]` cannot be identified or accessed collectively using just the name `h`.

E PARTIAL FRACTIONS

The partial fraction decomposition of a quotient of two polynomials yields a form that is often useful for mathematical analysis. Consider as an initial example the equation

$$\frac{1}{1-x^2} = \frac{1}{2}\left[\frac{1}{1-x} + \frac{1}{1+x}\right] . \tag{E.1}$$

The right-hand side of this equation is identified as the set of **partial fractions** equivalent to the left-hand side. Equation (E.1) is easily checked by combining the two terms on the right-hand side over a common denominator.

One reason the decomposition of Eq. (E.1) is useful is that it makes it easy to evaluate the integral

$$\int \frac{dx}{1-x^2} = \frac{1}{2}\int \left[\frac{1}{1-x} + \frac{1}{1+x}\right] dx = \frac{1}{2}\left[-\ln(1-x) + \ln(1+x)\right] = \frac{1}{2}\ln\left(\frac{1+x}{1-x}\right) . \tag{E.2}$$

The general problem of which the above is an example is to obtain the partial fraction decomposition of $f(x)/g(x)$, where $f(x)$ and $g(x)$ are polynomials, with the degree of $f(x)$ less than that of $g(x)$ (a so-called **proper fraction**). Note that if the numerator were originally of degree equal to or higher than the denominator, we could carry out the indicated division to reduce $f(x)/g(x)$ to a polynomial plus a proper fraction. For example,

$$\frac{x^3}{1-x^2} = -x + \frac{x}{1-x^2} .$$

Our first step in obtaining the decomposition is to bring the denominator to factored form. If we cannot do so, equivalent to obtaining all the roots of $g(x) = 0$, we cannot proceed. Thus, we assume that $f(x)/g(x)$ has been brought to the form

$$\frac{f(x)}{g(x)} = \frac{a_0 + a_1 x + \cdots + a_m x^m}{(x-r_1)(x-r_2)\cdots(x-r_n)} , \tag{E.3}$$

where $m < n$.

If all the r_n are distinct, the partial fraction decomposition we seek has the general form

$$\frac{f(x)}{g(x)} = \frac{c_1}{x-r_1} + \frac{c_2}{x-r_2} + \cdots + \frac{c_n}{x-r_n} , \tag{E.4}$$

and we can determine the c_i by requiring that Eq. (E.4) be consistent with Eq. (E.3), in which the a_i and r_i are known. We note that if we multiply the expansion in Eq. (E.4) by some $x - r_i$, the i term will become simply c_i and all the other terms of the expansion will have $x - r_i$ as a common factor. If we then take the limit $x \to r_i$, all but the ith term will vanish, leaving c_i as the limit. We therefore have the formula

$$c_i = \lim_{x\to r_i}\left[\frac{(x-r_i)g(x)}{h(x)}\right] . \tag{E.5}$$

That limit amounts to writing Eq. (E.3) without the factor $x - r_i$ in the denominator and then setting $x = r_i$.

If the r_n are not all distinct, we generalize Eq. (E.4) when the factor $(x - r_j)$ of Eq. (E.3) occurs p times by the replacement

$$\frac{c_j}{x-r_j} \longrightarrow \frac{c_{j,1}}{x-r_j} + \frac{c_{j,2}}{(x-r_j)^2} + \cdots + \frac{c_{j,p}}{(x-r_j)^p} . \tag{E.6}$$

It is now often easiest to proceed by putting the expansion of Eq. (E.4) over a common denominator and equating the numerator to the numerator $a_0 + \cdots + a_m x^m$ of Eq. (E.3). Because power series expansions are unique, we equate the coefficients of individual powers of x, thereby obtaining a systems of linear equations that we can solve for the c_i. In typical cases, the c_i are usually found easily.

Sometimes, in order to avoid the introduction of complex quantities from the factorization of the denominator $g(x)$, we leave as a partial fraction a quadratic denominator, of the general form

$$\frac{c + c'x}{x^2 + rx + r'} .$$

Here also it may be useful to find the coefficients by the method described after Eq. (E.6).

Example E.1. Partial Fraction Expansion

Let

$$f(x) = \frac{k^2}{x(x^2 + k^2)} = \frac{c}{x} + \frac{ax + b}{x^2 + k^2} .$$

We have written the form of the partial fraction expansion, but have not yet determined the values of a, b, and c. Putting the right side of the equation over a common denominator, we have

$$\frac{k^2}{x(x^2 + k^2)} = \frac{c(x^2 + k^2) + x(ax + b)}{x(x^2 + k^2)} .$$

Expanding the right-side numerator and equating it to the left-side numerator, we get

$$0(x^2) + 0(x) + k^2 = (c + a)x^2 + bx + ck^2 ,$$

which we solve by requiring the coefficient of each power of x to have the same value on both sides of this equation. We get $b = 0$, $c = 1$, and then $a = -1$. The final result is therefore

$$f(x) = \frac{1}{x} - \frac{x}{x^2 + k^2} . \tag{E.7}$$

■

Example E.2. More Partial Fractions

We desire the partial fraction decomposition of

$$F \equiv \frac{3x^2 - 1}{\mathcal{D}} \equiv \frac{3x^2 - 1}{x(x^2 + 4)(x - 2)^2} = \frac{c_1}{x} + \frac{c_2 x + c_3}{x^2 + 4} + \frac{c_4}{x - 2} + \frac{c_5}{(x - 2)^2} , \tag{E.8}$$

which according to the preceding discussion can be written in the form shown here. Putting the right-hand side of Eq. (E.8) over a common denominator, we have

$$F = \frac{c_1(x^2 + 4)(x - 2)^2 + x(c_2 x + c_3)(x - 2)^2 + x(x^2 + 4)c_4(x - 2) + x(x^2 + 4)c_5}{\mathcal{D}} . \tag{E.9}$$

Expanding the numerator and collecting terms with the same power of x, we reach

$$F = \frac{3x^2 - 1}{\mathcal{D}} = \frac{C_4 x^4 + C_3 x^3 + C_2 x_2 + C_1 x_1 + C_0}{\mathcal{D}}, \tag{E.10}$$

with

$$C_4 = c_1 + c_2 + c_4, \qquad C_3 = -4c_1 - 4c_2 + c_3 - 2c_4 + c_5,$$

$$C_2 = 8c_1 + 4c_2 - 4c_3 + 4c_4, \qquad C_1 = -16c_1 + 4c_3 - 8c_4 + 4c_5,$$

$$C_0 = 16c_1.$$

Invoking the linear independence of different powers of x, we now set each C_i equal to its value as specified by the left member of the equation:

$$C_0 = -1, \quad C_1 = 0, \quad C_2 = 3, \quad C_3 = 0, \quad C_4 = 0\,.$$

We have thereby obtained a system of linear equations which we can solve for the c_i. The result is

$$F = -\frac{x}{16} - \frac{13}{16}\frac{1}{(x^2+4)} + \frac{1}{16}\frac{1}{(x-2)} + \frac{11}{16}\frac{1}{(x-2)^2}\,.$$

■

SYMBOLIC COMPUTATION

Symbolic computation systems know how to find partial fraction decompositions. If the expression to be decomposed contains only a single variable, the commands involved are the following:

`convert(`*expr*`,parfrac);` (MAPLE) `Apart[`*expr*`]` (MATHEMATICA)

If it is necessary to identify the variable x to be separated into partial fractions, those commands become

`convert(`*expr*`,parfrac,x);` (MAPLE) `Apart[`*expr*`,x]` (MATHEMATICA)

Exercises

E.1. Find partial fraction decompositions of the following fractions. Do so first by hand, and then verify your result using a symbolic computation system.

(a) $\dfrac{1}{(x+1)(x+2)(x+3)}$. (b) $\dfrac{1}{x^2 - a^2}$, assuming x to be the variable.

(c) Repeat part (b), with a the variable. (d) $\dfrac{x^2}{(x+a)^3}$; (the variable is x).

E.2. Prove the partial fraction expansion (with p a positive integer)

$$\frac{1}{n(n+1)\cdots(n+p)} =$$

$$\frac{1}{p!}\left[\binom{p}{0}\frac{1}{n} - \binom{p}{1}\frac{1}{n+1} + \binom{p}{2}\frac{1}{n+2} - \cdots + (-1)^p\binom{p}{p}\frac{1}{n+p}\right],$$

Hint. Use mathematical induction. Two binomial coefficient formulas of use here are

$$\frac{p+1}{p+1-j}\binom{p}{j} = \binom{p+1}{j}, \qquad \sum_{j=1}^{p+1}(-1)^{j-1}\binom{p+1}{j} = 1.$$

F MATHEMATICAL INDUCTION

We are occasionally faced with the need to establish a relation which is valid for a set of integer values, in situations where it may not initially be obvious how to proceed. However, it may be possible to show that if the relation is valid for an arbitrary value of some index n, then it is also valid if n is replaced by $n+1$. If we can also show that the relation is unconditionally satisfied for some initial value n_0, we may then conclude (unconditionally) that the relation is also satisfied for $n_0+1, n_0+2, \ldots$ That method of proof is known as **mathematical induction**. It is ordinarily most useful when we know (or suspect) the validity of a relation, but lack a more direct method of proof.

Example F.1. Sum of Integers

The sum of the integers from 1 through n, here denoted $S(n)$, is given by the formula $S(n) = n(n+1)/2$. An inductive proof of this formula proceeds as follows:

(1) Given the formula for $S(n)$, we calculate

$$S(n+1) = S(n)+(n+1) = \frac{n(n+1)}{2}+(n+1) = \left[\frac{n}{2}+1\right](n+1) = \frac{(n+1)(n+2)}{2}.$$

Thus, given $S(n)$, we can establish the validity of $S(n+1)$.

(2) It is obvious that $S(1) = 1(2)/2 = 1$, so our formula for $S(n)$ is valid for $n = 1$.

(3) The formula for $S(n)$ is therefore valid for all integer $n \geq 1$.

■

Exercises

F.1. Show that $\displaystyle\sum_{j=1}^{n} j^4 = \frac{n}{30}(2n+1)(n+1)(3n^2+3n-1)$.

F.2. Prove the Leibniz formula for the repeated differentiation of a product:

$$\left(\frac{d}{dx}\right)^n [f(x)g(x)] = \sum_{j=0}^{n}\binom{n}{j}\left[\left(\frac{d}{dx}\right)^j f(x)\right]\left[\left(\frac{d}{dx}\right)^{n-j} g(x)\right].$$

G CONSTRAINED EXTREMA

Lagrange's method of undetermined multipliers is a powerful approach for finding constrained minima or maxima. A typical situation to which the method is applicable is the minimization of a function $f(x,y,z)$ subject to a constraint of a form such as $g(x,y,z) =$ constant. One can, in principle, solve the constraint equation for

z as a function of x and y, then write $f(x, y, z)$ as $f(x, y, z(x, y))$, and finally solve the resulting unconstrained minimization problem in x and y by setting

$$\left(\frac{\partial f}{\partial x}\right)_y = \left(\frac{\partial f}{\partial y}\right)_x = 0\,.$$

However, that process may be unnecessarily cumbersome and, in cases where one cannot solve the constraint equation, impossible to carry out.

Lagrange's method can be developed starting from the observation that the constraint equation implies

$$dg = \left(\frac{\partial g}{\partial x}\right)_{yz} dx + \left(\frac{\partial g}{\partial y}\right)_{xz} dy + \left(\frac{\partial g}{\partial z}\right)_{zy} dz = 0\,. \tag{G.1}$$

If we set $dy = 0$, the above equation can be rearranged to the form

$$\left(\frac{\partial z}{\partial x}\right)_y = -\frac{\left(\frac{\partial g}{\partial x}\right)_{yz}}{\left(\frac{\partial g}{\partial z}\right)_{xy}}\,. \tag{G.2}$$

Next we use the chain rule to obtain $(\partial f/\partial x)_y$. The notation of this derivative indicates that z is to be viewed (because of the constraint) as a function of x and y.

$$\left(\frac{\partial f}{\partial x}\right)_y = \left(\frac{\partial f}{\partial x}\right)_{yz} + \left(\frac{\partial f}{\partial z}\right)_{xy}\left(\frac{\partial z}{\partial x}\right)_y\,. \tag{G.3}$$

Substituting from Eq. (G.2) and setting $(\partial f/\partial x)_y$ to zero, we can organize the result to read

$$(\partial f/\partial x)_y = \left(\frac{\partial f}{\partial x}\right)_{yz} - \frac{\left(\frac{\partial f}{\partial z}\right)_{xy}}{\left(\frac{\partial g}{\partial z}\right)_{xy}}\left(\frac{\partial g}{\partial x}\right)_{yz} = \left(\frac{\partial f}{\partial x}\right)_{yz} - \lambda\left(\frac{\partial g}{\partial x}\right)_{yz} = 0\,, \tag{G.4}$$

where

$$\lambda = \frac{\left(\frac{\partial f}{\partial z}\right)_{xy}}{\left(\frac{\partial g}{\partial z}\right)_{xy}}\,. \tag{G.5}$$

If the roles of x and y are reversed, we can obtain an equation analogous to Eq. (G.4) for $(\partial f/\partial y)_x$, with the **same** λ. And if we rearrange Eq. (G.5), we reach a third equation analagous to Eq. (G.4), but with z instead of x or y. We therefore have the symmetrical equation set

$$\begin{aligned}
\left(\frac{\partial f}{\partial x}\right)_{yz} - \lambda\left(\frac{\partial g}{\partial x}\right)_{yz} &= 0\,,\\
\left(\frac{\partial f}{\partial y}\right)_{xz} - \lambda\left(\frac{\partial g}{\partial y}\right)_{xz} &= 0\,,\\
\left(\frac{\partial f}{\partial z}\right)_{xy} - \lambda\left(\frac{\partial g}{\partial z}\right)_{xy} &= 0\,.
\end{aligned} \tag{G.6}$$

All the derivatives in these equations are straightforward to evaluate. Considering also the equation of constraint, we now have a total of four equations in the unknowns x, y, z, and λ, and we may solve them to find the desired minimum.

In general, the above procedure actually finds stationary points of f subject to the constraint; they can be minima, maxima, or saddle points.

A procedure similar to the above can be applied in the more general case where $f = f(x_1, x_2, \dots, x_n)$ and there are k constraints $g_i = C_i$, $i = 1, \dots, k$. There will then be as many multipliers λ_i as there are constraints, and the equations to be solved can be shown to take the form

$$\left(\frac{\partial f}{\partial x_j}\right) - \sum_{i=1}^{k} \lambda_i \left(\frac{\partial g_i}{\partial x_j}\right) = 0\,, \qquad j = 1, \dots, n, \tag{G.7}$$

$$g_i = C_i, \qquad i = 1, \dots, k. \tag{G.8}$$

Example G.1. Gradient

We want to maximize df with respect to dx_1, dx_2, dx_3, in

$$df = \left(\frac{\partial f}{\partial x_1}\right)_{x_2 x_3} dx_1 + \left(\frac{\partial f}{\partial x_2}\right)_{x_1 x_3} dx_2 + \left(\frac{\partial f}{\partial x_3}\right)_{x_1 x_2} dx_3$$

subject to the constraint $g = (dx_1)^2 + (dx_2)^2 + (dx_3)^2 = ds^2$, where ds is an infinitesimal constant. Here df corresponds to the change in f associated with a displacement of magnitude ds in the direction defined by dx, dy, dz in a Cartesian coordinate system.

Noting that the derivative of g with respect to dx_i is $2\,dx_i$, we see that the equations defining the constrained maximum of df are

$$\left(\frac{\partial f}{\partial x_i}\right) - 2\lambda x_i = 0\,, \qquad i = 1, 2, 3,$$

$$(dx_1)^2 + (dx_2)^2 + (dx_3)^2 = ds^2\,.$$

The first of those equations yields

$$dx_i = \frac{1}{2\lambda}\left(\frac{\partial f}{\partial x_i}\right), \qquad i = 1, 2, 3,$$

showing that the direction defined by the dx_i has components proportional to the derivatives $(\partial f/\partial x_i)$. Using the notation

$$\boldsymbol{\nabla} f = \left(\frac{\partial f}{\partial x_1}\right)_{x_2 x_3} \hat{\mathbf{e}}_1 + \left(\frac{\partial f}{\partial x_2}\right)_{x_1 x_3} \hat{\mathbf{e}}_2 + \left(\frac{\partial f}{\partial x_3}\right)_{x_1 x_2} \hat{\mathbf{e}}_3,$$

we see that we have shown that the maximum directional derivative of f (the gradient) is in the direction of $\boldsymbol{\nabla} f$.

It remains to verify that the magnitude of $\boldsymbol{\nabla} f$ is the magnitude of the gradient. To show this, we next solve the ds^2 equation for $1/2\lambda$ and write the result as follows:

$$\frac{1}{2\lambda} = \frac{ds}{\sqrt{\left(\frac{\partial f}{\partial x_1}\right)^2 + \left(\frac{\partial f}{\partial x_2}\right)^2 + \left(\frac{\partial f}{\partial x_3}\right)^2}} = \frac{ds}{|\boldsymbol{\nabla} f|}\,.$$

Finally, we insert the expressions for the dx_i into the equation for df, thereby finding its maximum value:

$$df = \frac{ds}{|\nabla f|}\left[\left(\frac{\partial f}{\partial x_1}\right)^2 + \left(\frac{\partial f}{\partial x_2}\right)^2 + \left(\frac{\partial f}{\partial x_3}\right)^2\right] = |\nabla f|\, ds\,.$$

Thus, ∇f is a vector whose direction is that of most rapid increase in f and whose magnitude is the directional derivative in the direction of that increase.

■

Example G.2. Boltzmann Distribution

In classical statistical mechanics, one frequently considers a system consisting of a very large number N of identical and independent subsystems (e.g., molecules). It is assumed that each molecule can be in one of a set of energy states ε_i, and the number of molecules in the ε_i state is n_i. Then N and the total energy E of the system satisfy

$$N = \sum_i n_i\,, \qquad E = \sum_i n_i \varepsilon_i\,.$$

The central assumption of classical (Boltzmann) statistics is that the most probable distribution of energies can be reached by assigning energies to individual molecules in the largest number of different ways; for the distribution $n_1, n_2, \ldots$ the number of different assignments is

$$W = \frac{N!}{n_1!\, n_2!\, n_3! \cdots}\,.$$

Finding the most probable (Boltzmann) distribution is a constrained maximization problem, and is most easily approached by maximizing

$$\ln W - \alpha \sum_i n_i - \beta \sum_i n_i \varepsilon_i$$

and imposing the constraints. We evaluate $\ln W$ using Stirling's formula, assuming all the n_i to be large, so we can use the approximation given as Eq. (9.28), $\ln N \approx N \ln N - N$ and $\ln n_i \approx n_i \ln n_i - n_i$. These simplifications lead us to

$$\ln W \approx N \ln N - \sum_i n_i \ln n_i\,,$$

so we need the unconstrained maximum with respect to the n_i of

$$N \ln N - \sum_i n_i \ln n_i - \alpha \sum_i n_i - \beta \sum_i n_i \varepsilon_i\,.$$

Setting the derivative of this expression with respect to n_i to zero, we get for each i

$$-\ln n_i - 1 - \alpha - \beta \varepsilon_i = 0\,.$$

Giving $-1 - \alpha$ the name $\ln C$ and exponentiating, we reach

$$n_i = C e^{-\beta \varepsilon_i}\,.$$

In statistical mechanics β is identified as $1/kT$, where T is the temperature and k is Boltzmann's constant. Typically, instead of directly enforcing the constraint on E one chooses a value of T (and hence also β), and then finds the value of C needed to make the sum of the n_i add to N. With the n_i in hand, it is straightforward to compute the total energy corresponding to the chosen temperature.

■

Exercises

G.1. Find the radius and height that will minimize the surface area of a right circular cylinder of unit volume.

H SYMBOLIC COMPUTING FOR VECTOR ANALYSIS

Both MAPLE and MATHEMATICA support computations in vector analysis through packages that extend those symbolic computation systems. While those packages share some common command features, their design philosophies are markedly different, and some understanding of these differences will be helpful to those who use the packages at a level that goes beyond the most elementary applications. The main issues become apparent when making computations in curvilinear coordinates; they arise because the unit basis vectors can then differ in direction when associated with different points. We review the situation separately for the two symbolic languages, but users of either language will benefit from reading both parts of this appendix.

MAPLE

MAPLE uses the `Vector` command in two distinct ways ("cases"):

(1) To create the set of three quantities (q_1, q_2, q_3) that are the coordinates of the head of a vector whose tail is at the origin, or

(2) To create the three functions $f_1(q_1, q_2, q_3)$, $f_2(q_1, q_2, q_3)$, $f_3(q_1, q_2, q_3)$ that are the coefficients of the respective unit vectors $\hat{\mathbf{e}}_1$, $\hat{\mathbf{e}}_2$, $\hat{\mathbf{e}}_3$ when at the coordinate point (q_1, q_2, q_3), with the intent of specifying the mathematical object

$$f_1(q_1, q_2, q_3)\,\hat{\mathbf{e}}_1 + f_2(q_1, q_2, q_3)\,\hat{\mathbf{e}}_2 + f_3(q_1, q_2, q_3)\,\hat{\mathbf{e}}_3\,.$$

When a set of data is entered for either of these cases, MAPLE records (as "attributes" of that individual vector) the coordinate system then active, the names of the coordinates, and the case to which the data correspond. Case (2) is identified by having the "vector field" attribute and is reached either by entering the vector using the command `VectorField` or by applying that command to an existing vector.

In Cartesian coordinates there isn't a lot of difference between Cases (1) and (2), because the coordinates of a vector with tail at the origin are equal to its respective components, and that remains true even if the vector is associated with a different coordinate point. But the same is not true in general curvilinear systems. To illustrate, consider the representation of a unit vector $\hat{\mathbf{z}}$ in spherical coordinates. If its tail is at the origin, its head is at $(r = 1,\ \theta = 0,\ \varphi = 0)$ and under Case (1) it would be represented by $[1, 0, 0]$. But in Case (2), we need the representation of $\hat{\mathbf{z}}$ when we are at a general point (r, θ, φ). Referring to Fig. H.1, we see that

$$\hat{\mathbf{z}} = \cos\theta\,\hat{\mathbf{e}_{\mathbf{r}}} - \sin\theta\,\hat{\mathbf{e}}_\theta\,,$$

so $\hat{\mathbf{z}}$ is represented as $[\cos\theta,\ -\sin\theta,\ 0]$.

Now suppose that we have two vectors $\mathbf{A}$ and $\mathbf{B}$, and that we have entered them under Case (1) (i.e., not designating them as vector fields). If these vectors are in Cartesian coordinates (i.e., entered when those coordinates were active), we can form $\mathbf{A} + \mathbf{B}$ by addition of corresponding elements of the arrays representing $\mathbf{A}$ and $\mathbf{B}$, and MAPLE lets us do so. But if either or both of these vectors were defined to be

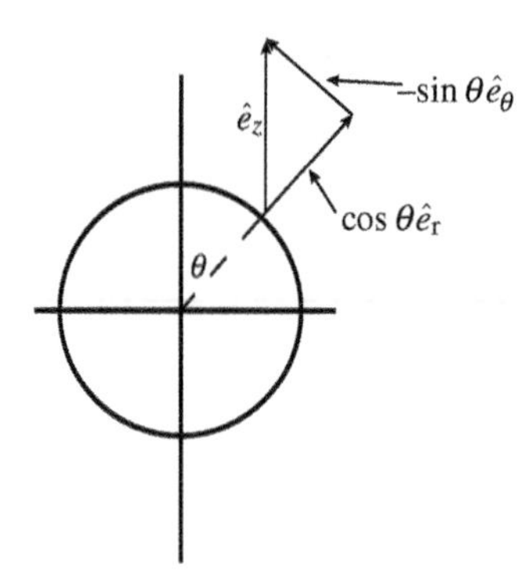

Figure H.1: Unit vector $\hat{\mathbf{z}}$, based at (r, θ, φ).

in a nonCartesian system, corresponding-element addition makes no sense; MAPLE detects this and sends an error signal. To illustrate this point, again using spherical coordinates, note that the Case (1) representation of $\hat{\mathbf{x}}$ is $[1,\ \pi/2,\ 0]$, but that $\hat{\mathbf{x}} + \hat{\mathbf{x}}$ is **not** represented by $[2, \pi, 0]$; it is actually $[2,\ \pi/2,\ 0]$. The basic problem here is that coordinate values are not components (and because of the curvature, not even proportional to components), and that the unit vectors are not in fixed directions.

On the other hand, under Case (2), component addition **is** legitimate, because the data **are** components and are relative to the **same** set of unit vectors. The only potential problem in Case (2) is that the vectors being added might have been defined in different coordinate systems, in which case MAPLE blocks the operation and sends an error signal.

Going forward, we follow MAPLE's usage, calling Case (1) quantities "vectors" and Case (2) quantities "vector fields." We have already noted that MAPLE blocks the addition of vectors in nonCartesian systems. It also blocks (for nonCartesian systems) the other operations of vector algebra (dot and cross products, multiplication by a scalar). However, all the algebraic vector operations are permitted (and are processed correctly) for both Cartesian and nonCartesian vector fields, providing that all the vector fields appearing in an expression have been defined in the same coordinate system.

It is possible to define a vector whose entries are functions of the variables used as coordinate names, but that does not automatically make the vector a vector field. Even in Cartesian systems MAPLE will not permit the use of vector differential operators unless the vector has been given the vector field attribute. But once a quantity has been identified as a vector field, MAPLE permits operation on it by the vector differential operators, doing so in a manner consistent with the coordinate system in which the vector field was defined.

MAPLE provides some tools to assist users in converting vector fields to desired coordinate systems. The most useful command for this purpose is `MapToBasis`, which can (1) convert the coordinates of a spatial point from one coordinate system to another, or (2) convert the expression of a vector field between coordinate systems. These features are illustrated in the following comprehensive example.

Example H.1. Coordinate Transformations in Maple

Let's set up some Maple vectors and transform them between coordinate systems.

```
> with(VectorCalculus):
> SetCoordinates(cartesian);                    cartesian
```

When not working with vector fields, we need not specify the names of the coordinates and MAPLE will use defaults. The names (whether default or user-supplied) are for convenience only. Their only function is to label the e_i, so, for example, if a vector A has been specified in the Cartesian coordinates (x, y, z), while B was specified in the Cartesian system (u, v, w) and the active system is Cartesian with variables (n, l, m), the coefficient of e_n in $(A + B)$ would be the sum of the coefficients of e_x from A and e_u from B. It is important to remember that default coordinates are not permitted for operations on vector fields. Continuing our MAPLE session,

```
> X := Vector([1,0,0]);
```
$$X := e_x$$

X is a unit vector in the x-direction.

```
> MapToBasis(X,spherical);
```
$$e_r + \tfrac{1}{2}\,\pi\, e_\phi$$

This illustration (which appears not to be correct) was prepared with Version 13 of MAPLE, for which the default spherical coordinates are reversed from their conventional order (probably a bug!). You can avoid such bugs or differences in convention by specifying the coordinate names (in this context valid only for the current command). The following agrees with the usual coordinate names:

```
> Xsph := MapToBasis(X,spherical[r,theta,phi]);
```
$$Xsph := e_r + \tfrac{1}{2}\,\pi\, e_\theta$$

As a check, we can transform back to a Cartesian representation:

```
> MapToBasis(Xsph,cartesian[x,y,z]);
```
$$e_x$$

Notice what happens if we transform a vector whose description contains the coordinates:

```
> W := Vector([z,0,0]);
```
$$W := z\, e_x$$

```
> MapToBasis(W,spherical[r,theta,phi]);
```
$$(\sqrt{z^2})\, e_r + (\arctan(\sqrt{z^2}, 0))e_\theta + \tfrac{1}{2}\,(1 - \mathrm{csgn}(z))\,\pi\, e_\phi$$

This fairly complicated expression indicates that the r coordinate of W is $|z|$, that the θ coordinate is $\pi/2$, and that the ϕ coordinate is zero if $z > 0$ and π if $z < 0$.[1] More important for the present discussion is that MAPLE does not treat z as a variable to be converted to spherical coordinates; for the present conversion it is regarded as a coordinate-independent parameter.

The basic vector conversion capability can be used to add vectors that have been defined in curvilinear coordinate systems. One can transform the vectors to Cartesian form, add them, and then transform back to the curvilinear system. For example, if U and V are in cylindrical coordinates, their cylindrical-coordinate sum $S = U + V$ can be obtained as follows:

```
> SetCoordinates(cylindrical[rho,phi,z]):
> U := Vector([1,0,0]);
```
$$U := e_\rho$$

```
> V := Vector([1,Pi/2,1]);
```
$$V := e_\rho + \tfrac{1}{2}\,\pi\, e_\phi + e_z$$

```
> Uc := MapToBasis(U,cartesian[x,y,z]);
```
$$Uc := e_x$$

```
> Vc := MapToBasis(V,cartesian[x,y,z]);
```
$$Vc := e_y + e_z$$

[1]Note that `arctan(y,x)` is the value of $\tan^{-1}(y/x)$ that corresponds to the individual values of x and y, so in the present case $y > 0$, $x = 0$, and the polar angle is 90°. The function `csgn(z)` is $+1$ when $z > 0$ and -1 when $z < 0$, so $(1 - \mathrm{csgn}(z))/2$ is zero for $z > 0$ and 1 for $z < 0$.

```
> Sc := Uc + Vc;
```
$$Sc := e_x + e_y + e_z$$

```
> S := MapToBasis(Sc,cylindrical[rho,phi,z]);
```
$$S := \sqrt{2}\, e_\rho + \tfrac{1}{4}\,\pi\, e_\phi + e_z$$

The situation becomes different if we make a coordinate conversion on a vector field W. Now the names of the coordinates in W need to have been specified when W was created, and any dependence on these named coordinates is converted upon application of the `MapToBasis` command. In addition, `MapToBasis` must include the names of the new coordinates. Thus, continuing our MAPLE session,

```
> SetCoordinates(cartesian[x,y,z]);
> WW := VectorField(W);
```
$$WW := z\,\overline{e}_x$$

```
> Wsph := MapToBasis(WW,spherical[r,theta,phi]);
```
$$Wsph := (r\cos(\theta)\sin(\theta)\cos(\phi))\,\overline{e}_r + (r\cos(\theta)^2\cos(\phi))\,\overline{e}_\theta - r\cos(\theta)\sin(\phi)\,\overline{e}_\phi$$

This result, which may seem less than obvious, indeed corresponds to the spherical-component decomposition of a vector $z\,\hat{\mathbf{x}}$ based at the point (r, θ, ϕ). We can at least easily check that $Wsph$ has length $|z|$:

```
> simplify(Wsph . Wsph);
```
$$r^2\cos(\theta)^2$$

which we recognize as $|z^2|$.

Finally, let's verify that there is consistency between the results of vector calculus in different coordinate systems. Define

```
> f := x^2 + y^2;
```
$$f := x^2 + y^2$$

The active coordinates are Cartesian, so

```
> A := Gradient(f);
```
$$A := 2x\,\overline{e}_x + 2y\,\overline{e}_y$$

Converting this result to cylindrical coordinates and simplifying,

```
> Acyl := simplify(MapToBasis(A,cylindrical[rho,phi,z]));
```
$$2\rho\,\overline{e}_\rho$$

We could also have obtained this result by writing f in cylindrical coordinates and taking its gradient in that coordinate system:

```
> SetCoordinates(cylindrical[rho,phi,z]):
> fcyl = rho^2;
```
$$fcyl = \rho^2$$

```
> Gradient(fcyl);
```
$$2\rho\,\overline{e}_\rho$$

Converting this result back to Cartesian coordinates, we further verify total consistency:

```
> MapToBasis(%,cartesian[x,y,z]);
```
$$2x\,\overline{e}_x + 2y\,\overline{e}_y$$

■

MATHEMATICA

MATHEMATICA does not create arrays with additional attributes associated with specific kinds of mathematical objects; a vector or vector field is represented simply as a three-element list enclosed in braces. Since such a list contains no information relevant to its purpose or meaning, it remains the responsibility of the user to use the array in ways consistent with what the data actually represent. In particular, if the user has created an array containing the three coordinate values that describe a vector with tail at the origin, as in MAPLE's Case (1), he/she should **not** add such data from two nonCartesian vectors in an attempt to form their vector sum. However, MATHEMATICA knows nothing about what the user is trying to do and willingly participates in making nonsense computations. On the other hand, if the three-element array contains the components of a vector field, an element-by-element sum is a relevant and meaningful operation.

MATHEMATICA determines how to carry out both algebraic and differential vector operations by referring to the currently active coordinate system and coordinates, which (if they have not been explicitly set) are taken to be the defaults `Cartesian[Xx,Yy,Zz]`. The processing assumes that all vector quantities are given as components in the active coordinate system and that the operations are evaluated using formulas appropriate to those coordinates, i.e., that the vector(s) involved are vector fields, as in MAPLE's Case (2). This protocol means that the user has the further responsibility to make sure that all vector quantities were originally entered as components of a vector field in the chosen coordinate system. MATHEMATICA is unable to determine whether that is the case.

MATHEMATICA provides limited support for coordinate-system conversions. The commands `CoordinatesToCartesian` and `CoordinatesFromCartesian` convert the coordinates of a spatial point between coordinate systems, but there is no single command that converts a vector field between coordinate systems.

Example H.2. Coordinate Conversions in Mathematica

Complete conversions of vector fields are not supported in a simple way in MATHEMATICA. MATHEMATICA does, however, provide support for the conversion of vectors that have been described by the coordinates of their points when their tails are placed at the coordinate origin. This capability is provided by commands that convert between the coordinates of a point (q_1, q_2, q_3) in curvilinear coordinates and those of the same point in Cartesian coordinates. The commands that accomplish this are `CoordinatesToCartesian` and `CoordinatesFromCartesian`. One use of these commands is to add vectors that have been specified in curvilinear coordinates. To obtain $S = U + V$, proceed as follows:

```
<<VectorAnalysis`
```

U and V are intended to be vectors in cylindrical coordinates, but that information is not needed until we make a conversion.

`U = {1, 0, 0}`	`{1, 0, 0}`
`V = {1, Pi/2, 1}`	$\{1, \frac{\pi}{2}, 1\}$
`Uc = CoordinatesToCartesian[U,Cylindrical]`	$\{1, 0, 0\}$
`Vc = CoordinatesToCartesian[V,Cylindrical]`	$\{0, 1, 1\}$
`Sc = Uc + Vc`	$\{1, 1, 1\}$
`CoordinatesFromCartesian[Sc,Cylindrical]`	$\{\sqrt{2}, \frac{\pi}{4}, 1\}$

Here is a check of the gradient operator in cylindrical coordinates:

`SetCoordinates[Cartesian[x,y,z]]`	`Cartesian[x,y,z]`
`f = x^2 + y^2`	$x^2 + y^2$
`Grad[f]`	$\{2\,x,\ 2\,y,\ 0\}$

We know (or can easily find) that in cylindrical coordinates,

`fc = rho^2`	$\mathtt{rho}^2$

We now make cylindrical coordinates active and take the gradient:

`SetCoordinates[Cylindrical[rho,phi,z]]`	`Cylindrical[rho,phi,z]`
`G = Grad[fc]`	$\{2\,\mathtt{rho},\ 0,\ 0\}$

This result corresponds to a vector of length 2ρ in the ρ-direction. When we are at (x, y, z), the unit vector $\hat{\mathbf{e}}_\rho$ has the Cartesian representation $(x/\rho)\hat{\mathbf{x}} + (y/\rho)\hat{\mathbf{y}}$, so $2\,\rho\,\hat{\mathbf{e}}_\rho = 2x\,\hat{\mathbf{x}} + 2y\,\hat{\mathbf{y}}$, in agreement with the original Cartesian computation.

Finally, note that if we are so foolish as to apply `CoordinatesToCartesian` to our cylindrical-coordinate gradient, we get an incorrect result:

`CoordinatesToCartesian[G]`	$\{2\,\mathtt{rho},\ 0,\ 0\}$

This result would still be wrong if we substituted for `rho` its Cartesian equivalent $\sqrt{x^2 + y^2}$. The problem is that a simple coordinate transformation does not take into account the change with position of the basis unit vectors.

■

I MAPLE TENSOR UTILITIES

In MAPLE, the Kronecker delta and the Levi-Civita symbol are not part of the basic language. To make these features available without invoking the packages in which they occur, the following procedures can be executed. Note that `KD` and `Eps` are defined here as **functions**. The MAPLE package definitions of these quantities defines them (with different names than used here) as **arrays** that can represent space-time tensors.

```
> KD := proc(i,j);
>   if i=j then 1 else 0 end if
> end proc;
```

```
> Eps := proc(i,j,k);
>   if not(member(i,{1,2,3}) and member(j,{1,2,3})
>           and member(k,{1,2,3})) then RETURN(0) end if;
>   if (i=j or i=k or j=k) then RETURN(0) end if;
>   if abs(j-i)=1 then j-i else (i-j)/2 end if
> end proc;
```

This coding uses the feature that `member` returns `true` if its first argument is a member of the set of items given as its second argument and returns `false` otherwise. Once it has been established that i, j, k is a permutation of $1, 2, 3$, the one-line test of $j - i$ suffices to determine its parity.

Checks of the above coding:

```
> KD(5,5);                          1
> KD(2,3);                          0
> Eps(2,3,4);                       0
> Eps(2,3,2);                       0
> Eps(2,3,1);                       1
> Eps(3,2,1);                      -1
```

Exercises

I.1. By explicit enumeration verify that the coding for `Eps` is correct.

I.2. Write symbolic code for `Eps` in dimension 4, i.e., `Eps(i,j,m,n)`.

I.3. Modify the code for `KD` to have it return zero if the two arguments are not integers 1, 2, or 3.

J WRONSKIANS IN ODE THEORY

The Wronskian is a valuable tool for the development of the theory of ordinary differential equations. As outlined in Section 4.8, it provides a test for linear dependence, which can be used to show that a linear homogeneous ODE of order n has no more than n linearly independent solutions. In addition, the Wronskian can be used to construct a second solution to a second-order linear homogeneous ODE if a first solution is known.

NUMBER OF SOLUTIONS

Given an ODE of order n, of the form

$$\mathcal{L}(x)\, y \equiv y^{(n)} + p_{n-1}(x)\, y^{(n-1)} + \cdots + p_0(x)\, y(x) = 0\,, \tag{J.1}$$

let $y_i(x)$, $i = 1, \ldots, n+1$ be any $n+1$ of its solutions. We form an $(n+1) \times (n+1)$ matrix $\mathsf{W}(x)$ of the $y_i(x)$ and their first n derivatives as follows:

$$\mathsf{W} = \begin{pmatrix} y_1 & y_1' & y_1'' & \cdots & y_1^{(n)} \\ y_2 & y_2' & y_2'' & \cdots & y_2^{(n)} \\ y_3 & y_3' & y_3'' & \cdots & y_3^{(n)} \\ \cdots & \cdots & \cdots & \cdots & \cdots \\ y_{n+1} & y_{n+1}' & y_{n+1}'' & \cdots & y_{n+1}^{(n)} \end{pmatrix}, \tag{J.2}$$

and note, by comparison with Eq. (4.77), that $\det(W)$ is the Wronskian of the $n+1$ ODE solutions. (The array constituting W is the transpose of that appearing in Eq. (4.77), but transposition does not change the value of a determinant.)

We next define a vector $\mathbf{a}$ as

$$\mathbf{a} = \begin{pmatrix} p_0(x) \\ p_1(x) \\ \cdots \\ p_{n-1}(x) \\ 1 \end{pmatrix}, \tag{J.3}$$

where the $p_j(x)$ are the coefficients of the ODE in Eq. (J.1). We now form the product Wa using the quantities in Eqs. (J.2) and (J.3), noting that the resulting expressions are instances of the ODE in Eq. (J.1). The result is

$$\mathsf{Wa} = \begin{pmatrix} \mathcal{L}(x)\, y_1 \\ \mathcal{L}(x)\, y_2 \\ \cdots \\ \mathcal{L}(x)\, y_{n+1} \end{pmatrix} = 0\,. \tag{J.4}$$

Equation (J.4) shows that W is singular. If W^{-1} were to exist, we could multiply that equation by it and obtain the inconsistent result $\mathbf{a} = 0$. We therefore conclude that $\det(\mathsf{W}) = 0$. Since $W(x) = \det(\mathsf{W})$ is the Wronskian of the ODE solutions, we have shown that every possible choice of the $n+1$ ODE solutions exhibits linear dependence, meaning that the ODE cannot have $n+1$ linearly independent solutions.

We already know from Section 10.1 that because our ODE is linear and homogeneous any linear combination of its solutions is also a solution, so a more complete statement of the present analysis is that the general solution of our ODE will be an arbitrary linear combination of n linearly independent solutions.

WRONSKIANS OF SECOND-ORDER ODEs

A second-order linear homogeneous ODE of the form

$$y''(x) + p(x)y'(x) + q(x)y(x) = 0 \tag{J.5}$$

has the Wronskian of its two linearly independent solutions $y_1(x)$ and $y_2(x)$

$$W(x) = \begin{vmatrix} y_1 & y_2 \\ y_1' & y_2' \end{vmatrix} = y_1 y_2' - y_1' y_2\,. \tag{J.6}$$

Differentiating this expression for W, we get initially

$$W' = y_1' y_2' + y_1 y_2'' - y_1'' y_2 - y_1' y_2' = y_1 y_2'' - y_1'' y_2\,. \tag{J.7}$$

Using Eq. (J.5) to remove the second derivatives from Eq. (J.7), we reach

$$W' = y_1(-py_2' - qy_2) - (-py_1' - qy_1)y_2 = p(y_1' y_2 - y_1 y_2') = -p\,W\,. \tag{J.8}$$

Equation (J.8) is a separable first-order ODE, which can be written

$$\frac{dW}{W} = -p(x)\,dx\,, \quad \text{with solution} \quad \ln W = -\int p(x)\,dx\,.$$

Solving for W,

$$W(x) = \exp\left(-\int^x p(x)\,dx\right)\,. \tag{J.9}$$

We do not need to retain a constant of integration because all it will affect is the scale chosen for y_1 and y_2.

The key result embodied in Eq. (J.9) is that the functional form of $W(x)$ depends only upon the coefficient $p(x)$ of the ODE.

FINDING A SECOND SOLUTION

We start by noticing that

$$\frac{W}{y_1^2} = \frac{y_1 y_2' - y_1' y_2}{y_1^2} = \left(\frac{y_2}{y_1}\right)'. \tag{J.10}$$

Integrating both sides of Eq. (J.10), we get

$$\frac{y_2}{y_1} = \int \frac{W}{y_1^2}\, dx\,. \tag{J.11}$$

Being careful with the naming of the variables of integration, Eq. (J.11) takes the explicit form

$$\frac{y_2(x)}{y_1(x)} = \int^x \frac{\exp(-\int^t p(u)\, du)}{y_1(t)^2}\, dt + C\,.$$

The constant of integration can be dropped because it only adds to y_2 a multiple of y_1, so our final formula for y_2 becomes

$$y_2(x) = y_1(x) \int^x \frac{\exp(-\int^t p(u)\, du)}{y_1(t)^2}\, dt\,. \tag{J.12}$$

K MAPLE CODE FOR ASSOCIATED LEGENDRE FUNCTIONS AND SPHERICAL HARMONICS

ASSOCIATED LEGENDRE FUNCTIONS

The default MAPLE environment setting for the associated Legendre functions causes them to contain a factor $(x^2-1)^{m/2}$ in place of the $(1-x^2)^{m/2}$ that occurs in the formulas of this text, and that setting makes these functions real (for all m) only when $|x| \geq 1$. Changing this setting (by the method provided in MAPLE) leads for some P_n^m to incorrect results, and produces sign differences relative to this text when $m < 0$.

The easiest way to avoid dealing with these inconsistences is to define a new MAPLE function that corresponds to the definitions in this text. We give that function the name `LegenP(n,m,x)`. Its coding is as follows:

```
> LegenP := proc(n,m,x) local P1,P2,P3,PX,M,j;
>   if (m = 0) then RETURN(simplify(LegendreP(n,x))) end if;
>   if not type(m,integer) then RETURN('procname(args)') end if;
>   if not type(n,integer) then RETURN('procname(args)') end if;
>   P3 := 1;
>   if (m > 0) then
>     M := m;
>     for j from 1 to M do
>       P3 := -P3*(2*j-1)*(1-x^2)^(1/2) end do;
>   else
```

```
>       M := -m;
>       for j from 1 to M do
>         P3 := P3/(2*j)*(1-x^2)^(1/2) end do;
>     end if;
>     P3:=simplify(P3);
>     P3:= eval(subs(csgn=ONE,P3));
>     if (M = n) then RETURN(P3) end if;
>     P1 := (M+m+1)*x;
>     P2:=1;
>     for j from 2 to n-M do
>        PX := P1;
>        P1 := ( (2*(M+j)-1)*x*P1-(M+m+j-1)*P2) /(M-m+j);
>        P2 := PX end do;
>     normal(P1*P3)
> end proc;
> ONE := proc(z); 1 end proc:
```

Here P3 contains $P_{|m|}^{m}(x)$, corresponding to Eq. (14.52) or (14.53) of the main text, and P1 contains the polynomial multiplying P3, built up using the recurrence formula, Eq. (14.47). Note the use of $M = |m|$ and m to give results that are valid for both positive and negative m.

Other comments regarding the coding details:

(1) If either m or n is not given a numerical value (and is therefore detected by MAPLE as "not an integer"), the procedure returns the calling command unevaluated. The quantities procname and args are special names; when they are encountered within a procedure they respectively stand for the procedure name and its arguments. By writing RETURN('procname(args)') we cause MAPLE to return this statement unevaluated. An attempt to evaluate procname(args) would create an infinite recursive loop.

(2) When MAPLE can simplify the quantity $(1 - x^2)$ it realizes that $(1 - x^2)^{1/2}$ can have a sign ambiguity, which it deals with by inserting in the output a sign-assignment factor using the function csgn. However, the intended use of LegenP is in contexts where $1 - x^2 = 1 - \cos^2\theta = \sin^2\theta$, with θ in the range $(0, \pi)$. In that range, $\sin\theta \geq 0$, and the csgn factor becomes unnecessary clutter. It is therefore removed by

(i) Changing csgn to ONE, a function that returns the result +1 for all input. The command subs does this by making the substitution indicated by its first argument in the expression designated by its second argument (P3), and then

(ii) Causing the altered P3 to be evaluated, making the factor ONE( ) disappear.

(3) The one-line procedure ONE is used when LegenP is called and both the procedures LegenP and ONE must have been executed in the current MAPLE session before LegenP is first used.

Checks vs. MATHEMATICA of the above coding:

LegenP(5,3,x);	$-\frac{105}{2}(1 - x^2)^{3/2}(9x^2 - 1)$
LegendreP[5,3,x]	$\frac{105}{2}\sqrt{1 - x^2}\,(-1 + x^2)(-1 + 9x^2)$

LegenP(5,-3,x); $\frac{1}{384}(1-x^2)^{3/2}(9x^2-1)$

LegendreP[5,-3,x] $-\frac{1}{384}\sqrt{1-x^2}\,(-1+x^2)(-1+9x^2)$

LegenP(4,2,x); $-\frac{15}{2}(7x^2-1)(-1+x^2)$

LegendreP[4,2,x] $-\frac{15}{2}(-1+x^2)(-1+7x^2)$

LegenP(4,-2,x); $-\frac{1}{48}(7x^2-1)(-1+x^2)$

LegendreP[4,-2,x] $-\frac{1}{48}(-1+x^2)(-1+7x^2)$

SPHERICAL HARMONICS

A MAPLE procedure for the spherical harmonics can now be constructed using the procedure LegenP of the preceding subsection. Making straightforward use of Eq. (15.91), the coding to produce $Y_L^M(\theta,\varphi)$ can be as follows:

```
> SphY := proc(L,M,theta,phi);
>    sqrt( (2*L+1)/(4*Pi) * (L-M)!/(L+M)! )
>                  * LegenP(L,M,cos(theta)) * exp(I*M*phi);
> end proc:
```

Use of SphY and SphericalHarmonicY for unevaluated arguments produces the following results:

```
> SphY(L,M,theta,phi);
```

$$\frac{1}{2}\sqrt{\frac{(2L+1)(L-M)!}{\pi(L+M)!}}\;LegenP(L,M,\cos(\theta))\,e^{I\,M\phi}$$

```
SphericalHarmonicY[L,M,theta,phi]
```

SphericalHarmonicY[L, M, theta, phi]

We see that SphY carries out the evaluation except for reporting the explicit form of P_l^m; MATHEMATICA returns the function call entirely unevaluated.

SphY and MATHEMATICA's SphericalHarmonicY are in complete agreement, as shown by the check values shown below.

> SphY(3, 1, Pi/2, 0); $\frac{1}{8}\frac{\sqrt{21}}{\sqrt{\pi}}$

SphericalHarmonicY[3, 1, Pi/2, 0] $\frac{\sqrt{\frac{21}{\pi}}}{8}$

> SphY(3, 1, Pi/2, Pi/4); $\frac{1}{8}\frac{\sqrt{21}\left(\frac{1}{2}\sqrt{2}+\frac{1}{2}\mathrm{I}\sqrt{2}\right)}{\sqrt{\pi}}$

```
> evalf(%);
```

$$0.2285228996 + 0.2285228996\,\mathrm{I}$$

```
SphericalHarmonicY[3, 1, Pi/2, Pi/4]
```

$$\frac{1}{8}(-1)^{1/4}\sqrt{\frac{21}{\pi}}$$

```
N[%]
```

$$-0.228523 - 0.228523\,i$$

Here MATHEMATICA does not resolve the complex exponential $e^{i\pi/4}$ into real and imaginary parts, instead making the somewhat ambiguous reduction to $(-1)^{1/4}$. However, MATHEMATICA interprets this notation consistently, as it gives the same fully reduced complex evaluation as MAPLE. One final check:

```
> SphY(3, 1, theta, phi);
```

$$-\frac{1}{8}\,\frac{\sqrt{21}\left(5\cos(\theta)^2-1\right)\sin(\theta)\,e^{\mathrm{I}\,\phi}}{\sqrt{\pi}}$$

```
SphericalHarmonicY[3, 1, theta, phi]
```

$$-\frac{1}{8}\,e^{i\,\mathrm{phi}}\sqrt{\frac{21}{\pi}}\,\left(-1+5\,\mathrm{Cos}[\mathrm{theta}]^2\right)\mathrm{Sin}[\mathrm{theta}]$$

MAPLE LANGUAGE SUMMARY

Commands that are underlined are not part of the basic language; code for them is given in this book.

Some Basics:

Arithmetic operations: `a+b`, `a-b`, `a*b`, `a/b`, `a^b` or `a**b` (a^b),

Logical/relational: `a or b`, `a and b`, `not a`, `a>b`, `a>=b`, `a=b`, `a<>b` ($a \neq b$), `a<b`, `a<=b`,

String: `"`*string*`"`, Assignment: `a :=` *expr*, Help: `?`*topic*, Comment: `#` *comment*,

Clear everything and start over: `restart`, `sort(`*polynomial*`)` (sorts terms),

Floating point precision: `Digits :=` *nn*, Command ending (print) `;` (do not print) `:`

Force evaluation: `eval`, Floating-point evaluation: `evalf`, Undefine x: `x := 'x'`,

Enable package: `with(`*package*`)`, Disable package: `unwith(`*package*`)`.

Simplify or Transform:

`simplify(`*expr*`)`, `normal(`*expr*`)`, `factor(`*expr*`)`, `expand(`*expr*`)`, `collect(`*expr*,*x*`)`,

`convert(`*expr*`, exp)`, (converts circular/hyperbolic trig functions to exponential form),

`convert(`*expr*`, trig)` (converts exponentials to circular/hyperbolic trig form),

`convert(`*expr*`, parfrac)` (converts to partial fractions), `op(n,A)` (get nth operand of `A`),

`round(x)` (round `x` to nearest integer), `lcoeff(`*polynomial*`,x)` (extract leading coefficient).

Substitution: `subs(x =` *value, expr*`)` (returns *expr* with x set to *value*, stored *expr* unchanged).

Impose Assumptions: `assume(`*assumption*`)` (use when needed to make computations valid).

Programming:

`for i from` *first* `to` *last* `do` *statement, statement,...* `end do`,

`break` (exit the `for` construct), `next` (go to next `i` value),

`if` *condition* `then` *statement, statement,...* `else` *statement, statement,...* `end if`,

`while` *condition* `do` *statement, statement,...* `end do`,

`F := proc(`*args*`) local` *local variables*; *body* `end proc`.

Constants: `I`, `Pi`, `true`, `false`, `infinity`, `gamma` (Euler-Mascheroni constant), `Catalan`.
Use `exp`: e is not defined.

Elementary Functions and Coefficients:

`n!`, `dfac(n)` $= n!!$, `binomial(n,m)` $= \binom{n}{m}$, `bernoulli(n)`, `sign(x)`,

`sqrt(x)`, `abs(x)`, `exp(x)`, `ln(x)` or `log(x)` (these are the same),

`sin(x)`, `sinh(x)`, `arcsin(x)`, `arcsinh(x)`.

In all four of the above `sin` can be replaced by `cos`, `tan`, `cot`, `sec`, or `csc`.

Special Functions:

`erf(x)`, `erfc(x)`, `GAMMA(s)`, `GAMMA(s,x)` (integral x to ∞), `Beta(p,q)`,

`Psi(s)`, `Psi(n,s)` $= \psi^{(n)}(s)$, `Ei(n,x)` $= E_n(x)$, `Ei(x)`, `Zeta(s)`,

`BesselJ(n,x)`, `BesselY(n,x)`, `BesselI(n,x)`, `BesselK(n,x)`,

`BesselJZeros(n,j)`, `BesselYZeros(n,j)`, `HankelH1(n,x)`, `HankelH2(n,x)`,

`SphBesselJ(n,x)`, `SphBesselY(n,x)`, `SphBesselI(n,x)`, `SphBesselK(n,x)`,

`SphHankelH1(n,x)`, `SphHankelH2(n,x)`, `SphY(L,M,theta,phi)`.

Orthogonal Polynomials:

`LegendreP(n,x)`, `LegendreQ(n,x)`, `LegenP(n,m,x)`, `HermiteH(n,x)`,

`LaguerreL(n,x)`, `LaguerreL(n,k,x)`, `ChebyshevT(n,x)`.

Complex Variables:

`I`, `Re(z)`, `Im(z)`, `conjugate(z)`, `abs(z)`, `argument(z)`, `residue(`*expr*`, z=z0)`,
`csgn(z)` (+1/–1, right/left half-plane), `evalc(z)` (attempts to find real and imaginary parts).

Sums, Integrals, Limits, Derivatives:

`sum(`*expr(m)*`, m = n1 .. n2)` (or `add`), `int(`*expr(x)*`, x = x1 .. x2)`, `limit(`*expr*`, x=x0)`,
`series(`*expr*`, x = a, n)` (expansion about a, which can be `infinity`; approximately n terms),
`convert(`*series*`, polynom)` (converts from series form to ordinary expression),
`diff(`*expr(x)*`, x)`, `diff(`*expr(x,y)*`, x, y)`, `diff(`*expr(x)*`, x, x)`.

Linear Algebra:

`V := Vector(n)` (column, length n, all zero), `VC := Vector([3, 1])` (column, elements 3, 1),
`VR := Vector[row]([2,9])` (row, elements 2, 9), `V[i]` (ith element), `M := Matrix(m,n)` (mxn matrix),
`M := Matrix([[1,2],[3,4]])` (matrix, rows 1,2 and 3,4), `M[i]` (ith row of `M`),
`Column(M,j)` (jth column of `M`), `M[i,j]` (element (i,j) of `M`), `VC.VR` (outputs a matrix),
`VR.VR`, `VC.VC`, `VR.VC` (scalar dot product), `M.M`, `M.VC`, `VR.M` valid, but not `VC.M` or `M.VR`.
If `Eq1 := a11*x1+a12*x2 = b1` etc, `solve([Eq1,Eq2,···],[x1,x2,···])` (solves equation system),
`SProd(`$F(\theta,\varphi)$, $G(\theta,\varphi)$`, theta, phi)` (angular scalar product in spherical coordinates).
After `with(LinearAlgebra)`: `Transpose(M)`, `HermitianTranspose(M)`, `MatrixInverse(M)`,
`Determinant(M)`, `Trace(M)`, `GramSchmidt(`*list of vectors*`, normalized)`, `Val := Eigenvalues(M)`,
`Val,Vec := Eigenvectors(M)` (eigenvalues in `Val`, vectors as columns in `Vec`).

Vector Analysis:

After `with(VectorCalculus)`: `A . B`, `CrossProduct(U,V)` or `U &x V`,
`SetCoordinates(cartesian[x,y,z])` or `spherical` or `cylindrical` to establish coordinates, then
`VV := VectorField(V)`, `Gradient(`*expr(x,y,z)*`)` or `Del(`*expr(x,y,z)*`)`, `Divergence(VV)`,
`Curl(VV)`, `Del . VV`, `Del &x VV`, `Laplacian(`*expr(x,y,z)*`)`, `Laplacian(VV)`,
`MapToBasis(VV, `*coords*`[q1,q2,q3])` (converts from prior system to *coords*),
`ScalarPotential(VV)`, `VectorPotential(VV)`,
`BasisFormat(true` *or* `false)` (`true`)$\longrightarrow e_x + 2e_y \cdots$, (`false`)$\longrightarrow [1, 2, \cdots]$.

Tensor Analysis:

`T := Array(1..3, 1..3, 1..3, `*nested list*`)` (build tensor), `ArrayElems(T)` (view elements),
`KD(i,j)` (Kronecker delta), `Eps(i,j,k)` (Levi-Civita symbol).

Solving Differential Equations:

`ODE :=` F`(diff(y(x),x,x), diff(y(x),x), y(x)) = g(x)`,
`dsolve({ODE,`*conditions*`})` (*conditions* are equations, e.g., `y(0)=0`).

Integral Transforms:

After `with(inttrans)`: `fourier(`*expr*`,t,w)`, `invfourier(`*expr*`,w,t)`, `fouriersin(`*expr*`,t,w)`,
`fouriercos(`*expr*`,t,w)`, `laplace(`*expr*`,t,s)`, `invlaplace(`*expr*`,s,t)`, `Heaviside(t)`, `Dirac(t)`.

Probability: `BinomialB(n,s,p)`, `PoissonP(n,nu)`, `GaussG(mu,sigma,x)`, `GaussPhi(mu,sigma,x)`.

Curve Fitting, Least Squares:

`data := [ [x1,y1],[x2,y2],···]` or `seq( [n,`*f(n)*`], n = n1 .. n2)`,
`LeastSquares(data, v, curve = `*form involving v*`)`

Making Tables and Plots: See Appendices A and B.

MATHEMATICA LANGUAGE SUMMARY

Commands that are underlined are not part of the basic language; code for them is given in this book.

Some Basics:

Arithmetic operations: `a+b, a-b, a*b, a/b, a^b` (a^b),

Logical/relational: `a||b, a&&b, !a,` (or, and, not); `a>b, a>=b, a<b, a<=b, a==b, a!=b` ($a \neq b$),

String: `"`*string*`"`, Assignment: `a =` *expr*, Comment: `(*` *comment* `*)`,

Command ending (print) (none) (do not print) `;` `FindRoot[`*eqn*`,{x,x0}]` (get root near `x0`),

Force evaluation: `Evaluate[`*expr*`]`, Floating-point evaluation: `N[`*expr*`]`, Undefine x: `Clear[x]`.

Simplify or Transform:

`Simplify[`*expr*`]`, `Together[`*expr*`]`, `Factor[`*expr*`]`, `FactorTerms[`*expr*`]`, `Expand[`*expr*`]`,

`Collect[`*expr*,*x*`]`, `Apart[`*expr*`]` (converts to partial fractions),

`TrigToExp[`*expr*`]` (converts circular/hyperbolic trig functions to exponential form),

`ExpToTrig[`*expr*`]` (converts exponentials to circular/hyperbolic trig form),

`Round[x]` (round `x` to nearest integer), `Coefficient[`*polynomial*`,x^n]` (coefficient of x^n),

`FunctionExpand[`*expr*`]` (tries to evaluate special functions).

Substitution: *expr* `/. x=`*value* (returns *expr* with x set to *value*, stored *expr* unchanged).

Programming:

`Do[`*statement;* ⋯ *; statement*`,{n,n1,n2}]`,

`If[`*condition, then-statement;* ⋯ *; then-statement, else-statement;* ⋯ *; else-statement*`]`,

`While[`*condition, statement;* ⋯ *; statement*`]`,

procname`[`*args*`_] := Module[{`*local variables*`},`*body*`]`,

`Break[ ]` (exit innermost loop), `Continue[ ]` (go to next case of innermost loop).

Constants:

`I, Pi, E, True, False, Infinity, EulerGamma` (Euler-Mascheroni constant), `Catalan`.

Elementary Functions and Coefficients:

`n!`, `dfac[n]` $= n!!$, `Binomial[n,m]` $= \binom{n}{m}$, `BernoulliB[n]`, `Sign[x]`,

`Sqrt[x]`, `Abs[x]`, `Exp[x]=e^x`, `Log[x]`,

`Sin[x]`, `Sinh[x]`, `ArcSin[x]`, `ArcSinh[x]`.

In all four of the above `Sin` can be replaced by `Cos`, `Tan`, `Cot`, `Sec`, or `Csc`.

Special Functions:

`Erf[x]`, `Erfc[x]`, `Gamma[s]`, `Gamma[s,x]` (integral x to ∞), `Beta[p,q]`,

`PolyGamma[s]`$= \psi(s)$, `PolyGamma[n,s]` $= \psi^{(n)}(s)$, `ExpIntegralE[n,x]` $= E_n(x)$,

`ExpIntegralEi[x]`, `Zeta[s]`, `BesselJ[n,x]`, `BesselY[n,x])`, `BesselI[n,x]`, `BesselK[n,x]`,

`BesselJZero[n,j]`, `BesselYZero[n,j]`, `HankelH1[n,x]`, `HankelH2[n,x]`,

`SphericalBesselJ[n,x]`, `SphericalBesselY[n,x]`, `SphericalBesselI[n,x]`,

`SphericalBesselK[n,x]`, `SphericalHankelH1[n,x]`, `SphericalHankelH2[n,x]`,

`SphericalHarmonicY[L,M,theta,phi]`.

Orthogonal Polynomials:

`LegendreP[n,x]`, `LegendreQ[n,x]`, `LegendreP[n,m,x]`, `HermiteH[n,x]`,

`LaguerreL[n,x]`, `LaguerreL[n,k,x]`, `ChebyshevT[n,x]`.

Complex Variables:

I, Re[z], Im[z], Conjugate[z], Abs[z], Arg[z], Residue[*expr*, {z,z0}],
ComplexExpand[z] (attempts to find real and imaginary parts)

Sums, Integrals, Limits, Derivatives:

Sum[*expr(m)*, {m,n1,n2}], Integrate[*expr(x)*, {x, x1, x2}], Limit[*expr*, x -> x0],
Series[*expr*, {x, a, n}] (expansion about a, which can be Infinity; approximately n terms),
Normal[*series*] (converts from series form to ordinary expression),
D[*expr(x)*, x], D[*expr(x,y)*, x, y], D[*expr(x)*, x, x]

Linear Algebra:

V = {v1, v2, ⋯ } (vector, row or column), V[[i]] (ith element of V), V1.V2 (dot product),
M = {{*row*},{*row*},⋯} (matrix), M[[i]] (ith row of M), Part[M,All,j] (jth column of M),
M[[i,j]] (element (i,j) of M), M.M, M.V, V.M valid, Transpose[M], ConjugateTranspose[M],
If Eq1 = a11*x1+a12*x2 == b1 etc, Solve[{Eq1,Eq2,⋯},{x1,x2,⋯}] (solves equation system),
Inverse[M], SProd[$F(\theta,\varphi)$, $G(\theta,\varphi)$] (angular scalar product in spherical coordinates),
Det[M], Tr[M] (trace), Orthogonalize[M] (vectors are rows of M), Val := Eigenvalues[M],
Vec := Eigenvectors[M], {Val,Vec} = Eigensystem[M] (eigenvalues in Val, vectors as rows in Vec).

Vector Analysis:

A . B, Cross[A,B],
After <<VectorAnalysis`:
SetCoordinates[Cartesian[x,y,z]] or Spherical or Cylindrical to establish coordinates, then
Grad[*expr(x,y,z)*], Div([*vector*], Curl[*vector*], Laplacian[*expr(x,y,z)*], Laplacian[*vector*],
CoordinatesFromCartesian[V,*coords*] (converts from Cartesian to *coords*),
CoordinatesToCartesian[V,*coords*] (converts to Cartesian from *coords*).

Tensor Analysis:

T := Array(1..3, 1..3, 1..3, *nested list*) (build tensor), MatrixForm(T) (view elements),
Transpose[*tensor*, {*new index order*}]
T1.T2 (contract last index of T1 with first of T2), Tr[T, plus, 2] (contract first two indices of T),
KroneckerDelta[i,j], Signature[i,j,k] (Levi-Civita symbol).

Solving Differential Equations:

ODE = *F*(diff(y(x),x,x), diff(y(x),x), y(x)) == g(x),
dsolve({ODE, *conditions*) (*conditions* are equations, e.g., y(0)==0).

Integral Transforms:

FourierTransform[*expr*,t,w], FourierSinTransform[*expr*,t,w], FourierCosTransform[*expr*,t,w],
InverseFourierTransform[*expr*,w,t], HeavisideTheta[t], DiracDelta[t],
LaplaceTransform[*expr*,t,s], InverseLaplaceTransform[*expr*,s,t].

Probability: BinomialB[n,s,p], PoissonP[n,nu], GaussG[mu,sigma,x], GaussPhi[mu,sigma,x].

Curve Fitting, Least Squares:

data = { {x1,y1},{x2,y2},⋯} or data = Table[{n, F[n]}, {n, n1, n2}],
Fit[data, {1, v, ⋯}, v].

Making Tables and Plots: See Appendices A and B.

INDEX OF MAPLE COMMANDS

INDEX OF MATHEMATICA COMMANDS

GENERAL INDEX

D

E

F

G

H

I

J

K

L

M

N

O

P

Q

R

S

T

U

V

W

X

Z

Printed and bound by CPI Group (UK) Ltd, Croydon, CR0 4YY
08/05/2025
01864939-0002